K-theory in Algebra, Analysis and Topology

CONTEMPORARY MATHEMATICS

749

K-theory in Algebra, Analysis and Topology

ICM 2018 Satellite School and Workshop
K-theory Conference, Argentina, 2018
July 16–20 and July 23–27, 2018
La Plata and Buenos Aires, Argentina

Guillermo Cortiñas
Charles A. Weibel
Editors

AMS AMERICAN MATHEMATICAL SOCIETY

Contemporary Mathematics
Volume **749**, 2020
https://doi.org/10.1090/conm/749/15067

Comparison of spaces associated to DGLA via higher holonomy

Paul Bressler, Alexander Gorokhovsky, Ryszard Nest, and Boris Tsygan

Abstract. For a differential graded Lie algebra $\mathfrak{g}$ whose components vanish in degrees below -1 we construct an explicit equivalence between the nerve of the Deligne 2-groupoid and the simplicial set of $\mathfrak{g}$-valued differential forms introduced by V. Hinich. The construction uses the theory of non-abelian multiplicative integration.

1. Introduction

The principal result of the present note is a construction of a weak equivalence of two spaces (simplicial sets) naturally associated with a nilpotent differential grade Lie algebra subject to certain restrictions.

To a nilpotent DGLA $\mathfrak{g}$ which satisfies the additional condition

$$\mathfrak{g}^i = 0 \text{ for } i < -1 \tag{1.1}$$

P. Deligne [**3**] and, independently, E. Getzler [**4**] associated a (strict) 2-groupoid which we denote $\mathrm{MC}^2(\mathfrak{g})$ and refer to as the Deligne 2-groupoid. In an earlier paper by the authors ([**2**]) it was shown that the simplicial nerve $\mathfrak{N}\,\mathrm{MC}^2(\mathfrak{g})$ of the 2-groupoid $\mathrm{MC}^2(\mathfrak{g})$, $\mathfrak{g}$ a nilpotent DGLA satisfying (1.1) is weakly equivalent to another simplicial set, denoted $\Sigma(\mathfrak{g})$, introduced by V. Hinich [**6**] by constructing natural maps from $\mathfrak{N}\,\mathrm{MC}^2(\mathfrak{g})$ and $\Sigma(\mathfrak{g})$ to a third simplicial set and showing that they are equivalences.

In this note we outline a construction of the map

$$\mathbb{I}(\mathfrak{g})\colon \Sigma(\mathfrak{g}) \to \mathfrak{N}\,\mathrm{MC}^2(\mathfrak{g}) \tag{1.2}$$

omitting technical details. The main tool in our construction is the theory of non-abelian multiplicative integration on surfaces.

In [**9**], Yekutieli constructed a theory of multiplicative integration on surfaces based on Riemann products(cf. also Aref'eva's work [**1**]; a sketch of a construction, also based on Riemann products, is due to Kontsevich, cf. [**8**]) and does not apply in the context at hand. In [**7**] Kapranov develops a theory of higher holonomy based on Chen's iterated integrals. In the present work we take the approach based on the fundamental theorem of calculus.

2010 *Mathematics Subject Classification.* Primary 18G50, 18G30; Secondary 53D55.

The third author was supported by the Danish National Research Foundation through the Centre for Symmetry and Deformation (DNRF92).

The paper is organized as follows. In Section 2 we recall some basic notions of differential and integral calculus. In Section 3 we review the theory of connections and holonomy. In Section 4 we define the two-dimensional holonomy of connection-curvature pairs and discuss some of its properties. In Section 5 we recall the definitions of the simplicial sets $\mathfrak{N}\mathrm{MC}^2(\mathfrak{g})$ and $\Sigma(\mathfrak{g})$, construct the map (1.2) and state its properties in Theorem 5.2 and Theorem 5.3.

1.1. Notation. Throughout the paper

- k is a field of characteristic zero
- for a variety X over k: $\mathbb{A}^n_X := X \times \mathbb{A}^n_k$; for a commutative k-algebra R: $\mathbb{A}^n_R := \mathbb{A}^n_{\operatorname{Spec} R}$
- for a variety $\mathrm{pr}_X \colon Y \to X$ over X and a $\mathcal{O}_X$-module $\mathcal{E}$ we denote by $\mathcal{E}_Y$ the $\mathcal{O}_Y$-module $\mathrm{pr}_X^*\mathcal{E}$
- for an X-variety $: Y \to X$ over X we denote by $\mathsf{d}_{/X} \colon \mathcal{O}_Y \to \Omega^1_{Y/X}$ the relative differential
- for an nilpotent $\mathcal{O}_X$-Lie algebra $\mathfrak{g}$ we denote by $\exp(\mathfrak{g})$ the algebraic group over X defined as $\exp(\mathfrak{g})(Y) = \mathfrak{g}_Y(Y)$ equipped with the group structure given by the Baker-Campbell-Hausdorff series
- for a group G and $g \in G$ we denote $\mathrm{Conj}(g)$ the automorphism $h \mapsto ghg^{-1}$

2. Calculus

2.1. Logarithmic derivatives. Suppose that $\mathfrak{g}$ is a nilpotent Lie algebra. For $D \in \mathrm{Der}(\mathfrak{g})$, $a = \exp(x) \in \exp(\mathfrak{g})$, $x \in \mathfrak{g}$. Define $D\log(a) \in \mathfrak{g}$ by

$$D\log(a) := (Da)\cdot a^{-1} = \left(\frac{\exp(\mathsf{ad}(x)) - 1}{\mathsf{ad}(x)}\right)(Dx).$$

For $D \in \mathrm{Der}(\mathfrak{g})$, $a, b \in \exp(\mathfrak{g})$, $x \in \mathfrak{g}$,

$$D\log(ab) = D\log(a) + \mathrm{Ad}_a(D\log(b)) \tag{2.1}$$

$$D\log(a^{-1}) = -\mathrm{Ad}_{a^{-1}}(D\log(a)) \tag{2.2}$$

$$\mathsf{ad}_x \log(a) = x - \mathrm{Ad}_a(x) \tag{2.3}$$

For $D_1, D_2 \in \mathrm{Der}(\mathfrak{g})$, $a \in \exp(\mathfrak{g})$

$$D_1(D_2\log(a)) - D_2(D_1\log(a)) = [D_1, D_2]\log(a) + [D_1\log(a), D_2\log(a)] \tag{2.4}$$

2.2. Initial value problems. Let $\mathrm{pr}_X \colon \mathbb{A}^1_X \to X$ denote the projection so that $\mathrm{pr}_{X*}\mathcal{O}_{\mathbb{A}^1_X} = \mathcal{O}_X[s]$.

Suppose that $\mathcal{E}$ is a vector bundle on X. For

- $A \in \mathrm{End}_{\mathcal{O}_{\mathbb{A}^1_X}}(\mathcal{E}_{\mathbb{A}^1_X})$ nilpotent
- a section $\sigma \colon X \to \mathbb{A}^1_X$ of the projection pr_X
- a section $e \in \Gamma(X; \mathcal{E})$

there exists a unique section $u \in \Gamma(\mathbb{A}^1_X; \mathcal{E}_{\mathbb{A}^1_X})$ which satisfies

$$\begin{cases} \dfrac{\partial u}{\partial s} + A(u) = 0 \\ \sigma^*(u) = e. \end{cases}$$

2.3. Integration. Let $x = s\otimes 1$ and $y = 1\otimes s$ denote the coordinate functions on $\mathbb{A}^1_X \times_X \mathbb{A}^1_X$. Let Δ denote the subvariety defined by the equation $x = y$.

For $\alpha \in \Omega^1_{\mathbb{A}^1_X/X}$ we denote by

$$\int_x^y \alpha \in \Gamma(\mathbb{A}^1_X \times_X \mathbb{A}^1_X; \mathcal{O}_{\mathbb{A}^1_X \times_X \mathbb{A}^1_X})$$

the solution of the initial value problem

$$\begin{cases} \dfrac{\partial u}{\partial y} = \mathrm{pr}_2^*\alpha \\ u|_\Delta = 0 \end{cases}$$

in the unknown $u = u(x, y) \in \Gamma(\mathbb{A}^1_X \times_X \mathbb{A}^1_X; \mathcal{O}_{\mathbb{A}^1_X \times_X \mathbb{A}^1_X})$.

3. Differential geometry

3.1. Connections and curvature. Suppose that $\mathfrak{g}$ is a nilpotent $\mathcal{O}_X$-Lie algebra. For $\mathrm{A} \in \Gamma(\mathbb{A}^k_X; \Omega^1_{\mathbb{A}^k_X/X} \otimes \mathfrak{g})$ we consider connection $\nabla := \mathsf{d}_{/X} + \mathrm{A}$. Its curvature is defined by $F = F(\nabla) := \mathsf{d}_{/X}\mathrm{A} + \frac{1}{2}[\mathrm{A}, \mathrm{A}] \in \Gamma(\mathbb{A}^k_X; \Omega^2_{\mathbb{A}^k_X/X} \otimes \mathfrak{g})$.

Suppose now that $\mathcal{E}$ is an $\mathcal{O}_X$-module and $\rho\colon \mathfrak{g} \to \underline{\mathrm{End}}_{\mathcal{O}_X}(\mathcal{E})$ is a representation of $\mathfrak{g}$. The connection ∇ gives rise to the connection $\nabla^{\mathcal{E}}$ on $\mathcal{E}_{\mathbb{A}^k_X}$.

The sheaf $\Omega^1_{\mathbb{A}^k_X/X}$ of relative differentials is equipped with the canonical relative connection. Hence, the connection ∇ induces a connection on the $\mathcal{O}_{\mathbb{A}^k_X}$-module $\Omega^1_{\mathbb{A}^k_X/X} \otimes \mathcal{E}$ which will be denoted $\nabla^{\mathcal{E}}$ as well.

An example of the above situation is given by the adjoint representation $\mathsf{ad}\colon \mathfrak{g} \to \mathrm{Der}(\mathfrak{g})$ giving rise to the connection $\nabla^{\mathfrak{g}}$.

3.2. Logarithmic covariant derivative. For $u = \exp(x) \in \exp(\mathfrak{g})(\mathbb{A}^k_X)$, let

$$d\log(u) := \left(\frac{\exp(\mathsf{ad}(x)) - 1}{\mathsf{ad}(x)}\right)(\mathsf{d}x)$$

and

$$\nabla \log(u) := d\log u + \mathrm{A}.$$

LEMMA 3.1. *For $\xi_1, \xi_2 \in \mathcal{T}_{\mathbb{A}^k_X/X}$ such that $[\xi_1, \xi_2] = 0$ the identity*

$$\nabla^{\mathfrak{g}}_{\xi_1}\nabla_{\xi_2}\log u - \nabla^{\mathfrak{g}}_{\xi_2}\nabla_{\xi_1}\log u = [\nabla_{\xi_1}\log u, \nabla_{\xi_2}\log u] + \iota_{\xi_1}\iota_{\xi_2}F \tag{3.1}$$

holds.

3.3. Holonomy. Suppose that $\mathfrak{g}$ is a nilpotent $\mathcal{O}_X$-Lie algebra. For $\mathrm{A} \in \Gamma(\mathbb{A}^1_X; \Omega^1_{\mathbb{A}^1_X/X} \otimes \mathfrak{g})$, $\nabla := \mathsf{d}_{/X} + \mathrm{A}$ we denote by

$$(x, y) \mapsto \oint_x^y \nabla \in \exp(\mathfrak{g})(\mathbb{A}^1_X \times_X \mathbb{A}^1_X)$$

the solution of the initial value problem

$$\begin{cases} \mathrm{pr}_2^*\nabla_{\frac{\partial}{\partial y}}\log(u) = 0 \\ u|_\Delta = 1, \end{cases}$$

where Δ is the subvariety defined by the equation $x = y$.

Suppose that $\mathcal{E}$ is an $\mathcal{O}_X$-module and $\rho\colon \mathfrak{g} \to \underline{\mathrm{End}}_{\mathcal{O}_X}(\mathcal{E})$ is a representation of $\mathfrak{g}$ which is nilpotent in the sense that there exists N such that $\mathrm{Im}(\rho)^N = 0$. The representation ρ induces the map of groups $\exp(\rho)\colon \exp(\mathfrak{g}) \to \underline{\mathrm{Aut}}(\mathcal{E})$. Since the two compositions $\mathbb{A}^1_X \times_X \mathbb{A}^1_X \xrightarrow{\mathrm{pr}_i} \mathbb{A}^1_X \xrightarrow{\mathrm{pr}_X} X$, where pr_i are the canonical projections, are equal it follows that $\underline{\mathrm{Hom}}_{\mathcal{O}_{\mathbb{A}^1_X \times_X \mathbb{A}^1_X}}(\mathrm{pr}_1^*\mathcal{E}_{\mathbb{A}^1_X}, \mathrm{pr}_2^*\mathcal{E}_{\mathbb{A}^1_X}) = \underline{\mathrm{End}}_{\mathcal{O}_{\mathbb{A}^1_X \times_X \mathbb{A}^1_X}}(\mathcal{E}_{\mathbb{A}^1_X \times_X \mathbb{A}^1_X})$ and there is a canonical map

$$\exp(\mathfrak{g})(\mathbb{A}^1_X \times_X \mathbb{A}^1_X) \to \mathrm{Isom}(\mathrm{pr}_1^*\mathcal{E}_{\mathbb{A}^1_X}, \mathrm{pr}_2^*\mathcal{E}_{\mathbb{A}^1_X}). \tag{3.2}$$

We denote by $\overset{y}{\underset{x}{\mathcal{P}}}\nabla^{\mathcal{E}}$ the image of $\overset{y}{\underset{x}{\mathcal{P}}}\nabla$ under (3.2).

LEMMA 3.2. *The following useful identities hold.*

$$\left(\overset{y}{\underset{x}{\mathcal{P}}}\nabla^{\mathcal{E}}\right)\frac{\partial}{\partial y}\left(\overset{y}{\underset{x}{\mathcal{P}}}\nabla^{\mathcal{E}}\right)^{-1} = \nabla^{\mathcal{E}}_{\frac{\partial}{\partial y}}$$

$$\left(\overset{y}{\underset{x}{\mathcal{P}}}\nabla\right)\cdot\left(\overset{z}{\underset{y}{\mathcal{P}}}\nabla\right) = \overset{z}{\underset{x}{\mathcal{P}}}\nabla$$

3.4. Holonomy and curvature. Suppose that $\mathfrak{g}$ is a nilpotent $\mathcal{O}_X$-Lie algebra, $\mathrm{A} \in \Gamma(\mathbb{A}^2_X; \Omega^1_{\mathbb{A}^2_X/X} \otimes \mathfrak{g})$, $\nabla := \mathsf{d}_{/X} + \mathrm{A}$, $F = F(\nabla) := \mathsf{d}_{/X}\mathrm{A} + \frac{1}{2}[\mathrm{A}, \mathrm{A}] \in \Gamma(\mathbb{A}^2_X; \Omega^2_{\mathbb{A}^2_X/X} \otimes \mathfrak{g})$.

The second projection $\mathbb{A}^2_X \to \mathbb{A}^1_X$ identifies $\mathbb{A}^2_X$ with $\mathbb{A}^1_{\mathbb{A}^1_X}$. Let

$$\overline{\mathrm{A}} \in \Gamma(\mathbb{A}^2_X; \Omega^1_{\mathbb{A}^2_X/\mathbb{A}^1_X} \otimes \mathfrak{g})$$

denote the image of A, $\overline{\nabla} := d_{/\mathbb{A}^1_X} + \overline{\mathrm{A}}$. Hence, the holonomy

$$((x_1, y_2), (y_1, y_2)) \mapsto \overset{y_1}{\underset{x_1}{\mathcal{P}}}\overline{\nabla} \in \exp(\mathfrak{g})(\mathbb{A}^2_X \times_{\mathbb{A}^1_X} \mathbb{A}^2_X)$$

is defined and will be denoted, by abuse of notation, by $\overset{y_1}{\underset{x_1}{\mathcal{P}}}\nabla$.

LEMMA 3.3.

$$\nabla_{\frac{\partial}{\partial y_2}} \log\left(\overset{y_1}{\underset{x_1}{\mathcal{P}}}\nabla\right) = \int_{t=x_1}^{t=y_1} \left(\overset{t}{\underset{y_1}{\mathcal{P}}}\nabla^{\mathfrak{g}}\right)^{-1}\left(\iota_{\frac{\partial}{\partial y_2}} F\right) \tag{3.3}$$

PROOF. We will show that both sides of (3.3) are solutions of the initial value problem

$$\begin{cases} \nabla^{\mathfrak{g}}_{\frac{\partial}{\partial y_1}}\Phi = \iota_{\frac{\partial}{\partial y_1}}\iota_{\frac{\partial}{\partial y_2}} F \\ \Phi|_{x_1=y_1} = 0. \end{cases}$$

It is clear that both sides of (3.3) satisfy the initial condition. By (3.1),

$$\nabla^{\mathfrak{g}}_{\frac{\partial}{\partial y_1}}\nabla_{\frac{\partial}{\partial y_2}}\log\left(\mathop{\mathcal{P}}\limits_{x_1}^{y_1}\nabla\right)-\nabla^{\mathfrak{g}}_{\frac{\partial}{\partial y_2}}\nabla_{\frac{\partial}{\partial y_1}}\log\left(\mathop{\mathcal{P}}\limits_{x_1}^{y_1}\nabla\right)=$$

$$[\nabla_{\frac{\partial}{\partial y_1}}\log\left(\mathop{\mathcal{P}}\limits_{x_1}^{y_1}\nabla\right),\nabla_{\frac{\partial}{\partial y_2}}\log\left(\mathop{\mathcal{P}}\limits_{x_1}^{y_1}\nabla\right)]+\iota_{\frac{\partial}{\partial y_1}}\iota_{\frac{\partial}{\partial y_2}}F$$

By definition, $\nabla_{\frac{\partial}{\partial y_1}}\log\left(\mathop{\mathcal{P}}\limits_{x_1}^{y_1}\nabla\right)=0$. Hence,

$$\nabla^{\mathfrak{g}}_{\frac{\partial}{\partial y_1}}\nabla_{\frac{\partial}{\partial y_2}}\log\left(\mathop{\mathcal{P}}\limits_{x_1}^{y_1}\nabla\right)=\iota_{\frac{\partial}{\partial y_1}}\iota_{\frac{\partial}{\partial y_2}}F$$

which is to say, the left-hand side satisfies the differential equation. Differentiating the right-hand side we obtain

$$\nabla^{\mathfrak{g}}_{\frac{\partial}{\partial y_1}}\int_{x_1}^{y_1}(\mathop{\mathcal{P}}\limits_{y_1}^{y}\nabla^{\mathfrak{g}})^{-1}(\iota_{\frac{\partial}{\partial y_2}}F)$$

$$=\left(\mathop{\mathcal{P}}\limits_{x_1}^{y_1}\nabla^{\mathfrak{g}}\right)\frac{\partial}{\partial y_1}\left(\mathop{\mathcal{P}}\limits_{x_1}^{y_1}\nabla^{\mathfrak{g}}\right)^{-1}\int_{x_1}^{y_1}(\mathop{\mathcal{P}}\limits_{y_1}^{y}\nabla^{\mathfrak{g}})^{-1}(\iota_{\frac{\partial}{\partial y_2}}F)$$

$$=\left(\mathop{\mathcal{P}}\limits_{x_1}^{y_1}\nabla^{\mathfrak{g}}\right)\frac{\partial}{\partial y_1}\int_{x_1}^{y_1}(\mathop{\mathcal{P}}\limits_{x_1}^{y}\nabla^{\mathfrak{g}})^{-1}(\iota_{\frac{\partial}{\partial y_2}}F)=\left(\mathop{\mathcal{P}}\limits_{x_1}^{y_1}\nabla^{\mathfrak{g}}\right)\iota_{\frac{\partial}{\partial y_1}}(\mathop{\mathcal{P}}\limits_{x_1}^{y_1}\nabla^{\mathfrak{g}})^{-1}(\iota_{\frac{\partial}{\partial y_2}}F)$$

$$=\iota_{\frac{\partial}{\partial y_1}}\iota_{\frac{\partial}{\partial y_2}}F$$

Thus, the right-hand side satisfies the differential equation as well. □

For a nilpotent $\mathcal{O}_X$-Lie algebra $\mathfrak{g}$, $\mathrm{A}\in\Gamma(\mathbb{A}^k_X;\Omega^1_{\mathbb{A}^k_X/X}\otimes\mathfrak{g})$, $\nabla:=\mathsf{d}_{/X}+\mathrm{A}$, a pair of points $(x_1,\dots,x_k),(y_1,\dots,y_k)\in\mathbb{A}^k_X(X)$ and an index $1\leqslant j\leqslant k$ such that $x_i=y_i$ for $i\neq j$ let

$$\mathop{\mathcal{P}}\limits_{x_j}^{y_j}\nabla:=\mathop{\mathcal{P}}\limits_{x_j}^{y_j}\overline{\nabla}^j,$$

where $Y=\mathbb{A}^{k-1}_X$ and $\mathbb{A}^k_X\to Y$ is the projection along the jth coordinate, $\overline{\nabla}^j=\mathsf{d}_{/Y}+\overline{\mathrm{A}}^j$, $\overline{\mathrm{A}}^j\in\Gamma(\mathbb{A}^1_Y;\Omega^1_{\mathbb{A}^k_Y/Y}\otimes\mathfrak{g})$ is the image of A.

3.5. Special broken lines.

DEFINITION 3.4. A *special broken line (of length n)* γ in $\mathbb{A}^k_X$ is a sequence of points $(p^{(i)})_{i=0}^n$, $p^{(i)}=(p_1^{(i)},\dots,p_k^{(i)})$ such that for any $0\leqslant i<n-1$ there exists a $1\leqslant j_i\leqslant k$ such that $p_l^{(i)}=p_l^{(i+1)}$ for all $l\neq j_i$.

A special broken line is *closed* if $p^{(0)}=p^{(n)}$.

For a nilpotent $\mathcal{O}_{\mathbb{A}^k_X}$-Lie algebra $\mathfrak{g}$, $\mathrm{A} \in \Gamma(\mathbb{A}^k_X; \Omega^1_{\mathbb{A}^k_X/X} \otimes \mathfrak{g})$, $\nabla := \mathtt{d}_{/X} + \mathrm{A}$ and a special broken line $\gamma = (p^{(i)})_{i=0}^n$ in $\mathbb{A}^k_X$, we define the *holonomy of* ∇ *along* γ as the product

$$\prod_{\gamma} \nabla := \left(\prod_{p^{(n-1)}_{j_{n-1}}}^{p^{(n)}_{j_{n-1}}} \nabla \right) \cdot \left(\prod_{p^{(n-1)}_{j_{n-2}}}^{p^{(n-1)}_{j_{n-2}}} \nabla \right) \cdot \dots \cdot \left(\prod_{p^{(0)}_{j_0}}^{p^{(1)}_{j_0}} \nabla \right)$$

For a special broken line $\gamma = (p^{(i)})_{i=0}^n$ we denote by $-\gamma$ the special broken line $(q^{(i)})_{i=0}^n$ with $q^{(i)} = p^{(n-i)}$.

LEMMA 3.5.

$$\prod_{-\gamma} \nabla = (\prod_{\gamma} \nabla)^{-1}$$

For a closed special broken line $\gamma = (p^{(i)})_{i=0}^n$ we denote by $\tau\gamma$ the closed special broken line $(q^{(i)})_{i=0}^n$ with $q^{(i)} = p^{(i+1 \pmod{n+1})}$.

LEMMA 3.6.

$$\prod_{\tau\gamma} \nabla = \mathrm{Conj}(\prod_{p^{(0)}_{j_0}}^{p^{(1)}_{j_0}} \nabla)(\prod_{\gamma} \nabla)$$

where j_0 *is as in Definition* 3.4.

4. Two-dimensional holonomy

4.1. Connection-curvature pairs. Data of a *connection-curvature pair* on $\mathbb{A}^k_X$ consists, by definition, of

- a nilpotent crossed module[1] in $\mathcal{O}_X$-Lie algebras $\mathfrak{h} \xrightarrow{\delta} \mathfrak{g} \xrightarrow{\rho} \mathrm{Der}(\mathfrak{h})$
- $\mathrm{A} \in \Gamma(\mathbb{A}^k_X; \Omega^1_{\mathbb{A}^k_X/X} \otimes \mathfrak{g})$
- $\beta \in \Gamma(\mathbb{A}^k_X; \Omega^2_{\mathbb{A}^k_X/X} \otimes \mathfrak{h})$

Let $\nabla := \mathtt{d}_{/X} + \mathrm{A}$, $F = F(\nabla) := \mathtt{d}_{/X}\mathrm{A} + \dfrac{1}{2}[\mathrm{A}, \mathrm{A}]$.

These data are required to satisfy

- $\rho(F) = \mathtt{ad}(\beta)$,
- the Bianchi identity $\nabla\beta = 0$.

4.2. Two-dimensional holonomy. For a connection-curvature pair (∇, β) on $\mathbb{A}^2_X$ we denote by

$$\prod_{x_2}^{y_2} \prod_{x_1}^{y_1} (\nabla, \beta) \in \exp(\mathfrak{h})(\mathbb{A}^2_X \times_X \mathbb{A}^2_X)$$

[1]A crossed module is the same thing as a dgla concentrated in degrees 0 and -1. A *nilpotent* crossed module is one which corresponds to a nilpotent dgla.

the solution of the initial value problem

$$\begin{cases} \nabla_{\frac{\partial}{\partial y_2}} \log(u) = \int\limits_{x_1}^{y_1} (\prod\limits_{y_1}^{y} \nabla^{\mathfrak{h}})^{-1}(\iota_{\frac{\partial}{\partial y_2}}\beta) \\ u|_{\Delta_1} = 1 \end{cases}$$

where Δ_1 is given by the equation $x_1 = y_1$.

For a connection-curvature pair (∇, β) on $\mathbb{A}^k_X$ and

- points $(x_1, \ldots, x_k), (y_1, \ldots, y_k) \in \mathbb{A}^k_X(X)$;
- indices $1 \leqslant j_1, j_2 \leqslant k$ such that $x_i = y_i$ for $i \neq j_1$ and $i \neq j_2$

let

$$\prod_{x_{j_2}}^{y_{j_2}} \prod_{x_{j_1}}^{y_{j_1}} (\nabla, \beta) := \prod_{x_{j_2}}^{y_{j_2}} \prod_{x_{j_1}}^{y_{j_1}} (\overline{\nabla}^{(j_1,j_2)}, \overline{\beta}^{(j_1,j_2)})$$

where $Y = \mathbb{A}^{k-2}_X$ and $\mathbb{A}^k_X \to Y$ is the projection along the j_1^{th} and j_2^{th} coordinates, $\overline{\nabla}^{(j_1,j_2)} = \mathsf{d}_{/Y} + \overline{\mathrm{A}}^{(j_1,j_2)}$, $\overline{\mathrm{A}}^{(j_1,j_2)} \in \Gamma(\mathbb{A}^2_Y; \Omega^1_{\mathbb{A}^k_Y/Y} \otimes \mathfrak{g})$ is the image of A, $\overline{\beta}^{(j_1,j_2)} \in \Gamma(\mathbb{A}^2_Y; \Omega^1_{\mathbb{A}^k_Y/Y} \otimes \mathfrak{h})$ is the image of β.

4.3. Rectangles. A *rectangle* R in $\mathbb{A}^2_X$ is a sequence of points $(p^{(i)})_{i=0}^3$, $p^{(i)} = (p_1^{(i)}, p_2^{(i)}) \in \mathbb{A}^2_X(X)$, such that

$$p_1^{(i)} = p_1^{(i+1)} \Leftrightarrow p_2^{(i+1)} = p_2^{(i+2)} \qquad \forall i \pmod 4$$

Rectangles are in one-to-one correspondence with pairs of points in $\mathbb{A}^2_X(X)$. The pair $((x_1, x_2), (y_1, y_2))$ determines the rectangle $((x_1, x_2), (y_1, x_2)(y_1, y_2)(x_1, y_2))$.

For a rectangle $R = ((x_1, x_2), (y_1, x_2)(y_1, y_2)(x_1, y_2))$ the closed special broken line of length four

$$\partial R := ((x_1, x_2), (y_1, x_2)(y_1, y_2)(x_1, y_2), (x_1, x_2)),$$

will be called the *boundary of* R.

For a rectangle $R = ((x_1, x_2), (y_1, x_2)(y_1, y_2)(x_1, y_2))$ in $\mathbb{A}^2_X$ let

$$\prod_R (\nabla, \beta) := \prod_{x_2}^{y_2} \prod_{x_1}^{y_1} (\nabla, \beta).$$

4.4. Chains of rectangles. A *chain of rectangles* $S = (\gamma, x, y)$ in $\mathbb{A}^{k+1}_X$ is given by the following data:

- A special broken line $\gamma = (p^{(i)})_{i=0}^n$, $p^{(i)} = (p_1^{(i)}, \ldots, p_k^{(i)})$ in $\mathbb{A}^k_{\mathbb{A}^1_X}$.
- Points $x, y \in \mathbb{A}^1_X(X)$.

The chain of rectangles S consists of rectangles

$$R_i := \left((x, p_{j_{i-1}}^{(i-1)}), (y, p_{j_{i-1}}^{(i-1)}), (y, p_{j_{i-1}}^{(i)}), (x, p_{j_{i-1}}^{(i)})\right),$$

$1 \leqslant i \leqslant n$, in the notations of Definition 3.4. The *boundary* of S, denoted ∂S is the closed special broken line

$$\partial S := ((p^{(0)}, x), (p^{(0)}, y), (p^{(1)}, y), \ldots, (p^{(n)}, y), (p^{(n)}, x), \ldots, (p^{(0)}, x))$$

in $\mathbb{A}^{k+1}_X$.

For $1 \leqslant i \leqslant n$ let $\gamma_{\leqslant i} = (p^{(l)})_{l=0}^{i}$, $S_{\leqslant i} = (\gamma_{\leqslant i}, x, y)$. Note that $\gamma_{\leqslant n} = \gamma$ and $S_{\leqslant n} = S$. Define

$$\prod_{S_{\leqslant 1}} (\nabla, \beta) := \prod_{R_1} (\nabla, \beta)$$

and for $2 \leqslant i \leqslant n$ let

$$\prod_{S_{\leqslant i}} (\nabla, \beta) := \mathrm{Conj}\left(\prod_{\gamma_{\leqslant i-1}} \nabla \right) \left(\prod_{R_i} (\nabla, \beta) \right) \cdot \prod_{S_{\leqslant i-1}} (\nabla, \beta)$$

4.5. Green's theorem for chains of rectangles.

THEOREM 4.1 (Green's Theorem). *Suppose that S is a chain of rectangles in $\mathbb{A}_X^{k+1}$ and (∇, β) is a connection-curvature pair as in* 4.1. *Then,*

$$\prod_{\partial S} \nabla = \exp(\delta)\Big(\prod_{S} (\nabla, \beta)\Big).$$

4.6. Parallelepipeds. A parallelepiped is a pair of points in $\mathbb{A}_X^3$.

In what follows we use that first coordinate projection to identify $\mathbb{A}_X^3$ with $\mathbb{A}^2_{\mathbb{A}^1_X}$. A parallelepiped $Q = ((x_1, x_2, x_3), (y_1, y_2, y_3)) \in \mathbb{A}_X^3 \times_X \mathbb{A}_X^3(X)$ gives rise to the rectangle $R = ((x_2, x_3), (y_2, x_3)(y_2, y_3)(x_2, y_3))$ in $\mathbb{A}^2_{\mathbb{A}^1_X}$ and the chain of rectangles $S = (\partial R, x_1, y_1)$.

THEOREM 4.2 (Gauss-Ostrogradsky Theorem). *Suppose that Q is a parallelepiped in $\mathbb{A}_X^3$ and (∇, β) is a connection-curvature pair as in* 4.1.

$$\prod_{S} (\nabla, \beta) = \mathrm{Conj}\Big(\prod_{x_1}^{y_1} \nabla \Big|_{(x_2, x_3)} \Big) \left(\prod_{R} (\nabla, \beta) \Big|_{y_1} \right) \cdot \left(\prod_{R} (\nabla, \beta) \Big|_{x_1} \right)^{-1}$$

4.7. Parametrization of simplexes. Recall that the n-simplex Δ^n is defined as the hyperplane in $\mathbb{A}^{n+1}$ given by the equation $t_0 + \cdots + t_n = 1$.

We parameterize Δ^1, Δ^2 and Δ^3, respectively, by

Δ^1: $t_0 = 1 - x$, $t_1 = x$;
Δ^2: $t_0 = 1 - x_1$, $t_1 = x_1(1 - x_2)$, $t_2 = x_1 x_2$;
Δ^3: $t_0 = 1 - x_1$,$t_1 = x_1(1 - x_2)$, $t_2 = x_1 x_2 (1 - x_3)$, $t_3 = x_1 x_2 x_3$.

Let Φ^i denote the parameterization of Δ^i, $i = 1, 2, 3$.

$$\prod_{\Delta^1} \nabla := \prod_{0}^{1} \Phi^{1*} \nabla$$

$$\prod_{\Delta^2} (\nabla, \beta) := \prod_{0}^{1} \prod_{0}^{1} (\Phi^{2*} \nabla, \Phi^{2*} \beta)$$

With these notations, for a connection-curvature pair (∇, β) on $\mathbb{A}^3_X$ Gauss-Ostrogradsky Theorem 4.2 says

$$\mathrm{Conj}\Big(\prod_{\Delta^1} f^*\nabla\Big)\left(\prod_{\Delta^2} \partial_0^*(\nabla,\beta)\right) = \left(\prod_{\Delta^2} \partial_2^*(\nabla,\beta)\right)^{-1} \cdot \left(\prod_{\Delta^2} \partial_1^*(\nabla,\beta)\right) \cdot \left(\prod_{\Delta^2} \partial_3^*(\nabla,\beta)\right), \tag{4.1}$$

where $f = \partial_3 \circ \partial_2 = \partial_2 \circ \partial_2$ is the map $\Delta^1 \to \Delta^3$ induced by $0 \mapsto 0$, $1 \mapsto 1$.

5. From Σ to $\mathfrak{N}\,\mathrm{MC}^2$

Throughout this section we assume that $\mathfrak{g}$ is a nilpotent DGLA which satisfies $\mathfrak{g}^i = 0$ for $i \leqslant -2$. Recall that in [**2**] we showed that the simplicial nerve $\mathfrak{N}\,\mathrm{MC}^2(\mathfrak{g})$ of the 2-groupoid $\mathrm{MC}^2(\mathfrak{g})$ is weakly equivalent to another simplicial set $\Sigma(\mathfrak{g})$. In this section we apply the theory of multiplicative integration outlined above to construct the map

$$\mathbb{I}(\mathfrak{g})\colon \Sigma(\mathfrak{g}) \to \mathfrak{N}\,\mathrm{MC}^2(\mathfrak{g})$$

omitting technical details. We will rely heavily on results and notations from [**2**].

In what follows we denote by Ω_n, $n = 0, 1, 2, \ldots$ the commutative differential graded algebra over $\mathbb{Q}$ with generators $t_0, \ldots, t_n$ of degree zero and $dt_0, \ldots, dt_n$ of degree one subject to the relations $t_0 + \cdots + t_n = 1$ and $dt_0 + \cdots + dt_n = 0$. The differential $d\colon \Omega_n \to \Omega_n[1]$ is defined by $t_i \mapsto dt_i$ and $dt_i \mapsto 0$. The assignment $[n] \mapsto \Omega_n$ extends in a natural way to a simplicial commutative differential graded algebra.

Recall that the set of Maurer-Cartan elements of $\mathfrak{g}$, denoted $\mathrm{MC}(\mathfrak{g})$ is defined as

$$\mathrm{MC}(\mathfrak{g}) := \left\{\gamma \in \mathfrak{g}^1 \mid \delta\gamma + \frac{1}{2}[\gamma,\gamma] = 0\right\}.$$

5.1. The simplicial set $\Sigma(\mathfrak{g})$. For a nilpotent L_∞-algebra $\mathfrak{g}$ and a non-negative integer n let

$$\Sigma_n(\mathfrak{g}) = \mathrm{MC}(\mathfrak{g} \otimes \Omega_n).$$

Equipped with structure maps induced by those of $\Omega_\bullet$ the assignment $n \mapsto \Sigma_n(\mathfrak{g})$ defines a simplicial set denoted $\Sigma(\mathfrak{g})$.

The simplicial set $\Sigma(\mathfrak{g})$ was introduced by V. Hinich in [**6**] for DGLA and used by E. Getzler in [**4**] (where it is denoted $\mathrm{MC}_\bullet(\mathfrak{g})$) for general nilpotent L_∞-algebras.

5.2. The Deligne 2-groupoid. We denote by $\mathrm{MC}^2(\mathfrak{g})$ the Deligne 2-groupoid as defined by P. Deligne [**3**] and independently by E. Getzler, [**4**]. Below we review the construction of Deligne 2-groupoid of a nilpotent DGLA following [**4, 5**] and references therein.

Since $\mathfrak{g}$ is a nilpotent DGLA it follows that $\mathfrak{g}^0$ is a nilpotent Lie algebra. The unipotent group $\exp\mathfrak{g}^0$ acts on the space $\mathfrak{g}^1$ by affine transformations. The action of $\exp X$, $X \in \mathfrak{g}^0$, on $\gamma \in \mathfrak{g}^1$ is given by the formula

$$(\exp X)\cdot\gamma = \gamma - \sum_{i=0}^{\infty} \frac{(\operatorname{ad} X)^i}{(i+1)!}(\delta X + [\gamma, X]). \tag{5.1}$$

The action of gauge transformations (5.1) preserves the subset of Maurer-Cartan elements $\mathrm{MC}(\mathfrak{g}) \subset \mathfrak{g}^1$.

We denote by $\mathrm{MC}^1(\mathfrak{g})$ the Deligne groupoid defined as the groupoid associated with the action of the group $\exp \mathfrak{g}^0$ by gauge transformations on the set $\mathrm{MC}(\mathfrak{g})$.

We denote by $\mathrm{MC}^2(\mathfrak{g})$ the Deligne 2-groupoid whose objects and the 1-morphisms of $\mathrm{MC}^2(\mathfrak{g})$ are those of $\mathrm{MC}^1(\mathfrak{g})$. The 2-morphisms of $\mathrm{MC}^2(\mathfrak{h})$are defined as follows.

For $\gamma \in \mathrm{MC}(\mathfrak{g})$ let $[\cdot,\cdot]_\gamma$ denote the Lie bracket on $\mathfrak{g}^{-1}$ defined by

$$[a, b]_\gamma = [a, \delta b + [\gamma, b]]. \tag{5.2}$$

Equipped with this bracket, $\mathfrak{g}^{-1}$ becomes a nilpotent Lie algebra. We denote by $\exp_\gamma \mathfrak{g}^{-1}$ the corresponding unipotent group, and by

$$\exp_\gamma \colon \mathfrak{g}^{-1} \to \exp_\gamma \mathfrak{g}^{-1}$$

the corresponding exponential map. If γ_1, γ_2 are two Maurer-Cartan elements, then the group $\exp_{\gamma_2} \mathfrak{g}^{-1}$ acts on $\mathrm{Hom}_{\mathrm{MC}^1(\mathfrak{g})}(\gamma_1, \gamma_2)$. For $\exp_{\gamma_2} t \in \exp_{\gamma_2} \mathfrak{g}^{-1}$ and $\mathrm{Hom}_{\mathrm{MC}^1(\mathfrak{g})}(\gamma_1, \gamma_2)$ the action is given by

$$(\exp_{\gamma_2} t) \cdot (\exp X) = \exp(\delta t + [\gamma_2, t]) \exp X \in \exp \mathfrak{g}^0.$$

By definition, $\mathrm{Hom}_{\mathrm{MC}^2(\mathfrak{g})}(\gamma_1, \gamma_2)$ is the groupoid associated with the above action.

5.3. The simplicial nerve of the Deligne 2-groupoid. We will need the following explicit description of the simplicial nerve of $\mathrm{MC}^2(\mathfrak{g})$. See [**2**] for details.

LEMMA 5.1. *The simplicial nerve $\mathfrak{N}\,\mathrm{MC}^2(\mathfrak{g})$ admits the following explicit description:*

1. $\mathfrak{N}_0\,\mathrm{MC}^2(\mathfrak{g}) = \mathrm{MC}(\mathfrak{g})$
2. *For $n \geq 1$ there is a canonical bijection between $\mathfrak{N}_n\,\mathrm{MC}^2(\mathfrak{g})$ and the set of data of the form*

$$((\mu_i)_{0\leq i\leq n}, (g_{ij})_{0\leq i<j\leq n}, (c_{ijk})_{0\leq i<j<k\leq n}),$$

where
- *(μ_i) is an $(n+1)$-tuple of objects of Maurer-Cartan elements of $\mathfrak{g}$,*
- *(g_{ij}) is a collection of 1-morphisms (gauge transformations) $g_{ij} \colon \mu_j \to \mu_i$*
- *(c_{ijk}) is a collection of 2-morphisms $c_{ijk} : g_{ij}g_{jk} \to g_{ik}$ which satisfies*

$$c_{ijl}c_{jkl} = c_{ikl}c_{ijk} \tag{5.3}$$

(in the set of 2-morphisms $g_{ij}g_{jk}g_{kl} \to g_{il}$).

For a morphism $f \colon [m] \to [n]$ in Δ the induced structure map $f^* : \mathfrak{N}_n\,\mathrm{MC}^2(\mathfrak{g}) \to \mathfrak{N}_m\,\mathrm{MC}^2(\mathfrak{g})$ is given (under the above bijection) by $f^*((\mu_i), (g_{ij}), (c_{ijk})) = ((\nu_i), (h_{ij}), (d_{ijk}))$, where $\nu_i = \mu_{f(i)}$, $h_{ij} = g_{f(i),f(j)}$, $d_{ijk} = c_{f(i),f(j),f(k)}$.

5.4. From Σ to $\mathfrak{N}\,\mathrm{MC}^2$. Recall that, for $n = 0, 1, 2, \ldots$, $\Sigma_n(\mathfrak{g}) = \mathrm{MC}(\Omega_n \otimes \mathfrak{g}) \subset (\Omega_n \otimes \mathfrak{g})^1 = \Omega_n^0 \otimes \mathfrak{g}^1 \oplus \Omega_n^1 \otimes \mathfrak{g}^0 \oplus \Omega_n^2 \otimes \mathfrak{g}^{-1}$. Thus, every element $\mu \in \Sigma_n(\mathfrak{g})$ is a triple: $\mu = (\mu^{0,1}, \mu^{1,0}, \mu^{2,-1})$.

For $n = 0, 1, 2, \dots$ the map

$$\mathbb{I}_n(\mathfrak{g})\colon \Sigma_n(\mathfrak{g}) \to \mathfrak{N}_n \operatorname{MC}^2(\mathfrak{g}) \tag{5.4}$$

is defined as follows:

$n = 0$ The map $\mathbb{I}_0(\mathfrak{g})\colon \Sigma_0(\mathfrak{g}) = \operatorname{MC}(\mathfrak{g}) \to \operatorname{MC}(\mathfrak{g}) = \mathfrak{N}_0 \operatorname{MC}^2(\mathfrak{g})$ is defined to be the identity.

$n = 1$ For $\mu = (\mu^{0,1}, \mu^{1,0}) \in \Sigma_1(\mathfrak{g})$,

$$\mathbb{I}_1(\mathfrak{g})(\mu) = ((\partial_i^* \mu^{0,1})_{i=0,}, \int_{\Delta^1} (d + \mu^{1,0})).$$

$n \geq 2$ For $\mu = (\mu^{0,1}, \mu^{1,0}, \mu^{2,-1}) \in \Sigma_n(\mathfrak{g})$,

$$\mathbb{I}_n(\mathfrak{g})(\mu) = ((\mu_i^{0,1})_{0 \leqslant i \leqslant n}, (\int_{\Delta^1} (d + \mu_{ij}^{1,0})_{0 \leqslant i < j \leqslant n}, (\int_{\Delta^2} (d + \mu_{ijk}^{1,0}, \mu_{ijk}^{2,-1}))_{0 \leqslant i < j < k \leqslant n}),$$

where

- $\mu_i^{0,1} = f^* \mu^{0,1}$, where $f\colon \Delta^0 \to \Delta^n$ is the map induced by $[0] \to [n]\colon 0 \mapsto i$;
- $\mu_{ij}^{1,0} = f^* \mu^{1,0}$, where $f\colon \Delta^1 \to \Delta^n$ is the map induced by $[1] \to [n]\colon 0 \mapsto i, 1 \mapsto j$;
- $\mu_{ijk}^{2,-1} = f^* \mu^{2,-1}$, where $f\colon \Delta^2 \to \Delta^n$ is the map induced by $[1] \to [n]\colon 0 \mapsto i, 1 \mapsto j, 2 \mapsto k$.

THEOREM 5.2. *Suppose that $\mathfrak{g}$ is a nilpotent DGLA which satisfies $\mathfrak{g}^i = 0$ for $i \leqslant -2$. The collection of maps $\mathbb{I}_n(\mathfrak{g})$, $n = 0, 1, 2, \dots$, defines a morphism of simplicial sets*

$$\mathbb{I}(\mathfrak{g})\colon \Sigma(\mathfrak{g}) \to \mathfrak{N} \operatorname{MC}^2(\mathfrak{g}) \tag{5.5}$$

natural in $\mathfrak{g}$ which satisfies

(1) *$\mathbb{I}_0(\mathfrak{g})\colon \Sigma_0(\mathfrak{g}) \to \mathfrak{N}_0 \operatorname{MC}^2(\mathfrak{g})$ is the identity map after the identification $\Sigma_0(\mathfrak{g}) = \operatorname{MC}(\mathfrak{g}) = \mathfrak{N}_0 \operatorname{MC}^2(\mathfrak{g})$.*

(2) *The restriction of $\mathbb{I}$ to the subcategory of abelian algebras (a.k.a. complexes) coincides with the integration map*

$$\int \colon \Sigma(\mathfrak{a}) \to K(\mathfrak{a}[1]) = \mathfrak{N} \operatorname{MC}^2(\mathfrak{a}), \tag{5.6}$$

$\mathfrak{a}$ abelian. Note that the integration map is a morphism of simplicial groups.

(3) *If $\mathfrak{a} \hookrightarrow \mathfrak{g}$ is a central subalgebra the diagram*

$$\begin{array}{ccc} \Sigma(\mathfrak{g}) & \xrightarrow{\mathbb{I}} & \mathfrak{N} \operatorname{MC}^2(\mathfrak{g}) \\ \downarrow & & \downarrow \\ \Sigma(\mathfrak{g}/\mathfrak{a})_0 & \xrightarrow{\mathbb{I}} & \mathfrak{N} \operatorname{MC}^2(\mathfrak{g}/\mathfrak{a})_0 \end{array}$$

is a morphism of principal fibrations relative to the morphism of groups (5.6), where $\Sigma(\mathfrak{g}/\mathfrak{a})_0$ (respectively, $\mathfrak{N} \operatorname{MC}^2(\mathfrak{g}/\mathfrak{a})_0$) denotes the image of the map $\Sigma(\mathfrak{g}) \to \Sigma(\mathfrak{g}/\mathfrak{a})$ (respectively, $\mathfrak{N} \operatorname{MC}^2(\mathfrak{g}) \to \mathfrak{N} \operatorname{MC}^2(\mathfrak{g}/\mathfrak{a})$).

Theorem 5.2 and induction on the nilpotency length imply the following statement.

THEOREM 5.3. *The map* $\mathbb{I}(\mathfrak{g})$ *is an equivalence.*

References

[1] I. Ja. Aref′eva, *Nonabelian Stokes formula* (Russian, with English summary), Teoret. Mat. Fiz. **43** (1980), no. 1, 111–116. MR570948

[2] Paul Bressler, Alexander Gorokhovsky, Ryszard Nest, and Boris Tsygan, *Deligne groupoid revisited*, Theory Appl. Categ. **30** (2015), Paper No. 29, 1001–1017. MR3370863

[3] Pierre Deligne, Letter to L. Breen, 1994.

[4] Ezra Getzler, *A Darboux theorem for Hamiltonian operators in the formal calculus of variations*, Duke Math. J. **111** (2002), no. 3, 535–560, DOI 10.1215/S0012-7094-02-11136-3. MR1885831

[5] Ezra Getzler, *Lie theory for nilpotent L_∞-algebras*, Ann. of Math. (2) **170** (2009), no. 1, 271–301, DOI 10.4007/annals.2009.170.271. MR2521116

[6] Vladimir Hinich, *Descent of Deligne groupoids*, Internat. Math. Res. Notices **5** (1997), 223–239, DOI 10.1155/S1073792897000160. MR1439623

[7] Mikhail Kapranov, *Membranes and higher groupoids*, arXiv:1502.06166 (2015).

[8] Maxim Kontsevich, *Deformation quantization of algebraic varieties*, Lett. Math. Phys. **56** (2001), no. 3, 271–294, DOI 10.1023/A:1017957408559. EuroConférence Moshé Flato 2000, Part III (Dijon). MR1855264

[9] Amnon Yekutieli, *Nonabelian multiplicative integration on surfaces*, World Scientific Publishing Co. Pte. Ltd., Hackensack, NJ, 2016. MR3381955

DEPARTAMENTO DE MATEMÁTICAS, UNIVERSIDAD DE LOS ANDES, BOGOTÁ, COLOMBIA
Email address: paul.bressler@gmail.com

DEPARTMENT OF MATHEMATICS, UCB 395, UNIVERSITY OF COLORADO, BOULDER, COLORADO 80309-0395
Email address: Alexander.Gorokhovsky@colorado.edu

DEPARTMENT OF MATHEMATICS, COPENHAGEN UNIVERSITY, UNIVERSITETSPARKEN 5, 2100 COPENHAGEN, DENMARK
Email address: rnest@math.ku.dk

DEPARTMENT OF MATHEMATICS, NORTHWESTERN UNIVERSITY, EVANSTON, ILLINOIS 60208-2730
Email address: b-tsygan@northwestern.edu

Contemporary Mathematics
Volume **749**, 2020
https://doi.org/10.1090/conm/749/15068

Equivariant coarse homotopy theory and coarse algebraic K-homology

Ulrich Bunke, Alexander Engel, Daniel Kasprowski, and Christoph Winges

ABSTRACT. We study equivariant coarse homology theories through an axiomatic framework. To this end we introduce the category of equivariant bornological coarse spaces and construct the universal equivariant coarse homology theory with values in the category of equivariant coarse motivic spectra.

As examples of equivariant coarse homology theories we discuss equivariant coarse ordinary homology and equivariant coarse algebraic K-homology.

Moreover, we discuss the cone functor, its relation with equivariant homology theories in equivariant topology, and assembly and forget-control maps. This is a preparation for applications in subsequent papers aiming at split-injectivity results for the Farrell–Jones assembly map.

CONTENTS

2010 *Mathematics Subject Classification.* Primary 18F25.

The first and second authors were supported by the SFB 1085 "Higher Invariants" funded by the Deutsche Forschungsgemeinschaft DFG.

The second author was furthermore supported by the Research Fellowship EN 1163/1-1 "Mapping Analysis to Homology", also funded by the DFG.

The third and fourth authors acknowledge support by the Max Planck Society.

The fourth author was also supported by Wolfgang Lück's ERC Advanced Grant "KL2MG-interactions" (no. 662400).

1. Introduction

In this paper we study equivariant coarse homology theories. We start with the equivariant generalization of the coarse homotopy theory developed by Bunke–Engel [**8**]. To this end we introduce the category of equivariant bornological coarse spaces and construct the universal equivariant coarse homology theory with values in the category of equivariant coarse motivic spectra.

As examples of equivariant coarse homology theories we discuss equivariant ordinary coarse homology and equivariant coarse algebraic K-homology of an additive category.

An important application of equivariant coarse homotopy theory is in the study of assembly maps which appear in isomorphism conjectures of Farrell–Jones or Baum–Connes type. The main tools for the transition between equivariant homology theories and equivariant coarse homology theories are the cone functor and the process of coarsification. In this paper we give a detailed account of the cone functor and the construction of equivariant homology theories from equivariant coarse homology theories. Then we introduce the coarsification functor and the forget-control map, and discuss its relation with the assembly maps.

The third part of the present paper provides the technical background for subsequent papers:

(1) In [**11**] we show that a certain large scale geometric condition called finite decomposition complexity implies that a motivic version of the forget-control map is an equivalence.

(2) In [**10**] we study the descent principle. It states that under certain conditions the fact that the forget-control map becomes an equivalence after restriction of the action to all finite subgroups implies that it is split injective for the original (in general infinite) group. This idea goes originally back to Carlsson and Pedersen [**15**] and was used to show injectivity of the K-theoretic assembly map in several cases [**5, 15, 23, 34, 35**].

In [**10**] we use this idea in combination with the results of [**11**] and the technical results of the present paper to deduce split injectivity results in a more general framework. These yield (and slightly generalize) previous injectivity results for the original Farrell–Jones assembly map when applied to coarse K-homology. In [**12**] we apply the results from [**10**] to deduce injectivity results about the assembly map for Waldhausen's A-theory.

(3) In [**9**] we study more formal aspects of the process of coarsification of homology theories and provide a general account for coarse assembly maps. Further, we compare the forget-control map for coarse K-homology with the coarse analytic assembly map featuring in the coarse Baum–Connes conjecture.

A bornological coarse space is a set equipped with a coarse and a bornological structure such that these structures are compatible with each other. The category of bornological coarse spaces **BC** was introduced in [**8**], where it was denoted **BornCoarse**, as a general framework for coarse geometry and coarse topology. Interesting invariants of bornological coarse spaces up to coarse equivalence are coarse homology theories. In [**8**] the examples of coarse ordinary homology and various coarse versions of topological K-homology were discussed. In order to study general properties of coarse homology theories the category of motivic coarse spectra $\mathbf{Sp}\mathcal{X}$ was constructed as the target of the universal coarse homology theory

$$\mathrm{Yo}^s\colon \mathbf{BC} \to \mathbf{Sp}\mathcal{X} \ .$$

One of the motivations to consider equivariant coarse algebraic topology is that it appears as a building block of proofs that certain assembly maps are equivalences (Farrell–Jones conjecture [**2**], [**1**] and Baum–Connes conjecture [**42**]).

In Section 2.1 we define for every group Γ the category $\Gamma\mathbf{BC}$ of Γ-bornological coarse spaces. We provide various constructions of Γ-bornological coarse spaces from Γ-sets, metric spaces with isometric Γ-action, or Γ-simplicial complexes. We further show that the category of Γ-bornological coarse spaces admits coproducts and fiber products. It also has an interesting symmetric monoidal structure $\otimes$.

In Section 3 we introduce the notion of an equivariant coarse homology theory, and the terminology necessary to state its defining properties:

(1) coarse invariance,
(2) coarse excision,
(3) vanishing on flasques, and
(4) u-continuity.

Strongness (introduced in Section 4.4) is an additional property which an equivariant coarse homology might have. It is important in order to interpret the forget-control map as a transformation between equivariant coarse homology theories.

In the present paper the most important additional property is continuity. We introduce this notion in Section 5. Continuity is crucial if one wants to relate the forget-control map with the assembly map for the family of finite subgroups.

Following the line of thought of [**8**], in Section 4 we construct the universal equivariant coarse homology theory

$$\mathrm{Yo}^s\colon \Gamma\mathbf{BC} \to \Gamma\mathbf{Sp}\mathcal{X}$$

with values in the category of equivariant coarse motivic spectra. Similarly, in Section 5 we construct a universal continuous equivariant coarse homology theory

$$\mathrm{Yo}^s_c\colon \Gamma\mathbf{BC} \to \Gamma\mathbf{Sp}\mathcal{X}_c \ .$$

As examples of equivariant coarse homology theories, in Part 2 we introduce equivariant coarse ordinary homology $H\mathcal{X}^\Gamma$ and equivariant coarse algebraic K-homology $K\mathbf{A}\mathcal{X}^\Gamma$ of an additive category $\mathbf{A}$. Our main results are the verification that the definitions indeed satisfy the four defining properties of an equivariant coarse homology theory. We also show that these examples have the additional properties of being strong, strongly additive and continuous. We calculate the evaluations of these equivariant homology theories on simple Γ-bornological coarse spaces. These calculations are important if one wants to understand which equivariant homology theories they induce after pull-back with the cone functor.

In most of the present paper we consider the theory for a fixed group Γ. But in Section 6 we consider a homomorphism of groups $H \to \Gamma$ and provide various transitions from H-bornological coarse spaces to Γ-bornological coarse spaces and back. The most important examples are restriction and induction. Our examples of equivariant coarse homology theories are defined for all groups, so in particular for H and Γ. The group-change construction on the level of bornological coarse spaces are accomplished by natural transformations relating the evaluations of the H- and Γ-equivariant versions of the equivariant homology theories.

In Section 9 we introduce the category of Γ-uniform bornological coarse spaces $\Gamma\mathbf{UBC}$ and the cone functor

$$\mathcal{O}\colon \Gamma\mathbf{UBC} \to \Gamma\mathbf{BC}\ .$$

The construction of the cone is motivated by Bartels–Farrell–Jones–Reich [**1**] and Mitchener [**31**,**32**], and its main ingredient in the construction is the hybrid coarse structure first introduced by Wright [**41**] and studied in detail in [**8**]. Our main technical results are homotopy invariance and excisiveness of the cone. While the cone $\mathcal{O}(X)$ depends on the coarse structure on X, its germ at ∞, denoted by $\mathcal{O}^\infty(X)$, is essentially independent of the coarse structure. So $\mathcal{O}^\infty$ is very close to an equivariant homology theory. But it is still defined on $\Gamma\mathbf{UBC}$ and does not satisfy a wedge axiom.

In Section 10 we first review some general features of equivariant homotopy theory and then derive an equivariant homology theory

$$\mathcal{O}^\infty_{\mathrm{hlg}}\colon \Gamma\mathbf{Top} \to \Gamma\mathbf{Sp}\mathcal{X}$$

from the functor $\mathcal{O}^\infty$. We then introduce the classifying space $E_{\mathcal{F}}\Gamma$ for a family $\mathcal{F}$ of subgroups of Γ and define the motivic assembly map as the map

$$\alpha_{E_{\mathcal{F}}\Gamma}\colon \mathcal{O}^\infty_{\mathrm{hlg}}(E_{\mathcal{F}}\Gamma) \to \mathcal{O}^\infty_{\mathrm{hlg}}(*)$$

induced by the morphism $E_{\mathcal{F}}\Gamma \to *$. We discuss some conditions on a Γ-bornological coarse space Q implying that the twisted version $\alpha_{E_{\mathcal{F}}} \otimes \mathrm{Yo}^s(Q)$ of the assembly map becomes an equivalence.

In Section 11 we introduce the universal coarsification functor

$$F^\infty\colon \Gamma\mathbf{BC} \to \Gamma\mathbf{Sp}\mathcal{X}$$

and the forget-control map

$$\beta\colon F^\infty \to \Sigma F^0\ ,$$

a natural transformation of functors from $\Gamma\mathbf{BC}$ to $\Gamma\mathbf{Sp}\mathcal{X}$. The coarse geometry approach to the isomorphism conjectures provides conditions on a Γ-bornological coarse space X implying that the forget-control map $\beta_X\colon F^\infty(X) \to \Sigma F^0(X)$ is an equivalence, or becomes an equivalence after application of a suitable equivariant coarse homology theory. Since one is also interested in the assembly maps of equivariant homotopy theory like $\alpha_{E_{\mathcal{F}}\Gamma}$ we provide a comparison between this assembly map and the forget-control map.

Part 1. General constructions

2. Equivariant bornological coarse spaces

2.1. Basic notions. In this section we introduce the equivariant version of the category of bornological coarse spaces introduced in [**8**]. We assume familiarity with [**8**, Section 2].

Let Γ be a group. If Γ acts on a bornological coarse space X by automorphisms, then it acts on the set of coarse entourages $\mathcal{C}$ of X. We let $\mathcal{C}^\Gamma$ denote the partially ordered subset of $\mathcal{C}$ of entourages of X which are fixed set-wise.

DEFINITION 2.1. A Γ-*bornological coarse space* is a bornological coarse space X together with an action of Γ by automorphisms such that $\mathcal{C}^\Gamma$ is cofinal in $\mathcal{C}$.

A morphism between Γ-bornological coarse spaces is a morphism of bornological coarse spaces which is in addition Γ-equivariant.

We let Γ**BC** denote the category of Γ-bornological coarse spaces and morphisms. By considering a bornological coarse space as a Γ-bornological coarse space with the trivial action we get a fully faithful functor

$$C\colon \mathbf{BC} \to \Gamma\mathbf{BC} . \tag{2.1}$$

EXAMPLE 2.2. Let X be a set with an action of Γ and $A \subseteq \mathcal{P}(X \times X)^\Gamma$ be a family of Γ-invariant subsets. Then we can form the coarse structure $\mathcal{C} := \mathcal{C}\langle A\rangle$ generated by A. In this case $\mathcal{C}^\Gamma$ is cofinal in $\mathcal{C}$. Hence a Γ-coarse structure can be generated by a family of Γ-invariant entourages.

REMARK 2.3. Let **Coarse** denote the category of coarse spaces (i.e., sets with coarse structures) and controlled morphisms. We can consider Γ equipped with the minimal coarse structure $\mathcal{C}\langle \mathrm{diag}_\Gamma\rangle$ as a group object in **Coarse**.

Let X be a set with an action of a group Γ and $\mathcal{C}$ be a coarse structure on X. The following conditions are equivalent:

(1) $\mathcal{C}^\Gamma$ is cofinal in $\mathcal{C}$.
(2) For every entourage U in $\mathcal{C}$ the set $\bigcup_{\gamma\in\Gamma}(\gamma\times\gamma)(U)$ also belongs to $\mathcal{C}$.
(3) The action is a morphism $\Gamma \times X \to X$ in **Coarse**.

The proof is straightforward.

We denote the category of Γ-coarse spaces consisting of coarse spaces with a Γ-action satisfying the above conditions and equivariant and controlled maps by Γ**Coarse**.

EXAMPLE 2.4. We consider Γ as a Γ-set with the left action. We furthermore let $\mathcal{B}_{min}$ be the minimal bornology on Γ consisting of the finite subsets. Finally, we let the coarse structure $\mathcal{C}_{can}$ on Γ be generated by the Γ-invariant sets $\Gamma(B \times B)$ for all B in $\mathcal{B}_{min}$.

Then $(\Gamma, \mathcal{C}_{can}, \mathcal{B}_{min})$ is a Γ-bornological coarse space called the canonical Γ-bornological coarse space associated to Γ. We will denote it by $\Gamma_{can,min}$.

EXAMPLE 2.5. Let X be a set with an action of Γ. It gives rise to the Γ-bornological coarse space $X_{min,min}$ with the minimal structures $\mathcal{B}_{min}$ consisting of the finite subsets of X and $\mathcal{C}_{min} := \mathcal{C}\langle \mathrm{diag}_X\rangle$. We will use the notation $X_{min,min}$ for this Γ-bornological coarse space.

For example, the identity of the underlying set of Γ is a morphism $\Gamma_{min,min} \to \Gamma_{can,min}$ of Γ-bornological coarse spaces.

EXAMPLE 2.6. Let X again be a set with an action of Γ. Then we can equip X with the maximal bornological and coarse structures. In this way we get a Γ-bornological coarse space $X_{max,max}$.

Our notation convention is such that the first subcript indicates the coarse structure, while the second subscript reflects the bornological structure.

For example, we also have a Γ-bornological coarse space $X_{min,max}$ and morphisms of Γ-bornological coarse spaces $X_{min,max} \to X_{max,max}$ and $X_{min,max} \to X_{min,min}$ given by the identity of the underlying set of X.

Note that in general $X_{max,min}$ does not make sense since the minimal bornology is not compatible with the maximal coarse structure.

EXAMPLE 2.7. Let (X, d) be a metric space with an isometric Γ-action. For r in $(0, \infty)$ we consider the invariant entourages

$$U_r := \{(x,y) \in X \times X | d(x,y) \le r\} .$$

The coarse structure associated to the metric is defined by these entourages, i.e., given by

$$\mathcal{C}_d := \mathcal{C}\langle\{U_r | r \in (0,\infty)\}\rangle .$$

Furthermore, the bornology associated to the metric is generated by the metrically bounded subsets, i.e., given by

$$\mathcal{B}_d := \mathcal{B}\langle\{B(x,r) | x \in X , r \in (0,\infty)\}\rangle ,$$

where $B(x, r)$ denotes the metric ball of radius r centered at x.

The associated bornological coarse space $X_d := (X, \mathcal{C}_d, \mathcal{B}_d)$ is a Γ-bornological coarse space.

The identity of the underlying set of X is a morphism $X_{min,max} \to X_d$ of Γ-bornological coarse spaces.

REMARK 2.8. Let Γ be a countable group equipped with a proper left invariant metric d. Then the Γ-bornological coarse spaces Γ_d and $\Gamma_{can,min}$ are equal.

EXAMPLE 2.9. Let X be a Γ-bornological coarse space with the coarse structure $\mathcal{C}$ and the bornology $\mathcal{B}$. For every invariant entourage U in $\mathcal{C}$ we consider the coarse structure $\mathcal{C}_U := \mathcal{C}\langle\{U\}\rangle$. The coarse structure $\mathcal{C}_U$ is compatible with $\mathcal{B}$. We let

$$X_U := (X, \mathcal{C}_U, \mathcal{B})$$

denote the resulting Γ-bornological coarse space. The identity of the underlying set is a morphism of Γ-bornological coarse spaces $X_U \to X$. If U' is a second invariant entourage such that $U \subseteq U'$, then we also have a morphism $X_U \to X_{U'}$. This construction is important for the formulation of the u-continuity condition in Definition 3.10.

EXAMPLE 2.10. Let X be a Γ-bornological coarse space with coarse structure $\mathcal{C}$ and bornology $\mathcal{B}$, and let Z be a Γ-invariant subset of X. Then we define the induced coarse structure and bornology on Z as follows:

(1) $\mathcal{C}_Z := \{(Z \times Z) \cap U | U \in \mathcal{C}\}$
(2) $\mathcal{B}_Z := \{Z \cap B | B \in \mathcal{B}\}$.

Then $Z_X := (Z, \mathcal{C}_Z, \mathcal{B}_Z)$ is a Γ-bornological coarse space. The inclusion $Z_X \to X$ is a morphism of Γ-bornological coarse spaces.

EXAMPLE 2.11. If Γ acts on the underlying set of a bornological coarse space $(X, \mathcal{C}, \mathcal{B})$, then we can define a Γ-bornological coarse space $\Gamma X := (X, \Gamma\mathcal{C}, \Gamma\mathcal{B})$, where

$$\Gamma\mathcal{C} := \mathcal{C}\left\langle\left\{\bigcup_{\gamma\in\Gamma}(\gamma \times \gamma)(U) | U \in \mathcal{C}\right\}\right\rangle$$

and

$$\Gamma\mathcal{B} := \mathcal{B}\langle\{U[\gamma B] | U \in \Gamma\mathcal{C}, \gamma \in \Gamma \text{ and } B \in \mathcal{B}\}\rangle . \tag{2.2}$$

In general we must enlarge the bornology $\mathcal{B}$ to $\Gamma\mathcal{B}$ as described above in order to keep it compatible with the new coarse structure.

If X was a Γ-bornological coarse spaces, then $\Gamma X = X$.

DEFINITION 2.12. Let Γ denote a group and let X be a set with an action of Γ. If $\mathcal{B}$ is a bornology on X, then we let $\mathcal{B}_\Gamma$ denote the bornology on X which is generated by the sets ΓB for all B in $\mathcal{B}$.

Let $(X,\mathcal{C},\mathcal{B})$ be a Γ-bornological coarse space. Then the new bornology $\mathcal{B}_\Gamma$ is compatible with the original coarse structure. The Γ-*completion* of $(X,\mathcal{C},\mathcal{B})$ is defined to be the Γ-bornological coarse space $(X,\mathcal{C},\mathcal{B}_\Gamma)$.

Let $(X,\mathcal{C},\mathcal{B})$ be a Γ-bornological coarse space and Y be a subset of X. The subset Y is called Γ-*bounded* if it belongs to $\mathcal{B}_\Gamma$.

EXAMPLE 2.13. Let $(X,\mathcal{B})$ be a bornological space with an action of Γ by proper maps. We say that Γ acts *properly* if for every B in $\mathcal{B}$ the set $\{\gamma\in\Gamma|\gamma B\cap B\neq\emptyset\}$ is finite.

We define the coarse structure $\mathcal{C}_\mathcal{B}$ on X to be generated by the Γ-invariant entourages $U_B:=\Gamma(B\times B)$ for all B in $\mathcal{B}$. If Γ acts properly, then the bornological structure is compatible with this coarse structure and we get a Γ-bornological coarse space $(X,\mathcal{C}_\mathcal{B},\mathcal{B})$.

EXAMPLE 2.14. Let X be a Γ-complete (i.e., for every bounded subset B the set ΓB is again bounded) Γ-bornological coarse space with coarse structure $\mathcal{C}$ and bornology $\mathcal{B}$. We equip the quotient set $\bar{X}:=\Gamma\backslash X$ with the maximal bornology $\bar{\mathcal{B}}$ such that the projection $q\colon X\to\bar{X}$ is proper, i.e.,

$$\bar{\mathcal{B}}=\mathcal{B}\langle\{\bar{B}\subseteq\bar{X}|q^{-1}(\bar{B})\in\mathcal{B}\}\rangle\ .$$

We furthermore equip $\bar{X}$ with the minimal coarse structure $\bar{\mathcal{C}}$ such that q is controlled, i.e.,

$$\bar{\mathcal{C}}:=\mathcal{C}\langle\{(q\times q)(U)|U\in\mathcal{C}\}\rangle\ .$$

Then $\bar{\mathcal{C}}$ and $\bar{\mathcal{B}}$ are compatible and we obtain a bornological coarse space $(\bar{X},\bar{\mathcal{C}},\bar{\mathcal{B}})$.

This construction produces a functor

$$Q\colon\Gamma\mathbf{BC}\to\mathbf{BC}\ ,\quad Q(X,\mathcal{C},\mathcal{B}):=(\bar{X},\bar{\mathcal{C}},\bar{\mathcal{B}})\ .$$

It is easy to check that the morphism of bornological coarse spaces $q\colon(X,\mathcal{C},\mathcal{B})\to(\bar{X},\bar{\mathcal{C}},\bar{\mathcal{B}})$ can be interpreted as the unit of an adjunction

$$Q\colon\Gamma\mathbf{BC}\leftrightarrows\mathbf{BC}:C\ ,$$

where C is the inclusion (2.1).

LEMMA 2.15. *The category* $\Gamma\mathbf{BC}$ *admits arbitrary coproducts and cartesian products.*

PROOF. The coproduct of a family of Γ-bornological coarse spaces is represented by the coproduct of the underlying bornological coarse spaces with the induced Γ-action.

Similarly, the cartesian product of a family of Γ-bornological coarse spaces is represented by the cartesian product of the family of the underlying bornological coarse spaces with the induced Γ-action. □

For more information about limits and colimits in $\Gamma\mathbf{BC}$ we refer to Section 2.2.

EXAMPLE 2.16. We consider a family $(X_i)_{i\in I}$ of Γ-bornological coarse spaces. We define the free union $\bigsqcup_{i\in I}^{\mathrm{free}}X_i$ as follows:

(1) The underlying Γ-set of the free union is the disjoint union of Γ-sets $\bigsqcup_{i\in I}X_i$.

(2) The coarse structure of the free union is generated by entourages $\bigsqcup_{i \in I} U_i$ for all families $(U_i)_{i \in I}$, where U_i is an entourage of X_i for every i in I.
(3) The bornology is generated by the set $\{B \mid B \in \mathcal{B}(X_i), i \in I\}$ of subsets of $\bigsqcup_{i \in I} X_i$.

If I is finite, then the free union is the coproduct of the family. In general, we have a morphism of Γ-bornological coarse spaces

$$\coprod_{i \in I} X_i \to \bigsqcup_{i \in I}^{\text{free}} X_i$$

induced by the identity of the underlying sets.

EXAMPLE 2.17. Let X and X' be two Γ-bornological coarse spaces. Then we can form the Γ-bornological coarse space $X' \otimes X$ whose coarse structure is the one of the cartesian product and the bornology is generated by the products $B' \times B$ for bounded subsets B' of X' and B of X. This construction defines a symmetric monoidal structure

$$- \otimes - \colon \Gamma\mathbf{BC} \times \Gamma\mathbf{BC} \to \Gamma\mathbf{BC}$$

with tensor unit given by the one-point space.

Let Y be a Γ-set. We can form the space $Y_{min,min} \otimes X$. For a second Γ-set Y' we have a canonical isomorphism

$$(Y' \times Y)_{min,min} \otimes X \cong Y'_{min,min} \otimes (Y_{min,min} \otimes X) \ .$$

2.2. Limits and colimits in ΓBC. In this section we show that the category $\Gamma\mathbf{BC}$ admits all limits of non-empty diagrams and various colimits. We furthermore discuss some special cases.

Let Γ be a group. In the following arguments we let

$$\iota \colon \Gamma\mathbf{BC} \to \Gamma\mathbf{Set} \tag{2.3}$$

be the forgetful functor. For a Γ-bornological coarse space X we let $\mathcal{B}_X$ and $\mathcal{C}_X$ denote its bornology and coarse structure.

PROPOSITION 2.18. (1) *The category* $\Gamma\mathbf{Coarse}$ *admits all small limits.*
(2) *The category* $\Gamma\mathbf{BC}$ *admits all limits of diagrams indexed by non-empty small categories.*

PROOF. We give the proof for $\Gamma\mathbf{BC}$. The statement for $\Gamma\mathbf{Coarse}$ can be obtained by ignoring all comments pertaining to bornologies and allowing in addition the index set I below to be empty.

Note that the non-emptiness assumption on I only enters into the part of the proof concerning bornologies. See also Remark 2.19.

Let I be a non-empty small category and

$$X \colon I \to \Gamma\mathbf{BC}$$

be a functor. We will show that $\lim_I X$ exists.

The category $\Gamma\mathbf{Set}$ is complete. In a first step we form the Γ-set

$$\widetilde{Y} := \lim_I \iota X \ .$$

For every i in I we have a map of Γ-sets $e_i \colon \widetilde{Y} \to \iota X(i)$.

On $\widetilde{Y}$ we define the bornology

$$\mathcal{B}_Y := \mathcal{B}\langle\{e_i^{-1}(B) | i \in I \text{and} B \in \mathcal{B}_{X(i)}\}\rangle \ .$$

Since $\mathcal{B}_{X(i)}$ is Γ-invariant for every i in I, we see that $\mathcal{B}_Y$ is Γ-invariant.

We can view $\widetilde{Y}$ as a subset of $\prod_{i\in I}\iota X(i)$. On $\widetilde{Y}$ we define the coarse structure

$$\mathcal{C}_Y := \mathcal{C}\langle\{(\prod_{i\in I} U_i)\cap(\widetilde{Y}\times\widetilde{Y})|(U_i)_{i\in I}\in\prod_{i\in I}\mathcal{C}_{X(i)}\}\rangle \ .$$

Using that $\mathcal{C}_{X(i)}$ has a cofinal subset of invariant entourages for every i in I, we see that the coarse structure $\mathcal{C}_Y$ is generated by invariant entourages and is hence a Γ-coarse structure.

Finally, the relation

$$(\prod_{i\in I} U_i)[e_j^{-1}(B)] = e_j^{-1}(U_j[B])$$

for j in I and B a subset of X_j shows that the compatibility of $\mathcal{B}_{X(j)}$ with $\mathcal{C}_{X(j)}$ for all j in I implies that $\mathcal{B}_Y$ and $\mathcal{C}_Y$ are compatible.

We therefore have defined an object $Y := (\widetilde{Y}, \mathcal{C}_Y, \mathcal{B}_Y)$ in $\Gamma\mathbf{BC}$. We now show that $e_j \colon Y \to X(j)$ is a morphism of Γ-bornological coarse spaces for all j in I. For every family $(U_i)_{i\in I}$ we have $e_j((\prod_{i\in I} U_i)\cap(\widetilde{Y}\times\widetilde{Y}))\subseteq U_j$. This implies that e_j is controlled. Furthermore, $\mathcal{B}_Y$ is defined such that e_j is proper for every j in I.

The morphisms $e_i \colon Y \to X(i)$ give a transformation $e \colon \underline{Y} \to X$ in $\mathbf{Fun}(I, \Gamma\mathbf{BC})$, where $\underline{Y}$ is the constant functor with value Y. We now show that (Y, e) has the universal property of the limit .

We consider a pair (Z, f) with Z in $\Gamma\mathbf{BC}$ and with $f \colon \underline{Z} \to X$ a morphism in $\mathbf{Fun}(I, \Gamma\mathbf{BC})$. Because the underlying Γ-set of Y is the limit of the diagram ιX there is a unique map $h \colon \iota Z \to \iota Y$ of Γ-sets such that $\iota f = \iota e \circ \underline{h}$.

It suffices to show that h is a morphism in $\Gamma\mathbf{BC}$. Let U be an entourage of Z. Then $f_i(U)$ is an entourage of $X(i)$ for every i in I. We have

$$h(U) \subseteq (\prod_{i\in I}(f_i\times f_i)(U))\cap(\widetilde{Y}\times\widetilde{Y}) \ ,$$

i.e., $h(U)$ is contained in one of the generating entourages of $\mathcal{C}_Y$. Hence h is a controlled map.

In order to show that h is proper, we consider a generating bounded subset $e_i^{-1}(B)$ for i in I and B in $\mathcal{B}_{X(i)}$. Then $h^{-1}(e_i^{-1}(B)) = f_i^{-1}(B)$ is bounded in Z. Since I is non-empty, the set of subsets $e_i^{-1}(B)$ for all i in I and B in $\mathcal{B}_{X(i)}$ cover Y. Therefore, every element of $\mathcal{B}_Y$ is contained in a finite union of such subsets. We conclude that h is proper. □

REMARK 2.19. Note that the last piece of the argument goes wrong if the index category of the diagram is empty. Then $Y = *$, and we need the subset $\{*\}$ (which is not of the form $e_i^{-1}(B)$) to generate the bornology. The empty limit does not exist since the category $\Gamma\mathbf{BC}$ does not have a final object.

We now turn to colimits. Let I be a small category and

$$X \colon I \to \Gamma\mathbf{BC}$$

be a functor. The category $\Gamma\mathbf{Set}$ is cocomplete. We form the Γ-set

$$\widetilde{Y} := \operatorname*{colim}_I \iota X \ ,$$

where ι is the forgetful functor from (2.3). For every i in I we have a map $e(i) \colon \iota X(i) \to \widetilde{Y}$ of Γ-sets.

DEFINITION 2.20. We say that the diagram X is colim-admissible if we have

$$e_i^{-1}((e_{j_1}(U_1) \circ \cdots \circ e_{j_k}(U_k))[\{y\}]) \in \mathcal{B}_{X(i)}$$

for every y in $\widetilde{Y}$, every i in I, every r in $\mathbb{N}$, every family $(j_1, \ldots, j_r)$ of objects of I, and every family of entourages U_k in $\mathcal{C}_{X(j_k)}$ for k in $\{1, \ldots, r\}$.

PROPOSITION 2.21. (1) *The category* Γ**Coarse** *admits all small colimits.*
(2) *The category* Γ**BC** *admits colimits for all colim-admissible diagrams.*

PROOF. We give the proof for Γ**BC**. The statement for Γ**Coarse** can be obtained by ignoring all comments pertaining to bornologies. Note that the assumption of colim-admissibility only enters to see that the bornology and the coarse structure on the colimit are compatible.

Assume that

$$X : I \to \Gamma\mathbf{BC}$$

is a colim-admissible diagram. We show that $\operatorname{colim}_I X$ exists.

In a first step we form the Γ-set

$$\widetilde{Y} := \operatorname*{colim}_I \iota X \ .$$

On $\widetilde{Y}$ we define the coarse structure

$$\mathcal{C}_Y := \mathcal{C}\langle\{(e_i \times e_i)(U) | i \in I \text{ and } U \in \mathcal{C}_{X(i)}\}\rangle \ .$$

Using the fact that $\mathcal{C}_{X(i)}$ has a cofinal subset of invariant entourages we see that the coarse structure $\mathcal{C}_Y$ is generated by invariant entourages and is hence a Γ-coarse structure.

We define the bornology $\mathcal{B}_Y$ to be the subset of $\mathcal{P}(\widetilde{Y})$ consisting of the sets B satisfying

$$e_i^{-1}((e_{j_1}(U_1) \circ \cdots \circ e_{j_k}(U_k))[B]) \in \mathcal{B}_{X(i)}$$

for every i in I, every r in $\mathbb{N}$, every family $(j_1, \ldots, j_r)$ of objects of I, and every family of entourages U_k in $\mathcal{C}_{X(j_k)}$ for k in $\{1, \ldots, r\}$. Since the diagram is colim-admissible, all one-point sets belong to $\mathcal{B}_Y$. Furthermore $\mathcal{B}_Y$ is obviously closed under forming finite unions and subsets. Consequently, $\mathcal{B}_Y$ is a bornology on Y. Since $\mathcal{B}_{X(i)}$ and $\mathcal{C}_{X(i)}$ are Γ-invariant for every i in I we see that $\mathcal{B}_Y$ is Γ-invariant. We finally observe that $\mathcal{C}_Y$ and $\mathcal{B}_Y$ are compatible by construction.

We define now the object $Y := (\widetilde{Y}, \mathcal{C}_Y, \mathcal{B}_Y)$ of Γ**BC**. By construction the maps $e_i : X(i) \to Y$ are morphisms in Γ**BC** for all i in I. The family of morphisms $(e_i)_{i \in I}$ provides a morphism $e : X \to \underline{Y}$ in **Fun**$(I, \Gamma\mathbf{BC})$, where $\underline{Y}$ is the constant functor with value Y.

We now show that (Y, e) has the universal property of a colimit.

Consider a pair (Z, f) with Z in Γ**BC** and $f : X \to \underline{Z}$. Since the underlying Γ-set of Y is the colimit of the diagram ιX there is a unique map $h : \iota Y \to \iota Z$ of Γ-sets such that $\iota f = \underline{h} \circ \iota e$.

It suffices to show that h is a morphism.

Let i in be I and U be in $\mathcal{C}_{X(i)}$. Then $h(e_i(U)) = f_i(U_i)$ is an entourage of Z. This implies that h is controlled.

Let now B be a bounded subset of Z. Since $\mathcal{C}_Z$ and $\mathcal{B}_Z$ are compatible and f_j is controlled, $f_j(U)[B]$ is bounded for every j in I and U in $\mathcal{C}_{X_j}$. We now have[1]

$$e_i^{-1}(e_j(U)[h^{-1}(B)]) \subseteq e_i^{-1}(h^{-1}(h(e_j(U))[B])) = f_i^{-1}(f_j(U)[B]) \ .$$

Since f_i is proper, we conclude that $e_i^{-1}(e_j(U)[h^{-1}(B)])$ is bounded. Since i, j in I and U in $\mathcal{C}_{X(j)}$ were arbitrary this shows that $h^{-1}(B)$ is bounded in Y. We conclude that h is a proper map. □

EXAMPLE 2.22. If $(\mathcal{Y}, Z)$ is an equivariant complementary pair (see Definition 3.7) on X, then we have a push-out

$$\begin{array}{ccc} \operatorname{colim} \mathcal{Y} \cap Z & \longrightarrow & Z \\ \downarrow & & \downarrow \\ \operatorname{colim} \mathcal{Y} & \longrightarrow & X \end{array}$$

We first note that $\operatorname{colim} \mathcal{Y}$ is admissible since it is a filtered colimit of inclusions. It is straightforward to check that the diagram is colim-admissible and that the bornology on the space X is the one of the colimit. The only non-trivial fact to check is that the coarse structure on X given by the colimit is not too small. Let U be an entourage of X. Assume that i is in I such that $Y_i \cup Z = X$. Since the family is big, there exists j in I such that $U[Y_i] \subseteq Y_j$. Let $e \colon Y_j \to X$ and $f \colon Z \to X$ be the inclusions. Then

$$U \subseteq e(U \cap (Y_j \times Y_j)) \cup f(U \cap (Z \times Z)) \ .$$

EXAMPLE 2.23. By Proposition 2.18 the category $\Gamma\mathbf{BC}$ admits fiber products. Here we give an explicit description. We consider a diagram

$$\begin{array}{ccc} & & X \\ & & \downarrow{\scriptstyle a} \\ Y & \xrightarrow{b} & Z \end{array}$$

of Γ-bornological coarse spaces. We then form the cartesian product $X \times Y$ in the category $\Gamma\mathbf{BC}$. We define the Γ-bornological coarse space $X \times_Z Y$ to be the subset of $X \times Y$ of pairs (x, y) with $a(x) = b(y)$ with the induced bornological and coarse structure. It is straightforward to check that the square

$$\begin{array}{ccc} X \times_Z Y & \longrightarrow & X \\ \downarrow & & \downarrow{\scriptstyle a} \\ Y & \xrightarrow{b} & Z \end{array}$$

is cartesian in $\Gamma\mathbf{BC}$.

EXAMPLE 2.24. Assume that (Y, Z) is a coarsely excisive pair (see Definition 4.13) on X. Then we have a push-out

$$\begin{array}{ccc} Y \cap Z & \longrightarrow & Z \\ \downarrow & & \downarrow \\ Y & \longrightarrow & X \end{array}$$

[1] Here we use the general relation $U[f^{-1}(B)] \subseteq f^{-1}(f(U)[B])$ for a map $f \colon X \to Y$, entourage U of X and subset B of Y.

It is straightforward to check that the diagram is colim-admissible and that the bornology on the space X is the one of the colimit. The only non-trivial fact to check is that the coarse structure on X given by the colimit is not too small. Let U be an entourage of X. Then there is an entourage W of X such that $U \subseteq W$ and $U[Z] \cap U[Y] \subseteq W[Z \cap Y]$. Let $e\colon Y \to X$ and $f\colon Z \to X$ be the inclusions. Then

$$U \subseteq e(W^2 \cap (Y \times Y)) \cup f(W^2 \cap (Z \times Z)) .$$

EXAMPLE 2.25. Let H be a group acting on a Γ-bornological coarse space X such that the set of H-invariant entourages of X is cofinal in all entourages.

We let $B_H(X)$ denote H-completion of X obtained from X by replacing the bornology $\mathcal{B}$ of X by the bornology $B_H(\mathcal{B})$ generated by the subsets HB for all B in $\mathcal{B}$. Then the coequalizer

$$(H_{min,min} \otimes X)_{max-\mathcal{B}} \rightrightarrows B_H(X) \xrightarrow{\pi} X/H$$

for the H-action exists. Here the index $-_{max-\mathcal{B}}$ indicates that we replaced the bornology by the maximal bornology. The two arrows are given by $(h,x) \mapsto x$ and $(h,x) \mapsto hx$. One checks easily that they are both morphisms of Γ-bornological coarse spaces: they are both proper since their domain has the maximal bornology, and if U is a H-invariant entourage of X then both maps send $\mathrm{diag}(H) \times U$ to U and hence the maps are controlled in view of our assumption on the coarse structure of X.

Finally, we check that the coequalizer diagram is colim-admissible. It suffices to check that for every H-invariant entourage U of X and point Hx in X/H the set $U[\pi^{-1}(Hx)]$ is bounded in $B_H(X)$. This is the case since $U[x]$ belongs to $\mathcal{B}$ and so $U[\pi^{-1}(Hx)] = HU[x]$ belongs to $B_H(\mathcal{B})$.

EXAMPLE 2.26. Let

$$M\colon \Gamma\mathbf{BC} \to \Gamma\mathbf{Set}$$

be the functor which sends a Γ-bornological coarse space to its underlying Γ-set. In view of Proposition 2.18, it preserves all limits over non-empty small index categories. It is in fact the right-adjoint of an adjunction

$$(-)_{min,max}\colon \Gamma\mathbf{Set} \leftrightarrows \Gamma\mathbf{BC} : M ,$$

where $(-)_{min,max}$ sends a Γ-set S to the Γ-bornological coarse space obtained by equipping S with the minimal coarse structure and maximal bornology. Indeed, for a Γ-bornological coarse space X we have a natural identification

$$\mathrm{Hom}_{\Gamma\mathbf{BC}}(S_{min,max}, X) \cong \mathrm{Hom}_{\Gamma\mathbf{Set}}(S, M(X)) .$$

3. Equivariant coarse homology theories

The following notions are the obvious generalizations from the non-equivariant situation considered in [**8**].

Two morphisms $f, f'\colon X \to X'$ between Γ-bornological coarse spaces are *close* to each other if the subset $\{(f(x), f'(x)) \mid x \in X\}$ of $X' \times X'$ is an entourage, i.e. if they are close as morphisms between the underlying bornological coarse spaces.

DEFINITION 3.1. A morphism between Γ-bornological coarse spaces is an *equivalence* if it admits an inverse morphism up to closeness.

EXAMPLE 3.2. We consider a Γ-bornological coarse space X, a Γ-invariant subset A of X, and a Γ-invariant entourage U of X. Then we can form the U-thickening $U[A]$ which is again Γ-invariant. We now assume that U contains the diagonal. Then we have a natural inclusion $i\colon A \to U[A]$. This inclusion is in general not an equivalence of Γ-bornological coarse spaces.

For example, let the group $\mathbb{Z}$ act on $\mathbb{C}$ by $(n, z) \mapsto e^{2\pi i\theta n}z$, where θ is an irrational real number. Then the subset $\mathbb{C} \setminus \{0\}$ of $\mathbb{C}$ is $\mathbb{Z}$-invariant. Every non-trivial thickening of this subset contains the point 0. This point is fixed by the action, but $\mathbb{C} \setminus \{0\}$ does not contain any fixed point which could serve as the image of 0 under a potential inverse of the inclusion $\mathbb{C} \setminus \{0\} \to \mathbb{C}$.

Let X be a Γ-bornological coarse space and A be a Γ-invariant subset of X.

DEFINITION 3.3. The subset A is called *nice* if for every invariant entourage U of X containing the diagonal the inclusion $A \to U[A]$ is an equivalence.

EXAMPLE 3.4. Let X be a Γ-bornological coarse space and let Y be a bornological coarse space considered as a Γ-bornological coarse space with the trivial Γ-action.

For every subset A of Y the subset $A \times X$ of the Γ-bornological coarse space $Y \times X$ (or of $Y \otimes X$) is nice.

A filtered family of subsets of a set X is a family $(Y_i)_{i\in I}$ of subsets indexed by a filtered partially ordered set I such that the map $I \to \mathcal{P}(X)$ given by $i \mapsto Y_i$ is order-preserving.

Let X be a Γ-bornological coarse space. Recall from [**8**] that a *big family* on X is a filtered family of subsets $(Y_i)_{i\in I}$ of X such that for every entourage U of X and i in I there exists j in I such that $U[Y_i] \subseteq Y_j$.

DEFINITION 3.5. An *equivariant big family* on X is a big family consisting of Γ-invariant subsets.

EXAMPLE 3.6. Let X be a Γ-bornological coarse space and A be a Γ-invariant subset of X. Then the family

$$\{A\} := (U[A])_{U\in\mathcal{C}^\Gamma}$$

is an equivariant big family.

Let X be a Γ-bornological coarse space. Recall from [**8**] that a complementary pair $(Z, \mathcal{Y})$ on X is a pair of a subset Z of X and a big family $\mathcal{Y} = (Y_i)_{i\in I}$ on X such that there exists i in I with $Z \cup Y_i = X$.

DEFINITION 3.7. An *equivariant complementary pair* on X is a complementary pair $(Z, \mathcal{Y})$ such that Z is a Γ-invariant subset and $\mathcal{Y}$ is an equivariant big family.

DEFINITION 3.8. A Γ-bornological coarse space X is *flasque* if it admits a morphism $f\colon X \to X$ such that

(1) f is close to id_X.
(2) For every entourage U of X the subset $\bigcup_{n\in\mathbb{N}}(f^n \times f^n)(U)$ is an entourage of X.
(3) For every bounded subset B of X there exists an integer n such that $\Gamma B \cap f^n(X) = \emptyset$.

We say that flasqueness of X is *implemented* by f.

REMARK 3.9. In Condition 3 above one could require the weaker condition $B \cap f^n(X) = \emptyset$ instead of $\Gamma B \cap f^n(X) = \emptyset$. Then much of the theory would go through, but we lose the possibility of descending the group change functors "H-completion", "quotient" and "induction" to the motivic level.

Let $\mathbf{C}$ be a cocomplete stable ∞-category. We consider a functor

$$E\colon \Gamma\mathbf{BC} \to \mathbf{C} .$$

If $\mathcal{Y} = (Y_i)_{i\in I}$ is a filtered family of Γ-invariant subsets of X, then we set

$$E(\mathcal{Y}) := \operatorname*{colim}_{i\in I} E(Y_i) . \tag{3.1}$$

In this formula we consider the subsets Y_i as Γ-bornological coarse spaces with the structures induced from X.

The set $\{0,1\}_{max,max}$ is a Γ-bornological coarse space with the trivial Γ-action.

DEFINITION 3.10. A *Γ-equivariant $\mathbf{C}$-valued coarse homology theory* is a functor

$$E\colon \Gamma\mathbf{BC} \to \mathbf{C}$$

with the following properties:

(1) (Coarse invariance) The projection $\{0,1\}_{max,max} \otimes X \to X$ is sent by E to an equivalence for all $X \in \Gamma\mathbf{BC}$.
(2) (Excision) $E(\emptyset) \simeq 0$ and for every equivariant complementary pair $(Z,\mathcal{Y})$ on a Γ-bornological coarse space X the square

$$\begin{array}{ccc} E(Z\cap\mathcal{Y}) & \longrightarrow & E(Z) \\ \downarrow & & \downarrow \\ E(\mathcal{Y}) & \longrightarrow & E(X) \end{array}$$

is a push-out.
(3) (Flasqueness) If a Γ-bornological coarse space X is flasque, then $E(X) \simeq 0$.
(4) (u-Continuity) For every Γ-bornological coarse space X the natural map

$$\operatorname*{colim}_{U\in\mathcal{C}^\Gamma} E(X_U) \xrightarrow{\simeq} E(X)$$

is an equivalence (see Example 2.9 for notation).

If the group Γ is clear from the context, then we will often just speak of an equivariant coarse homology theory.

REMARK 3.11. Condition 1 in the above definition is equivalent to the condition that E sends equivalences (Definition 3.1) of Γ-bornological coarse spaces to equivalences in $\mathbf{C}$:

The projection $\{0,1\}_{max,max} \otimes X \to X$ is an equivalence of Γ-bornological coarse spaces. If E preserves equivalences, then it sends this morphism to an equivalence.

Vice versa, if E satisfies Condition 1, then it sends pairs of close maps to equivalent maps. If f is an equivalence with inverse g up to closeness, then $E(f\circ g)$ and $E(g\circ f)$ are equivalent to $E(\mathrm{id})$ and therefore themselves equivalences. This implies by functoriality that $E(f)$ is an equivalence.

Let the cocomplete stable ∞-category $\mathbf{C}$ have all small products. Let $(X_i)_{i\in I}$ be a family of Γ-bornological coarse spaces. If E is a $\mathbf{C}$-valued equivariant coarse homology theory, then by excision for every index i in I we have a projection $E(\bigsqcup_{i\in I}^{\mathrm{free}} X_i) \to E(X_i)$. The collection of these projections induces a morphism

$$E(\bigsqcup_{i\in I}^{\mathrm{free}} X_i) \to \prod_{i\in I} E(X_i) \tag{3.2}$$

DEFINITION 3.12. E is called *strongly additive* if (3.2) is an equivalence for every family $(X_i)_{i\in I}$ of Γ-bornological coarse spaces.

Let E be an equivariant coarse homology theory and let S be a Γ-set.

LEMMA 3.13. *If E is strongly additive, then the twist $E(- \otimes S_{max,max})$ is strongly additive.*

PROOF. This follows from the fact that for every family $(X_i)_{i\in I}$ of Γ-bornological coarse spaces we have an isomorphism

$$(\coprod_{i\in I}^{\mathrm{free}} X_i) \otimes S_{max,max} \cong \coprod_{i\in I}^{\mathrm{free}} (X_i \otimes S_{max,max})$$

of Γ-bornological coarse spaces. □

4. Equivariant coarse motivic spectra

4.1. Construction. In this section we define the stable ∞-category of coarse motives $\Gamma\mathbf{Sp}\mathcal{X}$. This is completely analogous to [**8**, Sec. 3 & 4]. The category $\Gamma\mathbf{Sp}\mathcal{X}$ is designed such that equivariant $\mathbf{C}$-valued coarse homology theories (Definition 3.10) are the same as colimit-preserving functors $\Gamma\mathbf{Sp}\mathcal{X} \to \mathbf{C}$. The precise formulation is Corollary 4.10.

Let $\mathbf{Spc}$ be the ∞-category of spaces, i.e., the universal presentable ∞-category generated by $*$. We start with the category

$$\mathbf{PSh}(\Gamma\mathbf{BC}) := \mathbf{Fun}(\Gamma\mathbf{BC}^{op}, \mathbf{Spc})$$

of $\mathbf{Spc}$-valued presheaves on $\Gamma\mathbf{BC}$. Let

$$\mathrm{yo}\colon \Gamma\mathbf{BC} \to \mathbf{PSh}(\Gamma\mathbf{BC}) \tag{4.1}$$

be the Yoneda embedding. Precomposition with it induces an equivalence

$$\mathbf{Fun}^{\lim}(\mathbf{PSh}(\Gamma\mathbf{BC})^{op}, \mathbf{Spc}) \simeq \mathbf{PSh}(\Gamma\mathbf{BC}) ,$$

where $\mathbf{Fun}^{\lim}$ denotes limit-preserving functors [**26**, Thm. 5.1.5.6] (see also [**8**, Rem. 3.9]). We use this equivalence in order to evaluate presheaves on other presheaves.

For a filtered family $\mathcal{Y} = (Y_i)_{i\in I}$ of invariant subsets on some Γ-bornological coarse space X we write

$$\mathrm{yo}(\mathcal{Y}) := \operatorname*{colim}_{i\in I} \mathrm{yo}(Y_i) \in \mathbf{PSh}(\Gamma\mathbf{BC}) . \tag{4.2}$$

For E in $\mathbf{PSh}(\Gamma\mathbf{BC})$ we set

$$E(\mathcal{Y}) := E(\mathrm{yo}(\mathcal{Y})) .$$

REMARK 4.1. For a Γ-bornological coarse space X we have the equivalence

$$E(X) \simeq E(\mathrm{yo}(X)) .$$

Furthermore we have

$$E(\mathcal{Y}) \simeq \lim_{i\in I} E(Y_i) \tag{4.3}$$

for a filtered family $\mathcal{Y} = (Y_i)_{i\in I}$ of invariant subsets on X.

Let E be an object of $\mathbf{PSh}(\Gamma\mathbf{BC})$.

DEFINITION 4.2. We say that E *satisfies descent* if

(1) $E(\emptyset) \simeq *$, and
(2) for every equivariant complementary pair $(Z, \mathcal{Y})$ on a Γ-bornological coarse space X the square

$$\begin{array}{ccc} E(X) & \longrightarrow & E(Z) \\ \downarrow & & \downarrow \\ E(\mathcal{Y}) & \longrightarrow & E(Z\cap\mathcal{Y}) \end{array} \tag{4.4}$$

is cartesian.

Presheaves which satisfy descent are called *sheaves*.

REMARK 4.3. One can show that there is a subcanonical Grothendieck topology τ_χ on $\Gamma\mathbf{BC}$ such that the τ_χ-sheaves are exactly the presheaves which satisfy descent for equivariant complementary pairs. Since we will not be using this fact in this paper we will omit the arguments.

We let $\mathbf{Sh}(\Gamma\mathbf{BC})$ denote the full subcategory of $\mathbf{PSh}(\Gamma\mathbf{BC})$ of sheaves. We can characterize sheaves as presheaves which are local with respect to the morphisms

$$\begin{array}{rcl} \mathrm{yo}(\mathcal{Y}) \sqcup_{\mathrm{yo}(Z\cap\mathcal{Y})} \mathrm{yo}(Z) & \to & \mathrm{yo}(X) \\ \mathrm{yo}(\emptyset) & \to & * . \end{array} \tag{4.5}$$

REMARK 4.4. In order to fix set-theoretic issues we assume that all Γ-bornological coarse spaces and the index sets I for the big families belong to some Grothendieck universe of small sets. The class of local objects is then generated by a small set of morphisms. The category $\mathbf{Sh}(\Gamma\mathbf{BC})$ then belongs to a bigger universe.

We have a sheafification adjunction

$$\mathbf{PSh}(\Gamma\mathbf{BC}) \leftrightarrows \mathbf{Sh}(\Gamma\mathbf{BC}) : \mathit{inclusion} .$$

REMARK 4.5. For a Γ-bornological coarse space X the presheaf $\mathrm{yo}(X)$ is a compact object of $\mathbf{PSh}(\Gamma\mathbf{BC})$. If $\mathcal{Y}$ is a big family, then $\mathrm{yo}(\mathcal{Y})$ is an in general infinite colimit of compact objects and hence not compact anymore. Consequently, the morphisms (4.5) are not morphisms between compact objects. The localization $\mathbf{Sh}(\Gamma\mathbf{BC})$ is therefore a presentable ∞-category, but it is not compactly generated.

Let E be an object of $\mathbf{Sh}(\Gamma\mathbf{BC})$.

DEFINITION 4.6. (1) E is *coarsely invariant* if it is local with respect to the morphisms

$$\mathrm{yo}(\{0,1\}_{max,max} \otimes X) \to \mathrm{yo}(X)$$

induced by the projection for every Γ-bornological coarse space X.

(2) E *vanishes on flasques* if it is local with respect to the morphisms

$$\mathrm{yo}(\emptyset) \to \mathrm{yo}(X)$$

for every flasque Γ-bornological coarse space X.

(3) E is *u-continuous* if it is local for the morphism

$$\operatorname*{colim}_{U \in \mathcal{C}^\Gamma} \mathrm{yo}(X_U) \to \mathrm{yo}(X)$$

for every Γ-bornological coarse space X (where $\mathcal{C}^\Gamma$ denotes the invariant entourages of the space X).

The above notions are just the equivariant analogues of the corresponding notions from [**8**, Sec. 3].

DEFINITION 4.7. We define the ∞-category of Γ-*equivariant motivic coarse spaces* $\Gamma\mathbf{Spc}\mathcal{X}$ as the full localizing subcategory of $\mathbf{Sh}(\Gamma\mathbf{BC})$ of coarsely invariant, u-continuous sheaves which vanish on flasques.

The locality condition is generated by a small set of morphisms. Therefore we have a localization adjunction

$$\mathcal{L}\colon \mathbf{PSh}(\Gamma\mathbf{BC}) \leftrightarrows \Gamma\mathbf{Spc}\mathcal{X} : \mathit{inclusion}\ . \tag{4.6}$$

We define

$$\mathrm{Yo} := \mathcal{L} \circ \mathrm{yo}\colon \Gamma\mathbf{BC} \to \Gamma\mathbf{Spc}\mathcal{X}\ .$$

The ∞-category $\Gamma\mathbf{Spc}\mathcal{X}$ is a presentable ∞-category.

DEFINITION 4.8. We define the category of *equivariant motivic coarse spectra* as the stabilization

$$\Gamma\mathbf{Sp}\mathcal{X} := \Gamma\mathbf{Spc}\mathcal{X}_*[\Sigma^{-1}]$$

in the realm of presentable ∞-categories.

Then $\Gamma\mathbf{Sp}\mathcal{X}$ is a stable presentable ∞-category which fits into an adjunction

$$\Sigma^{mot}_{+}\colon \Gamma\mathbf{Spc}\mathcal{X} \leftrightarrows \Gamma\mathbf{Sp}\mathcal{X} : \Omega^{mot}\ .$$

We further define the Yoneda functor

$$\mathrm{Yo}^s := \Sigma^{mot}_{+} \circ \mathrm{Yo}\colon \Gamma\mathbf{BC} \to \Gamma\mathbf{Sp}\mathcal{X}\ .$$

DEFINITION 4.9. We call $\mathrm{Yo}^s\colon \Gamma\mathbf{BC} \to \Gamma\mathbf{Sp}\mathcal{X}$ the universal equivariant coarse homology theory.

For a Γ-bornological coarse space X we consider the object $\mathrm{Yo}^s(X)$ of $\Gamma\mathbf{Sp}\mathcal{X}$ as the motive of X.

Let $\mathbf{C}$ be a cocomplete stable ∞-category. Let $\Gamma\mathbf{CHT}_{\mathbf{C}}$ denote the full subcategory of $\mathbf{Fun}(\Gamma\mathbf{BC},\mathbf{C})$ of functors which are Γ-equivariant $\mathbf{C}$-valued coarse homology theories in the sense of Definition 3.10. We denote by $\mathbf{Fun}^{\mathrm{colim}}(\Gamma\mathbf{Sp}\mathcal{X},\mathbf{C})$ the full subcategory of $\mathbf{Fun}(\Gamma\mathbf{Sp}\mathcal{X},\mathbf{C})$ of colimit preserving functors.

The construction of $\Gamma\mathbf{Sp}\mathcal{X}$ has the following consequence (see [**8**, Cor. 4.6]):

COROLLARY 4.10. *The functor* Yo^s *is a* $\Gamma\mathbf{Sp}\mathcal{X}$*-valued equivariant coarse homology theory. Furthermore, precomposition with* Yo^s *induces an equivalence of* ∞*-categories*

$$\mathbf{Fun}^{\mathrm{colim}}(\Gamma\mathbf{Sp}\mathcal{X},\mathbf{C})\to\Gamma\mathbf{CHT}_{\mathbf{C}}\ .$$

4.2. Properties. If $\mathcal{Y}=(Y_i)_{i\in I}$ is an equivariant big family on a Γ-bornological coarse space X, then we define the equivariant motivic coarse spectrum

$$\mathrm{Yo}^s(\mathcal{Y}):=\Sigma^{mot}_{+}\circ\mathcal{L}\circ\mathrm{yo}(\mathcal{Y})\ . \tag{4.7}$$

Note that we have $\Sigma^{mot}_{+}\circ\mathcal{L}\circ\mathrm{yo}(\mathcal{Y})\simeq\mathrm{colim}_{i\in I}\mathrm{Yo}^s(Y_i)$.

We will use the notation

$$\mathrm{Yo}^s(X,\mathcal{Y}):=\mathrm{Cofib}(\mathrm{Yo}^s(\mathcal{Y})\to\mathrm{Yo}^s(X))\ . \tag{4.8}$$

By construction we have the following properties:

COROLLARY 4.11. (1) *We have a fiber sequence*

$$\mathrm{Yo}^s(\mathcal{Y})\to\mathrm{Yo}^s(X)\to\mathrm{Yo}^s(X,\mathcal{Y})\to\Sigma\mathrm{Yo}^s(\mathcal{Y})$$

(2) *For an equivariant complementary pair* $(Z,\mathcal{Y})$ *on* X *the natural morphism*

$$\mathrm{Yo}^s(Z,Z\cap\mathcal{Y})\to\mathrm{Yo}^s(X,\mathcal{Y})$$

is an equivalence.

(3) *If* $X\to X'$ *is an equivalence of* Γ*-bornological coarse spaces, then the induced morphism* $\mathrm{Yo}^s(X)\to\mathrm{Yo}^s(X')$ *is an equivalence in* $\Gamma\mathbf{Sp}\mathcal{X}$.

(4) *If* X *is a flasque* Γ*-bornological coarse space, then* $\mathrm{Yo}^s(X)\simeq 0$.

(5) *For every* Γ*-bornological coarse space* X *with coarse structure* $\mathcal{C}$ *the natural map*

$$\operatorname*{colim}_{U\in\mathcal{C}^\Gamma}\mathrm{Yo}^s(X_U)\xrightarrow{\simeq}\mathrm{Yo}^s(X)$$

is an equivalence.

Let X be a Γ-bornological coarse space and A be a Γ-invariant subset of X. Recall that $\{A\}$ denotes the equivariant big family generated by A (Example 3.6).

COROLLARY 4.12. *If* A *is nice, then we the natural map* $\mathrm{Yo}^s(A)\to\mathrm{Yo}^s(\{A\})$ *is an equivalence.*

PROOF. Since A is nice, for every invariant entourage U of X the inclusion $A\to U[A]$ is an equivalence. The assertion now follows since Yo^s preserves equivalences. □

Let X be a Γ-bornological coarse space and Y,Z be invariant subsets such that $Y\cup Z=X$.

DEFINITION 4.13. We say that (Y,Z) is a *coarsely excisive pair*, if:

(1) For every entourage U of X there exists an entourage W of X such that

$$U[Y]\cap U[Z]\subseteq W[Y\cap Z]\ .$$

(2) There exists a cofinal set of invariant entourages V of X such that $V[Y]\cap Z$ is nice.

Note that Condition 2 is a new aspect of the equivariant theory.

Let X be a Γ-bornological coarse space and Y, Z be invariant subsets such that $Y \cup Z = X$.

COROLLARY 4.14. *If (Y, Z) is a coarsely excisive pair, then we have a cocartesian square*

$$\begin{array}{ccc} \mathrm{Yo}^s(Y \cap Z) & \longrightarrow & \mathrm{Yo}^s(Z) \\ \downarrow & & \downarrow \\ \mathrm{Yo}^s(Y) & \longrightarrow & \mathrm{Yo}^s(X) \end{array}$$

PROOF. The proof of [**8**, Lem. 3.38] goes through literally. In the proof we need the equivalence

$$\mathrm{Yo}^s(V[Y] \cap Z) \simeq \mathrm{Yo}^s(\{V[Y] \cap Z\})$$

for sufficiently large invariant entourages V of X. This is ensured by Condition 2 in the Definition 4.13 of coarse excisiveness. $\square$

Let X be a Γ-bornological coarse space. Given two bornological and Γ-invariant maps $p = (p_-, p_+)$ with $p_- : X \to (-\infty, 0]$ and $p_+ : X \to [0, \infty)$ we can form the coarse cylinder I_pX as in the non-equivariant case [**8**, Sec. 4.3]. With its natural Γ-action it is a Γ-bornological coarse space. The projection $I_pX \to X$ is a morphism. We will call it an equivariant cylinder in order to stress that the datum p was Γ-invariant.

Let X be a Γ-bornological coarse space and I_pX be an equivariant coarse cylinder.

COROLLARY 4.15. *The projection $I_pX \to X$ induces an equivalence* $\mathrm{Yo}^s(I_pX) \to \mathrm{Yo}^s(X)$.

PROOF. We observe that the proof of [**8**, Prop. 4.16] goes through. At all places in the argument where Corollary 4.12 is used the corresponding subset is nice, see Example 3.4. $\square$

We say that two morphisms $f_+, f_- : X \to X'$ between Γ-bornological coarse spaces are *homotopic* if there exists a cylinder I_pX such that $p_\pm$ are Γ-invariant, bornological and in addition controlled, and if there exists a morphism $h : I_pX \to X'$ such that $f_\pm = h \circ i_\pm$. This leads to an extension of the notion of coarse invariance.

COROLLARY 4.16. *If f_+ and f_- are homotopic, then* $\mathrm{Yo}^s(f_+)$ *and* $\mathrm{Yo}^s(f_-)$ *are equivalent.*

4.3. Symmetric monoidal structure. Recall that the category $\Gamma\mathbf{BC}$ has a symmetric monoidal structure

$$- \otimes - : \Gamma\mathbf{BC} \times \Gamma\mathbf{BC} \to \Gamma\mathbf{BC} \tag{4.9}$$

with tensor unit $*$.

LEMMA 4.17. *$\Gamma\mathbf{Sp}\mathcal{X}$ has an induced closed symmetric monoidal structure $\otimes$ such that the functor $\mathrm{Yo}^s : \Gamma\mathbf{BC} \to \Gamma\mathbf{Sp}\mathcal{X}$ is symmetric monoidal. The functor $\otimes$ commutes with colimits in each variable separately.*

PROOF. We get an induced symmetric monoidal structure on $\mathbf{PSh}(\Gamma\mathbf{BC})$ by the Day convolution product. The unit is given by $\mathrm{yo}(*)$, the Yoneda embedding is a strong symmetric monoidal functor, and $\mathbf{PSh}(\Gamma\mathbf{BC})$ is closed symmetric monoidal.

For a Γ-bornological coarse space Q the functor

$$- \otimes Q \colon \Gamma\mathbf{BC} \to \Gamma\mathbf{BC}$$

maps big families to big families and complementary pairs to complementary pairs. So the monoidal structure of $\mathbf{PSh}(\Gamma\mathbf{BC})$ restricts to one on $\mathbf{Sh}(\Gamma\mathbf{BC})$ and the sheafification adjunction is a symmetric monoidal adjunction.

The functor $- \otimes Q$ furthermore respects closeness of morphisms and therefore coarse equivalences, and it respects flasqueness and u-continuity. So we get an induced symmetric monoidal structure on $\Gamma\mathbf{Spc}\mathcal{X}$ and $\mathrm{Yo}\colon \Gamma\mathbf{BC} \to \Gamma\mathbf{Spc}\mathcal{X}$ is symmetric monoidal.

Since $\Gamma\mathbf{Spc}\mathcal{X}$ is presentable, we can equip its stabilization $\Gamma\mathbf{Sp}\mathcal{X}$ with a unique symmetric monoidal structure $\otimes$ such that stabilization $\Sigma^{mot}_{+}\colon \Gamma\mathbf{Spc}\mathcal{X} \to \Gamma\mathbf{Sp}\mathcal{X}$ is symmetric monoidal [**20**, Thm. 5.1].

It follows from the construction that $\otimes$ commutes with colimits in each variable separately. □

4.4. Strong version. In this section we discuss an additional property (strongness) which an equivariant coarse homology theory can have. Another condition (continuity) will be discussed in Section 5.

By definition, a flasque Γ-bornological coarse space X admits a morphism $f\colon X \to X$ satisfying the conditions listed in Definition 3.8. The first condition is the condition that f is close to the identity. This fact is usually used in order to deduce that $\mathrm{Yo}^s(f) \simeq \mathrm{id}_{\mathrm{Yo}^s(X)}$. In the following we will use this weaker condition in order to define a more general notion of flasqueness.

DEFINITION 4.18. X is called *weakly flasque* if it admits a morphism $f\colon X \to X$ satisfying

(1) $\mathrm{Yo}^s(f) \simeq \mathrm{id}_{\mathrm{Yo}^s(X)}$.
(2) For every entourage U of X the subset $\bigcup_{n\in\mathbb{N}}(f^n \times f^n)(U)$ is again an entourage of X.
(3) For every bounded subset B of X there exists an integer n such that $\Gamma B \cap f^n(X) = \emptyset$.

We say that f *implements weak flasqueness* of X.

Let $\mathbf{C}$ be a cocomplete stable ∞-category and consider a $\mathbf{C}$-valued equivariant coarse homology theory E.

DEFINITION 4.19. E is called *strong* if $E(X) \simeq 0$ for all weakly flasque Γ-bornological coarse spaces X.

Let us incorporate now the condition of strongness on the motivic level.

DEFINITION 4.20. We define the version of equivariant motivic spectra $\Gamma\mathbf{Sp}\mathcal{X}_{\mathrm{wfl}}$ as the localization of the category $\Gamma\mathbf{Sp}\mathcal{X}$ at the set of morphisms $0 \to \mathrm{Yo}^s(X)$ for all weakly flasque Γ-bornological coarse spaces X.

The corresponding Yoneda functor is denoted by

$$\mathrm{Yo}^s_{\mathrm{wfl}}\colon \Gamma\mathbf{BC} \to \Gamma\mathbf{Sp}\mathcal{X}_{\mathrm{wfl}} .$$

We consider now the ∞-category of strong Γ-equivariant coarse homology theories. The construction of $\Gamma\mathbf{Sp}\mathcal{X}_{\mathrm{wfl}}$ has the following immediate consequence:

COROLLARY 4.21. *The functor* $\mathrm{Yo}^s_{\mathrm{wfl}}$ *is a* $\Gamma\mathbf{Sp}\mathcal{X}_{\mathrm{wfl}}$*-valued equivariant coarse homology theory. Furthermore, precomposition with* $\mathrm{Yo}^s_{\mathrm{wfl}}$ *induces an equivalence of* ∞*-categories*

$$\mathbf{Fun}^{\mathrm{colim}}(\Gamma\mathbf{Sp}\mathcal{X}_{\mathrm{wfl}}, \mathbf{C}) \to \mathbf{strong}\Gamma\mathbf{CHT}_{\mathbf{C}} \ .$$

5. Continuity

The purpose of this section is to introduce the notion of continuity for equivariant coarse homology theories. This property will be crucially needed in Section 11.2. We will first introduce the notion of trapping exhaustions in Section 5.1. Section 5.2 contains the actual definition of continuous equivariant coarse homology theories, and Section 5.3 incorporates continuity motivically. In the last Section 5.4 we will show how one can force continuity for an equivariant coarse homology theory.

5.1. Trapping exhaustions. In this section we will introduce the notion of a trapping exhaustion of a Γ-bornological space and discuss some examples and basic properties of this notion. We will also introduce the stronger notion of a co-Γ-bounded exhaustion.

Let X be a bornological space and let F be a subset of X.

DEFINITION 5.1. The subset F is called *locally finite* if $B \cap F$ is finite for every bounded subset B of X.

EXAMPLE 5.2. Every finite subset of X is locally finite.

EXAMPLE 5.3. If X has the minimal bornology on X, i.e., a subset is bounded if and only if it is finite, then every subset of X is locally finite.

EXAMPLE 5.4. If X has the maximal bornology on X, i.e., every subset of X is bounded, then the locally finite subsets of X are exactly the finite subsets.

Let $f\colon X \to X'$ be a proper map between bornological spaces. Let F be a subset of X.

LEMMA 5.5. *If* F *is locally finite, then* $f(F)$ *is locally finite.*

PROOF. We use the relation $f(F) \cap B \subseteq f(F \cap f^{-1}(B))$. □

We consider in the following a Γ-bornological space X and a filtered family of invariant subsets $\mathcal{Y} = (Y_i)_{i \in I}$.

DEFINITION 5.6. The family $\mathcal{Y}$ is called a *trapping exhaustion* if for every locally finite, invariant subset F of X there exists i in I such that $F \subseteq Y_i$.

EXAMPLE 5.7. The family consisting of all locally finite, invariant subsets is a trapping exhaustion.

It might happen that a Γ-bornological coarse space does note admit any non-empty invariant locally finite subset. Consider e.g. Γ with the maximal bornology. In this case the empty family is a trapping exhaustion.

In the following we will introduce a particular kind of trapping exhaustions which we call co-Γ-bounded exhaustions.

We consider a Γ-bornological space X and a filtered family of invariant subsets $\mathcal{Y} = (Y_i)_{i \in I}$. We use the notation and terminology introduced in Definition 2.12.

DEFINITION 5.8. The family $\mathcal{Y}$ is called a *co-Γ-bounded exhaustion* if

(1) $\mathcal{Y}$ is an exhaustion, i.e., $\bigcup_{i\in I} Y_i = X$, and
(2) $\mathcal{Y}$ is co-Γ-bounded, i.e., there exists i in I such that $X \setminus Y_i$ is Γ-bounded.

We consider Γ-bornological spaces X and Z and a filtered family $\mathcal{Y} := (Y_i)_{i\in I}$ of invariant subsets of X. In the following we denote by $Z \otimes X$ the Γ-bornological space whose bornology is generated by products $A \times B$ for all bounded subsets A of Z and B of X.

LEMMA 5.9. *If Z is bounded and $\mathcal{Y}$ is a co-Γ-bounded (resp., trapping) exhaustion of X, $(Z \times Y_i)_{i\in I}$ is a co-Γ-bounded (resp., trapping) exhaustion of $Z \otimes X$.*

PROOF. The co-Γ-bounded case is straightforward.

For the trapping case assume that F is an invariant, locally finite subset of $Z \otimes X$. Since the projection $p\colon Z \otimes X \to X$ is proper, by Lemma 5.5 the subset $p(F)$ of X is locally finite. Hence there exists i in I such that $p(F) \subseteq Y_i$, and therefore $F \subseteq Z \times Y_i$. □

LEMMA 5.10. *If $\mathcal{Y}$ is a co-Γ-bounded exhaustion of a Γ-bornological space X then it is a trapping exhaustion.*

PROOF. Let F be an invariant, locally finite subset of X. Since $\mathcal{Y}$ is a co-Γ-bounded exhaustion of X there exists an index i in I and a bounded subset B of X such that $\Gamma B \cup Y_i = X$. Since the union of $\mathcal{Y}$ is X and $F \cap B$ is finite there exists an index j in I such that $i \leq j$ and $F \cap B \subseteq Y_j$. Since Y_j is invariant, then also $F \subseteq Y_j$. □

EXAMPLE 5.11. Let Z be a Γ-bounded Γ-bornological space and let $\mathcal{Z} = (Z_i)_{i\in I}$ be an exhaustion by not necessarily Γ-invariant subsets. For every i in I we consider the subset

$$D_i := \Gamma(Z_i \times \{1\})$$

of $Z \times \Gamma$. We consider the Γ-bornological space $Z \otimes \Gamma$, where Γ has any Γ-invariant bornology. The family

$$\mathcal{D} := (D_i)_{i\in I}$$

is a co-Γ-bounded (and hence trapping) exhaustion of $Z \otimes \Gamma$.

We consider $[0, \infty)$ as a Γ-bornological space with the trivial action and the bornology generated by the subsets $[0, n]$ for all integers n. In the following we will construct an interesting trapping exhaustion of the Γ-bornological space

$$[0, \infty) \otimes Z \otimes \Gamma$$

which will play an important role in Section 11.2.

Note that in general $([0, \infty) \times D_i)_{i\in I}$ is not trapping.

We consider the set of functions $I^{\mathbb{N}}$ with its partial order induced from I. Then the partially ordered set $I^{\mathbb{N}}$ is filtered. For a function κ in $I^{\mathbb{N}}$ we define the set

$$Y_\kappa := \bigcup_{n\in\mathbb{N}} [n-1, n] \times D_{\kappa(n)} \ .$$

LEMMA 5.12. *If Z is Γ-bounded, then $\mathcal{Y} := (Y_\kappa)_{\kappa\in I^{\mathbb{N}}}$ is a trapping exhaustion of the space $[0, \infty) \otimes Z \otimes \Gamma$.*

Note that the exhaustion $\mathcal{Y}$ is not co-Γ-bounded.

PROOF. The members of $\mathcal{Y}$ are Γ-invariant subsets. For every integer n the family given by $([n-1,n]\times D_i)_{i\in I}$ is a co-Γ-bounded exhaustion of $[n-1,n]\otimes Z\otimes\Gamma$, since $[n-1,n]$ is bounded and $(D_i)_{i\in I}$ is a co-Γ-bounded exhaustion of $Z\otimes\Gamma$. So it is trapping.

Let F be a Γ-invariant locally finite subset of $[0,\infty)\times Z\times\Gamma$. Then $F\cap[n-1,n]\times Z\times\Gamma$ is also locally finite and Γ-invariant. For every integer n we can choose $\kappa(n)$ in I such that $(F\cap[n-1,n]\times Z\times\Gamma)\subseteq[n-1,n]\times D_{\kappa(n)}$. This describes a function κ in $I^{\mathbb{N}}$ such that by construction $F\subseteq Y_\kappa$. □

Let $f\colon X'\to X$ be a proper map between Γ-bornological spaces.

LEMMA 5.13. *If $\mathcal{Y}$ is a co-Γ-bounded (resp., trapping) exhaustion of X, then $f^{-1}\mathcal{Y}$ is a co-Γ-bounded (resp., trapping) exhaustion of X'.*

PROOF. The co-Γ-bounded case is a direct consequence of properness and the fact that forming preimages commutes with forming complements.

For the trapping case one uses in addition Lemma 5.5. □

5.2. Continuous equivariant coarse homology theories. In this section we will introduce an additional continuity condition on an equivariant coarse homology theory. We then verify that a continuous equivariant coarse homology theory preserves coproducts.

REMARK 5.14. In Section 7.3 we show continuity of equivariant coarse ordinary homology theory and in Proposition 8.17 continuity of equivariant coarse algebraic K-homology. In Example 8.31 we describe a version of the coarse algebraic K-homology theory which is not continuous.

We can extend the notions of locally finite subsets and trapping / co-Γ-bounded exhaustions to Γ-bornological coarse spaces by just considering the underlying Γ-bornological spaces.

Let $\mathbf{C}$ be a stable cocomplete ∞-category and

$$E\colon \Gamma\mathbf{BC}\to\mathbf{C}$$

an equivariant coarse homology theory. We use the convention (3.1) for the evaluation $E(\mathcal{Y})$ on a filtered family $\mathcal{Y}$ of invariant subsets of a Γ-bornological coarse space. We have a natural morphism

$$E(\mathcal{Y})\to E(X)\ . \tag{5.1}$$

DEFINITION 5.15. E is called *continuous* if for every trapping exhaustion $\mathcal{Y}$ of some Γ-bornological coarse space X the morphism (5.1) is an equivalence.

REMARK 5.16. A continuous equivariant coarse homology theory E is determined by its values on locally finite, invariant spaces. More precisely, let $\mathcal{F}(X)$ be the filtered partially ordered set of locally finite, invariant subsets of X. In view of Example 5.7 we have a trapping exhaustion $\mathcal{Y}:=(F)_{F\in\mathcal{F}(X)}$ of X. So we get

$$E(X)\simeq E(\mathcal{Y})=\operatorname*{colim}_{F\in\mathcal{F}(X)}E(F)$$

showing the claim.

If $\mathcal{F}(X)$ is empty, i.e., X does not admit non-empty locally finite Γ-invariant subsets (see Example 5.7), and E is continuous, then $E(X)\simeq 0$.

By excision an equivariant coarse homology theory preserves coproducts of finite families of Γ-bornological coarse spaces. By the following lemma, for a continuous equivariant coarse homology theory, we can drop the word *finite*.

LEMMA 5.17. *A continuous equivariant coarse homology theory preserves coproducts.*

PROOF. Let $(X_i)_{i\in I}$ be a family of Γ-bornological coarse spaces. We must show that the natural map

$$\bigoplus_{i\in I} E(X_i) \to E(\coprod_{i\in I} X_i)$$

is an equivalence. All invariant subsets of $\coprod_{i\in I} X_i$ have the induced bornological coarse structures.

Consider the following diagram, where the horizontal maps are equivalences by continuity and where $\mathcal{F}(-)$ is the trapping exhaustion consisting of all locally finite, invariant subsets.

$$\begin{array}{ccc} \bigoplus_{i\in I} \operatorname{colim}_{F_i\in\mathcal{F}(X_i)} E(F_i) & \xrightarrow{\simeq} & \bigoplus_{i\in I} E(X_i) \\ \downarrow ! & & \downarrow \\ \operatorname{colim}_{F\in\mathcal{F}(\coprod_{i\in I} X_i)} E(F) & \xrightarrow{\simeq} & E(\coprod_{i\in I} X_i) \end{array}$$

It remains to show that the map marked with ! is an equivalence. The bornology of the coproduct is described in [**8**, Lemma 2.24]. A subset of the coproduct is bounded if and only if its intersection with X_i for every i in I is bounded. This implies that for every invariant, locally finite subset F of $\coprod_{i\in I} X_i$ there exists a minimal finite subset $J(F)$ of I such that $F \subseteq \coprod_{i\in J(F)} X_i$. We write $F_i := F \cap X_i$. Then

$$F = \bigcup_{i\in J(F)} F_i$$

is a finite, coarsely disjoint decomposition. Hence

$$E(F) \simeq \bigoplus_{i\in J(F)} E(F_i) \simeq \bigoplus_{i\in I} E(F_i)\ .$$

That the map marked with ! is an equivalence now follows from the equivalence

$$\bigoplus_{i\in I} \operatorname*{colim}_{F_i\in\mathcal{F}(X_i)} E(F_i) \simeq \operatorname*{colim}_{(F_i)_i\in\mathcal{F}(\coprod_{i\in I} X_i)} \bigoplus_{i\in I} E(F_i)\ .$$

□

5.3. Continuous coarse motives. In this section we will explain how to incorporate continuity on the motivic level and we will discuss basic properties of this procedure. Recall the Yoneda embedding (4.1) and the notation (4.2). We have a natural morphism

$$\mathrm{yo}(\mathcal{Y}) \to \mathrm{yo}(X)\ . \tag{5.2}$$

Let E be an object of $\mathbf{PSh}(\Gamma\mathbf{BC})$.

DEFINITION 5.18. We call E *continuous* if it is local with respect to the morphisms (5.2) for all trapping exhaustions $\mathcal{Y}$ of Γ-bornological coarse spaces X.

REMARK 5.19. Let E be an object of $\mathbf{PSh}(\Gamma\mathbf{BC})$ and recall (4.3). The collection of restriction morphisms $E(X) \to E(Y_i)$ for all i in I induce a natural morphism

$$E(X) \to E(\mathcal{Y}) \ . \tag{5.3}$$

Then E is continuous if and only if the morphism (5.3) is an equivalence for every trapping exhaustion $\mathcal{Y}$ of a Γ-bornological coarse space X.

We now incorporate continuity on the motivic level by adding this relation to the list in Definition 4.7.

DEFINITION 5.20. We define the ∞-category of continuous Γ-equivariant motivic coarse spaces $\Gamma\mathbf{Spc}\mathcal{X}_c$ to be the full localizing subcategory of $\mathbf{Sh}(\Gamma\mathbf{BC})$ of coarsely invariant, continuous and u-continuous sheaves which vanish on flasques.

The locality condition is generated by a small set of morphisms. Therefore we have a localizing adjunction

$$\mathcal{L}_c \colon \mathbf{PSh}(\Gamma\mathbf{BC}) \leftrightarrows \Gamma\mathbf{Spc}\mathcal{X}_c : \mathit{inclusion}\ . \tag{5.4}$$

We define

$$\mathrm{Yo}_c := \mathcal{L}_c \circ \mathrm{yo} \colon \Gamma\mathbf{BC} \to \Gamma\mathbf{Spc}\mathcal{X}_c \ .$$

We furthermore have a localizing adjunction

$$C \colon \Gamma\mathbf{Spc}\mathcal{X} \leftrightarrows \Gamma\mathbf{Spc}\mathcal{X}_c : \mathit{inclusion}$$

and the relations

$$\mathcal{L}_c \simeq C \circ \mathcal{L}\ , \quad \mathrm{Yo}_c \simeq C \circ \mathrm{Yo}\ .$$

where $\mathcal{L}$ is as in (4.6).

The ∞-category $\Gamma\mathbf{Spc}\mathcal{X}$ is a presentable ∞-category.

DEFINITION 5.21. We define the category of continuous equivariant motivic coarse spectra as the stabilization

$$\Gamma\mathbf{Sp}\mathcal{X}_c := \Gamma\mathbf{Spc}\mathcal{X}_{c,*}[\Sigma^{-1}]$$

in the realm of presentable ∞-categories.

Then $\Gamma\mathbf{Sp}\mathcal{X}_c$ is a stable presentable ∞-category which fits into an adjunction

$$\Sigma^{mot}_{c,+} \colon \Gamma\mathbf{Spc}\mathcal{X}_c \leftrightarrows \Gamma\mathbf{Sp}\mathcal{X}_c : \Omega^{mot}_c \ .$$

We further define the following stable continuous version of the Yoneda functor:

$$\mathrm{Yo}^s_c := \Sigma^{mot}_{c,+} \circ \mathrm{Yo}_c \colon \Gamma\mathbf{BC} \to \Gamma\mathbf{Sp}\mathcal{X}_c \ . \tag{5.5}$$

We obtain the functor C^s in the following commuting square from the universal property of the stabilization:

$$\begin{array}{ccc} \Gamma\mathbf{Spc}\mathcal{X} & \xrightarrow{\Sigma^{mot}_{+}} & \Gamma\mathbf{Sp}\mathcal{X} \\ \downarrow C & & \downarrow C^s \\ \Gamma\mathbf{Spc}\mathcal{X}_c & \xrightarrow{\Sigma^{mot}_{c,+}} & \Gamma\mathbf{Sp}\mathcal{X}_c \end{array} \tag{5.6}$$

We furthermore have the relation

$$\mathrm{Yo}^s_c \simeq C^s \circ \mathrm{Yo}^s \ . \tag{5.7}$$

Let $\mathbf{C}$ be a cocomplete stable ∞-category. We denote the full subcategory of $\mathbf{Fun}(\Gamma\mathbf{BC}, \mathbf{C})$ of functors which are continuous Γ-equivariant coarse homology theories in the sense of Definition 3.10 by $\mathbf{Cont}\Gamma\mathbf{CHT}_{\mathbf{C}}$.

The construction of $\Gamma\mathbf{Sp}\mathcal{X}_c$ has the following immediate consequence:

COROLLARY 5.22. *Precomposition with* Yo^s_c *induces an equivalence of* ∞*-categories*

$$\mathbf{Fun}^{\mathrm{colim}}(\Gamma\mathbf{Sp}\mathcal{X}_c, \mathbf{C}) \to \mathbf{Cont}\Gamma\mathbf{CHT}_{\mathbf{C}} .$$

The Yoneda functor Yo^s_c has all the properties listed in Section 4.2. In addition it satisfies:

COROLLARY 5.23. *We have*

$$\mathrm{Yo}^s_c(\mathcal{Y}) \simeq \mathrm{Yo}^s_c(X)$$

for every trapping exhaustion $\mathcal{Y}$ *of a* Γ*-bornological coarse space* X.

So in particular, $\mathrm{Yo}^s_c(X)$ is determined by the collection of invariant, locally finite subsets of X:

$$\mathrm{Yo}^s_c(X) \simeq \operatorname*{colim}_{F\in\mathcal{F}(X)} \mathrm{Yo}^s_c(F) .$$

Note that Yo^s_c is a $\Gamma\mathbf{Sp}\mathcal{X}_c$-valued continuous Γ-equivariant coarse homology theory. Hence Lemma 5.17 implies:

COROLLARY 5.24. *The functor* $\mathrm{Yo}^s_c\colon \Gamma\mathbf{BC} \to \Gamma\mathbf{Spc}\mathcal{X}_c$ *preserves coproducts.*

If $\mathcal{Y} = (Y_i)_{i\in I}$ is a co-Γ-bounded (or trapping, respectively) exhaustion of a Γ-bornological space X and Z is a second Γ-bornological coarse space, then $\mathcal{Y}\times Z := (Y_i \times Z)_{i\in I}$ is not necessarily trapping in $X \otimes Z$. As a consequence the symmetric monoidal structure $\otimes$ does not descend to continuous motivic coarse spectra. But if Z is bounded, then $\mathcal{Y}\times Z$ is again a trapping exhaustion of $X\otimes Z$ by Lemma 5.9. We can conclude:

COROLLARY 5.25. *If* Z *is a bounded* Γ*-bornological coarse space, then the functor*

$$-\otimes Z\colon \Gamma\mathbf{BC} \to \Gamma\mathbf{BC}$$

descends to a functor

$$-\otimes^{mot} Z\colon \Gamma\mathbf{Sp}\mathcal{X}_c \to \Gamma\mathbf{Sp}\mathcal{X}_c$$

such that

$$\mathrm{Yo}^s_c(X\otimes Z) \simeq \mathrm{Yo}^s_c(X) \otimes^{mot} Z . \tag{5.8}$$

5.4. Forcing continuity. To every $\mathbf{C}$-valued equivariant coarse homology theory E we can naturally associate a continuous version E_{cont}. This is actually the best approximation of E by some continuous $\mathbf{C}$-valued equivariant coarse homology theory. We will show that there is an adjunction

$$(-)_{cont}\colon \Gamma\mathbf{CHT}_{\mathbf{C}} \leftrightarrows \mathbf{Cont}\Gamma\mathbf{CHT}_{\mathbf{C}} : \mathit{inclusion} .$$

Let us construct the functor $(-)_{cont}$. Denote by $\Gamma\mathbf{BC}^{mb}$ the full subcategory of $\Gamma\mathbf{BC}$ spanned by the Γ-bornological coarse spaces which carry the minimal bornology. Given a functor

$$E\colon \Gamma\mathbf{BC} \to \mathbf{C}$$

with target a stable cocomplete ∞-category, we define the functor E_{cont} by the left Kan extension

$$\begin{array}{ccc} \Gamma\mathbf{BC}^{mb} & \xrightarrow{E} & \mathbf{C} \\ \downarrow & \nearrow_{E_{cont}} & \\ \Gamma\mathbf{BC} & & \end{array}$$

Since $\mathbf{C}$ is cocomplete, this left Kan extension exists. Because the image of every morphism originating in a Γ-bornological coarse space with minimal bornology is locally finite, the point-wise formula for the left Kan-extension implies that the canonical map

$$\operatorname*{colim}_{F\in\mathcal{F}(X)} E(F_X) \xrightarrow{\sim} E_{cont}(X) \tag{5.9}$$

is an equivalence.

LEMMA 5.26. *If E is an equivariant coarse homology theory, then E_{cont} is a continuous equivariant coarse homology theory.*

PROOF. We start with showing that E_{cont} is coarsely invariant. Let X be a Γ-bornological space. Since the subsets $\{0,1\}\times F$ of $\{0,1\}\times X$ for all locally finite invariant subsets F of X are cofinal in all locally finite invariant subsets of $\{0,1\}_{max,max}\otimes X$ we get the second equivalence in the chain

$$\begin{aligned} E_{cont}(\{0,1\}_{max,max}\otimes X) &\simeq \operatorname*{colim}_{F'\in\mathcal{F}(\{0,1\}\otimes X)} E(F'_{\{0,1\}_{max,max}\otimes X}) \\ &\simeq \operatorname*{colim}_{F\in\mathcal{F}(X)} E(\{0,1\}_{max,max}\otimes F_X) \\ &\simeq \operatorname*{colim}_{F\in\mathcal{F}(X)} E(F_X) \\ &\simeq E_{cont}(X)\ . \end{aligned}$$

See Example 2.10 for the notation F_X (the induced structures on the subset F). The third equivalence in the above chain of equivalences follows from the coarse invariance of E.

Next we show that E_{cont} satisfies excision. If $(Z,\mathcal{Y})$ is an invariant complementary pair on the Γ-bornological coarse space X, then $(F\cap Z, F\cap\mathcal{Y})$ is an invariant complementary pair on F_X. Hence

$$\begin{array}{ccc} E_{cont}(Z\cap\mathcal{Y}) & \longrightarrow & E_{cont}(\mathcal{Y}) \\ \downarrow & & \downarrow \\ E_{cont}(Z) & \longrightarrow & E_{cont}(X) \end{array}$$

is the colimit of the push-out diagrams over F in $\mathcal{F}(X)$:

$$\begin{array}{ccc} E(F\cap Z\cap\mathcal{Y}) & \longrightarrow & E(F\cap\mathcal{Y}) \\ \downarrow & & \downarrow \\ E(F\cap Z) & \longrightarrow & E(F_X) \end{array}$$

(all subsets of F are equipped with the bornological coarse structures induced from F_X) and hence itself a push-out diagram.

We now show that E_{cont} vanishes on flasques. Assume that X is a flasque Γ-bornological coarse space and that flasqueness is implemented by the morphism $f\colon X \to X$. If F is an invariant, locally finite subset of X, then $\widetilde{F} := \bigcup_{n\in\mathbb{N}} f^n(F)$ is again an invariant, locally finite subset of X. Furthermore, $\widetilde{F}_X$ is flasque with flasqueness implemented by the restriction $f|_{\widetilde{F}}$. The inclusion $F \to \widetilde{F}$ belongs to the structure maps for the colimit over $\mathcal{F}(X)$. Since $E(\widetilde{F}) \simeq 0$ this finally implies that $E_{cont}(X) \simeq 0$.

Finally, we have u-continuity by

$$\begin{aligned}\operatorname*{colim}_{U\in\mathcal{C}} E_{cont}(X_U) &\simeq \operatorname*{colim}_{U\in\mathcal{C}} \operatorname*{colim}_{F\in\mathcal{F}(X)} E(F_{X_U})\\ &\simeq \operatorname*{colim}_{U\in\mathcal{C}} \operatorname*{colim}_{F\in\mathcal{F}(X)} E(F_{(F\times F)\cap U})\\ &\simeq \operatorname*{colim}_{F\in\mathcal{F}(X)} \operatorname*{colim}_{U\in\mathcal{C}} E(F_{(F\times F)\cap U})\\ &\simeq \operatorname*{colim}_{F\in\mathcal{F}(X)} E(F_X)\\ &\simeq E_{cont}(X)\ .\end{aligned}$$

This finishes the proof that E_{cont} is an equivariant coarse homology theory.

We now argue that E_{cont} is continuous. Let X be a Γ-bornological coarse space and $\mathcal{Y} = (Y_i)_{i\in I}$ be a trapping exhaustion of X. For every invariant, locally finite subset F of X exists an index i in I such that $F \subseteq Y_i$. Furthermore, for every i in I we have an inclusion $\mathcal{F}(Y_i) \subseteq \mathcal{F}(X)$. This implies that

$$E_{cont}(\mathcal{Y}) \simeq \operatorname*{colim}_{i\in I} \operatorname*{colim}_{F\in\mathcal{F}(Y_i)} E(F_{Y_i}) \simeq \operatorname*{colim}_{F\in\mathcal{F}(X)} E(F_X) \simeq E_{cont}(X)\ .$$

This finishes the proof of Lemma 5.26. □

PROPOSITION 5.27. *There exists an adjunction*

$$(-)_{cont}\colon \Gamma\mathbf{CHT_C} \leftrightarrows \mathbf{Cont}\Gamma\mathbf{CHT_C} : \mathit{inclusion}\ .$$

PROOF. The collection of maps $F \to X$ for all locally finite invariant subsets F of X induces a transformation of functors

$$\eta\colon (-)_{cont} \to \mathrm{id}$$

on $\Gamma\mathbf{CHT_C}$. By Remark 5.16, if E is continuous, then the transformation $\eta_E\colon E_{cont} \to E$ is an equivalence. Moreover the transformations $\eta_{E_{cont}}$ and $(\eta_E)_{cont}$ (i.e., the functor $(-)_{cont}$ applied to η_E) are equivalences. By [**26**, Prop. 5.2.7.4] we get the desired adjunction. □

Let the cocomplete stable ∞-category $\mathbf{C}$ admit all small products. Let E be a $\mathbf{C}$-valued equivariant coarse homology theory.

LEMMA 5.28. *If E is strongly additive, then so is E_{cont}.*

PROOF. Let $(X_i)_{i\in I}$ be a family of Γ-bornological coarse spaces and set $X := \bigsqcup^{\mathrm{free}}_{i\in I} X_i$ (Example 2.16). For a subset F of X and i in I we write $F_i := F \cap X_i$. Then F is locally finite in X if and only if F_i is locally finite in X_i for every i in I. Furthermore, we have an isomorphism of Γ-bornological coarse spaces $F_X \cong \bigsqcup^{\mathrm{free}}_{i\in I} F_{i,X_i}$. This implies

$$\operatorname*{colim}_{F\in\mathcal{F}(X)} \prod_{i\in I} E(F_{i,X_i}) \simeq \prod_{i\in I} \operatorname*{colim}_{F_i\in\mathcal{F}(X_i)} E(F_{i,X_i})\ .$$

We must show that the right vertical map in the diagram

$$\begin{array}{ccc} \operatorname{colim}_{F\in\mathcal{F}(X)} E(F_X) & \xrightarrow{\simeq} & E_{cont}(X) \\ \downarrow\simeq & & \\ \operatorname{colim}_{F\in\mathcal{F}(X)} \prod_{i\in I} E(F_{i,X_i}) & & \downarrow \\ \downarrow\simeq & & \\ \prod_{i\in I} \operatorname{colim}_{F_i\in\mathcal{F}(X_i)} E(F_{i,X_i}) & \xrightarrow{\simeq} & \prod_{i\in I} E_{cont}(X_i) \end{array}$$

is an equivalence. Indeed, the horizontal maps are equivalences by the definition of E_{cont} and the upper vertical map is an equivalence since E is strongly additive. □

6. Change of groups

In this section we describe various change of groups constructions. They induce adjunctions between the corresponding categories of equivariant motivic coarse spectra which are very similar to the base change functors in motivic homotopy theory.

Assume that we are given an equivariant coarse homology theory defined for every group (as it is the case for all our examples). The compatibility of the equivariant coarse homology theory with the change of groups functors is expressed by natural transformations. These transformations are additional data and will be discussed for every example of equivariant coarse homology theory separately.

What we describe here is the beginning of a story which should finally capture all group-change functors in a sort of spectral Mackey functor formalism (see, e.g., Barwick [**6**] and Barwick–Glasman–Shah [**7**]). The obvious task here is to capture the relations between these functors on the motivic level (like iterated restrictions or inductions and the Mackey relation, Lemma 6.8) together with all their higher coherences in a proper way.

All change of groups transformations are associated to a homomorphism of groups

$$\iota\colon H\to\Gamma\ .$$

6.1. Restriction. Every Γ-bornological coarse space gives rise to an H-bornological coarse space, where the action of H is induced from the action of Γ via ι. In this way we get a restriction functor

$$\mathrm{Res}^{\Gamma}_{H}\colon \Gamma\mathbf{BC}\to H\mathbf{BC}\ .$$

If the homomorphism ι is not clear from the context, then we add it to the notation and write $\mathrm{Res}^{\Gamma}_{H}(\iota)$.

The functor $\mathrm{Res}^{\Gamma}_{H}$ induces a pull back functor $\mathrm{Res}^{\Gamma}_{H,pre}$ for presheaves, which preserves all limits and colimits. Since $\mathbf{PSh}(\Gamma\mathbf{BC})$ is a presentable ∞-category, by Lurie [**26**, Cor. 5.5.2.9] the functor $\mathrm{Res}^{\Gamma}_{H,pre}$ is the right-adjoint of an adjunction

$$\mathrm{Res}^{\Gamma,pre}_{H}\colon \mathbf{PSh}(\Gamma\mathbf{BC})\leftrightarrows\mathbf{PSh}(H\mathbf{BC}) : \mathrm{Res}^{\Gamma}_{H,pre}\ .$$

The functor $\mathrm{Res}^{\Gamma}_{H}$ sends equivariant complementary pairs on a Γ-bornological coarse space X to equivariant complementary pairs on the H-bornological coarse space $\mathrm{Res}^{\Gamma}_{H}(X)$. Consequently, the restriction functor $\mathrm{Res}^{\Gamma}_{H,pre}$ preserves sheaves. In the following, we decorate the Yoneda functors by the relevant group. Using $\mathrm{Res}^{\Gamma,pre}_{H}\circ$

$\mathrm{yo}_\Gamma \simeq \mathrm{yo}_H \circ \mathrm{Res}^\Gamma_H$, we see that $\mathrm{Res}^{\Gamma,pre}_H$ sends the generators of the localization listed in Definition 4.6 for the group Γ to corresponding generators for H. We conclude that $\mathrm{Res}^\Gamma_{H,pre}$ preserves coarsely invariant sheaves, sheaves which vanish on flasque spaces, and u-continuous sheaves. Hence we get an adjunction

$$\mathrm{Res}^{\Gamma,*}_H : \Gamma\mathbf{Spc}\mathcal{X} \leftrightarrows H\mathbf{Spc}\mathcal{X} : \mathrm{Res}^\Gamma_{H,*} \ .$$

Here $\mathrm{Res}^\Gamma_{H,*}$ is given by the restriction of $\mathrm{Res}^\Gamma_{H,pre}$ to coarse motivic spaces, and its left adjoint satisfies

$$\mathrm{Res}^{\Gamma,*}_H \simeq \mathcal{L}_H \circ \mathrm{Res}^{\Gamma,pre}_H \ ,$$

where $\mathcal{L}_H$ is the localization as in (4.6) (we have added a subscript H in order to indicate the relevant group).

Hence by passing to the stabilizations we get an adjunction

$$\mathrm{Res}^{\Gamma,Mot}_H : \Gamma\mathbf{Sp}\mathcal{X} \leftrightarrows H\mathbf{Sp}\mathcal{X} : \mathrm{Res}^\Gamma_{H,Mot} \ ,$$

where $\mathrm{Res}^\Gamma_{H,Mot}$ is defined by the extension of the functor $\mathrm{Res}^\Gamma_{H,pre}$ to stable objects. The obvious equivalence

$$\mathrm{Res}^{\Gamma,pre}_H \circ \mathrm{yo}_\Gamma \simeq \mathrm{yo}_H \circ \mathrm{Res}^\Gamma_H$$

implies the equivalence

$$\mathrm{Res}^{\Gamma,Mot}_H \circ \mathrm{Yo}^s_\Gamma \simeq \mathrm{Yo}^s_H \circ \mathrm{Res}^\Gamma_H \ .$$

where we have decorated the Yoneda functors by the relevant group.

6.2. Completion. We consider a Γ-bornological coarse space X. Let $\mathcal{C}$ and $\mathcal{B}$ denote the coarse structure and the bornology of X. We define a new compatible bornology $\mathcal{B}_H$ on X generated by the $\iota(H)$-completions $\iota(H)B$ of the bounded subsets B of X. We observe that $\mathcal{C}$ and $\mathcal{B}_H$ are compatible and that $\mathcal{B}_H$ is $N_\Gamma(\iota(H))$-invariant (as a subset of $\mathcal{P}(X)$), where $N_\Gamma(\iota(H))$ denotes the normalizer of the subgroup $\iota(H)$ in Γ.

DEFINITION 6.1. The *H-completion* of X is the $N_\Gamma(\iota(H))$-bornological coarse space defined by $B_H(X) := (X, \mathcal{C}, \mathcal{B}_H)$.

In this way we define a functor

$$B_H : \Gamma\mathbf{BC} \to N_\Gamma(\iota(H))\mathbf{BC} \ .$$

The pull-back along B_H induces an adjunction

$$B^{pre}_H : \mathbf{PSh}(\Gamma\mathbf{BC}) \leftrightarrows \mathbf{PSh}(N_\Gamma(\iota(H))\mathbf{BC}) : B_{H,pre} \ .$$

It is easy to see that $B_{H,pre}$ preserves sheaves, u-continuity and coarse invariance, since its left-adjoint adjoint B^{pre}_H preserves the corresponding generating morphisms. See Section 6.1 for a similar argument.

If X is a flasque Γ-bornological coarse space with flasqueness implemented by $f : X \to X$, then $B_H(X)$ is flasque with flasqueness implemented by the same map. Here it is important to define flasqueness with Condition 3 in Definition 3.8 and not the weaker one discussed in Remark 3.9. Consequently, $B_{H,pre}$ preserves presheaves which vanish on flasques.

Similarly as in the case of the restriction, we get an adjunction

$$B^{Mot}_H : \Gamma\mathbf{Sp}\mathcal{X} \leftrightarrows N_\Gamma(\iota(H))\mathbf{Sp}\mathcal{X} : B_{H,Mot}$$

and an equivalence

$$B_H^{Mot} \circ \mathrm{Yo}_\Gamma^s \simeq \mathrm{Yo}_{N_\Gamma(\iota(H))}^s \circ B_H \ .$$

The identity of the underlying set induces a morphism $b_H \colon B_H(X) \to \mathrm{Res}^\Gamma_{N_\Gamma(\iota(H))}(X)$. The transformation b_H induces a natural transformation of functors

$$b_H \colon B_H \to \mathrm{Res}^\Gamma_{N_\Gamma(\iota(H))} \colon \Gamma\mathbf{BC} \to N_\Gamma(\iota(H))\mathbf{BC} \ .$$

By functoriality, we get a transformation

$$b_H \colon E \circ B_H \to E \circ \mathrm{Res}^\Gamma_{N_\Gamma(\iota(H))}$$

for any equivariant coarse homology theory E.

REMARK 6.2. If H is a finite group, then $B_H \cong \mathrm{Res}^\Gamma_{N_\Gamma(\iota(H))}$.

6.3. Quotients. We consider a Γ-bornological coarse space X. We form the quotient set $H\backslash X$ which carries an action of the group

$$W_\Gamma(H) := N_\Gamma(\iota(H))/\iota(H) \ ,$$

where $N_\Gamma(\iota(H))$ denotes the normalizer of the subgroup $\iota(H)$ in Γ.

Let $\pi \colon X \to H\backslash X$ denote the projection. We equip $H\backslash X$ with the maximal bornology such that the projection $\pi \colon B_H(X) \to H\backslash X$ is proper. Furthermore, we equip $H\backslash X$ with the minimal coarse structure such that $\pi \colon X \to H\backslash X$ is controlled. In this way we define a functor

$$Q_H \colon \Gamma\mathbf{BC} \to W_\Gamma(H)\mathbf{BC} \ .$$

The projection maps define a natural transformation $\pi_H \colon B_H \to \mathrm{Res}^{W_\Gamma(H)}_{N_\Gamma(\iota(H))}(Q_H)$ of $N_\Gamma(\iota(H))\mathbf{BC}$-valued functors.

For X in $\Gamma\mathbf{BC}$ one can interpret $\mathrm{Res}^{W_\Gamma(H)}_{N_\Gamma(\iota(H))} Q_H(X)$ as a coequalizer. For a bornological coarse space Z let $Z_{max-\mathcal{B}}$ denote the bornological coarse space obtained by replacing its bornology by the maximal bornology. The two maps $H \times X \to X$ in the lemma below are given by the projection $(h,x) \mapsto x$ and the action $(h,x) \mapsto hx$. The diagram and the colimit are considered in $N_\Gamma(\iota(H))\mathbf{BC}$, where the action of σ in $N_\Gamma(\iota(H))$ on $H \times X$ is given by $\sigma(h,x) := (\sigma h \sigma^{-1}, \sigma x)$.

LEMMA 6.3. *We have an isomorphism*

$$\mathrm{Res}^{W_\Gamma(H)}_{N_\Gamma(\iota(H))} Q_H(X) \cong \mathrm{colim}\left((H_{min,max} \otimes X)_{max-\mathcal{B}} \rightrightarrows B_H(X)\right) \ .$$

PROOF. One checks that the two morphisms in the coequalizer diagram are controlled. Since we replaced the bornology on the domain by the maximal one they are obviously proper. The coequalizer diagram is colim-admissible, see Example 2.25. One checks that the description of the structures on $Q_H(X)$ given above coincides with the explicit description of the structures of the colimit given in the proof of Proposition 2.21. □

The functor Q_H induces a pull-back in presheaves $Q_{H,pre}$. This functor preserves all limits and colimits. Again by Lurie [**26**, Cor. 5.5.2.9] it fits into an adjunction

$$Q_H^{pre} \colon \mathbf{PSh}(\Gamma\mathbf{BC}) \leftrightarrows \mathbf{PSh}(W_\Gamma(H)\mathbf{BC}) : Q_{H,pre} \ .$$

Note that the functor Q_H induces a bijection between equivariant complementary pairs on a Γ-bornological coarse space X and on the $W_\Gamma(H)$-bornological coarse space $Q_H(X)$. Consequently, $Q_{H,pre}$ preserves sheaves.

It is furthermore clear that $Q_{H,pre}$ preserves coarsely invariant sheaves and u-continuous sheaves. If the Γ-bornological coarse space X is flasque with flasqueness implemented by $f\colon X \to X$, then the induced map $\bar{f}\colon Q_H(X) \to Q_H(X)$ implements flasqueness of $Q_H(X)$. It follows that $Q_{H,pre}$ preserves coarse motivic spaces.

Hence $Q_{H,pre}$ restricts to equivariant coarse motivic spaces. Similar as before we get an adjunction

$$Q_H^{Mot}\colon \Gamma\mathbf{Sp}\mathcal{X} \leftrightarrows W_\Gamma(H)\mathbf{Sp}\mathcal{X} : Q_{H,Mot} .$$

We have the relation

$$\mathrm{Yo}^s_{W_\Gamma(H)} \circ Q_H \simeq Q_H^{Mot} \circ \mathrm{Yo}^s_\Gamma .$$

6.4. Products. We consider two groups Γ and Γ'. For a Γ-bornological coarse space X and a Γ'-bornological coarse space X' we can form the product $X \otimes X'$ which is a $(\Gamma\times\Gamma')$ bornological coarse space. We fix the Γ-bornological coarse space X and consider the functor

$$P_X := X \otimes -\colon \Gamma'\mathbf{BC} \to (\Gamma \times \Gamma')\mathbf{BC} .$$

As in the preceding cases one can check that the restriction

$$P_{X,pre}\colon \mathbf{PSh}((\Gamma \times \Gamma')\mathbf{BC}) \to \mathbf{PSh}(\Gamma'\mathbf{BC})$$

along P_X preserves coarse motivic spaces and induces an adjunction

$$P_X^{Mot}\colon \Gamma'\mathbf{Sp}\mathcal{X} \leftrightarrows (\Gamma \times \Gamma')\mathbf{Sp}\mathcal{X} : P_{X,Mot}$$

such that

$$P_X^{Mot} \circ \mathrm{Yo}^s_{\Gamma'} \simeq \mathrm{Yo}^s_{(\Gamma\times\Gamma')} \circ P_X .$$

6.5. Induction. We now come back to our original situation and consider the homomorphism $\iota\colon H \to \Gamma$ of groups. We define a $(\Gamma \times H)$-bornological coarse space $\hat{\Gamma}$ as follows:

(1) The underlying bornological coarse space of $\hat{\Gamma}$ is $\Gamma_{min,min}$.
(2) The group $(\Gamma \times H)$ acts on the set $\hat{\Gamma}$ by

$$(\gamma, h)\gamma' := \gamma\gamma'\iota(h)^{-1} .$$

We define the functor

$$\hat{P}_\Gamma := \mathrm{Res}^{\Gamma\times H\times H}_{\Gamma\times H} \circ P_{\hat{\Gamma}}\colon H\mathbf{BC} \to (\Gamma \times H)\mathbf{BC} ,$$

where the restriction is along the homomorphism

$$\mathrm{id}_\Gamma \times \mathrm{diag}_H\colon \Gamma \times H \to \Gamma \times H \times H .$$

This functor extends to motives

$$\hat{P}_\Gamma^{Mot} := \mathrm{Res}^{\Gamma\times H\times H,Mot}_{\Gamma\times H} \circ P^{Mot}_{\hat{\Gamma}} .$$

By construction we have an adjunction

$$\hat{P}_\Gamma^{Mot}\colon H\mathbf{Sp}\mathcal{X} \leftrightarrows (\Gamma \times H)\mathbf{Sp}\mathcal{X} : \hat{P}_{\Gamma,Mot}$$

and the relation

$$\mathrm{Yo}^s_{\Gamma\times H} \circ \hat{P}_\Gamma \simeq \hat{P}_\Gamma^{Mot} \circ \mathrm{Yo}^s_H .$$

We consider the canonical embedding $\kappa\colon H \to \Gamma \times H$ into the second factor. Note that then we have $\Gamma \times H = N_{\Gamma\times H}(H)$ and hence $W_{\Gamma\times H}(H) \cong \Gamma$. We thus have the quotient functor

$$Q_H(\kappa)\colon (\Gamma \times H)\mathbf{BC} \to \Gamma\mathbf{BC}$$

(it is useful to add the embedding κ as an argument since there is also the other obvious homomorphism $(\iota, \mathrm{id})\colon H \to \Gamma \times H$). We define the induction functor as the composition

$$\mathrm{Ind}_H^\Gamma\colon H\mathbf{BC} \to \Gamma\mathbf{BC}\ , \quad \mathrm{Ind}_H^\Gamma := Q_H(\kappa) \circ \hat{P}_\Gamma\ .$$

The induction functor extends to motives

$$\mathrm{Ind}_H^{\Gamma,Mot} := Q_H^{Mot}(\kappa) \circ \hat{P}_\Gamma^{Mot} : H\mathbf{Sp}\mathcal{X} \to \Gamma\mathbf{Sp}\mathcal{X}$$

such that

$$\mathrm{Ind}_H^{\Gamma,Mot} \circ \mathrm{Yo}_H^s \simeq \mathrm{Yo}_\Gamma^s \circ \mathrm{Ind}_H^\Gamma\ .$$

REMARK 6.4. The underlying Γ-set of $\mathrm{Ind}_H^\Gamma(X)$ is $\Gamma \times_H X$ with the left-action of Γ on the left factor. Here $\Gamma \times_H X$ stands for the quotient set $H\backslash(\Gamma \times X)$ with respect to the action given by

$$h(\gamma, x) = (\gamma\iota(h^{-1}), hx)\ .$$

The bornology on $\Gamma \times_H X$ is generated by the images of the subsets $\{\gamma\} \times B$ for all bounded subsets of X, and the coarse structure is generated by the images of $\mathrm{diag}_\Gamma \times U$ for all entourages U of X.

Let $K := \ker(\iota\colon H \to \Gamma)$.

LEMMA 6.5. *We have* $\mathrm{Ind}_H^\Gamma \cong \mathrm{Ind}_H^\Gamma \circ B_K$.

PROOF. Straightforward. □

REMARK 6.6. Using Lemma 6.3 one can easily check that we have an isomorphism

$$\mathrm{Ind}_H^\Gamma(X) \cong \mathrm{colim}\left((H_{min,max} \otimes \Gamma_{min,min} \otimes X)_{max-\mathcal{B}} \rightrightarrows B_H(\Gamma_{min,min} \otimes X)\right)$$

in $\Gamma\mathbf{BC}$. The two arrows are given by $(h, \gamma, x) \mapsto (\gamma, x)$ and $(h, \gamma, x) \mapsto (\gamma h^{-1}, hx)$. The group Γ acts on the factor $\Gamma_{min,min}$ by left multiplication. We used the more complicated description of the induction as a composition of various previously defined functors in order to deduce that induction descends to motives.

PROPOSITION 6.7. *If K is finite and the image of ι has finite index in Γ, then we have an adjunction*

$$\mathrm{Ind}_H^\Gamma\colon H\mathbf{BC} \leftrightarrows \Gamma\mathbf{BC} : \mathrm{Res}_H^\Gamma\ .$$

PROOF. As we observed earlier, the underlying set of $\mathrm{Ind}_H^\Gamma(X)$ is given by $\Gamma \times_H X$. Recall that the functions [2]

$$X \to \Gamma \times_H X, \quad x \mapsto [e, x]$$

and

$$\Gamma \times_H \mathrm{Res}_H^\Gamma(Y) \to Y, \quad [\gamma, y] \mapsto \gamma y$$

[2] We use the word "function" in order to denote maps between underlying sets

define the unit and counit, respectively, of an adjunction $H\mathbf{Set} \rightleftarrows \Gamma\mathbf{Set}$. We must check that these functions define morphisms of equivariant bornological coarse spaces.

Since K is finite and normal in H, we have $B_K X \cong X$ for every H-bornological coarse space X. We claim that the function

$$X \to \Gamma \times X\ , \quad x \mapsto (e, x)$$

defines a natural morphism $B_K X \to B_H(\Gamma_{min,min} \otimes X)$. This function is obviously controlled. To see that it is proper, we note that the generating bounded subsets of $B_H(\Gamma_{min,min} \otimes X)$ are of the form $\iota(H)(\{\gamma\} \times B)$. The intersection of such a subset with $\{e\} \times X$ is equal to $\{e\} \times KB$, hence bounded because K is finite. Then the unit is given by the composition

$$X \cong B_K(X) \to \mathrm{Res}^\Gamma_H(B_H(\Gamma_{min,min} \otimes X)) \xrightarrow{\pi} \mathrm{Res}^\Gamma_H(Q_H(\Gamma_{min,min} \otimes X))$$

using $\mathrm{Res}^\Gamma_H(Q_H(\Gamma_{min,min} \otimes X)) = \mathrm{Res}^\Gamma_H(\mathrm{Ind}^\Gamma_H(X))$. Consider now the composition

$$F \colon \Gamma \times \mathrm{Res}^\Gamma_H Y \to \Gamma \times_H \mathrm{Res}^\Gamma_H Y \to Y$$

of the projection map with the function which will provide the desired counit. It suffices to show that this composition is a morphism of bornological coarse spaces $B_H(\Gamma_{min,min} \otimes \mathrm{Res}^\Gamma_H(Y)) \to Y$. If U is an entourage of Y, then $F(\mathrm{diag}_\Gamma \times U) = \bigcup_{\gamma\in\Gamma}(\gamma\times\gamma)(f\times f)(U)$ is an entourage of Y, see Remark 2.3. If B is a bounded subset of Y, then $F^{-1}(B) = \bigcup_{\gamma\in\Gamma}\{\gamma^{-1}\} \times \gamma B$ is bounded since $\bigcup_{h\in H}\{\gamma\iota(h)^{-1}\} \times \iota(h)B$ is bounded for every γ in Γ and the image of ι has finite index in Γ. This shows that F, and hence the counit, is a morphism. □

We now consider homomorphisms $H \to \Gamma$ and $H' \to \Gamma$, and set $K := \ker(H \to \Gamma)$ and $K' := \ker(H' \to \Gamma)$. Note that $W_H(K) \cong H/K =: \bar{H}$ and $W_{H'}(K') \cong H'/K' =: \bar{H}'$.

LEMMA 6.8. *Assume that $H'\backslash\Gamma/H$ is finite.*
For an H-bornological coarse space X we have the relation

$$\mathrm{Res}^\Gamma_{H'} \circ \mathrm{Ind}^\Gamma_H(X) \cong \coprod_{[\gamma]\in\bar{H}'\backslash\Gamma/\bar{H}} \mathrm{Res}^{\bar{H}'}_{H'} \circ \mathrm{Ind}^{\bar{H}'}_{\bar{H}\cap\gamma^{-1}\bar{H}'\gamma} \circ \mathrm{Res}^{\bar{H}}_{\bar{H}\cap\gamma^{-1}\bar{H}'\gamma}(c_\gamma) \circ \mathrm{Ind}^{\bar{H}}_H(X)\ .$$

where $c_\gamma \colon \bar{H} \cap \gamma^{-1}\bar{H}'\gamma \to \bar{H}$ is given by $\bar{h} \mapsto \gamma^{-1}\bar{h}\gamma$.

PROOF. One just makes all definitions explicit. □

The induction on the motivic level is given by

$$\mathrm{Ind}^{\Gamma,Mot}_H := Q^{Mot}_H \circ \hat{P}^{Mot}_\Gamma\ .$$

This functor fits into the adjunction

$$\mathrm{Ind}^{\Gamma,Mot}_H \colon H\mathbf{Sp}\mathcal{X} \leftrightarrows \Gamma\mathbf{Sp}\mathcal{X} : \mathrm{Ind}^\Gamma_{H,Mot}$$

and is compatible with the Yoneda functor:

$$\mathrm{Ind}^{\Gamma,Mot}_H \circ \mathrm{Yo}^s_H \simeq \mathrm{Yo}^s_\Gamma \circ \mathrm{Ind}^\Gamma_H\ .$$

Since the Yoneda functor preserves finite coproducts (since it is excisive), the Lemma 6.8 implies:

COROLLARY 6.9. *If $H'\backslash\Gamma/H$ is finite, then*

$$\mathrm{Res}^{\Gamma,Mot}_{H'}\circ\mathrm{Ind}^{\Gamma,Mot}_{H} \simeq \bigoplus_{[\gamma]\in\bar{H}'\backslash G/\bar{H}} \mathrm{Res}^{\bar{H}',Mot}_{H'}\circ\mathrm{Ind}^{\bar{H}',Mot}_{\bar{H}\cap\gamma^{-1}\bar{H}'\gamma}\circ\mathrm{Res}^{\bar{H},Mot}_{\bar{H}\cap\gamma^{-1}\bar{H}'\gamma}(c_\gamma)\circ Q^{Mot}_{K}\ .$$

Part 2. Examples

7. Equivariant coarse ordinary homology

7.1. Construction. In this section we introduce equivariant coarse ordinary homology theory. Its construction is completely analogous to the non-equivariant case [**8**, Sec. 6.3].

Let X be a Γ-bornological coarse space and let n be a natural number. An *n-chain on X* is a function $c\colon X^{n+1}\to\mathbb{Z}$. Its *support* is defined by $\mathrm{supp}(c):=\{x\in X^{n+1}\,|\,c(x)\neq 0\}$. We typically think of n-chains as infinite linear combinations of points in X^{n+1}.

Let U be a coarse entourage of X and let B be a bounded subset. A point $(x_0,\dots,x_n)$ in X^{n+1} is *U-controlled* if $(x_i,x_j)\in U$ for every $0\le i,j\le n$. The point $(x_0,\dots,x_n)$ in X^{n+1} *meets B* if there exists $0\le i\le n$ such that x_i lies in B.

An n-chain c is *U-controlled* if every point of its support is U-controlled. The n-chain c is *controlled* if it is U-controlled for some coarse entourage U of X. Furthermore, c is *locally finite* if for every bounded subset B of X the set of points in $\mathrm{supp}(c)$ which meet B is finite. We let $C\mathcal{X}_n(X)$ denote the abelian group of controlled locally finite n-chains.

The boundary operator $\partial\colon C\mathcal{X}_n(X)\to C\mathcal{X}_{n-1}(X)$ (for all $n\ge 1$) is defined to be $\partial:=\sum_{i=0}^{n}(-1)^i\partial_i$, where ∂_i is the linear extension of the operator $X^{n+1}\to X^n$ omitting the i'th entry. One checks that ∂ is well-defined and a differential of a chain complex. By $C\mathcal{X}(X)$ we denote the chain complex of locally finite and controlled chains on X.

DEFINITION 7.1. For every natural number n we let $C\mathcal{X}^\Gamma_n(X)$ denote the subgroup of $C\mathcal{X}_n(X)$ of locally finite and controlled n-chains which are in addition Γ-invariant.

For every natural number n the boundary operator $\partial\colon C\mathcal{X}_{n+1}(X)\to C\mathcal{X}_n(X)$ restricts to a boundary operator $\partial\colon C\mathcal{X}^\Gamma_{n+1}(X)\to C\mathcal{X}^\Gamma_n(X)$ between the subgroups of Γ-invariants. Hence we have defined a subcomplex $C\mathcal{X}^\Gamma(X)$ of $C\mathcal{X}(X)$. If $f\colon X\to X'$ is a morphism between Γ-bornological coarse spaces, then the induced map $C\mathcal{X}(f)\colon C\mathcal{X}(X)\to C\mathcal{X}(X')$ of chain complexes preserves the subcomplexes of Γ-invariants. Therefore, we obtain a functor

$$C\mathcal{X}^\Gamma\colon \Gamma\mathbf{BC}\to\mathbf{Ch}\ ,$$

where $\mathbf{Ch}$ denotes the category of chain complexes.

In order to go from chain complexes to spectra we use the Eilenberg–MacLane correspondence

$$\mathcal{EM}\colon \mathbf{Ch}\to\mathbf{Sp}\ . \tag{7.1}$$

One way to define this functor is as the composition

$$\mathcal{EM}\colon \mathbf{Ch}\to\mathbf{Ch}[W^{-1}]\to\mathbf{Sp}\ ,$$

where the first functor is the ∞-categorical localization of the category of chain complexes at quasi-isomorphisms, and the second functor is the mapping spectrum functor $\mathrm{map}(\mathbb{Z}[0],\dots)$ of the stable ∞-category $\mathbf{Ch}[W^{-1}]$, where $\mathbb{Z}[0]$ is the chain complex with $\mathbb{Z}$ placed in degree zero.

DEFINITION 7.2. We define the functor $H\mathcal{X}^\Gamma\colon \mathbf{BC}\to\mathbf{Sp}$ by

$$H\mathcal{X}^\Gamma := \mathcal{EM}\circ C\mathcal{X}^\Gamma\ .$$

THEOREM 7.3. *$H\mathcal{X}^\Gamma$ is an equivariant coarse homology theory.*

PROOF. We observe that the arguments given in the proof of the [**8**, Thm. 6.15] extend word-by-word to the equivariant case. □

7.2. Calculations for spaces of the form $\Gamma_{can,min}\otimes S_{min,max}$. In this section we will do some computations of equivariant coarse ordinary homology groups. In particular, we will relate it to ordinary group homology.

EXAMPLE 7.4. If the Γ-bornological coarse space X has the trivial Γ-action, then we have an isomorphism $H\mathcal{X}^\Gamma(X)\cong H\mathcal{X}(X)$.

In order to provide more examples we consider the group homology functor

$$H(\Gamma,-)\colon \mathbf{Mod}(\mathbb{Z}[\Gamma])\to\mathbf{Sp}$$

which can be defined as the composition

$$\mathbf{Mod}(\mathbb{Z}[\Gamma])\to\mathbf{Ch}_{\mathbb{Z}[\Gamma]}\to\mathbf{Ch}_{\mathbb{Z}[\Gamma]}[W^{-1}]\xleftarrow{\simeq}\mathbf{Ch}^{\mathit{free}}_{\mathbb{Z}[\Gamma]}[W^{-1}]\xrightarrow{\mathbb{Z}\otimes_{\mathbb{Z}[\Gamma]}-}\mathbf{Ch}_{\mathbb{Z}}[W^{-1}]\xrightarrow{\mathcal{EM}}\mathbf{Sp}\ .$$

The first functor sends a $\mathbb{Z}[\Gamma]$-module to a chain complex of $\mathbb{Z}[\Gamma]$-modules concentrated in degree 0. The second functor is the localization at quasi-isomorphisms. The equivalence in the third step is induced by the inclusion of the full subcategory of chain complexes of free $\mathbb{Z}[\Gamma]$-modules. It is essentially surjective by the existence of free resolutions. Finally, the functor $\mathbb{Z}\otimes_{\mathbb{Z}[\Gamma]}-$ is well-defined since it preserves quasi-isomorphisms between chain complexes of free $\mathbb{Z}[\Gamma]$-modules.

Since the above definition involves the inverse of an equivalence it does not directly provide an explicit formula. But for calculations it is useful to choose an explicit model for $H(\Gamma,-)$. The standard choice is as follows. We consider the chain complex of $\mathbb{Z}[\Gamma]$-modules $C(\Gamma)$ given by

$$\dots\to\mathbb{Z}[\Gamma^{n+1}]\to\mathbb{Z}[\Gamma^n]\to\dots\to\mathbb{Z}[\Gamma]\ .$$

The differential $C(\Gamma)_{n+1}\to C(\Gamma)_n$ is defined as the linear extension of the map

$$(\gamma_0,\dots,\gamma_{n+1})\to\sum_{i=0}^{n+1}(-1)^i(\gamma_0,\dots,\widehat{\gamma}_i,\dots,\gamma_{n+1})\ ,$$

where $\widehat{\gamma}_i$ indicates that this component gets omitted. The group Γ acts diagonally on the products Γ^n and this induces the $\mathbb{Z}[\Gamma]$-module structure on $C(\Gamma)$. We now consider the functor

$$C(\Gamma,-)\colon\mathbf{Mod}(\mathbb{Z}[\Gamma])\to\mathbf{Ch}_{\mathbb{Z}}\ ,\quad V\mapsto\mathbb{Z}\otimes_{\mathbb{Z}[\Gamma]}(C(\Gamma)\otimes_{\mathbb{Z}}V)\ ,$$

which sends a $\mathbb{Z}[\Gamma]$-module V to its standard complex $C(\Gamma,V)$. Here Γ acts diagonally on $C(\Gamma)\otimes_{\mathbb{Z}}V$. Then we have an equivalence of functors

$$H(\Gamma,-)\simeq\mathcal{EM}\circ C(\Gamma,-)\ .$$

If S is a Γ-set, then we form the Γ-bornological coarse space $S_{min,max}$ given by the Γ-set S with the maximal bornological and the minimal coarse structure. In this way we get a functor

$$(-)_{min,max}\colon \Gamma\mathbf{Set} \to \Gamma\mathbf{BC} .$$

We have furthermore a functor

$$\Gamma\mathbf{Set} \to \mathbf{Mod}(\mathbb{Z}[\Gamma]) , \quad S \mapsto \mathbb{Z}[S] .$$

We now have two functors $\Gamma\mathbf{Set} \to \mathbf{Sp}$ given by

$$S \mapsto H\mathcal{X}^{\Gamma}(\Gamma_{can,min} \otimes S_{min,max}) \quad \text{and} \quad S \mapsto H(\Gamma, \mathbb{Z}[S]) .$$

PROPOSITION 7.5 (cf. [**19**, Prop. 3.8]). *There is a natural equivalence*

$$H\mathcal{X}^{\Gamma}(\Gamma_{can,min} \otimes S_{min,max}) \simeq H(\Gamma, \mathbb{Z}[S]) .$$

PROOF. We claim that there is a natural isomorphism between $C\mathcal{X}^{\Gamma}(\Gamma_{can,min} \otimes S_{min,max})$ and the standard complex $C(\Gamma, \mathbb{Z}[S])$. To do so, we identify $\mathbb{Z}[\Gamma^{n+1}] \otimes_{\mathbb{Z}} \mathbb{Z}[S] \cong \mathbb{Z}[\Gamma^{n+1} \times S]$, where $\Gamma^{n+1} \times S$ carries the diagonal Γ-action. Then we define the homomorphism

$$\phi_n\colon C_n(\Gamma, \mathbb{Z}[S]) \cong \mathbb{Z} \otimes_{\mathbb{Z}[\Gamma]} \mathbb{Z}[\Gamma^{n+1} \times S] \to C\mathcal{X}^{\Gamma}(\Gamma_{can,min} \otimes S_{min,max}) \tag{7.2}$$

as the linear extension of

$$1 \otimes (\gamma_0, \gamma_1, \dots, \gamma_n, s) \mapsto \sum_{\gamma \in \Gamma} ((\gamma\gamma_0, \gamma s), \dots, (\gamma\gamma_n, \gamma s)) . \tag{7.3}$$

Note that all summands are different points on $(\Gamma \times S)^{n+1}$ so that the infinite sum makes sense, and it is Γ-invariant by construction. Every $((\gamma\gamma_0, \gamma s), \dots, (\gamma\gamma_n, \gamma s))$ is controlled by the entourage $\Gamma\{(\gamma_i, \gamma_j) \mid 0 \le i, j \le n\} \times \mathrm{diag}_S$ of the Γ-bornological coarse space $\Gamma_{can,min} \otimes S_{min,max}$. To show that this chain is also locally finite, it suffices to check that there are only finitely many points in the support of the chain (7.3) which meet bounded sets of the form $B \times S$, where B is some finite subset of Γ. This is clear since Γ acts freely on Γ^{n+1}. This finishes the argument for the assertion that (7.2) is well-defined.

It is straightforward to check that the collection $\{\phi_n\}_n$ is a chain map.

We now argue that the map (7.2) is an isomorphism. To this end we define an inverse

$$\psi\colon C\mathcal{X}_n(\Gamma_{can,min} \otimes S_{min,max}) \to \mathbb{Z} \otimes_{\mathbb{Z}} \mathbb{Z}[\Gamma^{n+1} \times S] \cong C_n(\Gamma, \mathbb{Z}[S]) .$$

Let

$$c = \sum_{x \in (\Gamma \times S)^{n+1}} n_x x$$

be an invariant, controlled and locally finite n-chain on $\Gamma_{can,min} \otimes S_{min,max}$. We now define

$$\psi(c) := \sum_{(\gamma_1, \dots, \gamma_n, s) \in \Gamma^n \times S} n_{((1,s),(\gamma_1,s)\dots,(\gamma_n,s))} \otimes (1, \gamma_1, \dots, \gamma_n, s) .$$

Assume that c is U-controlled. Then only summands with $\{\gamma_1, \dots, \gamma_n\} \subseteq U[\{1\}]$ contribute to the sum. Since $U[\{1\}]$ is bounded and c is locally finite we see that the number of non-trivial summands is finite. This implies that $\psi(c)$ is well-defined.

It is straightforward to check that ϕ and ψ are inverse to each other: To see that $\psi \circ \phi = \mathrm{id}$, use that

$$1 \otimes (\gamma_0, \gamma_1, \dots, \gamma_n, s) = 1 \otimes (1, \gamma_0^{-1}\gamma_1, \dots, \gamma_0^{-1}\gamma_n, \gamma_0^{-1}s) \ .$$

The equality $\phi \circ \psi = \mathrm{id}$ follows from the Γ-invariance of a chain $c = \sum_{x \in (\Gamma \times S)^{n+1}} n_x x$ together with the observation that $n_{((\gamma_0,s_0),\dots,(\gamma_n,s_n))} = 0$ unless $s_0 = \dots = s_n$. The latter fact is due to S carrying the minimal coarse structure.

One easily checks that ϕ is natural for maps between Γ-sets. □

REMARK 7.6. Assume that S is a transitive Γ-set. If we fix a point s in S and let Γ_s denote the stabilizer subgroup of s, then we have an isomorphism of $\mathbb{Z}[\Gamma]$-modules

$$\mathbb{Z}[S] \cong \mathrm{Ind}_{\Gamma_s}^{\Gamma} \mathbb{Z} \ .$$

The induction isomorphism in group homology now gives the chain of equivalences

$$H(\Gamma, \mathbb{Z}[S]) \simeq H(\Gamma, \mathrm{Ind}_{\Gamma_s}^{\Gamma} \mathbb{Z}) \simeq H(\Gamma_s, \mathbb{Z}) \ .$$

Since this identification involves the choice of the base point s it is not reasonable to state any naturality (for morphisms of Γ-sets) of this equivalence.

The following definition and proposition depend on definitions and results of the following sections. They are not needed later and can safely be skipped on first reading.

We can apply Proposition 7.5 in order to study the equivariant homology theory $H\mathbb{Z}^{\Gamma}$ on Γ-topological spaces induced by the equivariant coarse homology $H\mathcal{X}^{\Gamma}$ and the twist $\mathrm{Yo}^s(\Gamma_{can,min})$; see Definition 10.29 and Equation (10.17) for the notation.

DEFINITION 7.7. We define the equivariant ordinary homology theory by

$$H\mathbb{Z}^{\Gamma} := H\mathcal{X}^{\Gamma}_{\mathrm{Yo}^s(\Gamma_{can,min})} \mathcal{O}^{\infty}_{\mathrm{hlg}} \colon \Gamma\mathbf{Top} \to \mathbf{Sp} \ .$$

By the results from Section 10.1 the associated equivariant homology theory is determined on Γ-CW complexes by the restriction of the functor $H\mathbb{Z}^{\Gamma}$ to transitive Γ-sets. The following result describes this functor explicitly.

We consider the following two functors $\mathbf{Orb}(\Gamma) \to \mathbf{Sp}$ given by

$$S \mapsto H\mathbb{Z}^{\Gamma}(S) \ , \quad S \mapsto H(\Gamma, \mathbb{Z}[S]) \ .$$

PROPOSITION 7.8. *For every transitive Γ-set S we have a natural equivalence*

$$H\mathbb{Z}^{\Gamma}(S) \simeq \Sigma H(\Gamma, \mathbb{Z}[S])$$

of spectra.

PROOF. By definition we have a natural equivalence

$$\begin{aligned} H\mathbb{Z}^{\Gamma}(S) &\simeq H\mathcal{X}^{\Gamma}(\mathcal{O}^{\infty}(\mathcal{M}(\mathcal{U}(S))) \otimes \mathrm{Yo}^s(\Gamma_{can,min})) \\ &\simeq H\mathcal{X}^{\Gamma}(\mathcal{O}^{\infty}(S_{disc,max,max}) \otimes \mathrm{Yo}^s(\Gamma_{can,min})) \ . \end{aligned}$$

By Proposition 9.35, $\mathcal{O}^{\infty}(S_{disc,max,max}) \simeq \Sigma\mathrm{Yo}^s(S_{min,max})$. This gives the equivalence

$$H\mathbb{Z}^{\Gamma}(S) \simeq \Sigma H\mathcal{X}^{\Gamma}(S_{min,max} \otimes \Gamma_{can,min}) \ .$$

By Proposition 7.5 we have a natural equivalence $H\mathcal{X}^{\Gamma}(S_{min,max} \otimes \Gamma_{can,min}) \simeq H(\Gamma, \mathbb{Z}[S])$ which finishes this proof. □

7.3. Additional properties. Recall Definition 5.15 of continuity of an equivariant coarse homology theory.

LEMMA 7.9. *The equivariant coarse homology theory $H\mathcal{X}^\Gamma$ is continuous.*

PROOF. Let X be a Γ-bornological coarse space and $\mathcal{Y} := (Y_i)_{i\in I}$ be a trapping exhaustion. If c is a chain in $C\mathcal{X}^\Gamma_n(X)$, then $\mathrm{supp}(c)$ is a Γ-invariant subset of X^{n+1} which meets every bounded subset of X in a finite set. For i in $\{0,\dots,n\}$ let $p_i\colon X^{n+1}\to X$ be the projection. Then we consider the Γ-invariant subset

$$F := \bigcup_{i=0}^{n} p_i(\mathrm{supp}(c)) \ .$$

Note that c belongs to the image of the map $C\mathcal{X}^\Gamma_n(F_X)\to C\mathcal{X}^\Gamma_n(X)$ induced by the inclusion of F into X.

Observe that F is locally finite. Hence there exists an index i in I such that $F\subseteq Y_i$. We conclude that

$$C\mathcal{X}^\Gamma_n(X)\cong \operatorname*{colim}_{i\in I} C\mathcal{X}^\Gamma_n(Y_i) \ .$$

The argument above implies that we have an isomorphism of chain complexes

$$C\mathcal{X}^\Gamma(X)\cong \operatorname*{colim}_{i\in I} C\mathcal{X}^\Gamma(Y_i) \ .$$

Since the Eilenberg–MacLane correspondence (7.1) preserves filtered colimits we conclude that

$$H\mathcal{X}^\Gamma(X)\simeq \operatorname*{colim}_{i\in I} H\mathcal{X}^\Gamma(Y_i)$$

which was to be shown. □

Recall Definition 4.19 of strongness of an equivariant coarse homology theory.

LEMMA 7.10. *The equivariant coarse homology theory $H\mathcal{X}^\Gamma$ is strong.*

PROOF. We can essentially repeat the proof of [**8**, Prop. 6.18].

Let $f\colon X\to X$ implement weak flasqueness of a Γ-bornological coarse space X. Then we can define the chain map

$$S := \sum_{n=0}^{\infty} C\mathcal{X}(f^n)\colon C\mathcal{X}(X)\to C\mathcal{X}(X) \ .$$

We refer to [**8**, Prop. 6.18] for the verification that this map is well-defined. We then have the identity of endomorphisms of $C\mathcal{X}^\Gamma(X)$

$$\mathrm{id}_{C\mathcal{X}(X)} + C\mathcal{X}(f)\circ S = S \ .$$

Applying the Eilenberg–MacLane correspondence $\mathcal{EM}$ this gives

$$\mathrm{id}_{H\mathcal{X}^\Gamma(X)} + H\mathcal{X}^\Gamma(f)\circ \mathcal{EM}(S) = \mathcal{EM}(S) \ .$$

Since we already know that $H\mathcal{X}^\Gamma$ is a coarse homology theory we have the equivalence $H\mathcal{X}^\Gamma(f)\simeq \mathrm{id}_{H\mathcal{X}^\Gamma(X)}$. Hence we get $\mathrm{id}_{H\mathcal{X}^\Gamma(X)} + \mathcal{EM}(S)\simeq \mathcal{EM}(S)$, and this implies that we must have $H\mathcal{X}^\Gamma(X)\simeq 0$. □

Recall the definition of the free union of a family of Γ-bornological coarse spaces which was given in Example 2.16 and the Definition 3.12 of the notion of strong additivity for an equivariant coarse homology theory.

LEMMA 7.11. *The equivariant coarse homology theory $H\mathcal{X}^\Gamma$ is strongly additive.*

PROOF. Let $(X_i)_{i\in I}$ be a family of Γ-bornological coarse spaces. By inspection of the definitions,

$$C\mathcal{X}^\Gamma\Big(\coprod_{i\in I}^{\text{free}} X_i\Big) \cong \prod_{i\in I} C\mathcal{X}^\Gamma(X_i)\ .$$

We then use that the Eilenberg–MacLane correspondence (7.1) preserves products. □

7.4. Change of groups. In this section we provide natural transformations which relate the equivariant coarse homology theory with the change of group functors considered in Section 6.

Let $\iota\colon H\to\Gamma$ be a homomorphism of groups. Let X be a Γ-bornological coarse space. Since every Γ-invariant chain is H-invariant we have an inclusion

$$C\mathcal{X}^\Gamma(X)\hookrightarrow C\mathcal{X}^H(\mathrm{Res}^\Gamma_H X)\ .$$

This gives a natural transformation between equivariant coarse homology theories

$$\mathrm{res}^\Gamma_H\colon H\mathcal{X}^\Gamma\to H\mathcal{X}^H\circ\mathrm{Res}^\Gamma_H\ . \tag{7.4}$$

We let the subset X' of X be a set of representatives of H-orbits in X. Then we have a restriction map

$$r\colon C\mathcal{X}_n(X)\to\mathbb{Z}^{X'\times X^n}\ ,\quad c\mapsto c|_{X'\times X^n}\ .$$

We want to define a homomorphism

$$C\mathcal{X}^\Gamma_n(X)\to C\mathcal{X}^{W_\Gamma(H)}_n(Q_H(X)) \tag{7.5}$$

by linear extension of the projection map $X'\times X^n\to(H\backslash X)^{n+1}$ composed with the restriction r. We will argue now that this is well-defined, i.e., that the sums appearing in this extension are finite. For x in X we denote by $[x]$ its orbit in $H\backslash X$. We consider a point $([x_0],\dots,[x_n])$ in $(H\backslash X)^{n+1}$. Then $[x_0]\cap X'$ consists of a unique point x_0.

Let c be in $C\mathcal{X}_n(X)$ and assume that c is U-controlled for some entourage U of X. Then we have

$$\mathrm{supp}(c)\cap\{x_0\}\times X^n\subseteq(U[x_0])^{n+1}\ .$$

Since $U[x_0]$ is bounded the number of points of $\mathrm{supp}(c)$ which meet $U[x_0]$ is finite.

One checks that the homomorphism (7.5) does not depend on the choice of the set of representatives X' and has values in $W_\Gamma(H)$-invariant chains. Furthermore, it is compatible with the differential and takes values in controlled and locally finite chains.

We therefore get a natural transformation

$$q_H(\iota)\colon H\mathcal{X}^\Gamma\to H\mathcal{X}^{W_\Gamma(H)}\circ Q_H(\iota)\ . \tag{7.6}$$

Recall from Section 6.5 that $P_{\hat\Gamma}(X)=\Gamma_{min,min}\otimes X$ with the action of $\Gamma\times H$ given by $(\gamma,h)(\gamma',x)=(\gamma\gamma'\iota(h)^{-1},hx)$. We define a morphism of chain complexes

$$C\mathcal{X}^H_n\to C\mathcal{X}^{\Gamma\times H}_n(P_{\hat\Gamma}(X))$$

by linear extension of the map

$$(x_0,\dots,x_n)\mapsto\sum_{\gamma'\in\Gamma}((\gamma',x_0),(\gamma',x_1),\dots,(\gamma',x_n))\ .$$

In this way we get a transformation

$$\hat{p}_\Gamma\colon H\mathcal{X}^H \to H\mathcal{X}^{\Gamma\times H} \circ \hat{P}_\Gamma\ . \tag{7.7}$$

We finally get a natural transformation

$$\mathrm{ind}_H^\Gamma \simeq q_H(\kappa)\circ \hat{p}_\Gamma\colon H\mathcal{X}^H \to H\mathcal{X}^\Gamma \circ \mathrm{Ind}_H^\Gamma\ , \tag{7.8}$$

where $\kappa\colon H\to \Gamma\times H$ is the inclusion of the second factor.

PROPOSITION 7.12. *If $\iota\colon H\to\Gamma$ is injective, then the transformation (7.8) is an equivalence of H-equivariant coarse homology theories.*

PROOF. By Remark 6.4 the map $X\to \Gamma\times_H X = \mathrm{Ind}_H^\Gamma(X)$ given by $x\mapsto [1,x]$ is an embedding of an H-invariant coarse component. Restriction along this map gives a map of chain complexes $C\mathcal{X}^\Gamma(\mathrm{Ind}_H^\Gamma(X)) \to C\mathcal{X}^H(X)$ which induces the inverse to (7.8). □

8. Equivariant coarse algebraic K-homology

In this section, we define for every additive category $\mathbf{A}$ with Γ-action its Γ-equivariant coarse algebraic K-homology

$$K\mathbf{A}\mathcal{X}^\Gamma\colon \Gamma\mathbf{BC}\to\mathbf{Sp}\ .$$

The construction associates to a Γ-bornological coarse space X an additive category of equivariant X-controlled $\mathbf{A}$-objects $\mathbf{V}_{\mathbf{A}}^\Gamma(X)$, and defines $K\mathbf{A}\mathcal{X}^\Gamma$ to be the (non-connective) algebraic K-theory spectrum of this category.

8.1. The algebraic K-theory functor. We describe the properties of the K-theory functor that we will use subsequently. See [**1**, Sec. 2.1] for similar statements. Let **Add** denote the 1-category of small additive categories and exact functors. In the following, all additive categories will be small so that we can omit this adjective safely.

The K-theory functor is a functor

$$K\colon \mathbf{Add}\to\mathbf{Sp}$$

which has the following properties (we will recall the occurring notions further below):

(1) (Normalization) It sends (a skeleton of) the additive category of finitely generated free modules over a ring R to the non-connective K-theory (see e.g. [**36**]) of that ring.
(2) (Invariance) It sends isomorphic exact functors to equivalent maps.
(3) (Colimits) If $\mathbf{A} = \mathrm{colim}_i\,\mathbf{A}_i$ is a filtered colimit of additive subcategories, then the natural map $\mathrm{colim}_i\,K(\mathbf{A}_i)\xrightarrow{\simeq} K(\mathbf{A})$ is an equivalence.
(4) (Additivity) If $\Phi,\Psi\colon \mathbf{A}\to\mathbf{A}'$ are exact functors between additive categories, then we have an equivalence $K(\Phi)+K(\Psi)\simeq K(\Phi\oplus\Psi)$ of morphisms between K-theory spectra.
(5) (Exactness) If $\mathbf{A}$ is a Karoubi filtration of $\mathbf{C}$, then we have a fiber sequence

$$K(\mathbf{A})\to K(\mathbf{C})\to K(\mathbf{C}/\mathbf{A})\xrightarrow{\partial} K(\mathbf{A})\ .$$

(6) (Products) If we have a family $(\mathbf{A}_i)_{i\in I}$ of additive categories, then the natural map $K(\prod_{i\in I}\mathbf{A}_i) \to \prod_{i\in I} K(\mathbf{A}_i)$ is an equivalence.
(7) (Flasqueness) It sends flasque additive categories to zero.

Let us recall the definition of some notions appearing above.

A category is additive if admits a zero object and biproducts such that the operation $\mathrm{Hom}(A,B)\times\mathrm{Hom}(A,B)\to\mathrm{Hom}(A,B)$ which sends f,g to

$$f+g\colon A \xrightarrow{\Delta} A\oplus A \xrightarrow{f\oplus g} B\oplus B \xrightarrow{\mathrm{id}+\mathrm{id}} B$$

defines an abelian group structure on morphism sets. In applications, it is useful to consider the equivalent characterization of additive categories as categories which are enriched over abelian groups, have a zero object and admit finite coproducts Lurie [**27**, Sec. 1.1.2].

A morphism between additive categories is a functor between the underlying categories which preserves the zero object and finite coproducts. Equivalently, one can require that the functor is compatible with the enrichement in abelian groups, see Mac Lane [**28**, Prop. VIII.2.4].

Given morphisms $\Phi,\Psi\colon \mathbf{A}\to\mathbf{A}'$ between additive categories, we define a new morphism $\Phi\oplus\Psi\colon\mathbf{A}\to\mathbf{A}'$ by choosing for every object A of $\mathbf{A}$ an object of $\mathbf{A}'$ representing the sum $\Phi(A)\oplus\Psi(A)$. Since the sum of two functors is unique up to unique isomorphism, there is an essentially unique map $K(\Phi\oplus\Psi)$ by virtue of Property (2).

DEFINITION 8.1. An additive category $\mathbf{A}$ is called *flasque* if there exists a functor $\Sigma\colon\mathbf{A}\to\mathbf{A}$ such that $\mathrm{id}_{\mathbf{A}}\oplus\Sigma\cong\Sigma$.

Note that Property (4) implies Property (7): Assume that $\mathbf{A}$ is flasque and that $\Sigma\colon\mathbf{A}\to\mathbf{A}$ is a functor satisfying $\mathrm{id}_{\mathbf{A}}\oplus\Sigma\cong\Sigma$. By Property 4 we then have an equivalence $K(\Sigma)+\mathrm{id}_{K(\mathbf{A})}\simeq K(\Sigma)$, which implies that $K(\mathbf{A})\simeq 0$.

Let $\mathbf{A}\subseteq\mathbf{C}$ be a full additive subcategory. For C,D in $\mathbf{C}$, let $\mathrm{Hom}_{\mathbf{C}}(C,\mathbf{A},D)$ denote the set of all morphisms in $\mathrm{Hom}_{\mathbf{C}}(C,D)$ which factor through some object in $\mathbf{A}$. Then $\mathrm{Hom}_{\mathbf{C}}(C,\mathbf{A},D)$ is a subgroup of $\mathrm{Hom}_{\mathbf{C}}(C,D)$. We let $\mathbf{C}/\mathbf{A}$ be the category with the same objects as $\mathbf{C}$ and whose morphisms are given by

$$\mathrm{Hom}_{\mathbf{C}/\mathbf{A}}(C,D) := \mathrm{Hom}_{\mathbf{C}}(C,D)/\,\mathrm{Hom}_{\mathbf{C}}(C,\mathbf{A},D)\ .$$

Note that $\mathbf{C}/\mathbf{A}$ is again an additive category. It has the universal property that exact functors on $\mathbf{C}/\mathbf{A}$ correspond bijectively to exact functors on $\mathbf{C}$ which vanish on $\mathbf{A}$.

Let $\mathbf{C}$ be an additive category.

DEFINITION 8.2. The inclusion $\mathbf{A}\subseteq\mathbf{C}$ of a full additive subcategory is a *Karoubi filtration* if every diagram

$$A\xrightarrow{f} C\xrightarrow{g} B$$

in $\mathbf{C}$, where A,B are objects of $\mathbf{A}$, admits an extension to a commutative diagram

$$\begin{array}{ccccc} A & \xrightarrow{f} & C & \xrightarrow{g} & B \\ \downarrow & & \downarrow\cong & & \uparrow \\ D & \xrightarrow{inc} & D\oplus D^{\perp} & \xrightarrow{\mathrm{pr}} & D \end{array}$$

for some object D of $\mathbf{A}$.

In [**23**, Lem. 5.6] it is shown that Definition 8.2 is equivalent to the standard definition of a Karoubi filtration as considered in [**13**].

An algebraic K-theory functor as described here can be furnished by restricting the K-theory functor constructed by Schlichting [**37**] to additive categories. For Property (1), see [**37**, Thms. 5 & 8]. Property (2) is so elementary that it is rarely stated explicitly, but can be easily read off from the construction in [**37**]. Property (3) is a combination of [**33**, Eq. (9)] and [**37**, Cor. 5]. Property (4) follows from Property (5) (cf. also [**40**, Prop. 1.3.2]); the latter is proved in [**36**, Thm. 2.10]. Property (6) is shown in [**24**, Thm 1.2]; for connective K-theory, this is originally due to Carlsson [**14**].

8.2. X-controlled A-objects. Let X be a bornological coarse space with the bornology $\mathcal{B}$ and the coarse structure $\mathcal{C}$, and let $\mathbf{A}$ be an additive category with a (strict) Γ-action. The poset $\mathcal{B}$ is ordered by the subset inclusion and will be regarded as a category. For a functor $A\colon \mathcal{B} \to \mathbf{A}$ we define $\gamma A\colon \mathcal{B} \to \mathbf{A}$ to be the functor sending a bounded set B to $\gamma(A(\gamma^{-1}(B)))$.

DEFINITION 8.3. An *equivariant X-controlled* $\mathbf{A}$*-object* is a pair (A, ρ) consisting of a functor $A\colon \mathcal{B} \to \mathbf{A}$ and a family $\rho = (\rho(\gamma))_{\gamma\in\Gamma}$ of natural isomorphisms $\rho(\gamma)\colon A \to \gamma A$ satisfying the following conditions:

(1) $A(\emptyset) \cong 0$.
(2) For all B, B' in $\mathcal{B}$, the commutative square

$$\begin{array}{ccc} A(B\cap B') & \longrightarrow & A(B) \\ \downarrow & & \downarrow \\ A(B') & \longrightarrow & A(B\cup B') \end{array}$$

is a push-out.
(3) For all B in $\mathcal{B}$, there exists a finite subset F of B such that the inclusion $F \subseteq B$ induces an isomorphism $A(F) \xrightarrow{\cong} A(B)$.
(4) For all pairs of elements γ, γ' of Γ we have $\rho(\gamma\gamma') = \gamma\rho(\gamma') \circ \rho(\gamma)$, where $\gamma\rho(\gamma')$ is the natural transformation from γA to $\gamma\gamma' A$ induced from $\rho(\gamma')$.

Let (A, ρ) be an equivariant X-controlled $\mathbf{A}$-object.

LEMMA 8.4. (1) (a) *The canonical morphism*

$$\bigoplus_{x\in F} A(\{x\}) \xrightarrow{\sum_{x\in F} A(\{x\}\subseteq F)} A(F)$$

is an isomorphism for every finite subset F of X.

(b) *For two finite subsets F, F' of X with $F \subseteq F'$ we have a commuting square*

$$\begin{array}{ccc} \bigoplus_{x\in F} A(\{x\}) & \xrightarrow{\sum_{x\in F} A(\{x\}\subseteq F)} & A(F) \\ \downarrow & & \downarrow \\ \bigoplus_{x\in F'} A(\{x\}) & \xrightarrow{\sum_{x\in F'} A(\{x\}\subseteq F')} & A(F') \end{array}$$

(2) (a) *For a bounded subset B of X there exists a unique minimal finite subset F_B of B such that $A(F_B) \to A(B)$ is an isomorphism.*

(b) *If B and F_B are as in 2a, then for any subset B' of X with $F_B \subseteq B' \subseteq B$, the morphisms $A(F_B) \to A(B')$ and $A(B') \to A(B)$ are isomorphisms.*

PROOF. Item 1 is a direct consequence of Definition 8.3(1) and (2).

Suppose now that F, F' are two finite subsets of B as in Definition 8.3(3). Then we obtain a push-out square

$$\begin{array}{ccc} A(F \cap F') & \longrightarrow & A(F) \\ \downarrow & & \downarrow \cong \\ A(F') & \xrightarrow{\cong} & A(B) \end{array}$$

in which all arrows are inclusions of direct summands. It follows that $A(F \cap F') \to A(B)$ is also an isomorphism. This implies the existence of F_B.

For a bounded subset B' with $F_B \subseteq B' \subseteq B$, inspection of a similar push-out square implies that $A(F_B) \to A(B')$ is an isomorphism, and the claim follows. □

Let (A, ρ) be an equivariant X-controlled $\mathbf{A}$-object.

DEFINITION 8.5. The function σ which sends a bounded subset B of X to the finite subset F_B from Lemma 8.4 is called the *support function* of (A, ρ).

The support function is an order preserving, equivariant function from $\mathcal{B}$ to the set of finite subsets of X with the property that $\sigma(\sigma(B)) = \sigma(B)$ for every bounded subset B.

Let $(A, \rho), (A', \rho')$ be equivariant X-controlled $\mathbf{A}$-objects and let U be an invariant entourage of X.

DEFINITION 8.6. An *equivariant U-controlled morphism* $f\colon (A, \rho) \to (A', \rho')$ is a natural transformation

$$f\colon A(-) \to A'(U[-]) ,$$

such that $\rho'(\gamma) \circ f = (\gamma f) \circ \rho(\gamma)$ for all elements γ of Γ.

We let $\mathrm{Mor}_U((A, \rho), (A', \rho'))$ denote the set of equivariant U-controlled morphisms. Furthermore, we define the set of controlled morphisms from A to A' by

$$\mathrm{Hom}_{\mathbf{V}^\Gamma_\mathbf{A}(X)}((A, \rho), (A', \rho')) := \operatorname*{colim}_{U \in \mathcal{C}^\Gamma} \mathrm{Mor}_U((A, \rho), (A', \rho')) .$$

We denote the resulting category of equivariant X-controlled $\mathbf{A}$-objects and equivariant controlled morphisms by $\mathbf{V}^\Gamma_\mathbf{A}(X)$.

We observe that the composition of a U-controlled and a U'-controlled morphism is a $U \circ U'$-controlled morphism. We conclude that composition in $\mathbf{V}^\Gamma_\mathbf{A}(X)$ is well-defined.

LEMMA 8.7. *The category $\mathbf{V}^\Gamma_\mathbf{A}(X)$ is additive.*

PROOF. Let $(A, \rho), (A', \rho')$ be equivariant X-controlled $\mathbf{A}$-objects. Denote by $A \oplus A'$ their direct sum in $\mathbf{Fun}(\mathcal{B}, \mathbf{A})$. Note that $(A \oplus A', \rho \oplus \rho')$ is an X-controlled $\mathbf{A}$-object because finite unions of bounded sets are bounded. Since finite unions of coarse entourage are coarse entourages and there are canonical isomorphisms of Γ-sets

$$\mathrm{Nat}(A \oplus A', C \circ U[-]) \cong \mathrm{Nat}(A, C \circ U[-]) \times \mathrm{Nat}(A', C \circ U[-])$$

and

$$\mathrm{Nat}(C,(A\oplus A')\circ U[-])\cong \mathrm{Nat}(C,A\circ U[-])\times \mathrm{Nat}(C,A'\circ U[-])$$

(we use the symbol Nat to denote the morphism sets in $\mathbf{Fun}(\mathcal{B},\mathbf{A})$) for all U in $\mathcal{C}^\Gamma$ and equivariant X-controlled $\mathbf{A}$-objects C, it follows that $(A\oplus A',\rho\oplus\rho')$ is also a direct sum in $\mathbf{V}^\Gamma_{\mathbf{A}}(X)$.

Similarly, addition of morphisms in $\mathbf{Fun}(\mathcal{B},\mathbf{A})$ induces the addition operation of morphisms in $\mathbf{V}^\Gamma_{\mathbf{A}}(X)$. □

Let us discuss functoriality of $\mathbf{V}^\Gamma_{\mathbf{A}}(X)$ in the variable $\mathbf{A}$. For a Γ-equivariant exact functor $\Phi\colon \mathbf{A}\to\mathbf{A}'$ of additive categories, there is an induced exact functor

$$\mathbf{V}^\Gamma_{\Phi}(X)\colon \mathbf{V}^\Gamma_{\mathbf{A}}(X)\to \mathbf{V}^\Gamma_{\mathbf{A}'}(X)$$

which sends an object (A,ρ) to $(\Phi\circ A,\Phi(\rho))$. Therefore, we have a functor

$$\mathbf{V}^\Gamma_{-}(X)\colon \mathbf{Fun}(B\Gamma,\mathbf{Add})\to\mathbf{Add}\ .$$

Next we discuss functoriality of $\mathbf{V}^\Gamma_{\mathbf{A}}(X)$ in the variable X. Let $\phi\colon (X,\mathcal{B},\mathcal{C})\to (X',\mathcal{B}',\mathcal{C}')$ be a morphism of Γ-bornological coarse spaces, and let (A,ρ) be an equivariant X-controlled $\mathbf{A}$-object. Since ϕ is proper, we can define a functor $\phi_*A\colon \mathcal{B}'\to\mathbf{A}$ by

$$\phi_*A(B):=A(\phi^{-1}(B))\ ,$$

and we define

$$\phi_*\rho(\gamma)(B)=\rho(\gamma)(\phi^{-1}(B))\ .$$

All properties of Definition 8.3 except (3) are immediate. To see that (3) also holds, we note that $\sigma(\phi^{-1}(B))\subseteq\phi^{-1}(\phi(\sigma(\phi^{-1}(B))))\subseteq\phi^{-1}(B)$ and apply Lemma 8.4 to see that $\phi_*A(\phi(\sigma(\phi^{-1}(B))))\to\phi_*A(B)$ is an isomorphism.

Let $f\colon (A,\rho)\to(A',\rho')$ be an equivariant U-controlled morphism. Then there exists some V in $\mathcal{C}'^{,\Gamma}$ such that $(\phi\times\phi)(U)\subseteq V$. Then $U[\phi^{-1}(B)]\subseteq\phi^{-1}(V[B])$ for all bounded subsets B of X, so we obtain an induced V-controlled morphism

$$\phi_*f=\{f_{\phi^{-1}(B)}\colon \phi_*A(B)\to\phi_*A(V[B])\}_{B\in\mathcal{B}'}\ .$$

This defines a functor

$$\phi_*\colon \mathbf{V}^\Gamma_{\mathbf{A}}(X)\to\mathbf{V}^\Gamma_{\mathbf{A}}(X')\ .$$

We thus have constructed a functor

$$\mathbf{V}^\Gamma_{\mathbf{A}}\colon \Gamma\mathbf{BC}\to\mathbf{Add}\ .$$

8.3. Coarse algebraic K-homology. Let Γ be a group and $\mathbf{A}$ be an additive category with a Γ-action.

DEFINITION 8.8. We define the *coarse algebraic K-homology $K\mathbf{A}\mathcal{X}^\Gamma$* associated to $\mathbf{A}$ as

$$K\mathbf{A}\mathcal{X}^\Gamma:=K\circ\mathbf{V}^\Gamma_{\mathbf{A}}\colon \Gamma\mathbf{BC}\to\mathbf{Sp}\ .$$

This section discusses the homological properties of $K\mathbf{A}\mathcal{X}^\Gamma$. Our first goal is to prove the following theorem.

THEOREM 8.9. *The functor $K\mathbf{A}\mathcal{X}^\Gamma$ is an equivariant coarse homology theory.*

We divide the proof of Theorem 8.9 into a sequence of lemmas.

LEMMA 8.10. *The functor $K\mathbf{A}\mathcal{X}^\Gamma$ is u-continuous.*

PROOF. Let X be a Γ-bornological coarse space, and let U be an invariant entourage of X. The natural map $X_U \to X$ induces a functor $\Phi_U \colon \mathbf{V}^\Gamma_{\mathbf{A}}(X_U) \to \mathbf{V}^\Gamma_{\mathbf{A}}(X)$. Since the definition of equivariant X-controlled $\mathbf{A}$-objects is independent of the coarse structure, Φ_U is the identity on objects. Additionally, since inclusions of direct summands are monomorphisms, Φ_U is faithful.

This allows us to view $\mathbf{V}^\Gamma_{\mathbf{A}}(X_U)$ as a subcategory of $\mathbf{V}^\Gamma_{\mathbf{A}}(X)$, and we have

$$\mathbf{V}^\Gamma_{\mathbf{A}}(X) = \bigcup_{U \in \mathcal{C}^\Gamma} \mathbf{V}^\Gamma_{\mathbf{A}}(X_U)$$

since every morphism in $\mathbf{V}^\Gamma_{\mathbf{A}}(X)$ is U-controlled for some U in $\mathcal{C}^\Gamma$.

Since the algebraic K-theory functor is compatible with filtered colimits (Property (3)), the claim of the lemma follows. □

Let $\phi, \psi \colon X \to X'$ be morphisms of Γ-bornological coarse spaces.

LEMMA 8.11. *If ϕ and ψ are close, then ϕ_* and ψ_* are isomorphic.*

PROOF. Let U' be a symmetric entourage of X' containing the diagonal such that $(\phi(x), \psi(x))$ lies in U' for all x in X. Note that this implies $\phi^{-1}(B') \subseteq \psi^{-1}(U'[B'])$ and $\psi^{-1}(B') \subseteq \phi^{-1}(U'[B'])$ for all bounded subsets B' of X'.

Let (A, ρ) be an equivariant X-controlled $\mathbf{A}$-object. The maps

$$A(\phi^{-1}(B')) \to A(\psi^{-1}(U'[B']))$$

define a natural morphism $f \colon \phi_* A \to \psi_* A$, and similarly we have a natural morphism $g \colon \psi_* A \to \phi_* A$. Since the composition $g \circ f$ is given by the natural transformation

$$\{A(\phi^{-1}(B') \subseteq \phi^{-1}((U')^2[B'])) \colon \phi_* A \to \phi_* A \circ (U')^2[-]\}_{B' \in \mathcal{B}'} ,$$

we have $g \circ f = \mathrm{id}_{\phi_* A}$. Similarly, $f \circ g = \mathrm{id}_{\psi_* A}$. It follows that $\phi_* \cong \psi_*$. □

COROLLARY 8.12. *The functor $K\mathbf{A}\mathcal{X}^\Gamma$ is coarsely invariant.*

PROOF. This is a direct consequence of Lemma 8.11 together with the Property (2) of the algebraic K-theory functor. □

LEMMA 8.13. *The functor $K\mathbf{A}\mathcal{X}^\Gamma$ vanishes on flasque Γ-bornological coarse spaces.*

PROOF. Let X be a Γ-bornological coarse space with flasqueness implemented by $\phi \colon X \to X$. We claim that the functor

$$\Sigma := \bigoplus_{n \in \mathbb{N}} (\phi^n)_* \colon \mathbf{V}^\Gamma_{\mathbf{A}}(X) \to \mathbf{V}^\Gamma_{\mathbf{A}}(X)$$

is well-defined (up to canonical isomorphism).

For every bounded subset B of X, there exists n in $\mathbb{N}$ such that $(\phi^n)^{-1}(B) = \emptyset$, so the direct sum $\bigoplus_{n \in \mathbb{N}} (\phi^n)_* A$ exists for every equivariant X-controlled object (A, ρ). Let $f \colon (A, \rho) \to (A', \rho')$ be a U-controlled morphism. Then $\bigoplus_{n \in \mathbb{N}} (\phi^n)_* f$ is V-controlled, where $V := \bigcup_{n \in \mathbb{N}} (\phi \times \phi)^n(U)$ is again a coarse entourage of X by assumption on ϕ. So Σ is an exact functor.

Since ϕ is close to id_X, we conclude from Lemma 8.11 that $\phi_* \circ \Sigma$ and Σ are isomorphic. Hence,

$$\mathrm{id}_{\mathbf{V}^\Gamma_{\mathbf{A}}(X)} \oplus \Sigma \cong \mathrm{id}_{\mathbf{V}^\Gamma_{\mathbf{A}}(X)} \oplus (\phi_* \circ \Sigma) \cong \Sigma ,$$

and we deduce the lemma from Property (7) of the algebraic K-theory functor. □

Let X be a Γ-bornological coarse space and Z a Γ-invariant subset of X. For an equivariant X-controlled object (A, ρ), we denote by $(A|_Z, \rho|_Z)$ the restriction to Z, that is $A|_Z$ is the restriction of A to $\mathcal{B} \cap Z$ and $\rho|_Z$ the appropriate restriction of ρ.

Let X be a Γ-bornological coarse space and let $\mathcal{Y} = (Y_i)_{i \in I}$ be an equivariant big family in X. For each i in I, the canonical exact functor $\mathbf{V}^\Gamma_{\mathbf{A}}(Y_i) \to \mathbf{V}^\Gamma_{\mathbf{A}}(X)$ is injective on objects and fully faithful, so we can regard $\mathbf{V}^\Gamma_{\mathbf{A}}(Y_i)$ as a full subcategory of $\mathbf{V}^\Gamma_{\mathbf{A}}(X)$. Define

$$\mathbf{V}^\Gamma_{\mathbf{A}}(\mathcal{Y}) := \bigcup_{i \in I} \mathbf{V}^\Gamma_{\mathbf{A}}(Y_i) \,,$$

considered as a full subcategory of $\mathbf{V}^\Gamma_{\mathbf{A}}(X)$.

LEMMA 8.14. *The inclusion $\mathbf{V}^\Gamma_{\mathbf{A}}(\mathcal{Y}) \to \mathbf{V}^\Gamma_{\mathbf{A}}(X)$ is a Karoubi filtration.*

PROOF. Let $(A, \rho), (A', \rho')$ be objects in $\mathbf{V}^\Gamma_{\mathbf{A}}(\mathcal{Y})$, (C, ρ_C) be an object in $\mathbf{V}^\Gamma_{\mathbf{A}}(X)$, and let $f \colon A \to C$ and $g \colon C \to A'$ be morphisms. Choose i in I such that both A and A' are objects in $\mathbf{V}^\Gamma_{\mathbf{A}}(Y_i)$, and pick an invariant and symmetric entourage U which contains the diagonal such that f and g are U-controlled. Let j in I be such that $U[Y_i] \subseteq Y_j$. Since f is a natural transformation $A \to C \circ U[-]$ and $g|_{C|_{X \setminus Y_j}} = 0$, the following diagram commutes:

$$\begin{array}{ccccc}
A & \xrightarrow{\quad f \quad} & C & \xrightarrow{\quad g \quad} & A' \\
\downarrow & & \downarrow \cong & & \uparrow \\
C|_{Y_j} & \xrightarrow{inc} & C|_{Y_j} \oplus C|_{X \setminus Y_j} & \xrightarrow{\mathrm{pr}} & C|_{Y_j}
\end{array}$$

Hence, the inclusion $\mathbf{V}^\Gamma_{\mathbf{A}}(\mathcal{Y}) \to \mathbf{V}^\Gamma_{\mathbf{A}}(X)$ is a Karoubi filtration. $\square$

PROPOSITION 8.15. *The functor $K\mathbf{A}\mathcal{X}^\Gamma$ is excisive.*

PROOF. The category of $\emptyset$-controlled $\mathbf{A}$-objects is the zero category, which has trivial K-theory by Property (1).

Let X be a Γ-bornological coarse space, and let $(Z, \mathcal{Y})$ be an equivariant complementary pair on X. Both inclusions $\mathbf{V}^\Gamma_{\mathbf{A}}(Z \cap \mathcal{Y}) \to \mathbf{V}^\Gamma_{\mathbf{A}}(Z)$ and $\mathbf{V}^\Gamma_{\mathbf{A}}(\mathcal{Y}) \to \mathbf{V}^\Gamma_{\mathbf{A}}(X)$ are Karoubi filtrations by Lemma 8.14. Therefore, we obtain by Property (5) of the algebraic K-theory functor a map of fiber sequences

$$\begin{array}{ccccccc}
K(\mathbf{V}^\Gamma_{\mathbf{A}}(Z \cap \mathcal{Y})) & \longrightarrow & K(\mathbf{V}^\Gamma_{\mathbf{A}}(Z)) & \longrightarrow & K(\mathbf{V}^\Gamma_{\mathbf{A}}(Z)/\mathbf{V}^\Gamma_{\mathbf{A}}(Z \cap \mathcal{Y})) & \xrightarrow{\partial} & K(\mathbf{V}^\Gamma_{\mathbf{A}}(Z \cap \mathcal{Y})) \\
\downarrow & & \downarrow & & \downarrow & & \downarrow \\
K(\mathbf{V}^\Gamma_{\mathbf{A}}(\mathcal{Y})) & \longrightarrow & K(\mathbf{V}^\Gamma_{\mathbf{A}}(X)) & \longrightarrow & K(\mathbf{V}^\Gamma_{\mathbf{A}}(X)/\mathbf{V}^\Gamma_{\mathbf{A}}(\mathcal{Y})) & \xrightarrow{\partial} & K(\mathbf{V}^\Gamma_{\mathbf{A}}(\mathcal{Y}))
\end{array}$$

Consider the induced exact functor $\Phi \colon \mathbf{V}^\Gamma_{\mathbf{A}}(Z)/\mathbf{V}^\Gamma_{\mathbf{A}}(Z \cap \mathcal{Y}) \to \mathbf{V}^\Gamma_{\mathbf{A}}(X)/\mathbf{V}^\Gamma_{\mathbf{A}}(\mathcal{Y})$.

Let (A, ρ) in $\mathbf{V}^\Gamma_{\mathbf{A}}(X)$ and consider the natural morphisms $f \colon (A|_Z, \rho|_Z) \to (A, \rho)$ and $p \colon (A, \rho) \to (A|_Z, \rho|_Z)$. Clearly, $pf = \mathrm{id}_{(A|_Z, \rho|_Z)}$. Pick i in I such that $X \setminus Z \subseteq Y_i$. Then $\mathrm{id}_{(A,\rho)} - fp$ factors through $(A|_{Y_i}, \rho|_{Y_i})$, so f and p define mutually inverse isomorphisms in $\mathbf{V}^\Gamma_{\mathbf{A}}(X)/\mathbf{V}^\Gamma_{\mathbf{A}}(\mathcal{Y})$. We conclude that $\Phi \circ \Psi \cong \mathrm{id}_{\mathbf{V}^\Gamma_{\mathbf{A}}(X)/\mathbf{V}^\Gamma_{\mathbf{A}}(\mathcal{Y})}$, so Φ is an equivalence of categories.

It follows from Property (2) of the algebraic K-theory functor that

$$\begin{array}{ccc} K(\mathbf{V}^{\Gamma}_{\mathbf{A}}(Z\cap\mathcal{Y})) & \longrightarrow & K(\mathbf{V}^{\Gamma}_{\mathbf{A}}(Z)) \\ \downarrow & & \downarrow \\ K(\mathbf{V}^{\Gamma}_{\mathbf{A}}(\mathcal{Y})) & \longrightarrow & K(\mathbf{V}^{\Gamma}_{\mathbf{A}}(X)) \end{array}$$

is a push-out. By Property (3) of algebraic K-theory we have $K(\mathbf{V}^{\Gamma}_{\mathbf{A}}(\mathcal{Y}))\simeq K\mathbf{A}\mathcal{X}^{\Gamma}(\mathcal{Y})$. This proves excision. □

REMARK 8.16. Let $(X,\mathcal{B},\mathcal{C})$ be a Γ-bornological coarse space and let Y be a Γ-invariant subspace of X with the property that $U[Y]=Y$ for every U in $\mathcal{C}$. Then $(Y,X\setminus Y)$ is a coarsely excisive pair. Inspecting the proof of Proposition 8.15, we obtain the following commutative diagram:

$$\begin{array}{ccccc} 0 & \longrightarrow & K(\mathbf{V}^{\Gamma}_{\mathbf{A}}(Y)) & \xrightarrow{\simeq} & K(\mathbf{V}^{\Gamma}_{\mathbf{A}}(Y)) \\ \downarrow & & \downarrow & & \downarrow{\simeq} \\ K(\mathbf{V}^{\Gamma}_{\mathbf{A}}(X\setminus Y)) & \longrightarrow & K(\mathbf{V}^{\Gamma}_{\mathbf{A}}(X)) & \longrightarrow & K(\mathbf{V}^{\Gamma}_{\mathbf{A}}(X)/\mathbf{V}^{\Gamma}_{\mathbf{A}}(X\setminus Y)) \end{array}$$

In addition, we observe that an inverse to the right vertical equivalence is induced by the functor Ψ which is given by $\Psi(A,\rho)=(A|_Y,\rho|_Y)$. Since this functor is already well-defined as a functor $\Psi\colon \mathbf{V}^{\Gamma}_{\mathbf{A}}(X)\to\mathbf{V}^{\Gamma}_{\mathbf{A}}(Y)$, we see that the projection map

$$K(\mathbf{V}^{\Gamma}_{\mathbf{A}}(X))\simeq K(\mathbf{V}^{\Gamma}_{\mathbf{A}}(X\setminus Y))\oplus K(\mathbf{V}^{\Gamma}_{\mathbf{A}}(Y))\to K(\mathbf{V}^{\Gamma}_{\mathbf{A}}(Y))$$

arising from excision coincides with $K(\Psi)$.

Theorem 8.9 follows now by combining Lemma 8.10, Corollary 8.12, Lemma 8.13 and Proposition 8.15. In the remainder of this section, we establish some additional properties of the equivariant coarse homology theory $K\mathbf{A}\mathcal{X}^{\Gamma}$. For the next two propositions recall the notions of continuity (Definition 5.15) and strongness (Definition 4.19).

PROPOSITION 8.17. *The equivariant coarse homology theory $K\mathbf{A}\mathcal{X}^{\Gamma}$ is continuous.*

PROOF. Let (A,ρ) be an equivariant X-controlled object. Set $S:=\{x\in X\mid A(\{x\})\not\cong 0\}$. By definition, we have $S\cap B=\sigma(B)$, where σ is the support function of A. Hence, S is a locally finite subset of X, so (A,ρ) lies in the full subcategory $\mathbf{V}^{\Gamma}_{\mathbf{A}}(S)$ of $\mathbf{V}^{\Gamma}_{\mathbf{A}}(X)$. This shows that $\mathbf{V}^{\Gamma}_{\mathbf{A}}(X)=\bigcup_{S\subseteq X\text{ locally finite}}\mathbf{V}^{\Gamma}_{\mathbf{A}}(S)$. By Property (3) of the algebraic K-theory functor, it follows that $K\mathbf{A}\mathcal{X}^{\Gamma}$ is continuous. □

PROPOSITION 8.18. *The equivariant coarse homology theory $K\mathbf{A}\mathcal{X}^{\Gamma}$ is strong.*

PROOF. Let X be a Γ-bornological coarse space with weak flasqueness implemented by $\phi\colon X\to X$. As in the proof of Lemma 8.13, the functor $\Sigma\colon\mathbf{V}^{\Gamma}_{\mathbf{A}}(X)\to\mathbf{V}^{\Gamma}_{\mathbf{A}}(X)$ given by

$$\Sigma:=\bigoplus_{n\in\mathbb{N}}(\phi^n)_*$$

is well-defined. By assumption, we have $\mathrm{id}_{K\mathbf{A}\mathcal{X}^{\Gamma}(X)}=K\mathbf{A}\mathcal{X}^{\Gamma}(\phi)$. Now apply

Property (4) of algebraic K-theory to deduce that

$$\mathrm{id}_{K\mathbf{A}\mathcal{X}^\Gamma(X)} + K(\Sigma) \simeq \mathrm{id}_{K\mathbf{A}\mathcal{X}^\Gamma(X)} + K\mathbf{A}\mathcal{X}^\Gamma(\phi)\circ K(\Sigma) \simeq K(\mathrm{id}_{\mathbf{V}^\Gamma_\mathbf{A}(X)} \oplus \phi_*\circ\Sigma) \simeq K(\Sigma)\ ,$$

so $\mathrm{id}_{K\mathbf{A}\mathcal{X}^\Gamma(X)} \simeq 0$. □

Recall the definition of the free union of a family of Γ-bornological coarse spaces which was given in Example 2.16 and Definition 3.12 of the notion of strong additivity for an equivariant coarse homology theory.

PROPOSITION 8.19. *The equivariant coarse homology theory $K\mathbf{A}\mathcal{X}^\Gamma$ is strongly additive.*

PROOF. Let $(X_i)_{i\in I}$ be a family of Γ-bornological coarse spaces. The functors

$$\Phi_j\colon \mathbf{V}^\Gamma_\mathbf{A}(\bigsqcup_{i\in I}^{\mathrm{free}} X_i) \to \mathbf{V}^\Gamma_\mathbf{A}(X_j)$$

sending a $\bigsqcup_{i\in I}^{\mathrm{free}} X_i$-controlled $\mathbf{A}$-object (A,ρ) to $(A|_{X_j},\rho|_{X_j})$ for j in I assemble to a functor

$$\Phi\colon \mathbf{V}^\Gamma_\mathbf{A}(\bigsqcup_{i\in I}^{\mathrm{free}} X_i) \to \prod_{i\in I}\mathbf{V}^\Gamma_\mathbf{A}(X_i)\ .$$

For j in I, let $\iota_j\colon X_j \to \bigsqcup_{i\in I}^{\mathrm{free}} X_i$ denote the inclusion. We claim that the functor

$$\Psi\colon \prod_{i\in I}\mathbf{V}^\Gamma_\mathbf{A}(X_i) \to \mathbf{V}^\Gamma_\mathbf{A}(\bigsqcup_{i\in I}^{\mathrm{free}} X_i)$$

which sends a sequence $(A_i,\rho_i)_i$ to $\bigoplus_{i\in I}(\iota_i)_*(A_i,\rho_i)$ is well-defined (up to canonical isomorphism). We only have to check that the direct sum exists. This follows from the fact that for every bounded subset B of $\bigsqcup_{i\in I}^{\mathrm{free}} X_i$ the subset $\{i\in I | B\cap X_i \neq \emptyset\}$ is finite, and that $B\cap X_i$ is bounded for all i in I.

Clearly, $\Psi\circ\Phi$ is isomorphic to the identity. The composition $\Psi\circ\Phi$ is also isomorphic to the identity since $(A,\rho)\cong\bigoplus_{i\in I}(A|_{X_i},\rho|_{X_i})$ for all objects (A,ρ).

Using Properties (2) and (6), we conclude that

$$K(\mathbf{V}^\Gamma_\mathbf{A}(\bigsqcup_{i\in I}^{\mathrm{free}} X_i)) \xrightarrow{K(\Phi)} K(\prod_{i\in I}\mathbf{V}^\Gamma_\mathbf{A}(X_i)) \xrightarrow{\simeq} \prod_{i\in I} K(\mathbf{V}^\Gamma_\mathbf{A}(X_i))$$

is an equivalence. Note that the j-th component of this equivalence is given by the map $K(\Phi_j)$. Now apply Remark 8.16 to see that $K(\Phi_j)$ agrees with the projection map coming from excision. □

8.4. Calculations.

8.4.1. *Examples of the form $(\Gamma/H)_{min,min}$.* Let H be a subgroup of Γ. Let $\mathbf{A}$ be an additive category with trivial Γ-action.

LEMMA 8.20. *We have an equivalence*

$$K\mathbf{A}\mathcal{X}^\Gamma((\Gamma/H)_{min,min}) \simeq K(\mathbf{Fun}(BH,\mathbf{A}))\ .$$

PROOF. In view of the Property 2 of the K-theory functor it suffices to construct an equivalence of additive categories

$$\Phi\colon \mathbf{V}^\Gamma_\mathbf{A}((\Gamma/H)_{min,min}) \to \mathbf{Fun}(BH,\mathbf{A})\ .$$

This functor sends an object (A,ρ) of $\mathbf{V}^\Gamma_{\mathbf{A}}((\Gamma/H)_{min,min})$ to the functor sending γ to $\rho(\gamma)(\{eH\})\colon A(\{eH\})\to A(\{eH\})$.

Furthermore, the functor Φ sends a morphism

$$f\colon (A,\rho)\to (A',\rho')$$

in $\mathbf{V}^\Gamma_{\mathbf{A}}((\Gamma/H)_{min,min})$ to the transformation $f(\{eH\})\colon A(\{eH\})\to A'(\{eH\})$.

In order to define an inverse functor we choose a section $s\colon G/H\to G$ of the projection $G\to G/H$. Then

$$\Psi\colon \mathbf{Fun}(BH,\mathbf{A})\to \mathbf{V}^\Gamma_{\mathbf{A}}((\Gamma/H)_{min,min})$$

sends a functor $F\colon BH\to\mathbf{A}$ to the following object (A,ρ) of $\mathbf{V}^\Gamma_{\mathbf{A}}((\Gamma/H)_{min,min})$: we choose $A(B)=\bigoplus_{b\in B}F(*)$ and ρ is defined on an element γ of Γ such that $\rho(\gamma)(B)$ is the morphism $\bigoplus_{b\in B}F(*)\to\bigoplus_{b\in\gamma^{-1}(B)}F(*)$ sending the summand with index $b=gH$ to the summand with index $\gamma^{-1}gH$ via $F(h)$, where h is the element of H which is uniquely determined by the equation $\gamma^{-1}s(gH)=s(\gamma^{-1}gH)h$.

It is an easy exercise to construct the isomorphisms from the compositions $\Psi\circ\Phi$ and $\Phi\circ\Psi$ to the respective identity functors. □

8.4.2. *Examples of the form $X_{min,max}\otimes\Gamma_{can,min}$.* We consider the group Γ as a Γ-bornological coarse space $\Gamma_{can,min}$. In applications of coarse homotopy theory to proofs of the Farrell–Jones conjecture the coarse algebraic K-homology $K\mathbf{A}\mathcal{X}^\Gamma$ twisted by $\mathrm{Yo}^s(\Gamma_{can,min})$ plays an important role. Therefore it is relevant to calculate the spectra

$$K\mathbf{A}\mathcal{X}^\Gamma_{\mathrm{Yo}^s(\Gamma_{can,min})}((\Gamma/H)_{min,max})\simeq K\mathbf{A}\mathcal{X}^\Gamma((\Gamma/H)_{min,max}\otimes\Gamma_{can,min})\ .$$

More generally, we will replace Γ/H by any Γ-set X.

DEFINITION 8.21 ([**4**, Def. 2.1]). Let $\mathbf{A}$ be an additive category with a Γ-action and let X be a Γ-set. We define a new additive category denoted $\mathbf{A}*_\Gamma X$ as follows. An object A in $\mathbf{A}*_\Gamma X$ is a family $A=(A_x)_{x\in X}$ of objects in $\mathbf{A}$ where we require that $\{x\in X\mid A_x\neq 0\}$ is a finite set. A morphism $\phi\colon A\to B$ is a collection of morphisms $\phi=(\phi_{x,g})_{(x,g)\in X\times\Gamma}$, where $\phi_{x,g}\colon A_x\to g(B_{g^{-1}x})$ is a morphism in $\mathbf{A}$. We require that the set of pairs (x,g) in $X\times\Gamma$ with $\phi_{x,g}\neq 0$ is finite. Addition of morphisms is defined componentwise. Composition of morphisms is defined as the convolution product.

REMARK 8.22. In [**4**, Def. 2.1] additive categories with right Γ-action are used. For us it is more convenient to consider left Γ-actions.

Let H be a subgroup of Γ.

DEFINITION 8.23. We will denote $\mathbf{A}*_\Gamma(\Gamma/H)$ by $\mathbf{A}[H]$.

If $\mathbf{A}$ is the category of finitely generated, free R-modules for some ring R, then $\mathbf{A}[H]$ is equivalent to the category of finitely generated, free $R[H]$-modules.

The following calculation closely follows Bartels–Farrell–Jones–Reich [**1**, Sec. 6.1 and Proof of Prop. 6.2].

PROPOSITION 8.24. *For every Γ-set X we have an equivalence*

$$\mathbf{V}^\Gamma_{\mathbf{A}}(X_{min,max}\otimes\Gamma_{can,min})\simeq\mathbf{A}*_\Gamma X\ .$$

PROOF. The desired equivalence is given by an exact functor

$$\Phi\colon \mathbf{V}^{\Gamma}_{\mathbf{A}}(X_{min,max} \otimes \Gamma_{can,min}) \to \mathbf{A} *_{\Gamma} X \ .$$

We define Φ as follows:

(1) For an object (A, ρ) in $\mathbf{V}^{\Gamma}_{\mathbf{A}}(X_{min,max} \otimes \Gamma_{can,min})$, we define $\Phi(A, \rho)_x$ as $A(\{(x, 1)\})$.
(2) For a morphism $f\colon (A, \rho) \to (A', \rho')$ we define $\Phi(f)_{x,g}$ as the composition

$$A(\{x,1\}) \xrightarrow{f} A'(\{x\} \times F) \xrightarrow{p_g} A'(\{x,g\}) \xrightarrow{\rho'(g)} gA'(\{g^{-1}x,1\}) \ ,$$

where F is a finite subset of Γ containing g and p_g is the projection arising from the identification $\bigoplus_{f\in F} A(\{x,f\}) \xrightarrow{\cong} A(\{x\} \times F)$.

Note that $\Phi(f)_{x,g}$ is independent of the choice of F.

We will first show that Φ is fully faithful. A morphism $f\colon (A, \rho) \to (A', \rho')$ is determined by its values on $A(\{x, 1\})$ by equivariance. Since $X_{min,max}$ has the minimal coarse structure, the family $(\Phi(f)_{x,g})_{(x,g)\in X\times\Gamma}$ determines f. Hence Φ is faithful.

Since the Γ-action on $X_{min,max} \otimes \Gamma_{can,min}$ is free, for every finite subset F of Γ and every family of morphisms $A(\{x, 1\}) \to A'(\{x\} \times F)$ indexed by points x in X there exists a unique equivariant extension to a morphism $f\colon (A, \rho) \to (A, \rho')$. Let $\phi\colon \Phi(A, \rho) \to \Phi(A', \rho')$ be any morphism in $A *_{\Gamma} X$. Let $F := \{g \in \Gamma \mid \exists x \in X : \phi_{g,x} \neq 0\}$, then F is a finite subset of Γ. The family of morphisms

$$\left((A(\{x,1\}) \xrightarrow{\bigoplus_{g\in F} \rho'_{g^{-1}} \circ \phi_{x,g}} \bigoplus_{g\in F} A'(\{x,g\}) \xrightarrow{\cong} A'(\{x\} \times F) \right)_{x\in X}$$

extends to a morphism $f\colon (A, \rho) \to (A', \rho')$ with $\Phi(f) = \phi$. This shows that Φ is fully faithful.

We now show that Φ is essentially surjective. Every finitely supported family $(A_x)_{x\in X}$ of objects of A extends essentially uniquely to an equivariant object (A, ρ) with $A(\{x, 1\}) = A_x$ for all x in X. This uses the choice of finite sums of the objects A_x. □

Let X be a Γ-set.

COROLLARY 8.25. *We have an equivalence*

$$K\mathbf{A}\mathcal{X}^{\Gamma}(X_{min,max} \otimes \Gamma_{can,min}) \simeq K(\mathbf{A} *_{\Gamma} X) \ .$$

8.4.3. *Examples of the form* $(\Gamma/H)_{min,min} \otimes \Gamma_{?,max}$. Let H be a subgroup of Γ and let $\mathbf{A}$ be an additive category with Γ-action.

LEMMA 8.26. *If H is finite, then*

$$K\mathbf{A}\mathcal{X}^{\Gamma}((\Gamma/H)_{min,min} \otimes \Gamma_{max,max}) \simeq K\mathbf{A}\mathcal{X}^{\Gamma}((\Gamma/H)_{min,min} \otimes \Gamma_{can,max}) \simeq K(\mathbf{A}[H]) \ .$$

Otherwise

$$K\mathbf{A}\mathcal{X}^{\Gamma}((\Gamma/H)_{min,min} \otimes \Gamma_{max,max}) \simeq K\mathbf{A}\mathcal{X}^{\Gamma}((\Gamma/H)_{min,min} \otimes \Gamma_{can,max}) \simeq 0 \ .$$

PROOF. We argue as in the proof of Proposition 8.24 for $X = (\Gamma/H)_{min,min}$. If H is finite, then the set F appearing in the proof is still finite, but for a different reason.

If H is infinite, then there are no non-trivial Γ-invariant $(\Gamma/H)_{min,min} \otimes \Gamma_{?,max}$ controlled modules (this does not depend on the coarse structures). □

8.5. Change of groups. Let H be a subgroup of Γ.

THEOREM 8.27. *There is an equivalence of H-equivariant coarse homology theories*

$$\mathrm{ind}_H^\Gamma \colon K\mathbf{A}\mathcal{X}^H \xrightarrow{\simeq} K\mathbf{A}\mathcal{X}^\Gamma \circ \mathrm{Ind}_H^\Gamma \ .$$

In the proof we will construct a equivalence in the other direction. We state the theorem in this form since this is the more common direction.

PROOF. Let $(X, \mathcal{B}, \mathcal{C})$ be a bornological coarse space. Recall from the Remark 6.4 that the bornological coarse space $\mathrm{Ind}_H^\Gamma X$ is given by the set $\Gamma \times_H X$ with bornology generated by the subsets $\{g\} \times B$ for B in $\mathcal{B}$ and coarse structure generated by the entourages $\mathrm{diag}_\Gamma \times U$ for U in $\mathcal{C}$.

Note that $H \times_H X \simeq X$ is an H-invariant coarse component of $\mathrm{Ind}_H^\Gamma X$. Hence, restricting an object (A, ρ) of $\mathbf{V}_\mathbf{A}^\Gamma(\mathrm{Ind}_H^\Gamma X)$ to $(A|_X, \rho_X)$ yields a functor

$$\mathbf{V}_\mathbf{A}^\Gamma(\mathrm{Ind}_H^\Gamma X) \to \mathbf{V}_\mathbf{A}^H(X) \ .$$

Similarly to the proof of Lemma 8.20, one checks that this functor is an equivalence. □

Let H be a subgroup of Γ. Sending a Γ-equivariant X-controlled object (A, ρ) to the object $(A, \{\rho(h)\}_{h \in H})$ yields a natural transformation

$$\mathrm{res}_H^\Gamma \colon K\mathbf{A}\mathcal{X}^\Gamma \to K\mathbf{A}\mathcal{X}^H \circ \mathrm{Res}_H^\Gamma \ . \tag{8.1}$$

8.6. Variations on the definition. Let again $\mathbf{A}$ be an additive category with a Γ-action and $(X, \mathcal{B}, \mathcal{C})$ be a Γ-bornological coarse space. Intuitively, an equivariant X-controlled $\mathbf{A}$-object (A, ρ) is some (infinite) sum of objects in $\mathbf{A}$ parametrized by points in X together with an action of Γ. One may want to keep track of the "global" object associated to an X-controlled object explicitly. The purpose of this section is to give an alternative definition of $\mathbf{V}_\mathbf{A}^\Gamma(X)$ which accomplishes precisely this, and discuss a variation of this definition which leads to an example of a non-continuous coarse homology theory.

Since an equivariant X-controlled $\mathbf{A}$-object usually involves an infinite number of objects in $\mathbf{A}$, we need to enlarge our coefficient category appropriately. Therefore, let $\mathbf{A} \to \widehat{\mathbf{A}}$ be a fully faithful and Γ-equivariant embedding of $\mathbf{A}$ into an additive category $\widehat{\mathbf{A}}$ with Γ-action which admits sufficiently large direct sums. The sum completion of $\mathbf{A}$ is a canonical choice for $\widehat{\mathbf{A}}$.

For an object A in $\widehat{\mathbf{A}}$, let $\Pi(A)$ denote the set of idempotents on A whose image splits off as a direct sum and is isomorphic to an object in $\mathbf{A}$.

DEFINITION 8.28. A *global equivariant X-controlled $\mathbf{A}$-object* is a triple (A, ϕ, ρ) consisting of

(1) an object A in $\widehat{\mathbf{A}}$;
(2) a function $\phi \colon \mathcal{B} \to \Pi(A)$;
(3) a morphism $\rho(\gamma) \colon A \to \gamma A$ for every element γ in Γ;

such that the following conditions are satisfied:

(1) The function ϕ satisfies the following relations:
  (a) $\phi(\emptyset) = 0$;
  (b) $\phi(B_1 \cup B_2) = \phi(B_1) + \phi(B_2) - \phi(B_1 \cap B_2)$;
  (c) $\phi(B_1 \cap B_2) = \phi(B_1) \circ \phi(B_2)$.

(2) For every bounded subset B of X, there exists some finite subset F of B such that $\phi(B) = \phi(F)$.
(3) For all pairs of elements γ, γ' in Γ, we have $\rho(\gamma'\gamma) = \gamma\rho(\gamma') \circ \rho(\gamma)$.
(4) For all elements γ of Γ and bounded subsets B of X, we have the equality

$$\phi(\gamma^{-1}B) = \rho(\gamma)^{-1} \circ \gamma\phi(B) \circ \rho(\gamma)\ .$$

In contrast to the data specifying an equivariant X-controlled object, this definition does not fix a chosen image for each of the idempotents $\phi(B)$.

DEFINITION 8.29. A morphism $f\colon (A, \phi, \rho) \to (A', \phi', \rho')$ of global equivariant X-controlled $\mathbf{A}$-objects is a morphism $f\colon A \to A'$ in $\widehat{\mathbf{A}}$ satisfying the following conditions:

(1) f is equivariant in the sense that $(\gamma f) \circ \rho(\gamma) = \rho'(\gamma) \circ f$;
(2) f is controlled in the sense that the set

$$\bigcap\{U \subseteq X \times X \mid \forall\, B, B' \in \mathcal{B}\colon U[B] \cap B' = \emptyset \Rightarrow \phi'(B') f \phi(B) = 0\}$$

is an entourage of X.

Morphisms of global equivariant X-controlled $\mathbf{A}$-objects can be composed. We denote the resulting category by $\mathbf{V}^{\Gamma}_{\mathbf{A}\subseteq\widehat{\mathbf{A}}}(X)$. Similar to the discussion in Section 8.2, one shows that $\mathbf{V}^{\Gamma}_{\mathbf{A}\subseteq\widehat{\mathbf{A}}}(X)$ is additive. If $f\colon X \to X'$ is a morphism of Γ-bornological coarse spaces, then we define a functor

$$f_*\colon \mathbf{V}^{\Gamma}_{\mathbf{A}\subseteq\widehat{\mathbf{A}}}(X) \to \mathbf{V}^{\Gamma}_{\mathbf{A}\subseteq\widehat{\mathbf{A}}}(X')$$

which sends (A, ϕ, ρ) to $(A, f_*\phi, \rho)$, were $f_*\phi(B') := \phi(f^{-1}(B'))$ for all B' in $\mathcal{B}'$. Furthermore, the construction is functorial with respect to commutative squares of additive functors

$$\begin{array}{ccc} \mathbf{A} & \longrightarrow & \widehat{\mathbf{A}} \\ {\scriptstyle\Phi}\downarrow & & \downarrow{\scriptstyle\widehat{\Phi}} \\ \mathbf{A}' & \longrightarrow & \widehat{\mathbf{A}}' \end{array}$$

in which the horizontal arrows are fully faithful and Γ-equivariant embeddings.

We have already indicated how objects in $\mathbf{V}^{\Gamma}_{\mathbf{A}}(X)$ correspond to objects in $\mathbf{V}^{\Gamma}_{\mathbf{A}\subseteq\widehat{\mathbf{A}}}(X)$ and vice versa, namely by summing up all values of an equivariant X-controlled object and choosing images of idempotents, respectively. In fact, it is not difficult to show the following.

PROPOSITION 8.30. *There is a zig-zag of equivalence between the functors* $\mathbf{V}^{\Gamma}_{\mathbf{A}}$ *and* $\mathbf{V}^{\Gamma}_{\mathbf{A}\subseteq\widehat{\mathbf{A}}}$ *from* $\Gamma\mathbf{BC}$ *to* $\mathbf{Add}$.

EXAMPLE 8.31. We now provide a modification of the definition of $\mathbf{V}^{\Gamma}_{\mathbf{A}\subseteq\widehat{\mathbf{A}}}$ which leads to a non-continuous coarse homology theory.

Let (A, ϕ, ρ) be a global equivariant X-controlled $\mathbf{A}$-object. Condition (2) of Definition 8.28 implies, in the presence of the Condition (1), that the object (A, ϕ) is essentially determined by its restriction to the poset of finite subsets of X, and that the support of (A, ϕ), i.e. the set $\{x \in X \mid \phi(\{x\}) \neq 0\}$, is a locally finite subset of X.

Dropping the Condition (2), we obtain another additive category $\mathbf{V}^{\Gamma}_{\mathbf{A},\Psi}(X)$ which is also functorial in Γ-bornological coarse spaces. Taking non-connective algebraic K-theory of this category also gives rise to an equivariant coarse homology theory $K\mathbf{A}\mathcal{X}^{\Gamma}_{\Psi}$ since the proofs of Lemma 8.10, Lemma 8.11, Lemma 8.13 and Proposition 8.15 go through without change. However, the proof of continuity (Proposition 8.17) does not apply to $K\mathbf{A}\mathcal{X}^{\Gamma}_{\Psi}$ since the condition ensuring that the support of an object is locally finite has been omitted.

In fact, the following example shows that the coarse homology theory $K\mathbf{A}\mathcal{X}_{\Psi}$ is not continuous: Suppose that there exists an object A in $\mathbf{A}$ whose class in K_0 is non-trivial. Consider the bornological coarse space $\mathbb{N}_{min,max}$. Choose an ultrafilter $\mathcal{F}$ on $\mathbb{N}$ and define the function $\phi\colon \mathcal{P}(\mathbb{N}) \to \Pi(A)$ by

$$\phi(B) := \begin{cases} \mathrm{id}_A & B \in \mathcal{F}, \\ 0 & B \notin \mathcal{F}. \end{cases}$$

Then (A, ϕ) is a global $\mathbb{N}_{min,max}$-controlled $\mathbf{A}$-object since $\mathcal{F}$ is an ultrafilter. Moreover, the morphism of bornological coarse spaces $\mathbb{N}_{min,max} \to *$ induces a homomorphism

$$\pi_0 K\mathbf{A}\mathcal{X}_{\Psi}(\mathbb{N}_{min,max}) \to \pi_0 K\mathbf{A}\mathcal{X}_{\Psi}(*) \cong K_0(\mathbf{A})$$

which maps the class $[(A, \phi)]$ to $[A] \neq 0$.

On the other hand, locally finite subsets of $\mathbb{N}_{min,max}$ are precisely the finite subsets of $\mathbb{N}$. Since each finite subset F is a union of coarse components of $\mathbb{N}_{min,max}$, the pair of subsets $(F, \mathbb{N} \setminus F)$ is a complementary pair. From excision, we obtain a direct sum decomposition

$$K\mathbf{A}\mathcal{X}_{\Psi}(\mathbb{N}_{min,max}) \simeq K\mathbf{A}\mathcal{X}_{\Psi}(F_{min,max}) \oplus K\mathbf{A}\mathcal{X}_{\Psi}((\mathbb{N} \setminus F)_{min,max}) ,$$

in which the projection $K\mathbf{A}\mathcal{X}_{\Psi}(\mathbb{N}_{min,max}) \to K\mathbf{A}\mathcal{X}_{\Psi}(F_{min,max})$ is induced by the functor restricting $\mathbb{N}_{min,max}$-controlled objects to F (cf. Remark 8.16). Since the restriction of (A, ϕ) to any finite subset of $\mathbb{N}$ is the zero object, we conclude that the class $[(A, \phi)]$ does not lie in the image of the comparison map

$$\operatorname*{colim}_{F \subseteq \mathbb{N} \text{ finite}} \pi_0 K\mathbf{A}\mathcal{X}_{\Psi}(F_{min,max}) \to \pi_0 K\mathbf{A}\mathcal{X}_{\Psi}(\mathbb{N}_{min,max}) .$$

In particular, $K\mathbf{A}\mathcal{X}_{\Psi}$ is not continuous.

Part 3. Cones and assembly maps

9. Cones

In this section we will introduce the cone and the 'cone at infinity' functors, and discuss their properties. The cone at infinity is later used in Section 10.3 to define the universal assembly map, and in Section 10.4 to transform coarse homology theories into topological homology theories. In conjunction with the Rips complex construction the cone at infinity will be used in Section 11.3 to construct the universal coarse assembly map.

The first three Sections 9.1–9.3 are technical preparations. The cone functor is then defined and discussed in Section 9.4 and Section 9.5.

9.1. Γ-uniform bornological coarse spaces. In this section we introduce the category $\Gamma\mathbf{UBC}$ of Γ-uniform bornological coarse spaces. Its objects are Γ-bornological coarse spaces with an additional Γ-uniform structure. The additional datum of a Γ-uniform structure is needed in order to define hybrid structures, see Section 9.2.

We start with recalling some basics on uniform spaces. Let X be a set and $\mathcal{T}$ be a subset of $\mathcal{P}(X \times X)$, the power set of $X \times X$.

DEFINITION 9.1. The set $\mathcal{T}$ is a uniform structure if it is non-empty, closed under composition, inversion, supersets, finite intersection, every element of $\mathcal{T}$ contains the diagonal of X, and if for every U in $\mathcal{T}$ there exists V in $\mathcal{T}$ such that $V \circ V \subseteq U$.

REMARK 9.2. Note that any subset S of $\mathcal{P}(X \times X)$ with the property that for every U in S there exists V in X such that $V \circ V \subseteq U$ generates a uniform structure on X by taking the closure of S under composition, inversion, supersets and finite intersection.

The elements of $\mathcal{T}$ are called *uniform entourages*. We will consider $\mathcal{T}$ as a filtered partially ordered set whose order relation is the opposite of the inclusion relation.

DEFINITION 9.3. A *uniform space* is a pair $(X, \mathcal{T})$ of a set X together with a uniform structure $\mathcal{T}$.

Let $(X, \mathcal{T})$ and $(X', \mathcal{T}')$ be uniform spaces and $f \colon X \to X'$ be a map of sets.

DEFINITION 9.4. The map f is called a *uniform map* if for every uniform entourage U' of X' we have $(f \times f)^{-1}(U') \in \mathcal{T}$.

If a group Γ acts on a uniform space $(X, \mathcal{T})$, then it acts on the set of uniform entourages $\mathcal{T}$. We let $\mathcal{T}^\Gamma$ denote the subset of $\mathcal{T}$ of Γ-invariant uniform entourages.

DEFINITION 9.5. A Γ*-uniform space* is a uniform space $(X, \mathcal{T})$ with an action of Γ by automorphisms such that $\mathcal{T}^\Gamma$ is cofinal in $\mathcal{T}$ (i.e., for every U in $\mathcal{T}$ there exists V in $\mathcal{T}^\Gamma$ with $V \subseteq U$).

A uniform structure $\mathcal{T}$ on a Γ-set X such that X is a Γ-uniform space will be called a Γ*-uniform structure.*

EXAMPLE 9.6. The uniform structure $\mathcal{T}_d$ of a metric space (X, d) is generated by the uniform entourages

$$U_r := \{(x, y) \in X \times X | d(x, y) < r\} \tag{9.1}$$

for all r in $(0, \infty)$.

If Γ acts isometrically on a metric space (X, d), then the associated uniform space $X_{u,d} := (X, \mathcal{T}_d)$ is a Γ-uniform space.

If the metric is implicitly clear, then we will also write X_u instead of $X_{u,d}$.

A uniformly continuous map between metric spaces induces a uniform map between the associated uniform spaces.

The standard metric turns $\mathbb{R}$ into a uniform space $\mathbb{R}_u$. The action of $\mathbb{Z}$ on $\mathbb{R}$ by dilatations $(n, x) \mapsto 2^n x$ is an action on $\mathbb{R}_u$ by automorphisms of uniform spaces, but $\mathbb{R}_u$ is not a $\mathbb{Z}$-uniform space.

Let $\Gamma\mathbf{U}$ be the category of Γ-uniform spaces and uniform equivariant maps.

EXAMPLE 9.7. Let K be a simplicial complex with a simplicial Γ-action. On K we consider the path quasi-metric induced by the spherical metric on the simplices. This quasi-metric is preserved by Γ and the associated uniform structure is a Γ-uniform structure. Therefore a simplicial complex with a simplicial Γ-action gives rise to a Γ-uniform space K_u.

If $K \to K'$ is an equivariant simplicial map, then it is a uniform map $K_u \to K'_u$ between the Γ-uniform spaces.

We now consider the combination of uniform and bornological coarse structures. We consider a Γ-set X with a Γ-coarse structure $\mathcal{C}$ and a Γ-uniform structure $\mathcal{T}$.

DEFINITION 9.8. We say that $\mathcal{C}$ and $\mathcal{T}$ are *compatible* if $\mathcal{C}^\Gamma \cap \mathcal{T}^\Gamma \neq \emptyset$.

In words, the coarse and the uniform structures are compatible if there exists an invariant entourage which is both a coarse entourage and a uniform entourage.

DEFINITION 9.9. We define the category $\mathbf{\Gamma UBC}$ of Γ-uniform bornological coarse spaces as follows:

(1) The objects of $\mathbf{\Gamma UBC}$ are Γ-bornological coarse spaces with an additional compatible Γ-uniform structure.
(2) The morphisms of $\mathbf{\Gamma UBC}$ are morphisms of Γ-bornological coarse spaces which are in addition uniform.

EXAMPLE 9.10. Let (X, d) be a quasi-metric space with an action of Γ by isometries and a Γ-invariant bornology $\mathcal{B}$. Assume that the metric and the bornology are compatible in the sense that for every r in $(0, \infty)$ and B in $\mathcal{B}$ we have $U_r[B] \in \mathcal{B}$, where U_r is as in (9.1). Then we get a Γ-uniform bornological coarse space X_{du} with the following structures:

(1) The coarse structure is generated by the coarse entourages U_r for all r in $(0, \infty)$.
(2) The uniform structure is generated by the uniform entourages U_r for all r in $(0, \infty)$.
(3) The bornology is $\mathcal{B}$.

If (X', d') is a second quasi-metric space with isometric Γ-action and $f \colon X \to X'$ is a proper (this refers to the bornologies), Γ-equivariant contraction, then $f \colon X_{du} \to X'_{du}$ is a morphism of Γ-uniform bornological coarse spaces.

EXAMPLE 9.11. Let K be a Γ-simplicial complex equipped with the spherical quasi-metric. Then we can equip K with the bornology of metrically bounded subsets and obtain a Γ-uniform bornological coarse space K_{du}. Alternatively we can equip it with the maximal bornology and get the Γ-uniform bornological coarse space $K_{du,max}$.

A morphism of Γ-simplicial complexes $f \colon K \to K'$ always induces the two morphisms $f \colon K_{du,max} \to K'_{du,max}$ and $K_{du,max} \to K'_{du}$ of Γ-uniform bornological coarse spaces. If f is proper (in the sense that preimages of simplices are finite complexes), then it also induces a morphism $f \colon K_{du} \to K'_{du}$.

EXAMPLE 9.12. For a Γ-uniform bornological coarse space X let $F_{\mathcal{T}}(X)$ denote the underlying Γ-bornological coarse space obtained by forgetting the datum of the uniform structure. Let X and Y be Γ-uniform bornological coarse spaces. Then we define the Γ-uniform bornological coarse space $X \otimes Y$ such that $\mathcal{F}_{\mathcal{T}}(X \otimes Y) = F_{\mathcal{T}}(X) \otimes F_{\mathcal{T}}(Y)$ and the Γ-uniform structure on $X \otimes Y$ is generated by the products $U \times V$ for all pairs of uniform entourages U of X and V of Y.

9.2. Hybrid structures. In this section we will define hybrid coarse structures, which will feature in the definition of cones in Section 9.4.

We consider a Γ-uniform bornological coarse space X with coarse structure $\mathcal{C}$, bornology $\mathcal{B}$, and uniform structure $\mathcal{T}$. Let furthermore a Γ-invariant big family

$\mathcal{Y} = (Y_i)_{i\in I}$ be given (recall from [**8**] that a *big family* on X is a filtered family of subsets $(Y_i)_{i\in I}$ of X such that for every entourage U of X and i in I there exists j in I such that $U[Y_i] \subseteq Y_j$).

DEFINITION 9.13. The pair $(X, \mathcal{Y})$ is called *hybrid data.*

In this situation we can define the hybrid coarse structure $\mathcal{C}_h$ as follows.

Note that $\mathcal{P}(X \times X)^\Gamma$ is a filtered poset with the opposite of the inclusion relation. We consider a function $\phi\colon I \to \mathcal{P}(X \times X)^\Gamma$, i.e., an order-preserving map.

DEFINITION 9.14. The function ϕ is called $\mathcal{T}^\Gamma$-admissible, if for every U in $\mathcal{T}^\Gamma$ there exists i in I such that $\phi(i) \subseteq U$.

Given a $\mathcal{T}^\Gamma$-admissible function $\phi\colon I \to \mathcal{P}(X \times X)^\Gamma$ we define the entourage

$$U_\phi := \big\{(x,y) \in X \times X | \big(\forall i \in I | (x,y) \in Y_i \times Y_i \text{ or } (x,y) \in \phi(i)\big)\big\} .$$

Note that U_ϕ is Γ-invariant.

DEFINITION 9.15. The *hybrid coarse structure* $\mathcal{C}_h$ is the coarse structure generated by the entourages $U \cap U_\phi$ for all U in $\mathcal{C}^\Gamma$ and $\mathcal{T}^\Gamma$-admissible functions $\phi\colon I \to \mathcal{P}(X \times X)^\Gamma$.

DEFINITION 9.16. The *hybrid space* X_h is defined to be the Γ-bornological coarse space with underlying set X, the hybrid coarse structure $\mathcal{C}_h$ and the bornological structure $\mathcal{B}$.

A morphism of hybrid data $f\colon (X, \mathcal{Y}) \to (X', \mathcal{Y}')$ is a morphism of Γ-uniform bornological coarse spaces which is compatible with the big families $\mathcal{Y} = (Y_i)_{i\in I}$ and $\mathcal{Y}' = (Y'_{i'})_{i'\in I'}$ in the sense that for every i in I there exists i' in I' such that $f(Y_i) \subseteq Y'_{i'}$.

LEMMA 9.17. *If f is a morphism of hybrid data, then the underlying map of sets is a morphism $f\colon X_h \to X'_h$ of Γ-bornological coarse spaces.*

PROOF. [**8**, Lem. 5.15]. □

REMARK 9.18. One could set up a category of hybrid data and understand the construction of the hybrid structure as a functor from hybrid data to Γ-bornological coarse spaces.

9.3. Decomposition theorem and homotopy theorem. In this section we discuss the Decomposition Theorem and the Homotopy Theorem for hybrid coarse structures. These two theorems constitute important technical results which are needed to prove crucial properties of the cone functor in Section 9.5. Let A, B be Γ-invariant subsets of a Γ-uniform space Y with uniform structure $\mathcal{T}$. For an entourage U of Y we set

$$\mathcal{T}^\Gamma_{\subseteq U} := \{V \in \mathcal{T}^\Gamma | V \subseteq U\} .$$

DEFINITION 9.19. The pair (A, B) is an *equivariant uniform decomposition* if

(1) $Y = A \cup B$, and
(2) there is an invariant uniform entourage U of Y and a function

$$s\colon \mathcal{P}(Y \times Y)^\Gamma_{\subseteq U} \to \mathcal{P}(Y \times Y)^\Gamma$$

such that for every W in $\mathcal{P}(Y \times Y)^\Gamma_{\subseteq U}$ we have the inclusion

$$W[A] \cap W[B] \subseteq s(W)[A \cap B]$$

and the restriction $s_{|\mathcal{T}^\Gamma_{\subseteq U}}$ is $\mathcal{T}^\Gamma$-admissible.

According to Definition 9.14, the function $s_{|\mathcal{T}^\Gamma_{\subseteq U}}$ is $\mathcal{T}^\Gamma$-admissible if for every entourage V in $\mathcal{T}^\Gamma$ there is an entourage W in $\mathcal{T}^\Gamma_{\subseteq U}$ such that $s(W) \subseteq V$.

EXAMPLE 9.20. Assume that K is a Γ-simplicial complex, and A and B are Γ-invariant subcomplexes such that $K = A \cup B$. Then (A, B) is an equivariant uniform decomposition of K_u. This follows from [**8**, Ex. 5.19].

Let Y be a Γ-uniform bornological coarse space with an invariant big family $\mathcal{Y} = (Y_i)_{i \in I}$. We further assume that (A, B) is an equivariant uniform decomposition of Y. We let Y_h denote the associated bornological coarse space with the hybrid structure.

We write A_h for the Γ-bornological coarse space obtained from the Γ-uniform bornological coarse structure on A induced from Y by first restricting the hybrid data to A and then forming the hybrid structure. By A_{Y_h} we denote the Γ-bornological coarse space obtained by restricting the structures of Y_h to the subset A. It was shown in [**8**, Lem. 5.17] that $A_h = A_{Y_h}$.

DEFINITION 9.21. A Γ-uniform space $(Y, \mathcal{T})$ is called Hausdorff, if $\bigcap_{U \in \mathcal{T}} U = \mathrm{diag}_Y$.

THEOREM 9.22. *(Decomposition Theorem) If $I = \mathbb{N}$ and Y is Hausdorff, then the following square in $\Gamma\mathbf{Sp}\mathcal{X}$ is cocartesian:*

$$\begin{array}{ccc} \mathrm{Yo}^s((A \cap B)_h, A \cap B \cap \mathcal{Y}) & \longrightarrow & \mathrm{Yo}^s(A_h, A \cap \mathcal{Y}) \\ \downarrow & & \downarrow \\ \mathrm{Yo}^s(B_h, B \cap \mathcal{Y}) & \longrightarrow & \mathrm{Yo}^s(Y_h, \mathcal{Y}) \end{array} \tag{9.2}$$

PROOF. The proof of [**8**, Thm. 5.20] goes through word-for-word. One just works with invariant entourages or invariant uniform neighbourhoods everywhere. □

We consider a Γ-uniform bornological coarse space Y with an invariant big family $\mathcal{Y} = (Y_n)_{n \in \mathbb{N}}$. We consider the unit interval $[0, 1]$ with the trivial Γ-action as a Γ-uniform bornological coarse space $[0, 1]_{du}$ with the structures induced from the metric. On the tensor product $[0, 1]_{du} \otimes Y$ (Example 9.12) we consider the big family $([0, 1] \times Y_n)_{n \in \mathbb{N}}$. Let $\mathcal{B}$ denote the bornology of Y.

THEOREM 9.23 (Homotopy Theorem). *Assume that for every B in $\mathcal{B}$ there exists n in $\mathbb{N}$ such that $B \subseteq Y_n$. Then the projection induces an equivalence*

$$\mathrm{Yo}^s(([0, 1]_{du} \otimes Y)_h) \to \mathrm{Yo}^s(Y_h) \ .$$

PROOF. The proof of [**8**, Thm. 5.25] goes through in the present equivariant case. □

9.4. The cone functor. In this section we define the cone functor

$$\mathcal{O} \colon \Gamma\mathbf{UBC} \to \Gamma\mathbf{BC} \tag{9.3}$$

and prove that decompositions of the spaces which are simultaneously uniformly and coarsely excisive lead to corresponding coarsely excisive decompositions of the cones.

We consider the metric space $[0, \infty)$ with the metric induced from the inclusion into $\mathbb{R}$ and the trivial Γ-action. We get a Γ-uniform bornological coarse space

$[0,\infty)_{du}$. For a Γ-uniform bornological coarse space Y we form the Γ-uniform bornological coarse space

$$[0,\infty)_{du}\otimes Y$$

(Example 9.12). This Γ-uniform bornological coarse space has a canonical big family given by

$$\mathcal{Y}(Y) := ([0,n]\times Y)_{n\in\mathbb{N}} \;. \tag{9.4}$$

A morphism $f\colon Y\to Y'$ of Γ-uniform bornological coarse spaces induces a morphism of hybrid data

$$([0,\infty)_{du}\otimes Y,\mathcal{Y}(Y))\to([0,\infty)_{du}\otimes Y',\mathcal{Y}(Y'))$$

given by the map $\mathrm{id}_{[0,\infty)}\times f$ on the underlying sets.

DEFINITION 9.24. The *cone functor* (9.3) is defined such that it sends the Γ-uniform bornological coarse space Y to the Γ-bornological coarse space

$$\mathcal{O}(Y) := ([0,\infty)_{du}\otimes Y)_h$$

and a morphism $f\colon Y\to Y'$ of Γ-uniform bornological coarse spaces to the morphism

$$\mathcal{O}(f)\colon \mathcal{O}(Y)\to\mathcal{O}(Y')$$

of Γ-bornological coarse spaces given by the map $\mathrm{id}_{[0,\infty)}\times f$ of the underlying sets.

EXAMPLE 9.25. If the Γ-uniform bornological coarse space Y is discrete as a uniform and as a coarse space, then $\mathcal{O}(Y)$ is flasque: Flasqueness of $\mathcal{O}(Y)$ can be implemented by the map $(t,y)\mapsto(t+\frac{1}{t+1},y)$.

If the uniform structure of Y is discrete, but the coarse structure of Y is strictly larger than the discrete one, then the above morphism f does not implement flasqueness of $\mathcal{O}(Y)$, because Condition 2 in Definition 3.8 is violated.

Let X and Y be Γ-uniform bornological coarse spaces. Recall that $F_{\mathcal{T}}$ is the functor from Γ-uniform bornological coarse spaces to Γ-bornological coarse spaces which forgets the uniform structure.

LEMMA 9.26. *If Y is discrete as a coarse space, then*

$$\mathcal{O}(X)\otimes F_{\mathcal{T}}(Y)\cong\mathcal{O}(X\otimes Y)\;.$$

PROOF. Immediate from the definitions. □

Let Y be a Γ-uniform bornological coarse space with coarse and uniform structures $\mathcal{C}$, $\mathcal{T}$, and let A,B be Γ-invariant subsets of Y.

LEMMA 9.27. *If (A,B) is an equivariant uniform (Definition 9.19) and coarsely excisive decomposition of Y, then*

$$([0,\infty)\times A,[0,\infty)\times B)$$

is a coarsely excisive pair on $\mathcal{O}(Y)$.

PROOF. We have $([0,\infty)\times A)\cup([0,\infty)\times B)=[0,\infty)\times Y$.

Let s and U be as in Definition 9.19, let $\phi\colon\mathbb{N}\to\mathcal{P}(Y\times Y)^\Gamma$ be a $\mathcal{T}^\Gamma$-admissible function and let $\kappa\colon[0,\infty)\to[0,\infty)$ be monotoneously decreasing such that $\lim_{u\to\infty}\kappa(u)=0$. The pair $\psi:=(\phi,\kappa)$ determines the invariant entourage

$$U_\psi := \{((a,x),(b,y))\in([0,\infty)\times Y)^{\times 2}\mid |a-b| \\ \le\kappa(\max\{a,b\}),(x,y)\in\phi(\lceil a\rceil)\cap\phi(\lceil b\rceil)\}\;.$$

For W in $\mathcal{C}^\Gamma$ and r in $(0,\infty)$ we consider the entourage $W_r := U_r \times W$ of $[0,\infty)_d \otimes Y$. The entourages of the form $U_\psi \cap W_r$ for all ψ as above, r in $(0,\infty)$ and W in $\mathcal{C}^\Gamma$ are cofinal in the hybrid coarse structure of $\mathcal{O}(Y)$.

We now fix ψ, W and r as above. We must show that there exist r' in $(0,\infty)$, W' in $\mathcal{C}^\Gamma$, and ψ' such that

$$(9.5)\quad (U_\psi \cap W_r)[[0,\infty)\times A] \cap (U_\psi \cap W_r)[[0,\infty)\times B] \subseteq (U_{\psi'} \cap W'_{r'})[[0,\infty)\times (A\cap B)]\ .$$

Using coarse excisiveness of (A,B) we can choose an invariant entourage W' of X such that $W[A]\cap W[B] \subseteq W'[A\cap B]$. We further set $r' := r$. Then

$$(9.6)\qquad W_r[[0,\infty)\times A] \cap W_r[[0,\infty)\times B] \subseteq W'_r[[0,\infty)\times (A\cap B)].$$

By $\mathcal{T}^\Gamma$-admissibility of ϕ there is u_0 in $\mathbb{N}$ such that $\phi(u) \subseteq U$ for all u in $\mathbb{N}$ with $u \ge u_0$. We define

$$\phi' \colon \mathbb{N} \to \mathcal{P}(Y\times Y)^\Gamma\ ,\quad \phi'(u) := \begin{cases} W' & u < u_0 \\ s(\phi(u)) & u \ge u_0 \end{cases}$$

Then ϕ' is $\mathcal{T}^\Gamma$-admissible. We further set $\psi' := (\phi', \kappa)$. We claim that

$$(9.7)\quad (U_\psi \cap W_r)[[0,\infty)\times A] \cap (U_\psi \cap W_r)[[0,\infty)\times B] \subseteq U_{\psi'}[[0,\infty)\times (A\cap B)]\ .$$

Consider a point (u,z) in $(U_\psi \cap W_r)[[0,\infty)\times A] \cap (U_\psi \cap W_r)[[0,\infty)\times B]$. Then there exists (a,x) in $[0,\infty)\times A$ such that we have

$$|a-u| \le \kappa(\max\{a,u\}) \quad\text{and}\quad (z,x)\in \phi(\lceil u\rceil)\cap\phi(\lceil a\rceil)\ ,$$

and there exist (b,y) in $[0,\infty)\times B$ such that

$$|b-u| \le \kappa(\max\{b,u\}) \quad\text{and}\quad (z,y)\in \phi(\lceil u\rceil)\cap\phi(\lceil b\rceil)\ .$$

In particular, we have $z \in (\phi(\lceil u\rceil)\cap W)[A] \cap (\phi(\lceil u\rceil)\cap W)[B]$. If $u \ge u_0$, then we have $z \in \phi'(\lceil u\rceil)[A\cap B]$ by the corresponding property of s (see Definition 9.19). Let w in $A\cap B$ be such that we have $(z,w)\in\phi'(\lceil u\rceil)$. Then $((u,z),(u,w)) \in U_{\psi'}[[0,\infty)\times(A\cap B)]$. If $u<u_0$, then again $((u,z),(u,w))\in U_{\psi'}[[0,\infty)\times(A\cap B)]$ because of the choice of W'.

The relations (9.6) and (9.7) together imply (9.5). □

REMARK 9.28. In the above proof, in contrast to the general Decomposition Theorem 9.22 for hybrid structures, we do not use that Y is Hausdorff.

9.5. The cone at infinity. In this section we will define the "cone at infinity" functor. It fits into the cone fiber sequence. We discuss invariance under coarsenings and calculate it for discrete spaces. Furthermore, we show that the "cone at infinity" is excisive and homotopy invariant.

If Y is a Γ-uniform bornological coarse space Y, then $\mathcal{O}(Y)$ has a canonical big family $\mathcal{Y}(Y)$ given by (9.4). Recall the notation (4.8).

DEFINITION 9.29. We define the functor

$$\mathcal{O}^\infty \colon \Gamma\mathbf{UBC} \to \Gamma\mathbf{Sp}\mathcal{X}$$

by

$$\mathcal{O}^\infty(Y) := \mathrm{Yo}^s(\mathcal{O}(Y), \mathcal{Y}(Y))\ .$$

Recall that $F_{\mathcal{T}}$ is the functor from Γ-uniform bornological coarse spaces to Γ-bornological coarse spaces which forgets the uniform structure. For n in $\mathbb{N}$ let $([0,n]\times Y)_{\mathcal{O}(Y)}$ denote the Γ-bornological coarse space given by the subset $[0,n]\times Y$ of $\mathcal{O}(Y)$ with the induced structures. The inclusion

$$F_{\mathcal{T}}(Y) \to ([0,n]\times Y)_{\mathcal{O}(Y)}\ , \quad y\mapsto (0,y)$$

is an equivalence of Γ-bornological coarse spaces for every integer n. Hence we have an equivalence

$$\mathrm{Yo}^s(Y)\simeq \mathrm{Yo}^s(\mathcal{Y}(Y)_{\mathcal{O}(Y)})\ .$$

COROLLARY 9.30. *For every Γ-uniform bornological coarse space Y we have a natural fiber sequence*

$$\mathrm{Yo}^s(F_{\mathcal{T}}(Y))\to \mathrm{Yo}^s(\mathcal{O}(Y))\to \mathcal{O}^\infty(Y)\xrightarrow{\partial}\Sigma\mathrm{Yo}^s(F_{\mathcal{T}}(Y)) \tag{9.8}$$

in $\Gamma\mathbf{Sp}\mathcal{X}$.

PROOF. The fiber sequence is associated to the pair $(\mathcal{O}(Y),\mathcal{Y}(Y))$, see Corollary 4.11.1. □

Let Y be a Γ-uniform bornological coarse space. Then we consider the Γ-bornological coarse space $\mathcal{O}(Y)_-$ obtained from the Γ-uniform bornological coarse space $\mathbb{R}\otimes Y$ by taking the hybrid coarse structure 9.15 associated to the big family $((-\infty,n]\times Y)_{n\in\mathbb{N}}$. Note that the subset $[0,\infty)\times Y$ of $\mathcal{O}(Y)_-$ with the induced structures is the cone $\mathcal{O}(Y)$. We then have maps of Γ-bornological coarse spaces

$$F_{\mathcal{T}}(Y)\xrightarrow{i}\mathcal{O}(Y)\xrightarrow{j}\mathcal{O}(Y)_-\xrightarrow{d}F_{\mathcal{T}}(\mathbb{R}_{du}\otimes Y)\ .$$

The first two maps i and j are the inclusions, and the last map d is given by the identity of the underlying sets.

PROPOSITION 9.31. *We have a commutative diagram in $\Gamma\mathbf{Sp}\mathcal{X}$*

$$\begin{array}{ccccc}
\mathrm{Yo}^s(\mathcal{O}(Y)) & \xrightarrow{j} & \mathrm{Yo}^s(\mathcal{O}(Y)_-) & \xrightarrow{d} & \mathrm{Yo}^s(F_{\mathcal{T}}(\mathbb{R}_{du}\otimes Y))\\
\| & & \downarrow\simeq & & \downarrow\simeq\\
\mathrm{Yo}^s(\mathcal{O}(Y)) & \longrightarrow & \mathcal{O}^\infty(Y) & \xrightarrow{\partial} & \Sigma\mathrm{Yo}^s(F_{\mathcal{T}}(Y))
\end{array} \tag{9.9}$$

REMARK 9.32. This proposition identifies a segment of the cone sequence (9.8) with a sequence represented by maps between Γ-bornological coarse spaces. It in particular shows that the cone $\mathcal{O}^\infty(Y)$ is represented by the Γ-bornological coarse space $\mathcal{O}(Y)_-$.

PROOF OF PROPOSITION 9.31. We consider the diagram of motivic coarse spectra

$$\begin{array}{ccccc}
\mathrm{Yo}^s(F_{\mathcal{T}}(Y)) & \xrightarrow{i} & \mathrm{Yo}^s(\mathcal{O}(Y)) & \longrightarrow & \mathrm{Yo}^s(F_{\mathcal{T}}([0,\infty)\otimes Y))\\
\downarrow & & \downarrow j & & \downarrow\\
\mathrm{Yo}^s(F_{\mathcal{T}}((-\infty,0]\otimes Y)) & \longrightarrow & \mathrm{Yo}^s(\mathcal{O}(Y)_-) & \xrightarrow{d} & \mathrm{Yo}^s(F_{\mathcal{T}}(\mathbb{R}_{du}\otimes Y))
\end{array} \tag{9.10}$$

The left and right vertical and the lower left horizontal map are given by the canonical inclusions. The upper right horizontal map is the identity map of the underlying sets. This diagram commutes since it is obtained by applying Yo^s to a commuting diagram of bornological coarse spaces.

The left square in (9.10) is cocartesian since the pair $((-\infty,0]\times Y, \mathcal{O}(Y))$ in $\mathcal{O}(Y)_-$ is coarsely excisive. Furthermore, since $((-\infty,0]\times Y,[0,\infty)\times Y)$ is coarsely excisive in $F_{\mathcal{T}}(\mathbb{R}_{du}\otimes Y)$ the outer square is cocartesian. It follows that the right square is cocartesian.

Since the upper right and the lower left corners in (9.10) are trivial by flasqueness of the rays the diagram is equivalent to the composition

$$\begin{array}{ccccc} \mathrm{Yo}^s(F_{\mathcal{T}}(Y)) & \xrightarrow{i} & \mathrm{Yo}^s(\mathcal{O}(Y)) & \longrightarrow & 0 \\ \downarrow & & \downarrow{\scriptstyle j} & & \downarrow \\ 0 & \longrightarrow & \mathrm{Yo}^s(\mathcal{O}(Y)_-) & \xrightarrow{d} & \mathrm{Yo}^s(F_{\mathcal{T}}(\mathbb{R}_{du}\otimes Y)) \end{array}$$

of cocartesian squares. Note that $\mathcal{O}^\infty(Y)$ is defined as the cofiber of the left upper horizontal map i. Hence the left square yields the middle vertical equivalence in (9.9).

The outer square yields the equivalence $\mathrm{Yo}^s(F_{\mathcal{T}}(\mathbb{R}_{du}\otimes Y))\simeq\Sigma\mathrm{Yo}^s(Y)$. The right square then identifies d with the boundary map ∂ of the cone sequence. □

Next we will observe that $\mathcal{O}^\infty(Y)$ is essentially independent of the coarse structure on Y. Let Y be a Γ-uniform bornological coarse space with coarse structure $\mathcal{C}$, bornology $\mathcal{B}$ and uniform structure $\mathcal{T}$. Let $\mathcal{C}'$ be a Γ-coarse structure on Y such that $\mathcal{C}\subseteq\mathcal{C}'$ and $\mathcal{C}'$ is still compatible with the bornology. We write Y' for the Γ-uniform bornological coarse space obtained from Y by replacing the coarse structure $\mathcal{C}$ by $\mathcal{C}'$. Then the identity map of the underlying sets is a morphism $Y\to Y'$ of Γ-uniform bornological coarse spaces. We will call such a morphism a *coarsening.*

Let Y be a Γ-uniform bornological coarse space.

PROPOSITION 9.33. *If $Y\to Y'$ is a coarsening, then the induced map*

$$\mathcal{O}^\infty(Y)\to\mathcal{O}^\infty(Y')$$

is an equivalence.

PROOF. By definition of $\mathcal{O}^\infty(Y)$ we have

$$\mathcal{O}^\infty(Y)\simeq\operatorname*{colim}_{n\in\mathbb{N}}\mathrm{Yo}^s(([0,\infty)_{du}\otimes Y)_h,[0,n]\times Y)\ , \tag{9.11}$$

where the subsets $[0,n]\times Y$ of $([0,\infty)_{du}\otimes Y)_h$ have the induced bornological coarse structure. By u-continuity of Yo^s we have

$$\mathcal{O}^\infty(Y)\simeq\operatorname*{colim}_{n\in\mathbb{N}}\operatorname*{colim}_{U}\mathrm{Yo}^s(([0,\infty)_{du}\otimes Y)_U,([0,n]\times Y)_U)\ , \tag{9.12}$$

where U runs over the Γ-invariant entourages of $([0,\infty)_{du}\otimes Y)_h$. Here for a subset X' of X the notation X'_U denotes the set X' with the structures induced from X_U, i.e., X'_U is a short-hand notation for X'_{X_U}. For every integer n, there is a cofinal set of entourages U such that the pair

$$([0,n]\times Y,[n,\infty)\times Y)$$

is coarsely excisive on $([0,\infty)\times Y)_U$. In fact, this is a coarsely excisive pair for any entourage U that allows propagation from $\{n\}\times Y$ in the direction of the ray.

Since the Yoneda functor Yo^s is excisive we get the equivalence

$$\mathcal{O}^\infty(Y) \simeq \operatorname*{colim}_{n\in\mathbb{N}} \operatorname*{colim}_{U} \mathrm{Yo}^s(([n,\infty)\times Y)_U, (\{n\}\times Y)_U) \ . \tag{9.13}$$

In general, for a Γ-bornological coarse space X with coarse structure $\mathcal{C}$ and an invariant subset Z we have an equivalence

$$\operatorname*{colim}_{U\in\mathcal{C}^\Gamma} \mathrm{Yo}^s(Z_U) \simeq \operatorname*{colim}_{U\in\mathcal{C}^\Gamma} \mathrm{Yo}^s(Z_{(Z\times Z)\cap U}) \tag{9.14}$$

(here we must not omit the colimit). We insert this into (9.13) and get

$$\mathcal{O}^\infty(Y) \simeq \operatorname*{colim}_{n\in\mathbb{N}} \operatorname*{colim}_{U} \mathrm{Yo}^s(([n,\infty)\times Y)_{U_n}, (\{n\}\times Y)_{U_n}) \ , \tag{9.15}$$

where we use the abbreviation $U_n := (([n,\infty)\times Y)\times([n,\infty)\times Y))\cap U$. We can now interchange the order of the colimits and get

$$\mathcal{O}^\infty(Y) \simeq \operatorname*{colim}_{U} \operatorname*{colim}_{n\in\mathbb{N}} \mathrm{Yo}^s(([n,\infty)\times Y)_{U_n}, (\{n\}\times Y)_{U_n}) \ . \tag{9.16}$$

We argue now that in this formula we can replace the colimit over the invariant entourages U of $([0,\infty)_{du}\otimes Y)_h$ by the colimit over all invariant entourages U' of $([0,\infty)_{du}\otimes Y')_h$. To this end we consider the generating entourages $U_\psi\cap W'$ of $\mathcal{O}(Y')$ (see the proof of Lemma 9.27 for notation). Since $\mathcal{T}$ and $\mathcal{C}$ are compatible, there exists an integer n_0 sufficiently large such that $\phi(n_0)\in\mathcal{C}$. But then, since ϕ is monotoneous, we have $\phi(x)\in\mathcal{C}$ for every x in $[n_0,\infty)$. We conclude that for every integer n with $n\geq n_0$ we have

$$(U_\psi\cap W')\cap\big(([n,\infty)\times Y)\times([n,\infty)\times Y)\big)\subseteq U_\psi\cap(W'\cap\phi(n))$$

and $W'\cap\phi(n)\in\mathcal{C}$.

This gives

$$\mathcal{O}^\infty(Y) \simeq \operatorname*{colim}_{U'} \operatorname*{colim}_{n\in\mathbb{N}} \mathrm{Yo}^s(([n,\infty)\times Y)_{U'_n}, (\{n\}\times Y)_{U'_n}) \ , \tag{9.17}$$

where now U' runs over the invariant entourages of $([0,\infty)_{du}\otimes Y')_h$. Going the argument above backwards with Y replaced by Y' we end up with

$$\mathcal{O}^\infty(Y) \simeq \operatorname*{colim}_{n\in\mathbb{N}} \mathrm{Yo}^s(([0,\infty)_{du}\otimes Y')_h, [0,n]\times Y) \simeq \mathcal{O}^\infty(Y') \tag{9.18}$$

and this completes the proof. □

In the following proposition we use the invariance under coarsening in order to calculate the value of the $\mathcal{O}^\infty$-functor on Γ-uniform bornological coarse spaces whose underlying uniform structure is discrete.

For a Γ-uniform bornological coarse space X which is discrete as a uniform space let X_{disc} denote the Γ-uniform bornological coarse space obtained by replacing the coarse structure by the discrete coarse structure.

REMARK 9.34. If X is not discrete as a uniform space, then the discrete coarse structure is not compatible with the uniform structure.

Let X be a Γ-uniform bornological coarse space.

PROPOSITION 9.35. *If X is discrete as a uniform space, then we have an equivalence*

$$\mathcal{O}^\infty(X)\simeq\Sigma\mathrm{Yo}^s(F_{\mathcal{T}}(X_{disc}))$$

in $\Gamma\mathbf{Sp}\mathcal{X}$.

PROOF. Since $X_{disc} \to X$ is a coarsening, by Proposition 9.33 we have an equivalence

$$\mathcal{O}^\infty(X_{disc}) \xrightarrow{\simeq} \mathcal{O}^\infty(X) .$$

By Example 9.25 we know that $\mathcal{O}(X_{disc})$ is flasque and hence $\mathrm{Yo}^s(\mathcal{O}(X_{disc})) \simeq 0$. The fiber sequence obtained in Corollary 9.30 yields an equivalence $\mathcal{O}^\infty(X_{disc}) \simeq \Sigma\mathrm{Yo}^s(F_{\mathcal{T}}(X_{disc}))$ as desired. □

Next we discuss excision and homotopy invariance for $\mathcal{O}^\infty$.

Let Y be a Γ-uniform bornological coarse space and A, B be Γ-invariant subsets of Y.

COROLLARY 9.36. *If (A, B) is an equivariant uniformly and coarsely excisive decomposition, then the following square in $\Gamma\mathbf{Sp}\mathcal{X}$ is cocartesian:*

$$\begin{array}{ccc} \mathcal{O}^\infty(A\cap B) & \longrightarrow & \mathcal{O}^\infty(B) \\ \downarrow & & \downarrow \\ \mathcal{O}^\infty(A) & \longrightarrow & \mathcal{O}^\infty(Y) \end{array} \tag{9.19}$$

PROOF. Since (A, B) is coarsely excisive the square

$$\begin{array}{ccc} \mathrm{Yo}^s(\mathcal{F}_{\mathcal{T}}(A)\cap \mathcal{F}_{\mathcal{T}}(B)) & \longrightarrow & \mathrm{Yo}^s(\mathcal{F}_{\mathcal{T}}(B)) \\ \downarrow & & \downarrow \\ \mathrm{Yo}^s(\mathcal{F}_{\mathcal{T}}(A)) & \longrightarrow & \mathrm{Yo}^s(\mathcal{F}_{\mathcal{T}}(Y)) \end{array}$$

is cocartesian. Furthermore, by Lemma 9.27 the square

$$\begin{array}{ccc} \mathcal{O}(A\cap B) & \longrightarrow & \mathcal{O}(B) \\ \downarrow & & \downarrow \\ \mathcal{O}(A) & \longrightarrow & \mathcal{O}(Y) \end{array}$$

is cocartesian. Now it just remains to use the cone sequence (9.8) in order to conclude that the square (9.19) is cocartesian. □

REMARK 9.37. If we assume that the underlying uniform space of Y is Hausdorff, then we could drop the assumption that (A, B) is coarsely excisive. In this case it will follow from the Decomposition Theorem 9.22 applied to the equivariant uniform decomposition $([0,\infty)\times A, [0,\infty)\times B)$ of $[0,\infty)_{du}\otimes Y$ that (9.19) is cocartesian.

COROLLARY 9.38. *The functor $\mathcal{O}^\infty\colon \Gamma\mathbf{UBC} \to \Gamma\mathbf{Sp}\mathcal{X}$ is homotopy invariant.*

PROOF. Let Y be a Γ-uniform bornological coarse space. Let $h\colon [0,1]_{du}\otimes Y \to Y$ be the projection. By functoriality of the cone we get the morphism

$$\mathcal{O}(h)\colon \mathcal{O}([0,1]_{du}\otimes Y) \to \mathcal{O}(Y) .$$

We now observe that

$$\mathcal{O}([0,1]_{du}\otimes Y) \cong ([0,\infty)_{du}\otimes [0,1]_{du}\otimes Y)_h \cong ([0,1]_{du}\otimes [0,\infty)_{du}\otimes Y)_h .$$

By the Homotopy Theorem 9.23 we get an equivalence

$$\begin{aligned}\mathrm{Yo}^s(\mathcal{O}([0,1]_{du}\otimes Y)) &\simeq \mathrm{Yo}^s(([0,1]_{du}\otimes[0,\infty)_d\otimes Y)_h)\simeq \mathrm{Yo}^s(([0,\infty)_{du}\otimes Y)_h)\\ &\simeq \mathrm{Yo}^s(\mathcal{O}(Y))\ .\end{aligned}$$

Since the projections $[0,1]_{du}\otimes[0,n]_{du}\otimes Y\to[0,n]_{du}\otimes Y$ induce equivalences of underlying Γ-bornological coarse spaces we conclude that the projection h induces an equivalence

$$\mathcal{O}^\infty([0,1]_{du}\otimes Y)\to\mathcal{O}^\infty(Y)$$

in $\Gamma\mathbf{Sp}\mathcal{X}$. □

10. Topological assembly maps

The overall theme of this section is the interplay between equariant coarse homology theories and equivariant homology theories.

We start in Section 10.1 with the general discussion of equivariant homology theories, and then in Section 10.2 we modify the "cone at infinity" functor $\mathcal{O}^\infty$ to get an equivariant $\Gamma\mathbf{Sp}\mathcal{X}$-valued homology theory

$$\mathcal{O}^\infty_{\mathrm{hlg}}\colon\Gamma\mathbf{Top}\to\Gamma\mathbf{Sp}\mathcal{X}\ .$$

We introduce classifying spaces $E_{\mathcal{F}}\Gamma$ for families $\mathcal{F}$ of subgroups and define the motivic assembly map

$$\alpha_{\mathcal{F}}\colon\mathcal{O}^\infty_{\mathrm{hlg}}(E_{\mathcal{F}}\Gamma)\to\mathcal{O}^\infty_{\mathrm{hlg}}(*)$$

in Section 10.3. Using the cone sequence we define further versions $\alpha_{X,Q}$ of the motivic assembly map with twist Q and discuss some instances where it is an equivalence. Finally, in Section 10.4 we use the functor $\mathcal{O}^\infty_{\mathrm{hlg}}$ in order to derive equivariant homology theories from coarse homology theories.

10.1. Equivariant homology theories. In this section we will recall the notion of a (strong) equivariant homology theory on Γ-topological spaces.

Our basic category of topological spaces is the convenient category **Top** of compactly generated weakly Hausdorff spaces. A map between topological spaces is a weak equivalence if it induces an isomorphism between the sets of connected components and isomorphisms of homotopy groups in all positive degrees and for all choices of base points. The ∞-category obtained from (the nerve of) **Top** by inverting these weak equivalences is a model for the presentable ∞-category **Spc** of spaces. In particular, we have the localization functor

$$\kappa\colon\mathbf{Top}\to\mathbf{Spc}\ . \tag{10.1}$$

A Γ-topological space is a topological space with an action of the group Γ by automorphisms. We denote the category of Γ-topological spaces and equivariant continuous maps by $\Gamma\mathbf{Top}$. A weak equivalence between Γ-topological spaces is a Γ-equivariant map which induces weak equivalences on fixed-point spaces for all subgroups of Γ. We will model this homotopy theory by presheaves on the orbit category of Γ.

The orbit category $\mathbf{Orb}(\Gamma)$ of Γ is the category of transitive Γ-sets and equivariant maps. A Γ-set can naturally be considered as a discrete Γ-topological space. In this way we get a fully faithful functor $\mathbf{Orb}(\Gamma)\to\Gamma\mathbf{Top}$. For a transitive Γ-set S and Γ-topological space X we consider the topological space $\mathrm{Map}_{\Gamma\mathbf{Top}}(S,X)$ of equivariant maps from S to X.

REMARK 10.1. We consider a transitive Γ-set S. If we fix a base point s in S and denote the stabilizer of s by Γ_s, then we get an identification $\mathrm{Map}(S,X) \simeq X^{\Gamma_s}$, where X^{Γ_s} is the subspace of Γ_s-fixed points.

We define a functor

(10.2)
$$\ell\colon \Gamma\mathbf{Top} \to \mathbf{PSh}(\mathbf{Orb}(\Gamma)) \text{ by } \ell(X)(S) := \kappa(\mathrm{Map}_{\Gamma\mathbf{Top}}(S,X)) \text{ for } S \in \mathbf{Orb}(\Gamma)\ .$$

REMARK 10.2. A map between topological spaces is a weak equivalence if and only if its image under κ is an equivalence. Consequently, a map between Γ-topological spaces is a weak equivalence if and only if its image under ℓ is an equivalence. By Elmendorf's theorem [**29**, Thm. VI.6.3] (which boils down to the assertion that ℓ is essentially surjective) the functor ℓ induces an equivalence

$$\Gamma\mathbf{Top}[W^{-1}] \xrightarrow{\simeq} \mathbf{PSh}(\mathbf{Orb}(\Gamma))\ , \tag{10.3}$$

where $\Gamma\mathbf{Top}[W^{-1}]$ denotes the ∞-category obtained from $\Gamma\mathbf{Top}$ by inverting the weak equivalences. Occasionally we will use the fact that a weak equivalence in $\Gamma\mathbf{Top}$ between Γ-CW-complexes is actually a homotopy equivalence in $\Gamma\mathbf{Top}$.

Let $\mathbf{C}$ be a cocomplete ∞-category. By the universal property of the presheaf category we have an equivalence of ∞-categories

$$\mathbf{Fun}^{\mathrm{colim}}(\mathbf{PSh}(\mathbf{Orb}(\Gamma)),\mathbf{C}) \simeq \mathbf{Fun}(\mathbf{Orb}(\Gamma),\mathbf{C})\ , \tag{10.4}$$

where the superscript colim stands for colimit-preserving. The localization functor (10.2) induces a faithful restriction functor

$$\mathbf{Fun}^{\mathrm{colim}}(\mathbf{PSh}(\mathbf{Orb}(\Gamma),\mathbf{C}) \to \mathbf{Fun}(\Gamma\mathbf{Top},\mathbf{C})\ . \tag{10.5}$$

From now on we assume that $\mathbf{C}$ is cocomplete and stable.

Our preferred definition of the notion of an equivariant $\mathbf{C}$-valued homology theory would be the following.

Let $E\colon \Gamma\mathbf{Top} \to \mathbf{C}$ be a functor.

DEFINITION 10.3. E is called a *strong equivariant* $\mathbf{C}$*-valued homology theory* if it is in the essential image of (10.5).

Assume that we are given a functor E as above. If it sends weak equivalences to equivalences, then, using the equivalence (10.3), it extends essentially uniquely to a functor $\mathbf{PSh}(\mathbf{Orb}(\Gamma)) \to \mathbf{C}$. The functor E is an equivariant $\mathbf{C}$-valued homology theory if this extension preserves colimits. In general it seems to be complicated to check these conditions if E is given by some geometric construction. For this reason we add the adjective *strong* in order to distinguish this notion from the Definition 10.4 of an equivariant $\mathbf{C}$-valued homology theory that we actually work with.

Let $E\colon \Gamma\mathbf{Top} \to \mathbf{C}$ be a functor. We extend E to pairs (X,A) of Γ-topological spaces and subspaces by setting

$$E(X,A) := \mathrm{Cofib}\big(E(*) \to E(X \cup_A \mathrm{Cone}(A))\big)\ ,$$

where $\mathrm{Cone}(A)$ denotes the cone over A and $*$ is the base point of the cone.

DEFINITION 10.4. The functor E is called an *equivariant* $\mathbf{C}$*-valued homology theory* if it has the following properties:

(1) (Homotopy invariance) For every Γ-topological space X the projection induces an equivalence
$$E([0,1]\times X)\to E(X)\ .$$

(2) (Excision) If (X,A) is a pair of Γ-topological spaces and U is an invariant open subset of A such that $\overline{U}$ is contained in the interior of A, then the inclusion $(X\setminus U, A\setminus U)\to (X,A)$ induces an equivalence
$$E(X\setminus U, A\setminus U)\to E(X,A)\ .$$

(3) (Wedge axiom) For every family $(X_i)_{i\in I}$ of Γ-topological spaces the canonical map
$$\bigoplus_{i\in I} E(X_i)\xrightarrow{\simeq} E\Big(\coprod_{i\in I} X_i\Big)$$
is an equivalence.

REMARK 10.5. In order to verify that E satisfies excision one must show that E sends the square
$$\begin{array}{ccc}\mathrm{Cone}(A\setminus U) & \longrightarrow & (X\setminus U)\cup_{A\setminus U}\mathrm{Cone}(A\setminus U)\\ \downarrow & & \downarrow\\ \mathrm{Cone}(A) & \longrightarrow & X\cup_A \mathrm{Cone}(A)\end{array}$$
to a push-out square. For homotopy invariant functors E this is equivalent to the property that E sends the right vertical map in the square above to an equivalence.

For homotopy invariant functors E excision follows from the stronger condition of closed excision, i.e., that for every decomposition (A,B) of X into closed invariant subsets the diagram
$$\begin{array}{ccc}E(A\cap B) & \longrightarrow & E(A)\\ \downarrow & & \downarrow\\ E(B) & \longrightarrow & E(X)\end{array}$$
is a push-out square. This can be seen as follows. Assume that E is homotopy invariant and satisfies closed excision. Let X, A, and U be as above. Then we have a closed decomposition
$$(X\setminus U\cup_{A\setminus U}\mathrm{Cone}(A\setminus U), \mathrm{Cone}(\overline{U}))$$
of $X\cup_A \mathrm{Cone}(A)$ with intersection $\mathrm{Cone}(\overline{U}\setminus U)$. By closed excision we get the cocartesian square
$$\begin{array}{ccc}E(\mathrm{Cone}(\overline{U}\setminus U)) & \longrightarrow & E(X\setminus U\cup_{A\setminus U}\mathrm{Cone}(A\setminus U))\\ \downarrow & & \downarrow\\ E(\mathrm{Cone}(\overline{U})) & \longrightarrow & E(X\cup_A \mathrm{Cone}(A))\end{array}$$
Since E is homotopy invariant it sends the left vertical map to an equivalence since cones are contractible. Consequently, the right vertical map is an equivalence, too.

REMARK 10.6. Let E be an equivariant $\mathbf{C}$-valued homology theory. In general we can not expect that it factorizes over the localization (10.2).

Using the equivalence (10.4) the restriction of E to the orbit category gives rise to a strong equivariant $\mathbf{C}$-valued homology theory $E^{\%}$ which comes with a natural transformation $E^{\%} \to E$. Using the theory developed by Davis–Lück [**16**, Sec. 3] one can check that

$$E^{\%}(X) \xrightarrow{\simeq} E(X)$$

for all Γ-CW-complexes X.

10.2. The cone as an equivariant homology theory. In Section 9.5 we have seen that the "cone at infinity" functor $\mathcal{O}^\infty$ is a homotopy invariant and excisive functor from $\Gamma\mathbf{UBC}$ to $\mathbf{Sp}\mathcal{X}$. In this section we modify this functor in order to get an equivariant homology theory $\mathcal{O}^\infty_{\mathrm{hlg}} \colon \Gamma\mathbf{Top} \to \Gamma\mathbf{Sp}\mathcal{X}$.

If X is a Γ-uniform space, then we can consider X as a Γ-uniform bornological coarse space $X_{max,max}$ by equipping the uniform space X in addition with the maximal coarse structure and the maximal bornology. In this way we get a functor

$$\mathcal{M} \colon \Gamma\mathbf{U} \to \Gamma\mathbf{UBC}\ , \quad X \mapsto \mathcal{M}(X) := X_{max,max}\ . \tag{10.6}$$

Let Y be a Γ-set and Q be a Γ-bornological coarse space. The projection $Y_{max,max} \to *$ induces a morphism

$$\mathrm{Yo}^s(Y_{max,max}) \otimes \mathrm{Yo}^s(Q) \to \mathrm{Yo}^s(Q) \tag{10.7}$$

in $\Gamma\mathbf{Sp}\mathcal{X}$.

LEMMA 10.7. *If the underlying set of Q is a free Γ-set, then* (10.7) *is an equivalence.*

PROOF. Since Q is a free Γ-set we can choose a Γ-equivariant map of sets $\kappa \colon Q \to Y$. The map $(\kappa, \mathrm{id}) \colon Q \to Y_{max,max} \otimes Q$ is then a morphism of Γ-bornological coarse spaces. It is an inverse to the projection up to equivalence. Hence the projection is an equivalence of Γ-bornological coarse spaces and therefore $\mathrm{Yo}^s(Y_{max,max} \otimes Q) \to \mathrm{Yo}^s(Q)$ is an equivalence. We now use that Yo^s is a symmetric monoidal functor (see Section 4.3) in order to rewrite the domain of the morphism as in (10.7). □

The functor

$$\mathcal{O}^\infty \circ \mathcal{M} \colon \Gamma\mathbf{U} \to \Gamma\mathbf{Sp}\mathcal{X} \tag{10.8}$$

behaves very much like an equivariant homology theory. Of course, it is not defined on $\Gamma\mathbf{Top}$, but on $\Gamma\mathbf{U}$. On the other hand it is homotopy invariant by Corollary 9.38, and sends equivariant uniform decompositions to push-outs by Corollary 9.36. The drawback is that it in general only preserves finite coproducts.

In order to improve these points we define a new functor

$$\mathcal{O}^\infty_{\mathrm{hlg}} \colon \Gamma\mathbf{Top} \to \Gamma\mathbf{Sp}\mathcal{X}$$

by first restricting $\mathcal{O}^\infty \circ \mathcal{M}$ to the subcategory of Γ-compact Γ-metrizable spaces and then left-Kan extending the result to $\Gamma\mathbf{Top}$. In the following we describe the details.

Let X be a Γ-toplogical space. Recall that X is Γ-*compact* if there exists a compact subset K of X such that $\Gamma K = X$, and that X is Γ-*metrizable* if there exists a Γ-invariant metric on X which induces the topology of X.

We denote by $\Gamma\mathbf{Top}^{cm}$ the full sub-category of $\Gamma\mathbf{Top}$ spanned by all the Γ-compact and Γ-metrizable Γ-topological spaces. Associated to any X in $\Gamma\mathbf{Top}^{cm}$ we define

$$\mathcal{N}(X) := \{N \subseteq X \times X \mid N \text{ contains a } \Gamma\text{-invariant neighborhood of the diagonal}\} .$$

Furthermore we set $\mathcal{U}(X) := (X, \mathcal{N}(X))$. For a Γ-invariant metric on X which is compatible with the topology we let $\mathcal{T}_d$ denote the associated metric uniform structure on X.

LEMMA 10.8. *Assume that X is in $\Gamma\mathbf{Top}^{cm}$.*

(1) *$\mathcal{N}(X)$ is a Γ-uniform structure.*
(2) *If d is a Γ-invariant metric compatible with the topology, then $\mathcal{N}(X) = \mathcal{T}_d$.*
(3) *The assignment $X \mapsto \mathcal{U}(X)$ defines a functor $\mathcal{U}\colon \Gamma\mathbf{Top}^{cm} \to \Gamma\mathbf{U}$.*

PROOF. By assumption we can choose a Γ-invariant metric d which is compatible with the topology. We claim that for every Γ-invariant neighborhood N of the diagonal there exists some T in $\mathcal{T}_d$ such that $T \subseteq N$. The case $N = X \times X$ is trivial and we assume that N is a proper subset. We define a function

$$b\colon X \to (0,\infty), \quad x \mapsto \sup\{\epsilon \in (0,\infty) \mid B_\epsilon(x) \times B_\epsilon(x) \subseteq N\} ,$$

where $B_\epsilon(x)$ denotes the open ball of radius ϵ around x (with respect to d). Since d is compatible with the topology the argument of the sup is non-empty for every x in X and the value $b(x)$ is indeed positive. Furthermore, the supremum is attained. Moreover, the supremun can not be infinite since we assume that N is a proper subset. Finally, we observe that b is Γ-equivariant with respect to the trivial Γ-action on $(0,\infty)$.

We now show that the function b is 1-Lipschitz. Let x, y be two points in X and let δ be their distance. If both $b(x)$ and $b(y)$ are less than δ, then so is their distance. Therefore, we can assume that $b(x) \geq b(y)$ and $b(x) \geq \delta$. By the triangle inequality we have

$$B_{b(x)-\delta}(y) \times B_{b(x)-\delta}(y) \subseteq B_{b(x)}(x) \times B_{b(x)}(x) \subseteq N$$

This implies $b(x) \geq b(y) \geq b(x) - \delta$. In particular, $|b(x) - b(y)| \leq \delta$ and b is indeed 1-Lipschitz and thus continuous.

By assumption we can choose a compact subset K of X such that $\Gamma K = X$. Since b is continuous, the restriction $b|_K\colon K \to (0,\infty)$ attains a positive minimal value, which we denote by ϵ_0. By Γ-equivariance of b, we then have $b(x) \geq \epsilon_0$ for all x in X. We conclude that the metric uniform entourage $\{(x,y) \in X \times X \mid d(x,y) < \epsilon_0\}$ is contained in N.

Note that every metric uniform entourage is a neighbourhood of the diagonal since d is compatible with the topology. Since both $\mathcal{N}(X)$ and $\mathcal{T}_d$ are closed under taking supersets, we have shown that $\mathcal{N}(X) = \mathcal{T}_d$. This shows the first two assertions of the Lemma.

We now show the third assertion. Let $f\colon X \to X'$ be an equivariant continuous map between two Γ-compact and Γ-metrizable Γ-topological spaces. We must show that $f\colon (X, \mathcal{N}(X)) \to (X', \mathcal{N}(X'))$ is uniformly continuous. Let V' belong to $\mathcal{N}(X')$. Then V' contains a Γ-invariant neighbourhood of the diagonal U'. Since f is equivariant and continuous, $(f^{-1} \times f^{-1})(U)$ is a Γ-invariant neighbourhood of the diagonal of X contained in $(f^{-1} \times f^{-1})(V')$. Consequently,

$(f^{-1}\times f^{-1})(V')\in\mathcal{N}(X)$. This implies that $f\colon (X,\mathcal{N}(X))\to(X',\mathcal{N}(X'))$ is uniformly continuous. □

Let X be a Γ-topological space, and let A and B be closed Γ-invariant subsets of X such that $A\cup B=X$.

LEMMA 10.9. *If X is Γ-compact and Γ-metrizable, then (A,B) is an equivariant uniform decomposition of $\mathcal{U}(X)$ (Definition 9.19).*

PROOF. We choose a Γ-invariant metric d on X which is compatible with the topology. By Lemma 10.8, $(X,\mathcal{T}_d)=\mathcal{U}(X)$.

For a subset Z of Y and e in $(0,\infty)$ we consider the e-thickening

$$U_e[Z]:=\{y\in Y|d(y,Z)\le e\}$$

(note that U_e is defined in (9.1) with a $\le$-relation, too) of Z. If Z is invariant, then $U_e[Z]$ is again invariant.

If $A\cap B=\emptyset$, then by Γ-compactness of Y the subsets A and B are uniformly separated and (A,B) is an equivariant uniform decomposition.

Assume now that $A\cap B\neq\emptyset$. It suffices to define a monotoneous function $s\colon(0,\infty)\to(0,\infty)$ such that:

(1) $\lim_{e\to 0}s(e)=0$.
(2) For all e in $(0,\infty)$ we have
$$U_e[A]\cap U_e[B]\subseteq U_{s(e)}[A\cap B]\ .$$

We define a function $s\colon(0,\infty)\to[0,\infty]$ by

$$s(e):=\inf\{e'\in(0,\infty)|U_e[A]\cap U_e[B]\subseteq U_{e'}[A\cap B]\}\ .$$

By construction, this function is monotoneous. Since $A\cap B\neq 0$, and since by Γ-compactness of Y there exists R in $(0,\infty)$ such that $U_R[A\cap B]=Y$, the function s is finite. Condition 2 follows from this observation.

We claim that

$$A\cap B=\bigcap_{e>0}U_e[A]\cap U_e[B]\ . \tag{10.9}$$

It is clear that

$$A\cap B\subseteq\bigcap_{e>0}U_e[A]\cap U_e[B]\ .$$

On the other hand, assume that $y\in Y\setminus(A\cap B)$. Without loss of generality (interchange the roles of A and B, if necessary) we can assume that $y\in Y\setminus A$. Since A is closed there exists e in $(0,\infty)$ such that $y\notin U_e[A]$. Hence $y\in Y\setminus\bigcap_{e>0}U_e[A]\cap U_e[B]$. This shows the opposite inclusion

$$\bigcap_{e>0}U_e[A]\cap U_e[B]\subseteq A\cap B.$$

We now show Condition 1. Assume the contrary. Then there exists ϵ in $(0,\infty)$ such that $s(e)\ge\epsilon>0$ for all e in $(0,\infty)$. For every integer n there exists a point y_n in $U_{1/n}[A]\cap U_{1/n}[B]$ such that $y_n\notin U_\epsilon[A\cap B]$. We can assume (after replacing y_n by $\gamma_n y_n$ for suitable elements γ_n of Γ) by Γ-compactness that $y_n\to y$ for $n\to\infty$. Then $y\notin U_{\epsilon/2}[A\cap B]$ but $y\in A\cap B$ by (10.9). This is a contradiction, and so we have verified Condition 1. □

DEFINITION 10.10. We define the functor

$$\mathcal{O}^\infty_{\mathrm{hlg}}\colon \Gamma\mathbf{Top} \to \Gamma\mathbf{Sp}\mathcal{X}$$

as the left Kan extension

$$\begin{array}{ccc} \Gamma\mathbf{Top}^{cm} & \xrightarrow{\mathcal{O}^\infty\circ\mathcal{M}\circ\mathcal{U}} & \Gamma\mathbf{Sp}\mathcal{X} \\ \downarrow & \nearrow_{\mathcal{O}^\infty_{\mathrm{hlg}}} & \\ \Gamma\mathbf{Top} & & \end{array}$$

of $\mathcal{O}^\infty \circ \mathcal{M} \circ \mathcal{U}$ along the fully faithful functor $\Gamma\mathbf{Top}^{cm} \hookrightarrow \Gamma\mathbf{Top}$.

PROPOSITION 10.11. *$\mathcal{O}^\infty_{\mathrm{hlg}}$ is an equivariant $\Gamma\mathbf{Sp}\mathcal{X}$-valued homology theory.*

PROOF. We first note that the objectwise formula for the left Kan-extension gives

$$\mathcal{O}^\infty_{\mathrm{hlg}}(X) \simeq \operatorname*{colim}_{(Y\to X)\in\Gamma\mathbf{Top}^{cm}/X} \mathcal{O}^\infty(\mathcal{M}(\mathcal{U}(Y))) \ . \tag{10.10}$$

Let Y be a Γ-compact and Γ-metrizable Γ-topological space and $(\phi, f)\colon Y \to [0,1]\times X$ be an equivariant map. Then we have the factorization

$$(\phi, f)\colon Y \xrightarrow{(\phi,\mathrm{id}_Y)} [0,1]\times Y \xrightarrow{(\mathrm{id}_{[0,1]},f)} [0,1]\times X \ .$$

Note that $[0,1]\times Y$ is again a Γ-compact and Γ-metrizable Γ-topological space. This shows that the category of maps of the form $(\mathrm{id}_{[0,1]}, f)\colon [0,1]\times Y \to [0,1]\times X$ for morphisms of Γ-topological spaces $f\colon Y \to X$ with Y in $\Gamma\mathbf{Top}^{cm}$ is cofinal in $\Gamma\mathbf{Top}^{cm}/([0,1]\times X)$. By Lemma 10.8, we have an isomorphism of Γ-uniform spaces

$$\mathcal{U}([0,1]\times Y) \cong [0,1]_u \otimes \mathcal{U}(Y) \ .$$

By the homotopy invariance of $\mathcal{O}^\infty$ (Corollary 9.38), the projection $[0,1]_u\otimes\mathcal{U}(Y) \to \mathcal{U}(Y)$ induces an equivalence $\mathcal{O}^\infty(\mathcal{M}(\mathcal{U}([0,1]\times Y))) \simeq \mathcal{O}^\infty(\mathcal{M}(\mathcal{U}(Y)))$. Hence we conclude that $\mathcal{O}^\infty_{\mathrm{hlg}}([0,1]\times X) \simeq \mathcal{O}^\infty_{\mathrm{hlg}}(X)$, i.e., that $\mathcal{O}^\infty_{\mathrm{hlg}}$ is homotopy invariant.

Assume now that $(X_i)_{i\in I}$ is a filtered family of Γ-topological spaces and set $X := \operatorname{colim}_{i\in I} X_i$. Then every morphism $Y \to X$ for a Y in $\Gamma\mathbf{Top}^{cm}$ factorizes as $Y \to X_i \to X$ for some i in I since Y is Γ-compact. It follows that

$$\begin{aligned} \mathcal{O}^\infty_{\mathrm{hlg}}(X) &\simeq \operatorname*{colim}_{(Y\to X)\in\Gamma\mathbf{Top}^{cm}/X} \mathcal{O}^\infty(\mathcal{M}(\mathcal{U}(Y))) \\ &\simeq \operatorname*{colim}_{i\in I} \operatorname*{colim}_{(Y\to X_i)\in\Gamma\mathbf{Top}^{cm}/X_i} \mathcal{O}^\infty(\mathcal{M}(\mathcal{U}(Y))) \\ &\simeq \operatorname*{colim}_{i\in I} \mathcal{O}^\infty_{\mathrm{hlg}}(X_i) \ . \end{aligned}$$

This in particular implies the wedge axiom for $\mathcal{O}^\infty_{\mathrm{hlg}}$.

In order to show that $\mathcal{O}^\infty_{\mathrm{hlg}}$ satisfies excision, by Remark 10.5 it suffices to show the stronger result that for every decomposition (A, B) of a Γ-topological space X into two closed invariant subsets we have a push-out square

$$\begin{array}{ccc} \mathcal{O}^\infty_{\mathrm{hlg}}(A\cap B) & \longrightarrow & \mathcal{O}^\infty_{\mathrm{hlg}}(A) \\ \downarrow & & \downarrow \\ \mathcal{O}^\infty_{\mathrm{hlg}}(B) & \longrightarrow & \mathcal{O}^\infty_{\mathrm{hlg}}(X) \end{array} \ . \tag{10.11}$$

Let $r\colon Y \to X$ be an object of $\Gamma\mathbf{Top}^{cm}/X$. Then $(r^{-1}(A), r^{-1}(B))$ is a closed decomposition of Y into Γ-compact and Γ-metrizable subspaces. Note that the objects of the form $r^{-1}(A) \to A$ of $\Gamma\mathbf{Top}^{cm}/A$ for all $r\colon Y \to X$ are cofinal in $\Gamma\mathbf{Top}^{cm}/A$.

We conclude that (10.11) is a colimit of commuting squares

$$\begin{array}{ccc} \mathcal{O}^\infty(\mathcal{M}(\mathcal{U}(r^{-1}(A)\cap r^{-1}(B)))) & \longrightarrow & \mathcal{O}^\infty(\mathcal{M}(\mathcal{U}(r^{-1}(A)))) \\ \downarrow & & \downarrow \\ \mathcal{O}^\infty(\mathcal{M}(\mathcal{U}(r^{-1}(B)))) & \longrightarrow & \mathcal{O}^\infty(\mathcal{M}(\mathcal{U}(Y))) \end{array} \tag{10.12}$$

Since Y is a Γ-compact and Γ-metrizable Γ-topological space, by Lemma 10.9 the decomposition $(r^{-1}(A), r^{-1}(B))$ is an equivariant uniform decomposition of $\mathcal{U}(Y)$. Therefore the square (10.12) is a push-out square by Corollary 9.36. Being a colimit of push-out squares the square (10.11) is therefore also a push-out square. □

REMARK 10.12. Instead of $\mathcal{O}^\infty \circ \mathcal{M}$ we could consider any functor

$$A\colon \Gamma\mathbf{U} \to \mathbf{C}$$

for some cocomplete stable ∞-category $\mathbf{C}$, which is excisive for closed decompositions and homotopy invariant. The construction above produces an equivariant $\mathbf{C}$-valued homology theory $A_{\mathrm{hlg}}\colon \Gamma\mathbf{Top} \to \mathbf{C}$ as a left Kan-extension

$$\begin{array}{ccc} \Gamma\mathbf{Top}^{cm} & \xrightarrow{A\circ\mathcal{U}} & \mathbf{C} \\ \downarrow & \nearrow_{A_{\mathrm{hlg}}} & \\ \Gamma\mathbf{Top} & & \end{array}$$

The proof that this is indeed an equivariant $\mathbf{C}$-valued homology theory is word-for-word the same as the proof of Proposition 10.11.

Note that the proof also shows that for a filtered family of Γ-topological spaces $(X_i)_{i\in I}$ the following natural morphism is an equivalence

$$\operatorname*{colim}_{i\in I} A_{\mathrm{hlg}}(X_i) \xrightarrow{\simeq} A_{\mathrm{hlg}}(\operatorname*{colim}_{i\in I} X_i)\ . \tag{10.13}$$

We will use this equivalence later in this paper.

A Γ-uniform space X has an underlying Γ-topological space which we will denote by $\tau(X)$. The topology of $\tau(X)$ is generated by the sets $V[x]$ for all points x and uniform entourages V of X. We actually get a functor

$$\tau\colon \Gamma\mathbf{U} \to \Gamma\mathbf{Top}\ . \tag{10.14}$$

REMARK 10.13. Let Y be a Γ-compact and Γ-metrizable Γ-topological space, and let X be a Γ-uniform space. Then every continuous map $Y \to \tau(X)$ induces a uniform map $\mathcal{U}(Y) \to X$. Therefore, we obtain a natural morphism

$$A_{\mathrm{hlg}}(\tau(X)) \to A(X)\ . \tag{10.15}$$

Since the inclusion functor $\Gamma\mathbf{Top}^{cm} \hookrightarrow \Gamma\mathbf{Top}$ is fully faithful, the morphism (10.15) is an equivalence for $X := \mathcal{U}(Y)$ if Y is a Γ-compact and Γ-metrizable Γ-topological space.

10.3. Families of subgroups and the universal assembly map. In this section we will define the classifying space $E_{\mathcal{F}}\Gamma$ for a family $\mathcal{F}$ of subgroups of Γ, and we will define the corresponding universal assembly map

$$\alpha_{\mathcal{F}} \colon \mathcal{O}^{\infty}_{\mathrm{hlg}}(E_{\mathcal{F}}\Gamma) \to \mathcal{O}^{\infty}_{\mathrm{hlg}}(*) .$$

We will also introduce the motivic assembly map

$$\alpha_{X,Q} \colon \mathcal{O}^{\infty}_{\mathrm{hlg}}(X) \otimes \mathrm{Yo}^{s}(Q) \to \Sigma\mathrm{Yo}^{s}(Q)$$

for a Γ-topological space X twisted by a Γ-bornological coarse space Q.

DEFINITION 10.14. A *family of subgroups* $\mathcal{F}$ of Γ is a non-empty subset of the set of subgroups of Γ which is closed under conjugation and taking subgroups.

EXAMPLE 10.15. Examples of families of subgroups are:

(1) $\{\mathbf{1}\}$ – the family containing only the trivial subgroup
(2) **Fin** – the family of all finite subgroups
(3) **Vcyc** – the family of all virtually cyclic subgroups
(4) **All** – the family of all subgroups

In Section 11 we will mostly work with the family **Fin** since one can model the corresponding classifying space (see Definition 10.17) by the Rips complex.

For a family of subgroups $\mathcal{F}$ we let $\mathbf{Orb}_{\mathcal{F}}(\Gamma) \subseteq \mathbf{Orb}(\Gamma)$ be the full subcategory of transitive Γ-sets whose stabilizers belong to $\mathcal{F}$.

EXAMPLE 10.16. Note that $\mathbf{Orb}_{\mathbf{All}}(\Gamma) = \mathbf{Orb}(\Gamma)$.

There is furthermore an equivalence $B\Gamma \to \mathbf{Orb}_{\{1\}}(\Gamma)^{op}$ which sends the unique object $*$ of $B\Gamma$ to the Γ-set Γ (with the left action). On morphisms this equivalence sends the morphism γ in $\mathrm{Hom}_{B\Gamma}(*,*) = \Gamma$ to the automorphism of the Γ-set Γ given by right-multiplication with γ.

For two families of subgroups $\mathcal{F}, \mathcal{F}'$ satisfying $\mathcal{F} \subset \mathcal{F}'$ we have an inclusion

$$\mathbf{Orb}_{\mathcal{F}}(\Gamma) \hookrightarrow \mathbf{Orb}_{\mathcal{F}'}(\Gamma)$$

of orbit categories. On presheaves this inclusion of orbit categories defines a restriction functor which is the right-adjoint of an adjunction

$$\mathrm{Ind}_{\mathcal{F}}^{\mathcal{F}'} : \mathbf{PSh}(\mathbf{Orb}_{\mathcal{F}}(\Gamma)) \leftrightarrows \mathbf{PSh}(\mathbf{Orb}_{\mathcal{F}'}(\Gamma)) : \mathrm{Res}_{\mathcal{F}}^{\mathcal{F}'} . \tag{10.16}$$

We let $*_{\mathcal{F}}$ denote the final object in $\mathbf{Orb}_{\mathcal{F}}(\Gamma)$.

DEFINITION 10.17. The object

$$E_{\mathcal{F}}\Gamma := \mathrm{Ind}_{\mathcal{F}}^{\mathbf{All}}(*_{\mathcal{F}})$$

of $\mathbf{PSh}(\mathbf{Orb}(\Gamma))$ is called the classifying space of Γ for the family $\mathcal{F}$.

REMARK 10.18. Assume that X is a Γ-CW-complex with an equivalence $\ell(X) \simeq E_{\mathcal{F}}\Gamma$. Then the following Lemma 10.19 shows that for every subgroup H of Γ the fixed point set X^H is contractible or empty depending on whether H belongs to $\mathcal{F}$ or not. This property is the usual characterization of a classifying space of Γ for the family $\mathcal{F}$. We say that X is a model for $E_{\mathcal{F}}\Gamma$. A model for $E_{\mathcal{F}}\Gamma$ is unique up to contractible choice.

Let $y\colon \mathbf{Orb}(\Gamma) \to \mathbf{PSh}(\mathbf{Orb}(\Gamma))$ denote the Yoneda embedding.

LEMMA 10.19. *For $T \in \mathbf{Orb}(\Gamma)$ we have*

$$\mathrm{Map}(y(T), E_{\mathcal{F}}\Gamma) \simeq \begin{cases} \emptyset & \text{if } T \notin \mathbf{Orb}_{\mathcal{F}}(\Gamma) , \\ * & \text{if } T \in \mathbf{Orb}_{\mathcal{F}}(\Gamma) . \end{cases}$$

PROOF. For a presheaf E on $\mathbf{Orb}_{\mathcal{F}}(\Gamma)$ and object T in $\mathbf{Orb}(\Gamma)$ we have (we interpret the Hom-sets as discrete spaces)

$$\mathrm{Ind}_{\mathcal{F}}^{\mathbf{All}}(E)(T) \simeq \operatorname*{colim}_{(S \to T) \in \mathbf{Orb}_{\mathcal{F}}(\Gamma)^{op}/T} E(S) .$$

Consequently, we have

$$\mathrm{Map}(y(T), E_{\mathcal{F}}\Gamma) \simeq \operatorname*{colim}_{(S \to T) \in \mathbf{Orb}_{\mathcal{F}}(\Gamma)^{op}/T} *_{\mathcal{F}}(S) .$$

Note that $*_{\mathcal{F}}(S) \simeq *$ and $\mathbf{Orb}_{\mathcal{F}}(\Gamma)^{op}/T$ is empty if T is not in $\mathbf{Orb}_{\mathcal{F}}(\Gamma)$ and otherwise has the identity on T as final object. □

REMARK 10.20. We must consider models for $E_{\mathcal{F}}\Gamma$ since $\mathcal{O}^{\infty}_{\mathrm{hlg}}$ or the equivariant homology theory A_{hlg} constructed in Remark 10.12 are not expected to be strong equivariant homology theories (Definition 10.3) and therefore can not be applied to the object $E_{\mathcal{F}}\Gamma$ of the ∞-category $\mathbf{PSh}(\mathbf{Orb}(\Gamma))$. But on the other hand, since these functors are homotopy invariant, the evaluations $\mathcal{O}^{\infty}_{\mathrm{hlg}}(E_{\mathcal{F}}\Gamma^{cw})$ or $A_{\mathrm{hlg}}(E_{\mathcal{F}}\Gamma^{cw})$ on a model $E_{\mathcal{F}}\Gamma^{cw}$ for $E_{\mathcal{F}}\Gamma$ are well-defined up to equivalence.

From now on $E_{\mathcal{F}}\Gamma^{cw}$ will denote some choice of a model.

The projection $E_{\mathcal{F}}\Gamma^{cw} \to *$ is a morphism in $\Gamma\mathbf{Top}$.

If $F\colon \Gamma\mathbf{Top} \to \mathbf{C}$ is an equivariant $\mathbf{C}$-valued homology theory, then the Farrell–Jones and Baum–Connes type question is for which family $\mathcal{F}$ this projection induces an equivalence

$$F(E_{\mathcal{F}}\Gamma^{cw}) \to F(*) .$$

The coarse geometric approach to the question uses that this projection induces a morphism in $\Gamma\mathbf{Sp}\mathcal{X}$

$$\alpha_{E_{\mathcal{F}}\Gamma^{cw}}\colon \mathcal{O}^{\infty}_{\mathrm{hlg}}(E_{\mathcal{F}}\Gamma^{cw}) \to \mathcal{O}^{\infty}_{\mathrm{hlg}}(*) .$$

DEFINITION 10.21. The morphism $\alpha_{E_{\mathcal{F}}\Gamma^{cw}}$ is called the *universal assembly map* for the family of subgroups $\mathcal{F}$.

One could ask if there is an interesting family $\mathcal{F}$ (i.e., not the family of all subgroups of Γ) for which the universal assembly map is an equivalence.

Most of the study of the assembly map is based on an identification of this map with a forget-control map, or equivalently, a boundary operator of a cone sequence. We develop this point of view below.

The following diagram is one of the starting points of [**3**]. We consider a Γ-topological space X and the projection $X \to *$. It induces the vertical maps in the diagram in $\Gamma\mathbf{Sp}\mathcal{X}$ whose horizontal parts are segments of the cone sequence Corollary 9.30:

$$\begin{array}{ccccc} \operatorname*{colim}_{(Y \to X) \in \Gamma\mathbf{Top}^{cm}/X} \mathrm{Yo}^s(Y_{max,max}) & \longrightarrow & \operatorname*{colim}_{(Y \to X) \in \Gamma\mathbf{Top}^{cm}/X} \mathrm{Yo}^s(\mathcal{O}(\mathcal{M}(\mathcal{U}(Y)))) & \longrightarrow & \mathcal{O}^{\infty}_{\mathrm{hlg}}(X) \\ \downarrow & & \downarrow & & \downarrow{\scriptstyle \alpha_X} \\ \mathrm{Yo}^s(*) & \longrightarrow & \mathrm{Yo}^s(\mathcal{O}(*)) & \longrightarrow & \mathcal{O}^{\infty}_{\mathrm{hlg}}(*) \end{array} .$$

Note that $\mathrm{Yo}^s(\mathcal{O}(*)) \simeq 0$ since $\mathcal{O}(*)$ is flasque. Furthermore, if we tensor-multiply (with respect to $-\otimes-$) the diagram with $\mathrm{Yo}^s(Q)$ for a Γ-bornological coarse space Q whose underlying Γ-set is free, then by Lemma 10.7 the left vertical map becomes an equivalence.

DEFINITION 10.22. We define the *obstruction motive* to be the object

$$M(X) := \operatorname*{colim}_{(Y\to X)\in\Gamma\mathbf{Top}^{cm}/X} \mathrm{Yo}^s(\mathcal{O}(\mathcal{M}(\mathcal{U}(Y))))$$

of the category $\Gamma\mathbf{Sp}\mathcal{X}$.

The discussion above has the following corollary. Let Q be a Γ-bornological coarse space.

COROLLARY 10.23. *If the underlying Γ-set of Q is free, then we have a fiber sequence*

$$M(X)\otimes \mathrm{Yo}^s(Q) \to \mathcal{O}^\infty_{\mathrm{hlg}}(X)\otimes \mathrm{Yo}^s(Q) \xrightarrow{\alpha_{X,Q}} \Sigma\mathrm{Yo}^s(Q) \to \Sigma(M(X)\otimes \mathrm{Yo}^s(Q)) \ .$$

DEFINITION 10.24. The morphism $\alpha_{X,Q} := \alpha_X \otimes \mathrm{id}_{\mathrm{Yo}^s(Q)}$ is called the *motivic assembly map* with twist Q.

Let Q be a Γ-bornological coarse space.

COROLLARY 10.25. *Let the underlying Γ-set of Q be free. Then the motivic assembly map with Q-twist $\alpha_{X,Q}$ is an equivalence if and only if $M(X)\otimes\mathrm{Yo}^s(Q)\simeq 0$.*

A typical example for Q is $\Gamma_{can,min}$, and variants like $\Gamma_{can,max}$, $\Gamma_{max,max}$, $\Gamma_{min,min}$, etc.

In general we expect the assembly map $\alpha_{E_{\mathcal{F}}\Gamma,Q}$ to become an equivalence only after applying suitable equivariant coarse homology theories. But the following is an example where the assembly map is an equivalence already motivically.

Let $\mathcal{F}$ be a family of subgroups of Γ and consider a Γ-bornological coarse space Q.

LEMMA 10.26. *Assume that:*

1. *$E_{\mathcal{F}}\Gamma^{cw}$ is Γ-compact and Γ-metrizable.*
2. *Q is discrete as a coarse space.*
3. *Q has stabilizers in $\mathcal{F}$.*
4. *Q is Γ-finite.*

Then we have

$$M(E_{\mathcal{F}}\Gamma^{cw})\otimes \mathrm{Yo}^s(Q)\simeq 0\ .$$

REMARK 10.27. Note that if the family $\mathcal{F}$ is **VCyc**, i.e., the family of all virtually cyclic subgroups, then the condition of $E_{\mathcal{F}}\Gamma^{cw}$ being Γ-compact is very restrictive: by a conjecture of Juan-Pineda–Leary [**22**] this implies that Γ is virtually cyclic itself.

This conjecture is proven for hyperbolic groups (Juan-Pineda–Leary [**22**]), elementary amenable groups (Kochloukova–Martinez-Perez–Nucinkis [**25**] and Groves–Wilson [**21**]) and for one-relator groups, acylindrically hyperbolic groups, 3-manifold groups, CAT(0) cube groups, linear groups (von Puttkamer–Wu [**38, 39**]).

PROOF. Since $E_{\mathcal{F}}\Gamma^{cw}$ is Γ-compact and Γ-metrizable the colimit in the Definition 10.22 of $M(E_{\mathcal{F}}\Gamma)^{cw}$ stabilizes. We consider Q as a Γ-uniform bornological coarse space Q_{disc} with the discrete uniform structure. Using Lemma 9.26, we get the equivalence

$$M(E_{\mathcal{F}}\Gamma^{cw}) \otimes \mathrm{Yo}^s(Q) \simeq \mathrm{Yo}^s(\mathcal{O}(\mathcal{M}(\mathcal{U}(E_{\mathcal{F}}\Gamma^{cw}))) \otimes Q_{disc}) \ .$$

The space $E_{\mathcal{F}}\Gamma^{cw}$ has the following universal property: the space $\mathrm{Map}_{\Gamma}(X, E_{\mathcal{F}}\Gamma^{cw})$ is contractible for every Γ-space X with stabilizers in $\mathcal{F}$ which is homotopy equivalent to a CW-complex. It follows from the universal property of $E_{\mathcal{F}}\Gamma^{cw}$ that there is a unique homotopy class of maps $\iota\colon Q \to E_{\mathcal{F}}\Gamma^{cw}$. Furthermore, the maps $\iota \circ \mathrm{pr}_Q, \mathrm{pr}_{E_{\mathcal{F}}\Gamma^{cw}}\colon E_{\mathcal{F}}\Gamma^{cw} \times Q \to E_{\mathcal{F}}\Gamma^{cw}$ are homotopic. Since Q is Γ-finite and $E_{\mathcal{F}}\Gamma^{cw}$ is Γ-compact Γ-metrizable these maps and homotopies are automatically uniform maps when we consider $E_{\mathcal{F}}\Gamma^{cw}$ as a uniform space by applying $\mathcal{U}$. Hence $\mathrm{id}_{\mathcal{M}(\mathcal{U}(E_{\mathcal{F}}\Gamma^{cw}))\otimes Q_{disc}}$ is homotopic to the composition

$$\mathcal{M}(\mathcal{U}(E_{\mathcal{F}}\Gamma^{cw})) \otimes Q_{disc} \xrightarrow{\mathrm{pr}_Q} Q_{disc} \xrightarrow{(\iota, \mathrm{id}_Q)} \mathcal{M}(\mathcal{U}(E_{\mathcal{F}}\Gamma^{cw})) \otimes Q_{disc}$$

of morphisms between Γ-uniform bornological coarse spaces. Using that $\mathcal{O}$ is homotopy invariant we conclude that $\mathrm{id}_{M(E_{\mathcal{F}}\Gamma^{cw})\otimes \mathrm{Yo}^s(Q)}$ factorizes over $\mathrm{Yo}^s(\mathcal{O}(Q_{disc}))$. We now use Example 9.25 in order to conclude that $\mathrm{Yo}^s(\mathcal{O}(Q_{disc})) \simeq 0$. This finishes this proof. □

For the following we just combine Lemma 10.26 and Corollary 10.25.

COROLLARY 10.28. *If $E_{\mathcal{F}}\Gamma^{cw}$ is Γ-compact and Γ-metrizable, then the motivic assembly map $\alpha_{E_{\mathcal{F}}\Gamma^{cw},\Gamma_{min,?}}$ is an equivalence for $? \in \{\min, \max\}$.*

10.4. Homology theories from coarse homology theories. Let E be an equivariant $\mathbf{C}$-valued coarse homology theory, i.e., a colimit preserving functor $E\colon \Gamma\mathbf{Sp}\mathcal{X} \to \mathbf{C}$. By Proposition 10.11 we can use the cone $\mathcal{O}^{\infty}_{\mathrm{hlg}}$ in order to pull-back E to an equivariant homology theory:

DEFINITION 10.29. Let us define

$$E\mathcal{O}^{\infty}_{\mathrm{hlg}} := E \circ \mathcal{O}^{\infty}_{\mathrm{hlg}}\colon \Gamma\mathbf{Top} \to \mathbf{C} \ ,$$

which is an equivariant $\mathbf{C}$-valued homology theory.

Recall that for an equivariant coarse motivic spectrum L the functor

$$- \otimes L\colon \Gamma\mathbf{Sp}\mathcal{X} \to \Gamma\mathbf{Sp}\mathcal{X}$$

preserves colimits. If E is a $\mathbf{C}$-valued equivariant coarse homology theory, then we can define a new equivariant coarse homology theory

$$E_L(-) := E(- \otimes L)\colon \Gamma\mathbf{Sp}\mathcal{X} \to \mathbf{C} \ . \tag{10.17}$$

We will refer to L as a *twist.*

The Farrell–Jones/Baum–Connes type question for $E_{\mathrm{Yo}^s(Q)}\mathcal{O}^{\infty}_{\mathrm{hlg}}$ is now the question for which family $\mathcal{F}$ of subgroups the morphism $E_{\mathrm{Yo}^s(Q)}(\alpha_{E_{\mathcal{F}}\Gamma})$ is an equivalence of spectra. The next corollary is a consequence of Corollary 10.25.

Let Q be a Γ-bornological coarse space.

COROLLARY 10.30. *Assume that the underlying Γ-set of Q is free. Then the assembly map $E_{\mathrm{Yo}^s(Q)}(\alpha_{E_{\mathcal{F}}\Gamma})$ is an equivalence if and only if*

$$E_{\mathrm{Yo}^s(Q)}(M(E_{\mathcal{F}}\Gamma)) \simeq 0 \ .$$

Given the coarse homology theory E and a twist L it is of particular interest to characterize the homology theory $E_L\mathcal{O}^{\infty}_{\mathrm{hlg}}$. As explained in Section 10.1, the restriction of an equivariant homology theory from $\Gamma\mathbf{Top}$ to the full subcategory $\Gamma\mathbf{CW}$ of Γ-CW complexes is determined by its restriction to the orbit category $\mathbf{Orb}(\Gamma)$. So we must understand the evaluations $E_L\mathcal{O}^{\infty}_{\mathrm{hlg}}(S)$ for all transitive Γ-sets S. The main tool is Proposition 9.35. It gives

$$E_L\mathcal{O}^{\infty}_{\mathrm{hlg}}(S) \simeq \Sigma E_L(\mathrm{Yo}^s(S_{min,max})) \ .$$

We have calculated these evaluations for equivariant coarse ordinary homology and for equivariant coarse algebraic K-homology explicitly:

(1) see Section 7.2 for $H\mathcal{X}^{\Gamma}_{\mathrm{Yo}^s(\Gamma_{can,min})}$, and
(2) see Section 8.4 for $K\mathbf{A}\mathcal{X}^{\Gamma}_{\mathrm{Yo}^s(\Gamma_{?,?})}$.

11. Forget-control and assembly maps

11.1. The forget-control map. Let X be a coarse space and U an entourage of X. Let $\mu\colon \mathcal{P}(X) \to [0,1]$ be a probability measure on the measurable space $(X, \mathcal{P}(X))$.

DEFINITION 11.1. The measure μ is called finite U-bounded if there is a finite U-bounded subset F of X with $\mu(F) = 1$. We then define the support $\mathrm{supp}(\mu)$ to be the smallest subset of X with measure one.

We let $P_U(X)$ denote the topological space of finite U-bounded probability measures on X equipped with the topology induced by the evaluation against finitely supported functions.

Every point x in X gives rise to a Dirac measure δ_x at x. If U contains the diagonal of X, then we get a map $X \to P_U(X)$, $x \mapsto \delta_x$, of sets. A probability measure μ which is finite U-bounded can be written as a finite convex combination of Dirac measures. More concretely,

$$\mu = \sum_{x \in \mathrm{supp}(\mu)} \mu(\{x\})\delta_x \ .$$

The set $P_U(X)$ has a natural structure of a simplicial complex. In other words, $P_U(X)$ is the simplicial complex with vertex set X such that a subset $\{x_1, \ldots, x_m\}$ spans a simplex if and only if $(x_i, x_j) \in U$ for all i, j in $\{1, \ldots, m\}$. Let $P_U(X)_u$ denote the uniform space with the uniform structure induced by the canonical path quasi-metric on the simplicial complex, see Example 9.7. Note that the topology induced by this uniform structure is the topology on $P_U(X)$ induced by the evaluation against finitely supported functions, i.e., we have

$$\tau(P_U(X)_u) = P_U(X) \ ,$$

where τ is the functor associating the underlying topological space to a uniform space, cf. (10.14). Eilenberg–Steenrod [**18**, p. 75] called this the metric topology. In general this topology differs from the weak topology on the simplicial complex in the sense of Eilenberg–Steenrod [**18**, p. 75].

If X' is a second coarse space with entourage U' and $f\colon X \to X'$ is a map such that $(f \times f)(U) \subseteq U'$, then we get a map of simplicial complexes

$$f_*\colon P_U(X) \to P_{U'}(X') \ , \quad \mu \mapsto f_*\mu \ ,$$

where the measure $f_*\mu$ is the push-forward of the measure μ.

If X is a Γ-coarse space and the entourage U is Γ-invariant, then Γ acts on the simplicial complex $P_U(X)$ such that the latter becomes a Γ-simplicial complex, so in particular it becomes a Γ-topological space. Furthermore we obtain a Γ-uniform space $P_U(X)_u$. If X' is a second Γ-coarse space, U' is a Γ-invariant entourage, and $f\colon X \to X'$ is equivariant with $(f \times f)(U) \subseteq U'$, then $f_*\colon P_U(X) \to P_{U'}(X')$ is equivariant.

Let X be a Γ-coarse space.

DEFINITION 11.2. The Γ-topological space

$$\mathrm{Rips}(X) := \operatorname*{colim}_{U \in \mathcal{C}^\Gamma} P_U(X)$$

is called the *Rips complex* of X. Note that the colimit is taken in $\Gamma\mathbf{Top}$.

REMARK 11.3. Note that in general the simplicial complex $P_U(X)$ is not locally finite. In this case its topology does not exhibit it as a CW-complex (this happens if and only if the simplicial complex is locally finite).

But by a result of Dowker [**17**, Thm. 1 on P. 575] $P_U(X)$ has the homotopy type of a CW-complex (see also Milnor [**30**, Thm. 2] for a short proof of this). Concretely, the underlying identity map of the set induces a homotopy equivalence $\mathrm{CW}(P_U(X)) \to P_U(X)$, where $\mathrm{CW}(P_U(X))$ denotes $P_U(X)$ retopologized as a CW-complex.

Let Γ_{can} be the group Γ considered as a Γ-coarse space with its canonical coarse structure. Recall the localization (10.2).

LEMMA 11.4. *We have an equivalence* $\ell(\mathrm{Rips}(\Gamma_{can})) \simeq E_{\mathbf{Fin}}\Gamma$.

PROOF. By definition, $\ell(\mathrm{Rips}(\Gamma))$ is the presheaf on $\mathbf{Orb}(\Gamma)$ given by

$$S \mapsto \kappa(\mathrm{Map}_{\Gamma\mathbf{Top}}(S, \mathrm{Rips}(\Gamma_{can}))) ,$$

where κ is as in (10.1). We now observe that the stabilizers of the points of $\mathrm{Rips}(\Gamma_{can})$ are finite subgroups of Γ. Consequently, the presheaf belongs to the subcategory of presheaves supported on the orbits with finite stabilizers. Since, by definition, $E_{\mathbf{Fin}}\Gamma$ corresponds to the final object in this subcategory there exists a morphism $\ell(\mathrm{Rips}(\Gamma_{can})) \to E_{\mathbf{Fin}}\Gamma$. In order to show that it is an equivalence it suffices to show that the spaces $\mathrm{Map}_{\Gamma\mathbf{Top}}(\Gamma/H, \mathrm{Rips}(X))$ for all finite subgroups H of Γ are equivalent to $*$, i.e., that these spaces are connected and that their homotopy groups are trivial.

We now study these homotopy groups. Since the spheres S^n are compact for all n in $\mathbb{N}$ and the structure maps

$$\mathrm{Map}_{\Gamma\mathbf{Top}}(\Gamma/H, P_U(\Gamma_{can})) \to \mathrm{Map}_{\Gamma\mathbf{Top}}(\Gamma/H, P_{U'}(\Gamma_{can}))$$

for $U \subseteq U'$ are inclusions of CW-complexes, we have

$$\pi_*(\mathrm{Map}_{\Gamma\mathbf{Top}}(\Gamma/H, \mathrm{Rips}(\Gamma_{can}))) \cong \operatorname*{colim}_{U \in \mathcal{C}^\Gamma} \pi_*(\mathrm{Map}_{\Gamma\mathbf{Top}}(\Gamma/H, P_U(\Gamma_{can}))) .$$

We have a homeomorphism

$$\mathrm{Map}_{\Gamma\mathbf{Top}}(\Gamma/H, P_U(\Gamma_{can})) \cong P_U(\Gamma_{can})^H .$$

We fix an integer n and consider a map

$$f\colon S^n \to P_U(\Gamma_{can})^H .$$

The image $f(S^n)$ is a compact subset of $P_U(\Gamma_{can})^H$. In the case $n = 0$, since Γ_{can} is coarsely connected, we can increase the entourage U such that the image $f(S^0)$ belongs to a connected component of $P_U(\Gamma_{can})$. Let now n be arbitrary. We can now assume that $f(S^n)$ belongs to a connected component of $P_U(\Gamma_{can})$. It is hence bounded in diameter by some integer N. We conclude that under the map $P_U(\Gamma_{can}) \to P_{U^N}(\Gamma_{can})$ the image $f(S^n)$ is mapped to a subset of a single simplex of $P_{U^N}(\Gamma_{can})$. The intersection of the H-fixed points with this simplex is a convex subset and hence itself contractible. We conclude that f is homotopic to a constant map. □

DEFINITION 11.5. If X is a Γ-bornological coarse space and U an invariant entourage of X, then we equip the Γ-simplicial complex $P_U(X)$ with the bornology generated by the subsets $P_U(B)$ for all bounded subsets B of X. Equipped with this bornology and the coarse structure induced by the metric we obtain a Γ-bornological coarse space which we denote by $P_U(X)_{bd}$. We furthermore write $P_U(X)_{bdu}$ for the corresponding Γ-uniform bornological coarse space. Note in contrast that the notation $P_U(X)_d$ (or $P_U(X)_{du}$, respectively) would mean the Γ-bornological coarse space (or Γ-uniform bornological coarse space, respectively) whose bornological and coarse (and uniform, respectively) structures are induced from the metric, see Example 9.11.

Let X be a Γ-bornological coarse space and U be an invariant entourage of X. The Dirac measures provide a morphism of Γ-bornological coarse spaces

$$\delta\colon X_U \to P_U(X)_{bd}\ . \tag{11.1}$$

LEMMA 11.6. *Assume:*

1. *Γ is torsion-free.*
2. *The underlying Γ-set of X is free.*

Then the morphism (11.1) *is an equivalence of Γ-bornological coarse spaces.*

PROOF. We first observe that Γ acts freely on $P_U(X)$. Indeed, for γ in Γ and μ in $P_U(X)$ satisfying $\gamma\mu = \mu$ the subgroup of Γ generated by γ has a finite orbit contained in $\mathrm{supp}(\mu)$. Since Γ is torsion-free and X is a free Γ-set this can only happen if $\gamma = 1$.

To define an inverse morphism $g\colon P_U(X)_{bd} \to X_U$ we first choose representatives for the orbits $P_U(X)/\Gamma$. Then we define $g(\mu)$ for every chosen representative μ to be a point in $\mathrm{supp}(\mu)$, and extend equivariantly. Then $g \circ \delta = \mathrm{id}_{X_U}$ and $\delta \circ g$ is close to $\mathrm{id}_{P_U(X)}$. □

The following corollary follows immediately from the above lemma.

COROLLARY 11.7. *If Γ is a finitely generated torsion-free group, then $\Gamma_{can,min} \to P_U(\Gamma)_{bd}$ is an equivalence for every invariant generating entourage U of Γ.*

Assume that Q is a Γ-bornological coarse space. Let X be a Γ-bornological coarse space and U be an invariant entourage of X.

LEMMA 11.8. *If the underlying Γ-set of Q is free, then*

$$\delta \times \mathrm{id}_Q\colon X_U \otimes Q \to P_U(X)_{bd} \otimes Q$$

is an equivalence of Γ-bornological coarse spaces.

Note that we do not assume that X is a free Γ-set.

PROOF. First we note that Γ acts freely on the set $P_U(X) \times Q$. We choose representatives for the orbits $(P_U(X) \times Q)/\Gamma$. Then we choose for every representative $(\mu, q) \in P_U(X) \times Q$ a point $x \in \mathrm{supp}(\mu)$ and set $g(\mu, q) := (x, q)$. Then we extend this to an equivariant map $g \colon P_U(X) \times Q \to X \times Q$. This map of sets is a morphism

$$g \colon P_U(X)_{bd} \otimes Q \to X_U \otimes Q \ .$$

By construction $g \circ (\delta \times \mathrm{id}_Q) = \mathrm{id}_{X \times Q}$ and $g \circ (\delta \times \mathrm{id}_Q)$ is close to the identity. $\square$

For the following recall the cone functor $\mathcal{O}$ from Section 9.4 and the "cone at infinity" functor $\mathcal{O}^\infty$ from Section 9.5.

Let X be a Γ-bornological coarse space.

DEFINITION 11.9. We define the following equivariant coarse motivic spectra:

$$F(X) := \operatorname*{colim}_{U \in \mathcal{C}^\Gamma} \mathrm{Yo}^s(\mathcal{O}(P_U(X)_{bdu}))$$
$$F^\infty(X) := \operatorname*{colim}_{U \in \mathcal{C}^\Gamma} \mathcal{O}^\infty(P_U(X)_{bdu})$$
$$F^0(X) := \operatorname*{colim}_{U \in \mathcal{C}^\Gamma} \mathrm{Yo}^s(P_U(X)_{bd})$$

Using standard Kan-extension techniques one can refine the above description to functors

$$F, F^\infty, F^0 \colon \Gamma\mathbf{BC} \to \Gamma\mathbf{Sp}\mathcal{X} \ ,$$

see Remark 11.11 for details. The fiber sequence from Corollary 9.30 provides a natural fiber sequence of functors

$$F^0(X) \to F(X) \to F^\infty(X) \xrightarrow{\beta_X} \Sigma F^0(X) \ . \tag{11.2}$$

DEFINITION 11.10. We call β_X the *forget-control map.*

REMARK 11.11. We let $\Gamma\mathbf{BC}^{\mathcal{C}}$ denote the category of pairs (X, U), where X is a Γ-bornological coarse space and U is an invariant entourage of X containing the diagonal. A morphism $(X, U) \to (X', U')$ is a morphism $f \colon X \to X'$ in $\Gamma\mathbf{BC}$ such that $(f \times f)(U) \subseteq U'$. We have a forgetful functor

$$\Gamma\mathbf{BC}^{\mathcal{C}} \to \Gamma\mathbf{BC} \ , \quad (X, U) \mapsto X \ . \tag{11.3}$$

Let

$$\widetilde{E} \colon \Gamma\mathbf{BC}^{\mathcal{C}} \to \mathbf{C}$$

be a functor to some cocomplete target $\mathbf{C}$ and let E be the left Kan extension of $\widetilde{E}$ along (11.3). The evaluation of E on a Γ-bornological coarse space X is then given as follows:

LEMMA 11.12. *We have an equivalence*

$$E(X) \simeq \operatorname*{colim}_{U \in \mathcal{C}^\Gamma(X)} \widetilde{E}(X, U) \ .$$

PROOF. By the pointwise formula for the left Kan extension we have an equivalence

$$E(X) \simeq \operatorname*{colim}_{((X', U'), f \colon X' \to X) \in \Gamma\mathbf{BC}^{\mathcal{C}}/X} \widetilde{E}(X', U') \ .$$

If $((X', U'), f \colon X' \to X)$ belongs to $\Gamma\mathbf{BC}^{\mathcal{C}}/X$, then we have a morphism

$$(X', U') \to (X, f(U') \cup \mathrm{diag}(X))$$

in $\Gamma\mathbf{BC}^{\mathcal{C}}/X$. This easily implies that the full subcategory of objects of the form $((X,U),\mathrm{id}_X)$ of $\Gamma\mathbf{BC}^{\mathcal{C}}/X$ with U in $\mathcal{C}^\Gamma(X)$ is cofinal in $\Gamma\mathbf{BC}^{\mathcal{C}}/X$. □

We have a functor

$$P\colon \Gamma\mathbf{BC}^{\mathcal{C}} \to \Gamma\mathbf{UBC}\ , \quad (X,U)\mapsto P_U(X)_{bdu}\ .$$

We construct the fibre sequence (11.2) by applying the left Kan extension to the fibre sequence of functors $\Gamma\mathbf{BC}^{\mathcal{C}} \to \Gamma\mathbf{Sp}\mathcal{X}$

$$\mathrm{Yo}^s\circ\mathcal{F}_{\mathcal{T}}\circ P \to \mathrm{Yo}^s\circ\mathcal{O}\circ P \to \mathcal{O}^\infty\circ P \to \Sigma\mathrm{Yo}^s\circ\mathcal{F}_{\mathcal{T}}\circ P$$

obtained by precomposing the sequence from Corollary 9.30 with P.

Let X be a Γ-bornological space. In the following two corollaries we identify the Γ-coarse motivic spectrum $F^0(X)$.

COROLLARY 11.13. *If Γ is torsion-free and the underlying Γ-set of X is free, then*

$$F^0(X)\simeq \mathrm{Yo}^s(X)\ .$$

PROOF. We have equivalences

$$F^0(X)=\operatorname*{colim}_{U\in\mathcal{C}^\Gamma}\mathrm{Yo}^s(P_U(X)_{bd}) \stackrel{Lemma\ 11.6}{\simeq} \operatorname*{colim}_{U\in\mathcal{C}^\Gamma}\mathrm{Yo}^s(X_U) \stackrel{Corollary\ 4.11.5}{\simeq} \mathrm{Yo}^s(X)\ ,$$

which proves the claim. □

Let Q and X be Γ-bornological coarse spaces.

COROLLARY 11.14. *If the underlying Γ-set of Q is free, then*

$$F^0(X)\otimes\mathrm{Yo}^s(Q)\sim \mathrm{Yo}^s(X)\otimes\mathrm{Yo}^s(Q)\ .$$

PROOF. Here we use Lemma 11.8, that Yo^s is symmetric monoidal, and that the functor

$$-\otimes\mathrm{Yo}^s(Q)\colon \Gamma\mathbf{Sp}\mathcal{X}\to\Gamma\mathbf{Sp}\mathcal{X}$$

preserves colimits. Therefore we can write down an analogous sequence of equivalences as in the above proof of Corollary 11.13. □

11.2. Comparison of the assembly and the forget-control map. In this section we will compare the assembly map for the family of finite subgroups with the forget-control map.

For every two Γ-bornological coarse spaces X and L we have the forget-control morphism (Definition 11.10)

$$\beta_{X,L}\colon \operatorname*{colim}_{U\in\mathcal{C}^\Gamma}\mathcal{O}^\infty(P_U(X)_{bdu})\otimes\mathrm{Yo}^s(L)\to \operatorname*{colim}_{U\in\mathcal{C}^\Gamma}\Sigma\mathrm{Yo}^s(P_U(X)_{bd}\otimes L)\ .$$

We have furthermore the assembly map (see Definition 10.24)

$$\alpha_{\mathrm{Rips}(X),L}\colon \mathcal{O}^\infty_{\mathrm{hlg}}(\mathrm{Rips}(X))\otimes\mathrm{Yo}^s(L)\to\mathcal{O}^\infty_{\mathrm{hlg}}(*)\otimes\mathrm{Yo}^s(L)\simeq\Sigma\mathrm{Yo}^s(L)$$

induced by the morphism $\mathrm{Rips}(X)\to *$ of Γ-topological spaces.

Recall [**8**, Def. 2.28] that a coarse space $(X,\mathcal{C})$ is called *coarsely connected* if for any two points x,y in X there exists an entourage U in $\mathcal{C}$ such that $(x,y)\in U$.

DEFINITION 11.15. A Γ-bornological coarse space X is *eventually coarsely connected* if there exists a coarse entourage U such that X_U is coarsely connected.

Note that an eventually coarsely connected space is in particular coarsely connected.

While one is interested in $\alpha_{\mathrm{Rips}(X),\Gamma_{can,min}}$, using descent methods (like in [**10**]) we will only be able to derive split-injectivity of $\beta_{X,\Gamma_{max,max}}$ in several cases. The following theorem allows us to compare both maps. This comparison does not hold directly but only after forcing continuity.

Let X be a Γ-bornological coarse space. Recall [**8**, Def. 6.100] that X has *strongly bounded geometry* if it is equipped with the minimal compatible bornology and if for every entourage U of X there exists a uniform finite upper bound on the cardinalities of U-bounded subsets of X. Furthermore recall the functor C^s from (5.6).

THEOREM 11.16. *Assume:*

(1) *X has strongly bounded geometry.*
(2) *X is Γ-finite.*
(3) *The action of Γ on X is proper (Example 2.13).*
(4) *X is eventually coarsely connected.*

Then the morphisms $C^s(\alpha_{\mathrm{Rips}(X),\Gamma_{can,min}})$ and $C^s(\beta_{X,\Gamma_{max,max}})$ are equivalent.

REMARK 11.17. Note that X being Γ-finite and the action of Γ on X being proper implies that X has the minimal bornology.

Furthermore, if X is Γ-finite and U is a Γ-invariant entourage of X, then the assumption that every U-bounded subset is finite, already implies a uniform upper bound on the cardinality of U-bounded subsets. Hence Assumption 1 in Theorem 11.16 above could be equivalently replaced by the seemingly weaker assumption that every U-bounded subset is finite.

The rest of this section is devoted to the proof of Theorem 11.16.

We consider a Γ-simplicial complex K. Recall the notation introduced in Example 9.11:

(1) K_u denotes the Γ-uniform space associated to K.
(2) $K_{u,max,max}$ denotes the Γ-uniform bornological coarse space which has the uniform structure of K_u, but the maximal coarse and bornological structures.
(3) $K_{u,d,max}$ denotes the Γ-uniform bornological coarse space with the uniform structure of K_u, the metric coarse structure and the maximal bornological structure.
(4) $K_{d,max}$ denotes the Γ-bornological coarse space underlying $K_{u,d,max}$.

We denote by

$$\beta^{max,max}_{X,L}\colon \operatorname*{colim}_{U\in\mathcal{C}^\Gamma} \mathcal{O}^\infty(P_U(X)_{u,max,max})\otimes \mathrm{Yo}^s(L) \to \operatorname*{colim}_{U\in\mathcal{C}^\Gamma} \Sigma\mathrm{Yo}^s(P_U(X)_{max,max}\otimes L)$$

and

$$\beta^{d,max}_{X,L}\colon \operatorname*{colim}_{U\in\mathcal{C}^\Gamma} \mathcal{O}^\infty(P_U(X)_{u,d,max})\otimes \mathrm{Yo}^s(L) \to \operatorname*{colim}_{U\in\mathcal{C}^\Gamma} \Sigma\mathrm{Yo}^s(P_U(X)_{d,max}\otimes L)$$

the forget-control maps.

The proof of Theorem 11.16 consists of a sequence of lemmas.

Let X and L be Γ-bornological coarse spaces.

LEMMA 11.18. *Assume:*

(1) *X has strongly bounded geometry.*
(2) *X is Γ-finite.*
(3) *The underlying set of L is a free Γ-set.*

Then $\alpha_{\mathrm{Rips}(X),L}$ and $\beta^{max,max}_{X,L}$ are equivalent.

PROOF. The assumptions on X imply that $P_U(X)$ is Γ-compact and locally finite, hence Γ-metrizable, for every invariant coarse entourage U of X. Therefore, by (10.15), we have an equivalence

$$\mathcal{O}^\infty_{\mathrm{hlg}}(P_U(X)) \simeq \mathcal{O}^\infty(P_U(X)_{u,max,max}) \ .$$

Using Definition 11.2 of the Rips complex as a filtered colimit of Γ-topological spaces and the relation (10.13) we get

$$\begin{aligned}\mathcal{O}^\infty_{\mathrm{hlg}}(\mathrm{Rips}(X)) &\simeq \operatorname*{colim}_{U\in\mathcal{C}^\Gamma} \mathcal{O}^\infty_{\mathrm{hlg}}(P_U(X))\\ &\simeq \operatorname*{colim}_{U\in\mathcal{C}^\Gamma} \mathcal{O}^\infty(P_U(X)_{u,max,max}) \ .\end{aligned}$$

Hence we get the following commutative diagram:

$$\begin{array}{ccc}
\operatorname*{colim}_{U\in\mathcal{C}^\Gamma} \mathcal{O}^\infty(P_U(X)_{u,max,max})\otimes \mathrm{Yo}^s(L) & \xrightarrow{\beta^{max,max}_{X,L}} & \operatorname*{colim}_{U\in\mathcal{C}^\Gamma} \Sigma\mathrm{Yo}^s(P_U(X)_{max,max}\otimes L)\\
\downarrow{\simeq} & & \\
\mathcal{O}^\infty_{\mathrm{hlg}}(\mathrm{Rips}(X))\otimes \mathrm{Yo}^s(L) & & \downarrow{\simeq}\\
\downarrow{\alpha_{\mathrm{Rips}(X),L}} & & \\
\mathcal{O}^\infty_{\mathrm{hlg}}(*)\otimes \mathrm{Yo}^s(L) & \xrightarrow{\simeq} & \Sigma\mathrm{Yo}^s(L)
\end{array}$$

The vertical arrow on the right is an equivalence by Lemma 10.7. □

Recall the functor C^s from (5.6). Let X be a Γ-bornological coarse space.

LEMMA 11.19. *Assume:*

(1) *X has strongly bounded geometry.*
(2) *X is eventually coarsely connected.*

Then $C^s(\beta^{max,max}_{X,\Gamma_{can,min}})$ and $C^s(\beta^{d,max}_{X,\Gamma_{can,min}})$ are equivalent.

PROOF. The morphism $P_U(X)_{u,d,max} \to P_U(X)_{u,max,max}$ of Γ-uniform bornological coarse spaces induces a diagram

$$\begin{array}{ccc}
C^s(\mathcal{O}^\infty(P_U(X)_{u,d,max})\otimes \mathrm{Yo}^s(\Gamma_{can,min})) & \to & C^s(\mathcal{O}^\infty(P_U(X)_{u,max,max})\otimes \mathrm{Yo}^s(\Gamma_{can,min}))\\
\downarrow{\beta^{d,max}_{X,\Gamma_{can,min}}} & & \downarrow{\beta^{max,max}_{X,\Gamma_{can,min}}}\\
C^s(\Sigma\mathrm{Yo}^s(P_U(X)_{d,max}\otimes \Gamma_{can,min})) & \longrightarrow & C^s(\Sigma\mathrm{Yo}^s(P_U(X)_{max,max}\otimes \Gamma_{can,min}))
\end{array}$$

Since $P_U(X)_{u,d,max} \to P_U(X)_{u,max,max}$ is a coarsening, the left vertical arrow is an equivalence by Proposition 9.33. To show that the right vertical arrow is an equivalence for large entourages U we will need continuity.

For an invariant entourage U of X we denote the set of finite subcomplexes of $P_U(X)$ by $\mathcal{F}(P_U(X))$. It is a filtered partially ordered set with respect to the inclusion relation. For every finite subcomplex F we define a Γ-invariant subcomplex $D_F := \Gamma(F \times \{1\})$ of $P_U(X) \times \Gamma$. We consider the family of Γ-invariant subsets

$$\mathcal{D} := (D_F)_{F \in \mathcal{F}(P_U(X))}$$

of $P_U(X) \times \Gamma$. By Example 5.11 the family $\mathcal{D}$ is a co-Γ-bounded exhaustion of both spaces $P_U(X)_{max,max} \otimes \Gamma_{can,min}$ and $P_U(X)_{d,max} \otimes \Gamma_{can,min}$. By continuity, it suffices to show that the bornological coarse structures on $\mathcal{D}$ induced from $P_U(X)_{max,max} \otimes \Gamma_{can,min}$ and from $P_U(X)_{d,max} \otimes \Gamma_{can,min}$, respectively, agree.

Since the bornologies of $P_U(X)_{max,max} \otimes \Gamma_{can,min}$ and $P_U(X)_{d,max} \otimes \Gamma_{can,min}$ agree, we only have to care about the coarse structures. Let U be large enough, such that $P_U(X)$ is connected.

The coarse structure on D_F induced by $P_U(X)_{d,max} \otimes \Gamma_{can,min}$ is generated by the entourages

$$(D_F \times D_F) \cap (U_r \times V_B) ,$$

where U_r is a metric entourage of $P_U(X)$ of size r in $(0, \infty)$ and $V_B := \Gamma(B \times B)$ for a finite subset B is one of the generating entourages of the canonical structure of Γ.

The coarse structure on D_F induced by $P_U(X)_{max,max} \otimes \Gamma_{can,min}$ is generated by the entourages

$$(D_F \times D_F) \cap ((P_U(X) \times P_U(X)) \times V_B)$$

for finite subsets B of Γ. It is clear that the coarse structure of the latter is larger than the one of the first, and it remains to show to other inclusion.

We have

$$(D_F \times D_F) \cap ((P_U(X) \times P_U(X)) \times V_B) \cong \bigcup_{(\gamma,\gamma') \in V_B} (\gamma F \times \gamma' F) \times \{(\gamma, \gamma')\} .$$

Since F is a finite subcomplex and $P_U(X)$ is connected, there exists an r in $(0, \infty)$ such that $BF \times BF \subseteq U_r$. This implies that

$$(\gamma F \times \gamma' F) \times \{(\gamma, \gamma')\} \subseteq U_r \times V_B$$

for all pairs (γ, γ') in V_B. We conclude that

$$(D_F \times D_F) \cap ((P_U(X) \times P_U(X)) \times V_B) \subseteq (D_F \times D_F) \cap (U_r \times V_B) .$$

This finishes the proof. □

Let X be a Γ-bornological coarse space.

LEMMA 11.20. *Assume:*

1. *X is Γ-finite.*
2. *X has strongly bounded geometry.*
3. *the Γ-action on X is proper.*

Then the maps $C^s(\beta^{d,max}_{X,\Gamma_{can,min}})$ and $C^s(\beta_{X,\Gamma_{can,max}})$ are equivalent.

PROOF. Recall from Definition 11.5 that $P_U(X)_{bd}$ denotes the Γ-bornological coarse space whose coarse structure is induced from the metric and whose bornology is generated by the subsets $P_U(B)$ for all bounded subsets B of X. Recall furthermore that we write $P_U(X)_{bdu}$ for the corresponding Γ-uniform bornological coarse space.

Recall from (5.7) that $\mathrm{Yo}^s_c \simeq C^s \circ \mathrm{Yo}^s$. It suffices to produce diagrams

$$\begin{array}{ccc} \mathrm{Yo}^s_c([0,k]\otimes P_U(X)_{bd}\otimes \Gamma_{can,max}) & \longrightarrow & C^s(\mathcal{O}(P_U(X)_{bdu})\otimes \mathrm{Yo}^s(\Gamma_{can,max})) \\ \downarrow \simeq & & \downarrow \simeq \\ \mathrm{Yo}^s_c([0,k]\otimes P_U(X)_{d,max}\otimes \Gamma_{can,min}) & \longrightarrow & C^s(\mathcal{O}(P_U(X)_{u,d,max})\otimes \mathrm{Yo}^s(\Gamma_{can,min})) \end{array}$$

for every natural number k which are compatible with increasing k and U.

To produce the diagram we will use continuity with the exhaustion $\mathcal{Y} = (Y_\kappa)_{\kappa\in\mathcal{F}(P_U(X))^{\mathbb{N}}}$ from Lemma 5.12, where $\mathcal{F}(P_U(X))$ denotes the set of all finite subcomplexes of $P_U(X)$. Recall that for κ in $\mathcal{F}(P_U(X))^{\mathbb{N}}$ we set

$$Y_\kappa := \bigcup_{n\in\mathbb{N}} [n-1,n]\times D_{\kappa(n)} \ . \tag{11.4}$$

Since $P_U(X)_{bd}$ and $P_U(X)_{d,max}$ are Γ-bounded the exhaustion is trapping by Lemma 5.12 for both spaces

$$\mathcal{O}(P_U(X)_{u,d,max})\otimes \mathrm{Yo}^s(\Gamma_{can,min}) \quad \text{and} \quad \mathcal{O}(P_U(X)_{bdu})\otimes \mathrm{Yo}^s(\Gamma_{can,max}) \ .$$

Note that the hybrid coarse structure does not play a role here since trapping exhaustions are a bornological concept.

Since the definition of the exhaustion is independent of k and compatible with increasing U, it remains for us to show that the bornological coarse structures on Y_κ induced from $\mathcal{O}(P_U(X)_{bdu})\otimes \mathrm{Yo}^s(\Gamma_{can,max})$ and $\mathcal{O}(P_U(X)_{u,d,max})\otimes \mathrm{Yo}^s(\Gamma_{can,min})$, respectively, agree in order to obtain (by continuity) the equivalences in the above diagram. Since the coarse structures of $\mathcal{O}(P_U(X)_{bdu})\otimes \mathrm{Yo}^s(\Gamma_{can,max})$ and $\mathcal{O}(P_U(X)_{u,d,max})\otimes \mathrm{Yo}^s(\Gamma_{can,min})$ agree, we only have to consider the bornologies.

Every bounded subset of $\mathcal{O}(P_U(X)_{u,d,max})\otimes \mathrm{Yo}^s(\Gamma_{can,min})$ or $\mathcal{O}(P_U(X)_{bdu})\otimes \mathrm{Yo}^s(\Gamma_{can,max})$ is contained in $[0,n]\times P_U(X)\times\Gamma$ for some n. It therefore suffices to see that the induced bornologies on $([0,n]\times P_U(X)\times\Gamma)\cap Y_\kappa$ coincide. We can now further finitely decompose

$$([0,n]\times P_U(X)\times\Gamma)\cap Y_\kappa \subseteq \bigcup_{i=1}^{n+1}[i-1,i]\times D_{\kappa(i)} \ .$$

It suffices to show that the induced bornologies on $[i-1,i]\times D_{\kappa(i)}$ coincide. For this we have to show that for every $F\in\mathcal{F}(P_U(X))$ the bornologies on D_F induced from $P_U(X)_{bd}$ and $P_U(X)_{d,max}$, respectively, agree.

Since X is Γ-finite and the Γ-action on X is proper, X carries the minimal bornology. Consequently, every bounded subset of $P_U(X)_{bd}$ is contained in a finite subcomplex. Hence, the bornology on D_F induced by $P_U(X)_{bd}$ is generated by the sets $D_F\cap(F'\times\Gamma)$ for all F' in $\mathcal{F}(P_U(X))$. This set is equal to

$$\Big(\bigcup_{\gamma\in\Gamma}\gamma F\times\{\gamma\}\Big)\cap(F'\times\Gamma) = \bigcup_{\{\gamma\in\Gamma\mid\gamma F\cap F'\neq\emptyset\}}(\gamma F\cap F')\times\{\gamma\} \ .$$

Note that the index set of the union on the right hand side is finite since the Γ-action is proper.

The bornology induced by $P_U(X)_{d,max}$ is generated by the sets $D_F \cap (P_U(X) \times B)$ for all finite subsets B of Γ. This set can be written in the form

$$\bigcup_{\gamma \in B} \gamma F \times \{\gamma\} .$$

The families of subsets

$$\Big(\bigcup_{\{\gamma\in\Gamma \mid \gamma F \cap F' \neq \emptyset\}} (\gamma F \cap F') \times \{\gamma\} \Big)_{F' \in \mathcal{F}(P_U(X))} \quad \text{and} \quad \Big(\bigcup_{\gamma\in B} \gamma F \times \{\gamma\} \Big)_{B \subseteq \Gamma, |B| < \infty}$$

generate the same bornologies. This finishes the proof of the lemma. □

Let X be a Γ-bornological coarse space.

LEMMA 11.21. *Assume:*

(1) *X has strongly bounded geometry.*
(2) *X is Γ-finite.*

Then the morphisms $C^s(\beta_{X,\Gamma_{can,max}})$ and $C^s(\beta_{X,\Gamma_{max,max}})$ are equivalent.

PROOF. The morphism of Γ-bornological coarse spaces $\Gamma_{can,max} \to \Gamma_{max,max}$ induces the commutative diagram

$$\begin{array}{ccc} C^s(\mathcal{O}^\infty(P_U(X)_{bdu}) \otimes \mathrm{Yo}^s(\Gamma_{can,max})) & \longrightarrow & \Sigma\mathrm{Yo}^s_c(P_U(X)_{bd} \otimes \Gamma_{can,max}) \\ \big\downarrow & & \big\downarrow \\ C^s(\mathcal{O}^\infty(P_U(X)_{bdu}) \otimes \mathrm{Yo}^s(\Gamma_{max,max})) & \longrightarrow & \Sigma\mathrm{Yo}^s_c(P_U(X)_{bd} \otimes \Gamma_{max,max}) \end{array} \tag{11.5}$$

The functor $\mathcal{O}^\infty$ from $\Gamma\mathbf{UBC}$ to $\mathbf{Sp}\mathcal{X}$ is homotopy invariant and excisive for equivariant uniform decompositions. Since X is Γ-finite and has strongly bounded geometry, for every invariant entourage U the complex $P_U(X)$ is a Γ-finite simplicial complex. Using excision and homotopy invariance we conclude that the left vertical map in the above diagram is an equivalence if

$$C^s(\mathcal{O}^\infty(S) \otimes \mathrm{Yo}^s(\Gamma_{can,max})) \to C^s(\mathcal{O}^\infty(S) \otimes \mathrm{Yo}^s(\Gamma_{max,max}))$$

is an equivalence for every Γ-uniform bornological coarse space S which is a transitive Γ-set that has the minimal bornology and the discrete uniform structure. Note that in this case

$$\mathcal{O}^\infty(S) \simeq \mathcal{O}^\infty(S_{disc,min,min}) \simeq \Sigma\mathrm{Yo}^s(S_{min,min}) ,$$

since $S_{disc,min,min} \to S$ is a coarsening and $\mathcal{O}(S_{disc,min,min})$ is flasque.

If we can show that for every Γ-bounded Γ-bornological coarse space X the map

$$\mathrm{Yo}^s_c(X \otimes \Gamma_{can,max}) \to \mathrm{Yo}^s_c(X \otimes \Gamma_{max,max})$$

is an equivalence, then we can conclude that the right vertical map in (11.5) is an equivalence, and by the above argument the left vertical map is an equivalence, too. The lemma then follows by taking the colimit over all invariant entourages U of X.

By continuity, it suffices to show that if F is a locally finite, invariant subset of $X \times \Gamma_{?,max}$ (the coarse structure on Γ does not matter since local finiteness is a bornological concept), then the bornological coarse structures on F induced by $X \times \Gamma_{can,max}$ and $X \times \Gamma_{max,max}$, respectively, agree. Since the bornologies are the same, we only have to care about the coarse structures.

Every entourage of $F_{\Gamma_{can,max}}$ is an entourage of $F_{\Gamma_{max,max}}$. So it remains to show the other inclusion. We choose a bounded subset A of X such that $\Gamma A = X$. Let U be an invariant entourage of X containing the diagonal. Then $U[A]$ is bounded. Furthermore, we have $U \subseteq \Gamma(A \times U[A])$. The set $W' := F \cap (U[A] \times \Gamma)$ is finite since F is locally finite and $U[A] \times \Gamma$ is bounded. We let W denote the projection of W' to Γ. Then we have

$$(U \times \Gamma \times \Gamma) \cap (F \times F) \subseteq (U \times \Gamma(W \times W)) \cap (F \times F) .$$

Now note that $\Gamma(W \times W)$ is an entourage of $\Gamma_{can,max}$. This shows that every entourage of $F_{\Gamma_{max,max}}$ is an entourage of $F_{\Gamma_{can,max}}$. □

Theorem 11.16 follows from combining Lemma 11.18 (with $L = \Gamma_{can,min}$), Lemma 11.19, Lemma 11.20 and Lemma 11.21.

11.3. Homological properties of pull-backs by the cone. Recall Definition 11.9 of the three functors F, F^0, and F^∞. In this section we analyze the homological properties of the functor F^∞. It turns out that this functor is almost a coarse homology theory. The only problematic axiom is vanishing on flasques. In order to improve on this point recall the definition of $\Gamma\mathbf{Sp}\mathcal{X}_{\mathrm{wfl}}$ from Definition 4.20 and consider the composition

$$F^\infty_{\mathrm{wfl}} \colon \Gamma\mathbf{BC} \xrightarrow{F^\infty} \Gamma\mathbf{Sp}\mathcal{X} \to \Gamma\mathbf{Sp}\mathcal{X}_{\mathrm{wfl}} .$$

In a similar manner, we derive functors F^0_{wfl} and F_{wfl} from F^0 and F, respectively. For every Γ-bornological coarse space X we have a fiber sequence in $\Gamma\mathbf{Sp}\mathcal{X}_{\mathrm{wfl}}$

$$F^0_{\mathrm{wfl}}(X) \to F_{\mathrm{wfl}}(X) \to F^\infty_{\mathrm{wfl}}(X) \xrightarrow{\beta_{X,\mathrm{wfl}}} \Sigma F^0_{\mathrm{wfl}}(X) . \tag{11.6}$$

The morphism $\beta_{X,\mathrm{wfl}}$ is a version of the forget-control morphism from Definition 11.10.

PROPOSITION 11.22. *The functor F^∞_{wfl} is an equivariant $\Gamma\mathbf{Sp}\mathcal{X}_{\mathrm{wfl}}$-valued coarse homology theory.*

PROOF. We verify the axioms.

(Coarse invariance) We consider a Γ-bornological coarse space X. For i in $\{0,1\}$ let

$$\iota_i \colon X \to \{0,1\}_{max,max} \otimes X$$

denote the corresponding inclusions. It suffices to show that $F^\infty_{\mathrm{wfl}}(\iota_0)$ and $F^\infty_{\mathrm{wfl}}(\iota_1)$ are equivalent. For every invariant entourage U of X we consider the invariant entourage $\widetilde{U} := \{0,1\}^2 \times U$ of $\{0,1\}_{max,max} \otimes X$. Then the map

$$[0,1]_{du} \otimes P_U(X)_{bdu} \to P_{\widetilde{U}}(\{0,1\}_{max,max} \otimes X)_{bdu}$$

given by

$$(t,\mu) \mapsto (1-t)\iota_{0,*}\mu + t\iota_{1,*}\mu$$

is a homotopy between the morphisms of Γ-uniform bornological coarse spaces

$$P_U(X)_{bdu} \to P_{\widetilde{U}}(\{0,1\}_{max,max} \otimes X)_{bdu}$$

induced by ι_0 and ι_1.

By the homotopy invariance of the functor $\mathcal{O}^\infty$ (Corollary 9.38) we conclude that the morphisms

$$\mathcal{O}^\infty(P_U(X)_{bdu}) \to \mathcal{O}^\infty(P_{\widetilde{U}}(\{0,1\}_{max,max} \otimes X)_{bdu})$$

induced by ι_0 and ι_1 are equivalent. Since the entourages of the form $\widetilde{U}$ for all U in $\mathcal{C}^\Gamma$ are cofinal in the entourages of $\{0,1\} \otimes X$, we get the equivalence of $F^\infty_{\mathrm{wfl}}(\iota_0)$ and $F^\infty_{\mathrm{wfl}}(\iota_1)$ as desired.

(Excision) Let X be a Γ-bornological coarse space and Z an invariant subset. For an invariant entourage U of X the subset $P_U(Z)$ of $P_U(X)_{bdu}$ is invariant and closed.

Let $(\mathcal{Y}, Z)$ be an equivariant complementary pair on X with $\mathcal{Y} = (Y_i)_{i \in I}$. Let i_0 in I be such that $Y_{i_0} \cup Z = X$. Let i_1 in I be such that $U[Y_{i_0}] \subseteq Y_{i_1}$. Then for every i in I with $i \geq \max\{i_0, i_1\}$ we have $P_U(Y_i) \cup P_U(Z) = P_U(X)$. The pair of invariant subsets $(P_U(Y_i), P_U(Z))$ is then an equivariant uniform decomposition of $P_U(X)_{bdu}$. By Corollary 9.36 and Remark 9.37 the functor $\mathcal{O}^\infty$ sends equivariant uniform decompositions to push-outs. We conclude that for i in I with $i \geq \max\{i_0, i_1\}$ we have a push-out

$$\begin{array}{ccc} \mathcal{O}^\infty(P_U(Z \cap Y_i)_{bdu}) & \longrightarrow & \mathcal{O}^\infty(P_U(Y_i)_{bdu}) \\ \downarrow & & \downarrow \\ \mathcal{O}^\infty(P_U(Z)_{bdu}) & \longrightarrow & \mathcal{O}^\infty(P_U(X)_{bdu}) \end{array}$$

Since colimits of push-out squares are push-out squares, we now take the colimits over the invariant entourages U in $\mathcal{C}^\Gamma$ and over i in I to get the push-out square

$$\begin{array}{ccc} F^\infty(Z \cap \mathcal{Y}) & \longrightarrow & F^\infty(\mathcal{Y}) \\ \downarrow & & \downarrow \\ F^\infty(Z) & \longrightarrow & F^\infty(X) \end{array}$$

We get the desired push-out

$$\begin{array}{ccc} F^\infty_{\mathrm{wfl}}(Z \cap \mathcal{Y}) & \longrightarrow & F^\infty_{\mathrm{wfl}}(\mathcal{Y}) \\ \downarrow & & \downarrow \\ F^\infty_{\mathrm{wfl}}(Z) & \longrightarrow & F^\infty_{\mathrm{wfl}}(X) \end{array}$$

(Flasqueness) We assume that X is flasque with the flasqueness implemented by the equivariant map $f\colon X \to X$. For an invariant entourage U of X with the property $(\mathrm{id}, f)(\mathrm{diag}_X) \subseteq U$ we form the entourage of X

$$\widetilde{U} := \bigcup_{n \in \mathbb{N}} (f^n \times f^n)(U) .$$

Note that $(f \times f)(\widetilde{U}) \subseteq \widetilde{U}$. Therefore we have a morphism

$$P_{\widetilde{U}}(f)\colon P_{\widetilde{U}}(X)_{bdu} \to P_{\widetilde{U}}(X)_{bdu} .$$

Like every simplicial map it is distance decreasing. Moreover, for every $\mu \in P_{\widetilde{U}}(X)$ we have

$$d(\mu, P_{\widetilde{U}}(f)(\mu)) \leq 2 .$$

Finally, if B is a bounded subset of X and n is an integer such that $\widetilde{U}[B] \cap f^n(X) = \emptyset$, then

$$P_{\widetilde{U}}(f^n)(P_{\widetilde{U}}(X)) \cap P_{\widetilde{U}}(B) = \emptyset .$$

We conclude that $P_{\widetilde{U}}(f)$ implements flasqueness of the bornological coarse space $P_{\widetilde{U}}(X)_{bd}$. The set of invariant entourages of the form $\widetilde{U}$ as above is cofinal in all invariant entourages of X. Therefore, we get $F^0(X) \simeq 0$ and hence $F^0_{\mathrm{wfl}}(X) \simeq 0$ by taking the colimit over these entourages.

We now claim that $\mathcal{O}(P_{\widetilde{U}}(f))$ implements weak flasqueness of $\mathcal{O}(P_{\widetilde{U}}(X)_{bdu})$. In the following we verify the conditions stated in Definition 4.18.

Since f is U-close to id_X, as in the proof of coarse invariance we can conclude that the map $P_{\widetilde{U}}(f)_{bdu}$ is uniformly homotopic to $\mathrm{id}_{P_{\widetilde{U}}(X)_{bdu}}$. By the homotopy invariance of $\mathcal{O}$ we conclude that

$$\mathrm{Yo}^s(\mathcal{O}(P_{\widetilde{U}}(f))) \simeq \mathrm{id}_{\mathrm{Yo}^s(\mathcal{O}(P_{\widetilde{U}}(f)_{bdu}))}$$

as required in Definition 4.18.1.

In order to save notation we define the map

$$Q\colon \mathcal{P}([0,\infty) \times P_{\widetilde{U}}(X) \times [0,\infty) \times P_{\widetilde{U}}(X)) \to \mathcal{P}([0,\infty) \times P_{\widetilde{U}}(X) \times [0,\infty) \times P_{\widetilde{U}}(X))$$

by

$$Q(V) := \bigcup_{n \in \mathbb{N}} (([0,\infty) \times P_{\widetilde{U}}(f))^n \times ([0,\infty) \times P_{\widetilde{U}}(f))^n)(V)\ .$$

Let now V be an entourage of $\mathcal{O}(P_{\widetilde{U}}(X)_{bdu})$. We must show that $Q(V)$ is again an entourage of $\mathcal{O}(P_{\widetilde{U}}(X)_{bdu})$. After enlarging V we can assume that it is of the form $V = U_\psi \cap W_r$ as in the proof of Lemma 9.27, where the function ϕ (which is the first component of ψ) is such that $\phi(i)$ is a uniform entourage of the form $U_{r(i)}$ for every i in $\mathbb{N}$, see (9.1). Since $P_{\widetilde{U}}(f)$ is distance decreasing we see that $Q(W_r) \subseteq W_r$. Since $P_{\widetilde{U}}(f)$ preserves the first coordinate of the cone and is distance decreasing we also see that $Q(U_\psi) \subseteq U_\psi$. Hence we actually get $Q(V) \subseteq V$.

Finally, for every bounded subset A of $\mathcal{O}(P_{\widetilde{U}}(X)_{bdu})$ there exists r in $(0,\infty)$ and a bounded subset B of X such that $A \subseteq [0,r] \times P_{\widetilde{U}}(B)$. We can choose an integer n such that $f^n(X) \cap B = \emptyset$. Then $\mathcal{O}(P_{\widetilde{U}}(f))^n(\mathcal{O}(P_{\widetilde{U}}(X)_{bdu})) \cap A = \emptyset$.

We conclude that

$$\mathrm{Yo}^s_{\mathrm{wfl}}(\mathcal{O}(P_{\widetilde{U}}(X)_{bdu})) \simeq 0\ .$$

Taking the colimit over the invariant entourages U and again using the cofinality of the resulting family of entourages $\widetilde{U}$ we get $F_{\mathrm{wfl}}(X) \simeq 0$.

From the fiber sequence (11.6) we now conclude that

$$F^\infty_{\mathrm{wfl}}(X) \simeq 0\ .$$

(u-continuity) This is just a cofinality check:

$$\operatorname*{colim}_{U \in \mathcal{C}^\Gamma} F^\infty_{\mathrm{wfl}}(X_U) \simeq \operatorname*{colim}_{U \in \mathcal{C}^\Gamma} \operatorname*{colim}_{V \in \mathcal{C}\langle U\rangle^\Gamma} \mathcal{O}^\infty_{\mathrm{wfl}}(P_V(X)_{bdu}) \simeq \operatorname*{colim}_{V \in \mathcal{C}^\Gamma} \mathcal{O}^\infty_{\mathrm{wfl}}(P_V(X)_{bdu}) \simeq F^\infty_{\mathrm{wfl}}(X)\ .$$

This finishes the proof of Proposition 11.22. □

REMARK 11.23. Let X be flasque. In the above proof we have shown that $F_{\mathrm{wfl}}(X) \simeq 0$. Note that we do not expect that $F(X) \simeq 0$.

Let E be a strong Γ-equivariant $\mathbf{C}$-valued coarse homology theory. Then we have an essentially unique factorization $E_{\mathrm{wfl}}\colon \Gamma\mathbf{Sp}\mathcal{X}_{\mathrm{wfl}} \to \mathbf{C}$. The composition

$$E_{\mathrm{wfl}} \circ F^\infty_{\mathrm{wfl}}\colon \Gamma\mathbf{BC} \to \mathbf{C}$$

is then a Γ-equivariant $\mathbf{C}$-valued coarse homology theory. We have an equivalence

$$E \circ F^\infty \simeq E_{\mathrm{wfl}} \circ F^\infty_{\mathrm{wfl}}\ .$$

COROLLARY 11.24. *If E is a strong equivariant $\mathbf{C}$-valued coarse homology theory, then*

$$E \circ F^{\infty} \colon \Gamma\mathbf{BC} \to \mathbf{C}$$

is a Γ-equivariant coarse homology theory.

Let E be an equivariant coarse homology theory and Q be a Γ-bornological coarse space.

LEMMA 11.25. *If E is strong, then E_Q is also strong.*

PROOF. If X is a weakly flasque Γ-bornological coarse space with weak flasqueness implemented by $f \colon X \to X$, then $f \otimes \mathrm{id}_Q$ implements weak flasqueness of $X \otimes Q$. This implies the lemma. □

If the underlying Γ-set of Q is free, then by Corollary 11.14 we have

$$E_Q(F^0(X)) \simeq E(F^0(X) \otimes \mathrm{Yo}^s(Q)) \simeq E(X \otimes Q) \simeq E_Q(X)\ .$$

In particular, the functor

$$E_Q \circ F^0 \colon \Gamma\mathbf{BC} \to \mathbf{C}$$

is an equivariant coarse homology theory.

Let Q be a Γ-bornological coarse space and E be an equivariant coarse homology theory.

COROLLARY 11.26. *If E is strong and the underlying Γ-set of Q is free, then the forget-control map*

$$\beta \colon E_Q \circ F^{\infty} \to \Sigma E_Q \circ F^0$$

is a transformation between equivariant coarse homology theories.

This aspect of the theory (in the case of a trivial group Γ) is further studied in [**9**].

Acknowledgements

Parts of the present work were obtained during the Junior Hausdorff Trimester Program "Topology" at the Hausdorff Research Institute for Mathematics (HIM) in Bonn.

We would also like to thank Mark Ullmann for helpful discussions.

References

[1] Arthur Bartels, Tom Farrell, Lowell Jones, and Holger Reich, *On the isomorphism conjecture in algebraic K-theory*, Topology **43** (2004), no. 1, 157–213, DOI 10.1016/S0040-9383(03)00032-6. MR2030590

[2] A. Bartels and W. Lück, *The Farrell-Hsiang method revisited*, Math. Ann. **354** (2012), no. 1, 209–226, DOI 10.1007/s00208-011-0727-3. MR2957625

[3] Arthur Bartels, Wolfgang Lück, and Holger Reich, *The K-theoretic Farrell-Jones conjecture for hyperbolic groups*, Invent. Math. **172** (2008), no. 1, 29–70, DOI 10.1007/s00222-007-0093-7. MR2385666

[4] Arthur Bartels and Holger Reich, *Coefficients for the Farrell-Jones conjecture*, Adv. Math. **209** (2007), no. 1, 337–362, DOI 10.1016/j.aim.2006.05.005. MR2294225

[5] Arthur Bartels and David Rosenthal, *On the K-theory of groups with finite asymptotic dimension*, J. Reine Angew. Math. **612** (2007), 35–57, DOI 10.1515/CRELLE.2007.083. MR2364073

[6] Clark Barwick, *Spectral Mackey functors and equivariant algebraic K-theory (I)*, Adv. Math. **304** (2017), 646–727, DOI 10.1016/j.aim.2016.08.043. MR3558219

[7] Clark Barwick, *Spectral Mackey functors and equivariant algebraic K-theory (I)*, Adv. Math. **304** (2017), 646–727, DOI 10.1016/j.aim.2016.08.043. MR3558219

[8] U. Bunke and A. Engel, *Homotopy theory with bornological coarse spaces*, arXiv:1607.03657v3, 2016.

[9] ———, *Coarse assembly maps*, arXiv:1706.02164v2, 2017.

[10] U. Bunke, A. Engel, D. Kasprowski, and Ch. Winges, *Injectivity results for coarse homology theories*, arXiv:1809.11079.

[11] Ulrich Bunke, Alexander Engel, Daniel Kasprowski, and Christoph Winges, *Coarse homology theories and finite decomposition complexity*, Algebr. Geom. Topol. **19** (2019), no. 6, 3033–3074, DOI 10.2140/agt.2019.19.3033. MR4023335

[12] U. Bunke, D. Kasprowski, and Ch. Winges, *Split injectivity of A-theoretic assembly maps*, arXiv:1811.11864.

[13] M. Cárdenas and E. K. Pedersen, *On the Karoubi filtration of a category*, K-Theory **12** (1997), no. 2, 165–191, DOI 10.1023/A:1007726201728. MR1469141

[14] Gunnar Carlsson, *On the algebraic K-theory of infinite product categories*, K-Theory **9** (1995), no. 4, 305–322, DOI 10.1007/BF00961467. MR1351941

[15] Gunnar Carlsson and Erik Kjær Pedersen, *Controlled algebra and the Novikov conjectures for K- and L-theory*, Topology **34** (1995), no. 3, 731–758, DOI 10.1016/0040-9383(94)00033-H. MR1341817

[16] James F. Davis and Wolfgang Lück, *Spaces over a category and assembly maps in isomorphism conjectures in K- and L-theory*, K-Theory **15** (1998), no. 3, 201–252, DOI 10.1023/A:1007784106877. MR1659969

[17] C. H. Dowker, *Topology of metric complexes*, Amer. J. Math. **74** (1952), 555–577, DOI 10.2307/2372262. MR48020

[18] Samuel Eilenberg and Norman Steenrod, *Foundations of algebraic topology*, Princeton University Press, Princeton, New Jersey, 1952. MR0050886

[19] Alexander Engel, *Wrong way maps in uniformly finite homology and homology of groups*, J. Homotopy Relat. Struct. **13** (2018), no. 2, 423–441, DOI 10.1007/s40062-017-0187-x. MR3802801

[20] David Gepner, Moritz Groth, and Thomas Nikolaus, *Universality of multiplicative infinite loop space machines*, Algebr. Geom. Topol. **15** (2015), no. 6, 3107–3153, DOI 10.2140/agt.2015.15.3107. MR3450758

[21] J. R. J. Groves and John S. Wilson, *Soluble groups with a finiteness condition arising from Bredon cohomology*, Bull. Lond. Math. Soc. **45** (2013), no. 1, 89–92, DOI 10.1112/blms/bds065. MR3033956

[22] Daniel Juan-Pineda and Ian J. Leary, *On classifying spaces for the family of virtually cyclic subgroups*, Recent developments in algebraic topology, Contemp. Math., vol. 407, Amer. Math. Soc., Providence, RI, 2006, pp. 135–145, DOI 10.1090/conm/407/07674. MR2248975

[23] Daniel Kasprowski, *On the K-theory of groups with finite decomposition complexity*, Proc. Lond. Math. Soc. (3) **110** (2015), no. 3, 565–592, DOI 10.1112/plms/pdu062. MR3342098

[24] D. Kasprowski and Ch. Winges, *Shortening binary complexes and commutativity of K-theory with infinite products*, arXiv:1705.09116, 2017.

[25] D. H. Kochloukova, C. Martínez-Pérez, and B. E. A. Nucinkis, *Cohomological finiteness conditions in Bredon cohomology*, Bull. Lond. Math. Soc. **43** (2011), no. 1, 124–136, DOI 10.1112/blms/bdq088. MR2765556

[26] Jacob Lurie, *Higher topos theory*, Annals of Mathematics Studies, vol. 170, Princeton University Press, Princeton, NJ, 2009. MR2522659

[27] ———, *Higher algebra*, Available at www.math.harvard.edu/lurie, 2014.

[28] Saunders Mac Lane, *Categories for the working mathematician*, 2nd ed., Graduate Texts in Mathematics, vol. 5, Springer-Verlag, New York, 1998. MR1712872

[29] J. P. May, *Equivariant homotopy and cohomology theory*, CBMS Regional Conference Series in Mathematics, vol. 91, Published for the Conference Board of the Mathematical Sciences, Washington, DC; by the American Mathematical Society, Providence, RI, 1996. With contributions by M. Cole, G. Comezaña, S. Costenoble, A. D. Elmendorf, J. P. C. Greenlees, L. G. Lewis, Jr., R. J. Piacenza, G. Triantafillou, and S. Waner. MR1413302

[30] John Milnor, *On spaces having the homotopy type of a* CW*-complex*, Trans. Amer. Math. Soc. **90** (1959), 272–280, DOI 10.2307/1993204. MR100267

[31] Paul D. Mitchener, *Coarse homology theories*, Algebr. Geom. Topol. **1** (2001), 271–297, DOI 10.2140/agt.2001.1.271. MR1834777

[32] Paul D. Mitchener, *The general notion of descent in coarse geometry*, Algebr. Geom. Topol. **10** (2010), no. 4, 2419–2450, DOI 10.2140/agt.2010.10.2419. MR2748337

[33] Daniel Quillen, *Higher algebraic K-theory. I*, Algebraic K-theory, I: Higher K-theories (Proc. Conf., Battelle Memorial Inst., Seattle, Wash., 1972), Springer, Berlin, 1973, pp. 85–147. Lecture Notes in Math., Vol. 341. MR0338129

[34] Daniel A. Ramras, Romain Tessera, and Guoliang Yu, *Finite decomposition complexity and the integral Novikov conjecture for higher algebraic K-theory*, J. Reine Angew. Math. **694** (2014), 129–178, DOI 10.1515/crelle-2012-0112. MR3259041

[35] David Rosenthal, *Splitting with continuous control in algebraic K-theory*, K-Theory **32** (2004), no. 2, 139–166, DOI 10.1023/B:KTHE.0000037563.35102.0d. MR2083578

[36] Marco Schlichting, *Delooping the K-theory of exact categories*, Topology **43** (2004), no. 5, 1089–1103, DOI 10.1016/j.top.2004.01.005. MR2079996

[37] Marco Schlichting, *Negative K-theory of derived categories*, Math. Z. **253** (2006), no. 1, 97–134, DOI 10.1007/s00209-005-0889-3. MR2206639

[38] Timm von Puttkamer and Xiaolei Wu, *On the finiteness of the classifying space for the family of virtually cyclic subgroups*, Groups Geom. Dyn. **13** (2019), no. 2, 707–729, DOI 10.4171/GGD/502. MR3950648

[39] Timm von Puttkamer and Xiaolei Wu, *Linear groups, conjugacy growth, and classifying spaces for families of subgroups*, Int. Math. Res. Not. IMRN **10** (2019), 3130–3168, DOI 10.1093/imrn/rnx215. MR3952560

[40] Friedhelm Waldhausen, *Algebraic K-theory of spaces*, Algebraic and geometric topology (New Brunswick, N.J., 1983), Lecture Notes in Math., vol. 1126, Springer, Berlin, 1985, pp. 318–419, DOI 10.1007/BFb0074449. MR802796

[41] Nicholas John Wright, *C(0) coarse geometry*, ProQuest LLC, Ann Arbor, MI, 2002. Thesis (Ph.D.)–The Pennsylvania State University. MR2703848

[42] Guo Liang Yu, *Baum-Connes conjecture and coarse geometry*, K-Theory **9** (1995), no. 3, 223–231, DOI 10.1007/BF00961665. MR1344139

Fakultät für Mathematik, Universität Regensburg, 93040 Regensburg, Germany
Email address: `ulrich.bunke@mathematik.uni-regensburg.de`

Fakultät für Mathematik, Universität Regensburg, 93040 Regensburg, Germany
Email address: `alexander.engel@mathematik.uni-regensburg.de`

Rheinische Friedrich-Wilhelms-Universität Bonn, Mathematisches Institut, Endenicher Allee 60, 53115 Bonn, Germany
Email address: `kasprowski@uni-bonn.de`

Rheinische Friedrich-Wilhelms-Universität Bonn, Mathematisches Institut, Endenicher Allee 60, 53115 Bonn, Germany
Email address: `winges@math.uni-bonn.de`

Contemporary Mathematics
Volume **749**, 2020
https://doi.org/10.1090/conm/749/15069

Injectivity, crossed products, and amenable group actions

Alcides Buss, Siegfried Echterhoff, and Rufus Willett

Abstract. This paper is motivated primarily by the question of when the maximal and reduced crossed products of a G-C^*-algebra agree (particularly inspired by results of Matsumura and Suzuki), and the relationships with various notions of amenability and injectivity. We give new connections between these notions. Key tools in this include the natural equivariant analogues of injectivity, and of Lance's weak expectation property: we also give complete characterizations of these equivariant properties, and some connections with injective envelopes in the sense of Hamana.

Contents

1. Introduction

This paper studies the relationship between various notions of amenability of actions of a group G on a C^*-algebras A and the 'weak containment' question of whether $A \rtimes_{\max} G = A \rtimes_r G$. We were motivated by trying to elucidate relationships between the following works:

- recent interesting examples of Suzuki [**25**] showing a disconnect between notions of amenable actions and the equality $A \rtimes_{\max} G = A \rtimes_r G$;

2010 *Mathematics Subject Classification.* Primary 46L55, 43A35.

The authors were supported by Deutsche Forschungsgemeinschaft (DFG, German Research Foundation) under Germany 's Excellence Strategy – EXC 2044 – 390685587, Mathematics Münster: Dynamics – Geometry – Structure and SFB 878, Groups, Geometry & Actions, by CNPq/CAPES/Humboldt - Brazil, and by the US NSF (DMS 1401126 and DMS 1564281).

- Matsumura's work [**20**] relating the weak containment property $A \rtimes_{\max} G = A \rtimes_r G$ to amenability (at least in the presence of exactness of the acting group);
- the extensive work of Anantharaman-Delaroche on amenable actions, most relevantly for this paper in [**4**] and [**5**];
- the seminal theorem of Guentner-Kaminker [**12**] and Ozawa [**21**] relating exactness to existence of amenable actions;
- our study of the so-called maximal injective crossed product [**9**] and the connections to equivariant versions of injectivity;
- Hamana's classical study of equivariant injective envelopes [**15**] and the recent important exploitation of these ideas by Kalantar and Kennedy in their work on the Furstenberg boundary [**16**].

Our goal was to try to bridge connections between some of this in a way that we hope systematises some of the existing literature a little better, as well as solving some open problems. Our results are perhaps most easily explained by discussing the contents of the paper, which we now do.

In Section 2, we recall two notions of amenable action on a C^*-algebra due to Claire Anantharaman-Delaroche, and recall the theorems of Guentner-Kaminker and Ozawa on exactness. We also make a simple observation based on this (Proposition 2.6) that will be used over and over again in one form or another throughout the paper: roughly, this says that given an exact group G, for A to be amenable, it is sufficient that there exists an equivariant ucp map from $\ell^\infty(G)$ to the centre of the multiplier algebra $\mathcal{M}(A)$ (or the center of the double dual A^{**}). In Section 3, we briefly recall Suzuki's examples. We then show that while they do not have Anantharaman-Delaroche's strong amenability property, they do satisfy a version of Exel's approximation property and are therefore amenable in the sense of Anantharaman-Delaroche. The key idea here is to drop a precise centrality condition in favour of some form of 'quasi-centrality'.

Motivated by Suzuki's examples, in Section 4 we try to find reasonable conditions on A that characterize when $A \rtimes_{\max} G = A \rtimes_r G$ in general. We are able to do this (at least in the presence of exactness) in terms of injectivity-type conditions on representations (Corollary 4.11), and in terms of a weak amenability-like condition that we call commutant amenability (Theorem 4.17). We should point out that while these results seem theoretically useful, the conditions involved seem difficult to check without knowing something a priori stronger. In Section 5, we continue our study of weak containment. Following ideas of Matsumura, we apply the Haagerup standard form of the double dual to get more precise results on weak containment. The most satisfactory of these are in the setting of actions of exact groups on commutative C^*-algebras (Theorem 5.2), but we also give partial results for noncommutative algebras, and non-exact groups.

In Section 6, we go back to the relationship with exactness. Thanks to the above-mentioned work of Guentner-Kaminker and Ozawa, it is known that a group G is exact if and only if it admits an amenable action on a compact space. It is natural to ask whether the analogous result holds for amenable actions on unital, possibly noncommutative C^*-algebras, i.e. is it true that G is exact if and only if it admits an amenable action on a unital C^*-algebra? The answer is (clearly) 'yes' if 'amenable' in this statement means what Anantharaman-Delaroche calls strong amenability, but this is less clear in general. We show in fact (see Theorem 6.1) the answer is 'yes' in the strong sense that a group G is exact if and only if it admits a commutant amenable action on a unital C^*-algebra; commutant amenability is the weakest reasonable notion of amenability that we know of.

Section 7 studies equivariant versions of injectivity, and of Lance's weak expectation property; these are used throughout the paper, but here we study them in their own right. We give complete characterizations of when a (unital) G-algebra has these properties in terms of amenability and of the underlying non-equivariant versions (Theorems 7.2 and 7.3); again, exactness turns out to play a fundamental role. Finally, in Section 8 we show that our notion of injectivity is the same as that introduced by Hamana, and discuss relationships with his injective envelopes. We give applications of this material to a conjecture of Ozawa on nuclear subalgebras of injective envelopes (Corollary 8.4), and to the existence of amenable actions on injective envelopes (Theorem 8.3). In both cases, our results generalize work of Kalantar and Kennedy.

The initial motivation for this paper grew out of the relation of injectivity and the weak expectation property with the maximal injective crossed product functor as studied in [**9**]. It is interesting to note that this functor is, in a sense, 'dual' to the minimal exact crossed product which was studied by the authors in [**10**] and has a close relation to the Baum-Connes conjecture. The present paper shows that the maximal injective crossed product nicely relates to amenability.

2. Notation, basic definitions and preliminaries

We use the following notation. The abbreviations ucp and ccp stand for 'unital completely positive' and 'contractive completely positive' respectively. Throughout the paper, G always refers to a *discrete* group. A *G-algebra* will always mean a C^*-algebra A equipped with an action of G by $*$-automorphisms. A *G-space* will always mean a locally compact Hausdorff space X equipped with an action of G by homeomorphisms. Note that if X is a G-space, then $A = C_0(X)$ is a G-algebra and vice versa.

Given a G-algebra A, we denote the multiplier algebra, double dual, opposite algebra, and center by $\mathcal{M}(A)$, A^{**}, A^{op}, and $Z(A)$ respectively. These are always equipped with the canonically induced G-actions, which we will generally denote using the same letter as for the action on A unless this might cause confusion. Given G-algebras A and B, the spatial and maximal tensor products are denoted by $A \otimes B$ and $A \otimes_{\max} B$ respectively, and are equipped with the associated diagonal G-actions.

A map $\phi : A \to B$ between sets with G-actions α and β is *equivariant* if $\phi(\alpha_g(a)) = \beta_g(\phi(a))$ for all $a \in A$ and $g \in G$. We will also call equivariant maps *G-maps* and allow other similar modifiers as appropriate (for example 'ucp G-map', 'G-embedding', ...). Relatedly, a G-subalgebra A of B will be a C*-subalgebra that is invariant under the G-action (and is therefore a G-algebra in its own right).

The C^*-algebra $\ell^\infty(G)$ will play a special role in this paper. It is always considered as a G-algebra via the (left) translation action defined by

$$(\gamma_g f)(h) := f(g^{-1}h).$$

The following definitions are (to the best of our knowledge) due to Claire Anantharaman-Delaroche: the definition of positive type functions is from [**4**, Definition 2.1] and that of amenability of an action is from [**4**, Definition 4.1]. If A is commutative or unital, then our notion of strong amenability as defined below is equivalent to a notion of strong amenability due to Ananthraman-Delaroche [**5**, Definition 6.1] (see Lemma 2.4 below). In general our notion of strong amenability could possibly be weaker than the one introduced by Anantharaman-Delaroche.

DEFINITION 2.1. Let A be a G-algebra with associated action α.

A function $\theta : G \to A$ is *positive type* if for any finite subset $\{g_1, ..., g_n\}$ of G, the matrix

$$\left(\alpha_{g_i}(\theta(g_i^{-1}g_j))\right)_{i,j} \in M_n(A)$$

is positive.

We say that A is *amenable* if there exists a net $(\theta_i : G \to Z(A^{**}))_{i\in I}$ of positive type functions such that:

(i) each θ_i is finitely supported;
(ii) for each i, $\theta_i(e) \leqslant 1$;
(iii) for each $g \in G$, $\theta_i(g) \to 1$ ultraweakly as $i \to \infty$.

And we call A *strongly amenable* if there exists a net $(\theta_i : G \to Z\mathcal{M}(A))_{i\in I}$ of positive type functions such that:

(i) each θ_i is finitely supported;
(ii) for each i, $\theta_i(e) \leqslant 1$;
(iii) for each $g \in G$, $\theta_i(g) \to 1$ strictly as $i \to \infty$.

REMARK 2.2. The terminology for amenable actions is unfortunately not completely consistent in the literature. The notion of strong amenability also appears as [**7**, Definition 4.3.1] in the text of Brown and Ozawa (although only in the special case of unital G-algebras). There it is just called *amenability*. The notion of amenability as defined above is equivalent to what is called *weak amenability* in [**5**, Definition 6.1] (see [**5**, Proposition 6.4]) for nuclear G-algebras. However, it is not clear if this is true in the non-nuclear case, so we will keep to the terminology 'amenability'. Observe that strong amenability always implies amenability, since the canonical inclusion $\mathcal{M}(A) \hookrightarrow A^{**}$ is strict to ultraweak continuous.

For commutative G-algebras $A = C_0(X)$, the notions of amenability and strong amenability are the same, and both are equivalent to amenability of the G-space X by [**4**, Théorème 4.9 and Remarque 4.10]. However, we will see in Section 3 below that they are *not* the same in general, even on the class of nuclear, unital G-algebras.

The following lemma will get used many times in the paper.

LEMMA 2.3. *Let A and B be G-algebras and suppose there exists a strictly continuous ucp G-map $\phi : Z\mathcal{M}(A) \to Z\mathcal{M}(B)$. Then, if A is strongly amenable, so is B.*

PROOF. If (θ_i) is a net with the properties required to show strong amenability of A, then $(\phi \circ \theta_i)$ has the properties required to show strong amenability of B. □

In [**5**, Definition 6.1], Anantharaman-Delaroche defines a (possibly noncommutative) G-algebra A to be *strongly amenable* if there exists an amenable G-space X and a nondegenerate G-equivariant $*$-homomorphism $\Phi : C_0(X) \to Z\mathcal{M}(A)$.

LEMMA 2.4. *Every strongly amenable G-algebra in the sense of* [**5**, Definition 6.1] *is strongly amenable in the sense of Definition* 2.1 *above. If the G-algebra A is unital or commutative, then both notions of strong amenability coincide.*

PROOF. The first assertion follows from Lemma 2.3 as every nondegenerate *-homomorphism $C_0(X) \to Z\mathcal{M}(A)$ extends to a strictly continuous *-homomorphism $\mathcal{M}(C_0(X)) \to Z\mathcal{M}(A)$. If A is commutative, both definitions coincide by [**5**, Proposition 6.3], so let us assume now that A is unital and let X be the Gelfand dual of $Z(A)$. Then strong amenability in the sense of Definition 2.1 implies that $C(X)$ is an amenable G-algebra, thus X is an amenable G-space by [**4**, Théorème 4.9 and Remarque 4.10]. Since the inclusion $C(X) \cong Z(A) \hookrightarrow A$ is unital (hence nondegenerate), A is strongly amenable in the sense of [**5**, Definition 6.1]. □

The following important theorem is due to Ozawa [**21**] and (partially) Guentner-Kaminker [**12**]. For background on exact groups, see for example [**26**] or [**7**, Chapter 5].

THEOREM 2.5. *For a discrete group G, the following are equivalent:*

(i) *G is exact;*

(ii) *the canonical G-action on $\ell^\infty(G)$ is strongly amenable.* □

PROPOSITION 2.6. *Let G be an exact group and A be a G-algebra. The following hold.*

(i) *If there exists a ucp G-map $\ell^\infty(G) \to Z(\mathcal{M}(A))$, then A is strongly amenable.*

(ii) *There exists a ucp G-map $\ell^\infty(G) \to Z(A^{**})$ if and only if A is amenable.*

As a consequence, A is amenable if and only if A^{**} is strongly amenable, and the ultraweak convergence in item (iii) of the definition of an amenable G-algebra (see Definition 2.1) can be replaced by norm convergence.

PROOF. The first part follows immediately from Lemma 2.3 and Theorem 2.5.

For the second part, let α be the action on A^{**}, and equip $\ell^\infty(G, Z(A^{**}))$ with the G-action defined by

$$(\tilde{\alpha}_g f)(h) := \alpha_g(f(g^{-1}h)).$$

If A is amenable, we have by [**4**, Théorème 3.3] that there is a ucp G-map $P\colon \ell^\infty(G, Z(A^{**})) \to Z(A^{**})$. Composing this with the canonical unital G-embedding $\ell^\infty(G) \hookrightarrow \ell^\infty(G, Z(A^{**}))$ gives the desired ucp G-map $\ell^\infty(G) \to Z(A^{**})$.

Conversely, assume that there is a ucp G-map $\ell^\infty(G) \to Z(A^{**})$. Since G is exact, its translation action on $\ell^\infty(G)$ is strongly amenable, whence A^{**} is strongly amenable by Lemma 2.3. As norm convergence implies ultraweak convergence, this clearly implies amenability of A. □

We will need some equivalent versions of amenability and strong amenability.

DEFINITION 2.7. Let A be a G-algebra with action α. Define $\ell^2(G, A)$ to be the collection of all functions $\xi : G \to A$ such that the sum

$$\sum_{g \in G} \xi(g)^* \xi(g)$$

converges in the norm of A. Equip $\ell^2(G, A)$ with the A-valued inner product defined by

$$\langle \xi, \eta \rangle := \sum_{g \in G} \xi(g)^* \eta(g),$$

the norm defined by

$$\|\xi\|_2 := \sqrt{\|\langle \xi, \xi \rangle\|_A}$$

and the G-action $\widetilde{\alpha}$ defined by

$$(\widetilde{\alpha}_h\xi)(g) := \alpha_h(\xi(h^{-1}g)).$$

LEMMA 2.8. *Let A be a G-algebra.*

(i) *A is amenable if and only if there exists a net $(\xi_i : G \to Z(A^{**}))_{i\in I}$ of functions such that:*
  (i) *each ξ_i is finitely supported;*
  (ii) *for each i, $\langle \xi_i, \xi_i\rangle \leqslant 1$;*
  (iii) *for each $g \in G$, $\langle \xi_i, \widetilde{\alpha}_g\xi_i\rangle \to 1$ ultraweakly as $i \to \infty$.*

(ii) *A is strongly amenable if and only if there exists a net $(\xi_i : G \to Z\mathcal{M}(A))_{i\in I}$ of functions such that:*
  (i) *each ξ_i is finitely supported;*
  (ii) *for each i, $\langle \xi_i, \xi_i\rangle \leqslant 1$;*
  (iii) *for each $g \in G$, $\langle \xi_i, \widetilde{\alpha}_g\xi_i\rangle \to 1$ strictly as $i \to \infty$.*

PROOF. The proofs are essentially contained in that of [**4**, Théorème 3.3], so we just sketch the idea in the amenable case. If (ξ_i) is a net as in the statement of the lemma, then

$$\theta_i : G \to Z(A^{**}), \quad g \mapsto \langle \xi_i, \widetilde{\alpha}_g\xi_i\rangle$$

satisfies the properties needed to check amenability. Conversely, if (θ_i) is as in the definition of amenability, then the fact that each θ_i is positive type and finitely supported means we can apply a GNS-type construction for each i as in [**4**, Proposition 2.5] to get a vector $\eta_i \in \ell^2(G, Z(A^{**}))$ such that $\theta_i(g) = \langle \eta_i, \widetilde{\alpha}_g\eta_i\rangle$ for all $g \in G$. Using that the collection of finitely supported elements is norm dense in $\ell^2(G, Z(A^{**}))$ and approximating each η_i by some finitely supported ξ_i gives the result. □

3. Suzuki's examples

In [**25**], Yuhei Suzuki produces (among other things) a very striking class of examples. Let G be a countable, exact, non-amenable group. Suzuki shows in [**25**, Proposition B] that there exists a simple, unital, separable, nuclear G-algebra A such that $A \rtimes_r G = A \rtimes_{\max} G$. As A is simple and unital, its center is just scalar multiples of the unit. Hence A cannot be strongly amenable as if it were, G would necessarily be amenable. It is well known that strong amenability implies equality of the maximal and reduced crossed product C^*-algebras, but the converse had been an open question until these examples.

Now, it is also known [**4**, Proposition 4.8] that if A is an amenable G-algebra, then $A \rtimes_r G = A \rtimes_{\max} G$. The converse is again open, and so it is natural to ask if Suzuki's examples are amenable. The answer turns out to be yes: one way to see this is to note that Suzuki's examples arise as a direct limit

$$A \rtimes_r G = \lim_n (A_n \rtimes_r G)$$

with each A_n a strongly amenable nuclear G-subalgebra of A; as A_n is (strongly) amenable, $A_n \rtimes G$ is nuclear by [**4**, Théorème 4.5], whence $A \rtimes G$ is nuclear, and so A is amenable by [**4**, Théorème 4.5] again.

While we guess Suzuki (and others) are aware of this, it does not seem to have been explicitly recorded in his paper. Suzuki's examples seem to be the first known examples of G-algebras with an amenable action that is not strongly amenable.

In the rest of this section, we will give another approach to amenability of Suzuki's algebras, partly as it is more concrete, and partly as we suspect it will be useful in other contexts. This involves another variant of amenability in the form of an approximation property. Although this variant will not be used in the rest of the paper, it seems a natural notion so worth including. It is also the strongest 'amenability-type' condition that we could show that Suzuki's examples satisfy, and so seems worthwhile from that point of view.

DEFINITION 3.1. A G-algebra A has the *quasi-central approximation property* (QAP) if there is a net $(\xi_i\colon G \to \mathcal{M}(A))_{i\in I}$ of finitely supported functions satisfying

(i) $\langle \xi_i, \xi_i\rangle \leqslant 1$ for all i;
(ii) $\langle \xi_i, \tilde{\alpha}_g\xi_i\rangle$ converges strictly to 1 in $\mathcal{M}(A)$ for all $g \in G$;
(iii) $\|\xi_i a - a\xi_i\|_2 \to 0$ for all $a \in A$.

One can actually assume that the functions ξ_i in the definition of the QAP take their values in A:

LEMMA 3.2. *Suppose that the G-algebra A has the QAP. Then the functions $(\xi_i)_{i\in I}$ in the definition of the QAP can be chosen to take their values in A, i.e., there exists a net $(\xi_i\colon G \to A)_{i\in I}$ which satisfies conditions (i), (ii), and (iii) of Definition* 3.1.

PROOF. Let $(\eta_i : G \to \mathcal{M}(A))_{i\in I}$ be a net as in Definition 3.1 and let $(e_j)_{j\in J}$ be an approximate unit of A. Define $\xi_{i,j} : G \to A$ by $\xi_{i,j}(g) := \eta_i(g)e_j$. Then $\langle \xi_{i,j}, \xi_{i,j}\rangle = e_j\langle \eta_i, \eta_i\rangle e_j \leqslant 1$ for all $(i, j) \in I \times J$ and

$$\langle \xi_{i,j}, \tilde{\alpha}_g\xi_{i,j}\rangle = e_j\langle \eta_i, \tilde{\alpha}_g\eta_i\rangle\alpha_g(e_j) \to 1$$

in the strict topology of A, using the fact that multiplication is strictly continuous on bounded subsets of $\mathcal{M}(A)$. Thus $(\xi_{i,j})_{(i,j)\in I\times J}$ satisfies conditions (i) and (ii) of Definition 3.1. To check condition (iii) let $a \in A$ be fixed. Then

$$\begin{aligned}\|\xi_{i,j}a - a\xi_{i,j}\|_2 &= \|\eta_i e_j a - a\eta_i e_j\|_2\\ &\leqslant \|\eta_i e_j a - \eta_i a e_j\|_2 + \|\eta_i a e_j - a\eta_i e_j\|_2\\ &\leqslant \|\eta_i\|_2 \cdot \|e_j a - a e_j\| + \|\eta_i a - a\eta_i\|_2 \cdot \|e_j\|\\ &\leqslant \|e_j a - a e_j\| + \|\eta_i a - a\eta_i\|_2 \to 0.\end{aligned}$$

This finishes the proof. □

It follows from Lemma 2.8 that if A is strongly amenable, then it has the QAP. On the other hand, it follows from Lemma 3.2 that the QAP implies the so-called *approximation property* of Exel (see [**11**, Definition 20.4]), and therefore that the QAP implies amenability by the results of [**1**, Theorem 6.11 and Corollary 6.16]. To summarize, we have the following implications in general:

$$\text{strong amenability} \;\Rightarrow\; \text{QAP} \;\Rightarrow\; \text{amenability}.$$

We will soon show that some of Suzuki's examples have the QAP (but are not strongly amenable), whence the first implication above is not reversible. We do not know if the second is reversible. The key point for showing that Suzuki's examples have the QAP is as follows.

PROPOSITION 3.3. *Assume the G-algebra A is the inductive limit of a sequence (or net) of G-algebras $(A_n)_{n\in N}$. If all A_n have the QAP, then so does A. In particular, if all A_n are strongly amenable, then A is amenable.*

PROOF. Since the QAP passes to quotients by G-invariant ideals, we may assume without loss of generality that $(A_n)_{n\in N}$ is an increasing net of G-algebras such that $\cup_{n\in N} A_n$ is dense in A. For each n, let $(\xi_{i,n})$ be a net of functions $\xi_{i,n} : G \to A_n$ satisfying the conditions of Definition 3.1 for the QAP of A_n. Let $\eta_{i,n} : G \to A$ denote the composition of $\xi_{i,n}$ with the inclusion $A_n \hookrightarrow A$. It is clear that the net $(\eta_{i,n})$ satisfies condition (i) in the definition of the QAP. Moreover, conditions (ii) and (iii) for $(\xi_{i,n})$ imply that

$$a\langle \eta_{i,n}, \tilde{\alpha}_g \eta_{i,n}\rangle \to a, \ \langle \eta_{i,n}, \tilde{\alpha}_g \eta_{i,n}\rangle a \to a \quad \text{and} \quad \|\eta_{i,n} a - a\eta_{i,n}\|_2 \to 0$$

for all $a \in A_n$, and hence for all $a \in \cup_{n\in N} A_n$. Since $(\eta_{i,n})$ is uniformly bounded (by 1) with respect to the ℓ^2-norm, it follows that (ii) and (iii) hold for all $a \in A$. $\square$

Now, let us briefly describe Suzuki's examples as given in [**25**, Proposition B] in order to see how they fit into the above discussion. Let G be any countable exact group. Then it is always possible to choose a second countable, compact, amenable G-space X such that the G-action is also free and minimal (see for example [**24**, Section 6]). The crossed product $A_0 := C(X) \rtimes G$ is therefore a simple, separable, unital and nuclear C^*-algebra. Consider A_0 as a G-algebra endowed with the inner action implemented by the canonical unitaries $u_g \in C_r^*(G) \subseteq A_0$. Observe that an inner action can never be amenable unless G is amenable (or the algebra is zero). However, the diagonal G-action on the infinite tensor product $A := A_0^{\otimes \mathbb{N}}$ has the QAP (and is therefore amenable).

Indeed, as observed by Suzuki, it is enough to realise A as the direct limit of the G-subalgebras $A_n := A_0^{\otimes n} \otimes C(X)$, which are $X \rtimes G$-algebras in the obvious way. It follows that all A_n are strongly amenable, hence A has the QAP by Proposition 3.3.

Note that Suzuki's examples also show that (unlike the QAP), strong amenability does not behave well with respect to limits. It is not clear to us whether amenability passes to inductive limits of G-algebras.

We close this section with a brief discussion on the amenability of some other examples of G-algebras A with $A \rtimes G = A \rtimes_r G$ constructed by Suzuki in [**25**]. In [**25**, Theorem A] he produces examples of actions $\alpha : G \to \mathrm{Aut}(A)$ of non-amenable second countable locally compact groups on simple C^*-algebras A such that the full and reduced crossed products coincide. Since A is simple, we have $Z\mathcal{M}(A) \cong C_b(\mathrm{Prim}(A)) = \mathbb{C}$, hence, since G is not amenable, A cannot be a strongly amenable G-algebra.

To see that A is amenable if G is discrete, recall that Suzuki constructs A as a crossed product $A = C_0(X \times G) \rtimes \Gamma$ with respect to a certain minimal diagonal action, say β, of a free group Γ such that X is a compact amenable Γ-space and β commutes with the G-action $\tilde{\rho} := \mathrm{id}_X \otimes \rho$ on $C_0(X \times G)$, where ρ denotes the right translation action of G on itself. Then the action β on $C_0(X \times G)$ is amenable and therefore $A = C_0(X \times G) \rtimes \Gamma$ is nuclear by [**4**, Théorème 4.5]. Since $\tilde{\rho}$ commutes with β, it induces an action, say γ, of G on $A = C_0(X \times G) \rtimes \Gamma$, which is the action considered by Suzuki. To see that this action is amenable, by [**4**, Théorème 4.5] it suffices to show that $A \rtimes G$ is nuclear. Since the action of Γ on the nuclear algebra $C(X) \otimes \mathbb{K}(\ell^2(G))$ is amenable this follows from the isomorphisms

$$\begin{aligned} A \rtimes_\gamma G &= \big(C_0(X \times G) \rtimes_\beta \Gamma\big) \rtimes_\gamma G \\ &= \big(C_0(X \times G) \rtimes_{\tilde{\rho}} G\big) \rtimes \Gamma \cong \big(C(X) \otimes \mathbb{K}(\ell^2(G))\big) \rtimes \Gamma. \end{aligned}$$

A similar argument shows that Suzuki's examples of [**25**, Proposition C] are also amenable if G is discrete (but not strongly amenable). We believe that all these examples should also have the QAP, but so far we did not succeed to give a proof.

4. Weak containment

In this section, motivated in part by Suzuki's examples from [**25**], and partly by issues that came up in our earlier work on exotic crossed products [**9**], we study the question of characterizing when $A \rtimes_{\max} G$ is equal to $A \rtimes_r G$. If A satisfies this property, we say that A has *weak containnment.*

This question seems difficult in general: while amenability of the action is a sufficient condition by [**4**, Proposition 4.8], finding a 'good' necessary condition that works in complete generality has proven elusive, even in the case when A is commutative.

It turns out to be easier to characterize when $A \rtimes_{\max} G$ equals the so-called *maximal injective* crossed product $A \rtimes_{\text{inj}} G$, which was introduced by the current authors in [**9**]. As proved in [**9**, Proposition 4.2], $A \rtimes_{\text{inj}} G = A \rtimes_r G$ whenever G is exact, so characterizing when $A \rtimes_{\max} G = A \rtimes_{\text{inj}} G$ is the same as characterizing when $A \rtimes_{\max} G = A \rtimes_r G$ for 'most' groups that come up in 'real life'.

We first recall the definition of the maximal injective crossed product from [**9**, Section 3].

DEFINITION 4.1. For a G-algebra A, the *injective crossed product* $A \rtimes_{\text{inj}} G$ is defined as the completion of $C_c(G, A)$ for the norm defined on $a \in C_c(G, A)$ by

$$\|a\| := \inf\{\|a \circ \phi\|_{B \rtimes_{\max} G} \mid \phi : A \to B \text{ an injective equivariant } *\text{-homomorphism}\}.$$

It is not immediate from the definition, but this is a C^*-norm. Moreover, it defines a crossed product functor that takes injective equivariant $*$-homomorphisms to injective $*$-homomorphisms. The following definitions are important for establishing the basic properties of $\rtimes_{\text{inj}}$, and will be fundamental to our work in this paper.

DEFINITION 4.2. A G-algebra A is *G-injective* if for any G-embedding $A \subseteq B$, there exists an equivariant conditional expectation $P : B \to A$.

A G-algebra A has the *G-WEP* if for any G-embedding $\iota : A \hookrightarrow B$, there exists a ccp map $\phi : B \to A^{**}$ such that $\phi \circ \iota : A \to A^{**}$ coincides with the canonical embedding $A \hookrightarrow A^{**}$.

The above definition of G-injectivity is maybe a little non-standard. We will see in Section 8 below (see Proposition 8.1) that it is equivalent to the more usual definition due to Hamana [**15**].

We have the following basic lemma: this will get used several times below.

LEMMA 4.3. *If B is an injective G-algebra in the sense of Definition 4.2, then it is unital.*

PROOF. Let $\widetilde{B}$ be the unitisation of B, equipped with the unique extension of the G-action. By G-injectivity of B the natural inclusion $B \hookrightarrow \widetilde{B}$ admits an equivariant ccp splitting $E : \widetilde{B} \to B$. This splitting is necessarily a conditional expectation (see for example [**7**, Theorem 1.5.10]), whence for any $b \in B$,

$$E(1)b = E(1b) = E(b) = b = E(b) = E(b1) = bE(1),$$

so $E(1)$ is a unit for B. □

REMARK 4.4. A G-injective G-algebra clearly has the G-WEP. Moreover, if A^{**} is G-injective, then A has the G-WEP. To see this, let $A \hookrightarrow B$ be a G-embedding, and extend it canonically to an embedding of double duals $A^{**} \hookrightarrow B^{**}$. As A^{**} is G-injective, this admits a splitting $P : B^{**} \to A^{**}$, and the restriction of this splitting to B is a map with the properties required by the G-WEP. The converse is false in general: indeed, $A = \ell^\infty(G)$ is G-injective by Example 4.5 just below, so always has the G-WEP; however, A^{**} is G-injective if and only if G is exact as will follow from Theorems 2.5 and 7.3 below.

EXAMPLE 4.5. Perhaps the simplest example of a G-injective algebra is $\ell^\infty(G)$. Indeed, let $\ell^\infty(G) \hookrightarrow B$ be any G-embedding. Choose a state ϕ on B that extends the Dirac mass at the identity δ_e, considered as a multiplicative linear functional $\delta_e : \ell^\infty(G) \to \mathbb{C}$. Then define

$$P : B \to \ell^\infty(G), \quad P(b) : g \mapsto \phi(\beta_{g^{-1}}(b)).$$

It is not too difficult to see that this is a ccp G-map splitting the original inclusion (compare [**9**, Proposition 2.2] for a more general result).

Noting that $C_0(G)^{**} = \ell^\infty(G)$, Remark 4.4 gives that $C_0(G)$ has the G-WEP. It is never G-injective for infinite G, as G-injectivity implies unitality by Lemma 4.3.

More generally (and with essentially the same proof: see [**9**, Proposition 2.2 and Remark 2.3]) we have the following example.

EXAMPLE 4.6. Let B be a C^*-algebra that is injective in the usual sense (i.e. G-injective where G is the trivial group). Then $\ell^\infty(G, B)$ equipped with the translation action defined by

$$(\gamma_g(f))(h) := f(g^{-1}h)$$

is G-injective. More generally, if B is equipped with a G-action β (but is still only assumed non-equivariantly injective) and $\ell^\infty(G, B)$ is equipped with the diagonal type action

$$(\widetilde{\beta}_g f)(h) := \beta_g(f(g^{-1}h)),$$

then $\ell^\infty(G, B)$ is G-injective.

The following lemma, proved in [**9**, Proposition 3.12], is key to establishing these properties.

LEMMA 4.7. *Let A be a G-algebra with the G-WEP (in particular, A could be G-injective). Then $A \rtimes_{\max} G = A \rtimes_{\text{inj}} G$.* □

The following lemma treats integration of 'ccp covariant pairs' to the maximal crossed product and it is probably well known to the experts. We thank the anonymous referee for providing us with the short proof given below, which replaces a much longer one we had before.

LEMMA 4.8. *Let A be a G-algebra, and $(\sigma, u) : (A, G) \to \mathbb{B}(H)$ be a pair consisting of a ccp map σ and a unitary representation u satisfying the usual covariance relation*

$$\sigma(\alpha_g(a)) = u_g \sigma(a) u_g^*$$

for all $a \in A$ and $g \in G$. Then the integrated form

$$\sigma \rtimes u : C_c(G, A) \to \mathbb{B}(H), \quad f \mapsto \sum_{g \in G} \sigma(f(g)) u_g$$

extends to a ccp map $\sigma \rtimes u : A \rtimes_{\max} G \to \mathbb{B}(H)$.

PROOF. By [**7**, Exercise 4.1.4] (or [**8**, Theorem 4.9, (5) $\Rightarrow$ (6)]) the ccp G-map $\sigma : A \to \mathbb{B}(H)$ descends to a ccp map $\sigma \rtimes_{\max} G : A \rtimes_{\max} G \to \mathbb{B}(H) \rtimes_{\max} G$, with respect to the action $\mathrm{Ad}_u : G \to \mathrm{Aut}(\mathbb{B}(H))$ given by conjugation with u. Composing $\sigma \rtimes_{\max} G$ with the integrated form $\mathrm{id} \rtimes u : \mathbb{B}(H) \rtimes_{\max} G \to \mathbb{B}(H)$ of the covariant pair (id, u) gives the ccp map $\sigma \rtimes u$. □

The statement and proof of the following result are inspired by Lance's *tensor product trick*: see for example the exposition in [**7**, Proposition 3.6.6], or the original article [**18**].

THEOREM 4.9. *Let $\iota \colon A \hookrightarrow B$ be a faithful G-embedding. The following are equivalent:*

(i) *$\iota \rtimes_{\max} G \colon A \rtimes_{\max} G \to B \rtimes_{\max} G$ is injective;*
(ii) *for any covariant representation $(\pi, u) \colon (A, G) \to \mathbb{B}(H)$, there is a ccp G-map $\varphi \colon B \to \mathbb{B}(H)$ with $\varphi \circ \iota = \pi$;*
(iii) *there exists a covariant representation $(\pi, u) : (A, G) \to \mathbb{B}(H)$ such that the integrated form $\pi \rtimes u : A \rtimes_{\max} G \to \mathbb{B}(H)$ is faithful and for which there is a ccp G-map $\varphi \colon B \to \mathbb{B}(H)$ with $\varphi \circ \iota = \pi$.*

PROOF. Assume (i), and let $(\pi, u) : (A, G) \to \mathbb{B}(H)$ be a covariant representation. We must show that the dashed arrow below can be filled in with a ccp G-map

(1)

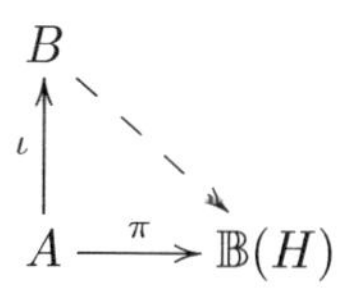

Let $\widetilde{A}$ and $\widetilde{B}$ be the unitizations of A and B and let $\widetilde{\pi} : \widetilde{A} \to \mathbb{B}(H)$ and $\widetilde{\iota} : \widetilde{A} \to \widetilde{B}$ be the canonical (equivariant) unital extensions. It will suffice to prove that the dashed arrow below

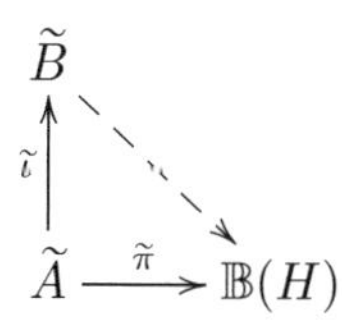

can be filled in with an equivariant ccp map; indeed, if we can do this, then the restriction of the resulting equivariant ccp map $\widetilde{B} \to \mathbb{B}(H)$ to B will have the desired property.

Since the descent $\iota \rtimes G : A \rtimes_{\max} G \to B \rtimes_{\max} G$ of ι is injective by assumption, it follows from this and the commutative diagram

$$\begin{array}{ccccccccc} 0 & \longrightarrow & B \rtimes_{\max} G & \longrightarrow & \widetilde{B} \rtimes_{\max} G & \longrightarrow & \mathbb{C} \rtimes_{\max} G & \longrightarrow & 0 \\ & & \uparrow & & \uparrow & & \| & & \\ 0 & \longrightarrow & A \rtimes_{\max} G & \longrightarrow & \widetilde{A} \rtimes_{\max} G & \longrightarrow & \mathbb{C} \rtimes_{\max} G & \longrightarrow & 0 \end{array}$$

of short exact sequences that the map

$$\widetilde{\iota} \rtimes G : \widetilde{A} \rtimes_{\max} G \to \widetilde{B} \rtimes_{\max} G$$

is injective as well. From now on, to avoid cluttered notation, we will assume that A, B, π and ι are unital, and that the map $\iota \rtimes G : A \rtimes_{\max} G \to B \rtimes_{\max} G$ is injective; our goal is to fill in the dashed arrow in line (1) under these new assumptions.

Now, as we are assuming that $\iota \rtimes G : A \rtimes_{\max} G \to B \rtimes_{\max} G$ is injective, in the diagram below

$$\begin{array}{ccc} B \rtimes_{\max} G & & \\ {\scriptstyle \iota \rtimes G}\uparrow & \overset{\widetilde{\pi \rtimes u}}{\searrow} & \\ A \rtimes_{\max} G & \xrightarrow{\pi \rtimes u} & \mathbb{B}(H) \end{array}$$

we may thus use injectivity of $\mathbb{B}(H)$ (i.e. Arveson's extension theorem as in for example [**7**, Theorem 1.6.1]) to show that the dashed arrow can be filled in with a ucp map. Any operator of the form u_g is in the multiplicative domain of $\widetilde{\pi \rtimes u}$, from which it follows that the restriction ϕ of $\widetilde{\pi \rtimes u}$ to B is equivariant. This restriction ϕ is the desired map.

The implication (ii)$\Rightarrow$(iii) is clear, so it remains to show (iii)$\Rightarrow$(i). Let $\pi \rtimes u : A \rtimes_{\max} G \to \mathbb{B}(H)$ be a faithful representation such that there is a ccp G-map $\tilde{\pi} : B \to \mathbb{B}(H)$ that extends π as in (iii). Lemma 4.8 implies that this ccp map integrates to a ccp map $\tilde{\pi} \rtimes u : B \rtimes_{\max} G \to \mathbb{B}(H)$. As the diagram

$$\begin{array}{ccc} B \rtimes_{\max} G & & \\ {\scriptstyle \iota \rtimes G}\uparrow & \overset{\tilde{\pi} \rtimes u}{\searrow} & \\ A \rtimes_{\max} G & \xrightarrow{\pi \rtimes u} & \mathbb{B}(H) \end{array}$$

commutes and the horizontal map is injective, the vertical map is injective too. □

Notice that $A \rtimes_{\max} G = A \rtimes_{\text{inj}} G$ if and only if every G-embedding $\iota \colon A \hookrightarrow B$ satisfies the equivalent conditions in Proposition 4.9. Hence we get the following immediate consequence, for which we need one additional definition.

DEFINITION 4.10. A covariant representation $(\pi, u) : (A, G) \to \mathbb{B}(H)$ is *G-injective* if for any G-embedding $A \subseteq B$ there exists a ccp map $\sigma : B \to \mathbb{B}(H)$ that extends π, and satisfies the covariance relation for u.

COROLLARY 4.11. *For a G-algebra A, the following are equivalent:*

(i) $A \rtimes_{\max} G = A \rtimes_{\text{inj}} G$;
(ii) *every covariant representation (π, u) is G-injective;*
(iii) *there is a G-injective covariant representation that integrates to a faithful representation of $A \rtimes_{\max} G$.*

Moreover, if G is exact, $\rtimes_{\text{inj}}$ may be replaced by $\rtimes_r$ in the above. □

Our next goal is to develop this to get a characterization in terms of an amenability property of a more traditional 'approximation property' form.

LEMMA 4.12. *Let A be a G-algebra and let $(\pi, u) \colon (A, G) \to \mathbb{B}(H)$ be a nondegenerate G-injective covariant pair. Then for any unital G-algebra C there exists a ucp G-map $\phi : C \to \pi(A)' \subseteq \mathbb{B}(H)$.*

PROOF. Consider the canonical G-embedding

$$\iota \colon A \hookrightarrow C \otimes A, \quad a \mapsto 1 \otimes a.$$

Then G-injectivity of π yields a ccp G-map $\varphi\colon C\otimes A \to \mathbb{B}(H)$ with $\varphi\circ\iota = \pi$. Since π is nondegenerate, so is φ, that is, $\varphi(e_i) \to 1$ strongly if (e_i) is an approximate unit for A. It follows that φ extends to a ucp map $\bar{\varphi}\colon \mathcal{M}(C\otimes A) \to \mathbb{B}(H)$, see [**19**, Corollary 5.7]. Moreover, this extension is G-equivariant as can be seen from the construction of $\bar{\varphi}$ in [**19**]. We now consider the canonical G-embedding $j\colon C \to \mathcal{M}(C\otimes A)$, $c \mapsto c\otimes 1$, and then define $\phi\colon C \to \mathbb{B}(H)$ by $\phi(c) := \bar{\varphi}(j(c))$. It remains to show that $\phi(C) \subseteq \pi(A)'$. But since $\bar{\varphi}\circ\iota = \varphi\circ\iota = \pi$ is a homomorphism, the image of ι lies in the multiplicative domain of $\bar{\varphi}$, so that

$$\begin{aligned}\phi(c)\pi(a) = \bar{\varphi}(j(c))\varphi(\iota(a)) = \bar{\varphi}(j(c)\iota(a))\\ = \bar{\varphi}(\iota(a)j(c)) = \varphi(\iota(a))\bar{\varphi}(j(c)) = \pi(a)\phi(c).\end{aligned}$$

□

Here is the version of amenability we will use. To state it, if A is a G-algebra and $(\pi, u) : (A, G) \to \mathbb{B}(H)$ a covariant pair, then $\pi(A)'$ will be equipped with the G-action $\beta = \mathrm{Ad}_u$ defined by conjugation by u.

DEFINITION 4.13. Let A be a G-algebra, and $(\pi, u) : (A, G) \to \mathbb{B}(H)$ a covariant pair. The pair (π, u) is *commutant amenable* (C-amenable) if there exists a net $(\theta_i : G \to \pi(A)')$ of positive type functions (with respect to $\beta = \mathrm{Ad}_u$) such that:

(i) each θ_i is finitely supported;
(ii) for each i, $\theta_i(e) \leqslant 1$;
(iii) for each $g \in G$, $\theta_i(g) \to 1$ ultraweakly as $i \to \infty$.

The G-algebra A is *commutant amenable* (C-amenable) if every covariant pair is C-amenable.

REMARK 4.14. If a G-algebra A is amenable, then it is C-amenable. This follows as any covariant representation (π, u) of (A, G) extends to a covariant representation of (A^{**}, G), and as the image of $Z(A^{**})$ under this extension is necessarily contained in the commutant $\pi(A)'$.

We give the above definition to make the analogy with amenability clearer. However, it will be more convenient to work with the following reformulation. To state it, recall that if $(\pi, u)\colon (A, G) \to \mathbb{B}(H)$ is a covariant representation of the G-algebra A, we use the action $\beta = \mathrm{Ad}_u$ on the commutant $\pi(A)'$ to define $\ell^2(G, \pi(A)')$ as in Definition 2.7. The proof of the next lemma is essentially the same as that of Lemma 2.8, and so omitted.

LEMMA 4.15. *Let A be a G-algebra with action α, and let (π, u) be a covariant pair. Then (π, u) is C-amenable if and only if there exists a net (ξ_i) in $\ell^2(G, \pi(A)')$ such that:*

(1) *each ξ_i is finitely supported;*
(2) *for each i, $\langle \xi_i, \xi_i\rangle \leqslant 1$;*
(3) *for each $g \in G$, $\langle \xi_i, \widetilde{\beta}_g(\xi_i)\rangle \to 1$ ultraweakly as $i \to \infty$.* □

PROPOSITION 4.16. *Let A be a G-algebra, and say there exists a C-amenable covariant pair (π, u) which integrates to a faithful representation of $A \rtimes_{\max} G$. Then $A \rtimes_{\max} G = A \rtimes_r G$.*

PROOF. Let $(\pi, u) : (A, G) \to \mathbb{B}(H)$ be a covariant pair as in the statement. Let $(\xi_i : G \to \pi(A)')$ be a net as in Lemma 4.15. For each i, define

$$T_i : H \to \ell^2(G, H), \quad v \mapsto \big(g \mapsto \xi_i(g)v\big).$$

A direct computation shows that

$$\|T_i v\|^2 = \langle v, \langle \xi_i, \xi_i \rangle v \rangle.$$

As $\langle \xi_i, \xi_i \rangle \leqslant 1$ for all i, the term on the right is bounded above by $\|v\|^2$, and thus $\|T_i\| \leqslant 1$. The adjoint of T_i is easily seen to be given by

$$T_i^*(\eta) = \sum_G \xi_i(g)^* \eta(g)$$

for all $\eta \in C_c(G, H)$.

Now, via Fell's trick, the covariant pair $(\pi \otimes 1, u \otimes \lambda) : (A, G) \to \mathbb{B}(\ell^2(G, H))$ integrates to $A \rtimes_r G$. Consider now the ccp map

$$\phi_i : \mathbb{B}(\ell^2(G, H)) \to \mathbb{B}(H), \quad b \mapsto T_i^* b T_i.$$

We compute for $f \in C_c(G, A)$ and $v \in H$:

$$\begin{aligned} \phi_i \circ (\pi \otimes 1) \rtimes (u \otimes \lambda)(f) &= T_i^*((\pi \otimes 1) \rtimes (u \otimes \lambda)(f)) T_i \\ &= \sum_{g,h \in G} \xi_i(h)^* \pi(f(g)) u_g \xi_i(g^{-1}h) \end{aligned}$$

Using that ξ_i takes values in $\pi(A)'$, this equals

$$\sum_{g \in G} \pi(f(g)) \left(\sum_{h \in G} \xi_i(h)^* u_g \xi_i(g^{-1}h) u_g^* \right) u_g = \sum_G \pi(f(g)) \langle \xi_i, \widetilde{\beta}_g \xi_i \rangle u_g.$$

As $\langle \xi_i, \widetilde{\beta}_g \xi_i \rangle$ converges ultraweakly to 1 and as multiplication is separately ultraweakly continuous, we get ultraweak convergence

$$\phi_i \circ (\pi \otimes 1) \rtimes (u \otimes \lambda)(f) \to (\pi \rtimes u)(f) \text{ as } i \to \infty.$$

As ultraweak limits do not increase norms and as each ϕ_i is ccp, we get

$$\|(\pi \rtimes u)(f)\| \leqslant \limsup_{i \to \infty} \|\phi_i \circ (\pi \otimes 1) \rtimes (u \otimes \lambda)(f)\| \leqslant \|(\pi \otimes 1) \rtimes (u \otimes \lambda)(f)\|.$$

Hence as $(\pi \otimes 1) \rtimes (u \otimes \lambda)$ extends to $A \rtimes_r G$, we get

$$\|(\pi \rtimes u)(f)\| \leqslant \|f\|_{A \rtimes_r G}.$$

As $\pi \rtimes u$ is faithful on $A \rtimes_{\max} G$, however, we are done. □

Finally in this section, we are able to give a characterization of weak containment in terms of commutant amenability, at least for exact groups.

THEOREM 4.17. *Let G be an exact discrete group, and let A be a G-algebra. Then the following are equivalent:*

(i) *A is commutant amenable;*
(ii) *$A \rtimes_{\max} G = A \rtimes_r G$;*
(iii) *$A \rtimes_{\max} G = A \rtimes_{\text{inj}} G$.*

PROOF. The implication (i)⇒(ii) is Proposition 4.16. The implication (ii)⇒(iii) is trivial, so it remains to show that $A \rtimes_{\max} G = A \rtimes_{\text{inj}} G$ implies C-amenability. Let then (π, u) be a covariant pair for (A, G). We may apply Corollary 4.11 and Lemma 4.12 to get an equivariant ucp map $\phi : \ell^\infty(G) \to \pi(A)'$. As G is exact, $\ell^\infty(G)$ is strongly amenable; postcomposing a net $(\theta_i : G \to \ell^\infty(G))$ that shows $\ell^\infty(G)$ is strongly amenable with ϕ gives C-amenability. □

REMARK 4.18. Similarly to Proposition 2.6, it follows from the above proof that for exact G, we can replace ultraweak convergence of the net $\theta_i : A \to \pi(A)'$ in the definition of C-amenability by norm convergence.

5. Matsumura's characterisations of weak containment

In this section, we connect the ideas in the previous section to other forms of amenability, and related results. The key ideas here are due to Matsumura [**20**], and the results are essentially fairly mild generalizations of Matsumura's. Nonetheless, our proofs are somewhat different from those of [**20**]. We also think some of the generalizations are worthwhile in their own right: for example, we remove some unitality and nuclearity assumptions, and have some applications to actions of non-exact groups.

The key technical tool in this section is a seminal theorem of Haagerup [**13**] on the existence of standard forms. We summarize what will be the key points for us in the next theorem (see [**20**, Theorem 2.3] for a brief discussion how the following theorem follows from [**13**]).

THEOREM 5.1. *Let A be a G-algebra. There exist* standard form representa-tions

$$\pi : A^{**} \to \mathbb{B}(H), \quad \textit{and} \quad \pi^{\text{op}} : (A^{\text{op}})^{**} \to \mathbb{B}(H)$$

on the same Hilbert space H together with a unitary representation $u : G \to \mathcal{U}(H)$ with the following properties:

(i) *π and π^{op} are normal, unital, and faithful;*
(ii) *(π, u) and (π^{op}, u) are covariant with respect to the canonical G-actions on A^{**} and $(A^{\text{op}})^{**}$;*
(iii) *having identified A^{**} and $(A^{\text{op}})^{**}$ with their images under π and π^{op}, we get $\pi(A)' = (A^{\text{op}})^{**}$ and $\pi^{\text{op}}(A^{\text{op}})' = A^{**}$;*
(iv) *if A is commutative, then $\pi(A)' = A^{**}$.*

The cleanest results we can prove on weak containment are in the case when G is exact and the G-algebra A is commutative, so we turn to this first.

THEOREM 5.2. *Let G be an exact group, and A a commutative G-algebra. The following are equivalent:*

(i) *A is strongly amenable;*
(ii) *A is amenable;*
(iii) *A is C-amenable;*
(iv) *$A \rtimes_{\max} G = A \rtimes_r G$;*
(v) *$A \rtimes_{\max} G = A \rtimes_{\text{inj}} G$;*
(vi) *A^{**} is strongly amenable;*
(vii) *A^{**} is amenable;*
(viii) *A^{**} is C-amenable;*
(ix) *$A^{**} \rtimes_{\max} G = A^{**} \rtimes_r G$;*
(x) *$A^{**} \rtimes_{\max} G = A^{**} \rtimes_{\text{inj}} G$.*

PROOF. That (i) implies (ii) is trivial, and that (ii) implies (iii) follows from Remark 4.14. That (iii) implies (iv) is Proposition 4.16, and (iv) implies (v) is immediate. Assume condition (v); we will show condition (vi). Indeed, let $\pi : A \to \mathbb{B}(H)$ be the restriction of a standard form of A^{**} as in Theorem 5.1 to A, so π

is covariant for some unitary representation u on H, and $\pi(A)'$ identifies naturally with A^{**}. Corollary 4.11 and Lemma 4.12 give us an equivariant ucp map

$$\phi : \ell^\infty(G) \to A^{**}.$$

It follows from Proposition 2.6 and exactness of G that A^{**} is strongly amenable.

Continuing, (vi) implies (vii) is trivial and (vii) implies (viii) is Remark 4.14 again, while (viii) implies (ix) is Proposition 4.16 again. Also, (ix) implies (x) is again immediate. Finally, it remains to show that (x) implies (i). For this, note that Lemma 4.12 applies to a standard form representation (π, u) to give us an equivariant ucp map

$$\phi : \ell^\infty(G) \to A^{**}.$$

Using Proposition 2.6 and exactness again, this implies that A is amenable. The proof is completed by appealing to [**4**, Théorème 4.9] (and commutativity of A) to get from there back to strong amenability of A. □

Many of the equivalences of Theorem 5.2 were known before this paper: notably, Matsumura [**20**] showed that (iv) and (ii) are equivalent when A is unital, and the equivalence of (i) and (ii) (and similarly of (vi) and (vii)) is due to Anantharaman-Delaroche [**4**, Théorème 4.9]. Many of the other implications are either trivial, or probably known to at least some experts, so overall we certainly don't claim much profundity!

Nonetheless, we hope collecting these conditions in one place clarifies the theory somewhat. It is perhaps also interesting that the only real way the proof goes beyond the classical results of Anantharaman-Delaroche from her seminal 1987 paper [**4**] are in the existence of standard forms [**13**] from 1979 and the adaptations in Corollary 4.11 and Lemma 4.12 of well-known tricks involving injectivity and multiplicative domains due originally to Lance [**18**] (see also the exposition in [**7**, Section 3.6]), and dating to 1973. Some of this already appears in Matsumura's work.

In the case where G is not necessarily exact, some of Theorem 5.2 still holds; however, some of it becomes false, and other parts are unclear.

THEOREM 5.3. *Let G be a discrete group, and let A be a commutative G-algebra. Consider the following conditions:*

(i) *A^{**} is strongly amenable;*
(ii) *A^{**} is amenable;*
(iii) *A^{**} is C-amenable;*
(iv) *$A^{**} \rtimes_{\max} G = A^{**} \rtimes_r G$;*
(v) *$A^{**} \rtimes_{\max} G = A^{**} \rtimes_{\mathrm{inj}} G$;*
(vi) *A is strongly amenable;*
(vii) *A is amenable;*
(viii) *A is C-amenable.*

We have in general that

$$\text{(i)} \Leftrightarrow \text{(ii)} \Leftrightarrow \text{(iii)}, \tag{2}$$

that

$$\text{(v)} \Leftrightarrow \text{(vi)} \Leftrightarrow \text{(vii)} \Leftrightarrow \text{(viii)} \tag{3}$$

and that the conditions in line (2) *imply condition* (iv), *which in turn implies the conditions in line* (3).

Moreover, if A is unital, all the conditions listed above are equivalent, and if $A \neq 0$, they force G to be exact.

PROOF. The implication (i) $\Rightarrow$ (ii) is trivial, and the converse is part of [**4**, Théorème 4.9]. The implication (ii) $\Rightarrow$ (iii) is likewise trivial, and the converse is a consequence of C-amenability of a standard form representation (see Theorem 5.1), and the fact that the bidual of a commutative C*-algebra is again commutative. Thus we have the equivalences in line (2).

The implication (iii) $\Rightarrow$ (iv) is Proposition 4.16, and (iv) $\Rightarrow$ (v) is trivial.

Look finally at the equivalences in line (3). First note that condition (v) implies via Corollary 4.11 that a standard form representation (Theorem 5.1) π of A^{**} is an injective representation. In particular, we can fill in the dashed line below with a ucp G-map making the diagram commute

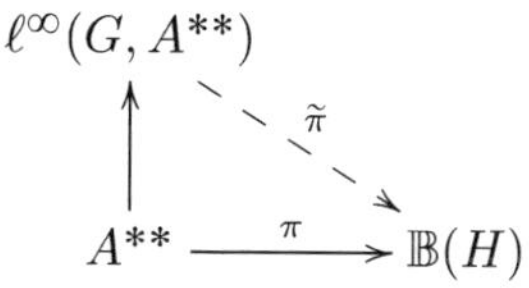

where the vertical arrow is the canonical inclusion of A^{**} in $\ell^\infty(G, A^{**})$ as constant functions. As A^{**} is in the multiplicative domain of $\widetilde{\pi}$, it follows that $\widetilde{\pi}(\ell^\infty(G, A^{**}))$ commutes with $\pi(A^{**})$; however, as π is a standard form, this implies that the image of $\widetilde{\pi}$ is exactly $\pi(A^{**}) \cong A^{**}$. In other words, $\widetilde{\pi}$ identifies with an equivariant conditional expectation $\ell^\infty(G, A^{**}) \to A^{**}$ that splits the constant inclusion. The existence of such a conditional expectation is equivalent to amenability by [**4**, Théorème 3.3]. Conversely, assuming amenability, we have such an equivariant conditional expectation. As A is commutative, A^{**} is injective, whence $\ell^\infty(G, A^{**})$ is G-injective by Example 4.6. Hence A^{**} is also G-injective, whence we get condition (v) by Lemma 4.7.

To complete the equivalences in line (3), note that the implications from (vi) to (vii), and from (vii) to (viii) are straightforward. On the other hand, we have implications from (viii) to (vii) by C-amenability of a standard form representation, and from (vii) to (vi) by part of [**4**, Théorème 4.9].

In the unital case, the implication from (vi) to (i) is trivial, so we are done. □

We also get the following observation in general.

THEOREM 5.4. *Let A be a commutative G-algebra. The following are equivalent:*

(i) $A \rtimes_{\max} G = A \rtimes_{\text{inj}} G$;
(ii) *A has the G-WEP.*

PROOF. The implication from (i) to (ii) follows as if $A \to \ell^\infty(G, A^{**})$ is the equivariant inclusion of A as constant functions from G to A and $\pi : A \to A^{**} \subseteq \mathbb{B}(H)$ is the restriction of a standard form of A^{**} from Theorem 5.1 to A, then Corollary 4.11 gives us a commutative diagram

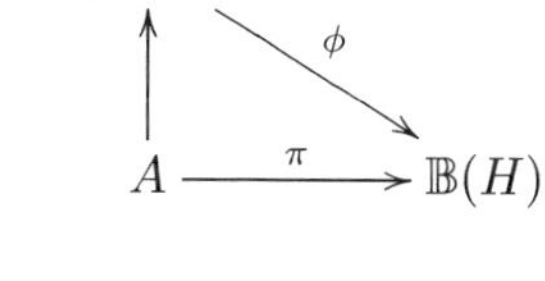

with ϕ a ccp G-map. As A is in the multiplicative domain of ϕ and the algebra $\ell^\infty(G, A^{**})$ is commutative, ϕ takes image in $\pi(A)' = A^{**}$. In other words we have factored the canonical inclusion $A \hookrightarrow A^{**}$ through the G-injective algebra (Example 4.6) $\ell^\infty(G, A^{**})$, which easily implies the G-WEP.

The converse is Lemma 4.7. $\square$

REMARK 5.5. It follows from Proposition 4.16 that

$$\text{(Theorem 5.3 conditions)} \Rightarrow A \rtimes_{\max} G = A \rtimes_r G \Rightarrow \text{(Theorem 5.4 conditions)}.$$

On the other hand, the conditions in Theorem 5.4 are true for $A = \ell^\infty(G)$ and any G, while those from Theorem 5.3 are all false for any non-exact G and $A = \ell^\infty(G)$, so we do not have equivalence of the conditions in Theorems 5.3 and 5.4. The exact relationship between the conditions in Theorem 5.3 and 5.4 and the condition $A \rtimes_{\max} G = A \rtimes_r G$ is not at all clear in general.

Note that $A = C_0(G)$ satisfies the conditions in line (3) of Theorem 5.3 for any G, but never satisfies the conditions in line (2) when G is not exact. $A = C_0(G)$ is also known to fail (iv) for at least some non-exact groups (see [**9**, Lemma 4.7]), so (iv) cannot always be equivalent to the conditions in line (3). It is possible that condition (iv) is always equivalent to those in line (2) even for non-unital A; however, it also seems quite plausible that (iv) holds for some non-exact groups and $A = C_0(G)$. This issue seems quite open at the moment.

We now move on to the noncommutative case.

THEOREM 5.6. *Let G be exact and A a G-algebra. Let A^{op} be the opposite algebra of A, and let M be the von Neumann algebra generated by $\pi(A)$ and $\pi^{\mathrm{op}}(A^{\mathrm{op}})$ in standard forms π and π^{op} of A^{**} and $(A^{\mathrm{op}})^{**}$ (see Theorem 5.1). The following are equivalent:*

(i) *A is amenable;*
(ii) *$A \otimes_{\max} A^{\mathrm{op}}$ is C-amenable;*
(iii) *$(A \otimes_{\max} A^{\mathrm{op}}) \rtimes_{\max} G = (A \otimes_{\max} A^{\mathrm{op}}) \rtimes_r G$;*
(iv) *$(A \otimes_{\max} A^{\mathrm{op}}) \rtimes_{\max} G = (A \otimes_{\max} A^{\mathrm{op}}) \rtimes_{\mathrm{inj}} G$;*
(v) *A^{**} is strongly amenable;*
(vi) *M is strongly amenable;*
(vii) *M is amenable;*
(viii) *M is C-amenable;*
(ix) *$M \rtimes_{\max} G = M \rtimes_r G$;*
(x) *$M \rtimes_{\max} G = M \rtimes_{\mathrm{inj}} G$.*

PROOF. Assume first that A is amenable, let $B = A \otimes_{\max} A^{\mathrm{op}}$, and let $(\rho, v) : (B, G) \to \mathbb{B}(H)$ be any covariant pair for B. Then ρ 'restricts' to representations of A and A^{op} as in [**7**, Theorem 3.2.6], which we also denote ρ. Extending ρ to A^{**}, we have that $\rho(Z(A^{**})) \subseteq \rho(B)'$. It follows from this that amenability of A implies C-amenability of $A \otimes_{\max} A^{\mathrm{op}}$, so we get (i) implies (ii).

The implication from (ii) to (iii) is Proposition 4.16, and from (iii) to (iv) is trivial. Assuming (iv), let $\pi : A \to \mathbb{B}(H)$ be the restriction of a standard form (see Theorem 5.1) of A^{**} to A. Thanks to Theorem 5.1 and the universal property of the maximal tensor product we obtain the covariant representation $(\sigma := \pi \times_{\max} \pi^{\mathrm{op}}, u)$, of $A \otimes_{\max} A^{\mathrm{op}}$. Lemma 4.12 now gives us a ucp G-map $\phi : \ell^\infty(G) \to \sigma(A \otimes_{\max} A^{\mathrm{op}})'$. However,

$$\sigma(A \otimes_{\max} A^{\mathrm{op}})' = \pi(A)' \cap \pi^{\mathrm{op}}(A)' = \pi(A)' \cap \pi(A)'' = Z(\pi(A)') \cong Z(A^{**}). \tag{4}$$

As G is exact, the existence of an equivariant ucp G-map $\phi : \ell^\infty(G) \to Z(A^{**})$ implies that A^{**} is strongly amenable by Proposition 2.6. Hence we have shown that (iv) implies (v).

The implication from (v) to (vi) follows as a standard form $\pi : A^{**} \to \mathbb{B}(H)$ restricts to a unital equivariant $*$-homomorphism $\pi : Z(A^{**}) \to Z(M)$. The implications from (vi) to (vii) and (vii) to (viii) are straightforward, and that from (viii) to (ix) is Proposition 4.16 again.

The implication from (ix) to (x) is trivial, so it remains to get back from (x) to (i). Indeed, Lemma 4.12 gives an equivariant ucp map $\phi : \ell^\infty(G) \to M'$, and analogously to line (4), $M' \cong Z(A^{**})$. Proposition 2.6 again completes the proof. □

REMARK 5.7. The conditions in Theorem 5.6 are not equivalent to strong amenability of A. This follows from the properties of Suzuki's examples [**25**] as discussed in Section 3.

When G is not necessarily exact and A is a general C^*-algebra, little of Theorems 5.3 and 5.4 seems recoverable. Here are some implications that hold in general.

PROPOSITION 5.8. *Let A be a G-algebra. The following are equivalent:*

(i) *A is amenable;*
(ii) *$A \otimes_{\max} B$ is amenable for every G-algebra B;*
(iii) *$A \otimes_{\max} A^{\mathrm{op}}$ is amenable;*
(iv) *$A \otimes_{\max} A^{\mathrm{op}}$ is C-amenable.*

PROOF. The implication from (i) to (ii) is well-known, but maybe not explicit in the literature. It can be proved using Lemma 2.8 to get a net $(\xi_i \colon G \to Z(A^{**}))_{i\in I}$ with the properties states there, and the fact that there is a canonical G-embedding $A^{**} \hookrightarrow (A \otimes_{\max} B)^{**}$ inducing a G-embedding $Z(A^{**}) \to Z((A \otimes_{\max} B)^{**})$. The implication from (ii) to (iii) is trivial, and (iii) implies (iv) is Remark 4.14. Finally, if (iv) holds, we use the same technique from the previous proofs by making use of the standard representation $\pi\colon A^{**} \hookrightarrow \mathbb{B}(H)$ to build a representation

$$\sigma := \pi \times_{\max} \pi^{\mathrm{op}} \colon A \otimes_{\max} A^{\mathrm{op}} \to \mathbb{B}(H)$$

with

$$\sigma(A \otimes_{\max} A^{\mathrm{op}})' = \pi(A)' \cap \pi(A)'' \cong Z(A^{**}).$$

Then C-amenability for $A \otimes_{\max} A^{\mathrm{op}}$ implies the existence of an approximative net $(\xi_i \colon G \to \rho(A \otimes_{\max} A^{\mathrm{op}})' \cong Z(A^{**}))$ giving amenability for A. □

QUESTION 5.9. Is it true that C-amenability passes to (maximal) tensor products in the sense of the implication (i) $\Rightarrow$ (ii) from Proposition 5.8 above, even for trivial actions?

If so, we could conclude that C-amenability and amenability were equivalent for all nuclear G-algebras (and therefore also that amenability was equivalent to $A \rtimes_{\max} G = A \rtimes_r G$ in the nuclear case). Indeed, if A is a a C-amenable nuclear G-algebra and B is any C^*-algebra with trivial G-action, then

$$\begin{aligned}(A \rtimes_r G) \otimes_{\max} B &= (A \rtimes_{\max} G) \otimes_{\max} B = (A \otimes_{\max} B) \rtimes_{\max} G \\ &= (A \otimes B) \rtimes_{\max} G = (A \otimes B) \rtimes_r G = (A \rtimes_r G) \otimes B,\end{aligned}$$

where we have used Proposition 4.16 (twice), nuclearity of A to replace the maximal tensor product with the spatial one, and standard facts about commuting tensor products with trivial G-algebras with crossed products. This implies that $A \rtimes_r G$ is nuclear and hence that the action of G on A is amenable by [**4**, Théorème 4.5].

6. Can non-exact groups admit amenable actions on unital C^*-algebras?

It is well-known that a group admits a strongly amenable action on a unital commutative G-algebra if and only if the group is exact. From this, it is clear that a non-exact group cannot admit a strongly amenable action on a unital G-algebra: indeed, the action on the unital commutative C^*-algebra $Z(A)$ is then also strongly amenable.

However, as discussed in Section 3, Suzuki's examples show that amenability and strong amenability are not equivalent for unital noncommutative (even nuclear) G-algebras. It is therefore natural to ask whether a non-exact group can admit an amenable action on any unital G-algebra.

The purpose of this section is to show that the answer to this question is 'no', even if we replace amenability by the, *a priori,* weaker condition of C-amenability.

THEOREM 6.1. *Say G is a discrete group and A is a unital (nonzero) C-amenable G-algebra. Then the following hold:*

(i) *if A is nuclear, the inclusion*

$$A \rtimes_r G \to (A \otimes A^{\mathrm{op}}) \rtimes_r G$$

induced by the equivariant map $A \to A \otimes A^{\mathrm{op}}$, $a \mapsto a \otimes 1$, is nuclear;

(ii) *if A is exact, then $A \rtimes_r G$ is exact;*

(iii) *G is exact.*

COROLLARY 6.2. *For a discrete group G, the following assertions are equivalent:*

(i) *G is exact;*
(ii) *G admits a strongly amenable action on a unital nonzero C^*-algebra;*
(iii) *G admits an amenable action on a unital nonzero C^*-algebra;*
(iv) *G admits a C-amenable action on a unital nonzero C^*-algebra.*

PROOF. The implications (ii)⇒(iii)⇒(iv) are straightforward, and (iv)⇒(i) is Theorem 6.1. The implication from (i) to (ii) follows as if G is exact, then $\ell^\infty(G)$ is strongly amenable (see Theorem 2.5). □

REMARK 6.3. Exel [**11**, Definition 20.4] has introduced a different notion of amenability for G-algebras (and more generally for Fell bundles) under the name of the *approximation property.* He asked whether the existence of a unital G-algebra with his approximation property implies exactness of G. Since [**1**, Theorem 6.11 and Corollary 6.16] give that Exel's approximation property implies amenability, the answer is 'yes'.

For the proof of Theorem 6.1 we need some technical preparations. For a G-algebra A, see Definition 2.7 for the module $\ell^2(G, A)$. The proof of the following lemma is based partly on ideas of Anantharaman-Delaroche from the paper [**4**].

LEMMA 6.4. *Say G is a discrete group, and A is a unital C-amenable G-algebra. Then for any $\epsilon > 0$ and any finite subset $\mathcal{G}$ of G there exists a function*

$\xi \in \ell^2(G, A^{\mathrm{op}})$ *such that:*

(i) ξ *is finitely supported;*
(ii) $\langle \xi, \xi \rangle \leqslant 1$;
(iii) *for all* $g \in \mathcal{G}$, $\|1_{A^{\mathrm{op}}} - \langle \xi, \tilde{\alpha}_g^{\mathrm{op}} \xi \rangle\| < \epsilon$.

PROOF. Let $\pi : A \to \mathbb{B}(H)$ be the restriction of a standard form representation (see Theorem 5.1) of A^{**} to A. Fix $\epsilon > 0$, and finite subsets $\mathcal{G}$ of G, and Φ of the state space of A^{op} respectively. As A is C-amenable, there exists a finitely supported function $\xi : G \to \pi(A)'$ such that $\langle \xi, \xi \rangle \leqslant 1$ and such that

$$|\phi(\langle \xi, \tilde{\alpha}_g^{\mathrm{op}} \xi \rangle - 1)| < \frac{\epsilon}{3} \tag{5}$$

for all $g \in \mathcal{G}$ and $\phi \in \Phi$. As $\pi(A)'$ canonically identifies with $(A^{\mathrm{op}})^{**}$ we have from [**4**, Lemme 1.1] that the finitely supported elements in the unit ball of $\ell^2(G, A^{\mathrm{op}})$ are dense in the unit ball of $\ell^2(G, \pi(A)')$ for the topology defined by the seminorms

$$\|\eta\|_\psi := \sqrt{\psi(\langle \eta, \eta \rangle)}$$

as ψ ranges over the state space of A^{op}. Hence there exists a finitely supported function $\eta : G \to A^{\mathrm{op}}$ such that $\langle \eta, \eta \rangle \leqslant 1$, and such that

$$\phi(\langle \xi - \eta, \xi - \eta \rangle)^{1/2} < \frac{\epsilon}{3} \tag{6}$$

for all $\phi \in \Phi$. For each $\phi \in \Phi$ and $g \in \mathcal{G}$, we have

$$|\phi(\langle \eta, \tilde{\alpha}_g^{\mathrm{op}} \eta \rangle - \langle \xi, \tilde{\alpha}_g^{\mathrm{op}} \xi \rangle)| \leqslant |\phi(\langle \eta - \xi, \tilde{\alpha}_g^{\mathrm{op}} \eta \rangle)| + |\phi(\langle \xi, \tilde{\alpha}_g^{\mathrm{op}} (\xi - \eta) \rangle)|. \tag{7}$$

The Cauchy-Schwarz inequality for the state ϕ then implies

$$|\phi(\langle \eta - \xi, \tilde{\alpha}_g^{\mathrm{op}} \eta \rangle)| \leqslant \phi(\langle \eta - \xi, \tilde{\alpha}_g^{\mathrm{op}} \eta \rangle^* \langle \eta - \xi, \tilde{\alpha}_g^{\mathrm{op}} \eta \rangle)^{1/2}.$$

Positivity of ϕ and Cauchy-Schwarz for the inner product on $\ell^2(G, \pi(A)')$ gives

$$\phi(\langle \eta - \xi, \tilde{\alpha}_g^{\mathrm{op}} \eta \rangle^* \langle \eta - \xi, \tilde{\alpha}_g^{\mathrm{op}} \eta \rangle)^{1/2} \leqslant \|\tilde{\alpha}_g^{\mathrm{op}} \eta\|_{\ell^2(G,A)} \phi(\langle \eta - \xi, \eta - \xi \rangle)^{1/2},$$

and hence we get from line (6) that

$$|\phi(\langle \eta - \xi, \tilde{\alpha}_g^{\mathrm{op}} \eta \rangle)| \leqslant \|\tilde{\alpha}_g^{\mathrm{op}} \eta\|_{\ell^2(G,A)} \phi(\langle \eta - \xi, \eta - \xi \rangle)^{1/2} < \frac{\epsilon}{3}.$$

Similarly,

$$|\phi(\langle \xi, \tilde{\alpha}_g^{\mathrm{op}} (\xi - \eta) \rangle)| < \frac{\epsilon}{3}.$$

Hence from line (7) we get

$$|\phi(\langle \eta, \tilde{\alpha}_g^{\mathrm{op}} \eta \rangle - \langle \xi, \tilde{\alpha}_g^{\mathrm{op}} \xi \rangle)| < \frac{2\epsilon}{3}$$

and from this and line (5) we get

$$|\phi(\langle \eta, \tilde{\alpha}_g^{\mathrm{op}} \eta \rangle - 1)| < \epsilon$$

for all $\phi \in \Phi$ and $g \in \mathcal{G}$.

Now, as the finite subset Φ of the state space of A^{op} was arbitrary, and as the states span $(A^{\mathrm{op}})^*$, this implies that in the C^*-algebra

$$B := \bigoplus_{g \in \mathcal{G}} A^{\mathrm{op}}$$

we have that zero is in the weak closure of the set

$$\Big\{ \big(\langle \eta, \tilde{\alpha}_g^{\mathrm{op}} \eta \rangle - 1_{A^{\mathrm{op}}}\big)_{g \in \mathcal{G}} \in B \mid \eta : G \to A^{\mathrm{op}} \text{ finitely supported and } \langle \eta, \eta \rangle \leqslant 1 \Big\}.$$

Hence by the Hahn-Banach theorem, zero is in the norm closure of the convex hull of this set.

It follows that for some given $\epsilon > 0$ we can find a finite collection $\xi_1, \dots, \xi_n$ of finitely supported functions $G \to A^{\mathrm{op}}$ and $t_1, \dots, t_n \in [0,1]$ with $\sum t_i = 1$ such that $\langle \xi_i, \xi_i \rangle \leqslant 1$ for each i, and such that

$$\Big\| 1_{A^{\mathrm{op}}} - \sum_{i=1}^{n} t_i \langle \xi_i, \tilde{\alpha}_g^{\mathrm{op}} \xi_i \rangle \Big\| < \frac{\epsilon}{2} \tag{8}$$

for all $g \in \mathcal{G}$. Define now $h : G \to A^{\mathrm{op}}$ by

$$h(g) := \sum_{i=1}^{n} t_i \langle \xi_i, \tilde{\alpha}_g^{\mathrm{op}} \xi_i \rangle.$$

Then clearly h is finitely supported, and one checks directly that it is positive type. Hence a GNS-type construction as in [**4**, Proposition 2.5] gives $\xi_0 \in \ell^2(G, A^{\mathrm{op}})$ with

$$h(g) = \langle \xi_0, \tilde{\alpha}_g^{\mathrm{op}} \xi_0 \rangle$$

for all $g \in G$. Note that $\langle \xi_0, \xi_0 \rangle = h(e) \leqslant 1$, whence $\|\xi_0\|_{\ell^2(G, A^{\mathrm{op}})} \leqslant 1$. Hence we may find finitely supported $\xi : G \to A^{\mathrm{op}}$ with $\|\xi\|_{\ell^2(G,A^{\mathrm{op}})} \leqslant 1$ and $\|\xi_0 - \xi\|_{\ell^2(G,A^{\mathrm{op}})} < \epsilon/4$. This gives that $\langle \xi, \xi \rangle \leqslant 1$ and that for any $g \in \mathcal{G}$,

$$\|\langle \xi_0, \tilde{\alpha}_g^{\mathrm{op}} \xi_0 \rangle - \langle \xi, \tilde{\alpha}_g^{\mathrm{op}} \xi \rangle\|_{A^{\mathrm{op}}} \leqslant 2\|\xi_0 - \xi\|_{\ell^2(G,A^{\mathrm{op}})} < \frac{\epsilon}{2}$$

by the Cauchy-Schwarz inequality for Hilbert modules (twice); combined with line (8) above and the fact that $\sum_{i=1}^{n} t_i \langle \xi_i, \tilde{\alpha}_g^{\mathrm{op}} \xi_i \rangle = \langle \xi_0, \tilde{\alpha}_g^{\mathrm{op}} \xi_0 \rangle$ for all $g \in G$, this completes the proof. □

For a finite subset F of G, let M_F denote $\mathbb{B}(\ell^2(F))$, i.e. the 'F-by-F matrices', and for each $g, h \in F$, let $e_{g,h} \in M_F$ denote the corresponding matrix unit. The following result is contained in the proof of [**7**, Lemma 4.2.3].

LEMMA 6.5. *With notation as above, let F be a finite subset of G. Then there is a well-defined ccp map determined by the formula*

$$\psi_F : A \rtimes_r G \to A \otimes M_F, \quad a\delta_g \mapsto \sum_{h \in F \cap gF} \alpha_{g^{-1}}(a) \otimes e_{h, g^{-1}h}.$$

Moreover, there is a well-defined cp map determined by the formula

$$\phi_F : A \otimes M_F \to A \rtimes_r G, \quad a \otimes e_{g,h} \mapsto \alpha_g(a)\delta_{gh^{-1}}. \quad \square$$

Most of the computations in the proof of the following result are inspired by [**7**, Lemma 4.3.3].

LEMMA 6.6. *Let A be a unital C-amenable G-algebra and let $A \otimes_{\max} A^{\mathrm{op}}$ be equipped with the diagonal action $\alpha \otimes_{\max} \alpha^{\mathrm{op}}$. Then for all $\epsilon > 0$ and finite subsets $\mathcal{G} \subseteq G$ and $\mathcal{A} \subseteq A$ there is a finite subset $F \subseteq G$ and ccp maps*

$$\begin{array}{ccc} A \rtimes_r G & & (A \otimes_{\max} A^{\mathrm{op}}) \rtimes_r G \\ & \searrow^{\psi} \quad {}^{\phi}\nearrow & \\ & A \otimes M_F & \end{array}$$

such that $\|\phi\psi(a\delta_g) - (a \otimes 1)\delta_g\| < \epsilon$ for all $a \in \mathcal{A}$ and $g \in \mathcal{G}$.

PROOF. Let $d := \max_{a\in\mathcal{A}}\|a\|$ and let $\xi : G \to A^{\mathrm{op}}$ have the properties as in Lemma 6.4 with respect to the finite set $\mathcal{G}$, and the constant ϵ/d, so in particular

$$\|1_{A^{\mathrm{op}}} - \langle \xi, \tilde{\alpha}_g^{\mathrm{op}}\xi\rangle\| < \frac{\epsilon}{d} \tag{9}$$

for all $g \in \mathcal{G}$. Let $F \subseteq G$ be a finite set such that $F \cap gF$ contains the support of ξ for all $g \in \mathcal{G}$ (for example, $F = \mathrm{supp}(\xi) \cup \bigcup_{g\in\mathcal{G}} g^{-1}\mathrm{supp}(\xi)$ works). Define first

$$\psi : A \rtimes_r G \to A \otimes M_F$$

to be the map ψ_F from Lemma 6.5, which is ccp. Define $X \in A^{\mathrm{op}} \otimes M_F$ by

$$X := \sum_{h\in F} \alpha_{h^{-1}}^{\mathrm{op}}(\xi(h)) \otimes e_{hh}$$

and define

$$\kappa : A \otimes M_F \to A \otimes_{\max} A^{\mathrm{op}} \otimes M_F, \quad a \otimes e_{g,h} \mapsto (1_A \otimes X^*)(a \otimes 1_{A^{\mathrm{op}}} \otimes e_{g,h})(1_A \otimes X)$$

Clearly κ is cp. Finally, let $\phi_F : A \otimes_{\max} A^{\mathrm{op}} \otimes M_F \to (A \otimes_{\max} A^{\mathrm{op}}) \rtimes_r G$ be as in Lemma 6.5 applied to the G-algebra $A \otimes_{\max} A^{\mathrm{op}}$, and define

$$\phi : A \otimes M_F \to (A \otimes_{\max} A^{\mathrm{op}}) \rtimes_r G, \quad \phi := \phi_F \circ \kappa.$$

We show that ϕ and ψ have the claimed properties.

Indeed, we already have from Lemma 6.5 that ψ is ccp. As we also already know from the same lemma that ϕ is cp, to see that it is ccp it suffices to show that $\phi(1) \leqslant 1$. For this, we compute that

$$\begin{aligned}\phi(1) &= \phi_F\Big((1 \otimes X^*)1(1 \otimes X)\Big) = \phi_F\Big(\sum_{h\in F} 1_A \otimes \alpha_{h^{-1}}^{\mathrm{op}}(\xi(h)^*\xi(h)) \otimes e_{h,h}\Big) \\ &= 1_A \otimes \sum_{h\in F} \xi(h)^*\xi(h) = 1_A \otimes \langle \xi, \xi\rangle \leqslant 1\end{aligned}$$

as claimed. It remains to show that

$$\|\phi\psi(a\delta_g) - (a \otimes 1)\delta_g\| < \epsilon,$$

for all $a \in \mathcal{A}$ and $g \in \mathcal{G}$. For this, we compute that

$$\begin{aligned}\phi\psi(a\delta_g) &= \phi_F\kappa\Big(\sum_{h\in F\cap gF} \alpha_{h^{-1}}(a) \otimes e_{h,g^{-1}h}\Big) \\ &= \phi_F\Big(\Big(1_A \otimes \sum_{k\in F} \alpha_{k^{-1}}^{\mathrm{op}}(\xi(k)^*) \otimes e_{k,k}\Big) \cdot \Big(\sum_{h\in F\cap gF} \alpha_{h^{-1}}(a) \otimes 1_{A^{\mathrm{op}}} \otimes e_{h,g^{-1}h}\Big) \\ &\qquad \cdot \Big(1_A \otimes \sum_{l\in F} \alpha_{l^{-1}}^{\mathrm{op}}(\xi(l)) \otimes e_{l,l}\Big)\Big) \\ &= \phi_F\Big(\sum_{h\in F\cap gF} \alpha_{h^{-1}}(a) \otimes \alpha_{h^{-1}}^{\mathrm{op}}\big(\xi(h)^*\alpha_g^{\mathrm{op}}(\xi(g^{-1}h)\big) \otimes e_{h,g^{-1}h}\Big).\end{aligned}$$

Hence using the formula for ϕ_F in Lemma 6.5 we get

$$\phi\psi(a\delta_g) = \sum_{h\in F\cap gF} \big(a \otimes \xi(h)^*\alpha_g^{\mathrm{op}}(\xi(g^{-1}h))\big)\delta_g = \big(a \otimes \langle \xi, \tilde{\alpha}_g^{\mathrm{op}}\xi\rangle\big)\delta_g,$$

where the last equality follows from the fact that $F \cap gF$ contains the support of ξ for all $g \in \mathcal{G}$. Thus we get

$$\|\phi\psi(a\delta_g) - (a \otimes 1_{A^{\mathrm{op}}})(1 \otimes \delta_g)\| \leqslant (\max_{a\in\mathcal{A}}\|a\|)\|1 - \langle \xi, \tilde{\alpha}_g^{\mathrm{op}}\xi\rangle\| < \epsilon.$$

□

PROOF OF THEOREM 6.1. Suppose first that A is nuclear. The result follows then from Lemma 6.6 together with the fact that the identity map on $A \otimes M_F$ is nuclear, hence approximately factors through finite dimensional C^*-algebras via ccp maps.

Suppose now that A is exact. Then there exists a unital nuclear embedding $\tau_F : A \otimes M_F \hookrightarrow \mathbb{B}(H_F)$ for some Hilbert space H_F. Let $\sigma : (A \otimes_{\max} A^{\mathrm{op}}) \rtimes_r G \to \mathbb{B}(H)$ be any unital faithful representation. Then the diagram of Lemma 6.6 can be extended to a diagram

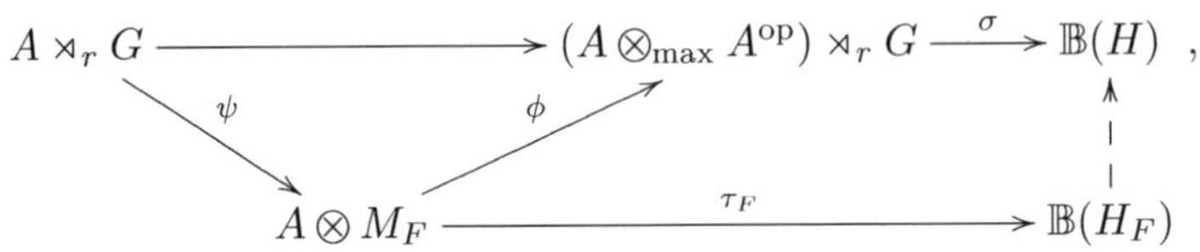

where the dashed arrow can be filled in with a ucp map by Arveson's extension theorem so that the right hand quadrilateral honestly commutes. As the map $A \otimes M_F \to \mathbb{B}(H_F)$ is nuclear, the existence of these diagrams gives that $A \rtimes_r G$ is nuclearly embeddable, so exact.

Finally, for exactness of G, we note that the above maps restricted to $C_r^*(G) \subseteq A \rtimes_r G$ show that $C_r^*(G)$ is nuclearly embedded in $\mathbb{B}(H)$ (whether or not A is exact), as the 'downwards' map ψ takes image in $M_F(\mathbb{C})$ when restricted to $C_r^*(G)$. Hence $C_r^*(G)$ is exact, and thus G is itself exact as it is discrete. □

QUESTION 6.7. Is it possible to improve the arguments for Theorem 6.1 to show that $A \rtimes_r G$ is nuclear for any C-amenable nuclear G-algebra A? As discussed in Question 5.9 above, this would imply that amenablity, C-amenability, and weak containment are equivalent for nuclear G-algebras.

7. Characterizing equivariant injectivity and the equivariant WEP

In this section, we study G-injectivity and the G-WEP in more detail. In particular, we give complete characterizations of both in terms of amenability and the respective non-equivariant versions. We also collect together many equivalent conditions in the special case of nuclear G-algebras and exact groups.

The following definition is partly inspired by work of Anantharaman-Delaroche [**3**, Section 2] in the setting of von Neumann algebras, and of Kirchberg [**17**, Proposition 3.1] in the non-equivariant setting.

DEFINITION 7.1. Let $\iota : A \hookrightarrow B$ be a G-embedding of G-algebras. The embedding is *relatively G-injective* (respectively, *weakly relatively G-injective*) if there exists a ccp G-map $P : B \twoheadrightarrow A$ splitting ι (respectively, a ccp G-map $P : B \to A^{**}$ such that $P \circ \iota : A \to A^{**}$ is the canonical inclusion).

Our main goals in this section are the following results. Note the different role played by exactness in the first two theorems: this is essentially due to the fact that if G admits an action on some A such that A^{**} is G-injective, then G must be exact; this is not true for the G-WEP, however.

THEOREM 7.2. *For a unital G-algebra A, the following are equivalent:*

(i) *A is amenable and has the WEP;*
(ii) *A has the G-WEP and G is exact.*

THEOREM 7.3. *For a G-algebra A, the following are equivalent:*

(i) *A is amenable and A^{**} is injective;*
(ii) *A^{**} is G-injective.*

Moreover, if there is a unital G-algebra satisfying these conditions, then G is exact.

The following theorem, which is essentially 'just' a compilation of our results and other results of Claire Anantharaman-Delaroche from [**2**] and [**4**], summarises some of the known facts about amenable actions of exact groups on nuclear C^*-algebras, and the relationships to G-injectivity and the G-WEP.

THEOREM 7.4. *If G is an exact group and A is a nuclear G-algebra, then the following assertions are equivalent:*

(i) *A is amenable;*
(ii) *A has the G-WEP;*
(iii) *A^{**} is G-injective;*
(iv) *A^{**} is strongly amenable;*
(v) *A^{**} is amenable;*
(vi) *$A \rtimes_r G$ is nuclear;*
(vii) *$A^{**} \bar{\rtimes} G$ is injective.*

Here $A^{**} \bar{\rtimes} G$ denotes the von Neumann algebra crossed product of A^{**} with G.

We cannot add the condition that A is strongly amenable to the equivalent conditions in Theorem 7.4 by Suzuki's examples in [**25**]. We do not know if we can add $A \rtimes_{\max} G = A \rtimes_r G$ to this list of equivalent conditions: compare Questions 5.9 and 6.7.

The proofs will proceed via a series of ancillary lemmas and propositions. The first few results compare G-injectivity and the G-WEP to the non-equivariant versions (sometimes also in the presence of amenability).

LEMMA 7.5. *Let G be a group, $H \subseteq G$ a subgroup and A a G-algebra. If A has the G-WEP (respectively, is G-injective), then A also has the H-WEP (respectively, is H-injective) with the restricted action.*

PROOF. Let $\pi : A \to \mathbb{B}(K)$ be a faithful representation (ignoring the G-action). Equip $B := \ell^\infty(G, \mathbb{B}(K))$ with the translation G-action defined by

$$(\gamma_g f)(h) := f(g^{-1}h)$$

and then consider the canonical G-embedding

$$\tilde{\pi} \colon A \hookrightarrow B, \quad \tilde{\pi}(a)(g) := \pi(\alpha_g^{-1}(a)).$$

If A has the G-WEP, then there exists a ccp G-map $P \colon B \to A^{**}$ with $P \circ \tilde{\pi}$ equal to the canonical embedding $A \hookrightarrow A^{**}$. Note that B is H-injective as an H-algebra with the restricted H-action: this is proved in [**9**, Remark 6.3]. Hence if A embeds H-equivariantly into some H-algebra C, then by the H-injectivity of B the dashed arrow below can be filled in

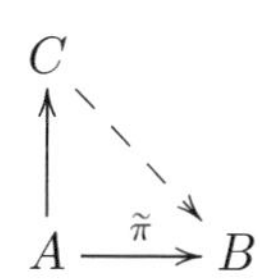

with a ccp H-map. Composing this with $P : B \to A^{**}$ yields the desired ccp H-map $C \to A^{**}$ extending the canonical inclusion $A \hookrightarrow A^{**}$.

The assertion on injectivity can be proved in essentially the same way. □

COROLLARY 7.6. *If a G-algebra has the G-WEP (respectively, is G-injective) then it has the WEP (respectively, is injective).*

PROOF. Take H to be the trivial group in Lemma 7.5. □

The following result is closely related to [**2**, Proposition 4.1]; we need something a little different, however, so give a direct proof. The statement essentially says that a non-equivariant ccp splitting can be bootstrapped up to an equivariant splitting in the presence of amenability.

LEMMA 7.7. *Let A be an amenable G-algebra. Let $A \subseteq B$ (respectively, $A^{**} \subseteq B$) be a G-embedding. Then if the embedding is relatively weakly injective (respectively, relatively injective), it is also relatively weakly G-injective (respectively, relatively G-injective).*

PROOF. We will just look at the case where $A^{**} \subseteq B$ is a relatively injective G-embedding; the other case is essentially the same. Relative injectivity gives us a ccp map $\phi : B \to A^{**}$ such that the restriction of ϕ to A^{**} is the identity; our task is to replace ϕ with a ccp G-map without changing it on A^{**}.

Let $\ell^2(G, A^{**})$ be as in Definition 2.7, equipped with the associated action $\tilde{\alpha}$ defined there. For each $b \in B$, define a multiplication-type operator $m(b)$ on $\ell^2(G, A^{**})$ by the formula

$$(m(b)\xi)(g) := \alpha_g(\phi(\beta_{g^{-1}}(b)))\xi(g).$$

Then it is not difficult to see that m defines a ccp map $m : B \to \mathbb{B}(\ell^2(G, A^{**}))$ from B to the adjointable operators on $\ell^2(G, A^{**})$. Let now $(\xi_i)_{i\in I}$ be a net as in the definition of amenability, so each ξ_i is a finitely supported function $\xi_i : G \to Z(A^{**})$ such that $\langle \xi_i, \xi_i\rangle \leqslant 1$ for all i, and so that $\langle \xi_i, \tilde{\alpha}_g(\xi_i)\rangle$ converges ultraweakly to 1 for all $g \in G$. For each i, define a map

$$\psi_i : B \to A^{**}, \quad b \mapsto \langle \xi_i, m(b)\xi_i\rangle.$$

One then checks that the net (ψ_i) consists of ccp maps, and so, by [**7**, Theorem 1.3.7] and after passing to a subnet if necessary, has an ultraweak limit point, which is also a ccp map $\psi : B \to A^{**}$. We claim that this limit has the right properties.

First, let us check that if a is an element of A^{**}, then $\psi(a) = a$. Indeed, in this case $m(a)$ is just the operator of left-multiplication by a, and so we have

$$\psi_i(a) = \langle \xi_i, a\xi_i\rangle = \sum_{g\in G} \xi_i(g)^* a\xi_i(g).$$

for all i. As ξ takes values in $Z(A^{**})$, this just equals $\langle \xi_i, \xi_i\rangle a$, however, which converges ultraweakly to a as i tends to infinity.

It remains to check that ψ is equivariant. Let then $b \in B$ and $h \in G$. Then

$$\psi_i(\beta_h(b)) = \langle \xi_i, m(\beta_h(b))\xi_i\rangle = \sum_{g\in G} \xi_i(g)^* \alpha_g\phi(\beta_{g^{-1}h}(b))\xi_i(g).$$

Making the substitution $k = h^{-1}g$, this equals

$$\begin{aligned}\sum_{k\in G} \xi_i(hk)^*\alpha_{hk}\phi(\beta_{k^{-1}}(b))\xi_i(hk) &= \alpha_h\Big(\sum_{k\in G}(\widetilde{\alpha}_{h^{-1}}\xi_i)(k)^*\alpha_k\phi(\beta_{k^{-1}}(b))(\widetilde{\alpha}_{h^{-1}}\xi_i)(k)\Big)\\ &= \alpha_h(\langle\widetilde{\alpha}_{h^{-1}}\xi_i, m(b)\widetilde{\alpha}_{h^{-1}}\xi_i\rangle).\end{aligned}$$

To prove equivariance, it thus suffices to show that

$$\langle\widetilde{\alpha}_{h^{-1}}\xi_i, m(b)\widetilde{\alpha}_{h^{-1}}\xi_i\rangle - \langle\xi_i, m(b)\xi_i\rangle \tag{10}$$

tends ultraweakly to zero. This follows as we have the identity

$$\langle\widetilde{\alpha}_{h^{-1}}\xi_i - \xi_i, \widetilde{\alpha}_{h^{-1}}\xi_i - \xi_i\rangle = \alpha_{h^{-1}}(\langle\xi_i,\xi_i\rangle) + \langle\xi_i,\xi_i\rangle - \langle\xi_i,\widetilde{\alpha}_{h^{-1}}\xi_i\rangle - \langle\widetilde{\alpha}_{h^{-1}}\xi_i,\xi_i\rangle,$$

and the right hand side tends ultraweakly to zero. The expression in line (10) therefore tends ultrweakly to zero using appropriate versions of the Cauchy-Schwarz inequality similarly to the proof of Lemma 6.4. □

COROLLARY 7.8. *Let A be an amenable G-algebra. Then if A has the WEP (respectively, if A^{**} is injective), then A has the G-WEP (respectively, A^{**} is G-injective).* □

The next lemma is closely related to Lemma 4.12.

LEMMA 7.9. *Let A be a G-algebra and assume A is G-injective (respectively, has the G-WEP). Let C be any unital G-algebra. Then there is a ucp G-map $C \to Z(A)$ (respectively, $C \to Z(A^{**})$).*

PROOF. Assume first that A has the G-WEP. Let B be the G-algebra $B := C \otimes A$ equipped with the diagonal action $\gamma\otimes\alpha$ where γ denotes the action on C, and α the action on A. Consider the canonical G-embedding

$$\iota\colon A \hookrightarrow B, \quad a \mapsto 1\otimes a.$$

Since A has the G-WEP, there is a ccp G-map $P\colon B \to A^{**}$ such that $P\circ\iota$ coincides with the canonical embedding $A \hookrightarrow A^{**}$. Fix an asymptotically G-invariant[1] approximate unit $(e_i)_{i\in I}$ for A, and for each i, define

$$P_i : C \to A^{**}, \quad c \mapsto P(c\otimes e_i).$$

The net (P_i) of ccp maps has a point-ultraweak limit, say $Q : C \to A^{**}$, which we claim is the required map. As Q is automatically ccp, we must check three things: that Q has image in $Z(A^{**})$; Q is unital; and that Q is equivariant.

Indeed, note first that the subalgebra $\{1\otimes a \mid a \in A\}$ of B is in the multiplicative domain of P, whence for each $a \in A$, $c \in C$, and $i \in I$,

$$aP_i(c) - P_i(c)a = aP(c\otimes e_i) - P(c\otimes e_i)a = P(c\otimes(ae_i - e_ia)),$$

which tends to zero (in norm) as i tends to infinity. Hence the image of Q commutes with A, and thus with all of A^{**}, so is central. To see that Q is unital, note that $P_i(1) = P(1\otimes e_i) = e_i$, and that any approximate unit for A converges ultraweakly to the unit of A^{**}. Finally, to see that Q is equivariant, we note that P is equivariant, whence if α denotes the G-actions on both A and A^{**}, and γ the

[1]This means that $\alpha_g(e_i) - e_i$ tends to zero in norm for each $g \in G$; such approximate units always exist.

action on C, then for any $c \in C$ and $i \in I$, we have

$$\alpha_g(P_i(c)) - P_i(\gamma_g(c)) = \alpha_g(P(c \otimes e_i)) - P(\gamma_g(c) \otimes e_i)) = P(\gamma_g(c) \otimes (\alpha_g(e_i) - e_i)),$$

which tends to zero (in norm) as i tends to infinity.

The case where A is G-injective is similar and easier because in this case A is now unital (see Lemma 4.3 above). Indeed, consider again the same embedding $\iota\colon A \hookrightarrow B$ as above. Notice that A, B and ι are unital. Since A is now G-injective, we get a ucp G-map $P\colon B \to A$ satisfying $P \circ \iota = \mathrm{id}_A$. The same argument as before shows that $P(C) \subseteq Z(A)$. Composing with the canonical embedding $C \hookrightarrow B$ yields the desired ucp G-map $Q : C \to Z(A)$. □

COROLLARY 7.10. *If G is an exact group and A is a G-algebra that is G-injective (respectively, has the G-WEP), then it is strongly amenable (respectively, amenable).*

PROOF. This follows directly from Proposition 2.6 and Lemma 7.9 in the special case $C = \ell^\infty(G)$. □

We are finally ready to prove our main theorems from the start of this section.

PROOF OF THEOREM 7.2. Assume first that A has the G-WEP and that G is exact. Then A has the WEP by Corollary 7.6, and is amenable by Corollary 7.10.

Conversely, if A has the WEP and is amenable, then it has the G-WEP by Corollary 7.8. As A is unital, the existence of an amenable action implies that G is exact by Theorem 6.1. □

REMARK 7.11. Say G is an exact group. Then the above proof shows that for any (not necessarily unital) G-algebra A, the following are equivalent:

(i) A is amenable and has the WEP;
(ii) A has the G-WEP.

In other words, if we are willing to assume exactness, we can drop the unitality assumption from Theorem 7.2. The above equivalences do not hold (for unital algebras) in the non-exact case: indeed, $A = \ell^\infty(G)$ is G-injective, so in particular has the G-WEP, but it is not amenable if G is not exact. On the other hand, the equivalence of

(i) A is amenable and has the WEP, and
(ii) A has the G-WEP and G is exact

from Theorem 7.2 do not hold in the non-unital case: $A = C_0(G)$ is amenable and has the WEP, so satisfies the first condition whether G is exact or not.

PROOF OF THEOREM 7.3. Assume first that A^{**} is G-injective. Then A^{**} is injective by Corollary 7.6. Moreover, G-injectivity of A^{**} gives an equivariant conditional expectation $\ell^\infty(G, A^{**}) \to A^{**}$ as in [**4**, Théorème 3.3, part (e)], which implies amenability.

Conversely, say A^{**} is injective and amenable. Then A^{**} is G-injective by Corollary 7.8.

Finally if A is unital, then amenability of A implies exactness of G by Theorem 6.1. □

PROOF OF THEOREM 7.4. As A is nuclear, it has the WEP (see for example [**7**, Corollary 3.6.8]). Hence the equivalence of (i) and (ii) follows from Theorem 7.2.

Similarly, if A is nuclear than A^{**} is injective, and so the equivalence of (iii) and (i) follows from Theorem 7.3.

The fact that (iii) implies (iv) follows from Corollary 7.10, and (iv) implies (v) is trivial. Assuming that A^{**} is amenable, note that the universal property of A^{****} gives a normal equivariant surjective $*$-homomorphism $A^{****} \twoheadrightarrow A^{**}$ splitting the canonical inclusion $A^{**} \hookrightarrow A^{****}$. This restricts to a normal $*$-homomorphism $Z(A^{****}) \twoheadrightarrow Z(A^{**})$, from which it follows that A is amenable, giving (i). We now have that conditions (i) through (v) are equivalent.

Finally, note that the equivalence of (i) to both (vi) and (vii) was established by Anantharaman-Delaroche in [**4**, Théorème 4.5]. □

REMARK 7.12. In Theorem 7.2, we have compared the G-WEP for A to the WEP for A and amenability type conditions. It is also natural to compare the G-WEP for A to the WEP for the crossed products $A \rtimes_r G$ and $A \rtimes_{\max} G$.

We first note that if A is amenable and has the WEP, then $A \rtimes_{\max} G = A \rtimes_r G$ has the WEP. This was proved by Bhattarcharya and Farenick in [**6**]. One can also give a short argument using that a C^*-algebra B has the WEP if and only if

$$B \otimes_{\max} C^*(\mathbb{F}_\infty) = B \otimes C^*(\mathbb{F}_\infty)$$

(see for example [**7**, Corollary 13.2.5]). If G is exact, we already know from Theorem 7.2 that A has the G-WEP if and only if A is amenable and has the WEP. In particular, if A has the G-WEP and G is exact, then $A \rtimes_{\max} G = A \rtimes_r G$ has the WEP.

On the other hand, is is shown in [**9**, Proposition 5.4] that if $A \rtimes_{\mathrm{inj}} G$ has the WEP, then $A \rtimes_{\max} G = A \rtimes_{\mathrm{inj}} G$. Moreover, if $A \rtimes_{\mathrm{inj}} G$ has the WEP, then so does A because A is a C^*-subalgebra of $A \rtimes_{\mathrm{inj}} G$ with a conditional expectation $A \rtimes_{\mathrm{inj}} G \twoheadrightarrow A$. Hence if G is exact and $A \rtimes_r G$ has the WEP, then we have that $A \rtimes_{\max} G = A \rtimes_{\mathrm{inj}} G = A \rtimes_r G$ and that A has the WEP.

Summarizing the above discussion, if G is an exact group, then we know that

$$A \text{ has } G\text{-WEP} \quad \Rightarrow \quad A \rtimes_r G \text{ has WEP}$$

and that

$$A \rtimes_r G \text{ has WEP} \quad \Rightarrow \quad A \text{ has WEP and } A \rtimes_{\max} G = A \rtimes_r G.$$

If A is commutative, the latter condition also implies that A is amenable by Theorem 5.2, and therefore that A has G-WEP by Theorem 7.2. The precise situation is not clear in general, however.

If G is not exact, then things are murky. For example, it is not clear whether the C^*-algebras $\ell^\infty(G) \rtimes_{\max} G$ or $\ell^\infty(G) \rtimes_r G$ could have the WEP if the G-action on $A = \ell^\infty(G)$ is not amenable.

8. Hamana's theory of injective envelopes

In this section, we discuss the relation of the notion of injectivity that we have been using with Hamana's from [**15**] (they turn out to be the same, fortunately). We also use some of our work above to address some questions about injective envelopes that seem to be of interest in their own right.

Hamana's definition of G-injectivity is as follows. Consider a diagram

(11)

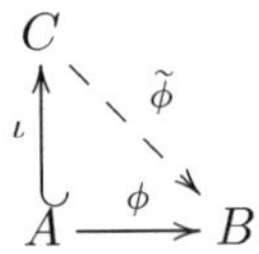

where B, C, and A are operator systems equipped with G-actions by complete order automorphisms, ι is a complete order injection, and ϕ is a ucp G-map. Then B is *G-injective* if the dashed arrow can be filled in with a ucp G-map.

On the other hand, in Definition 4.2, we say that a G-algebra is *injective* if in a diagram of the form (11) where C is a G-algebra, ι is an injective equivariant $*$-homomorphism, and ϕ is the identity map, the dashed arrow can be filled in with a ccp G-map.

Now, both Hamana's definition and our definition make sense for unital G-algebras. The two notions coincide, as we now show. In fact, they even coincide with respect to their domains of definition: this follows as injectivity of a G-algebra B in our sense forces B to be unital by Lemma 4.3, and injectivity of a G-operator system B in Hamana's sense forces B to admit a structure of a (unital) G-algebra by the proof of [**23**, Theorem 15.2].

PROPOSITION 8.1. *A unital G-algebra B is injective in the sense of Definition 4.2 if and only if it is injective in the sense of Hamana.*

PROOF. First assume that B satisfies Definition 4.2. In [**9**, Corollary 2.4] (compare also [**15**, Lemma 2.2]) it is shown that any (unital) G-algebra B admits a (unital) embedding $B \to B_H$ into a G-algebra B_H that is injective in Hamana's sense. Definition 4.2 gives an equivariant ucp map $E : B_H \to B$ splitting this inclusion. Consider now a diagram as in line (11) where ϕ and ι satisfy the conditions in Hamana's definition of injectivity. Consider the diagram

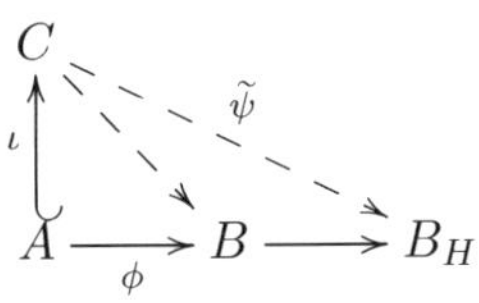

.

As B_H is injective in Hamana's sense, the long diagonal arrow can be filled in with an equivariant ucp map, say $\widetilde{\psi}$. The required map $\widetilde{\phi}$ can then be defined by $\widetilde{\phi} := E \circ \widetilde{\psi}$; it is not difficult to check that this works.

Conversely, say B is injective in Hamana's sense. We need to show that any injective equivariant $*$-homomorphism $\iota : B \hookrightarrow C$ admits an equivariant ccp splitting. Replacing C with the corner $\iota(1)C\iota(1)$, it is not difficult to see that we can assume that ι is unital, and the result is now straightforward. $\square$

We now turn to G-injective envelopes. Recall that in [**15**, Theorem 2.5], Hamana proves that every G-operator system (and in particular, every unital G-algebra) A has a *G-injective envelope* $I_G(A)$. This is a G-algebra $I_G(A)$ which is G-injective, equipped with a canonical unital G-embedding $A \to I_G(A)$, and has the universal property that whenever $A \to B$ is a ucp G-map into an injective operator system, there is a unique equivariant ucp extension $I_G(A) \to B$.

The following theorem provides a nice addition to the equivalent conditions in Theorem 7.4.

THEOREM 8.2. *Let G be an exact group, and let A be a unital nuclear G-algebra. The following are equivalent:*

(i) *A^{**} is G-injective;*
(ii) *A has the G-WEP;*
(iii) *there is a ucp G-map $I_G(A) \to A^{**}$ extending the inclusion $A \hookrightarrow A^{**}$;*
(iv) *there is a completely isometric ucp G-map $I_G(A) \hookrightarrow A^{**}$ extending the inclusion $A \hookrightarrow A^{**}$;*
(v) *the inclusion $A \hookrightarrow I_G(A)$ is relatively weakly G-injective in the sense of Definition 7.1.*

PROOF. The equivalence of (i) and (ii) is already proved in Theorem 7.4. Starting with (i), note that if A^{**} is G-injective, then we have a ucp G-map $I_G(A) \to A^{**}$ extending the canonical embedding $A \hookrightarrow A^{**}$. Hence (i) implies (iii). By G-injectivity of $I_G(A)$ we have a ucp G-map $A^{**} \to I_G(A)$ extending the identity on A. The composition with the map of (iii) is a G-ucp map $I_G(A) \to I_G(A)$ extending the identity on A. By the universal property of $I_G(A)$ it must be the identity. This gives (iii) implies (iv). It is clear that (iv) implies (v).

Finally, we claim that (v) implies (ii). Indeed, if A embeds into some G-algebra B, since $I_G(A)$ is G-injective, the inclusion $A \hookrightarrow I_G(A)$ extends to a ccp G-map $B \to I_G(A)$. Composing this with the map $I_G(A) \to A^{**}$ given by (v) yields the desired ccp G-map $B \to A^{**}$ extending the inclusion $A \hookrightarrow A^{**}$. □

We conclude this section with two results that can be seen as generalizations of results of Kalantar and Kennedy in their seminal work on the Furstenberg boundary [**16**]. The only new idea needed for the proofs in both cases is Corollary 7.10.

The first generalization of the work of Kalantar and Kennedy is as follows. In [**16**, Theorem 4.5], Kalantar and Kennedy prove that G is exact if only if its action on the Furstenberg boundary $\partial_F G$ is amenable. Recalling that $C(\partial_F G)$ is the G-injective envelope of $\mathbb{C}$ (see [**16**, Theorem 3.11]), the following result is a natural extension.

THEOREM 8.3. *The following are equivalent for a discrete group G.*

(i) *G is exact;*
(ii) *G acts strongly amenably (respectively, amenably, or C-amenably) on the injective envelope $I_G(A)$ of every nonzero G-algebra A.*
(iii) *G acts strongly amenably (respectively, amenably, or C-amenably) on the injective envelope $I_G(A)$ of some nonzero G-algebra A.*

PROOF. Assume G is exact, and let A be a G-algebra. Then as $I_G(A)$ is G-injective, Corollary 7.10 gives that $I_G(A)$ is strongly amenable. Hence (i) implies (ii). Since every injective C^*-algebra is unital by Lemma 4.3, (iii) implies (i) follows from Theorem 6.1. □

The second concerns a conjecture of Ozawa. In [**22**], Ozawa conjectures that every exact C^*-algebra B embeds into a nuclear C^*-algebra $N(B)$ with $B \subseteq N(B) \subseteq I(B)$. Here $I(B)$ denotes the *injective envelope* of B, see [**14**], which is the natural non-equivariant version of the G-injective envelope discussed above.

The above conjecture was established for $B = C^*_r(G)$ for any discrete group G by Kalantar and Kennedy in [**16**, Theorem 1.3] using the Furstenberg boundary $\partial_F G$.

Using the above observations we can prove Ozawa's conjecture for all crossed products of nuclear G-algebras by exact discrete groups. Note that in a previous version of this paper, we proved the following corollary only for the special case where A is commutative. We are grateful to Yuhei Suzuki for pointing out to us that the argument can be extended to cover all nuclear G-algebras.

COROLLARY 8.4. *Ozawa's conjecture holds for all C^*-algebras B of the form $B = A \rtimes_r G$, where A is a nuclear G-algebra and G is an exact group.*

PROOF. Let $I_G(A)$ be the G-injective envelope of A. By Theorem 8.3 (or Corollary 7.10), $I_G(A)$ is strongly amenable. Since $I_G(A)$ is unital, this implies that $Z(I_G(A))$ is (strongly) amenable as well. Let $D := C^*\big(A \cup Z(I_G(A))\big) \subseteq I_G(A)$ denote the C^*-subalgebra of $I_G(A)$ generated by A and $Z(I_G(A))$. Then D is a quotient of the nuclear algebra $A \otimes Z(I_G(A))$, and hence nuclear as well. Moreover, strong amenability of $Z(I_G(A)) \subseteq Z(D)$ implies strong amenability of D so that the crossed product $D \rtimes_r G$ is nuclear by [**4**, Théorème 4.5] (see also Theorem 7.4 above). On the other hand, by Hamana's results in [**15**, Theorem 3.4] we have that $I_G(A) \rtimes_r G \subseteq I(A \rtimes_r G)$, so that

$$A \rtimes_r G \subseteq D \rtimes_r G \subseteq I_G(A) \rtimes_r G \subseteq I(A \rtimes_r G),$$

so that Ozawa's conjecture holds with $N(B) = D \rtimes_r G$. $\square$

Acknowledgments

This paper started during visits of the first and third authors to the University of Münster. The first and third authors are grateful for the warm hospitality provided by the second author and the operator algebra group of that university. The authors thank Alex Bearden for pointing out a mistake in an earlier version, to Yuhei Suzuki for a very useful comment regarding Corollary 8.4 of this paper, and to the anonymous referee for several useful comments.

References

[1] Fernando Abadie, Alcides Buss, and Damián Ferraro, *Amenability and approximation properties for partial actions and Fell bundles*, 2019. arXiv:1907.03803.

[2] C. Anantharaman-Delaroche, *Action moyennable d'un groupe localement compact sur une algèbre de von Neumann* (French, with English summary), Math. Scand. **45** (1979), no. 2, 289–304, DOI 10.7146/math.scand.a-11844. MR580607

[3] C. Anantharaman-Delaroche, *Action moyennable d'un groupe localement compact sur une algèbre de von Neumann. II* (French, with English summary), Math. Scand. **50** (1982), no. 2, 251–268, DOI 10.7146/math.scand.a-11958. MR672928

[4] Claire Anantharaman-Delaroche, *Systèmes dynamiques non commutatifs et moyennabilité* (French), Math. Ann. **279** (1987), no. 2, 297–315, DOI 10.1007/BF01461725. MR919508

[5] Claire Anantharaman-Delaroche, *Amenability and exactness for dynamical systems and their C^*-algebras*, Trans. Amer. Math. Soc. **354** (2002), no. 10, 4153–4178, DOI 10.1090/S0002-9947-02-02978-1. MR1926869

[6] Angshuman Bhattacharya and Douglas Farenick, *Crossed products of C*-algebras with the weak expectation property*, New York J. Math. **19** (2013), 423–429. MR3104554

[7] Nathanial P. Brown and Narutaka Ozawa, *C^*-algebras and finite-dimensional approximations*, Graduate Studies in Mathematics, vol. 88, American Mathematical Society, Providence, RI, 2008. MR2391387

[8] Alcides Buss, Siegfried Echterhoff, and Rufus Willett, *Exotic crossed products and the Baum-Connes conjecture*, J. Reine Angew. Math. **740** (2018), 111–159, DOI 10.1515/crelle-2015-0061. MR3824785

[9] Alcides Buss, Siegfried Echterhoff, and Rufus Willett, *The maximal injective crossed product*, 2018. arXiv:1808.06804 (to appear in Ergodic Theory & Dyn. Systems).

[10] Alcides Buss, Siegfried Echterhoff, and Rufus Willett, *The minimal exact crossed product*, Doc. Math. **23** (2018), 2043–2077.

[11] Ruy Exel, *Partial dynamical systems, Fell bundles and applications*, Mathematical Surveys and Monographs, vol. 224, American Mathematical Society, Providence, RI, 2017. MR3699795

[12] Erik Guentner and Jerome Kaminker, *Exactness and the Novikov conjecture*, Topology **41** (2002), no. 2, 411–418, DOI 10.1016/S0040-9383(00)00036-7. MR1876896

[13] Uffe Haagerup, *The standard form of von Neumann algebras*, Math. Scand. **37** (1975), no. 2, 271–283, DOI 10.7146/math.scand.a-11606. MR407615

[14] Masamichi Hamana, *Injective envelopes of C^*-algebras*, J. Math. Soc. Japan **31** (1979), no. 1, 181–197, DOI 10.2969/jmsj/03110181. MR519044

[15] Masamichi Hamana, *Injective envelopes of C^*-dynamical systems*, Tohoku Math. J. (2) **37** (1985), no. 4, 463–487, DOI 10.2748/tmj/1178228589. MR814075

[16] Mehrdad Kalantar and Matthew Kennedy, *Boundaries of reduced C^*-algebras of discrete groups*, J. Reine Angew. Math. **727** (2017), 247–267, DOI 10.1515/crelle-2014-0111. MR3652252

[17] Eberhard Kirchberg, *On nonsemisplit extensions, tensor products and exactness of group C^*-algebras*, Invent. Math. **112** (1993), no. 3, 449–489, DOI 10.1007/BF01232444. MR1218321

[18] Christopher Lance, *On nuclear C^*-algebras*, J. Functional Analysis **12** (1973), 157–176, DOI 10.1016/0022-1236(73)90021-9. MR0344901

[19] E. C. Lance, *Hilbert C^*-modules*, London Mathematical Society Lecture Note Series, vol. 210, Cambridge University Press, Cambridge, 1995. A toolkit for operator algebraists. MR1325694

[20] Masayoshi Matsumura, *A characterization of amenability of group actions on C^*-algebras*, J. Operator Theory **72** (2014), no. 1, 41–47, DOI 10.7900/jot.2012sep07.1958. MR3246980

[21] Narutaka Ozawa, *Amenable actions and exactness for discrete groups* (English, with English and French summaries), C. R. Acad. Sci. Paris Sér. I Math. **330** (2000), no. 8, 691–695, DOI 10.1016/S0764-4442(00)00248-2. MR1763912

[22] Narutaka Ozawa, *Boundaries of reduced free group C^*-algebras*, Bull. Lond. Math. Soc. **39** (2007), no. 1, 35–38, DOI 10.1112/blms/bdl003. MR2303516

[23] Vern Paulsen, *Completely bounded maps and operator algebras*, Cambridge Studies in Advanced Mathematics, vol. 78, Cambridge University Press, Cambridge, 2002. MR1976867

[24] Mikael Rørdam and Adam Sierakowski, *Purely infinite C^*-algebras arising from crossed products*, Ergodic Theory Dynam. Systems **32** (2012), no. 1, 273–293, DOI 10.1017/S0143385710000829. MR2873171

[25] Yuhei Suzuki, *Simple equivariant C^*-algebras whose full and reduced crossed products coincide*, 2018. arXiv:1801.06949v1, to appear in JNCG.

[26] Rufus Willett, *Some notes on property A*, Limits of graphs in group theory and computer science, EPFL Press, Lausanne, 2009, pp. 191–281. MR2562146

Departamento de Matemática, Universidade Federal de Santa Catarina, 88.040-900 Florianópolis-SC, Brazil

Email address: alcides.buss@ufsc.br

Mathematisches Institut, Westfälische Wilhelms-Universität Münster, Einsteinstr. 62, 48149 Münster, Germany

Email address: echters@uni-muenster.de

Mathematics Department, University of Hawai'i at Mānoa, Keller 401A, 2565 McCarthy Mall, Honolulu, Hawaii 96822

Email address: rufus@math.hawaii.edu

Contemporary Mathematics
Volume **749**, 2020
https://doi.org/10.1090/conm/749/15070

Algebraic K-theory of planar cuspidal curves

Lars Hesselholt and Thomas Nikolaus

Introduction

The purpose of this paper is to evaluate the algebraic K-groups of a planar cuspidal curve over a perfect $\mathbb{F}_p$-algebra relative to the cusp point. A conditional calculation of these groups was given in [**4**, Theorem A], assuming a conjecture on the structure of certain polytopes. Our calculation here, however, is unconditional and illustrates the advantage of the new setup for topological cyclic homology by Nikolaus–Scholze [**10**], which we will be using throughout. The only input necessary for our calculation is the evaluation by the Buenos Aires Cyclic Homology group [**3**] and Larsen [**9**] of the structure of the Hochschild complex of the coordinate ring as a mixed complex, that is, as an object of the ∞-category of chain complexes with circle action.

We consider the planar cuspidal curve "$y^a = x^b$," where $a, b \geq 2$ are relatively prime integers. For $m \geq 0$, we define $\ell(a, b, m)$ to be the number of pairs (i, j) of positive integers such that $ai + bj = m$, and for $r \geq 0$, define $S(a, b, r)$ to be the set of positive integers m such that $\ell(a, b, m) \leq r$. The subset $S = S(a, b, r) \subset \mathbb{Z}_{>0}$ is a truncation set in the sense that if $m \in S$ and d divides m, then $d \in S$, and hence, the ring of (big) Witt vectors $\mathbb{W}_S(k)$ with underlying set k^S is defined. We refer to [**5**, Section 1] for a detailed introduction to Witt vectors.

THEOREM A. *Let k be a perfect $\mathbb{F}_p$-algebra, and let $a, b \geq 2$ be relatively prime integers. There is a canonical isomorphism*

$$K_j(k[x, y]/(y^a - x^b), (x, y)) \simeq \mathbb{W}_S(k)/(V_a\mathbb{W}_{S/a}(k) + V_b\mathbb{W}_{S/b}(k)),$$

if $j = 2r \geq 0$ with $S = S(a, b, r)$, and the remaining K-groups are zero.

We remark that recently Angeltveit [**1**] has given a different proof of this result that, unlike ours, employs equivariant homotopy theory. We also remark that the strategy employed in this paper was used by Speirs [**11**] to significantly simplify

---

2010 *Mathematics Subject Classification.* Primary 19D55; Secondary 55P43.

The first author was funded in part by the Isaac Newton Institute as a Rothschild Distinguished Visiting Fellow and by the Mathematical Sciences Research Institute as a Simons Visiting Professor.

The second author was funded in part by the Deutsche Forschungsgemeinschaft under Germany's Excellence Strategy EXC 2044 Ð390685587, Mathematics Münster: Dynamics–Geometry–Structure.

the calculation in [**6**] of the relative algebraic K-groups of a truncated polynomial algebra over a perfect $\mathbb{F}_p$-algebra.

We recall from [**4**, Section 1] that the group in the statement is a module over the ring $\mathbb{W}(k)$ of big Witt vectors in k of finite length $\frac{1}{2}(2r+1)(a-1)(b-1)$. The calculation of the length goes back to Sylvester [**12**]. Moreover, it admits a p-typical product decomposition indexed by positive integers m' not divisible by p. To state this, we let $s = s(a,b,r,p,m')$ be the unique integer such that

$$\ell(a,b,p^{s-1}m') \leq r < \ell(a,b,p^s m'),$$

if such an integer exists, and 0, otherwise, and assume (without loss of generality) that p does not divide b and write $a = p^u a'$ with a' not divisible by p. Then

$$\mathbb{W}_S(k)/(V_a\mathbb{W}_{S/a}(k) + V_b\mathbb{W}_{S/b}(k)) \simeq \textstyle\prod_{m'\in\mathbb{N}'} W_h(k)$$

by a canonical isomorphism, where

$$h = h(a,b,r,p,m') = \begin{cases} s, & \text{if neither } a' \text{ nor } b \text{ divides } m', \\ \min\{s,u\}, & \text{if } a' \text{ but not } b \text{ divides } m', \\ 0, & \text{if } b \text{ divides } m'. \end{cases}$$

Here we write $\mathbb{N}'$ for the set of positive integers not divisible by p.

1. Some recollections necessary for the proof

We first recall the Nikolaus–Scholze formula for topological cyclic homology from [**10**]; see also [**7**]. We write $\mathbb{T}$ for the circle group and $C_p \subset \mathbb{T}$ for the subgroup of order p. If R is a ring, then we write

$$\mathrm{TC}^-(R) \xrightarrow{\text{can}} \mathrm{TP}(R)$$

for the canonical map from the homotopy fixed points to the Tate construction of the spectrum with $\mathbb{T}$-action $\mathrm{THH}(R)$. The Frobenius map

$$\mathrm{THH}(R) \xrightarrow{\varphi} \mathrm{THH}(R)^{tC_p}$$

is $\mathbb{T}$-equivariant, provided that we let $\mathbb{T}$ act on the target through the isomorphism $\rho\colon \mathbb{T} \to \mathbb{T}/C_p$ given by the pth root, and therefore, it induces a map

$$\mathrm{TC}^-(R) = \mathrm{THH}(R)^{h\mathbb{T}} \xrightarrow{\varphi^{h\mathbb{T}}} (\mathrm{THH}(R)^{tC_p})^{h(\mathbb{T}/C_p)}.$$

The Tate-orbit lemma [**10**, I.2.1, II.4.2] identifies the p-completion of the target of this map with that of $\mathrm{TP}(R)$, and Nikolaus–Scholze show that, after p-completion, the topological cyclic homology of R is the equalizer

$$\mathrm{TC}(R) \longrightarrow \mathrm{TC}^-(R) \underset{\text{can}}{\overset{\varphi}{\rightrightarrows}} \mathrm{TP}(R)$$

of these two parallel maps. If p is nilpotent in R, as is the case in the situation that we consider, then the spectra in question are already p-complete.

The normalization of $A = k[x,y]/(y^a - x^b)$ is the k-algebra homomorphism to $B = k[t]$ that to x and y assigns t^a and t^b, respectively, and this homomorphism identifies A with sub-k-algebra $k[t^a, t^b] \subset k[t] = B$. In this situation, the square

$$\begin{array}{ccc} K(A) & \longrightarrow & \mathrm{TC}(A) \\ \downarrow & & \downarrow \\ K(B) & \longrightarrow & \mathrm{TC}(B) \end{array}$$

is cartesian. This follows from the birelative theorem, which has now been given a very satisfying conceptual proof by Land–Tamme [**8**]. (By contrast, the original proof in the rational case by Cortiñas [**2**] and the subsequent proof in the p-adic case by Geisser–Hesselholt both required rather elaborate calculational input.) Now, the map $K(B) \to K(k)$ induced by the k-algebra homomorphism that to t assigns 0 is an equivalence, and hence, the relative K-groups that we wish to determine are canonically identified with the homotopy groups of the common fibers of the vertical maps in the diagram above.

The k-algebras A and B are both monoid algebras. In general, if $k[\Pi]$ is the monoid algebra of an $\mathbb{E}_1$-monoid Π in spaces, then, as cyclotomic spectra,

$$\mathrm{THH}(k[\Pi]) \simeq \mathrm{THH}(k \otimes \mathbb{S}[\Pi]) \simeq \mathrm{THH}(k) \otimes B^{\mathrm{cy}}(\Pi)_+,$$

where $B^{\mathrm{cy}}(\Pi)$ denotes the unstable cyclic bar-construction of Π. In addition, on the right-hand side, the Frobenius map factors as a composition

$$\begin{aligned} \mathrm{THH}(k) \otimes B^{\mathrm{cy}}(\Pi)_+ &\xrightarrow{\varphi \otimes \tilde{\varphi}} \mathrm{THH}(k)^{tC_p} \otimes B^{\mathrm{cy}}(\Pi)_+^{hC_p} \\ &\xrightarrow{\mathrm{can}} (\mathrm{THH}(k) \otimes B^{\mathrm{cy}}(\Pi)_+)^{tC_p} \end{aligned}$$

of the map induced by the Frobenius $\varphi\colon \mathrm{THH}(k) \to \mathrm{THH}(k)^{tC_p}$ and the unstable Frobenius $\tilde{\varphi}\colon B^{\mathrm{cy}}(\Pi) \to B^{\mathrm{cy}}(\Pi)^{hC_p}$ and a canonical map. Importantly, we have a map of spectra with $\mathbb{T}$-action

$$\mathbb{Z} \longrightarrow \mathbb{Z}_p \simeq \tau_{\geq 0}\,\mathrm{TC}(k) \longrightarrow \mathrm{THH}(k)$$

from $\mathbb{Z}$ with (necessarily) trivial $\mathbb{T}$-action, and therefore, we can rewrite

$$\mathrm{THH}(k) \otimes B^{\mathrm{cy}}(\Pi)_+ \simeq \mathrm{THH}(k) \otimes_{\mathbb{Z}} \mathbb{Z} \otimes B^{\mathrm{cy}}(\Pi)_+.$$

Accordingly, we do not need to understand the homotopy type of the space with $\mathbb{T}$-action $B^{\mathrm{cy}}(\Pi)$. It suffices to understand the homotopy type of the chain complex with $\mathbb{T}$-action $\mathbb{Z} \otimes B^{\mathrm{cy}}(\Pi)_+$, which, in the case at hand, is exactly what the Buenos Aires Cyclic Homology group [**3**] and Larsen [**9**] have done for us.

To state their result, we let $\langle t^a, t^b \rangle \subset \langle t \rangle$ be the free monoid on a generator t and the submonoid generated by t^a and t^b, respectively, and set

$$B^{\mathrm{cy}}(\langle t \rangle, \langle t^a, t^b \rangle) = B^{\mathrm{cy}}(\langle t \rangle)/B^{\mathrm{cy}}(\langle t^a, t^b \rangle).$$

Counting powers of t gives a $\mathbb{T}$-equivariant decomposition of pointed spaces

$$B^{\mathrm{cy}}(\langle t \rangle, \langle t^a, t^b \rangle) \simeq \textstyle\bigvee_{m \in \mathbb{Z}_{>0}} B^{\mathrm{cy}}(\langle t \rangle, \langle t^a, t^b \rangle; m).$$

We can now state the result of the calculation by the Buenos Aires Cyclic Homology group [**3**] and by Larsen [**9**] as follows.

THEOREM 1. *Let $a, b \geq 2$ be relatively prime integers, and let $m \geq 1$ be an integer. In the ∞-category $D(\mathbb{Z})^{B\mathbb{T}}$ of chain complexes with $\mathbb{T}$-action, there is a canonical equivalence between*

$$\mathbb{Z} \otimes B^{\text{cy}}(\langle t \rangle, \langle t^a, t^b \rangle; m)$$

and the total cofiber of the square

$$\begin{array}{ccc} \mathbb{Z} \otimes (\mathbb{T}/C_{m/ab})_+[2\ell(a,b,m)] & \longrightarrow & \mathbb{Z} \otimes (\mathbb{T}/C_{m/a})_+[2\ell(a,b,m)] \\ \downarrow & & \downarrow \\ \mathbb{Z} \otimes (\mathbb{T}/C_{m/b})_+[2\ell(a,b,m)] & \longrightarrow & \mathbb{Z} \otimes (\mathbb{T}/C_{m})_+[2\ell(a,b,m)]. \end{array}$$

Here, all maps in the square are induced by the respective canonical projections, and if c does not divide m, then $\mathbb{T}/C_{m/c}$ is understood to be the empty space.

The result is not stated in this from in op. cit., and therefore, some explanation is in order. The ∞-category $D(\mathbb{Z})^{B\mathbb{T}}$ is equivalent to the ∞-category of modules over the $\mathbb{E}_1$-algebra $C_*(\mathbb{T}, \mathbb{Z})$ given by the singular chains on the circle. This $\mathbb{E}_1$-algebra, in turn, is a Postnikov section of a free $\mathbb{E}_1$-algebra over $\mathbb{Z}$, and therefore, it is formal in the sense that, as an $\mathbb{E}_1$-algebra over $\mathbb{Z}$, it is equivalent to the Pontryagin ring $H_*(\mathbb{T}, \mathbb{Z})$ given by its homology. As a model for the ∞-category of modules over the latter, we may use the dg-category of dg-modules over $H_*(\mathbb{T}, \mathbb{Z})$. But a dg-module over $H_*(\mathbb{T}, \mathbb{Z}) = \mathbb{Z}[d]/(d^2)$ is precisely what is called a mixed complex in op. cit. This shows that $D(\mathbb{Z})^{B\mathbb{T}}$ is equivalent to the ∞-category of mixed complexes. Some additional translation is necessary to bring the results in the above form for which we refer to [**4**, Section 5].

In the following, we will use the abbreviation

$$B_m = B^{\text{cy}}(\langle t \rangle, \langle t^a, t^b \rangle; m).$$

Theorem 1 shows, in particular, that the connectivity of B_m tends to infinity with m. Hence, tensoring with $\text{THH}(k)$, we obtain an equivalence

$$\text{THH}(k) \otimes B^{\text{cy}}(\langle t \rangle, \langle t^a, t^b \rangle) \simeq \textstyle\bigoplus_{m \in \mathbb{Z}_{>0}} \text{THH}(k) \otimes B_m \simeq \prod_{m \in \mathbb{Z}_{>0}} \text{THH}(k) \otimes B_m,$$

which, in turn, implies a product decomposition

$$(\text{THH}(k) \otimes B^{\text{cy}}(\langle t \rangle, \langle t^a, t^b \rangle))^{tC_p} \simeq \textstyle\prod_{m \in \mathbb{Z}_{>0}} (\text{THH}(k) \otimes B_m)^{tC_p}.$$

We will use the following result repeatedly below.

LEMMA 2. *The unstable Frobenius induces an equivalence*

$$\text{THH}(k)^{tC_p} \otimes B_{m/p} \xrightarrow{\text{id} \otimes \tilde{\varphi}} (\text{THH}(k) \otimes B_m)^{tC_p},$$

where the left-hand side is understood to be zero if p does not divide m.

PROOF. We recall two facts from [**4**, Section 3]. The first is that the pointed space $B_{m/p}$ is finite, and the second is that the cofiber of the composition

$$B_{m/p} \xrightarrow{\tilde{\varphi}} (B_m)^{hC_p} \longrightarrow B_m$$

of the unstable Frobenius map and the canonical "inclusion" is a finite colimit of free pointed C_p-cells. Here C_p acts trivially on the left-hand and middle terms.

Now, the map in the statement factors as the composition

$$\mathrm{THH}(k)^{tC_p} \otimes B_{m/p} \longrightarrow (\mathrm{THH}(k) \otimes B_{m/p})^{tC_p} \longrightarrow (\mathrm{THH}(k) \otimes B_m)^{tC_p}$$

of the canonical colimit interchange map, where $B_{m/p}$ is equipped with the trivial C_p-action, and the map of Tate spectra induced from the composite map above. The first fact implies that the left-hand map is an equivalence, and the second fact implies that the right-hand map is an equivalence. □

PROPOSITION 3. *Let G be a compact Lie group, let $H \subset G$ be a closed subgroup, let $\lambda = T_H(G/H)$ be the tangent space at $H = eH$ with the adjoint left H-action, and let S^λ the one-point compactification of λ. For every spectrum with G-action X, there are canonical natural equivalences*

$$(X \otimes (G/H)_+)^{hG} \simeq (X \otimes S^\lambda)^{hH},$$
$$(X \otimes (G/H)_+)^{tG} \simeq (X \otimes S^\lambda)^{tH}.$$

PROOF. We recall that for every map of spaces $f\colon S \to T$, the restriction functor $f^*\colon \mathrm{Sp}^T \to \mathrm{Sp}^S$ has both a left adjoint $f_!$ and a right adjoint f_*. In the case of the unique map $p\colon BG \to \mathrm{pt}$, we have $p_!(X) \simeq X_{hG}$ and $p_*(X) \simeq X^{hG}$. We now consider the following diagram of spaces.

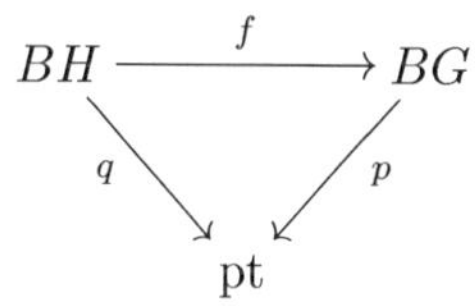

The top horizontal map is the map induced by the inclusion of H in G. It is a fiber bundle, whose fibers are compact manifolds.[1] Therefore, by parametrized Atiyah duality, its relative dualizing spectrum $Df \in \mathrm{Sp}^{BH}$ is given by the sphere bundle associated with the fiberwise normal bundle, which is $Df \simeq S^{-\lambda}$. By definition of the dualizing spectrum, we have for all $Y \in \mathrm{Sp}^{BH}$, a natural equivalence

$$f_!(Y \otimes S^{-\lambda}) \simeq f_*(Y)$$

in Sp^{BG}. It follows that for all $Y \in \mathrm{Sp}^{BH}$, we have a natural equivalence

$$p_* f_!(Y \otimes S^{-\lambda}) \simeq p_* f_*(Y) \simeq q_*(Y)$$

in Sp, which we also write as

$$((Y \otimes S^{-\lambda}) \otimes_H G_+)^{hG} \simeq Y^{hH}.$$

By [**10**, Theorem I.4.1 (3)], we further deduce a natural equivalence

$$((Y \otimes S^{-\lambda}) \otimes_H G_+)^{tG} \simeq Y^{tH}.$$

Indeed, the left-hand side vanishes for $Y = \Sigma^\infty H_+$, and the fiber of the map

$$((Y \otimes S^{-\lambda}) \otimes_H G_+)^{hG} \xrightarrow{\text{can}} ((Y \otimes S^{-\lambda}) \otimes_H G_+)^{tG}$$

preserves colimits in Y.

Finally, given $X \in \mathrm{Sp}^{BG}$, we set $Y = f^*(X) \otimes S^\lambda$ to obtain the equivalences in the statement. □

[1] Strictly speaking this statement does not make sense, since BH and BG are only defined as homotopy types. What we mean is that the map is classified by a map $BG \to B\,\mathrm{Diff}(G/H)$, where the latter is the diffeomorphism group of the compact manifold G/H.

2. Proof of Theorem A

We consider the diagram with horizontal equalizers

$$\begin{array}{ccccc} \mathrm{TC}(A) & \longrightarrow & \mathrm{TC}^-(A) & \overset{\varphi}{\underset{\mathrm{can}}{\rightrightarrows}} & \mathrm{TP}(A) \\ \downarrow & & \downarrow & & \downarrow \\ \mathrm{TC}(B) & \longrightarrow & \mathrm{TC}^-(B) & \overset{\varphi}{\underset{\mathrm{can}}{\rightrightarrows}} & \mathrm{TP}(B) \end{array}$$

and wish to evaluate the cofiber of the left-hand vertical map. We have

$$\mathrm{cofiber}(\mathrm{THH}(A) \to \mathrm{THH}(B)) \simeq \textstyle\bigoplus_{m\geq 1} \mathrm{THH}(k) \otimes B_m \simeq \prod_{m\geq 1} \mathrm{THH}(k) \otimes B_m.$$

The right-hand equivalence follows from the fact that the connectivity of B_m tends to infinity with m. Therefore, the cofiber of $\mathrm{TC}(A) \to \mathrm{TC}(B)$ is identified with the equalizer of the induced maps

$$\textstyle\prod_{m\geq 1}(\mathrm{THH}(k) \otimes B_m)^{h\mathbb{T}} \overset{\varphi}{\underset{\mathrm{can}}{\rightrightarrows}} \prod_{m\geq 1}(\mathrm{THH}(k) \otimes B_m)^{t\mathbb{T}}.$$

While the canonical map "can" preserves this product decomposition, the Frobenius map "φ" takes the factor indexed by m to the factor factor indexed pm. Therefore, we write $m = p^v m'$ with m' not divisible by p and rewrite the diagram as

$$\textstyle\prod_{m'}\prod_{v\geq 0}(\mathrm{THH}(k) \otimes B_{p^v m'})^{h\mathbb{T}} \overset{\varphi}{\underset{\mathrm{can}}{\rightrightarrows}} \prod_{m'}\prod_{v\geq 0}(\mathrm{THH}(k) \otimes B_m)^{t\mathbb{T}},$$

where both maps now preserve the outer product decomposition indexed by positive integers m' not divisible by p. We will abbreviate and write

$$\mathrm{TC}(m') \longrightarrow \mathrm{TC}^-(m') \overset{\varphi}{\underset{\mathrm{can}}{\rightrightarrows}} \mathrm{TP}(m')$$

for the equalizer diagram given by the factors indexed by m'. To complete the proof, we evaluate the induced diagram on homotopy groups.

We fix $m = p^v m'$. It follows from Theorem 1 that

$$\mathrm{THH}(k) \otimes B_m \simeq \mathrm{THH}(k) \otimes_{\mathbb{Z}} (\mathbb{Z} \otimes B_m)$$

agrees, up to canonical equivalence, with the total cofiber of the square[2]

$$\begin{array}{ccc} \mathrm{THH}(k) \otimes (\mathbb{T}/C_{m/ab})_+[2\ell(a,b,m)] & \longrightarrow & \mathrm{THH}(k) \otimes (\mathbb{T}/C_{m/a})_+[2\ell(a,b,m)] \\ \downarrow & & \downarrow \\ \mathrm{THH}(k) \otimes (\mathbb{T}/C_{m/b})_+[2\ell(a,b,m)] & \longrightarrow & \mathrm{THH}(k) \otimes (\mathbb{T}/C_m)_+[2\ell(a,b,m)]. \end{array}$$

By Proposition 3, the induced square of $\mathbb{T}$-homotopy fixed points takes the form

$$\begin{array}{ccc} \mathrm{THH}(k)^{hC_{m/ab}}[2\ell(a,b,m)+1] & \longrightarrow & \mathrm{THH}(k)^{hC_{m/a}}[2\ell(a,b,m)+1] \\ \downarrow & & \downarrow \\ \mathrm{THH}(k)^{hC_{m/b}}[2\ell(a,b,m)+1] & \longrightarrow & \mathrm{THH}(k)^{hC_m}[2\ell(a,b,m)+1] \end{array}$$

[2] Here we use in an essential way that, as a spectrum with $\mathbb{T}$-action, $\mathrm{THH}(k)$ is a $\mathbb{Z}$-module.

with the maps in the diagram given by the corestriction maps on homotopy fixed points. Indeed, the adjoint representation $\lambda = T_{C_r}(\mathbb{T}/C_r)$ is a trivial one-dimensional real C_r-representation. We now write $a = p^u a'$ with a' not divisible by p and assume that p does not divide b. If a and b both do not divide m, then

$$\begin{aligned}(\mathrm{THH}(k) \otimes B_m)^{h\mathbb{T}} &\simeq \mathrm{THH}(k)^{hC_m}[2\ell(a,b,m)+1] \\ &\simeq \mathrm{THH}(k)^{hC_{p^v}}[2\ell(a,b,m)+1],\end{aligned}$$

and if $a = p^u a'$ but not b divides m, then

$$\begin{aligned}(\mathrm{THH}(k) \otimes B_m)^{h\mathbb{T}} &\simeq \mathrm{cofiber}(\mathrm{THH}(k)^{hC_{m/a}} \to \mathrm{THH}(k)^{hC_m})[2\ell(a,b,m)+1] \\ &\simeq \mathrm{cofiber}(\mathrm{THH}(k)^{hC_{p^{v-u}}} \to \mathrm{THH}(k)^{hC_{p^v}})[2\ell(a,b,m)+1].\end{aligned}$$

Similarly, if b but not a divides m, then

$$\begin{aligned}(\mathrm{THH}(k) \otimes B_m)^{h\mathbb{T}} &\simeq \mathrm{cofiber}(\mathrm{THH}(k)^{hC_{m/b}} \to \mathrm{THH}(k)^{hC_m})[2\ell(a,b,m)+1] \\ &\simeq \mathrm{cofiber}(\mathrm{THH}(k)^{hC_{p^v}} \to \mathrm{THH}(k)^{hC_{p^v}})[2\ell(a,b,m)+1] \\ &\simeq 0,\end{aligned}$$

and if a and b both divide m, then

$$(\mathrm{THH}(k) \otimes B_m)^{h\mathbb{T}} \simeq 0,$$

since $B_m \simeq 0$. By the same reasoning, we find that

$$(\mathrm{THH}(k) \otimes B_m)^{t\mathbb{T}} \simeq \mathrm{THH}(k)^{tC_{p^v}}[2\ell(a,b,m)+1],$$

if a and b both do not divide m, that

$$(\mathrm{THH}(k) \otimes B_m)^{t\mathbb{T}} \simeq \mathrm{cofiber}(\mathrm{THH}(k)^{tC_{p^{v-u}}} \to \mathrm{THH}(k)^{tC_{p^v}})[2\ell(a,b,m)+1],$$

if $a = p^u a'$ divides m but b does not divide m, and that

$$(\mathrm{THH}(k) \otimes B_m)^{t\mathbb{T}} \simeq 0,$$

otherwise.

It follows from [**10**, Section IV.4] that, on homotopy groups, the diagram

$$\begin{array}{ccc} \mathrm{THH}(k)^{hC_{p^u}} & \xrightarrow{\mathrm{can}} & \mathrm{THH}(k)^{tC_{p^u}} \\ \downarrow{\scriptstyle \mathrm{cor}} & & \downarrow{\scriptstyle \mathrm{cor}} \\ \mathrm{THH}(k)^{hC_{p^v}} & \xrightarrow{\mathrm{can}} & \mathrm{THH}(k)^{tC_{p^v}} \end{array}$$

becomes

$$\begin{array}{ccc} W(k)[t,x]/(tx-p,p^u t) & \longrightarrow & W(k)[t^{\pm 1},x]/(tx-p,p^u t) \\ \downarrow & & \downarrow \\ W(k)[t,x]/(tx-p,p^v t) & \longrightarrow & W(k)[t^{\pm 1},x]/(tx-p,p^v t), \end{array}$$

where $\deg(t) = -2$ and $\deg(x) = 2$, where the horizontal maps are the unique graded $W(k)$-algebra homomorphisms that take t to t and x to $x = pt^{-1}$, and where the vertical maps are the unique maps of graded $W(k)[t,x]$-modules that take 1 to p^{v-u}.

We have now determined the diagram

$$\mathrm{TC}(m') \longrightarrow \mathrm{TC}^-(m') \underset{\mathrm{can}}{\overset{\varphi}{\rightrightarrows}} \mathrm{TP}(m')$$

at the level of homotopy groups, the Frobenius given by Lemma 2. Hence, it is merely a matter of bookkeeping to see that the statement of the theorem ensues. We recall the functions $s = s(a,b,r,p,m')$ and $h = h(a,b,r,p,m')$ from the p-typical decomposition recalled in the introduction,

$$\mathbb{W}_S(k)/(V_a\mathbb{W}_{S/a}(k) + V_b\mathbb{W}_{S/b}(k)) \simeq \prod_{m'\in\mathbb{N}'} W_h(k).$$

Suppose first that neither a' nor b divides m'. Then

$$\pi_{2r+1}((\mathrm{THH}(k)\otimes B_{p^v m'})^{h\mathbb{T}}) \simeq \begin{cases} W_{v+1}(k), & \text{if } 0\le v<s,\\ W_v(k), & \text{if } s\le v,\end{cases}$$
$$\pi_{2r+1}((\mathrm{THH}(k)\otimes B_{p^v m'})^{t\mathbb{T}}) \simeq W_v(k),$$

with $s = s(a,b,r,p,m')$. Also, we remark that the corresponding homotopy groups in even degree $2r$ are zero. The Frobenius map

$$\pi_{2r+1}((\mathrm{THH}(k)\otimes B_{p^v m'})^{h\mathbb{T}}) \xrightarrow{\varphi} \pi_{2r+1}((\mathrm{THH}(k)\otimes B_{p^{v+1} m'})^{t\mathbb{T}})$$

is an isomorphism for $0\le v<s$, and the canonical map

$$\pi_{2r+1}((\mathrm{THH}(k)\otimes B_{p^v m'})^{h\mathbb{T}}) \xrightarrow{\mathrm{can}} \pi_{2r+1}((\mathrm{THH}(k)\otimes B_{p^v m'})^{t\mathbb{T}})$$

is an isomorphism for $s\le v$. Hence, we have a map of exact sequences

$$\begin{array}{ccccccccc} 0 & \to & \prod_{s\le v} W_v(k) & \to & \mathrm{TC}^-_{2r+1}(m') & \to & \prod_{0\le v<s} W_{v+1}(k) & \to & 0\\ & & \downarrow{\scriptstyle \varphi-\mathrm{can}} & & \downarrow{\scriptstyle \varphi-\mathrm{can}} & & \downarrow{\scriptstyle \overline{\varphi-\mathrm{can}}} & & \\ 0 & \to & \prod_{s\le v} W_v(k) & \to & \mathrm{TP}_{2r+1}(m') & \to & \prod_{0\le v<s} W_v(k) & \to & 0,\end{array}$$

where the left-hand vertical map is an isomorphism, and where the right-hand vertical map is an epimorphism with kernel $W_s(k)$. Since $h = s$, we conclude that $\mathrm{TC}_{2r+1}(m') \simeq W_h(k)$ and $\mathrm{TC}_{2r}(m') \simeq 0$ as desired.

Suppose next that a' but not b divides m'. If $u\le s$, then

$$\pi_{2r+1}((\mathrm{THH}(k)\otimes B_{p^v m'})^{h\mathbb{T}}) \simeq \begin{cases} W_{v+1}(k), & \text{if } 0\le v<u,\\ W_u(k), & \text{if } u\le v,\end{cases}$$
$$\pi_{2r+1}((\mathrm{THH}(k)\otimes B_{p^v m'})^{t\mathbb{T}}) \simeq \begin{cases} W_v(k), & \text{if } 0\le v<u,\\ W_u(k), & \text{if } u\le v.\end{cases}$$

Moreover, we see as before that the Frobenius and canonical maps are isomorphisms for $0\le v<s$ and $s\le v$, respectively, so we have a map of exact sequences

$$\begin{array}{ccccccccc} 0 & \to & \prod_{s\le v} W_u(k) & \to & \mathrm{TC}^-_{2r+1}(m') & \to & \prod_{0\le v<s} W_c(k) & \to & 0\\ & & \downarrow{\scriptstyle \varphi-\mathrm{can}} & & \downarrow{\scriptstyle \varphi-\mathrm{can}} & & \downarrow{\scriptstyle \overline{\varphi-\mathrm{can}}} & & \\ 0 & \to & \prod_{s\le v} W_u(k) & \to & \mathrm{TP}_{2r+1}(m') & \to & \prod_{0\le v<s} W_d(k) & \to & 0,\end{array}$$

where $c = \min\{u, v+1\}$ and $d = \min\{u,v\}$. The left-hand vertical map is an isomorphism, and the right-hand vertical map is an epimorphism with kernel $W_u(k)$, so again $\mathrm{TC}_{2r+1}(m') \simeq W_h(k)$, since $h = u$, and $\mathrm{TC}_{2r}(m') \simeq 0$.

If $s < u$, then

$$\pi_{2r+1}((\mathrm{THH}(k) \otimes B_{p^v m'})^{h\mathbb{T}}) \simeq \begin{cases} W_{v+1}(k), & \text{if } 0 \leq v < s, \\ W_v(k), & \text{if } s \leq v < u, \\ W_u(k), & \text{if } u \leq v, \end{cases}$$

$$\pi_{2r+1}((\mathrm{THH}(k) \otimes B_{p^v m'})^{t\mathbb{T}}) \simeq \begin{cases} W_v(k), & \text{if } 0 \leq v < u, \\ W_u(k), & \text{if } u \leq v, \end{cases}$$

and a similar argument shows that $\mathrm{TC}_{2r+1}(m') \simeq W_h(k)$, since $h = s$, and that $\mathrm{TC}_{2r}(m') \simeq 0$.

Finally, if b divides m', then $\mathrm{TC}^{-}_{2r+1}(m')$ and $\mathrm{TP}_{2r+1}(m')$ are both zero, and therefore, so is $\mathrm{TC}_{2r+1}(m')$ and $\mathrm{TC}_{2r}(m')$. This completes the proof.

REMARK 4. As a pointed space with $\mathbb{T}$-action, the homotopy type of B_m was described conjecturally in [**4**, Conjecture B] and this conjecture was affirmed by Angeltveit in [**1**, Theorem 2.1]. The result is that B_m is equivalent to the total cofiber of a square of pointed spaces with $\mathbb{T}$-action

$$\begin{array}{ccc} (\mathbb{T}/C_{m/ab})_+ \wedge S^{\lambda(a,b,m)} & \longrightarrow & (\mathbb{T}/C_{m/a})_+ \wedge S^{\lambda(a,b,m)} \\ \downarrow & & \downarrow \\ (\mathbb{T}/C_{m/b})_+ \wedge S^{\lambda(a,b,m)} & \longrightarrow & (\mathbb{T}/C_{m})_+ \wedge S^{\lambda(a,b,m)}. \end{array}$$

So the trivial shift $[2\ell(a,b,m)]$ that appears in Theorem 1 is replaced by a shift by a non-trivial representation $\lambda(a,b,m)$ of real dimension $2\ell(a,b,m)$.

Acknowledgments

It is a pleasure to acknowledge the generous support that we have received while working on this paper. Finally, we are grateful to Tyler Lawson for pointing out that our arguments in an earlier version of this paper could be simplified significantly.

References

[1] V. Angeltveit, Picard groups and the K-theory of curves with cuspidal singularities. arXiv:1901.00264.

[2] Guillermo Cortiñas, *The obstruction to excision in K-theory and in cyclic homology*, Invent. Math. **164** (2006), no. 1, 143–173, DOI 10.1007/s00222-005-0473-9. MR2207785

[3] Buenos Aires Cyclic Homology Group, *Hochschild and cyclic homology of hypersurfaces*, Adv. Math. **95** (1992), no. 1, 18–60, DOI 10.1016/0001-8708(92)90043-K. MR1176152

[4] Lars Hesselholt, *On the K-theory of planar cuspical curves and a new family of polytopes*, Algebraic topology: applications and new directions, Contemp. Math., vol. 620, Amer. Math. Soc., Providence, RI, 2014, pp. 145–182, DOI 10.1090/conm/620/12368. MR3290091

[5] Lars Hesselholt, *The big de Rham-Witt complex*, Acta Math. **214** (2015), no. 1, 135–207, DOI 10.1007/s11511-015-0124-y. MR3316757

[6] Lars Hesselholt and Ib Madsen, *Cyclic polytopes and the K-theory of truncated polynomial algebras*, Invent. Math. **130** (1997), no. 1, 73–97, DOI 10.1007/s002220050178. MR1471886

[7] L. Hesselholt and T. Nikolaus, *Topological cyclic homology*, Handbook of Homotopy Theory, pp. 621–658, CRC Press/Chapman and Hall Handbook in Mathematical Series, Boca Raton, FL, 2019. DOI 10.1201/9781351251624.

[8] Markus Land and Georg Tamme, *On the K-theory of pullbacks*, Ann. of Math. (2) **190** (2019), no. 3, 877–930, DOI 10.4007/annals.2019.190.3.4. MR4024564

[9] M. Larsen, *Filtrations, mixed complexes, and cyclic homology in mixed characteristic*, K-Theory **9** (1995), no. 2, 173–198, DOI 10.1007/BF00961458. MR1340844

[10] Thomas Nikolaus and Peter Scholze, *On topological cyclic homology*, Acta Math. **221** (2018), no. 2, 203–409, DOI 10.4310/ACTA.2018.v221.n2.a1. MR3904731

[11] M. Speirs. *On the K-theory of truncated polynomial algebras, revisited.* Adv. Math. (to appear). DOI 10.1016/j.aim.2020.107083.

[12] J. J. Sylvester, *On Subvariants, i.e. Semi-Invariants to Binary Quantics of an Unlimited Order*, Amer. J. Math. **5** (1882), no. 1-4, 79–136, DOI 10.2307/2369536. MR1505319

Nagoya University, Japan; and University of Copenhagen, Denmark
Email address: `larsh@math.nagoya-u.ac.jp`

Universität Münster, Germany
Email address: `nikolaus@uni-muenster.de`

Contemporary Mathematics
Volume **749**, 2020
https://doi.org/10.1090/conm/749/15071

On NK_0 of the group of quaternions

Daniel Juan-Pineda

ABSTRACT. Let $\mathcal{Q}_8$ be the group of quaternions. We make a slight improvement on the calculation of the group $NK_0(\mathbb{Z}[\mathcal{Q}_8])$. We have in [**GJM** Proposition 52] an ambiguity about the possible values of this group. We solve this and prove that $NK_0(\mathbb{Z}[\mathcal{Q}_8])$ is isomorphic to an infinite direct sum of copies of $\mathbb{Z}/2$.

1. Introduction

Let $\mathcal{Q}_8$ be the group of quaternions. J. Guaschi, D. Juan Pineda and S. Millán performed, in [**GJM**] extensive computations of the lower algebraic K groups of the group rings of the virtually cyclic groups that appear as subgroups of the braid groups of the two dimensional sphere, then a more detailed computation for the case of the braid group on four strands of the 2-dimensional sphere. The case of the quaternion group is of particular interest, for example Martinet [**Mart**], and Reiner-Ullom [**R-U**] proved that $K_0(\mathbb{Z}[\mathcal{Q}_8]) \cong \mathbb{Z} \oplus \mathbb{Z}/2$. We found ambiguities in the calculations of both $NK_0(\mathbb{Z}[\mathcal{Q}_8])$ and $NK_1(\mathbb{Z}[\mathcal{Q}_8])$. We solve this for the case of $NK_0(\mathbb{Z}[\mathcal{Q}_8])$, unfortunately not for the case of $NK_1(\mathbb{Z}[\mathcal{Q}_8])$.

2. Nil group computations

In this section, we complete the computation of the the Bass Nil groups $NK_0(\mathbb{Z}[\mathcal{Q}_8])$.

THEOREM 1. *The group $NK_0(\mathbb{Z}[\mathcal{Q}_8])$ is isomorphic to a countable, infinite direct sum of copies of $\mathbb{Z}_2$.*

In order to prove Theorem 1, we recall some preliminaries that appeared in [**GJM**, Chapter 4]. We first consider the ring R of Lipschitz quaternions of the form $a + bi + cj + dk$, where $a, b, c, d \in \mathbb{Z}$ and i, j, k are the quaternionic roots of -1, and compute its NK_0 and NK_1 groups. Recall that the ring S of Hurwitz quaternions consists of the quaternions of the form $(a + bi + cj + dk)/2$ where a, b, c and d are integers that are either all even or all odd. Hence:

$$R = \mathbb{Z}[i, j, k] \text{ and } S = \mathbb{Z}[i, j, k] + \mathbb{Z}\Big[\frac{1 + i + j + k}{2}\Big].$$

2010 *Mathematics Subject Classification.* Primary 19A99.

This work was partially supported by grants PAPIIT-UNAM IN105318 and CONACYT CB 283988.

By [**Ch**, Example 5.1], S is a non-commutative principal ideal domain, and so is a regular ring. Let $M = (1+i)S$. Observe that $M \subset R \subset S$, and that R/M and S/M are the fields of two and four elements respectively. From this, it follows that S/R is the group with 2 elements. The calculations below involve the double relative term $K_1(R, S, M)$ for the injection $R \longrightarrow S$ and the two sided ideal $M = (1+i)S$ and is described as follows [**GW**, Remark 4.1.2, corollary to Theorem 4.1]:

(1)
$$K_1(R,S,M) \cong (S/R) \otimes_R (M/M^2)/\{\beta \otimes \gamma\xi + \gamma \otimes \xi\beta - \beta\gamma \otimes \xi \mid \beta,\gamma \in S/R,\ \xi \in M/M^2\}.$$

A straightforward computation yields:

$$M = \{a + bi + cj + dk \mid a+b \text{ and } c+d \text{ even}\} + \mathbb{Z}[i+k]$$
$$M^2 = (1+i)S(1+i)S = (1+i)^2S = 2iS = 2S.$$

Define $\beta \in S/R$ as the class of

$$s = \frac{1+i+j+k}{2},$$

it is immediate to verify that β is the only nontrivial element in S/R. This fact gives us that the set relations $\{\beta\otimes\gamma\xi+\gamma\otimes\xi\beta-\beta\gamma\otimes\xi \mid \beta,\gamma \in S/R,\ \xi \in M/M^2\}$ becomes $\{\beta \otimes (\beta\xi + \xi\beta) - \beta \otimes \xi | \xi \in M/M^2\}$.

Given an element $r = a + bi + cj + dk$ of M, a straightforward computation gives that $sr + rs = a_1 + b_1 i + c_1 j + d_1 k$, where all a_1, b_1, c_1 and d_1 are even. Thus, by the conditions defining the elements of M, they are in M^2, hence the class of $sr + rs$ is trivial in M/M^2 and it follows that $\beta \otimes \xi = 0$ for every $\xi \in M/M^2$ and therefore $K_1(R, S, M) = 0$.

We summarize these identities in the following:

LEMMA 2. *Let R be the ring of Lipschitz quaternions, and S be the Hurwitz quaternions. Let $R \longrightarrow S$ denote the natural inclusion and the two sided ideal $M = (1+i)S$. Then, the double relative term associated to this inclusion, $K_1(R, S, M)$, vanishes and the corresponding $NK_1(R, S, M)$ group vanishes as well.*

PROOF. The identities above imply that $K_1(R, S, M) = 0$, and by [**We**, Lemma 2.1], we obtain:

$$0 = NK_1(R,S,M) \cong K_1(R,S,M) \otimes x\mathbb{Z}[x].$$

□

REMARK 3. Here is another argument that shows that $K_1(R, S, M) = 0$, we thank the Referee for this, we saw that both $R/M \cong \mathbb{F}_2$ and $S/M \cong \mathbb{F}_4$ are commutative. Hence, by [**GW**, corollary to Theorem 0.2], we have that

$$K_1(R,S,M) \cong M/M^2 \otimes_{S/M} \Omega_{(S/M)/(R/M)}$$
$$\cong M/M^2 \otimes_{\mathbb{F}_2} \Omega_{\mathbb{F}_4/\mathbb{F}_2} = 0.$$

THEOREM 4. *The groups $NK_0(R)$ and $NK_1(R)$ are trivial.*

PROOF. As mentioned above, R/M and S/M are the fields of two and four elements respectively. Consider the following Cartesian square, where the right-hand vertical morphism is surjective:

$$\begin{array}{ccc} R & \longrightarrow & S \\ \big\downarrow & & \big\downarrow \\ R/M & \longrightarrow & S/M. \end{array}$$

Now R/M, S/M and S are all regular rings, so their corresponding Nil groups vanish, and it follows from the associated Mayer-Vietoris sequence that $NK_0(R) = 0$. On the other hand, using the fact that $0 = NK_1(R/M) = NK_1(S/M) = NK_1(S)$ and the description of the double relative group $K_1(R, S, M)$ given in [**GW**, Remark 4.1.2], and by Lemma 2, we have our result. □

To prove Theorem 1, we will require some general properties of Nil groups.

REMARK 5. We recall the following facts about the Nil groups NK_i:

(a) if NK_i is non trivial, it is an infinitely-generated group [**F-nil, R**].
(b) if $A \subset NK_i$ is a nontrivial finite subgroup then NK_i contains an infinite direct sum of copies of A [**LPW**, Theorem B].

In what follows, if $m \in \mathbb{N}$, C_m will denote the cyclic group of order m. Let $V = x\mathbb{F}_2[x]$ be the abelian group on countable many of copies of $\mathbb{F}_2 = \mathbb{Z}/2$ on generators x^i, $i > 0$.

We now recall some nil computations and comment on these, see [**LO1**, Lemmas 5.3 and 5.4], [**We**, Theorem 1.3] and [**WeK**, II 2.2].

$$NK_1(\mathbb{F}_2[C_2]) \cong (1 + x\varepsilon\mathbb{F}_2[x])^{\times} \cong V,\ NK_0(\mathbb{F}_2[C_2]) = 0, \tag{2}$$

$$NK_1(\mathbb{Z}[C_2 \times C_2]) \cong \Omega_{\mathbb{F}_2[x]} \cong \mathbb{F}_2[x]\,dx,\ NK_0(\mathbb{Z}[C_2 \times C_2]) \cong V = x\mathbb{F}_2[x], \text{ and} \tag{3}$$

$$NK_0(\mathbb{F}_2[C_2 \times C_2]) = 0. \tag{4}$$

As for the results of equations (2) and (4), asume $C_2 \times C_2$ has generators σ_1, σ_2. Let us define $\varepsilon_1 = \sigma_1 - 1$ and $\varepsilon_2 = \sigma_2 - 1$. Since $\mathbb{F}_2[C_2 \times C_2] \cong \mathbb{F}_2[\varepsilon_1, \varepsilon_2]/(\varepsilon_1^2, \varepsilon_2^2)$, the ideal $I = \langle \varepsilon_1, \varepsilon_2 \rangle$ is nilpotent in $\mathbb{F}_2[C_2 \times C_2]$, and it follows that $NK_0(\mathbb{F}_2[C_2 \times C_2]) \cong NK_0(\mathbb{F}_2) = 0$, which yields equation (4). In a similar fashion, we have $NK_0(\mathbb{F}_2[C_2]) = 0$. As Abelian groups, $\Omega_{\mathbb{F}_2[x]}$ and V are both countable infinite direct sums of copies of $\mathbb{Z}_2$. As we saw in Theorem 4, the ring R has trivial Nil groups in degrees 0 and 1. On the other hand, from the description of $\mathbb{F}_2[C_2 \times C_2]$ and by [**Ba**, Proposition XII.7.8] and [**Mar**, Theorem 3.3], we have:

$$\begin{aligned} NK_1(\mathbb{F}_2[C_2 \times C_2]) \cong (1 + x\varepsilon_1\mathbb{F}_2[x])^{\times} \times (1 + x\varepsilon_2\mathbb{F}_2[x])^{\times} \\ \times (1 + x\varepsilon_1\varepsilon_2\mathbb{F}_2[x])^{\times} \cong V^3. \end{aligned}$$

A better understanding of V and $\Omega_{\mathbb{F}_{2[x]}}$ is given by looking at the action of V_m (Verschiebung) and F_m (Frobenius) operators. In the first case, $V_m x^n = x^{nm}$, and $F_m(x^n) = 0$ if $(m, n) = 1$ $(m > 1)$ and $F_d(x^n) = dx^{n/d}$ when $d|n$. In the case of $\Omega_{\mathbb{F}_{2[x]}}$, $V_m(x^{n-1}dx) = mx^{mn-1}dx$ and $F_m(x^{n-1}dx) = x^{n/m-1}dx$ if $m|n$ and $F_m(x^{n-1}dx) = 0$ otherwise, see [**We**] for details.

Theverification of equation (3) goes as follows: start with the well known Cartesian square:

$$
\begin{array}{ccc}
\mathbb{Z}[C_2] & \xrightarrow{\sigma_1 \mapsto 1} & \mathbb{Z} \\
\downarrow{\scriptstyle \sigma_1 \mapsto -1} & & \downarrow \\
\mathbb{Z} & \longrightarrow & \mathbb{F}_2.
\end{array} \tag{5}
$$

The Mayer-Vietoris sequence associated to this, and the fact that $\mathbb{Z}$ and $\mathbb{F}_2$ are regular gives that $NK_1(\mathbb{Z}[C_2]) = NK_0(\mathbb{Z}[C_2]) = 0$. Next, tensor the above square with $\mathbb{Z}[C_2]$ to obtain the Cartesian square

$$
\begin{array}{ccc}
\mathbb{Z}[C_2 \times C_2] & \xrightarrow{\sigma_1 \mapsto 1} & \mathbb{Z}[C_2] \\
\downarrow{\scriptstyle \sigma_1 \mapsto -1} & & \downarrow \\
\mathbb{Z}[C_2] & \longrightarrow & \mathbb{F}_2[C_2].
\end{array} \tag{6}
$$

Note that in the above square, the second factor of $C_2 \times C_2$, the upper right, lower left and lower right group C_2 is generated by σ_2. Now, the Mayer Vietoris associated to the above square gives that $\partial : NK_1(\mathbb{F}_2[C_2]) \longrightarrow NK_0(\mathbb{Z}[C_2 \times C_2])$ is an isomorphism. Moreover, with our identification, $NK_1(\mathbb{F}_2[C_2]) \cong (1 + x\varepsilon_2\mathbb{F}_2[x])^{\times}$. Let $\mu = 1 + x\varepsilon_2 f(x)$, then we have

$$
\partial(\mu) = [N(\mu)[x]] - [\mathbb{Z}[C_2 \times C_2][x]], \tag{7}
$$

where $N(\mu)[x] = N(\mathbb{Z}[C_2][x], \mathbb{Z}[C_2][x], \mu^*)$ is a Milnor patching module, and the isomorphism $\mu^* : \mathbb{F}_2[C_2][x] \longrightarrow \mathbb{F}_2[C_2][x]$ is multiplication by μ. Observe that because μ has order 2, it follows that $\partial(\mu)$ has order two in $NK_0(\mathbb{Z}[C_2 \times C_2])$, see [**Mil**, Ch. 2], also [**CR2**, Lemma (42.10)].

Proof of Theorem 1. Let $\mathcal{Q}_8$ be equipped with the following presentation:

$$
\mathcal{Q}_8 = \langle x, y | x^2 = y^2, yxy^{-1} = x^{-1} \rangle.
$$

Note that the relations give $x^4 = y^4 = 1$. By [**CR2**, Theorem 50.36], the group ring $\mathbb{Z}[\mathcal{Q}_8]$ fits into the following Cartesian square:

$$
\begin{array}{ccc}
\mathbb{Z}[\mathcal{Q}_8] & \xrightarrow{f} & \mathbb{Z}[C_2 \times C_2] \\
\downarrow{\scriptstyle q} & & \downarrow{\scriptstyle p} \\
R & \xrightarrow{g} & \mathbb{F}_2[C_2 \times C_2],
\end{array} \tag{8}
$$

where q is defined on the generators of $\mathcal{Q}_8$ by $q(x) = i$ and $q(y) = j$, and f is induced by the homomorphism $\mathcal{Q}_8 \longrightarrow C_2 \times C_2$ given by taking the quotient of $\mathcal{Q}_8$ by its centre: $f(x) = \sigma_1, f(y) = \sigma_2$. So $gq = pf$. The Cartesian square (8) gives rise to the following Mayer-Vietoris sequence:

$$
\begin{gathered}
NK_2(\mathbb{Z}[\mathcal{Q}_8]) \longrightarrow NK_2(R) \oplus NK_2(\mathbb{Z}[C_2 \times C_2]) \longrightarrow NK_2(\mathbb{F}_2[C_2 \times C_2]) \xrightarrow{\partial} \\
NK_1(\mathbb{Z}[\mathcal{Q}_8]) \xrightarrow{(f_{*1})} NK_1(\mathbb{Z}[C_2 \times C_2]) \xrightarrow{p*} NK_1(\mathbb{F}_2[C_2 \times C_2]) \xrightarrow{\partial_1} \\
NK_0(\mathbb{Z}[\mathcal{Q}_8]) \xrightarrow{f_{*0}} NK_0(\mathbb{Z}[C_2 \times C_2]) \longrightarrow 0
\end{gathered} \tag{9}
$$

(the labelled homomorphisms are discussed below). The homomorphism

$$p_* : NK_1(\mathbb{Z}[C_2 \times C_2]) \longrightarrow NK_1(\mathbb{F}_2[C_2 \times C_2])$$

is trivial since $\mathbb{Z}[C_2 \times C_2]$ is reduced and $\mathbb{F}_2[C_2 \times C_2]$ is Artinian [**Mar**, Theorem 3.3].

Thus the part of (9) involving $NK_0(\mathbb{Z}[\mathcal{Q}_8])$ is:

$$0 \longrightarrow NK_1(\mathbb{F}_2[C_2 \times C_2]) \xrightarrow{\partial_1} NK_0(\mathbb{Z}[\mathcal{Q}_8]) \xrightarrow{f_{*0}} NK_0(\mathbb{Z}[C_2 \times C_2]) \longrightarrow 0. \tag{10}$$

Since both $NK_1(\mathbb{F}_2[C_2 \times C_2])$ and $NK_0(\mathbb{Z}[C_2 \times C_2])$ are non-trivial infinite sums of copies of $\mathbb{Z}_2$, it follows that $NK_0(\mathbb{Z}[\mathcal{Q}_8])$ is non trivial.

On the other hand, observe that ∂_1 is constructed in a similar fashion as in (7) with the data of the cartesian square (8). Next, f_{*0} is induced from the map $f : \mathbb{Z}[\mathcal{Q}_8] \longrightarrow \mathbb{Z}[C_2 \times C_2]$ in (8). Hence this is just tensoring with $\mathbb{Z}[C_2 \times C_2] \otimes_{\mathbb{Z}[\mathcal{Q}_8]} -$. Let $\tilde{N}(\mu)[x] = [N(\mu)[x]] - [\mathbb{Z}[C_2 \times C_2]]$ be an element of $NK_0(\mathbb{Z}[C_2 \times C_2])$, as f_{*0} is surjective, there exists a finitely generated projective $\mathbb{Z}[\mathcal{Q}_8][x]$-module $M[x]$ with $f_{*0}([M[x]]) = [\tilde{N}(\mu)[x]]$. On the other side by [**Mil**, Theorem 2.2], also [**CR2**, (42.6)], $M[x]$ fits in a cartesian square

$$\begin{array}{ccc} M[x] & \xrightarrow{f_*} & \tilde{N}(\mu)[x] \\ \Big\downarrow{\scriptstyle g_1} & & \Big\downarrow{\scriptstyle g_2} \\ R[x] & \longrightarrow & \mathbb{F}_2[C_2 \times C_2][x], \end{array} \tag{11}$$

where

$$M[x] = \{(a_1, a_2) \in \tilde{N}(\mu)[x] \times R[x] | \mu f_2(a_2) = g_2(a_1)\}.$$

Now, μ is a unit of order 2, hence $N(\mu)[x] \oplus N(\mu)[x]$ is free, it follows that $M[x]$ has order two in $NK_0(\mathbb{Z}[\mathcal{Q}_8])$ and is not of the form $\partial(\omega)$ because these elements are generated by elements of the form

$$[N(R[x], \mathbb{Z}[C_2 \times C_2][x], \nu^*)] - [\mathbb{Z}[\mathcal{Q}_8][x]],$$

for some unit $\nu \in NK_1(\mathbb{F}_2[C_2 \times C_2])$. This gives that $NK_0(\mathbb{Z}[\mathcal{Q}_8])$ is isomorphic to an infinite direct sum of copies of $\mathbb{Z}/2$. $\square$

As for the group $NK_1(\mathbb{Z}[\mathcal{Q}_8])$, by (9), this fits in an exact sequence

$$NK_2(\mathbb{F}_2[C_2 \times C_2]) \xrightarrow{\partial} NK_1(\mathbb{Z}[\mathcal{Q}_8]) \xrightarrow{(f_{*1})} NK_1(\mathbb{Z}[C_2 \times C_2]) \xrightarrow{p_*} 0.$$

At this point, we can only claim:

LEMMA 6. *$NK_1(\mathbb{Z}[\mathcal{Q}_8])$ is isomorphic to an infinite direct sum of copies of $\mathbb{Z}/2$, $\mathbb{Z}/4$ or $\mathbb{Z}/2 \oplus \mathbb{Z}/4$.*

PROOF. By exactness, and the fact that $NK_1(\mathbb{Z}[C_2 \times C_2])$ is a group of exponent 2, it suffices to prove that $NK_2(\mathbb{F}_2[C_2 \times C_2])$ is also a group of exponent 2. Let $S = \mathbb{F}_2[x]$, then the group we are interested in is $NK_2(\mathbb{F}_2[C_2 \times C_2]) = K_2(S \oplus I, I)$ where $I = S\varepsilon_1 \oplus S\varepsilon_2$. As $S \oplus I$ is commutative and $I^2 = 0$, $K_2(S \oplus I, I)$ has exponent 2 by [**Mag07**, 4. Relative K_2], see also [**We80**, Theorem 1.3]. $\square$

3. Twisted case

We now turn to the case of the semidirect product $\mathcal{Q}_8 \rtimes \mathbb{Z}$, where $\mathbb{Z} = \langle t \rangle$ is the infinite cyclic group with generator t.

PROPOSITION 7. [**GG**, Theorem 5] *There are, up to isomorphism, two non-trivial semi-direct products of the form $\mathcal{Q}_8 \rtimes \mathbb{Z}$, namely $\mathcal{Q}_8 \rtimes_\alpha \mathbb{Z}$, where $\alpha \in \{2,3\}$. These are:*

- $\mathbb{Z}\mathcal{Q}_8 \rtimes_2 \mathbb{Z}$, *by adding the relations* $txt^{-1} = y, \quad tyt^{-1} = x$,
- $\mathbb{Z}\mathcal{Q}_8 \rtimes_3 \mathbb{Z}$, *by adding the relations* $txt^{-1} = y, \quad tyt^{-1} = xy$.

The Bass *Nil*–groups are replaced by the *Farrell-Hsiang*–Nil groups, defined for the twisted polynomial ring $\mathbb{Z}[\mathcal{Q}_8]_\alpha[x]$. These are defined as follows, the natural augmentation map

$$\tau : \mathbb{Z}[\mathcal{Q}_8]_\alpha[x] \longrightarrow \mathbb{Z}[\mathcal{Q}_8], \text{ given by } \tau(t) = 0,$$

induces

$$\tau_* : K_i(\mathbb{Z}[\mathcal{Q}_8]_\alpha[x]) \longrightarrow K_i(\mathbb{Z}[\mathcal{Q}_8]).$$

Define $NK_i(\mathbb{Z}[\mathcal{Q}_8], \alpha) = ker(\tau_*)$, see [**FH, Gray**]. By [**GJM**, Proposition 55] and its proof, we have the following :

COROLLARY 8. *Let $\alpha \in \{2,3\}$, then*

$$NK_0(\mathbb{Z}[\mathcal{Q}_8], \alpha) \cong \bigoplus_{\infty} \mathbb{Z}/2.$$

Acknowledgment

We thank the Referee for the careful reading and pointing out key facts that we missed in the first draft of this paper.

References

[Ba] Hyman Bass, *Algebraic K-theory*, W. A. Benjamin, Inc., New York-Amsterdam, 1968. MR0249491

[Ch] A. W. Chatters, *Rings which are nearly principal ideal domains*, Glasgow Math. J. **40** (1998), no. 3, 343–351, DOI 10.1017/S0017089500032699. MR1660109

[CR2] Charles W. Curtis and Irving Reiner, *Methods of representation theory. Vol. II*, Pure and Applied Mathematics (New York), John Wiley & Sons, Inc., New York, 1987. With applications to finite groups and orders; A Wiley-Interscience Publication. MR892316

[F-nil] F. T. Farrell, *The nonfiniteness of Nil*, Proc. Amer. Math. Soc. **65** (1977), no. 2, 215–216, DOI 10.2307/2041893. MR450328

[FH] F. T. Farrell and W. C. Hsiang, *The Whitehead group of poly-(finite or cyclic) groups*, J. London Math. Soc. (2) **24** (1981), no. 2, 308–324, DOI 10.1112/jlms/s2-24.2.308. MR631942

[GW] Susan C. Geller and Charles A. Weibel, $K_1(A, B, I)$, J. Reine Angew. Math. **342** (1983), 12–34, DOI 10.1515/crll.1983.342.12. MR703484

[GG] Daciberg Lima Gonçalves and John Guaschi, *The classification of the virtually cyclic subgroups of the sphere braid groups*, SpringerBriefs in Mathematics, Springer, Cham, 2013. MR3112976

[GJM] John Guaschi, Daniel Juan-Pineda, and Silvia Millán López, *The lower algebraic K-theory of virtually cyclic subgroups of the braid groups of the sphere and of $\mathbb{Z}[B_4(\mathbb{S}^2)]$*, SpringerBriefs in Mathematics, Springer, Cham, 2018. MR3839305

[Gray] Daniel R. Grayson, *The K-theory of semilinear endomorphisms*, J. Algebra **113** (1988), no. 2, 358–372, DOI 10.1016/0021-8693(88)90165-2. MR929766

[LO1] Jean-François Lafont and Ivonne J. Ortiz, *Relative hyperbolicity, classifying spaces, and lower algebraic K-theory*, Topology **46** (2007), no. 6, 527–553, DOI 10.1016/j.top.2007.03.001. MR2363244

[LPW] Jean-François Lafont, Stratos Prassidis, and Kun Wang, *Revisiting Farrell's nonfiniteness of Nil*, Ann. K-Theory **1** (2016), no. 2, 209–225, DOI 10.2140/akt.2016.1.209. MR3514940

[Mart] Jacques Martinet, *Modules sur l'algèbre du groupe quaternionien* (French), Ann. Sci. École Norm. Sup. (4) **4** (1971), 399–408. MR0291208

[Mag07] Bruce A. Magurn, *Explicit K_2 of some finite group rings*, J. Pure Appl. Algebra **209** (2007), no. 3, 801–811, DOI 10.1016/j.jpaa.2006.08.001. MR2298858

[Mar] Robert Dawson Martin, *Nilgroups of Finite Abelian Groups*, ProQuest LLC, Ann Arbor, MI, 1976. Thesis (Ph.D.)–Columbia University. MR2625984

[Mil] John Milnor, *Introduction to algebraic K-theory*, Princeton University Press, Princeton, N.J.; University of Tokyo Press, Tokyo, 1971. Annals of Mathematics Studies, No. 72. MR0349811

[Q] Frank Quinn, *Ends of maps. II*, Invent. Math. **68** (1982), no. 3, 353–424, DOI 10.1007/BF01389410. MR669423

[R] R. Ramos, Grupos Nil torcidos en teoría K-algebraica, Doctoral Thesis, Universidad Nacional Autónoma de México, 2006.

[R-U] I. Reiner and S. Ullom, *Class groups of integral group rings*, Trans. Amer. Math. Soc. **170** (1972), 1–30, DOI 10.2307/1996291. MR304470

[We80] Charles A. Weibel, *K_2, K_3 and nilpotent ideals*, J. Pure Appl. Algebra **18** (1980), no. 3, 333–345, DOI 10.1016/0022-4049(80)90008-0. MR593622

[We] Charles Weibel, *NK_0 and NK_1 of the groups C_4 and D_4. Addendum to "Lower algebraic K-theory of hyperbolic 3-simplex reflection groups" by J.-F. Lafont and I. J. Ortiz [MR2495796]*, Comment. Math. Helv. **84** (2009), no. 2, 339–349, DOI 10.4171/CMH/164. MR2495797

[WeK] Charles A. Weibel, *The K-book*, Graduate Studies in Mathematics, vol. 145, American Mathematical Society, Providence, RI, 2013. An introduction to algebraic K-theory. MR3076731

Centro de Ciencias Matemáticas, Universidad Nacional Autónoma de México Campus Morelia, Morelia, Michoacán, Mexico 58089.

Email address: daniel@matmor.unam.mx

Contemporary Mathematics
Volume **749**, 2020
https://doi.org/10.1090/conm/749/15072

The Witt group of real surfaces

Max Karoubi and Charles Weibel

Abstract. Let V be an algebraic variety defined over $\mathbb{R}$, and V_{top} the space of its complex points. We compare the algebraic Witt group $W(V)$ of symmetric bilinear forms on vector bundles over V, with the topological Witt group $WR(V_{\text{top}})$ of symmetric forms on Real vector bundles over V_{top} in the sense of Atiyah, especially when V is 2-dimensional. To do so, we develop topological tools to calculate $WR(V_{\text{top}})$, and to measure the difference between $W(V)$ and $WR(V_{\text{top}})$.

If V is a smooth algebraic surface defined over the real numbers $\mathbb{R}$, the Witt group $W(V)$ is a finitely generated abelian group, whose rank equals the number ν of components of the 2-manifold $V(\mathbb{R})$ [**30**]. Sujatha has given formulas for the torsion subgroup in [**48**]; although the formulas she gives are not always easy to compute, they involve étale cohomology and are thus mostly topological in nature. Indeed, Cox' theorem [**14**] states that $H^n_{\text{et}}(V,\mathbb{Z}/2) \simeq \mathbb{H}^n_G(V_{\text{top}},\mathbb{Z}/2)$. Here V_{top} is the 4-manifold underlying the complex variety $V \times_{\mathbb{R}} \mathbb{C}$, G is the cyclic group $G = \text{Gal}(\mathbb{C}/\mathbb{R})$ of order 2, acting on V_{top} by complex conjugation, and $\mathbb{H}^*_G$ is Borel cohomology.

Together with Schlichting, we introduced the topological Witt group $WR(X)$ of a G-space X in [**23**] (see Definition 2.1 below), and showed that the natural map $W(V) \to WR(V_{\text{top}})$ is always an isomorphism modulo bounded 2-primary torsion; for curves it is an isomorphism; for surfaces, the kernel and cokernel of the map $W(V) \to WR(V_{\text{top}})$ have exponent 2 by [**23**, Thm. 8.7].

One of the main goals of this paper is to develop topological tools to compute $WR(X)$ when X has dimension ≤ 4, since if V is an algebraic surface then $WR(V_{\text{top}})$ is very close to $W(V)$. The second goal is to more precisely measure the difference between these two invariants in various examples.

Here are three important invariants of W that have analogues for WR. One is the classical signature on $W(V)$, which factors through the topological signature on $WR(V_{\text{top}})$, which is an invariant defined on the 1–skeleton of X for any G-space X (see Lemma 2.5). Another is the algebraic discriminant; its analogue defined on $WR(X)$ takes values in the group $\text{Pic}_G(X)$ of equivariant real line bundles; see Definition 4.1. The third is the Hasse invariant (see Theorem 6.5.)

2010 *Mathematics Subject Classification.* Primary 19G12; Secondary 55N15, 19L99, 19M05, 11E70, 19G38.

The second author was supported by NSA and NSF grants.

For any smooth projective surface V, $WR(V_{\text{top}})$ is a birational invariant of V (see Theorem 5.1). If V is either a complex projective surface, so that $V_{\text{top}} = G{\times}V(\mathbb{C})$, or a surface defined over $\mathbb{R}$ with no real points, (so that V_{top} is connected) the main theorems of [**26**] (see 7.1 and 7.2 below; cf. [**55**]) state that $W(V) \to WR(V_{\text{top}})$ is a split surjection, and an isomorphism if and only if the geometric genus $p_g = \dim H^0_{\text{an}}(X, \Omega^2_X)$ is 0. In particular, when V is a rational surface with G acting freely then $W(V) \cong WR(V_{\text{top}})$. To compute this group it suffices to compute $WR(X)$ when $X = S^2 \times S^2$. This is done in Theorem 7.4; there are 3 cases, corresponding to the 3 possible actions of G on $H^2(X, \mathbb{Z})$.

When G does not act freely, we determine the kernel and cokernel of $W(V) \to WR(V_{\text{top}})$ in Theorem 8.7; the kernel is 0 if and only if $p_g(V) = 0$. We also show in Theorem 8.6 that when $H^3_G(X; \mathbb{Z}(1))$ has no 2-torsion and X^G has $\nu > 0$ connected components then

$$WR(X) \cong \mathbb{Z}^\nu \oplus H^1(X/G, \mathbb{Z}/2).$$

The tools used to compute $WR(X)$ are markedly different from the tools used to compute $W(V)$, because WR is related to Atiyah's Real K-theory and to the equivariant K-group KO_G. Both KR and KO_G satisfy Bredon's axioms [**7**] for a G-equivariant cohomology theory: homotopy invariance, excision, and long exact sequences. In the Appendix, we recall the facts we need about the Bredon spectral sequence converging to equivariant cohomology.

Our paper is organized as follows. In Section 1, we review Sujatha's results on the Witt group of a surface, as a reference for our results. In Section 2, we define $WR(X)$ and review the construction of the signature map. In Section 3, we compute $WR(X)$ when $\dim(X) \leq 2$. We define the topological determinant in Section 4, relating it to the group $\text{Pic}_G(X)$ of symmetric forms on Real line bundles.

In Section 5, we show that WR is a birational invariant of projective surfaces over $\mathbb{R}$. In Section 6, we use the Stiefel–Whitney classes w_1 and w_2 to define maps on $WR(X)$ compatible with the discriminant and Hasse invariant on $W(V)$.

In Section 7, we recall and illustrate the calculations of [**26**] for $WR(X)$ when V has no real points. In Section 8 we study the map $W(V) \to WR(X)$ when V has real points.

In Section 9, we say a few things about 3–folds. The group $WR(V_{\text{top}})$ is easy to describe for complex 3–folds, and not much harder for 3–folds over $\mathbb{R}$ which have no real points. We did not look very hard at 3-folds with real points, because of Parimala's result in [**37**], showing that for a smooth algebraic 3-fold that $W(V)$ is finitely generated if and only if the mod-2 Chow group $CH^2(V)/2$ is finitely generated. (We give the short proof in Theorem 9.1 below.) Totaro has shown in [**50**] that $CH^2(V)/2$ is in fact not finitely generated for a very general abelian 3-fold, making computations of $W(V)$ for 3-folds problematic.

The Fundamental Theorem for Witt theory says that $W(V \times \mathbb{G}_m) \cong W(V) \oplus W(V)$ for a smooth variety V. This is true more generally when $K_{-1}(V) = 0$ (which is always true for smooth V); see [**24**]. In Section 10 of the present paper, we show that if $KR_{-1}(X) = 0$ then $WR(X \times S^{1,1}) \cong WR(X) \oplus WR(X)$ where $S^{1,1}$ is the circle $(\mathbb{G}_m)_{\text{top}}$.

NOTATION. Recall that G denotes $\mathbb{Z}/2$. If X is a space with involution, we write $\mathbb{H}^*_G(X, A)$ for the Borel cohomology of X with coefficients in a G-module A. Similarly, we write $H^*_G(X, h)$ for the Bredon cohomology of X, where h is a coefficient system; see the Appendix. We will occasionally write $H_{\text{et}}(V)$ for $H_{\text{et}}(V,\mathbb{Z}/2)$.

1. A filtration on $W(V)$

Let k be a field containing 1/2. A *symmetric form* (E,θ) on an algebraic variety V over k is an algebraic vector bundle E on V, with an isomorphism θ from E to its dual E^* such that $\theta=\theta^*$. Asssociated to any bundle E, there is a hyperbolic form $H(E)=(E\oplus E^*,\text{ev})$. The Grothendieck-Witt group $GW(V)$ is the Grothendick group of the category of symmetric forms on V, modulo the relation that $(E,\theta)=h(L)$ if E has a Lagrangian L (a subobject such that $L=L^{\perp}$). There is a hyperbolic map $H:K_0(V)\to GW(V)$, and the Witt group $W(V)$ is defined to be the cokernel of H. When V is connected, there is a canonical surjection $W(V)\to\mathbb{Z}/2$, and we define $I(V)$ to be its kernel.

When $\dim(V)\le 3$ and V is smooth, $W(V)$ injects into $W(F)$, where $F=k(V)$ is the function field of V over k, and there is a well known exact sequence

$$0\to W(V)\to W(F)\to\oplus_x W(k(x))$$

where x runs over all points of codimension 1. (See **[36]** and **[5]**, for example.) There is a natural filtration on $W(F)$ by the powers $I(F)^n$ of the maximal ideal $I(F)$, and $I(F)^n/I(F)^{n+1}\cong H^n_{\text{et}}(F,\mathbb{Z}/2)$ by **[35]**.

Still assuming that $\dim(V)\le 3$, we define $I_n=I_n(V)$ to be the ideal $W(V)\cap I(F)^n$ of $W(V)$; $I(V)=I_1(V)$. Parimala showed in **[37]** that above sequence restricts to an exact sequence

$$0\to I_n(V)\to I^n(F)\to\oplus_x\, I^{n-1}(k(x)) \tag{1.1}$$

which maps to the Bloch-Ogus sequence **[6]**

$$0\to H^0(V,\mathcal{H}^n)\to H^n_{\text{et}}(F,\mathbb{Z}/2)\to\oplus_x\, H^{n-1}_{\text{et}}(k(x),\mathbb{Z}/2). \tag{1.2}$$

Here $\mathcal{H}^n$ is the Zariski sheaf associated to the presheaf $U\mapsto H^n_{\text{et}}(U,\mathbb{Z}/2)$.

Recall that the discriminant of a diagonal form $(a_1,...,a_n)$ over a field F is the class of $(-1)^{[n/2]}\prod a_i$ in $F^\times/F^{\times 2}$. As the hyperbolic form $H(1)$ has discriminant $+1$, this induces a function disc: $W(F)\to F^\times/F^{\times 2}$.

DEFINITION 1.3. The map $I(V)\to I(F)\to I(F)/I^2(F)\cong F^\times/F^{\times 2}$ factors through the subgroup $H^0(V,\mathcal{H}^1)$ of $F^\times/F^{\times 2}$. The *algebraic discriminant* is the induced map $I(V)\to H^0(V,\mathcal{H}^1)\cong H^1_{\text{et}}(V,\mathbb{Z}/2)$.

Similarly, the *Hasse invariant* $I(V)\to H^0(V,\mathcal{H}^2)\cong {}_2\text{Br}(V)$ is the homomorphism induced by the morphism $I(V)\to I(F)$ followed by the Stiefel–Whitney class $w_2: I(F)\to H^2_{\text{et}}(F,\mathbb{Z}/2)$; see **[33**, 3.1**]**.

EXAMPLE 1.4. The algebraic discriminant $I(V)\to H^1_{\text{et}}(V,\mathbb{Z}/2)$ is onto. To see this, recall from Kummer Theory **[32**, III.4**]** that there is a split exact sequence

$$0\to\mathcal{O}^\times(V)/2\to H^1_{\text{et}}(V,\mathbb{Z}/2)\to {}_2\text{Pic}(V)\to 0.$$

If a is a global unit of V and $\mathcal{O}_V$ is the trivial line bundle then $(\mathcal{O}_V,a)$ is a symmetric form, representing an element of $W(V)$ whose discriminant is the class of a in $\mathcal{O}^\times(V)/2$. If L is a line bundle with $L\otimes L\cong\mathcal{O}_X$, the isomorphism $\theta:L\cong L^*$ defines a symmetric form, and the discriminant of (L,θ) maps to the class of L in ${}_2\text{Pic}(V)$.

If θ is any symmetric form on an algebraic vector bundle E, the discriminant and Hasse invariant of (E,θ) may be calculated by passing to an open subvariety U of V where $E \cong \mathcal{O}_U^n$ and θ is isomorphic to the diagonal form $(a_1, ..., a_n)$, where $a_i \in H^0(U, \mathcal{O}^\times)$; the discriminant of (E,θ) is the class of the product $(-1)^{[n/2]} \prod a_i$ in $H^0(U, \mathcal{O}^\times)/2 \subset F^\times/F^{\times 2}$, and the Hasse invariant of (E,θ) is the class of $\prod_{i<j}\{a_i, a_j\}$ in $H^2_{\text{et}}(F, \mathbb{Z}/2)$.

Let $\widetilde{H}^1_{\text{et}}(V, \mathbb{Z}/2)$ denote the cokernel of $H^1_{\text{et}}(k, \mathbb{Z}/2) \to H^1_{\text{et}}(V, \mathbb{Z}/2)$.

LEMMA 1.5. *If* $\dim(V) \leq 3$ *then* $I_n(V)/I_{n+1}(V) \to H^0(V, \mathcal{H}^n)$ *is an injection for all* $n \geq 0$*. For* $n = 1$*, this is the discriminant isomorphism*

$$I_1/I_2 \xrightarrow{\cong} H^0(V, \mathcal{H}^1) \cong H^1_{\text{et}}(V, \mathbb{Z}/2).$$

If -1 *is not a square in* $k(V)^\times$ *then* $W(V)/I_2(V) \cong \mathbb{Z}/4 \oplus \widetilde{H}^1_{\text{et}}(V, \mathbb{Z}/2)$.

PROOF. This follows from a diagram chase on the map between sequences (1.1) and (1.2), using the isomorphisms $I^n(F)/I^{n+1}(F) \cong H^n_{\text{et}}(F, \mathbb{Z}/2)$ (which are known to hold for all n by [**35**]). It is also well known that $W(F)/I(F)^2$ contains $\mathbb{Z}/4$ if and only if $\{-1,-1\} \neq 0$ in $H^2_{\text{et}}(k(V), \mathbb{Z}/2)$ (see, e.g., [**33**, 3.3]). □

REMARK 1.5.1. The injection $I_2/I_3 \to H^0(V, \mathcal{H}^2) \cong {}_2\text{Br}(V)$ is called the *Hasse invariant* [**37**]; it is not known if it is an isomorphism for non-affine V over k, even if $k = \mathbb{R}$.

Connection to signature

Now suppose that V is a variety over $\mathbb{R}$, and $V(\mathbb{R})$ has $\nu > 0$ connected components. The torsion subgroup of $W(V)$ is 2-primary (Pfister [**40**]), and Mahé [**30**] and Brumfiel [**8**] proved that the signature $W(V) \to \mathbb{Z}^\nu$ maps the torsionfree part of $W(V)$ isomorphically onto a subgroup of finite index in $\mathbb{Z}^\nu$. By [**23**, 2.4], the signature factors through $WR(V_{\text{top}})$.

Recall that the cup product with $\{-1\} \in H^0(V, \mathcal{H}^1)$ induces stabilization maps $H^0(V, \mathcal{H}^n) \to H^0(V, \mathcal{H}^{n+1})$; Colliot-Thélène and Parimala showed [**9**] that this map is an isomorphism for $n > d = \dim(V)$ and that this stable value is $(\mathbb{Z}/2)^\nu$, where ν is the number of real components of V. When $n \leq d$, the composite

$$H^0(V, \mathcal{H}^n) \to H^0(V, \mathcal{H}^{d+1}) \cong (\mathbb{Z}/2)^\nu$$

is called the *stabilization map*. Example 1.6 shows that $H^1_{\text{et}}(V, \mathbb{Z}/2) \cong H^0(V, \mathcal{H}^1)$ need not map onto $(\mathbb{Z}/2)^\nu$. In contrast, van Hamel proved in [**51**, 2.8] that the stabilization map $H^0(V, \mathcal{H}^d) \to H^0(V, \mathcal{H}^{d+1}) \cong (\mathbb{Z}/2)^\nu$ is a surjection. For surfaces, this becomes a surjection ${}_2\text{Br}(V) \cong H^0(V, \mathcal{H}^2) \xrightarrow{\text{onto}} (\mathbb{Z}/2)^\nu$.

EXAMPLE 1.6. A nonsingular cubic surface in $\mathbb{P}^3_{\mathbb{C}}$ is a special type of *Del Pezzo* surface, and is birationally equivalent to $\mathbb{P}^2_{\mathbb{C}}$ (see [**16**, V.4.7.1]). There are nonsingular cubic surfaces defined over $\mathbb{R}$ which are not birationally equivalent to $\mathbb{P}^2$ over $\mathbb{R}$; they have $\text{Pic}(V) \cong \mathbb{Z}^3$, and the real locus $V(\mathbb{R})$ has $\nu = 2$ components, homeomorphic to S^2 and $\mathbb{RP}^2$ respectively; see [**47**, V.5.4]. Since $H^1_{\text{et}}(V, \mathbb{Z}/2) \cong \mathbb{Z}/2$ by Kummer theory, it follows that $H^1_{\text{et}}(V, \mathbb{Z}/2) \to (\mathbb{Z}/2)^\nu$ is not onto.

The torsion subgroup

The torsion subgroup $I(V)_{\text{tors}}$ of $I(V)$ has an induced filtration by the ideals $I_n(V)_{\text{tors}} = I_n(V) \cap I(V)_{\text{tors}}$.

PROPOSITION 1.7. *If V is a smooth d-dimensional variety over $\mathbb{R}$, then*

(i) *$I_{d+1}(V)$ is torsionfree as an abelian group; $I_{d+1}(V)_{\text{tors}} = 0$.*
(ii) *the torsion subgroup $I(V)_{\text{tors}}$ of $I(V)$ has exponent 2^d.*
(iii) *If $V(\mathbb{R}) = \emptyset$ then $W(V)$ is an algebra over $\mathbb{Z}/2^{d+1}$.*

PROOF. It is classical [**15**, Thm. E] that $I^{d+1}(F)$ is a torsionfree abelian group. Hence the subgroup $I_{d+1}(V)$ is also torsionfree. Since the element '2' of $W(V)$ lies in $I(V)$ and $I^n(V) \subseteq I_n(V)$, $2^{d+1} \in 2I_d(V) \subseteq I_{d+1}(V)$. Finally, (iii) is immediate from (ii) and the fact that the torsionfree part has rank ν. □

When $\dim(V) = 2$, $I_3(V)$ is torsionfree and $I(V)_{\text{tors}}$ has exponent 4.

$$I(V)_{\text{tors}} \supseteq I_2(V)_{\text{tors}} \supseteq I_3(V)_{\text{tors}} = 0.$$

Following Sujatha [**48**], let $H^0_{\text{tors}}(V, \mathcal{H}^n)$ denote the kernel of the stable map $H^0(V, \mathcal{H}^n) \to H^0(V, \mathcal{H}^{d+1}) \cong (\mathbb{Z}/2)^\nu$. Sujatha described the torsion subgroup $I(V)_{\text{tors}}$ as follows.

PROPOSITION 1.8 (Sujatha). *When V is a smooth surface over $\mathbb{R}$, there is a short exact sequence*

$$0 \to H^0_{\text{tors}}(V, \mathcal{H}^2) \to I(V)_{\text{tors}} \to H^0_{\text{tors}}(V, \mathcal{H}^1) \to 0.$$

PROOF. Sujatha proves in [**48**, 2.1, 2.2] that the maps

$$I(V)_{\text{tors}}/I_2(V)_{\text{tors}} \to H^0_{\text{tors}}(V, \mathcal{H}^1), \quad I_2(V)_{\text{tors}}/I_3(V)_{\text{tors}} \to H^0_{\text{tors}}(V, \mathcal{H}^2)$$

are isomorphisms, and $I_3(V)_{\text{tors}} = 0$ by Proposition 1.7. □

REMARK 1.8.1. Set $j = \dim H^0_{\text{tors}}(V, \mathcal{H}^1)$ and $k = \dim H^0_{\text{tors}}(V, \mathcal{H}^2)$. It follows that (when $\nu > 0$) the torsion subgroup $W(V)_{\text{tors}}$ of $W(V)$ is a group of exponent 4 and order 2^{j+k}. (See [**48**, Lemma 3.2].) Note that $k + \nu = \dim {}_2\text{Br}(V)$ by van Hamel's result [**51**, 2.9].

Sujatha [**48**, 3.1] also defines a group N fitting into an exact sequence

$${}_2\operatorname{Pic}(V) \to {}_2\operatorname{Pic}(V_{\mathbb{C}}) \to N \to \operatorname{Pic}(V)/2 \to \operatorname{Pic}(V_{\mathbb{C}})/2.$$

Her invariant l is the dimension of N; she proves in [**48**, 3.3] that there are $k + l$ summands in $W(V)_{\text{tors}}$, and $j - l$ summands $\mathbb{Z}/4$ in $W(V)$. It follows by a counting argument that there are $k + 2l - j$ summands $\mathbb{Z}/2$ in $W(V)$. However, the numbers j, k and l are not always easy to determine; see Krasnov [**28**], [**29**].

Recall that a variety V over $\mathbb{R}$ is *geometrically connected* if $V \times_{\mathbb{R}} \mathbb{C}$ is connected, i.e., if the function field of V does not contain $\mathbb{C}$.

COROLLARY 1.9. *Let V be a geometrically connected surface over $\mathbb{R}$ with no real points. Then we have a short exact sequence*

$$0 \to {}_2\text{Br}(V) \to W(V) \to \mathbb{Z}/4 \times \widetilde{H}^1_{\text{et}}(V, \mathbb{Z}/2) \to 0,$$

$\widetilde{H}^1_{\text{et}}(V, \mathbb{Z}/2)$ denoting the cokernel of $H^1_{\text{et}}(\operatorname{Spec} \mathbb{R}, \mathbb{Z}/2) \to H^1_{\text{et}}(V, \mathbb{Z}/2)$.

PROOF. Since $\nu = 0$, Proposition 1.8 implies that $I_2(V) \cong {}_2\text{Br}(V)$. We conclude by Lemma 1.5. □

The group extension in Corollary 1.9 need not split, as illustrated by Example 1.11.

REMARK 1.9.1. Corollary 1.9 recovers Sujatha's result [**48**, 3.4] that if V is a geometrically connected surface with $V(\mathbb{R}) = \emptyset$ then $W(V)$ has order 2^{j+k+1}, because $j = \dim H^1_{\text{et}}(V, \mathbb{Z}/2)$ and $k = \dim {}_2\text{Br}(V)$.

Let Q_1 denote the projective Brauer-Severi curve $x^2 + y^2 + z^2 = 0$.

LEMMA 1.10. *Suppose that V is a smooth geometrically connected surface with no real points. Then $W(V)$ is a $\mathbb{Z}/8$-algebra and surjects onto $\mathbb{Z}/4$. It is a $\mathbb{Z}/4$-algebra if and only if there is a morphism $V \to Q_1$ defined over $\mathbb{R}$.*

PROOF. Since $W(V)$ injects into $W(F)$, it suffices to study the exponent of $W(F)$. When V is geometrically connected ($\mathbb{R}(V)$ does not contain $\mathbb{C}$), then -1 is not a square in F. Hence the element '2' of $W(F)$ is nonzero, because its discriminant is -1 in $F^\times/F^{\times 2}$. The group $W(F)$ has $I^3(F) = 0$ by Proposition 1.7(iii) (or [**15**, Thm. E]). Thus the ideal $I(F)$ has exponent 4, and '8' is zero in $W(F)$. Thus the image of $\mathbb{Z}$ in $W(F)$ is either $\mathbb{Z}/4$ or $\mathbb{Z}/8$.

It is well known [**33**, 3.3] that the element '4' in $W(F)$ corresponds to the symbol $\{-1, -1\}$ in $I^2(F) \cong H^2_{\text{et}}(F, \mathbb{Z}/2)$. This symbol vanishes if and only if there is a morphism $V \to Q_1$ defined over $\mathbb{R}$. $\square$

REMARK 1.10.1. Milnor showed in [**33**, 3.1, 3.3, 4.1] that $1 + w_1 + w_2$ is an additive isomorphism from $I(F)/I^3(F)$ to $1 + H^1_{\text{et}}(F) + H^2_{\text{et}}(F)$, which is a group under $(1+a_1+a_2)(1+b_1+b_2) = 1+(a_1+b_1)+(a_2+b_2+a_1 \cup b_1)$, with the inverse given by s_1 and s_2. For example, the element '2' of $I(F)$ maps to $1+\{-1\}$ and '4' maps to $(1+\{-1\})^2 = 1+\{-1,-1\}$.

EXAMPLE 1.11. There are real surfaces V for which the image of $\mathbb{Z} \to W(V)$ is $\mathbb{Z}/8$; since $W(V) \subset W(F)$, this is equivalent to the condition that the function field F of V has level 4 (i.e., -1 is a sum of 4 squares). As pointed out by Parimala in [**39**], this is the case if V is the hypersurface in $\mathbb{P}^3_{\mathbb{R}}$ defined by $z^2 - f$, where $f(x, y)$ is the Motzkin polynomial $x^4y^2 + x^2y^4 - 3x^2y^1 + 1$ (or certain other positive polynomials).

EXAMPLE 1.12. Consider the real surface $V = E \times E$, where E is an elliptic curve with 2 real connected components; this is a real form of the abelian variety $V_{\mathbb{C}} = E_{\mathbb{C}} \times E_{\mathbb{C}}$. To compute $W(V)$ using Remark 1.8.1, we first compute $H^*_{\text{et}}(V, \mathbb{Z}/2)$, using the Leray spectral sequence

$$H^p(G, H^q_{\text{et}}(V_{\mathbb{C}}, \mathbb{Z}/2)) \Rightarrow H^{p+q}_{\text{et}}(V, \mathbb{Z}/2).$$

Because $E(\mathbb{R})$ contains the exponent 2 points of the group $E(\mathbb{C})$, G acts trivially on $H^*_{\text{et}}(E_{\mathbb{C}}, \mathbb{Z}/2)$. Dropping the coefficients $\mathbb{Z}/2$ from the notation, the Künneth formula shows that G also acts trivially on $H^1_{\text{et}}(V_{\mathbb{C}}) \cong (\mathbb{Z}/2)^4$ and $H^2_{\text{et}}(V_{\mathbb{C}}) \cong (\mathbb{Z}/2)^6$. We get an exact sequence

$$0 \to (\mathbb{Z}/2) \to H^1_{\text{et}}(V) \to H^1(V_{\mathbb{C}})^G \to 0.$$

Hence $H^1_{\text{et}}(V) \cong (\mathbb{Z}/2)^5$. In the corresponding spectral sequence for $H^*_{\text{et}}(E)$, the differential $\mathbb{Z}/2 = E_2^{0,2}(E) \to E_2^{2,1}(E) = (\mathbb{Z}/2)^2$ is an injection. Hence the differential from $E_2^{0,2}(V) = H^2_{\text{et}}(V_{\mathbb{C}})$ to $E_2^{2,1}(V) = H^1_{\text{et}}(V_{\mathbb{C}})$ has rank 2, and we deduce that $H^2_{\text{et}}(V) \cong (\mathbb{Z}/2)^7$.

Since the stabilization map $H^1_{\text{et}}(V) \to H^2_{\text{et}}(V)$ is the cup product with $[-1]$, we also see that $H^1_{\text{tors}}(V) \cong (\mathbb{Z}/2)^3$ injects into $H^2_{\text{tors}}(V) \cong (\mathbb{Z}/2)^4$, so Sujatha's invariant l vanishes, and $j = 3$, $k = 4$. It follows from Remark 1.8.1 that

$$W(E \times E) \cong \mathbb{Z}^4 \oplus (\mathbb{Z}/4)^3 \oplus \mathbb{Z}/2.$$

2. WR and the signature

In this section, we define the signature map on $WR(X)$ and relate it to the edge map η^0 in the Bredon spectral sequence for $KO_G(X)$. First we recall the definition of $WR(X)$ from [**23**].

Let X be a topological space with involution. By a *Real vector bundle* on X we mean a complex vector bundle E with an involution σ compatible with the involution on X and such that for each $x \in X$ the isomorphism $\sigma : E_x \to E_{\sigma x}$ is $\mathbb{C}$-antilinear. Following Atiyah [**2**], we write $KR(X)$ for the Grothendieck group of Real vector bundles on a compact space X.

Recall from [**23**] that the Grothendieck-Witt group $GR(X)$ is the Grothendieck group of the category of symmetric forms (E, ϕ), where E is a Real vector bundle on X and ϕ is an isomorphism of Real vector bundles from E to its dual E^* such that $\phi = \phi^*$.

DEFINITION 2.1. The Real Witt group $WR(X)$ of a compact G-space X is the cokernel of the hyperbolic map $H : KR(X) \to GR(X)$.

By [**23**, Thm. 2.2], there is an isomorphism $KO_G(X) \cong GR(X)$ identifying the hyperbolic map with the map $KR(X) \to KO_G(X)$, sending a Real vector bundle to its underlying $\mathbb{R}$-linear bundle with involution. Thus $WR(X)$ is also the cokernel of $KR(X) \to KO_G(X)$.

EXAMPLE 2.1.1. It is easy to see that $KO_G(\text{pt}) \cong RO(G)$ and the signature $WR(\text{pt}) \cong \mathbb{Z}$ sends $[\mathbb{R}]$ and $[\mathbb{R}(1)]$ to 1 and -1. Suppose more generally that G acts trivially on X. Then $KO_G(X) \cong KO(X) \otimes R(G)$; we showed in [**23**, 2.4] that $KR(X) \cong KO(X)$ and $WR(X) \cong KO(X)$. In this case, the rank of a KO-bundle on each component of X defines a map $WR(X) \to \mathbb{Z}^\nu$, where $\nu = |\pi_0(X)|$.

In contrast, $WR(X)$ is a $\mathbb{Z}/2$-algebra when $X = G \times Y$. In this case, $WR(X)$ is the cokernel of the natural map $KU(Y) \to KO(Y)$ and hence a $\mathbb{Z}/2$-algebra; see [**23**, 2.4(b) and 2.6]. The following example illustrates this.

LEMMA 2.2. *When V is complex projective space $\mathbb{P}^d_{\mathbb{C}}$, $X = (\mathbb{P}^d_{\mathbb{C}})_{\text{top}}$ is $G \times \mathbb{CP}^d$ and*

$$W(\text{pt}) \cong W(\mathbb{P}^d_{\mathbb{C}}) \cong WR(X) \cong \mathbb{Z}/2.$$

PROOF. The isomorphism $\mathbb{Z}/2 = W(\text{pt}) \cong W(\mathbb{P}^d_{\mathbb{C}})$ is due to Arason [**1**]. As noted above, $WR(X)$ is the cokernel of $KU(\mathbb{CP}^d) \to KO(\mathbb{CP}^d)$. In this case, Karoubi and Mudrinski [**22**] proved that the forgetful map $KU(\mathbb{CP}^d, \text{pt}) \to KO(\mathbb{CP}^d, \text{pt})$ is onto, so $WR(\mathbb{CP}^d) = \mathbb{Z}/2$ as well. □

The topological signature. Now let X be any compact G-space, such that X^G has ν components. The signature of a symmetric form (E, ϕ) on X is an element of $\mathbb{Z}^\nu$; the signature on a component of X^G is the signature of the form ϕ_x on the vector space E_x, for any point x of the component (it is independent of the choice of x). This passes to a homomorphism $KO_G(X) \to \mathbb{Z}^\nu$.

Since hyperbolic forms have signature 0, it also defines a topological *signature* $\sigma : WR(X) \to \mathbb{Z}^\nu$; the signature of $w \in WR(X)$ is the signature of a quadratic form (E,ϕ) representing w. While σ may not be onto (see Remark 4.3), its image has finite 2-primary index in $\mathbb{Z}^\nu$ by [**23**, 7.4].

As observed in [**23**], there is a canonical map $W(V) \to WR(V_{\text{top}})$ and the composition with the topological signature is the classical algebraic signature.

On each component, the signature of an n-dimensional symmetric form ϕ is congruent to n modulo 2. If X is connected, it follows that the signature of ϕ lies in the subgroup Γ of $\mathbb{Z}^\nu$, defined as follows.

DEFINITION 2.3. Let Γ denote the subgroup of $\mathbb{Z}^\nu$ consisting of all $(c_1,...,c_\nu)$ with either all c_i even or all c_i odd; Γ contains $2\mathbb{Z}^\nu$ as an index 2 subgroup. The group Γ is free abelian of rank ν, but there is no isomorphism of Γ with $\mathbb{Z}^\nu$ which is natural in ν.

We now assume X is connected, and relate the signature to the canonical edge map

$$\eta^0 : KO_G(X) \to H^0_G(X;KO^0_G) \cong \mathbb{Z}\oplus\mathbb{Z}^\nu \tag{2.4}$$

in the Bredon spectral sequence (A.1); see Lemma A.5. The first coordinate of η^0 is the rank of the underlying vector bundle. To describe the other coordinates of η^0, let $\widetilde{KO_G(X)}$ denote the kernel of the rank map $KO_G(X)\to\mathbb{Z}$. For each point x of X^G, $\widetilde{KO}_G(x)$ is isomorphic to $\mathbb{Z}$ on $[\mathbb{R}]-[\mathbb{R}(1)]$. This yields a map $\widetilde{KO}_G(X)\to\widetilde{KO}_G(x)\cong\mathbb{Z}$; if E is a G-vector bundle of rank r on X, the map sends $[E]-r$ to $[E_x]-r$. Choosing a point on each of the ν components of X^G yields the edge map η^0 from $\widetilde{KO}_G(X)$ to $\mathbb{Z}^\nu$. It sends the line bundles $X\times\mathbb{R}$ and $X\times\mathbb{R}(1)$ to $(0,...,0)$ and $(-1,...,-1)$.

Both $H^0_G(X;KO_G)$ and ν are determined by the 1–skeleton of X.

LEMMA 2.5. *Let X be a connected finite G-CW complex whose fixed locus X^G has ν components. The signature $WR(X)\overset{\sigma}{\to}\mathbb{Z}^\nu$ is induced by*

$$KO_G(X) \xrightarrow{\eta^0} H^0_G(X;KO_G)\cong\mathbb{Z}\oplus\mathbb{Z}^\nu \xrightarrow{\rho} \Gamma\subseteq\mathbb{Z}^\nu,$$

where η^0 is the edge map in (A.1) *and ρ is given by*

$$\rho(r,a_1,...,a_\nu) = (r+2a_1,...,r+2a_\nu).$$

PROOF. If pt is a single point, we choose $[\mathbb{R}]$ and $\lambda=[\mathbb{R}]-[\mathbb{R}(1)]$ as a basis of $H^0_G(\text{pt},\mathbb{Z})\cong RO(G)\cong\mathbb{Z}^2$. The projection '+' in Example A.4(a) from $H^0_G(\text{pt},\mathbb{Z})\cong RO(G)$ to $H^0(\text{pt}/G,\mathbb{Z})\cong\mathbb{Z}$ is split by the map sending the generator to $[\mathbb{R}]$, and the kernel is generated by λ.

The image of $KR(\text{pt})\cong\mathbb{Z}$ in $H^0_G(\text{pt},\mathbb{Z})$ is generated by $[\mathbb{R}]+[\mathbb{R}(1)]=2[\mathbb{R}]-\lambda$; it is the kernel of the map $H^0_G(\text{pt},\mathbb{Z})\xrightarrow{s}\mathbb{Z}$ sending $(r[\mathbb{R}]+a\lambda)$ to its signature $r+2a$. Therefore s induces the signature $\sigma : WR(\text{pt})\xrightarrow{\cong}\mathbb{Z}$.

Now let Y denote ν discrete points. It follows that the signature $WR(Y)\overset{\cong}{\to}\mathbb{Z}^\nu$ is induced by the composition of the map

$$KO_G(Y)\xrightarrow{\cong} H^0_G(Y,\mathbb{Z})\xrightarrow{\cong} H^0(Y^G,\mathbb{Z})\oplus H^0(Y/G,\mathbb{Z})\cong\mathbb{Z}^\nu\oplus\mathbb{Z}^\nu,$$

with the map $\mathbb{Z}^{2\nu} \xrightarrow{s} \mathbb{Z}^\nu$, sending $(r_1, ..., a_1, ...)$ to $(r_1+2a_1, ..., r_\nu+2a_\nu)$. The general case now follows from the following diagram, whose rows are from Lemma A.5

$$\begin{array}{ccccccccc} 0 & \longrightarrow & H^0(X^G,\mathbb{Z}) & \longrightarrow & H^0_G(X;KO_G) & \xrightarrow{+} & H^0(X/G,\mathbb{Z}) & \longrightarrow & 0 \\ & & \downarrow\cong & & \downarrow\text{into} & & \downarrow\text{diag} & & \\ 0 & \longrightarrow & H^0(Y^G,\mathbb{Z}) & \xrightarrow{\lambda} & H^0_G(Y;KO_G) & \xrightarrow{+} & H^0(Y,\mathbb{Z}) & \longrightarrow & 0 \end{array}$$

and whose columns are associated to the evident map $Y \to X^G \subseteq X$, obtained by choosing a basepoint for each component of X^G. By inspection, the map $\rho : H^0_G(X;KO_G) \to \Gamma$ is the restriction of the map $s : H^0_G(Y;KO_G) \to \mathbb{Z}^\nu$. It follows that $\rho\eta^0_X$ is the restriction of $s\eta^0_Y$, which is the signature. $\square$

REMARK 2.5.1. The sequence $KR(X) \to \mathbb{Z}^{1+\nu} \xrightarrow{\rho} \Gamma$ of Lemma 2.5 is exact. Indeed, the kernel of ρ is the subgroup of $\mathbb{Z}^{1+\nu}$ generated by $(2, -1, ..., -1)$. This is the image of $\eta^0 \circ H : KR(X) \to \mathbb{Z} \oplus \mathbb{Z}^\nu$, because $\eta^0 \circ H$ maps the trivial Real line bundle $X \times \mathbb{C}$ to $(2, -1, ..., -1)$.

3. WR for low dimensional X

In this section, we determine $WR(X)$ when $\dim(X) \leq 2$. There is no harm in assuming that X is an irreducible G-space, so that X is either connected or has the form $X = G \times Y$ with Y connected. We saw in [**26**, 7.1] that $WR(G \times Y) \cong \mathbb{Z}/2 \oplus H^1(Y, \mathbb{Z}/2)$ when $\dim(Y) \leq 2$. Thus we are reduced to the case when X is connected.

In general, there is a map $w_1 : WR(X) \to \mathbb{H}^1_G(X, \mathbb{Z}/2)$, defined in 4.2 and 6.1 below. Let us write $[-1]$ for the element $w_1(X \times \mathbb{R}(1))$ of $\mathbb{H}^1_G(X, \mathbb{Z}/2)$. We will see in Remark 4.1.1 that $[-1]$ is always nonzero in $\mathbb{H}^1_G(X, \mathbb{Z}/2)$. Let $\widetilde{H}^1(X/G, \mathbb{Z}/2)$ denote the quotient of $H^1(X/G, \mathbb{Z}/2)$ by the subgroup generated by the element $[-1] = w_1(X \times \mathbb{R}(1))$.

LEMMA 3.1. *When X is a connected 1-dimensional G-complex, and X^G has ν components, we have*

$$WR(X) = \begin{cases} \mathbb{Z}^\nu \oplus H^1(X/G, \mathbb{Z}/2), & \nu > 0, \\ \mathbb{Z}/4 \oplus \widetilde{H}^1(X/G, \mathbb{Z}/2), & \nu = 0. \end{cases}$$

PROOF. The spectral sequences of (A.1) for KO_G and KR converge, having only two nonzero rows. By Lemma A.5, we have

$$KO_G(X) \cong \mathbb{Z} \oplus \mathbb{Z}^\nu \oplus H^1_G(X; KO_G^{-1}), \qquad KR(X) \cong \mathbb{Z} \oplus H^1_G(X; KR^{-1}),$$

and the cokernel of $H^1_G(X; KR^{-1}) \to H^1_G(X; KO_G^{-1})$ is $H^1(X/G, \mathbb{Z}/2)$. Thus it remains to consider the cokernel of the map of H^0_G-components $\mathbb{Z} \to \mathbb{Z} \oplus \mathbb{Z}^\nu$. It is described in Remark 2.5.1, and yields the displayed calculation. $\square$

EXAMPLE 3.2. There are three actions of G on the circle. Recall that $S^{p,q}$ denotes the sphere in $\mathbb{R}(1)^p \times \mathbb{R}^q$.

(i) $S^{0,2}$ is the circle with trivial G-action. Let μ denote the Möbius line bundle. Then $KO_G(S^{0,2})$ is $\mathbb{Z}^2 \oplus (\mathbb{Z}/2)^2$ with generators the line bundles 1, $X \times \mathbb{R}(1)$, μ and $\mu \otimes \mathbb{R}(1)$, while $WR(S^{0,2}) \cong KO(S^{0,2})$ is $\mathbb{Z} \oplus \mathbb{Z}/2$ by Example 2.1.1.

(ii) $S^{2,0}$ is the circle with the antipodal involution on S^1. Lemma 3.1 yields $WR(S^{2,0}) \cong \mathbb{Z}/4$.

(iii) $S^{1,1}$ is the unit circle in $\mathbb{C}$, with the induced complex conjugation. By Lemma 3.1,

$$WR(S^{1,1}) \cong \mathbb{Z}^2.$$

The Real line bundle $E = S^{1,1} \times \mathbb{C}$ carries a canonical Real symmetric form $\theta = t$, which is multiplication by t on the fiber over $t \in S^{1,1}$. By [**23**, Thm. 2.2], this corresponds to the G-subbundle of E whose fiber over t is $it \cdot \mathbb{R}$. It is a nontrivial element of $\mathrm{Pic}_G(S^{1,1}) \cong \{\pm 1\}^2$, and its discriminant $\iota = w_1(E,t)$ is nontrivial.

If $X \overset{a}{\to} S^{1,1}$ is an equivariant map, the pullback of (E,t) under the map a^* : $\mathrm{Pic}_G(S^{1,1}) \to \mathrm{Pic}_G(X)$ is the class of $(X \times \mathbb{C}, a)$ in $\mathrm{Pic}_G(X)$, and its discriminant is $w_1(a^*(E,t)) = a^*(\iota)$.

LEMMA 3.3. *Let X be a closed connected 2-manifold with involution, such that X^G is a union of ν circles. Then*

$$WR(X) = \begin{cases} \mathbb{Z}^\nu \oplus (\mathbb{Z}/2)^h, & \nu > 0 \\ \mathbb{Z}/4 \times (\mathbb{Z}/2)^{h-1}, & \nu = 0, \end{cases} \quad \textit{where } h = \dim H^1(X/G, \mathbb{Z}/2).$$

PROOF. If $\nu = 0$, this was proven in [**26**, 8.1], so suppose $\nu > 0$. Since X/G is a 2-manifold with boundary, $H^2(X/G, \mathbb{Z}/2) = 0$. (See the argument of [**41**, 1.9].) As $H^2(X^G, \mathbb{Z}/2) = 0$, Lemma A.5 yields $H^2_G(X; KO_G^{-2}) = 0$ and $H^1_G(X; KO_G^{-1}) = (\mathbb{Z}/2)^{h+\nu}$. By (A.1), we have $KO_G(X) = \mathbb{Z}^{\nu+1} \oplus (\mathbb{Z}/2)^{h+\nu}$. Now use the argument of Lemma 3.1. □

If C is a smooth projective curve over $\mathbb{R}$, C_{top} is a 2-manifold whose genus g is h when $\nu > 0$, and $h-1$ when $\nu = 0$. Thus we recover the following calculation of $WR(C_{\mathrm{top}})$, which was derived in [**23**, Thms. 4.6 and 4.7], using $H^2_{\mathrm{et}}(C, \mathbb{Z}/2) \cong (\mathbb{Z}/2)^{2\nu}$ and $\mathrm{Br}(C) \cong (\mathbb{Z}/2)^\nu$. (The fact that $W \cong WR$ for curves is also in *loc. cit.*)

COROLLARY 3.4. *If C is a smooth geometrically connected projective curve of genus g over $\mathbb{R}$, with ν real components, then*

$$W(C) \cong WR(C_{\mathrm{top}}) \cong \begin{cases} \mathbb{Z}^\nu \oplus (\mathbb{Z}/2)^g, & \nu > 0, \\ \mathbb{Z}/4 \oplus (\mathbb{Z}/2)^g, & \nu = 0. \end{cases}$$

REMARK 3.4.1. When $C(\mathbb{R}) = \emptyset$, and $U \subset C$ is the complement of $n > 0$ (complex) points, then $W(U) \cong W(C) \oplus (\mathbb{Z}/2)^{n-1}$. This follows from Corollary 3.4, (1.1), and the fact that $\mathrm{Pic}(C)/2 \cong \mathbb{Z}/2$. In addition, $W(U) \cong WR(U)$ by [**23**, 4.1]. We leave the details to the reader.

EXAMPLE 3.5. We can now calculate $WR(X)$ of a 2-sphere with involution. Writing $S^{p,q}$ for the sphere in $\mathbb{R}(1)^p \times \mathbb{R}^q$, we have:

$$WR(S^{0,3}) \cong \mathbb{Z} \oplus \mathbb{Z}/2; \quad WR(S^{1,2}) \cong \mathbb{Z}; \quad WR(S^{2,1}) \cong \mathbb{Z}^2 \oplus \mathbb{Z}/2;$$

and $WR(S^{3,0}) \cong \mathbb{Z}/4$. The calculation $WR(S^{0,3}) \cong KO(S^2) \cong \mathbb{Z} \oplus \mathbb{Z}/2$ is from Example 2.1.1. The formulas for the genus 0 curves $S^{1,2} = \mathbb{CP}^1$ and $S^{3,0} = Q_1$ are from Lemma 3.3.

For $X = S^{2,1}$, we note that X^G is 2 points and $X/G \simeq S^2$. Therefore $H^1_G(S^{2,1}, \mathbb{Z}/2) = 0$ and $H^2_G(X; KO^{-2}) \cong \mathbb{Z}/2$, using Lemma A.5. Thus $KO_G(X) \cong \mathbb{Z}^3 \oplus \mathbb{Z}/2$. We also have $\mathbb{H}^1_G(S^{2,1}, \mathbb{Z}/2) \cong \mathbb{Z}/2$. As in the proof of Lemma 3.3, we have $KR(X) = \mathbb{Z}$, and hence $WR(S^{2,1}) \cong \mathbb{Z}^2 \oplus \mathbb{Z}/2$.

More generally, if $q > 0$ then $KR^n(S^{p,q}, \mathrm{pt}) \cong KO^{n+p-q+1}(\mathrm{pt})$. This follows easily from Atiyah's sequence [**2**, (2.2)] for $(X, Y) = (B^{p,q}, S^{p,q})$. In addition, because $S^{p,q}$ is the 1–point compactification of $\mathbb{R}(1)^p \times \mathbb{R}^{q-1}$, $KO_G(S^{p,q}, \mathrm{pt})$ is isomorphic to the K-theory with compact supports $KO_G^{1-q}(\mathbb{R}(1)^p)$, a group isomorphic to the topological K-theory $K_{\mathrm{top}}^{1-q}(C^{0,p+1})$ of the Clifford algebra $C^{0,p+1}$; see [**19**].

If G acts trivially on Y then $KO_G(Y \times S^{p,q}, Y) \cong K_{\mathrm{top}}^{1-q}(C^{0,p+1}(Y))$, where $C^{0,p+1}(Y)$ is the Banach algebra of continuous functions from Y to $C^{0,p+1}$.

For completeness, we note that when $q = 0$ and $p > 0$ the group $KO_G(Y \times S^{p,0})$ is $KO(Y \times \mathbb{RP}^{p-1})$. This group was computed by Adams (when Y is a point) and in general by the first author in [**19**].

For later use in Example 8.9, we record that

$$KO_G(S^{1,3}, \mathrm{pt}) \cong \mathbb{Z}/2, \; KO_G(S^{2,2}, \mathrm{pt}) = 0 \text{ and } KO_G(S^{2,3}, \mathrm{pt}) = \mathbb{Z}.$$

Using this, we can check that $WR(S^{1,3}) \cong WR(S^{2,2}) \cong WR(S^{2,3}) \cong \mathbb{Z}$. (An alternative calculation of $WR(S^{2,2})$ was given in Example A.9.)

EXAMPLE 3.6. Let T be the torus $S^1 \times S^{1,1}$, where G acts trivially on S^1. Then T is a closed connected 2-manifold with involution, T^G is a union of 2 circles, and $T/G \cong S^1 \times [0,1]$. From Lemma 3.3 and Example 3.2(i), we see that $WR(S^1) \cong \mathbb{Z} \oplus \mathbb{Z}/2$ and

$$WR(T) \cong \mathbb{Z}^2 \oplus \mathbb{Z}/2 \ncong WR(S^1) \oplus WR(S^1).$$

If C is the affine circle $x^2 + y^2 = 1$ over $\mathbb{R}$ and $V = C \times \mathbb{G}_m$, then $W(C) \cong \mathbb{Z} \oplus \mathbb{Z}/2$ and $W(V) \cong \mathbb{Z}^2 \oplus (\mathbb{Z}/2)^2$ by the Fundamental Theorem for $W(C)$ [**20**, pp. 138–9]. Since C_{top} and V_{top} are equivariantly homotopic to S^1 and T, respectively, it follows that $W(V) \ncong WR(V_{\mathrm{top}})$.

EXAMPLE 3.7. The projective plane $\mathbb{P}^2$ over $\mathbb{R}$ has $\mathbb{P}^2_{\mathrm{top}} = \mathbb{CP}^2$ and

$$W(\mathbb{P}^2) \cong WR(\mathbb{CP}^2) = \mathbb{Z}.$$

Because W and WR are birational invariants of a surface (see Theorem 5.1), this also implies that $W(\mathbb{P}^1 \times \mathbb{P}^1) \cong WR(\mathbb{CP}^1 \times \mathbb{CP}^1) \cong \mathbb{Z}$.

Arason proved that $W(\mathbb{P}^d) = \mathbb{Z}$ for all d; see [**1**]. The G-space $X = \mathbb{P}^2_{\mathrm{top}}$ is $\mathbb{CP}^2$ with $X^G = \mathbb{RP}^2$. The quotient X/G has $H^*(X/G, \mathbb{Z}) \cong H^*(S^4, \mathbb{Z})$. In fact, X/G is homotopy equivalent to S^4; writing $\mathbb{CP}^2$ as the cone of the Hopf map, one sees that $X/G \simeq \Sigma(S^3/G) \simeq S^4$. By Lemma A.5,

$$E_2^{p,q} = H_G^p(X; KO_G^q) = H^p(\mathbb{RP}^2, KO^q) \oplus H^p(X/G, KO^q),$$

and the second factor vanishes unless $p = 0, 4$. Thus

$$H_G^p(X; KO_G^{-p}) = \begin{cases} \mathbb{Z}^2, & p = 0; \\ \mathbb{Z}/2, & p = 1, 2; \\ \mathbb{Z}, & p = 4; \\ 0, & \text{otherwise.} \end{cases}$$

Since $KO_G^n(X)$ maps onto $KO_G^n(\mathrm{pt})$ and $KO_G(X)$ maps to $KO(\mathbb{RP}^2) = \mathbb{Z} \oplus \mathbb{Z}/4$, the spectral sequence (A.1) degenerates to show that $KO_G(X)$ is the extension $\mathbb{Z}^3 \oplus \mathbb{Z}/4$ of $\mathbb{Z}^3 \oplus \mathbb{Z}/2$ by $\mathbb{Z}/2$. By Lemma A.5, Remark A.5.1, Lemma A.8 and Remark 2.5.1, $KR(X)$ maps onto the torsion, the final $\mathbb{Z}$, and the subgroup of $\mathbb{Z}^2 \oplus \mathbb{Z}/2 \oplus \mathbb{Z}$ generated by $(2, -1, [-1], 0)$. It follows that $WR(X) \cong \mathbb{Z}$.

REMARK 3.7.1. Atiyah proved in [**2**, 2.4] that $KR(\mathbb{CP}^2) \cong \mathbb{Z}[t]/t^3$. In the filtration arising from (A.1), t is detected by $H^1_G(X, KR^{-1}) \cong \mathbb{Z}/2$; t^2 and $2t$ are in filtration level 2.

4. The equivariant Picard group

In this section X will be a connected G-space, and ν will denote the number of components of X^G.

DEFINITION 4.1. Let $\mathrm{Pic}_G(X)$ denote the group of rank 1 real G-vector bundles on X with product $\otimes$; its unit '1' is the trivial line bundle $X \times \mathbb{R}$. If L is such a bundle, $L \otimes L$ is trivial, and the involution on the fibers of L over fixed points is multiplication by ± 1. This involution is locally constant, so if X^G has ν components it defines a natural map $\mathrm{Pic}_G(X) \to \{\pm 1\}^\nu$, called the *sign map*.

Let $\mathrm{Pic}^0_G(X)$ denote the kernel of the sign map. There is a natural isomorphism $\mathrm{Pic}^0_G(X) \overset{\cong}{\to} H^1(X/G, \mathbb{Z}/2)$, defined as follows. If L is a G-line bundle on X with trivial sign (so G acts trivially on L_x for all $x \in X^G$, then the identifications $L_x \cong L_{\sigma x}$ imply that L descends to a line bundle L/G on X/G. Since real line bundles on X/G are classified by the group $H^1(X/G, \mathbb{Z}/2)$, this defines a map $\mathrm{Pic}^0_G(X) \to H^1(X/G, \mathbb{Z}/2)$. Conversely, the pullback of such a line bundle along $X \to X/G$ is an equivariant line bundle on X with trivial sign, i.e., an element of $\mathrm{Pic}^0_G(X)$.

A natural isomorphism $\mathrm{Pic}_G(X) \overset{\omega}{\to} \mathbb{H}^1_G(X, \mathbb{Z}/2)$, was constructed in [**26**, 3.2]. It is compatible with the exact sequence

$$0 \to H^1(X/G, \mathbb{Z}/2) \to \mathbb{H}^1_G(X, \mathbb{Z}/2) \overset{\text{sign}}{\longrightarrow} H^0(X^G, \mathbb{Z}/2).$$

of Example A.3, as we see from the map $\mathbb{H}^1_G(X, \mathbb{Z}/2) \to \mathbb{H}^1_G(X^G, \mathbb{Z}/2)$. This exact sequence, combined with either Example 1.6 or Example 4.4, shows that the sign map is not always onto when $\nu \geq 2$.

REMARK 4.1.1. The G-line bundle $X \times \mathbb{R}(1)$ is nontrivial. This is clear if $X^G \neq \emptyset$, as $\mathbb{R}(1)$ has a nontrivial sign. When $X^G = \emptyset$, it is a nonzero element of $\mathrm{Pic}^0_G(X) \cong H^1(X/G, \mathbb{Z}/2)$. This is because $X \to X/G$ is a nontrivial covering space, covering spaces with group G are classified by elements of $H^1(X/G, \mathbb{Z}/2)$, and we showed in Example 2.7 of [**23**] that $\omega(X \times \mathbb{R}(1))$ is the element classifying this particular cover.

REMARK 4.1.2. If V is a geometrically ireducible variety over $\mathbb{R}$, and $X = V_{\mathrm{top}}$, Cox's Theorem [**14**] states that

$$\mathrm{Pic}_G(X) \cong \mathbb{H}^1_G(X, \mathbb{Z}/2) \cong H^1_{\mathrm{et}}(V, \mathbb{Z}/2) \cong \mathcal{O}^\times(V)/\mathcal{O}^{\times 2}(V) \oplus {}_2\mathrm{Pic}(V).$$

If V is projective then $\mathcal{O}^\times(V) = \mathbb{R}^\times$ and $\mathrm{Pic}_G(X) \cong \mathbb{Z}/2 \oplus {}_2\mathrm{Pic}(V)$.

DEFINITION 4.2. There is a determinant map $KO_G(X) \overset{\det}{\to} \mathrm{Pic}_G(X)$ satisfying $\det(E \oplus F) = \det(E) \otimes \det(F)$; $\det(L) = [L]$ if $\mathrm{rank}(L) = 1$. Composing det with the isomorphism of Definition 4.1 defines a surjection $w_1 : KO_G(X) \to \mathbb{H}^1_G(X, \mathbb{Z}/2)$, which we call the *discriminant*; we will relate the discriminant to the Stiefel–Whitney class w_1 in Section 6 below.

As in the algebraic setting, the discriminant does not factor through $WR(X)$, because the discriminant of $H(1)$ is -1. Instead, the discriminant factors through the subgroup $I(X)$ of forms in $WR(X)$ of even degree, as we shall see in Theorem 6.2 below.

REMARK 4.3. If E is a rank r G-bundle on X, then $e = [E] - r$ has rank 0 in $KO_G(X)$; if E has signature a_i on the ith component of X^G then the edge map η^0 of (2.4) sends e to $\eta^0(e) = (0, a_1 - r, ..., a_\nu - r)$, which is an element of $0 \times (2\mathbb{Z})^\nu$. Let $\widetilde{KO}_G(X)$ denote the kernel of $KO_G(X) \xrightarrow{\text{rank}} \mathbb{Z}$; since $\det(e) = \det(E)$ restricts to a line bundle with sign $(-1)^{(a_i-r)/2}$ on the ith component of X^G, we have a commutative diagram and a map from the kernel of $KO_G(X) \xrightarrow{\eta^0} \mathbb{Z}^{1+\nu}$ onto $\mathrm{Pic}^0_G(X)$.

$$\begin{array}{ccccc} \ker(\eta^0) & \longrightarrow & \widetilde{KO}_G(X) & \xrightarrow{\eta^0} & (2\mathbb{Z})^\nu \\ \downarrow{\scriptstyle\text{onto}} & & \downarrow{\scriptstyle\det} & & \downarrow{\scriptstyle\bmod 2} \\ \mathrm{Pic}^0_G(X) & \longrightarrow & \mathrm{Pic}_G(X) & \xrightarrow{\text{sign}} & \{\pm 1\}^\nu. \end{array}$$

As noted in Example 4.4, the sign map $\mathrm{Pic}_G(X) \to \{\pm 1\}^\nu$ need not be onto; it follows that the signature $KO_G(X) \to \Gamma \subseteq \mathbb{Z}^\nu$ of 2.5 need not be onto either.

EXAMPLE 4.4. The 2-sphere $X = S^{2,1}$ has $\nu = 2$, as X^G is two points, and we saw in Example 3.5 that $KO_G(S^{2,1}) \cong \mathbb{Z}^2 \oplus (\mathbb{Z}/2)$. It follows that $\mathrm{Pic}_G(X) \cong \mathbb{H}^1_G(X, \mathbb{Z}/2) = \mathbb{Z}/2$. From Remark 4.3, we see that the signature $KO_G(X) \to \Gamma \subseteq \mathbb{Z}^2$ is not onto.

Given an algebraic surface V defined over $\mathbb{R}$, such that $V(\mathbb{R})$ has ν real components, let S_a and S_t denote the images of the algebraic signature $W(V) \to \mathbb{Z}^\nu$ and the topological signature $WR(V_{\text{top}}) \to \mathbb{Z}^\nu$, respectively. Since the algebraic signature factors through the topological one, $S_a \subseteq S_t$. We do not know any examples where $S_a \neq S_t$.

As we noted in Lemma 2.5, $S_t \subseteq \Gamma$. Colliot-Thélène and Sansuc proved in [**10**, 4.1] that $I_3(V)$ maps onto $8\mathbb{Z}^\nu$, i.e, that S_a contains $8\mathbb{Z}^\nu$. (See also [**9**, proof of Thm. 2.3.2].) Therefore we have

$$8\mathbb{Z}^\nu \subseteq S_a \subseteq S_t \subseteq \Gamma \subseteq \mathbb{Z}^\nu.$$

We also know from [**23**, 8.7] that $2S_t \subseteq S_a$.

PROPOSITION 4.5. *The images of the algebraic and topological signatures in $\mathbb{Z}^\nu/4\mathbb{Z}^\nu$ are the same.*

PROOF. Set $X = V_{\text{top}}$; we may suppose X connected. Since both $W(V)/I(V)$ and $WR(X)/I(X)$ map isomorphically to $\Gamma/2\mathbb{Z}^\nu \cong \mathbb{Z}/2$, the signature maps $I(V)$ and $I(X)$ to $2\mathbb{Z}^\nu$. We need to consider their images modulo $4\mathbb{Z}^\nu$.

Let us write $\{\pm 1\}^\nu$ for $(2\mathbb{Z}/4\mathbb{Z})^\nu$. By Remark 4.3, the reduction modulo 2 of the topological signature, i.e., $I(X) \to (2\mathbb{Z})^\nu \to \{\pm 1\}^\nu$, factors as the discriminant $I(X) \to \mathrm{Pic}_G(X)$, which is onto, followed by the sign map $\mathrm{Pic}_G(X) \to \{\pm 1\}^\nu$. The same is true of the algebraic signature. Hence we have a commutative diagram:

$$\begin{array}{ccccc} I(V) & \xrightarrow{\text{onto}} & H^1_{\text{et}}(V, \mathbb{Z}/2) & \xrightarrow{\text{sign}} & \{\pm 1\}^\nu \\ \downarrow & & \downarrow{\scriptstyle\cong} & & \downarrow{\scriptstyle=} \\ I(X) & \xrightarrow{\text{onto}} & \mathbb{H}^1_G(X, \mathbb{Z}/2) & \xrightarrow{\text{sign}} & \{\pm 1\}^\nu. \end{array}$$

By inspection, the images of $I(V)$ and $I(X)$ in $\{\pm 1\}^\nu$ agree. $\square$

REMARK 4.5.1. Here are two consequences of Proposition 4.5:

i) If the signature $WR(X) \to \Gamma$ is onto, so is the signature $W(V) \to \Gamma$.

ii) If $4\mathbb{Z}^{\nu} \subseteq S_a$ then $S_a = S_t$.

5. WR is a birational invariant

Let V, V' be smooth projective surfaces defined over $\mathbb{R}$. We say that V and V' are *(real) birationally equivalent* if there exists a chain of blow-ups at smooth closed points (and their inverses, blow-downs) connecting them. This is equivalent to the assertion that their function fields are isomorphic (see [**47**, II.6.4]). By a *birational invariant* of smooth projective surfaces over $\mathbb{R}$ we mean a function that sends real birationally equivalent surfaces to isomorphic objects. For example, the number of real components is a birational invariant.

THEOREM 5.1. *WR is a birational invariant of smooth projective surfaces over $\mathbb{R}$.*

PROOF. Since every birational equivalence between smooth projective surfaces is a composition of blow-ups and blow-downs at smooth closed points, the result is immediate from Theorem 5.2 below. □

REMARK 5.1.1. We recall how to topologically construct the blow-up $V' \to V$ of a complex algebraic surface V at a smooth point x of V. Let D^4 denote a small 4-disk about x; the inverse image of D^4 in V'_{top} is a tubular neighborhood T of the exceptional fiber $\mathbb{P}^1_{\text{top}} \cong S^2$, and the boundary of T is the same as the boundary S^3 of D^4. (T has Euler class 1.) Thus V'_{top} is the union of $V_{\text{top}} - D^4$ and T along S^3.

THEOREM 5.2. *Let V be a smooth projective variety defined over $\mathbb{R}$, and $V' \to V$ the blowup at a smooth closed point p. Then $WR(V_{\text{top}}) \to WR(V'_{\text{top}})$ is an isomorphism.*

PROOF. Set $X = V_{\text{top}}$, $X' = V'_{\text{top}}$. There are two cases; we may blow up either a real point of V (a fixed point x of X), or a complex point of V (a conjugate pair of points on X).

Fixed points. Consider first the case where V' is the blow-up of V at an $\mathbb{R}$-point. That is, X' is obtained by removing an equivariant disk D^4 in X about the point and replacing it with a tubular neighborhood T of S^2. For any equivariant cohomology theory H^*, the excision $H^*(X, D^4) \cong H^*(X', T)$ yields a Mayer-Vietoris sequence.

Writing $\mathbb{P}$ for the G-space $\mathbb{CP}^1$, we have a commutative diagram, whose rows are Mayer–Vietoris sequences and whose columns are the exact Bott sequences (see [**44**, Thm. 6.1] and [**23**, (1.5)]).

(5.2.1)

$$\begin{array}{ccccccc}
KR_1(\mathbb{P},\mathrm{pt}) & \longrightarrow & KR(X) & \longrightarrow & KR(X') & \xrightarrow{\text{onto}} & KR(\mathbb{P},\mathrm{pt}) = \mathbb{Z} \\
\downarrow{\scriptstyle \cong} & & \downarrow{\scriptstyle H} & & \downarrow{\scriptstyle H} & & \downarrow{\scriptstyle \text{onto}} \\
GR_1(\mathbb{P},\mathrm{pt}) & \longrightarrow & GR(X) & \longrightarrow & GR(X') & \longrightarrow & GR(\mathbb{P},\mathrm{pt}) = \mathbb{Z}/2 \\
\downarrow{\scriptstyle 0} & & \downarrow{\scriptstyle \partial} & & \downarrow{\scriptstyle \partial} & & \downarrow \\
GR_0^{[-1]}(\mathbb{P},\mathrm{pt}) & \xrightarrow{\sigma} & GR_{-1}^{[-1]}(X) & \xrightarrow{u} & GR_{-1}^{[-1]}(X') & \longrightarrow & GR_{-1}^{[-1]}(\mathbb{P},\mathrm{pt}) = 0 \\
{\scriptstyle \tau}\uparrow\ \downarrow{\scriptstyle 2} & & {\scriptstyle \tau}\uparrow\ \downarrow{\scriptstyle F} & & {\scriptstyle \tau}\uparrow\ \downarrow{\scriptstyle F} & & \downarrow \\
KR(\mathbb{P},\mathrm{pt}) & \xrightarrow{0} & KR_{-1}(X) & \longrightarrow & KR_{-1}(X') & \longrightarrow & KR_{-1}(\mathbb{P},\mathrm{pt}) = 0
\end{array}$$

The Bott sequences in the first and last columns are well known, with $KR_1(\mathbb{P},\mathrm{pt}) \cong \mathbb{Z}/2$ mapping onto $GR_1(\mathbb{P},\mathrm{pt}) \cong \mathbb{Z}/2$, $KR(\mathbb{P},\mathrm{pt}) \cong \mathbb{Z}$ mapping onto $GR_1(\mathbb{P},\mathrm{pt}) \cong \mathbb{Z}/2$, and $GR_0^{[-1]}(\mathbb{P},\mathrm{pt}) = \mathbb{Z}$ mapping to $KR(\mathbb{P},\mathrm{pt}) = \mathbb{Z}$ by multiplication by 2.

Since the canonical bundle on $\mathbb{P}$ lifts to the Real bundle on X' associated to the invertible sheaf $\mathcal{O}(1)$ on V' [**16**, II.7], the top right horizontal map is onto, and the bottom left horizontal map is zero. It follows from a diagram chase that the map $WR(X) \to WR(X')$ is onto. To see that it is an injection, we will show that the map σ is zero, and hence that the map u is an injection; this suffices, since $WR(X)$ and $WR(X')$ are the images of the vertical maps ∂ in $GR_{-1}^{[-1]}(X)$ and $GR_{-1}^{[-1]}(X')$.

The map $\tau : KR_n(X) \to GR_n^{[-1]}(X)$ in (5.2.1) is defined as the composition of the map $KR_n(X) \to {}_{-}GR_n^{[1]}(X)$ in the Bott sequence with the isomorphism ${}_{-}GR_n^{[1]} \cong GR_n^{[-1]}$ of the Fundamental Theorem; see [**44**, 6.2]. Identifying $GR_0^{[-1]}(\mathbb{P},\mathrm{pt})$ with the relative group associated to the hyperbolic functor, τ sends the class of a Real bundle E to the hyperbolic bundle $E \oplus E^*$, equipped with its two Lagrangians $E \oplus 0$ and $0 \oplus E^*$. The composition $F \circ \tau$ sends $[E]$ to $[E] - [E^*]$. Since $[E^*] = -[E]$ in $KR(\mathbb{P},\mathrm{pt})$, the composition $F \circ \tau$ on the lower left group $KR(\mathbb{P},\mathrm{pt}) \cong \mathbb{Z}$ is multiplication by 2; it follows that $\tau : KR(\mathbb{P},\mathrm{pt}) \to GR_0^{[-1]}(\mathbb{P})$ is an isomorphism.

Conjugate points. We now consider the case where X' is the blow-up of X at a conjugate pair of points. In this case, we have a diagram like (5.2.1), with $(\mathbb{P},\mathrm{pt})$ replaced by $(G \times \mathbb{P}, G \times \mathrm{pt})$. As in [**23**, 2.4(b) and 2.6] (cf. Lemma 2.2 above), $GR_0^{[-1]}(G \times \mathbb{P}, G) \cong KO^{-2}(S^2,\mathrm{pt})$ and $KR(G \times \mathbb{P}, G) \cong KU(S^2,\mathrm{pt})$. Thus the bottom left and upper right vertical maps are the familiar maps:

$$\mathbb{Z} = KO^{-2}(S^2,\mathrm{pt}) \xrightarrow{2} KU(S^2,\mathrm{pt}) = \mathbb{Z}, \text{ and}$$

$$\mathbb{Z} = KU(S^2,\mathrm{pt}) \xrightarrow{\text{onto}} KO(S^2,\mathrm{pt}) = \mathbb{Z}/2.$$

Again, the Real bundle on X' associated to $\mathcal{O}(1)$ maps to the canonical complex bundle generating $KU(S^2,\mathrm{pt})$ so, as in (5.2.1), the upper right horizontal map is onto and the lower left horizontal map is zero. Again, as in diagram (5.2.1), this implies that the map σ is zero and hence that u is an injection. (In fact, u is an isomorphism as $KO^1(S^2,\mathrm{pt}) = 0$.) As before, this suffices to show that $WR(X) \to WR(X')$ is an isomorphism. $\square$

6. Stiefel–Whitney classes on KO_G

If X is a compact G-space, let us write X_G for $X \times_G EG$, so that the (Borel) equivariant cohomology group $H^n_G(X,\mathbb{Z}/2)$ is defined to be $H^n(X_G,\mathbb{Z}/2)$. Following [**3**], we can identify vector bundles on X_G with equivariant vector bundles on $X \times EG$; the pullback of a bundle on X_G is an equivariant vector bundle on $X \times EG$. As in [**3**], we define $KU(X_G)$ and $KO(X_G)$ to be representable K-theory: $KU(X_G) = [X_G, KU]$ and $KO(X_G) = [X_G, KO]$.

The map $X \times EG \to X$ induces natural maps $KU_G(X) \to KU(X_G)$ and $KO_G(X) \to KO(X_G)$, called *Atiyah-Segal maps*, since they were first studied by Atiyah and Segal in [**3**]. Atiyah and Segal showed in [**3**] that $KO(X_G)$ is the completion of $KO_G(X)$ (for finite CW complexes X) with respect to the augmentation ideal of $KO_G(\mathrm{pt}) \cong RO(G)$.

DEFINITION 6.1. The maps $w_n : KO_G(X) \to \mathbb{H}^n_G(X, \mathbb{Z}/2)$, obtained by composing the Atiyah–Segal map with the usual Stiefel–Whitney classes on X_G, are called the equivariant Stiefel–Whitney classes.

The equivariant Stiefel–Whitney class w_1 agrees on $KO_G(X)$ with the discriminant w_1 defined in Definition 4.2. To see this, note that $\mathrm{Pic}_G(X) \cong \mathrm{Pic}_G(X \times EG)$, and elements of $\mathrm{Pic}_G(X \times EG)$ correspond to rank 1 $\mathbb{R}$-linear vector bundles on X_G, which are classified by the group $H^1(X_G, \mathbb{Z}/2)$; see [**53**, I.4.11].

The equivariant Stiefel–Whitney class w_1 does not factor through $WR(X)$, as the example $X = \mathrm{pt}$ shows (see Example 2.1.1). However, we showed in [**26**, 5.2] that the restriction of w_1 to $I(X)$, the kernel of $KO_G(X) \to \mathbb{H}^0_G(X, \mathbb{Z}/2)$, does induce a map $I(X) \to \mathbb{H}^1_G(X, \mathbb{Z}/2)$, which we shall call w_1 by abuse of notation. The following result was proven in [**26**, Thm. 5.3].

THEOREM 6.2. *The algebraic discriminant of a smooth variety V factors as*

$$I(V) \to I(V_{\mathrm{top}}) \xrightarrow{w_1} \mathbb{H}^1_G(V_{\mathrm{top}}, \mathbb{Z}/2) \cong H^1_{\mathrm{et}}(V, \mathbb{Z}/2).$$

Write $I_2(X)$ for the kernel of $w_1 : I(X) \to \mathbb{H}^1_G(X, \mathbb{Z}/2)$.

COROLLARY 6.3. $W(V)/I_2(V) \xrightarrow{\cong} WR(X)/I_2(X)$, *where* $X = V_{\mathrm{top}}$.

PROOF. By Definition 4.2 and Theorem 6.2, the isomorphism of Lemma 1.5 is the composition

$$I(V)/I_2(V) \to I(X)/I_2(X) \xrightarrow{\cong} \mathbb{H}^1_G(X, \mathbb{Z}/2).$$

The extension from $I(V) \cong I(X)$ to $W(V) \cong W(X)$ is routine. □

As noted in Definition 4.2, the rank map and w_1 give a ring homomorphism $KO_G(X) \to \mathbb{Z} \times \mathbb{H}^1_G(X, \mathbb{Z}/2)$. It is onto, because $\mathrm{Pic}^0_G(X)$ maps onto $\mathbb{H}^1_G(X, \mathbb{Z}/2) \cong H^1(X/G, \mathbb{Z}/2)$ (see 4.1).

LEMMA 6.4. *Suppose that G acts freely on a connected space X. Then the rank and w_1 induce a non-trivial extension*

$$0 \to H^1(X/G, \mathbb{Z}/2) \to WR(X)/I_2(X) \xrightarrow{\mathrm{rank}} \mathbb{Z}/2 \to 0.$$

This is the extension such that $2 \cdot 1$ is $[-1]$, where '1' is the image of the unit of the ring $WR(X)$, and $[-1] \in H^1(X/G, \mathbb{Z}/2) \cong \mathrm{Pic}^0_G(X)$ is the class of the line bundle $X \times \mathbb{R}(1)$.

PROOF. The map $H : KR(X) \to KO_G(X)$ is compatible with the filtration F_* associated to the Bredon spectral sequence (A.1), and we have $KO_G(X)/F_2 = \mathbb{Z} \times \mathbb{H}^1_G(X, \mathbb{Z}/2)$. Since $H^1_G(X; KR^{-1}) = 0$ we have $KR(X)/F_2 \cong \mathbb{Z}$. By Remark 2.5.1, $(2, [-1])$ is the image under $w_1 \circ H$ of the trivial Real line bundle in $\mathbb{Z} \times \mathbb{H}^1_G(X, \mathbb{Z}/2)$. Since $[-1] \neq 0$ by Remark 4.1.1, the result follows. □

The Stiefel–Whitney class $w_2 : KO_G(X) \to \mathbb{H}^2_G(X, \mathbb{Z}/2)$ does not factor through $WR(X)$ either, because w_2 need not vanish on the image of $H : KR(X) \to KO_G(X)$. Consider the composition

$$\bar{w}_2 : WR(X) \to {}_2\mathbb{H}^3_G(X, \mathbb{Z}(1))$$

of the map $w_2 : KO_G(X) \to \mathbb{H}^2_G(X, \mathbb{Z}/2)$, followed by the Bockstein. We proved the following result in [**26**, 5.6 and 5.8].

THEOREM 6.5. *If V is a smooth variety, then the composition of*

$$GW(V) \to GR(V_{\mathrm{top}}) \cong KO_G(V_{\mathrm{top}}) \to WR(V_{\mathrm{top}})$$

with $\bar{w}_2$ agrees with the algebraic Hasse invariant 1.3*, followed by the Bockstein:*

$$GW(V) \xrightarrow{\text{Hasse}} H^2_{\mathrm{et}}(V,\mathbb{Z}/2)/(\operatorname{Pic} V/2) \xrightarrow{\beta} {}_2\mathbb{H}^3_G(V_{\mathrm{top}},\mathbb{Z}(1)).$$

That is, $\beta\,\mathrm{Hasse}(\theta) = \bar{w}_2(\theta_{\mathrm{top}})$ for all $\theta \in GW(V)$.

We end this section with a discussion of equivariant Chern classes.

DEFINITION 6.6. In [**17**], Bruno Kahn defined equivariant Chern classes $c_n : KR(X) \to \mathbb{H}^{2n}_G(X,\mathbb{Z}(n))$ for Real vector bundles, with the first Chern class c_1 inducing an isomorphism between the group of rank 1 Real vector bundles on X and $\mathbb{H}^2_G(X,\mathbb{Z}(1))$, where $\mathbb{Z}(1)$ is the sign representation. In particular, $c_1 : KR(X) \to \mathbb{H}^2_G(X,\mathbb{Z}(1))$ is a surjection.

The following calculation is taken from [**17**, Th. 1]. Recall from Definition 6.6 that the first Chern class $c_1(E)$ of a Real vector bundle E on X is an element of $H^2_G(X,\mathbb{Z}(1))$. If E has rank $d+1$ and $\mathbb{P}(E)$ is the associated projective bundle, then the fiber of $\mathbb{P}(E)$ over a point $x \in X$ is the copy of $\mathbb{CP}^d$ corresponding to $\pi^{-1}(x)$.

THEOREM 6.7 (Kahn). *Let $E \xrightarrow{\pi} X$ be a rank 2 Real bundle on a G-space X, and let $\mathbb{P}(E)$ denote the associated projective space. If $\xi = c_1(E) \in \mathbb{H}^2_G(X,\mathbb{Z}(1))$ is the first Chern class of E then for any coefficient system A*

$$\mathbb{H}^n_G(\mathbb{P}(E),A) = \mathbb{H}^n_G(X,A) \oplus \mathbb{H}^{n-2}_G(X,A(1))\cdot\xi.$$

In particular, $\mathbb{H}^1_G(\mathbb{P}(E),\mathbb{Z}/2) = \mathbb{H}^1_G(X,\mathbb{Z}/2)$ and

$$\mathbb{H}^2_G(\mathbb{P}(E),\mathbb{Z}(1)) = \mathbb{H}^2_G(X,\mathbb{Z}(1)) \oplus \mathbb{H}^0_G(X,\mathbb{Z}).$$

REMARK 6.7.1. Kahn's formula is more general. In particular, if X is a point and $E = \mathcal{O}^2_X$ so that $\mathbb{P}(E) = \mathbb{P}^2$, it yields

$$\mathbb{H}^n_G(\mathbb{P}^2,A) = \mathbb{H}^n_G(\mathrm{pt},A) \oplus \mathbb{H}^{n-2}_G(\mathrm{pt},A(1))\cdot\xi \oplus \mathbb{H}^{n-4}_G(\mathrm{pt},A)\cdot\xi^2.$$

7. Algebraic surfaces with no $\mathbb{R}$-points

Let V be an ireducible variety defined over $\mathbb{R}$. Recall [**16**, Ex. II.3.15] that V is said to be *geometrically connected* (or *geometrically integral*) if $V \times_{\mathbb{R}} \mathbb{C}$ is connected, i.e., the function field of V does not contain $\mathbb{C}$. In this case, V_{top} is connected.

If V is an irreducible surface defined over $\mathbb{R}$, and V_{top} is not connected, then V is also defined over $\mathbb{C}$, i.e., V is a complex variety. In this case, the G-space $V_{\mathrm{top}} = \mathrm{Hom}_{\mathbb{R}}(\mathrm{Spec}(\mathbb{C}),V)$ is $G \times V(\mathbb{C})$, where $V(\mathbb{C}) = \mathrm{Hom}_{\mathbb{C}}(\mathrm{Spec}(\mathbb{C}),V)$ is the usual space of complex points.

We proved the following theorem in [**26**, Thm. 7.4]; it is a refinement of a theorem of Zibrowius [**55**, 5.12]. The invariant ρ is the rank of the cokernel of the first Chern class $c_1 : NS(V) \to H^2(V(\mathbb{C}),\mathbb{Z})$, and p_g is the geometric genus; the inequality $\rho \ge 2p_g$ in Theorem 7.1 follows from Hodge theory.

THEOREM 7.1. *Suppose that V is a smooth projective surface over $\mathbb{C}$. Then there is a split exact sequence*

$$0 \to (\mathbb{Z}/2)^{\rho} \to W(V) \to WR(V_{\mathrm{top}}) \to 0,$$

where $\rho \geq 2p_g$, and $\rho = 0$ when $p_g = 0$. Thus $W(V) \to WR(V_{\rm top})$ is an isomorphism if and only if $p_g = 0$.

Surfaces with $p_g = 0$ include the projective plane $\mathbb{P}^2$, rational surfaces, ruled surfaces, K3 surfaces, and Enriques surfaces (see [**16**]). It also includes some surfaces of general type, such as Godeaux surfaces, Burniat surfaces and Mumford's fake projective plane.

When V is geometrically connected, i.e., $V_{\rm top}$ is connected, the following result was proven in [**26**, Thm. 8.5]. The invariant ρ_0 is the rank of $c_1 : \mathrm{Pic}(V) \to \mathbb{H}^2_G(X, \mathbb{Z}(1))$, and the relation $\rho_0 \geq p_g$ follows from Hodge theory.

THEOREM 7.2. *Let V be a smooth geometrically connected projective surface over $\mathbb{R}$ with no real points. Then there is a split exact sequence*

$$0 \to (\mathbb{Z}/2)^{\rho_0} \to W(V) \to WR(V_{\rm top}) \to 0,$$

where $\rho_0 \geq p_g$, and $W(V) \to WR(V_{\rm top})$ is an isomorphism if and only if V has geometric genus $p_g = 0$.

The following calculation was given in [**26**, 8.1]. As in Lemma 3.1, $\widetilde{H}^1(X/G, \mathbb{Z}/2)$ denotes the quotient of $H^1(X/G, \mathbb{Z}/2)$ by the subgroup generated by $[-1] = w_1(X \times \mathbb{R}(1))$.

THEOREM 7.3. *Let X be a connected 4-dimensional G-CW complex. If G acts freely on X, then $WR(X)$ is a $\mathbb{Z}/8$-algebra, and there is an extension:*

$$0 \to {}_2\mathbb{H}^3_G(X, \mathbb{Z}(1)) \to WR(X) \ \to \mathbb{Z}/4 \times \widetilde{H}^1(X/G, \mathbb{Z}/2) \to 0;$$

In the rest of this section, we describe $WR(V_{\rm top})$ for several surfaces with $p_g = 0$ and with no real points.

Forms of $\mathbb{P}^1 \times \mathbb{P}^1$. A variety V is a *form* of $\mathbb{P}^1 \times \mathbb{P}^1$ if $V \otimes_{\mathbb{R}} \mathbb{C} \cong \mathbb{P}^1 \times \mathbb{P}^1$. In this case, $p_g = 0$ and $V_{\rm top}$ is $S^2 \times S^2$, so $W(V) \cong WR(V_{\rm top})$ by Theorem 7.2. In this case, the calculation of WR is dictated by the structure of $H^2(V_{\rm top}, \mathbb{Z})$ as a G-module.

THEOREM 7.4. *Suppose that G acts freely on $X = S^2 \times S^2$.*

(i) *If G acts as -1 on $H^2(X, \mathbb{Z})$ then $WR(X) \cong \mathbb{Z}/4$.*
(ii) *If $H^2(X, \mathbb{Z}) \cong \mathbb{Z}[G]$ as a G-module then $WR(X) \cong \mathbb{Z}/4 \oplus \mathbb{Z}/2$.*
(iii) *If $H^2(X, \mathbb{Z}) \cong \mathbb{Z} \oplus \mathbb{Z}(1)$ as a G-module then $WR(X) \cong \mathbb{Z}/4 \oplus \mathbb{Z}/2$.*

EXAMPLES 7.4.1. Cases (i)–(iii) are the only possibilities, because the only G-module structures on $\mathbb{Z}^2$ are: $\mathbb{Z}^2$, $\mathbb{Z}(1)^2$, $\mathbb{Z}[G]$ and $\mathbb{Z} \oplus \mathbb{Z}(1)$ (see [**42**, 2.1]), and G cannot act trivially on $H^2(X, \mathbb{Z})$ (by the Lefschetz Fixed Point Theorem).

Let X be the space $S^2 \times S^2$, where $S^2 = \mathbb{C} \cup \{\infty\}$. The involutions sending (z_1, z_2) to $(-\bar{z}_1^{-1}, -\bar{z}_2^{-1})$, $(\bar{z}_2^{-1}, -\bar{z}_1^{-1})$ and $(-\bar{z}_1^{-1}, z_2)$ have no fixed points, and yield the G-space structures $\mathbb{Z}(1)^2$, $\mathbb{Z}[G]$ and $\mathbb{Z} \oplus \mathbb{Z}(1)$. These illustrate cases (i), (ii) and (iii) of Theorem 7.4.

PROOF OF THEOREM 7.4. Since $\pi_1(X) = 0$, we have $\pi_1(X/G) = \mathbb{Z}/2$. Hence $H^1(X/G, \mathbb{Z}/2)$ is $\mathbb{Z}/2$. It follows from Theorem 7.3 that $WR(X)$ is an extension of $\mathbb{Z}/4$ by ${}_2\mathbb{H}^3_G(X, \mathbb{Z}(1))$.

To determine ${}_2\mathbb{H}^3_G(X, \mathbb{Z}(1))$, we recall from A.3 that

$$\mathbb{H}^*_G(X, \mathbb{Z}) \cong H^*(X/G, \mathbb{Z}) \quad \text{and} \quad \mathbb{H}^*_G(X, \mathbb{Z}[G]) \cong H^*(X, \mathbb{Z}).$$

Hence the long exact cohomology sequence associated to the short exact sequence $0 \to \mathbb{Z} \to \mathbb{Z}[G] \to \mathbb{Z}(1) \to 0$ of G-modules becomes:

$$(7.4) \qquad 0 = H^3(X,\mathbb{Z}) \to \mathbb{H}^3_G(X,\mathbb{Z}(1)) \to H^4(X/G,\mathbb{Z}) \to H^4(X,\mathbb{Z}).$$

Now consider the spectral sequence $H^p(G, H^q(X,\mathbb{Z})) \Rightarrow H^{p+q}(X/G,\mathbb{Z})$ associated to $X_G \to BG$. For convenience, we omit the coefficient when it is $\mathbb{Z}$. If t is the nonzero element of $E_\infty^{2,0} = E_2^{2,0} = H^2(BG) \cong \mathbb{Z}/2$ then except for $p = 0$, cup product with t is an isomorphism $E_2^{p,q} \cong E_2^{p+2,q}$, commuting with the differentials.

We now consider the possible G-module structures on $H^2(X) \cong \mathbb{Z}^2$.

(i) If $H^2(X) \cong \mathbb{Z}(1)^2$ then $H^4(X) \cong \mathbb{Z}$ as a G-module. For even $p > 4$, the sequence

$$0 \to \mathbb{Z}/2 = E_3^{p-6,4} \xrightarrow{d_3} E_3^{p-3,2} = (\mathbb{Z}/2)^2 \xrightarrow{d_3} E_3^{p,0} = \mathbb{Z}/2 \to 0$$

must be exact. By t–periodicity, the map $d_3^{1,2} : E_3^{1,2} \to E_3^{4,0}$ must be onto. It follows that $H^4(X/G) \to H^4(X)$ is an injection (with cokernel $\mathbb{Z}/2$). The exact sequence (7.4) yields $\mathbb{H}^3_G(X,\mathbb{Z}(1)) = 0$.

(ii) If $H^2(X)$ is $\mathbb{Z}[G]$, we have $\mathbb{Z}/2 \cong H^4(BG) \xrightarrow{\cong} H^4(X/G)$. This is because $H^4(X) \cong \mathbb{Z}(1)$ so $H^4(X)^G = 0$, and $H^p(G, H^2(X)) = 0$ for $p > 0$. The result follows from (7.4).

(iii) If $H^2(X,\mathbb{Z}) \cong \mathbb{Z} \oplus \mathbb{Z}(1)$ then $H^4(X) \cong \mathbb{Z}(1)$ as a G-module. For even $p \geq 6$, the sequences

$$0 = E_3^{p-6,4} \xrightarrow{d_3} E_3^{p-3,2} = \mathbb{Z}/2 \xrightarrow{d_3} E_3^{p,0} = \mathbb{Z}/2 \to 0,$$

$$0 \to \mathbb{Z}/2 = E_3^{p-5,4} \xrightarrow{d_3} E_3^{p-2,2} = \mathbb{Z}/2 \xrightarrow{d_3} E_3^{p+1,0} = 0$$

must be exact. It follows that $H^4(X/G,\mathbb{Z}) \cong \mathbb{Z}/2$. □

The anisotropic quadric. Let Q_2 be the anisotropic quadric surface $x^2 + y^2 + z^2 + w^3 = 0$. It is well known that the topological G-space X underlying Q_2 is $S^{3,0} \times S^{3,0}$, where $S^{3,0}$ is the 2 sphere with the antipodal involution. (See [**47**, VI.5] or [**11**, Proof of 2.4] for example.)

COROLLARY 7.5. $W(Q_2) \cong WR(S^{3,0} \times S^{3,0}) \cong \mathbb{Z}/4$.

PROOF. This follows from Theorem 7.4(i) and Theorem 7.2. □

REMARK 7.5.1. The fact that $W(Q_2) \cong \mathbb{Z}/4$ is due to Parimala [**38**, p. 92]. See Theorem 7.6 for a short algebraic proof.

Rational surfaces. An algebraic surface defined over $\mathbb{R}$ is called a *real rational surface* if $V_\mathbb{C}$ is birational to the projective plane over $\mathbb{C}$. For example, Q_2 is a rational surface.

If V is a real rational surface with no real points, then $p_g = 0$, and $W(V) \cong WR(V_{\text{top}})$ by Theorem 7.2.

THEOREM 7.6. *Let V be a real rational surface with no real points. Then* $W(V) \xrightarrow{\cong} WR(V_{\text{top}}) \cong \mathbb{Z}/4$.

PROOF. We calculate $W(V)$ and $WR(V_{\text{top}})$ separately, for comparison purposes. We begin with the algebraic calculation of $W(V)$.

Sujatha proved in her thesis that $W(V) \cong \mathbb{Z}/4$. Unfortunately, the published version [**48**, 4.2] asserts that $W(V) = \mathbb{Z}/4 \oplus \mathbb{Z}/2$, due to a typo in the proof (on top of p. 100 in *loc. cit.*): the incorrect formula $\mathrm{Pic}(V_{\mathbb{C}})^G \cong \mathrm{Pic}(V) \oplus \mathbb{Z}/2$, should say that the torsionfree group $\mathrm{Pic}(V_{\mathbb{C}})^G$ is a nontrivial extension of $\mathrm{Br}(\mathbb{R}) \cong \mathbb{Z}/2$ by $\mathrm{Pic}(V)$:

$$0 \to \mathrm{Pic}(V) \to \mathrm{Pic}(V_{\mathbb{C}})^G \xrightarrow{d_3} \mathrm{Br}(\mathbb{R}) \to 0.$$

With this correction, Sujatha's proof of [**48**, 4.2] yields

$$\dim H^2(V, \mathbb{Z}/2) = \dim(\mathrm{Pic}\, V)/2 \quad \text{and} \quad {}_2\mathrm{Br}(V) = 0$$

(so $k = 0$ in the notation of [**48**]), and hence $W(V) \cong \mathbb{Z}/4$.

We turn to the topological calculation, setting $X = V_{\mathrm{top}}$. In this case, $H^1(X, \mathbb{Z}) = H^3(X, \mathbb{Z}) = 0$, and $H^2(X, \mathbb{Z})$ is given by Lemma 7.7 below. The spectral sequence $E_2^{p,q} = H^p(G, H^q(X, \mathbb{Z})) \Rightarrow \mathbb{H}_G^*(X, \mathbb{Z})$ agrees with the spectral sequence in the proof of Theorem 7.4(i), except for the $E_2^{0,2}$ term. As this term played no role in the proof of *loc. cit.*, the same argument (using 7.2) applies to yield $WR(X) \cong \mathbb{Z}/4$. □

LEMMA 7.7. *Let V be a real rational surface with no real points. Then there is an integer c such that $H^2(V_{\mathrm{top}}, \mathbb{Z}) \cong \mathbb{Z}(1)^2 \oplus \mathbb{Z}[G]^c$.*

PROOF. It is well known that $\mathrm{Pic}(V_{\mathbb{C}}) \cong \mathbb{Z}^2 \oplus \mathbb{Z}[G]^c$ as a G-module; see [**39**] or [**47**, p. 46]. Since $H^1_{\mathrm{an}}(X, \mathcal{O}) = H^2_{\mathrm{an}}(X, \mathcal{O}) = 0$, the equivariant exponential sequence $0 \to \mathbb{Z}(1) \to \mathcal{O}_{\mathrm{an}} \to \mathcal{O}^{\times}_{\mathrm{an}} \to 0$ on X implies that as a G-module:

$$\mathrm{Pic}(V_{\mathbb{C}}) \cong H^1_{\mathrm{an}}(X, \mathcal{O}^{\times}_{\mathrm{an}}) \cong H^2_{\mathrm{an}}(X, \mathbb{Z}(1)) \cong H^2(X, \mathbb{Z}) \otimes \mathbb{Z}(1). \qquad \square$$

REMARK 7.7.1. Every finitely generated torsionfree G-module A is isomorphic to $\mathbb{Z}^a \oplus \mathbb{Z}(1)^b \oplus \mathbb{Z}[G]^c$ for some integers a, b, c. The integer c is called the *Comesssatti characteristic* of A, since Comessatti used it to study abelian varieties over $\mathbb{R}$ in the 1920's; see [**12, 13**]. The integer c in Lemma 7.7 is called the *Comessatti characteristic* of $H^2(X)$. The Comessatti character of $H^d(V_{\mathrm{top}}, \mathbb{Z})$, $d = \dim V$, has played a major role in computations by Silhol [**47**, IV.1], Parimala–Sujatha [**39**] and the authors [**41**, 1.1], [**23**, 4.2].

Ruled surfaces. Recall that a *ruled surface* over a curve C is a variety of the form $V = \mathbb{P}(E)$, where E is a locally free sheaf of rank 2 over C. For example, if E is free then $V = C \times \mathbb{P}^1$. Ruled surfaces have $p_g = 0$, so Theorem 7.2 yields $W(V) \cong WR(V_{\mathrm{top}})$ if C has no $\mathbb{R}$-points, since in that case V has no $\mathbb{R}$-points either.

THEOREM 7.8. *Let C be a smooth, geometrically connected projective curve over $\mathbb{R}$ of genus g. If C has no real points and V is a ruled surface over C, then*

$$W(V) \cong WR(V_{\mathrm{top}}) \cong WR(C_{\mathrm{top}}) \cong \mathbb{Z}/4 \oplus (\mathbb{Z}/2)^g.$$

PROOF. It suffices to show that $WR(V_{\mathrm{top}}) \cong WR(C_{\mathrm{top}})$, because $W(C) \cong WR(C_{\mathrm{top}}) \cong \mathbb{Z}/4 \oplus (\mathbb{Z}/2)^g$ by Corollary 3.4. Because $\mathbb{H}^3_G(C_{\mathrm{top}}, \mathbb{Z}(1)) = 0$, we see from the Projective Bundle Theorem 6.7 that

$$\mathbb{H}^3_G(V_{\mathrm{top}}, \mathbb{Z}(1)) \cong \mathbb{H}^1_G(C_{\mathrm{top}}, \mathbb{Z}) \cong H^1(C_{\mathrm{top}}/G, \mathbb{Z}),$$

which is torsionfree. Theorem 7.3 yields $WR(V_{\mathrm{top}}) \cong WR(C_{\mathrm{top}})$. □

ALGEBRAIC PROOF. It suffices to show that $W(V) \cong W(C)$. By the algebraic Projective Bundle Theorem (which follows from 6.7 and [**14**]), $H^1_{\mathrm{et}}(V, \mathbb{Z}/2) = H^1_{\mathrm{et}}(C, \mathbb{Z}/2)$ and $H^2_{\mathrm{et}}(V, \mathbb{Z}/2) = H^2_{\mathrm{et}}(C, \mathbb{Z}/2) \oplus \mathbb{Z}/2$. In addition, $\mathrm{Pic}(V) \cong \mathrm{Pic}(C) \oplus \mathbb{Z}$. Hence $\mathrm{Br}(V) = \mathrm{Br}(C) = 0$, and $W(V) \cong W(C)$ by Corollary 1.9. □

REMARK 7.8.1. If C' is the (affine) curve obtained by removing $n \geq 1$ (complex) points from the projective curve C, and $V' = C' \times \mathbb{P}^1$, then we still have $W(V') \cong WR(V') \cong W(C')$, but now (using Remark 3.4.1) $W(V') \cong W(V) \oplus (\mathbb{Z}/2)^{n-1}$. This follows from Corollary 7.2, since $\mathrm{Br}(V') = \mathrm{Br}(C') = 0$ (see [**41**, 3.6]) and $H^1_{\mathrm{et}}(V', \mathbb{Z}/2) \cong H^1_{\mathrm{et}}(C', \mathbb{Z}/2)$.

Let Q_1 denote the projective curve $x^2 + y^2 + z^2 = 0$, which is a real form of $\mathbb{CP}^1$; its underlying G-space $Q_{1,\mathrm{top}}$ is $S^{3,0}$ (S^2 with the antipodal involution), and $Q_{1,\mathrm{top}}/G$ is $\mathbb{RP}^2$. Thus $H^1(Q_1, \mathbb{Z}/2) \cong \mathbb{Z}/2$.

Let C be a smooth (geometrically connected) projective curve over $\mathbb{R}$, and consider the real form $V = C \times Q_1$ of the ruled surface $C_{\mathbb{C}} \times \mathbb{P}^1_{\mathbb{C}}$.

PROPOSITION 7.9. *When $V = C \times Q_1$, and $C(\mathbb{R})$ has ν components,*

$$W(V) \xrightarrow{\cong} WR(V_{\mathrm{top}}) \cong \begin{cases} \mathbb{Z}/4 \oplus (\mathbb{Z}/2)^g, & \nu = 0; \\ \mathbb{Z}/4 \oplus (\mathbb{Z}/2)^g \oplus (\mathbb{Z}/4)^{\nu-1}, & \nu > 0. \end{cases}$$

PROOF. Since $p_g = 0$, this follows from Theorem 7.2. $\square$

EXAMPLES 7.10. Both $Q_1 \times Q_1$ and the ruled surface $Q_1 \times \mathbb{P}^1$ are surfaces with no real points, and $p_g = 0$. Their underlying G-space is $X = S^2 \times S^2$. By Theorem 7.2, $W(V) \cong WR(V_{\mathrm{top}}) \cong \mathbb{Z}/4$.

8. Real points

If $X^G \neq \emptyset$ and $\dim(X^G) \leq 3$ then $WR(X)$ is the sum of $\mathbb{Z}^\nu$ (given by the signature, as described in Lemma 2.5) and a finite 2–group. This follows from [**23**, 7.4], since $KO(X^G)$ has this structure. In particular, if V is a real algebraic surface and $V(\mathbb{R})$ has $\nu > 0$ components, then both $W(V)$ and $WR(V_{\mathrm{top}})$ are the sum of $\mathbb{Z}^\nu$ and a finite 2–group.

To describe the torsion, we use the cellular filtrations on KR and KO_G from the Bredon spectral sequence (A.1), defining $F_pWR(X)$ as the image of $F_pKO_G(X) \to WR(X)$. Thus $F_1WR(X)$ is the torsion subgroup and $WR(X)/F_1 \cong \mathbb{Z}^\nu$.

LEMMA 8.1. *Suppose that* $\dim(X) < 8$ *and* $\dim(X^G) \leq 3$. *Then* $F_3WR(X) = 0$.

PROOF. The groups $E_2^{p,-p}(KO_G)$ are zero if $p = 3, 5, 6, 7$. The map $E_2^{4,-4}(KR) \to E_2^{4,-4}(KO_G)$ is onto by Lemma A.8. Since $\dim X < 8$, $E_\infty^{4,-4}$ is a quotient of $E_2^{4,-4}$ for both the KR and KO_G cases, $E_\infty^{4,-4}(KR)$ maps onto $E_\infty^{4,-4}(KO_G)$, and hence $F_3WR(X) = 0$. $\square$

Since $F_1/F_2 \cong E_\infty^{1,-1}$ for both KO_G and KR, we have an exact sequence

$$E_\infty^{1,-1}(KR) \xrightarrow{H_\infty} E_\infty^{1,-1}(KO_G) \to F_1WR(X)/F_2WR(X) \to 0.$$

When $\dim(X) \leq 4$ and $\dim(X^G) \leq 2$, this fits into a diagram with exact rows, in which d_2 and $\bar{d}_2$ are the E_2 differentials:

(8.2)
$$\begin{array}{ccccc}
0 \to E_\infty^{1,-1}(KR) & \longrightarrow & H^1(X^G, \mathbb{Z}/2) & \xrightarrow{d_2} & {}_2H^3_G(X, \mathbb{Z}(1)) \\
\downarrow{\scriptstyle H_\infty} & & \downarrow{\scriptstyle \text{into}} & & \downarrow{\scriptstyle \text{mod } 2} \\
0 \to E_\infty^{1,-1}(KO_G) & \longrightarrow & H^1_G(X; KO_G^{-1}) & \xrightarrow{\bar{d}_2} & H^3(X/G, \mathbb{Z}/2) \\
\downarrow{\scriptstyle \text{onto}} & & \downarrow{\scriptstyle \text{onto}} & & \\
F_1WR(X)/F_2 & \xrightarrow{\text{onto}} & \mathrm{Pic}^0_G(X). & &
\end{array}$$

The bottom horizontal arrow in (8.2) is onto because, as pointed out in 4.1, $F_1WR(X)$ maps onto $\mathrm{Pic}^0_G(X)$. The top middle vertical is an injection with cokernel $H^1(X/G,\mathbb{Z}/2)\cong \mathrm{Pic}^0_G(X)$ by Lemma A.5 and Definition 4.1. It follows that H_∞ is also an injection.

By Example A.2, $H^3_G(X,\mathbb{Z}/2)\cong H^3(X/G,\mathbb{Z}/2)$; the right vertical map is reduction mod 2.

DEFINITION 8.3. Let Δ denote the kernel of the induced surjection $F_1WR(X)/F_2 \to \mathrm{Pic}^0_G(X)$ in the bottom row of (8.2).

By the snake lemma applied to the columns of (8.2), Δ is the image in $H^3(X,\mathbb{Z}(1))$ of the group of all a in $H^1_G(X^G,\mathbb{Z}/2)$ which vanish in $H^3(X/G,\mathbb{Z}/2)$ under the map $\overline{d}_2$ of (8.2), i.e., Δ is

$$\{a\in \mathrm{image}\big[H^1(X^G,\mathbb{Z}/2)\overset{d_2}{\to} H^3_G(X,\mathbb{Z}(1))\big] : \bar{a}=0 \text{ in } H^3(X/G,\mathbb{Z}/2)\}.$$

REMARK 8.3.1. Since the image of d_2 has exponent 2, Δ is also the intersection of image(d_2) and $2\cdot H^3_G(X,\mathbb{Z}(1))$.

Clearly, $\Delta=0$ if $H^3_G(X,\mathbb{Z}(1))$ has no 2-torsion. We also have $\Delta=0$ if the even Tate cohomology of $\mathbb{Z}/2$ acting on $H^3_G(X,\mathbb{Z}(1))$ is trivial.

LEMMA 8.4. *When* $\dim(X)\le 4$ *and* $\dim(X^G)\le 2$,

$$F_2WR(X)\cong F_2KO_G(X)/H(F_2KR(X))\cong E_\infty^{2,-2}(KR)/H(E_\infty^{2,-2}(KR)).$$

PROOF. Because H_∞ is an injection in diagram (8.2), the intersection

$$H(F_1KR(X))\cap F_2KO_G(X)$$

is the image of $F_2KR(X)$. □

LEMMA 8.5. *If* $\dim(X)\le 4$ *and* X^G *has* $\nu>0$ *components, of dimension* ≤ 2, *then there is an isomorphism* ${}_2H^3_G(X;\mathbb{Z}(1))\overset{\cong}{\to} F_2WR(X)$.

PROOF. By Lemma A.5 and Corollary A.7, and Lemma 8.4, we also have a commutative diagram with exact columns:

$$\begin{array}{ccccc}
(\mathbb{Z}/2)^\nu & \xrightarrow{d_2} & H^2_G(X;\mathbb{Z}(1))\oplus H^2(X^G,\mathbb{Z}/2) & \longrightarrow & E_\infty^{2,-2}(KR)\to 0\\
\downarrow & & \downarrow & & \downarrow{\scriptstyle H_\infty^{2,-2}}\\
(\mathbb{Z}/2)^{1+\nu} & \xrightarrow{d_2} & H^2(X/G,\mathbb{Z}/2)\oplus H^2(X^G,\mathbb{Z}/2) & \longrightarrow & E_\infty^{2,-2}(KO_G)\to 0\\
\downarrow{\scriptstyle\text{onto}} & & \downarrow{\scriptstyle\text{onto}} & & \downarrow{\scriptstyle\text{onto}}\\
(\mathbb{Z}/2) & \longrightarrow & {}_2H^3_G(X;\mathbb{Z}(1)) & \longrightarrow & F_2WR(X)\to 0.
\end{array}$$

A comparison with the corresponding diagram when X is a point shows that the bottom left horizontal map is zero and hence that we have ${}_2H^3_G(X;\mathbb{Z}(1))\cong F_2WR(X)$. □

Recall that the group Δ is defined in 8.3.

THEOREM 8.6. *If* X *is 4-dimensional and* X^G *has* $\nu>0$ *components, of dimension* ≤ 2, *the image* S_t *of the signature* $WR(X)\to\mathbb{Z}^\nu$ *is a subgroup of 2-primary index, and there is an exact sequence*

$$0\to {}_2H^3_G(X;\mathbb{Z}(1))\to WR(X)\to S_t\oplus H^1(X/G,\mathbb{Z}/2)\oplus\Delta\to 0.$$

In particular, if $H^3_G(X;\mathbb{Z}(1))$ *has no 2-torsion, then*

$$WR(X)\cong S_t\oplus H^1(X/G,\mathbb{Z}/2).$$

PROOF. Given Lemma 8.5, we need to determine the kernel of the surjection $F_1/F_2WR(X) \to \mathrm{Pic}^0_G(X)$ in (8.2). Since $H^1_G(X; KO_G^{-1})$ has exponent 2, the subgroup $E_\infty^{1,-1}(KO_G)$ and its quotient $F_1/F_2WR(X)$ also have exponent 2. This implies that the extension has a (non-canonical) splitting: $F_1/F_2WR(X) \cong H^1(X/G, \mathbb{Z}/2) \oplus \Delta$.

Finally, if $H^3_G(X; \mathbb{Z}(1))$ has no 2-torsion, then $\Delta = 0$. □

When V is a smooth projective variety over $\mathbb{R}$ and $X = V_{\mathrm{top}}$, we can use the isomorphisms $H^*_{\mathrm{et}}(V, \mathbb{Z}/2) \cong \mathbb{H}^*_G(X, \mathbb{Z}/2)$ to compare $W(V)$ and $WR(X)$. Recall from Section 4 that the image S_a of the algebraic signature $W(V) \to \mathbb{Z}^\nu$ is a subgroup of the image S_t of the topological signature $WR(X) \to \mathbb{Z}^\nu$.

Recall that ρ_0 denotes the rank of $c_1 : \mathrm{Pic}(V) \otimes \mathbb{Q} \to \mathbb{H}^2_G(X, \mathbb{Q}(1))$; we have $\rho_0 \geq p_g$ by Hodge theory, and if $p_g(V) = 0$ then $\rho_0 = 0$. The group Δ is defined in 8.3. (See also Remark 8.3.1.)

THEOREM 8.7. *Let V be a smooth geometrically connected projective surface defined over $\mathbb{R}$, with $V(\mathbb{R}) \neq \emptyset$. Then there is an exact sequence*

$$0 \to (\mathbb{Z}/2)^{\rho_0} \to W(V) \to WR(V_{\mathrm{top}}) \to \Delta \oplus (S_t/S_a) \to 0.$$

PROOF. As the signature on $W(V)$ factors through the signature on $WR(V_{\mathrm{top}})$, we are reduced to a comparison of their torsion subgroups $I(V)_{\mathrm{tors}}$ and $F_1WR(V_{\mathrm{top}})$. By Proposition 1.8, which is due to Sujatha,

$$I(V)_{\mathrm{tors}}/I_2(V)_{\mathrm{tors}} \cong H^0_{\mathrm{tors}}(V, \mathcal{H}^1) \text{ and } I_2(V)_{\mathrm{tors}} \cong H^0_{\mathrm{tors}}(V, \mathcal{H}^2).$$

By definition, $H^0_{\mathrm{tors}}(V, \mathcal{H}^1)$ and $H^0_{\mathrm{tors}}(V, \mathcal{H}^2)$ are the kernels of the stabilization maps from $H^0(V, \mathcal{H}^1) \cong H^1_{\mathrm{et}}(V, \mathbb{Z}/2)$ and $H^0(V, \mathcal{H}^2) \cong {}_2\mathrm{Br}(V)$ to $H^3_{\mathrm{et}}(V, \mathbb{Z}/2) \cong (\mathbb{Z}/2)^\nu$.

Recall from [**26**, 6.3] that when V is a smooth geometrically connected projective variety over $\mathbb{R}$ then there is a split exact sequence

$$0 \to (\mathbb{Q}/\mathbb{Z})^{\rho_0} \to \mathrm{Br}(V) \xrightarrow{\beta} {}_{\mathrm{tors}}\mathbb{H}^3_G(X, \mathbb{Z}(1)) \to 0, \tag{8.8}$$

where β is induced from the Bockstein on $\mathbb{H}^2_G(X, \mathbb{Z}/2)$, and $X = V_{\mathrm{top}}$. Hence $(\mathbb{Z}/2)^{\rho_0}$ is the kernel of $\beta : {}_2\mathrm{Br}(V) \to {}_2\mathbb{H}^3_G(X, \mathbb{Z}(1))$; by Theorem 6.5, $(\mathbb{Z}/2)^{\rho_0}$ is the kernel of $I_2(V)_{\mathrm{tors}} \to F_2WR(X)$.

It remains to show that the induced map from $I(V)_{\mathrm{tors}}/I_2(V)_{\mathrm{tors}} \cong H^0_{\mathrm{tors}}(V, \mathcal{H}^1)$ to $F_1WR(X)/F_2WR(X)$ is an injection with cokernel Δ. As we saw in Section 4, it follows from Definition 4.1 that

$$H^0_{\mathrm{tors}}(V, \mathcal{H}^1) \cong \mathrm{Pic}^0_G(X) \xrightarrow{\cong} H^1(X/G, \mathbb{Z}/2),$$

where the second map is w_1. By Theorem 6.2 and Corollary 6.3, this is the same as the discriminant on $I(V)$. It follows from Theorem 8.6 that the cokernel of $H^0_{\mathrm{tors}}(V, \mathcal{H}^1) \to WR(X)/F_2WR(X)$ is Δ. □

EXAMPLE 8.9. Consider the real surface $V = E \times E$, where E is an elliptic curve with 2 real connected components. We saw in Example 1.12 that $W(V) \cong \mathbb{Z}^4 \oplus (\mathbb{Z}/4)^3 \oplus \mathbb{Z}/2$. We claim that $WR(V_{\mathrm{top}}) \cong \mathbb{Z}^4 \oplus (\mathbb{Z}/2)^3$ so that $W(V) \to WR(V_{\mathrm{top}})$ is a surjection with kernel $(\mathbb{Z}/2)^4$.

To compute $WR(V_{\mathrm{top}})$ we choose a basepoint fixed by the involution, set $Y = E_{\mathrm{top}}$ and observe that since $V_{\mathrm{top}} = Y \times Y$ we have a split exact sequence

$$0 \to KR(Y \wedge Y) \to KR(Y \times Y) \to KR(Y, \mathrm{pt}) \oplus KR(Y, \mathrm{pt}),$$

and similarly for $KO_G(Y\times Y)$. Thus we have

$$WR(V_{\text{top}}) \cong WR(Y\wedge Y)\oplus WR(Y,\text{pt})\oplus WR(Y,\text{pt})$$

Finally, we use the identity $WR(Y,\text{pt})\cong\mathbb{Z}\oplus(\mathbb{Z}/2)$ of Corollary 3.4.

In the same way, $Y=S^1\times S^{1,1}$, where $S^{1,1}$ is the unit circle in $\mathbb{C}$. Therefore $WR(Y\wedge Y)$ decomposes as the sum of $WR(S^2)=\mathbb{Z}\oplus(\mathbb{Z}/2)$ and the groups $WR(S,\text{pt})$ for S on the ordered list

$$S^{2,3}, S^{2,2}, S^{2,2}, S^{1,3}, S^{1,3}, S^{1,2}, S^{1,2}, S^{2,1}.$$

Since $S^{1,2}\cong\mathbb{CP}^1$, we see from Corollary 3.4 that $WR(S^{1,2})=\mathbb{Z}$. We also have $WR(S^{2,1})=\mathbb{Z}^2$ by Example 3.5. A similar detailed computation, mentioned in Example 3.5, shows that

$$WR(S^{2,3})=WR(S^{2,2})=WR(S^{1,3})=\mathbb{Z}.$$

It follows that $WR(Y\times Y)\cong\mathbb{Z}^4\oplus(\mathbb{Z}/2)^3$. Thus $W(V)\to WR(V_{\text{top}})$ is a surjection with kernel $(\mathbb{Z}/2)^4$.

Ruled Surfaces

We return to *ruled surfaces* over a curve C, i.e., varieties of the form $\mathbb{P}=\mathbb{P}(E)$, where E is a rank 2 algebraic line bundle over C. Theorem 7.8 describes the case when $C(\mathbb{R})=\emptyset$.

THEOREM 8.10. *Let C be a curve defined over $\mathbb{R}$ with $\nu>0$ real components, and genus g. If $\mathbb{P}=\mathbb{P}(E)$ is a ruled surface over C, then*

$$WR(\mathbb{P}_{\text{top}}) \overset{\cong}{\leftarrow} WR(C_{\text{top}})\cong\mathbb{Z}^\nu\oplus H^1(C_{\text{top}}/G,\mathbb{Z}/2).$$

If C is smooth projective, then we also have $W(\mathbb{P})\overset{\cong}{\to} WR(\mathbb{P}_{\text{top}})$.

PROOF. Let X and Y denote the G-spaces underlying $\mathbb{P}=\mathbb{P}(E)$ and C, respectively. Theorem 8.6 yields $WR(Y)\cong\mathbb{Z}^\nu\oplus H^1(Y/G,\mathbb{Z}/2)$. By the Projective Bundle Theorem 6.7, $\mathbb{H}^3_G(X,\mathbb{Z}(1))\cong\mathbb{H}^1_G(Y,\mathbb{Z})$. This group is defined to be $H^1(Y_G,\mathbb{Z})$, which is torsionfree, so by Theorem 8.6 we have $WR(X)\cong\mathbb{Z}^\nu\oplus H^1(X/G,\mathbb{Z}/2)$. Since $H^1_{\text{et}}(\mathbb{P},\mathbb{Z}/2)\cong H^1_{\text{et}}(C,\mathbb{Z}/2)$, and $H^0(X^G)\cong H^0(Y^G)$, the exact sequence in Example A.3 implies that $H^1(X/G,\mathbb{Z}/2)\cong H^1(Y/G,\mathbb{Z}/2)$. Hence $WR(X)\cong WR(Y)$.

When C is smooth projective, $W(C)\cong WR(C_{\text{top}})$ by Corollary 3.4 and $W(\mathbb{P})\cong W(C)$ by Lemma 8.11. This gives the second assertion. □

LEMMA 8.11. *Let C be a smooth projective curve over $\mathbb{R}$, with $\nu>0$ real components, and genus g. If $\mathbb{P}=\mathbb{P}(E)$ is a ruled surface over C, then $W(\mathbb{P})\cong W(C)\cong\mathbb{Z}^\nu\oplus(\mathbb{Z}/2)^g$.*

PROOF. Now suppose that C is smooth and projective; we need to show that $W(C)\to W(\mathbb{P})$ is an isomorphism. We have $H^2_{\text{et}}(\mathbb{P},\mathbb{Z}/2)\cong H^2_{\text{et}}(C,\mathbb{Z}/2)\oplus\mathbb{Z}/2$ by the Projective Bundle Theorem [**32**, VI.10.1]. Since $\text{Pic}(\mathbb{P})\cong\text{Pic}(C)\oplus\mathbb{Z}$, we have ${}_2\text{Br}(\mathbb{P})={}_2\text{Br}(C)=\mathbb{Z}/2^\nu$. As in Corollary 3.4, $I_2(C)=\mathbb{Z}/2^\nu$; since $I_2(\mathbb{P})\subseteq{}_2\text{Br}(\mathbb{P})$, we see that $I_2(\mathbb{P})/I_2(C)$ is 2-torsionfree.

By Sujatha's formulas listed in Remark 1.8.1, $k=\dim H^0_{\text{tors}}(-,\mathcal{H}^2)$ is the same for C and $\mathbb{P}$. Similarly, $H^1_{\text{et}}(\mathbb{P},\mathbb{Z}/2)\cong H^1_{\text{et}}(C,\mathbb{Z}/2)$, and $j=\dim H^0_{\text{tors}}(-,\mathcal{H}^1)$ is the same for C and $\mathbb{P}$. Thus $W(C)=W(\mathbb{P})$, since both groups have order 2^{j+k} by Remark 1.8.1. □

REMARK 8.11.1. Charles Walter (unpublished [52]) proved that for Y smooth of dimension ≤ 2, and any projective bundle $\mathbb{P}$ over Y, we have $W(Y) \cong W(\mathbb{P})$.

9. 3-folds

In this section, we collect information about smooth 3-folds that follows from the techniques in this paper. We begin by giving a short proof of a result of Parimala [37]. Recall from (1.2) that $\mathcal{H}^n$ is the Zariski sheaf associated to the presheaf $U \mapsto H^n_{\text{et}}(U, \mathbb{Z}/2)$.

THEOREM 9.1 (Parimala). *Let V be a smooth 3-dimensional algebraic variety over $\mathbb{R}$ or $\mathbb{C}$. Then the group $W(V)$ is finitely generated if and only if $CH^2(V)/2$ is finitely generated.*

PROOF. By Cox' theorem, both $H^1_{\text{et}}(V, \mathbb{Z}/2)$ and $H^2_{\text{et}}(V, \mathbb{Z}/2)$ are finite. Hence Lemma 1.5 and Remark 1.5.1 show that $W(V)/I_3(V)$ is finite. In addition, $W(V)/$ torsion is $\mathbb{Z}^\nu$, where $V(\mathbb{R})$ has ν connected components, and $I_4(V)$ is torsionfree (by 1.7). Hence $W(V)$ is a finitely generated group if and only if $I_3(V)/I_4(V) \cong H^0(V, \mathcal{H}^3)$ is finite. Since $H^2(V, \mathcal{H}^2) \cong CH^2(V)/2$, and $H^p(V, \mathcal{H}^q) = 0$ for $p > q$ [6], the coniveau spectral sequence $H^p(V, \mathcal{H}^q) \Rightarrow H^{p+q}_{\text{et}}(V, \mathbb{Z}/2)$ yields the exact sequence

$$H^3_{\text{et}}(V, \mathbb{Z}/2) \to H^0(V, \mathcal{H}^3) \xrightarrow{d_2} CH^2(V)/2 \to H^4_{\text{et}}(V, \mathbb{Z}/2).$$

As the outer terms are finite, it follows that $H^0(V, \mathcal{H}^3)$ is finitely generated if and only if $CH^2(V)/2$ is finite. □

REMARK 9.1.1. If V is a 3-fold defined over $\mathbb{C}$, the Witt group $W(V)$ is an algebra over $W(\mathbb{C}) = \mathbb{Z}/2$, and no group extension issues arise.

For $WR(V_{\text{top}})$ of 3-folds, we need a new ingredient: the cohomology operation $\tilde{\beta}Sq^2$ from $H^2(X, \mathbb{Z}/2)$ to $H^5(X, \mathbb{Z})$, where $\tilde{\beta}$ is the integral Bockstein. By [26, 5.9], the kernel of $\tilde{\beta}Sq^2$ is the set of all Stiefel–Whitney classes $u = w_2(E) \in H^2(Y, \mathbb{Z}/2)$.

Since $H^5(\mathbb{CP}^\infty, \mathbb{Z}) = 0$ and $K(\mathbb{Z}, 2) = \mathbb{CP}^\infty$, $\tilde{\beta}Sq^2$ vanishes on the mod 2 reduction of $H^2(X, \mathbb{Z})$. Hence the following definition makes sense.

DEFINITION 9.2. We define $\bar{H}^2(Y)$ to be the kernel of the cohomology operation ${}_2H^3(Y, \mathbb{Z}) \to H^5(Y, \mathbb{Z})$ induced by $\tilde{\beta}Sq^2$.

The following result and its proof is a modification of the argument we gave in [26, 7.1] for 4-dimensional spaces.

THEOREM 9.3. *If Y is a connected 6-dimensional CW complex,*

$$WR(G \times Y) \cong \mathbb{Z}/2 \times H^1(Y, \mathbb{Z}/2) \times \bar{H}^2(Y).$$

PROOF. Recall that $WR(G \times Y)$ is the cokernel of $KU(Y) \to KO(Y)$. We compare the filtrations on $KU(Y)$ and $KO(Y)$ associated to their Atiyah-Hirzebruch spectral sequences. Since the quotient $KU(Y)/F_2KU(Y) = \mathbb{Z}$ injects into

$$KO(Y)/F_2KO(Y) = \mathbb{Z} \times H^1(Y, \mathbb{Z}/2),$$

we see that the cokernel of $F_2KU(Y) \to F_2KO(Y)$ is the kernel of the surjection $WR(G \times Y) \to \mathbb{Z}/2 \times H^1(Y, \mathbb{Z}/2)$. Since $E_2^{p,-p}(KO) = 0$ for $p = 3$ and $p > 4$, and the map

$$E_\infty^{4,-4}(KU) \cong H^4(Y, \mathbb{Z}) \to H^4(Y, \mathbb{Z}) \cong E_\infty^{4,-4}(KO)$$

is onto (because $KU^{-4} \to KO^{-4}$ is onto), we are left to consider $F_2/F_3 = E_\infty^{2,-2}$ for $KU(Y)$ and $KO(Y)$.

The differential $d_3 : E_2^{2,-2}(KU) \to E_2^{5,-4}(KU)$ is a cohomology operation $H^2(Y,\mathbb{Z}) \to H^5(Y,\mathbb{Z})$. As noted above, it must be zero, so $E_\infty^{2,-2}(KU) \cong H^2(Y,\mathbb{Z})$. Similarly, the differential $d_3 : H^2(Y,\mathbb{Z}/2) \to H^5(Y,\mathbb{Z})$ is the cohomology operation $\tilde{\beta}Sq^2$, so $E_\infty^{2,-2}(KO)$ is the subgroup of Stiefel–Whitney classes $w_2(E)$ in $H^2(Y,\mathbb{Z}/2)$, and the cokernel of $E_\infty^{2,-2}(KU) \to E_\infty^{2,-2}(KO_G)$ is the group $\bar{H}^2(Y)$ of Definition 9.2. □

COROLLARY 9.4. *Let V be a smooth 3-fold over $\mathbb{C}$, and set $Y = V(\mathbb{C})$. If $H^5(Y,\mathbb{Z})$ is a torsionfree group, then*

$$WR(V_{\text{top}}) \cong \mathbb{Z}/2 \times H^1(Y,\mathbb{Z}/2) \times {}_2H^3(Y,\mathbb{Z}).$$

PROOF. The image of $\tilde{\beta}Sq^2 : H^2(Y,\mathbb{Z}/2) \to H^5(Y,\mathbb{Z})$ has exponent 2. It must be zero when $H^5(Y,\mathbb{Z})$ is torsionfree. Now apply 9.3 to $G \times Y$. □

EXAMPLE 9.5. In particular, if V is an abelian 3-fold over $\mathbb{C}$, then $H^1(V(\mathbb{C}),\mathbb{Z}/2) \cong (\mathbb{Z}/2)^6$ and $WR(V_{\text{top}}) \cong (\mathbb{Z}/2)^7$. This contrasts with Totaro's result [**50**] that $W(V)$ can be infinite for very general V.

We now consider a connnected 6-dimensional complex over $\mathbb{R}$ on which G acts freely. We need the following modification of Definition 9.2. We will see in the proof of Theorem 9.7 below that the composition of $\mathbb{H}_G^2(X,\mathbb{Z}(1)) \cong H^2(X/G,\mathbb{Z}(1)) \to H^2(X/G,\mathbb{Z}/2)$ with $\tilde{\beta}Sq^2$ is zero.

DEFINITION 9.6. When G acts freely on X and $Y = X/G$, we define $\bar{H}_G^2(X)$ to be the kernel of the cohomology operation ${}_2H^3(Y,\mathbb{Z}(1)) \to H^5(Y,\mathbb{Z})$ induced by $\tilde{\beta}Sq^2$.

Here is the analogue of Theorem 7.2 for 3-folds over $\mathbb{R}$. It is a modification of the argument we gave in [**26**, 7.1] for 4-dimensional spaces; see Theorem 8.1 of [**26**]. As in Lemma 3.1, $\widetilde{H}^1(X/G,\mathbb{Z}/2)$ denotes the quotient of $H^1(X/G,\mathbb{Z}/2)$ by the subgroup generated by $[-1] = w_1(X \times \mathbb{R}(1))$.

THEOREM 9.7. *Let X be a connected 6-dimensional G-CW complex. If G acts freely on X, then $WR(X)$ is an extension:*

$$0 \to \bar{H}_G^2(X) \to WR(X) \to \mathbb{Z}/4 \times \widetilde{H}_G^1(X/G,\mathbb{Z}/2) \to 0,$$

In particular, $WR(X)$ is a $\mathbb{Z}/8$-algebra.

PROOF. Since X is a free G-space, $KO_G(X) = KO(X/G)$. To determine $KR(X)$, it is convenient to write Y for X/G. As in the proof of Theorem 9.3, we work with the filtration coming from the Bredon spectral sequence. By Lemma A.5, $E_2^{1,-1}(KR) \cong H^1(X^G,\mathbb{Z}/2) = 0$. Thus $KR(X)/F_2KR(X) = \mathbb{Z}$ injects into

$$KO_G(X)/F_2KO_G \cong KO(Y)/F_2KO(Y) = \mathbb{Z} \times H^1(Y,\mathbb{Z}/2),$$

and the kernel of the surjection $WR(X) \to \mathbb{Z}/4 \times \widetilde{H}^1(Y,\mathbb{Z}/2)$ is the cokernel of $F_2KR(X) \to F_2KO_G(X) \cong F_2KO(Y)$.

Since $E_2^{p,-p}(KO) = 0$ for $p > 4$, $F_5KO(Y) = 0$ and $F_4KO(Y) \cong E_\infty^{4,-4}$. For both KR and KO_G, $E_3^{3,-3} = 0$ and the group $E_\infty^{4,-4}$ is a quotient of $H^4(Y,\mathbb{Z})$. In addition, the map

$$E_2^{4,-4}(KR) \cong H^4(Y,\mathbb{Z}) \to H^4(Y,\mathbb{Z}) \cong E_2^{4,-4}(KO_G)$$

is an isomorphism (because $KU^{-4} \to KO^{-4}$ is). Thus $F_3KR(X)$ maps onto $F_3KO(Y)$, i.e., $F_3WR(X) = 0$. We are left to consider $F_2/F_3 = E_\infty^{2,-2}$ for $KR(X)$ and $KO(Y)$. We have a commutative diagram

$$\begin{array}{ccccc} 0 \to E_\infty^{2,-2}(KR) & \longrightarrow & H^2(Y,\mathbb{Z}(1)) & \xrightarrow{d_3} & H^5(Y,\mathbb{Z}) \\ \downarrow & & \downarrow & & \downarrow\cong \\ 0 \to E_\infty^{2,-2}(KO_G) & \longrightarrow & H^2(Y,\mathbb{Z}/2) & \xrightarrow{d_3} & H^5(Y,\mathbb{Z}). \end{array} \tag{9.8}$$

As in the proof of Theorem 9.3, the differential $d_3 : H^2(Y,\mathbb{Z}/2) \to H^5(Y,\mathbb{Z})$ is the cohomology operation $\tilde{\beta}Sq^2(u)$, so its kernel $E_\infty^{2,-2}(KO)$ is the subgroup $\{u \in H^2(Y,\mathbb{Z}/2) \mid \tilde{\beta}Sq^2(u) = 0\}$.

We claim that the differential $d_3 : H^2(Y,\mathbb{Z}(1)) \to H^5(Y,\mathbb{Z})$ is zero in the top row in (9.8), so that $E_\infty^{2,-2}(KR) \cong E_2^{2,-2}(KR) = H^2(Y,\mathbb{Z}(1))$. This will imply that the cokernel of

$$E_\infty^{2,-2}KR(X) \to E_\infty^{2,-2}KO_G(X)$$

is the subgroup $\bar{H}_G^2(Y)$ of Definition 9.6, finishing the proof.

The Real Chern class c_1 of Definition 6.6 is the composition of the map $F_2KR(X) \to E_\infty^{2,-2} \subseteq H_G^2(X,\mathbb{Z}(1)) \cong H^2(Y,\mathbb{Z}(1))$ with the isomorphism $H_G^2(X,\mathbb{Z}(1)) \cong \mathbb{H}_G^2(X,\mathbb{Z}(1))$ of A.4(c). Since c_1 is onto, the differential $d_3 : H^2(Y,\mathbb{Z}(1)) \to H^5(Y,\mathbb{Z})$ is zero, as claimed, □

REMARK 9.8.1. The fact that $WR(X)$ is a $\mathbb{Z}/8$–algebra can also be deduced from the considerations of Section 7 in [**23**], which are valid for any CW-complex with a free G-action, the torsion being a specific function of the dimension.

Now suppose that V is a 3-fold over $\mathbb{R}$ with no real points.

COROLLARY 9.9. *If V is a geometrically connected 3-fold over $\mathbb{R}$, with no $\mathbb{R}$-points, then $WR(V_{\text{top}})$ is an extension of $\mathbb{Z}/4 \times \widetilde{H}_{\text{et}}^1(V,\mathbb{Z}/2)$ by the subquotient $\bar{H}_G^2$ of $H_{\text{et}}^2(V,\mathbb{Z}/2)$ defined in 9.6.*

The kernel and cokernel of $W(V) \to WR(V_{\text{top}})$ have exponent 2, and are the same as the kernel and cokernel of $I_2(V) \to \bar{H}_G^2$.

PROOF. Setting $X = V_{\text{top}}$, the first assertion is Theorem 9.7. Combining this with Corollary 6.3, we get a commutative diagram with exact rows:

$$\begin{array}{ccccc} 0 \to I_2(V) & \longrightarrow & W(V) & \longrightarrow & \mathbb{Z}/4 \times \widetilde{H}_{\text{et}}^1(V,\mathbb{Z}/2) \to 0 \\ \downarrow & & \downarrow & & \downarrow\cong \\ 0 \to \bar{H}_G^2 & \longrightarrow & WR(X) & \longrightarrow & \mathbb{Z}/4 \times \widetilde{H}_G^1(X,\mathbb{Z}/2) \to 0. \end{array}$$

The result follows from the snake lemma. □

We now turn our attention to the anisotropic quadric 3-fold Q_3 defined by $\sum_{i=0}^4 x_i^2 = 0$. For notational simplicity, let X denote the G-manifold $Q_{3,\text{top}}$. Colliot-Thélène and Sujatha [**11**, 3.4] proved that $W(Q_3) \cong \mathbb{Z}/8$, with $I_2(Q_3) \cong \text{Br}(Q_3) \cong \mathbb{Z}/2$; we will show that $WR(X) \cong \mathbb{Z}/8$ as well.

LEMMA 9.10. *$\mathbb{Z}/2 \cong \text{Br}(Q_3) \to {}_2\mathbb{H}_G^3(X,\mathbb{Z}(1))$ is an isomorphism.*

PROOF. Colliot-Thélène and Sujatha observed in [**11**, p. 9] that $\text{Pic}(Q_3) = \mathbb{Z}$ and $\text{Br}(Q_3) \cong \text{Br}(\mathbb{R}) \cong \mathbb{Z}/2$, and observed that G acts trivially on $\text{Pic}(Q_3)$. Set

$m = 2^k$. From Kummer Theory, we deduce that $H^p_{\text{et}}(Q_3, \mu_m)$ is $\mathbb{Z}/2$ for $p = 0, 1$, and $\mathbb{Z}/m \oplus \mathbb{Z}/2$ for $p = 2$.

By Cox' theorem, we can identify $H^p_{\text{et}}(Q_3, \mu_m)$ with $\mathbb{H}^p_G(X, \mathbb{Z}(1)/m)$. Moreover, $\mathbb{H}^0_G(X, \mathbb{Z}(1)) = 0$, and the groups $\mathbb{H}^p_G(X, \mathbb{Z}(1))$ are finitely generated. Using the exact sequences

$$0 \to \mathbb{H}^p_G(X, \mathbb{Z}(1))/m \to H^p_{\text{et}}(Q_3, \mu_m) \to {}_m\mathbb{H}^{p+1}_G(X, \mathbb{Z}(1)) \to 0,$$

we see that $\mathbb{H}^1_G(X, \mathbb{Z}(1)) = \mathbb{Z}/2$ and $H^2_G(X, \mathbb{Z}(1)) = \mathbb{Z}$. From

$$0 \to \mathbb{H}^2_G(X, \mathbb{Z}(1))/2 \to \mathbb{H}^2_G(X, \mathbb{Z}/2) \xrightarrow{\partial} {}_2\mathbb{H}^3_G(X, \mathbb{Z}(1)) \to 0$$

we see that ${}_2\mathbb{H}^3_G(X, \mathbb{Z}(1)) = \mathbb{Z}/2$. Since ∂ factors through the surjection $H^2_{\text{et}}(Q_3, \mathbb{Z}/2) \to {}_2\operatorname{Br}(Q_3)$, we see that ${}_2\operatorname{Br}(Q_3) \cong {}_2\mathbb{H}^3_G(X, \mathbb{Z}(1))$ □

Combining Lemma 9.10 with Corollary 9.9, we conclude:

COROLLARY 9.11. $W(Q_3) \cong WR(Q_{3,\text{top}}) \cong \mathbb{Z}/8$.

When V is a geometrically connected 3-fold over $\mathbb{R}$, and $V(\mathbb{R}) \neq \emptyset$, we saw in Proposition 1.7 that the torsion subgroup of $I(V)$ has exponent 8. We have not yet thought about this case.

10. Fundamental theorem for Witt groups

The Fundamental Theorem for Witt groups concerns Laurent polynomial extensions. When $V = \operatorname{Spec}(A)$, and A is a regular $\mathbb{R}$-algebra, the formula $W(A[t, 1/t]) \cong W(A) \oplus W(A)$ goes back to the work of the first author [**20**, pp. 138–9] and Ranicki [**43**, 4.6] in the 1970s. More recently, Balmer and Gille proved in [**4**, 9.13] that for smooth schemes V we have $W(V \times \operatorname{Spec}(\mathbb{R}[t, 1/t])) \cong W(V) \oplus W(V)$.

Since the G-space underlying $\mathbb{G}_m$ is equivariantly homotopy equivalent to $S^{1,1}$, we consider $WR(X \times S^{1,1})$. Since $WR(S^{1,1}) \cong \mathbb{Z} \oplus \mathbb{Z}$ (by Lemma 3.1) we have a canonical external product map

$$WR(X) \oplus WR(X) \cong WR(X) \otimes WR(S^{1,1}) \xrightarrow{\mu} WR(X \times S^{1,1}). \tag{10.1}$$

When X is a finite G-set, it follows from Lemma 3.1 that the product (10.1) is an isomorphism. However, it is not true in general that $WR(X \times S^{1,1}) \cong WR(X) \oplus WR(X)$, even if $X = V_{\text{top}}$, as Example 3.6 shows.

Here is a sufficient condition. To state it, recall that sending a Real vector bundle E to its dual E^* defines an action of $\mathbb{Z}/2$ on $KR(X)$, and hence on $KR_n(X) = KR^{-n}(X)$ for all n.[1] Let $kr_n(X)$ (resp. $kr'_n(X)$) denote the even (resp. odd) Tate cohomology of $\mathbb{Z}/2$ acting on the group $KR_n(X)$. That is,

$$kr_n(X) = H^2(\mathbb{Z}/2, KR_n(X)), \quad kr'_n(X) = H^1(\mathbb{Z}/2, KR_n(X)).$$

THEOREM 10.2. *If $kr_{-1}(X) = kr'_{-1}(X) = 0$, (which is true in particular when $KR_{-1}(X) = 0$), then the product* (10.1) *is an isomorphism:*

$$W(X) \oplus WR(X) \xrightarrow{\mu} WR(X \times S^{1,1}).$$

More generally, the kernel of μ is a quotient of $kr_{-1}(X)$ and the cokernel of μ is a subgroup of $kr'_{-1}(X)$.

[1]The superscript notation KR^* is used by Atiyah [**2**], but the subscript notation is more appropriate for our uses.

We will prove Theorem 10.2 later in this section. We first record:

LEMMA 10.3. $KR_n(X \times S^{1,1}) \cong KR_n(X) \oplus KR_{n-1}(X)$.

PROOF. Since $B^{1,1}$ retracts equivariantly to a point in $S^{1,1}$, we have a (split) short exact sequence

$$0 \to KR_n(X \times B^{1,1}) \to KR_n(X \times S^{1,1}) \to KR_{n-1}(X \times B^{1,1}, X \times S^{1,1}) \to 0.$$

Since the Thom Isomorphism theorem [**2**, 2.4], identifies the final term with $KR_{n-1}(X^E, \text{pt}) \cong KR_{n-1}(X)$, $E = X \times \mathbb{R}^{1,1}$, the result follows. □

EXAMPLE 10.3.1. When $X = S^{1,1}$, Lemma 10.3 yields $KR_{-1}(X) = KR_{-1}(\text{pt}) \oplus KR_{-2}(\text{pt}) = 0$ and hence $WR(X \times S^{1,1}) \cong WR(X)^2 \cong \mathbb{Z}^4$. We also have $W(\mathbb{G}_m \times \mathbb{G}_m) \cong WR(S^{1,1} \times S^{1,1})$ in this case.

Since Real vector bundles on X form a Hermitian category, there is a classical Bott exact sequence (see [**44**, Thm. 6.1]):

$$(10.4) \qquad \to GR_n^{[i-1]}(X) \xrightarrow{F} KR_n(X) \xrightarrow{H} GR_n^{[i]}(X) \to GR_{n-1}^{[i-1]}(X) \xrightarrow{F} .$$

The groups $GR_0(X) = GR_n^{[0]}(X)$ are the usual Grothendieck–Witt groups of symmetric forms on Real vector bundles (see [**44**, 1.39]), and the groups $GR_n^{[2]}(X)$ are the Grothendieck–Witt groups of skew-symmetric forms.

For the rest of this section, we shall adopt the following terminology. If F is a functor on G-spaces, we'll write $F(\mathbb{R})$ and $F(X \times \mathbb{R})$ for $F(S^{1,1}, \text{pt})$ and $F(X \times S^{1,1}, X)$. Thus $KR_n(X \times \mathbb{R}) \cong KR_{n-1}(X)$ by 10.3, and Theorem 10.2 asserts that $WR(X \times \mathbb{R}) \cong WR(X)$.

We will use the following description of the group $GR_1^{[1]}(X)$, viewed as the Grothendieck group of the forgetful functor F (see [**18**, II.2.13]). An element is determined by a triple (E, g_1, g_2), up to a suitable homotopy equivalence. Here E is a Real vector bundle on X and the g_i are symmetric bilinear forms on E.

Taking $i = n = 1$ in (10.4), and using Lemma 10.3, we get an exact sequence

$$(10.5) \quad GR_1(X \times \mathbb{R}) \xrightarrow{F} KR_0(X) \xrightarrow{H} GR_1^{[1]}(X \times \mathbb{R}) \xrightarrow{\partial} GR(X \times \mathbb{R}) \to KR_{-1}(X).$$

EXAMPLE 10.6. When X is a point, the group $GR_1^{[1]}(\mathbb{R})$ has two special elements: $u = (E, \langle 1 \rangle, \langle t \rangle)$ and $u^- = (E, \langle -1 \rangle, \langle -t \rangle)$, where E is the bundle $S^{1,1} \times \mathbb{C}$ and $\langle \pm t \rangle$ (resp., $\langle \pm 1 \rangle$) is the quadratic form $x \mapsto \pm tx^2$ (resp., $x \mapsto \pm x^2$). The boundary map ∂ in (10.5) sends u to the generator $(E, \langle t \rangle)$ of $GR(\mathbb{R}) \cong \mathbb{Z}$, and $\partial(u^-) = -\partial(u)$, while the map H in (10.5) sends the generator of $KR_0(\text{pt}) \cong \mathbb{Z}$ to $u + u^-$. Since $GR_1(\mathbb{R}) \cong \mathbb{Z}/2$, the map $F : GR_1(\mathbb{R}) \to KR_0(\text{pt})$ is zero. Since $KR_{-1}(\text{pt}) = 0$, (10.5) reduces to:

$$0 \to KR_0(\text{pt}) \xrightarrow{H} GR_1^{[1]}(\mathbb{R}) \xrightarrow{\partial} GR(\mathbb{R}) \to 0.$$

It follows that $GR_1^{[1]}(\mathbb{R})$ is isomorphic to $\mathbb{Z}^2$, with basis $\{u, u^-\}$.

Similarly, $GR(\text{pt}) \cong RO(G) \cong \mathbb{Z}^2$; a basis is given by $e = (\mathbb{C}, \langle 1 \rangle)$ and $e^- = (\mathbb{C}, \langle -1 \rangle)$. Since the cup product $GR(\text{pt}) \xrightarrow{\cup u} GR_1^{[1]}(\mathbb{R})$ sends e to u and e^- to u^-, it is an isomorphism.

LEMMA 10.7. *Let $\varphi_X : KO(X) \to H(X)$ be a natural transformation of functors from finite CW complexes to abelian groups such that each $H(X)$ is naturally* *a $KO(X)$-module and each φ_X is a $KO(X)$-module map. Then φ_X is the cup product with $t = \varphi_{\rm pt}(1)$.*

If in addition $\varphi_{\rm pt} : KO(\mathrm{pt}) \to H(\mathrm{pt})$ is an isomorphism, and $H(X) \cong KO(X)$ for all X, then $\varphi : KO^* \to H^*$ is an isomorphism of cohomology theories.

PROOF. If $x \in KO(X)$ then $\varphi_X(x) = \varphi_X(x \cdot 1) = \varphi_X(x) \cdot t$. This proves the first assertion. Now assume that $H(X) \cong KO(X)$ for all X, and that $\varphi_{\rm pt}$ is an isomorphism; as $KO(\mathrm{pt}) \cong \mathbb{Z}$, $t = \pm 1$. To see that φ_X is an isomorphism, we may assume that X is connected. In this case, $KO(X) \cong KO(\mathrm{pt}) \oplus \widetilde{KO}(X)$, and the second factor is a nilpotent ideal in $KO(X)$. Therefore $t = \varphi_X(1)$ is a unit of $KO(X)$, and hence $\varphi_X : KO(X) \to KO(X)$ is an isomorphism. Taking products with spheres and using Bott periodicity, it follows that $\varphi_X : KO^*(X) \cong H^*(X)$ for all X, i.e., φ is an isomorphism of cohomology theories. □

REMARK 10.7.1. If G acts trivially on X, so $KO_G(X) \cong KO(X)[y]/(y^2-1)$ and $GR^{[1]}(X \times \mathbb{R})$ is a $GR(X)$-module, the lemma holds for $KO_G(X) \to GR^{[1]}(X \times \mathbb{R})$.

THEOREM 10.8. *The cup product with u defines isomorphisms*

$$GR_n(X) \xrightarrow{\cup u} GR^{[1]}_{n+1}(X \times \mathbb{R}).$$

COROLLARY 10.9. *(See* [**23**, C.6]*) We have an exact sequence*

$$\to GR_1(X \times \mathbb{R}) \to KR(X) \xrightarrow{H} GR(X) \to GR(X \times \mathbb{R}) \to,$$

where the map $GR(X) \to GR(X \times \mathbb{R})$ is the cup product with $\partial(u)$.

PROOF. Use Theorem 10.8 to replace the middle term in (10.5) with $GR(X)$. □

PROOF OF THEOREM 10.8. We first consider two extreme cases, and then the general case.

CASE 1. When $X = X^G$, the sequence of $GR(X)$-modules (10.5) reduces to

$$0 \to KO(X) \xrightarrow{H} GR^{[1]}_1(X \times \mathbb{R}) \xrightarrow{\partial} KO(X) \to 0.$$

We see from Example 10.6 and Lemma 10.7 that the map H is the cup product with $u + u^-$, and that the map ∂ is split by the cup product $KO(X) \to GR^{[1]}_1(X)$ with u. Hence $GR^{[1]}(X \times \mathbb{R}) \cong GR(X) \cong KO(X) \oplus KO(X)$ as a $KO(X)$-module.

We claim that $GR^{[1]}_1(X \times \mathbb{R}) \cong GR(X)$ as a $GR(X)$-module. Now $GR(X) \cong KO(X) \otimes RO(G)$, and $RO(G) = \mathbb{Z}[s]/(s^2-1)$, it suffices to check the action of the sign representation s. By inspection, $s = 1$ on $KO(X)$ and $s(u) = u^-$. Therefore H and ∂ are $RO(G)$-maps, establishing the claim.

By Remark 10.7.1, establishes Theorem 10.8 when $X = X^G$.

CASE 2. When $X = G \times Y$, the map $GR_n(X \times \mathbb{R}) \xrightarrow{F} KR_n(X \times \mathbb{R})$ in (10.5) is the complexification map $KO_{n-1}(Y) \to KU_{n-1}(Y)$, and (10.4) becomes

$$KO_{n+2}(Y) \xrightarrow{c} KU_n(Y) \xrightarrow{H} GR^{[1]}_{n+1}(X \times \mathbb{R}) \to KO_{n+1}(Y) \xrightarrow{c} KU_{n-1}(Y).$$

This shows that $GR^{[1]}_*(X \times \mathbb{R})$ is a cohomology theory on Y. Comparing (10.4) with the classical Bott sequence (see [**18**, III.5.18], [**2**, (3.4)])

$$KO_{n+2}(Y) \xrightarrow{c} KU_n(Y) \to KO_n(Y) \xrightarrow{\partial} KO_{n+1}(Y) \xrightarrow{c} KU_{n-1}(Y),$$

we see that $GR^{[1]}_{*+1}(X\times\mathbb{R}) \cong KO_*(Y)$ as $KO_*(Y)$-modules. Therefore it suffices to show that the cup product $\varphi : GR_0(X) \to GR^{[1]}_1(X\times\mathbb{R})$ sends 1 to a generator of the $KO(Y)$-module $GR^{[1]}_1(X) \cong KO(Y)$. By Lemma 10.7, we may assume that Y is a point. In this case, the above exact 5-term sequence reduces to

$$\mathbb{Z}/2 \xrightarrow{0} \mathbb{Z} \xrightarrow{H=2} \mathbb{Z} \xrightarrow{\partial} \mathbb{Z}/2 \to 0.$$

Since $KO(\mathrm{pt}) \cong \mathbb{Z}$ and $GR^{[1]}_1(\mathbb{R}) \cong \mathbb{Z}$, we need only show that the distinguished element $u = (\mathbb{C}, 1, \langle t\rangle)$ is a generator of $GR^{[1]}_1(\mathbb{R})$.

For this, we first note that $\partial(u)$ is nonzero and odd because, by construction, $\partial(u)$ is the class of $\langle t\rangle$ considered as a function $S^1 \to KO$, and this is the generator of $KO_1(S^1, \mathrm{pt}) \cong \mathbb{Z}/2$. Next, we compose the map $KO(Y) \to GR^{[1]}_1(X\times\mathbb{R})$ with the discriminant

$$GR^{[1]}_1(X\times\mathbb{R}) \xrightarrow{D} KR_1(X\times\mathbb{R}) \cong KU(Y).$$

Since D takes an element (E, g_1, g_2) of $GR^{[1]}_1(X\times\mathbb{R})$ to $[g_1 g_2^{-1}]$, we have $D(u) = [t]$, which is a generator of $KR(\mathbb{R}) \cong KU(\mathrm{pt}) \cong \mathbb{Z}$. Since $KU(\mathrm{pt}) \xrightarrow{H} KO(\mathrm{pt})$ is $\mathbb{Z} \xrightarrow{2} \mathbb{Z}$, we see that $|u| \leq 2$. Hence u is a generator of $GR^{[1]}_1(\mathrm{pt})$, establishing Theorem 10.8 when $X = G \times Y$.

CASE 3. When G acts freely on X, there is a finite cover of X by open subspaces of the form $\{G \times U_i\}$. By case 2), the cup product with u is an isomorphism for each of these opens, and for their interections. It follows from the 5–lemma that the cup product with u is an isomorphism for X.

CASE 4. For general X, let T be an equivariant closed neighborhood of the subcomplex X^G, so that X^G is in the interior of T and $X^G \subset T$ is a G-homotopy equivalence. Write Y (resp., A) for the closure of $X - T$ (resp., $T \cap Y$). Then the cup product with u determines a map of Mayer–Vietoris sequences

$$\begin{array}{ccccccccc}
GR_{n+1}(T \amalg Y) & \to & GR_{n+1}(A) & \to & GR_n(X) & \to & GR_n(T\amalg Y) & \to & GR_n(A) \\
\downarrow \cong & & \downarrow \cong & & \downarrow \cup u & & \downarrow \cong & & \downarrow \cong \\
GR^{[1]}_{n+1}(T \amalg Y) & \to & GR^{[1]}_{n+1}(A) & \to & GR^{[1]}_n(A) & \to & GR^{[1]}_n(T\amalg Y) & \to & GR^{[1]}_n(A).
\end{array}$$

The general case follows from the 5–lemma. □

Recall that the "co-Witt" group $WR'_0(X)$ is the kernel of the map $GR_0(X) \xrightarrow{F} KR_0(X)$ in (10.4); it is also the image of $GR^{[1]}_1(X) \to GR_0(X)$.

PROOF OF THEOREM 10.2. The relative group $WR(X\times S^{1,1}, X) = WR(X\times\mathbb{R})$ fits into the usual 12–term sequence [**21**, p. 278], an elementary part of which is:

$$kr'_0(X\times\mathbb{R}) \xrightarrow{j} WR'_0(X\times\mathbb{R}) \xrightarrow{\beta} WR_0(X\times\mathbb{R}) \xrightarrow{d} kr_0(X\times\mathbb{R}).$$

By Lemma 10.3, $kr_0(X\times\mathbb{R}) \cong kr_{-1}(X)$ and $kr'_0(X\times\mathbb{R}) \cong kr'_{-1}(X)$. If these groups are 0, for example if $KR_{-1}(X) = 0$, then $WR'_0(X\times\mathbb{R}) \cong WR_0(X\times\mathbb{R})$. In general, the kernel of β is a quotient of $kr_{-1}(X)$ and the cokernel is a subgroup of $kr'_{-1}(X)$.

On the other hand, by definition, Theorem 10.8, and Corollary 10.9:

$$\begin{aligned} WR'_0(X\times\mathbb{R}) &= \text{im } GR_1^{[1]}(X\times\mathbb{R}) \xrightarrow{H} GR_0(X\times\mathbb{R}) \\ &= \text{im } GR(X) \xrightarrow{\cup u} GR(X\times\mathbb{R}) \\ &= GR(X)/\text{im } KR(X) \cong WR(X). \end{aligned}$$ □

Appendix A. Equivariant cohomology

In this section, we recall basic facts about G-equivariant cohomology theories, and apply them to KO_G and KR. Recall from [**7**, I.2] that an *equivariant cohomology theory* is a sequence of functors h^q on pairs of G-complexes satisfying homotopy invariance, excision and the existence of long exact sequences, depending naturally on a pair of G-complex.

A *Bredon coefficient system* M for the cyclic group G of order 2 is a diagram $M(\text{pt}) \xrightarrow{a} M(G)$, together with an involution σ on $M(G)$ such that $\sigma a = a$ [**7**, I.4.1]. Any coefficient system M determines an equivariant cohomology theory $H^*_G(-;M)$ [**7**, I.6.4]. Conversely, any G-equivariant cohomology theory h^* defines a family of coefficient systems $M = h^q$, $h^q(\text{pt}) \to h^q(G)$, and there are canonical maps $\eta^q : h^q(X) \to H^0_G(X;h^q)$ for all $q \in \mathbb{Z}$.

If X is a finite dimensional G-CW complex, the (convergent) Bredon spectral sequence [**7**, IV.4] for an equivariant cohomology theory h^* is:

$$E_2^{p,q} = H^p_G(X;h^q) \Rightarrow h^{p+q}(X), \tag{A.1}$$

and the canonical maps η^q are the edge maps.

Any coefficient system M defines a sheaf $\mathcal{M}$ on X/G whose stalk at $\bar{x}$ is $M(\text{pt})$ or $M(G)$, depending on whether the inverse image of $\bar{x}$ in X is a fixed point or isomorphic to G. The sheaf cohomology $H^p(X/G;\mathcal{M})$ agrees with $H^p_G(X;M)$; this alternative definition is due to Segal [**45**]. We remark that (A.1) agrees with Segal's spectral sequence [**45**, 5.3].

Example A.2. For any G-module M, we have the coefficient system $M(\text{pt}) = M^G \to M(G) = M$, as well as a local system $\mathcal{M}$ on X/G, and $H^*_G(X;M)$ is $H^*\,\text{Hom}_G(C_*(X),M)$, the cohomology associated to the local system $\mathcal{M}$. If A is an abelian group, regarded as a trivial G-module, we get the *constant coefficient system* $A(0)$; since $\text{Hom}_G(C_*(X),A) = \text{Hom}(C_*(X/G),A)$, $H^p_G(X;A(0))$ is the usual cohomology group $H^*(X/G,A)$. Note that $H^p_G(X;A(0))$ is different from the Borel cohomology $\mathbb{H}^p_G(X,A)$; see Example A.3.

Example A.3. Let A be an abelian group. The Borel cohomology groups $\mathbb{H}^p_G(X,A)$ are defined to be $H^p(X_G,A)$, where $X_G = X\times_G EG$. The $\mathbb{H}^*_G(-,A)$ form an equivariant cohomology theory; the associated coefficient system has $H^q_G(\text{pt}) = H^q(BG,A)$ and $H^q_G(G) = 0$ for $q \neq 0$. Since $C_*(X\times EG)$ is a chain complex of free $\mathbb{Z}[G]$-modules, quasi-isomorphic to $C_*(X)$, we see that $\mathbb{H}^p_G(X,A)$ is the group hypercohomology of C_*X with coefficients in A.

$$\mathbb{H}^p_G(X,A) = H^p\,\text{Hom}(C_*(X_G),A) = H^p\,\text{Hom}_G(C_*(X\times EG),A).$$

More generally, if M is any G-module we will write $\mathbb{H}^*_G(X,M)$ for the group hypercohomology

$$\mathbb{H}^*_G(C_*X,M) = H^p\,\text{Hom}_G(C_*(X\times EG),M);$$

it is also an equivariant cohomology theory. For example, the reader may use the formula that $\mathrm{Hom}_{\mathbb{Z}}(A,\mathbb{Z}) \cong \mathrm{Hom}_G(A,\mathbb{Z}[G])$ for any G-module A to check that

$$\mathbb{H}^*_G(X,\mathbb{Z}[G]) \cong H^*(X,\mathbb{Z}).$$

Now suppose that $A = \mathbb{Z}/2$. The terms of the spectral sequence (A.1) are: $E_2^{p,q} = H^p(X^G,\mathbb{Z}/2)$ if $q > 0$, and $E_2^{p,0} = H^p(X/G,\mathbb{Z}/2)$. For example, if $X = X^G$ then $\mathbb{H}^n_G(X,\mathbb{Z}/2) \cong \bigoplus_{p=0}^n H^p(X,\mathbb{Z}/2)$.

The exact sequence of low degree terms is

$$0 \to H^1(X/G,\mathbb{Z}/2) \to \mathbb{H}^1_G(X,\mathbb{Z}/2) \xrightarrow{\eta^1} H^0(X^G,\mathbb{Z}/2) \xrightarrow{d_2} H^2(X/G,\mathbb{Z}/2).$$

The edge map $\mathbb{H}^1_G(X,\mathbb{Z}/2) \xrightarrow{\eta^1} H^0(X^G,\mathbb{Z}/2)$ in (A.1) need not be onto, as we see from Example 1.6.

The groups $\mathbb{H}^*_G(X,\mathbb{Z}/2)$ are graded modules over $\mathbb{H}^*_G(\mathrm{pt},\mathbb{Z}/2) = \mathbb{Z}/2[\beta]$, natural in X. It follows that $\beta \in \mathbb{H}^1_G(\mathrm{pt},\mathbb{Z}/2)$ acts on the Bredon spectral sequence. For $q > 0$, it sends $E_2^{p,q} \cong H^p(X^G,\mathbb{Z}/2)$ isomorphically to $E_2^{p,q+1}$. When $X = V_{\mathrm{top}}$ for a variety V, so that $H^*_{\mathrm{et}}(V,\mathbb{Z}/2) \cong \mathbb{H}^*_G(X,\mathbb{Z}/2)$, it follows from [**9**] that when $q > \dim V$ we have $E_2^{0,q} \cong H^0(V,\mathcal{H}^q) \cong (\mathbb{Z}/2)^\nu$.

EXAMPLES A.4. Let G be the group of order 2.

a) The equivariant cohomology theory KO^*_G determines coefficient systems KO^q_G:

$$KO^q_G(\mathrm{pt}) = KO^q(\mathrm{pt}) \otimes RO(G) \xrightarrow{+} KO^q_G(G) = KO^q(\mathrm{pt}).$$

Here $RO(G) \cong \mathbb{Z}^2$ is the real representation ring of G, with generators $[\mathbb{R}]$ and $[\mathbb{R}(1)]$, and '+' sends $a[\mathbb{R}] + b[\mathbb{R}(1)]$ to $a + b$.

b) Real K-theory KR^q determines the coefficient system $KO^q \to KU^q$: KR^0 is the constant system $\mathbb{Z}(0)$, KR^2 is $0 \to \mathbb{Z}(1)$, KR^4 is $\mathbb{Z} \xrightarrow{2} \mathbb{Z}$, $KR^6 = KR^{-2}$ is $\mathbb{Z}/2 \xrightarrow{0} \mathbb{Z}(1)$ and $KR^7 = KR^{-1}$ is $\mathbb{Z}/2 \to 0$. We also have $KR^q = 0$ for $q = 1,3,5$ and $KR^q = KR^{q+8}$.

c) The cohomology theory $h^q = \mathbb{H}^q_G(-,\mathbb{Z}(1))$ of Example A.3 has the coefficient systems $h^0 = \mathbb{Z}(1)$, $h^q = (\mathbb{Z}/2 \to 0)$ if $q > 0$ is odd, and $h^q = 0$ otherwise. In particular, there is an exact sequence

$$0 \to H^1_G(X;\mathbb{Z}(1)) \to \mathbb{H}^1_G(X,\mathbb{Z}(1)) \to (\mathbb{Z}/2)^\nu \xrightarrow{d_2}$$
$$H^2_G(X;\mathbb{Z}(1)) \to \mathbb{H}^2_G(X,\mathbb{Z}(1)) \to H^1(X^G;\mathbb{Z}/2) \xrightarrow{d_2} H^3_G(X;\mathbb{Z}(1)).$$

If G acts freely on X, then $H^n_G(X;\mathbb{Z}(1)) \cong \mathbb{H}^n_G(X,\mathbb{Z}(1))$ for all n.

REMARK A.4.1. If M is a G-module, $\mathbb{H}^*_G(X,M)$ denotes the Borel cohomology with coefficients in M (see A.3), while $H^*_G(X;M)$ denotes the Bredon cohomology of the coefficient system $(M^G \to M)$ (see A.2). In particular, $H^*_G(X;M) \to \mathbb{H}^*_G(X,M)$ need not be an isomorphism unless $X^G = \emptyset$.

LEMMA A.5. *For each p and q, there is a natural split exact sequence*

$$0 \to H^p(X^G,KO^q) \to H^p_G(X;KO^q_G) \xrightarrow{+} H^p(X/G,KO^q) \to 0.$$

If X/G is connected and X^G has ν components, then

$$H^0_G(X;KO_G) \cong \mathbb{Z} \oplus \mathbb{Z}^\nu.$$

Moreover, $H^1_G(X;KR^{-1})\cong H^1(X^G,\mathbb{Z}/2)$ *and*

$$H^1_G(X;KO^{-1}_G)\cong H^1_G(X;KR^{-1})\oplus H^1(X/G,\mathbb{Z}/2).$$

PROOF. There is a natural surjection from the coefficient system KO^q_G to the constant coefficient system $KO^q(0)=(KO^q\to KO^q)$; the map from $KO^q_G(\mathrm{pt})\cong KO^q\otimes R(G)$ to $KO^q(\mathrm{pt})$ is addition. It is split by the map $KO^q\to KO^q_G$ sending a to $a\otimes[\mathbb{R}]$.

Let E^q denote the kernel of this surjection; $E^q(\mathrm{pt})=KO^q$ and $E^q(G)=0$, so $H^p(X;E^q)\cong H^p(X^G,KO^q)$. Then $KO^q_G=E^q\oplus KO^q$ as coefficient systems. Applying $H^p_G(X;-)$ yields the exact sequence of the lemma. The last assertions follow from the isomorphisms $KR^0\cong\mathbb{Z}(0)$ and $KR^{-1}\cong E^{-1}$ of coefficient systems. □

REMARK A.5.1. The image of $H^p_G(X;KR^{-2})\to H^p_G(X;KO^{-2}_G)$ contains the summand $H^p(X^G,\mathbb{Z}/2)$ of Lemma A.5, since $E^{-2}=(\mathbb{Z}/2\to 0)$ is a summand of both coefficient systems KR^{-2} and KO^{-2}_G.

LEMMA A.6. *Let $\mathcal{A}$ be the coefficient system $(0\to A)$, where A is a G-module. Then for every G-CW complex X,*

$$H^*_G(X;\mathcal{A})\cong H^*_G(X,X^G;A)\cong H^*_G(X/X^G,\mathrm{pt};A).$$

PROOF. Recall that $C^q(X;\mathcal{A})=\mathrm{Hom}_G(C_q(X,X^G),A)$ is the group of functions f on the q-cells σ of $X-X^G$ satisfying $\overline{f(\sigma)}=f(\bar\sigma)$; see [**7**, I-14]. The usual differentials on $\mathrm{Hom}(C_*(X,X^G),A)$ make $\mathrm{Hom}_G(C_*(X,X^G),A)$ into a cochain complex, and $H^*_G(X;\mathcal{A})$ is the cohomology of this complex, i.e., the cohomology $H^*_G(X,X^G;A)\cong H^*_G(X/X^G,\mathrm{pt};A)$ of the local system A. (See Example A.2.) □

EXAMPLE A.6.1. When G acts trivially on X, $H^*_G(X;\mathcal{A})=0$. When G acts freely on X, $H^*_G(X;\mathcal{A})=H^*(X/G,A)$.

Recall that the coefficient system KR^2 is $0\to\mathbb{Z}(1)$.

COROLLARY A.7. $H^q_G(X;KR^{+2})\cong H^q_G(X;\mathbb{Z}(1))$, *and*

$$H^q_G(X;KR^{-2})\cong H^q(X^G,\mathbb{Z}/2)\oplus H^q_G(X;\mathbb{Z}(1)).$$

PROOF. By construction, $C^*(X^G,\mathbb{Z}(1))=0$, so $H^*_G(X^G;\mathbb{Z}(1))=0$. It follows from this and Lemma A.6 that $H^q_G(X;KR^{+2})\cong H^q_G(X;\mathbb{Z}(1))$.

The calculation of $H^q_G(X;KR^{-2})$ follows from this and the observation in Remark A.5.1 that the coefficient system KR^{-2} is the direct sum of $(\mathbb{Z}/2\to 0)$ and $(0\to\mathbb{Z}(1))$. □

LEMMA A.8. *If* $\dim X^G\le 3$, *then*

$$h^4: H^4_G(X;KR^{-4})\longrightarrow H^4_G(X;KO^{-4}_G)\cong H^4(X/G,\mathbb{Z}).$$

is a surjection. If $\dim X^G\le 2$, *then h^4 is an isomorphism.*

PROOF. The coefficient map $KO^{-4}_G\xrightarrow{+}\mathbb{Z}(0)$ of Lemma A.5 induces an isomorphism $H^4_G(X;KO^{-4}_G)\cong H^4_G(X;\mathbb{Z}(0))\cong H^4(X/G,\mathbb{Z})$, because $\dim X^G<4$. From the exact sequence of coefficient systems

$$0\to KR^{-4}\to\mathbb{Z}(0)\to(\mathbb{Z}/2\to 0)\to 0,$$

we get an exact sequence

$$H^3(X^G,\mathbb{Z}/2)\to H^4_G(X;KR^{-4})\to H^4_G(X;\mathbb{Z}(0))\to H^4(X^G,\mathbb{Z}/2).$$

The last term vanshes when $\dim X^G \le 3$, and the left term vanishes when $\dim X^G \le 2$, so the middle map $H^4_G(X; KR^{-4}) \to H^4_G(X; \mathbb{Z}(0))$ is onto (resp., an isomorphism) when $\dim X^G$ is at most 3 (resp., 2). As $KR^{-4} \to \mathbb{Z}(0)$ is the composite $KR^{-4} \to KO_G^{-4} \to \mathbb{Z}(0)$, the result follows. □

EXAMPLE A.9. Let X be $S^{2,2}$, the 3-sphere with $X^G = S^1$. It is not hard to show that $KO_G(S^{2,2}) = \mathbb{Z}^2$. Since $KR^1(B^{2,2}, S^{2,2}) \cong KO^1 = 0$ by [**2**, 2.3], we also have $KR(S^{2,2}) \cong \mathbb{Z}$ and hence $WR(S^{2,2}) \cong \mathbb{Z}$.

In this case, the differential $d_2 : H^1(X^G, \mathbb{Z}/2) \to H^3_G(X, \mathbb{Z}/2)$ is an isomorphism in the Bredon spectral sequence (A.1) for $KO_G(X)$. Indeed, since X/G is the suspension of $\mathbb{CP}^1$, we have $H^p(X/G, \mathbb{Z}/2) = 0$ for $p \ne 3$ and $H^3(X/G, \mathbb{Z}/2) = \mathbb{Z}/2$.

Acknowledgments

The authors thank the referee, and are grateful to Parimala, R. Sujatha and J.-L. Colliot-Thélène for several discussions. We are also grateful to Marco Schlichting for several discussions about Section 10.

References

[1] Jón Kr. Arason, *Der Wittring projektiver Räume* (German), Math. Ann. **253** (1980), no. 3, 205–212, DOI 10.1007/BF03219998. MR597829

[2] M. F. Atiyah, *K-theory and reality*, Quart. J. Math. Oxford Ser. (2) **17** (1966), 367–386, DOI 10.1093/qmath/17.1.367. MR206940

[3] M. F. Atiyah and G. B. Segal, *Equivariant K-theory and completion*, J. Differential Geometry **3** (1969), 1–18. MR0259946

[4] Paul Balmer and Stefan Gille, *Koszul complexes and symmetric forms over the punctured affine space*, Proc. London Math. Soc. (3) **91** (2005), no. 2, 273–299, DOI 10.1112/S0024611504015199. MR2167088

[5] Paul Balmer and Charles Walter, *A Gersten-Witt spectral sequence for regular schemes* (English, with English and French summaries), Ann. Sci. École Norm. Sup. (4) **35** (2002), no. 1, 127–152, DOI 10.1016/S0012-9593(01)01084-9. MR1886007

[6] Spencer Bloch and Arthur Ogus, *Gersten's conjecture and the homology of schemes*, Ann. Sci. École Norm. Sup. (4) **7** (1974), 181–201 (1975). MR0412191

[7] Glen E. Bredon, *Equivariant cohomology theories*, Lecture Notes in Mathematics, No. 34, Springer-Verlag, Berlin-New York, 1967. MR0214062

[8] G. W. Brumfiel, *Witt rings and K-theory*, Rocky Mountain J. Math. **14** (1984), no. 4, 733–765, DOI 10.1216/RMJ-1984-14-4-733. Ordered fields and real algebraic geometry (Boulder, Colo., 1983). MR773113

[9] J.-L. Colliot-Thélène and R. Parimala, *Real components of algebraic varieties and étale cohomology*, Invent. Math. **101** (1990), no. 1, 81–99, DOI 10.1007/BF01231498. MR1055712

[10] J.-L. Colliot-Thélène and J.-J. Sansuc, *Fibrés quadratiques et composantes connexes réelles* (French), Math. Ann. **244** (1979), no. 2, 105–134, DOI 10.1007/BF01420486. MR550842

[11] J.-L. Colliot-Thélène and R. Sujatha, *Unramified Witt groups of real anisotropic quadrics*, K-theory and algebraic geometry: connections with quadratic forms and division algebras (Santa Barbara, CA, 1992), Proc. Sympos. Pure Math., vol. 58, Amer. Math. Soc., Providence, RI, 1995, pp. 127–147. MR1327294

[12] Annibale Comessatti, *Sulle varietà abeliane reali* (Italian), Ann. Mat. Pura Appl. **2** (1925), no. 1, 67–106, DOI 10.1007/BF02409932. MR1553074

[13] Annibale Comessatti, *Sulle varietà abeliane reali* (Italian), Ann. Mat. Pura Appl. **3** (1926), no. 1, 27–71, DOI 10.1007/BF02418645. MR1553085

[14] David A. Cox, *The étale homotopy type of varieties over* **R**, Proc. Amer. Math. Soc. **76** (1979), no. 1, 17–22, DOI 10.2307/2042908. MR534381

[15] Richard Elman and T. Y. Lam, *Quadratic forms under algebraic extensions*, Math. Ann. **219** (1976), no. 1, 21–42, DOI 10.1007/BF01360856. MR401649

[16] Robin Hartshorne, *Algebraic geometry*, Springer-Verlag, New York-Heidelberg, 1977. Graduate Texts in Mathematics, No. 52. MR0463157

[17] Bruno Kahn, *Construction de classes de Chern équivariantes pour un fibré vectoriel réel* (French, with English summary), Comm. Algebra **15** (1987), no. 4, 695–711. MR877194

[18] Max Karoubi, *K-theory*, Classics in Mathematics, Springer-Verlag, Berlin, 2008. An introduction; Reprint of the 1978 edition; With a new postface by the author and a list of errata. MR2458205

[19] Max Karoubi, *Sur la K-théorie équivariante* (French), Séminaire Heidelberg-Saarbrücken-Strasbourg sur la K-théorie (1967/68), Lecture Notes in Mathematics, Vol. 136, Springer, Berlin, 1970, pp. 187–253. MR0268253

[20] Max Karoubi, *Localisation de formes quadratiques. I* (French), Ann. Sci. École Norm. Sup. (4) **7** (1974), 359–403 (1975). MR0384894

[21] Max Karoubi, *Le théorème fondamental de la K-théorie hermitienne* (French), Ann. of Math. (2) **112** (1980), no. 2, 259–282, DOI 10.2307/1971147. MR592292

[22] Max Karoubi and Vojislav Mudrinski, *K-théorie réelle de fibrés projectifs complexes* (French, with English summary), C. R. Acad. Sci. Paris Sér. I Math. **297** (1983), no. 6, 349–352. MR732504

[23] Max Karoubi, Marco Schlichting, and Charles Weibel, *The Witt group of real algebraic varieties*, J. Topol. **9** (2016), no. 4, 1257–1302, DOI 10.1112/jtopol/jtw024. MR3620457

[24] M. Karoubi, M. Schlichting and C. Weibel, The Witt groups of some singular schemes, preprint (2020).

[25] Max Karoubi and Charles Weibel, *Algebraic and Real K-theory of real varieties*, Topology **42** (2003), no. 4, 715–742, DOI 10.1016/S0040-9383(02)00069-1. MR1958527

[26] Max Karoubi and Charles Weibel, *The real graded Brauer group*, Q. J. Math. **70** (2019), no. 4, 1475–1503, DOI 10.1093/qmathj/haz037. MR4045110

[27] Manfred Knebusch, *Symmetric bilinear forms over algebraic varieties*, Conference on Quadratic Forms—1976 (Proc. Conf., Queen's Univ., Kingston, Ont., 1976), Queen's Univ., Kingston, Ont., 1977, pp. 103–283. Queen's Papers in Pure and Appl. Math., No. 46. MR0498378

[28] V. A. Krasnov, *On the Brauer group of a real algebraic surface* (Russian), Mat. Zametki **60** (1996), no. 6, 935–938, DOI 10.1007/BF02305166; English transl., Math. Notes **60** (1996), no. 5-6, 707–710 (1997). MR1619934

[29] V. A. Krasnov, *The Brauer and the Witt group of a real ruled surface* (Russian, with Russian summary), Mat. Zametki **72** (2002), no. 5, 706–714, DOI 10.1023/A:1021457005910; English transl., Math. Notes **72** (2002), no. 5-6, 652–659. MR1963133

[30] Louis Mahé, *Théorème de Pfister pour les variétés et anneaux de Witt réduits* (French), Invent. Math. **85** (1986), no. 1, 53–72, DOI 10.1007/BF01388792. MR842048

[31] A. S. Merkur′ev, *On the norm residue symbol of degree* 2 (Russian), Dokl. Akad. Nauk SSSR **261** (1981), no. 3, 542–547. MR638926

[32] James S. Milne, *Étale cohomology*, Princeton Mathematical Series, vol. 33, Princeton University Press, Princeton, N.J., 1980. MR559531

[33] John Milnor, *Algebraic K-theory and quadratic forms*, Invent. Math. **9** (1969/70), 318–344, DOI 10.1007/BF01425486. MR260844

[34] John W. Milnor and James D. Stasheff, *Characteristic classes*, Princeton University Press, Princeton, N. J.; University of Tokyo Press, Tokyo, 1974. Annals of Mathematics Studies, No. 76. MR0440554

[35] D. Orlov, A. Vishik, and V. Voevodsky, *An exact sequence for* $K_*^M/2$ *with applications to quadratic forms*, Ann. of Math. (2) **165** (2007), no. 1, 1–13, DOI 10.4007/annals.2007.165.1. MR2276765

[36] William Pardon, *A relation between Witt groups and zero-cycles in a regular ring*, Algebraic K-theory, number theory, geometry and analysis (Bielefeld, 1982), Lecture Notes in Math., vol. 1046, Springer, Berlin, 1984, pp. 261–328, DOI 10.1007/BFb0072028. MR750687

[37] R. Parimala, *Witt groups of affine three-folds*, Duke Math. J. **57** (1988), no. 3, 947–954, DOI 10.1215/S0012-7094-88-05742-0. MR975129

[38] R. Parimala, *Witt groups of conics, elliptic, and hyperelliptic curves*, J. Number Theory **28** (1988), no. 1, 69–93, DOI 10.1016/0022-314X(88)90120-5. MR925609

[39] R. Parimala and R. Sujatha, *Levels of nonreal function fields of real rational surfaces*, Amer. J. Math. **113** (1991), no. 4, 757–761, DOI 10.2307/2374845. MR1118459

[40] Albrecht Pfister, *Quadratische Formen in beliebigen Körpern* (German), Invent. Math. **1** (1966), 116–132, DOI 10.1007/BF01389724. MR200270

[41] C. Pedrini and C. Weibel, *Invariants of real curves*, Rend. Sem. Mat. Univ. Politec. Torino **49** (1991), no. 2, 139–173 (1993). MR1219265

[42] Claudio Pedrini and Charles Weibel, *The higher K-theory of real curves*, K-Theory **27** (2002), no. 1, 1–31, DOI 10.1023/A:1020865906645. MR1936583

[43] A. A. Ranicki, *Algebraic L-theory. IV. Polynomial extension rings*, Comment. Math. Helv. **49** (1974), 137–167, DOI 10.1007/BF02566724. MR414664

[44] Marco Schlichting, *Hermitian K-theory, derived equivalences and Karoubi's fundamental theorem*, J. Pure Appl. Algebra **221** (2017), no. 7, 1729–1844, DOI 10.1016/j.jpaa.2016.12.026. MR3614976

[45] Graeme Segal, *Equivariant K-theory*, Inst. Hautes Études Sci. Publ. Math. **34** (1968), 129–151. MR0234452

[46] P. Berthelot, A. Grothendieck and L. Illusie, *Théorie des intersections et Théorème de Riemann–Roch*, in SGA6, Lect. Notes Math. vol. 225, 1971.

[47] Robert Silhol, *Real algebraic surfaces*, Lecture Notes in Mathematics, vol. 1392, Springer-Verlag, Berlin, 1989. MR1015720

[48] R. Sujatha, *Witt groups of real projective surfaces*, Math. Ann. **288** (1990), no. 1, 89–101, DOI 10.1007/BF01444523. MR1070926

[49] R. W. Thomason, *Les K-groupes d'un schéma éclaté et une formule d'intersection excédentaire* (French), Invent. Math. **112** (1993), no. 1, 195–215, DOI 10.1007/BF01232430. MR1207482

[50] Burt Totaro, *Complex varieties with infinite Chow groups modulo 2*, Ann. of Math. (2) **183** (2016), no. 1, 363–375, DOI 10.4007/annals.2016.183.1.7. MR3432586

[51] Joost van Hamel, *Torsion zero-cycles and the Abel-Jacobi map over the real numbers*, The arithmetic and geometry of algebraic cycles (Banff, AB, 1998), CRM Proc. Lecture Notes, vol. 24, Amer. Math. Soc., Providence, RI, 2000, pp. 329–359. MR1738865

[52] C. Walter, Grothendieck-Witt groups of projective bundles, preprint, 2003. http://www.math.uiuc.edu/K-theory/644/

[53] Charles A. Weibel, *The K-book*, Graduate Studies in Mathematics, vol. 145, American Mathematical Society, Providence, RI, 2013. An introduction to algebraic K-theory. MR3076731

[54] Ernst Witt, *Zerlegung reeller algebraischer Funktionen in Quadrate. Schiefkörper über reellem Funktionenkörper* (German), J. Reine Angew. Math. **171** (1934), 4–11, DOI 10.1515/crll.1934.171.4. MR1581415

[55] Marcus Zibrowius, *Witt groups of curves and surfaces*, Math. Z. **278** (2014), no. 1-2, 191–227, DOI 10.1007/s00209-014-1312-8. MR3267576

Université Denis Diderot Paris 7, Institut Mathématique de Jussieu — Paris Rive Gauche, FRANCE
Email address: max.karoubi@gmail.com
URL: http://webusers.imj-prg.fr/~max.karoubi

Math. Dept., Rutgers University, New Brunswick, NJ 08901, USA
Email address: weibel@math.rutgers.edu
URL: http://math.rutgers.edu/~weibel

Contemporary Mathematics
Volume **749**, 2020
https://doi.org/10.1090/conm/749/15073

Negative K-theory and Chow group of monoid algebras

Amalendu Krishna and Husney Parvez Sarwar

ABSTRACT. We show, for a finitely generated partially cancellative torsion-free commutative monoid M, that $K_i(R) \cong K_i(R[M])$ whenever $i \leq -d$ and R is a quasi-excellent $\mathbb{Q}$-algebra of Krull dimension $d \geq 1$. In particular, $K_i(R[M]) = 0$ for $i < -d$. This is a generalization of Weibel's K-dimension conjecture to monoid algebras. We show that this generalization fails for $X[M]$ if X is not an affine scheme. We also show that the Levine-Weibel Chow group of 0-cycles $\mathrm{CH}_0^{LW}(k[M])$ vanishes for any finitely generated commutative partially cancellative monoid M if k is an algebraically closed field.

Introduction

The monoid algebras are natural generalizations of polynomial and Laurent polynomial algebras over commutative rings. One can often think of them as subalgebras of polynomial or Laurent polynomial algebras generated by monomials. These are as ubiquitous in the study of various properties of rings as the polynomial or Laurent polynomial algebras. A very natural question in algebraic K-theory is to find out to what extent various known facts about the K-theory of polynomial and Laurent polynomial algebras remain valid for more general monoid algebras.

Gubeladze proved several results on this subject in a series of many papers ([**11**], [**14**], [**15**] and [**16**] to name a few). Using the new direction provided by [**19**] and [**3**], Cortiñas, Haesemeyer, Walker and Weibel together have made significant advances in the study of algebraic K-theory of monoid algebras (see [**4**], [**5**], [**6**] and [**7**]). An old question on the K-theory of monoid algebras was recently settled in [**26**]. Gubeladze recently settled an old K-theoretic question about monoid algebras [**18**]. The message that comes out of these papers is that even though some properties of the algebraic K-theory of polynomial and Laurent polynomial algebras remain valid for monoid algebras, many of them do not directly extend.

This paper was motivated by our pursuit for those properties of the algebraic K-theory of polynomial and Laurent polynomial algebras which may extend to monoid algebras. Our intuition was that while the higher Quillen K-theory of monoid algebras may not resemble that of polynomial and Laurent polynomial algebras, the negative K-theory and part of K_0 should. In this paper, we attempt

2010 *Mathematics Subject Classification.* Primary 19D50; Secondary 13F15, 14F35.
Key words and phrases. Algebraic K-theory, Chow group, Monoid algebras.
The second author would also like to thank Govt. of India SERB fellowship for financial support during his stay at Warwick.

to understand two such properties, namely, Weibel's K-dimension conjecture and, vanishing of SK_0 and Levine-Weibel Chow group for monoid algebras.

0.1. Weibel's conjecture for monoid algebras. Recall that a famous conjecture of Weibel [**36**] asserts that if R is a commutative Noetherian ring of Krull dimension d, then $K_{-d}(R) \simeq K_{-d}(R[t_1, \cdots, t_n])$ and $K_i(R[t_1, \cdots, t_n]) = 0$ for $i < -d$ and $n \geq 0$. An affirmative answer to this conjecture was obtained recently by Kerz, Strunk and Tamme [**22**]. For Noetherian rings containing $\mathbb{Q}$, this was earlier solved by Cortiñas, Haesemeyer, Schlichting and Weibel [**3**] (see also [**10**], [**23**] and [**38**] for older results in positive characteristics).

The main technical tool that goes into the proof of Weibel's conjecture in [**22**] is a pro-cdh-descent theorem for algebraic K-theory. However, the final step in the proof of the conjecture using the pro-cdh-descent breaks down in the case of general monoids because it crucially uses the fact that the map $R \to R[M]$ is smooth.

In this paper, we shall use a combination of pro-cdh-descent techniques and the theory of monoid algebras to show that the assertion of Weibel's conjecture directly extends to more general monoid algebras over some rings. Broadly speaking, we prove the following. We refer to § 2 for the precise results including the relevant terms and notations.

THEOREM 0.1. *Let M be a monoid which is finitely generated, commutative, partially cancellative and torsion-free. Let R be a quasi-excellent $\mathbb{Q}$-algebra of Krull dimension d. Assume that one of the following conditions is satisfied.*

(1) $d \geq 1$.
(2) *M is cancellative and semi-normal.*

Then the map $K_i(R) \to K_i(R[M])$ is an isomorphism for all $i \leq -d$. In particular, $K_i(R[M]) = 0$ for all $i < -d$.

One can interpret this result as a generalization of Weibel's K_{-d}-regularity and $K_{<-d}$-vanishing conjectures to monoid algebras.

REMARK 0.2. We remark that the condition (2) of Theorem 0.1 is essential if $d = 0$. In this case, we can in fact assume that R is a field (see Lemma 1.4). Assuming this, Gubeladze has shown that $K_0(R) \cong K_0(R[M])$ if and only if M is semi-normal (see [**15**, Theorem 1.3]).

Even if we expect Theorem 0.1 to hold for $\mathbb{F}_p$-algebras, we do not know how to prove it yet. We hope to pursue this in a future work. However, if we allow ourselves to invert the characteristic of the ground field, then we can indeed show that $K_i(R) \cong K_i(R[M])$ for all $i \leq -d$ when M is as in Theorem 0.1. A proof of this is given in § 2.

For non-affine schemes, we can prove the following extension of Weibel's conjecture.

THEOREM 0.3. *Let M be a cancellative torsion-free monoid and X a quasi-excellent $\mathbb{Q}$-scheme of dimension d. Then $K_i(X[M]) = 0$ for $i < -d$. If M is semi-normal, then the map $K_i(X) \to K_i(X[M])$ is an isomorphism for all $i \leq -d$.*

If M is not semi-normal in Theorem 0.3, then we can show that the monoid extension of Weibel's K_{-d}-regularity conjecture to general monoids is not valid for non-affine schemes. We deduce this failure from the following precise result.

THEOREM 0.4. *Let X be a connected smooth projective curve over a field k of characteristic zero. Let $M \subset \mathbb{Z}_+$ be the submonoid generated by $\{2,3\}$. Then the map $K_{-1}(X) \to K_{-1}(X[M])$ is not an isomorphism if the genus of X is positive.*

0.2. Levine-Weibel Chow group and SK_0 of monoid algebras. Apart from the negative K-theory, we also wanted to look at other cohomological properties of polynomial and Laurent polynomial algebras which can generalize to monoid algebras and which are related to the non-negative K-theory. One of these cohomology groups is the Levine-Weibel Chow group of 0-cycles $\mathrm{CH}_0^{LW}(X)$ for a singular variety X [**27**]. This group is the singular analogue of the classical Chow group of 0-cycles on smooth varieties and it is directly related to the K-theory of locally free sheaves.

It is well known that the Levine-Weibel Chow group vanishes for polynomial and Laurent polynomial algebras over a field. On the other hand, even though affine toric varieties come up very naturally in algebraic geometry, it was not known yet if their Levine-Weibel Chow group is zero in positive characteristics. Our next result of this paper is the following.

THEOREM 0.5. *Let k be an algebraically closed field and M a finitely generated commutative partially cancellative monoid of rank at least two. Then $\mathrm{CH}_0^{LW}(k[M]) = 0$.*

Note that the lower bound on the rank of the monoid is essential because it is well known that $\mathrm{CH}_0^{LW}(k[t^2,t^3]) \cong k$. Combined with [**30**, Corollaries 2.7 and 3.4], the vanishing of the Levine-Weibel Chow group for monoid algebras has the following algebraic consequence.

COROLLARY 0.6. *Let k be an algebraically closed field and M a finitely generated commutative partially cancellative monoid of rank $n \geq 2$. Then every local complete intersection ideal of $k[M]$ of height n is a complete intersection.*

One of the two main ingredients in our proof of Theorem 0.5 is the following result of independent interest in the algebraic K-theory of monoid algebras. When the base ring is regular and the underlying monoid is cancellative and torsion-free, this result is due to Gubeladze [**15**, Theorem 1.3].

THEOREM 0.7. *Let R be a commutative Artinian ring and M a finitely generated commutative partially cancellative monoid. Then $SK_0(R[M]) = 0$.*

One may recall that in most of Gubeladze's works (except [**18**]) on monoid algebras, all monoids were cancellative and torsion-free. However, it turns out that for computing the K-theory of cancellative monoid algebras using various fundamental results like excision in algebraic K-theory, one needs to extend many K-theoretic results to partially cancellative torsion-free monoids. These monoids occur very naturally in affine geometry. The algebras over such monoids were first studied by Swan [**31**, § 15] under the name of discrete Hodge algebras. Higher K-theory analogue of Serre's problem on projective modules for such algebras was studied by Vorst [**35**] and Gubeladze [**13**]. More K-theoretic importance of these monoid algebras was later exhibited in several papers of Cortiñas, Haesemeyer, Walker and Weibel. See, for instance, [**6**] and [**7**].

REMARK 0.8. The reader may have observed that the class of monoids considered in Theorem 0.1 is more restrictive than the one in Theorem 0.5 in that we did not allow torsion monoids in Theorem 0.1. The reason for this is the following. If we let M be a finite torsion group of order $n \geq 2$ and R a quasi-excellent $\mathbb{Q}$-algebra of dimension $d \geq 1$ such that $K_{-d}(R) \neq 0$ (for instance, take R to be the coordinate ring of a product of affine nodal curves over $\mathbb{C}$), then we know that $K_{-d}(R[M]) = K_{-d}(R[M]_{\mathrm{red}})$ contains at least two copies of $K_{-d}(R)$ as direct summands one of which is the inclusion $K_{-d}(R) \subset K_{-d}(R[M])$. In particular, the canonical map $K_{-d}(R) \to K_{-d}(R[M])$ is not an isomorphism. This shows that the $i = -d$ case Theorem 0.1 fails for torsion monoids. We already saw that the map $K_{-d}(X) \to K_{-d}(X[M])$ is not an isomorphism even when M is torsion-free if X is not affine.

However, if we let $i < -d$, then the situation is likely to be different as some of the above results show. We expect this case of Theorem 0.1 to hold for a very general class of monoids and a general class of Noetherian schemes. So we end the discussion of our results with the following.

QUESTION 0.9. Let X be a Noetherian separated scheme of Krull dimension $d \geq 0$ and M a finitely generated commutative cancellative monoid. Is $K_i(X[M]) = 0$ for $i < -d$?

0.3. Outline of proofs. We give a brief outline of our proofs. Our main strategy for proving Theorem 0.1 is to eventually reduce the proof to the case of cancellative torsion-free semi-normal monoids with the help of several reductions. To prove the theorem in this restrictive case, we need to use the pro-cdh descent results of [**22**]. The pro-cdh-descent results and the weak resolution of singularities together allow us to reduce to the case when the base scheme X is regular. The proof of the vanishing in this case is done using some classical results of Gubeladze and a Zariski descent argument. In § 1, we recall the basic results about monoids and prove our reduction steps that are needed in the proof of Theorem 0.1. Gubeladze's Milnor square for monoid algebras plays a key role in these steps. The final proofs of Theorems 0.1 and 0.3 using pro-cdh descent are given in § 2.

The idea of proving Theorem 0.5 came to us from two known results. The first is a classical result of Gubeladze which says that $SK_0(k[M])$ vanishes if k is a field and M is cancellative torsion-free. The second is an old conjecture of Murthy [**30**] which says that the Levine-Weibel Chow group of an affine variety over an algebraically closed field is torsion-free. This conjecture was recently settled in [**25**]. In view of this positive answer to Murthy's conjecture, we are left with proving Theorem 0.7 which extends Gubeladze's result to a more general class of monoids.

We prove this extension in Sections 4 and 5. For the proof of Theorem 0.7, we have to establish several reduction steps to reduce the proof to a nicer class of monoids. This is done in § 4 using several Milnor squares. A crucial part of our reduction steps is a recent technique of Gubeladze [**18**], which tells us how one should deal with cancellative torsion monoids. This reduction technique plays a key role in our proofs. The final proof is obtained in § 5. The key step in the final proof is a generalization of Swan's result [**31**, Lemma 15.6] on the prime decomposition of radical ideals in cancellative torsion-free monoids to a more general class of monoids. This allows us to reduce the case of partially cancellative monoids to cancellative monoids.

One could now naturally ask if Theorems 0.5 and 0.7 could be valid for non-cancellative and other more general class of monoids. In the hope of dealing with this question in future, we came up with one more proof of Theorem 0.7 for partially cancellative torsion-free monoids. This proof is given in the end of § 5. It is substantially K-theoretic in flavor and hence has potential to generalize to a broader class of monoids.

1. Recollection and reduction steps

In this section, we fix our notations and provide a limited recollection of the basic definitions in the theory of monoids. We then establish our reduction steps for reducing the proof of Theorem 0.1 to the case of positive cancellative torsion-free semi-normal monoids.

Recall that a monoid is same as a semigroup with an identity element. Throughout this paper, we shall assume all monoids to be commutative and finitely generated. We shall assume all our rings to be commutative and Noetherian and all schemes to be separated and Noetherian. The dimension of a ring or a scheme will be its Krull dimension.

1.1. Recollection of monoids. Unless specified otherwise, we shall use the additive notation for the operation in a monoid but switch to multiplicative operation when we consider a monoid inside an algebra generated by it over a commutative ring.

For a monoid M, we shall let $\mathrm{gp}(M)$ denote the group completion of M. We have a natural monoid homomorphism $M \to \mathrm{gp}(M)$. We let $U(M)$ denote the largest submonoid of M which is also a group. We say that M is positive if $U(M) = 0$. The rank of the monoid M is the rank of free part of $\mathrm{gp}(M)$. We shall denote it by $\mathrm{rk}(M)$. Given a subset $S \subset M$, we shall let $\langle S \rangle$ denote the submonoid of M generated by S. It is the $\mathbb{Z}_+$-linear combination of elements of S.

Given a monoid M, we can construct another monoid M_+ by adding a base-point ∞ to M and by letting $m + \infty = \infty$ for all $m \in M_+$. We shall call M_+ the augmented monoid associated to M. It is a pointed monoid in the language of [**6**]. We have a canonical inclusion of monoids $M \subset M_+$. We shall usually avoid the usage of more general pointed monoids considered in [**6**] as we have no need for them. Every monoid homomorphism $f : M \to N$ uniquely extends to a monoid homomorphism $f : M_+ \to N_+$ which fixes the base-point.

A monoid M is called cancellative if for $a, b, c \in M$, the condition $a + b = a + c$ implies that $b = c$. This is equivalent to saying that the group completion map $M \to \mathrm{gp}(M)$ is injective. We say that M is torsion-free if for $a, b \in M$, the condition $na = nb$ for some $n \geq 1$ implies that $a = b$. If M is cancellative, then it is torsion-free if and only if $\mathrm{gp}(M) \simeq \mathbb{Z}^r$.

A subset $I \subset M$ is called an ideal in M if $I + M = I$. If $I \subset M$ is a proper ideal ($I \neq M$), then $I \cap U(M) = \emptyset$ and $M \setminus U(M)$ is the largest proper ideal of M. If $I \subset M$ is an ideal, we shall let I_* denote $I \cup \{0\}$. This is a submonoid of M.

If $I \subset M$ is an ideal, the quotient M/I is obtained as follows. We consider the inclusion $I \subset M_+$ and then take the quotient by the equivalence relation $m \sim \infty$ for all $m \in I$. There is a unique addition rule in M_+/I that turns the canonical surjection $M_+ \to M_+/I$ into a morphism of monoids (see [**6**, § 1]). We let M/I be the image of M under the quotient map $M_+ \to M_+/I$. There is an epimorphism of monoids $M \to M/I$ and a canonical isomorphism $(M/I)_+ \simeq M_+/I$. We allow $I \subset M$ to be the empty set, in which case, we identify M/I with M. If $I = M$, the quotient M/I is identified with the constant singleton monoid $\{+\}$.

A monoid is called partially cancellative (pc) if there is a cancellative monoid N and an ideal $I \subset N$ (possibly empty) such that $M = N/I$. A monoid M is called partially cancellative torsion-free (pctf) if there is a cancellative torsion-free monoid N and an ideal $I \subset N$ (possibly empty) such that $M = N/I$. Such monoids are called pctf monoids in [**7**]. We shall also use this notation in this paper.

For a monoid M and a ring R, we let $R[M]$ denote the free R-module on M. Then $R[M]$ is a commutative R-algebra in the usual way, with multiplication given by the addition rule for M. Since R is Noetherian and M is finitely generated by our assumption, the Hilbert basis theorem tells us that $R[M]$ is also Noetherian. One also checks that $R[M]$ is an integral domain if and only if R is an integral domain and M is cancellative torsion-free (see [**2**, Theorem 4.8]). $R[M]$ is reduced if R is reduced and M is cancellative torsion-free. It can also happen that $R[M]$ is reduced even if M is a torsion monoid. For example, $R[M]$ is reduced if R is reduced and M is cancellative whenever $\mathbb{Q} \subset R$ (see [**2**, Theorem 4.19]). If I is an ideal of the monoid M, then $R[I]$ is an ideal of the ring $R[M]$ and $R[M/I] = R[M]/R[I]$.

Recall that a cancellative monoid M is called normal if $M = \{a \in \mathrm{gp}(M) |\ na \in M \text{ for some } n > 0\}$. One says that M is semi-normal if $M = \{a \in \mathrm{gp}(M) |\ 2a, 3a \in M\}$. The semi-normalization of a cancellative monoid M is the submonoid $sn(M)$ consisting of all elements $a \in \mathrm{gp}(M)$ such that $2a, 3a \in M$. Given an inclusion of monoids $M \subset N \subset \mathrm{gp}(M)$ and an element $a \in N$, we let $M\langle a\rangle$ denote the submonoid of N generated by M and a. Given a finite set $F = \{a_1, \cdots, a_r\} \subset N$, we can inductively define $M\langle F\rangle \subset N$. It is then easy to see that $sn(M) = \operatorname*{colim}_{F \subset sn(M)} M\langle F\rangle$. One can also check that $sn(M) = \{a \in \mathrm{gp}(M) |\ na \in M \text{ for all } n \gg 0\}$ (see [**31**, § 1]). We shall let $n(M) = \{a \in \mathrm{gp}(M) |\ na \in M \text{ for some } n\}$ denote the normalization of M in $\mathrm{gp}(M)$. The following result is elementary.

LEMMA 1.1. *Let M be a cancellative monoid and let $M \subset M' \subset n(M)$ be inclusions of monoids. Then the following hold.*

(1) *M' is cancellative.*
(2) *M' is torsion-free if M is so.*
(3) *M' is positive torsion-free if M is so.*
(4) *M' is positive if M is so provided $M' \subset sn(M)$.*

PROOF. We have the inclusions $M \subset M' \subset n(M) \subset \mathrm{gp}(M)$. In particular, all these monoids have same group completions. The parts (1) and (2) of the lemma follow immediately from this. For (3), suppose that M is positive torsion-free and let $a \in U(M')$ be a non-zero element. We then have $a + b = 0$ for some $b \in M'$. We can choose some $n \gg 0$ such that $na, nb \in M$. Since M is torsion-free, we must have $na, nb \neq 0$. Since $na + nb = n(a + b) = 0$, it follows that $na \in U(M) \setminus \{0\}$. This contradicts our assumption that M is positive.

We now prove (4). Suppose M is positive and let $a \in U(M')$ be a non-zero element. We then have $a + b = 0$ for some $b \in M'$. Since $M' \subset sn(M)$, we can choose some $n_0 \gg 0$ such that $na, nb \in M$ for all $n \geq n_0$ (see above). If $n_0 a \neq 0$, then $n_0 b \neq 0$ and $n_0 a + n_0 b = 0$, so we get $0 \neq n_0 a \in U(M)$, which contradicts our hypothesis. Suppose that $n_0 a = 0$. Then $n_0 b = 0$. But this means that $(n_0 + 1)a = a \neq 0$ and $(n_0 + 1)b = b \neq 0$. On the other hand, we have $(n_0 + 1)a + (n_0 + 1)b = (n_0 + 1)(a + b) = 0$. Since $(n_0 + 1)a, (n_0 + 1)b \in M$, we get $0 \neq (n_0 + 1)a \in U(M)$. This again contradicts our hypothesis. We are therefore done. $\square$

REMARK 1.2. Note that the part (3) of Lemma 1.1 is not true in general if M is not torsion-free. In fact, it is easy to see in this case that $n(M)$ contains the whole torsion subgroup of $\mathrm{gp}(M)$.

1.2. Reduction of positive monoids to semi-normal monoids. In this subsection, we shall establish some reduction steps which will tell us how to replace a positive cancellative torsion-free monoid in Theorem 0.1 by the one which is positive cancellative torsion-free and semi-normal. Recall that a Cartesian square of commutative rings

$$\begin{array}{ccc} A_1 & \xrightarrow{\alpha} & A_2 \\ \phi\downarrow & & \downarrow\psi \\ B_1 & \xrightarrow[\beta]{} & B_2 \end{array} \tag{1.1}$$

called a *Milnor square* if one of ψ and β is surjective. We shall use the following consequence of the classical results of Bass [**1**] and Milnor [**29**] as one important tool.

PROPOSITION 1.3. *Given a Milnor square (1.1), the map $K_i(A_1, B_1) \to K_i(A_2, B_2)$ of relative K-groups is an isomorphism for $i \leq 0$. In particular, there is a long exact sequence sequence of algebraic K-groups*

$$K_i(A_1) \to K_i(A_2) \oplus K_i(B_1) \to K_i(B_2) \to K_{i-1}(A_1) \to \cdots$$

for $i \leq 1$.

If ψ and β are both surjective in the Milnor square, then $K_i(A_1, B_1) \to K_i(A_2, B_2)$ is an isomorphism for $i \leq 1$. In particular, the above sequence is exact for all $i \leq 2$.

LEMMA 1.4. *Let A be a ring and I a nilpotent ideal of A. Then $K_i(A) = K_i(A/I)$ for all $i \leq 0$.*

PROOF. See Bass' book [**1**, Chapter IX, Proposition 1.3]. □

LEMMA 1.5. *Let R be a ring and M a positive cancellative torsion-free monoid. Then $K_i(R[M]) \simeq K_i(R[sn(M)])$ for all $i \leq -1$.*

PROOF. We can assume R to be reduced by Lemma 1.4. It follows from [**2**, Theorem 4.19] that $R[M]$ is also reduced. As in § 1.1, we can write $sn(M) = \operatorname{colim}_{F \subset sn(M)} M\langle F\rangle$, where F is a finite set of elements $a \in \mathrm{gp}(M)$ such that $2a, 3a \in M$. Since K-theory commutes with direct limits, it is enough to prove that $K_i(R[M]) \simeq K_i(R[M\langle F\rangle])$ for $i \leq -1$. Since $M\langle F\rangle$ is an iterated extension of monoids of the form $N\langle\{a\}\rangle$ such that $a \in \mathrm{gp}(N)$ with $2a, 3a \in N$, it will suffice to prove inductively that $K_i(R[M]) \simeq K_i(R[M\langle\{a\}\rangle])$ for $i \leq -1$, where $a \in \mathrm{gp}(N)$ with $2a, 3a \in M$. Note here that $M\langle\{a\}\rangle$ is a positive cancellative torsion-free monoid if M is so and $a \in sn(M)$ by Lemma 1.1.

We let $A = R[M]$ and $B = R[M\langle\{a\}\rangle]$. Let $F(A)$ denote the total ring of quotients of A. Note that $F(A)$ is a product of fields. We can write $B = A[x] \subset$

$F(A)$, where $x^2, x^3 \in A$. We then get a conductor Milnor square

$$\begin{array}{ccc} A & \longrightarrow & B \\ \downarrow & & \downarrow \\ A/I & \longrightarrow & B/I \end{array} \tag{1.2}$$

with conductor ideal $I = (x^2, x^3)$.

Furthermore, one knows in this case that $(A/I)_{\text{red}} \to (B/I)_{\text{red}}$ is an isomorphism (see, for instance, [**31**, Proof of Theorem 14.1]). This gives a commutative diagram of long exact sequences

$$\begin{array}{ccccccccc} K_{i+1}(A/I) & \longrightarrow & K_i(A,I) & \longrightarrow & K_i(A) & \longrightarrow & K_i(A/I) & \longrightarrow & K_{i-1}(A,I) \\ \downarrow & & \downarrow & & \downarrow & & \downarrow & & \downarrow \\ K_{i+1}(B/I) & \longrightarrow & K_i(B,I) & \longrightarrow & K_i(B) & \longrightarrow & K_i(B/I) & \longrightarrow & K_{i-1}(B,I). \end{array}$$

Since $i \leq -1$ and $(A/I)_{\text{red}} \xrightarrow{\cong} (B/I)_{\text{red}}$, it follows from Lemma 1.4 that the first and the fourth (from the left) vertical arrows are isomorphisms. The second and the fifth vertical arrows are isomorphisms by Proposition 1.3. We conclude that the middle vertical arrow is also an isomorphism. □

1.3. Reduction to positive monoids. In this subsection, our goal is to prove a reduction step which will allow us to replace a cancellative torsion-free monoid in Theorem 0.1 by the one which is cancellative torsion-free and positive.

LEMMA 1.6. *Let A be a ring of dimension $d \geq 0$. Then the canonical map $K_i(A) \to K_i(A[X_1, \ldots, X_n, Y_1^{\pm 1}, \ldots, Y_m^{\pm 1}])$ is an isomorphism for all $i \leq -d$ and $n, m \geq 0$.*

PROOF. The proof is by induction on m. The case $m = 0$ is due to [**22**, Theorem B(ii)]. Let $m \geq 1$ and assume that $K_i(A) \to K_i(A[X_1, \ldots, X_n, Y_1^{\pm 1}, \ldots, Y_{m-1}^{\pm 1}])$ is an isomorphism for all $n \geq 0$.

We let $B = A[X_1, \ldots, X_n, Y_1^{\pm 1}, \ldots, Y_{m-1}^{\pm 1}]$. The induction hypothesis implies that $K_i(A) \cong K_i(B)$ for all $i \leq -d$ and all $n \geq 0$. Hence by [**22**, Theorem B(i)], we get $K_{i-1}(B) = 0$ for all $i \leq -d$.

We now consider the commutative diagram of K-groups

$$\begin{array}{ccccc} K_i(B) \oplus K_i(B) & \xrightarrow{\beta_i} & K_i(B) & & \\ \downarrow{\scriptstyle \xi_i} & & \downarrow{\scriptstyle \alpha_i} & & \\ K_i(B[Y_m]) \oplus K_i(B[Y_m^{-1}]) & \longrightarrow & K_i(B[Y_m^{\pm 1}]) & \longrightarrow & K_{i-1}(B) = 0, \end{array}$$

where the bottom row is the Bass fundamental exact sequence. The map β_i is defined by $\beta_i(a, b) = a - b$. This is clearly surjective. By induction, we have

$$K_i(B) \cong K_i(A) \cong K_i(A[X_1, \ldots, X_n, Y_m, Y_1^{\pm 1}, \ldots, Y_{m-1}^{\pm 1}]) = K_i(B[Y_m]),$$

where the second isomorphism holds by replacing n by $n+1$. By the same reason, we get $K_i(B) \cong K_i(A) \cong K_i(B[Y_m^{-1}])$ for all $i \leq -d$. Hence, ξ_i is an isomorphism for all $i \leq -d$. A diagram chase shows that α_i is surjective. Since this is already split injective, we are done. □

LEMMA 1.7. *Let M be a monoid and $N = M \setminus U(M)$. Let R be a ring of dimension $d \geq 0$. If $K_i(R) \xrightarrow{\cong} K_i(R[N_*])$ for all $i \leq -d$, then $K_i(R) \xrightarrow{\cong} K_i(R[M])$ for all $i \leq -d$.*

PROOF. By [**18**, Lemma 6.1] (the proof thereof), there is a Milnor square

$$\begin{array}{ccc} R[N_*] & \longrightarrow & R[M] \\ \downarrow & & \downarrow \\ R & \longrightarrow & R[U(M)], \end{array} \tag{1.3}$$

where ϕ is split surjective because the composite $R[U(M)] \to R[M] \to R[U(M)]$ is identity (see also [**31**, § 15]).

Since (1.3) is a commutative diagram of R-algebras, the trivial Milnor square

$$\begin{array}{ccc} R & \to & R \\ \downarrow & & \downarrow \\ R & \to & R \end{array} \tag{1.4}$$

(where all maps are identity) maps to this square. It follows from Proposition 1.3 that there is a commutative diagram of short exact sequences

$$\begin{array}{ccccccccc} 0 & \longrightarrow & K_i(R) & \longrightarrow & K_i(R) \oplus K_i(R) & \longrightarrow & K_i(R) & \longrightarrow & 0 \\ & & \downarrow & & \downarrow & & \downarrow & & \\ 0 & \to & K_i(R[N_*]) & \to & K_i(R) \oplus K_i(R[M]) & \to & K_i(R[U(M)]) & \to & 0. \end{array} \tag{1.5}$$

The left vertical arrow is an isomorphism by our assumption and the right vertical arrow is an isomorphism by Lemma 1.6 since $U(M)$ is torsion-free (and hence free). This implies the desired assertion. $\square$

1.4. Cancellative to partially cancellative monoids. We shall now show how we can reduce the proof of Theorem 0.1 to the case of cancellative torsion-free monoids. The result that we shall use is the following.

LEMMA 1.8. *Let M be a partially cancellative torsion-free monoid and R a ring of dimension $d \geq 0$. Assume that $K_i(R) \to K_i(R[N])$ is an isomorphism for all $i \leq -d$ and for all cancellative torsion-free monoids N. Then $K_i(R) \to K_i(R[M])$ is an isomorphism for all $i \leq -d$.*

PROOF. By definition of partially cancellative torsion-free monoids, we can write $M = N/I$, where N is a cancellative torsion-free monoid. This yields a Milnor square

$$\begin{array}{ccc} R[I_*] & \longrightarrow & R[N] \\ \downarrow & & \downarrow \\ R & \longrightarrow & R[N/I]. \end{array} \tag{1.6}$$

Mimicking the proof of Lemma 1.7, we get a commutative diagram of long exact sequences

(1.7)

$$\begin{array}{ccccccccc} K_i(R) & \to & K_i(R)\oplus K_i(R) & \to & K_i(R) & \to & K_{i-1}(R) & \to & K_{i-1}(R)\oplus K_{i-1}(R) \\ \downarrow & & \downarrow & & \downarrow & & \downarrow & & \downarrow \\ K_i(R[I_*]) & \to & K_i(R)\oplus K_i(R[N]) & \to & K_i(R[N/I]) & \to & K_{i-1}(R[I_*]) & \to & K_{i-1}(R)\oplus K_{i-1}(R[N]). \end{array}$$

All vertical arrows in this diagram except possibly the middle one are isomorphisms because N and I_* are cancellative torsion-free monoids. It follows that the middle vertical arrow is an isomorphism too. This finishes the proof. □

An identical argument also proves the following variant of Lemma 1.8. We shall not use this here but it may be useful in answering Question 0.9.

LEMMA 1.9. *Let M be a partially cancellative monoid and R a ring of dimension $d \geq 0$. Assume that $K_i(R) \to K_i(R[N])$ is an isomorphism for all $i \leq -d$ and for all cancellative monoids N. Then $K_i(R) \to K_i(R[M])$ is an isomorphism for all $i \leq -d$.*

We shall also use the following result in the proof of Theorem 0.1.

LEMMA 1.10. *Let R be a regular ring and M a partially cancellative torsion-free monoid. Then $K_i(R[M]) = 0$ for all $i \leq -1$.*

PROOF. If M is cancellative torsion-free, the lemma follows from [**15**, Theorem 1.3]. In this case, we can also prove it using Lemma 1.5. This lemma allows us to assume that M is semi-normal. Now, by the Gubeladze–Swan theorem (see [**31**, Corollary 1.4]), we have $K_0(R) \cong K_0(R[M \oplus \mathbb{Z}])$. Together with the fundamental exact sequence, this implies that $K_i(R[M]) = 0$ for all $i \leq -1$.

In case M is partially cancellative torsion-free, we can write $M = N/I$, where N is cancellative torsion-free. Hence, by similar arguments as in Lemma 1.8, we get a Milnor square

$$\begin{array}{ccc} R[I_*] & \to & R[N] \\ \downarrow & & \downarrow{\scriptstyle \phi} \\ R & \to & R[N/I]. \end{array} \tag{1.8}$$

Using Proposition 1.3, this yields the long exact sequence

$$\cdots \to K_i(R)\oplus K_i(R[N]) \to K_i(R[N/I]) \to K_{i-1}(R[I_*]) \to \cdots \tag{1.9}$$

for $i \leq -1$. Note that I_* is a cancellative torsion-free monoid. Hence, we have $K_i(R[I_*]) = 0$ for all $i \leq -1$. Since $K_i(R[N])$ and $K_i(R)$ are both zero for $i \leq -1$, we get $K_i(R[N/I]) = 0$ for $i \leq -1$. This finishes the proof. □

2. The monoid version of Weibel's conjecture

The goal of this section is to prove Theorem 0.1. We fix a field k. We let $\mathbf{Sch}_k$ denote the category of separated schemes which are essentially of finite type over k and let $\mathbf{Sm}_k$ denote the full subcategory of $\mathbf{Sch}_k$ consisting of schemes which are regular. We shall denote the product of $X, Y \in \mathbf{Sch}_k$ over k by $X \times Y$. Let $\widetilde{\mathbf{Sch}}_k$ denote the category of separated Noetherian schemes over k. We let $\widetilde{\mathbf{Sch}}_k/\mathrm{zar}$ denote the Grothendieck site on $\widetilde{\mathbf{Sch}}_k$ given by the Zariski topology. We shall consider all cohomology groups with respect to the Zariski topology in this paper.

2.1. Quasi-excellent schemes and resolution of singularities. Recall from [**28**, Chapter 32, p. 260] that a (Noetherian) ring A is called excellent if the following hold.

(1) The fibers of the map $A_{\mathfrak{m}} \to \widehat{A_{\mathfrak{m}}}$ are geometrically regular for every maximal ideal $\mathfrak{m} \in A$.
(2) The regular locus X_{reg} of every finite type affine scheme X over A is open in X.
(3) A is universally catenary.

If A satisfies only (1) and (2), it is called quasi-excellent. A Noetherian separated scheme X is called (quasi-) excellent if is it covered by the spectra of (quasi-) excellent rings. One important property of (quasi-) excellent schemes we shall use in this paper is that if X is a (quasi-) excellent scheme and $X' \to X$ is an essentially of finite type morphism, then X' is also (quasi-) excellent. In particular, all objects of $\mathbf{Sch}_k$ are excellent. This is also true if we replace k by any complete local ring.

Let $\mathcal{C}_k$ be a subcategory of $\widetilde{\mathbf{Sch}}_k$. We shall say that $\mathcal{C}_k$ admits weak resolution of singularities if the following hold.

(1) If $X \in \mathcal{C}_k$ and $Y \to X$ is a finite type morphism in $\widetilde{\mathbf{Sch}}_k$, then $Y \in \mathcal{C}_k$.
(2) Given a reduced scheme $X \in \mathcal{C}_k$, there exists a Cartesian square of schemes

$$\begin{array}{ccc} \widetilde{Y} & \longrightarrow & \widetilde{X} \\ \downarrow & & \downarrow \\ Y & \longrightarrow & X \end{array} \tag{2.1}$$

such that $\widetilde{X} \to X$ is a proper, the horizontal arrows are nowhere dense closed immersions, $\widetilde{X} \in \mathbf{Sm}_k$ and $\widetilde{X} \setminus \widetilde{Y} \to X \setminus Y$ is an isomorphism. Note that $\widetilde{X}$, Y and $\widetilde{Y}$ belong to $\mathcal{C}_k$ by (1).

We shall use the following celebrated result of Hironaka [**20**, Theorem 1*] and its extension to quasi-excellent schemes by Temkin [**32**, Theorem 1.1].

THEOREM 2.1. *The category of quasi-excellent schemes over $\mathbb{Q}$ admits weak resolution of singularities.*

2.2. The spectrum $C^M K$. We let $X \mapsto K(X)$ denote the non-connective Thomason-Trobaugh algebraic K-theory presheaf of spectra on $\widetilde{\mathbf{Sch}}_k$ (denoted by $K^B(X)$ in [**33**]). Given a monoid M, we have an augmented k-algebra $k[M]$. For any $X \in \widetilde{\mathbf{Sch}}_k$, we let $X[M] = X \times_k \mathrm{Spec}\,(k[M]) \cong X \times_{\mathrm{Spec}\,(\mathbb{Z})} \mathrm{Spec}\,(\mathbb{Z}[M])$. The augmentation of $k[M]$ yields natural maps $X \xrightarrow{\sigma_X} X[M] \xrightarrow{p_X} X$ whose composite is identity. In particular, there is functorial decomposition $K(X[M]) \simeq K(X) \amalg C^M K(X)$, where $C^M K(X) = \mathrm{hofiber}(K(X[M]) \xrightarrow{\sigma_X^*} K(X))$. It follows that $X \mapsto C^M K(X)$ is a presheaf of spectra on $\widetilde{\mathbf{Sch}}_k$. For any $i \in \mathbb{Z}$, we shall let $C^M K_i(X)$ denote the stable homotopy group of $\pi_i(C^M K(X))$. We let $C^M \mathcal{K}_{i,X}$ denote the Zariski sheaf on X associated to the presheaf $U \mapsto \pi_i(C^M K(U))$.

If $f : Y \to X$ is a morphism in $\widetilde{\mathbf{Sch}}_k$, we shall let $C^M K(X, Y)$ denote the homotopy fiber of the map $f^* : C^M K(X) \to C^M K(Y)$. We let $C^M K_i(X, Y)$ denote the homotopy groups of $C^M K(X, Y)$. We let $C^M \mathcal{K}_{i,(X,Y)}$ denote the Zariski sheaf on X associated to the presheaf $U \mapsto C^M K_i(U, U \times_X Y)$.

THEOREM 2.2. ([**33**, Theorem 10.3]) *Given $X \in \widetilde{\mathbf{Sch}}_k$ and a closed subscheme $Z \subset X$, there exists a strongly convergent spectral sequence*

$$E_2^{p,q} = H^p(X, C^M \mathcal{K}_{q,(X,Z)}) \Rightarrow C^M K_{q-p}(X, Z). \tag{2.2}$$

PROOF. This is proved by repeating the argument of [**33**, Theorem 10.3] verbatim with the aid of [**33**, Proposition 3.20.2] and the fact that if $U = \mathrm{Spec}\,(A)$ is an affine open in X with a prime ideal $\mathfrak{p} \subset A$ such that $A_{\mathfrak{p}} = \varinjlim_i A[f_i^{-1}]$, then $A_{\mathfrak{p}} \otimes_k k[M] \simeq \varinjlim_i (A[M])[f_i^{-1}]$. We leave out the details. □

LEMMA 2.3. *Let $X \in \widetilde{\mathbf{Sch}}_k$ be of dimension d and X_{red} denote the reduced subscheme of X. Let M be any monoid. Then $C^M K_i(X, X_{\mathrm{red}}) = 0$ for $i < -d$. If M is cancellative torsion-free, then $C^M K_i(X, X_{\mathrm{red}}) = 0$ for $i \le -d$. In particular, $C^M K_i(X) \simeq C^M K_i(X_{\mathrm{red}})$ for all $i \le -d$.*

PROOF. The first assertion is an immediate consequence of the spectral sequence (2.2) and Lemma 1.4. We show the second part.

Using the spectral sequence (2.2), it suffices to show that $C^M K_i(A, I) = 0$ for $i \le 0$ if A is the local ring of a Zariski point of X and I is its nil-radical. We shall prove the stronger result that $K_i(A[M], I[M]) = 0$ for $i \le 0$. Using Lemma 1.4 and the long exact sequence for relative K-theory, we only need to show that the map $K_1(A[M]) \to K_1(A/I[M])$ is surjective.

We now consider the commutative diagram of short exact sequences

$$\begin{array}{ccccccccc} 0 & \to & SK_1(A[M]) & \to & K_1(A[M]) & \to & (A[M])^\times & \to & 0 \\ & & \downarrow & & \downarrow & & \downarrow & & \\ 0 & \to & SK_1(A/I[M]) & \to & K_1(A/I[M]) & \to & (A/I[M])^\times & \to & 0. \end{array} \tag{2.3}$$

It follows from our assumption on M and [**2**, Theorem 4.19] that $A/I[M] = (A[M])_{\mathrm{red}}$. We conclude from [**1**, Chap. IX, Propositions 3.10 and 3.11] (see also [**39**, Chapter III, Lemma 2.4]) that the left vertical arrow in (2.3) is an isomorphism.

To prove that the right vertical arrow in (2.3) is surjective (which will finish the proof), we consider the commutative diagram

$$\begin{array}{ccc} A^\times \times U(M) & \longrightarrow & (A[M])^\times \\ \downarrow & & \downarrow \\ (A/I)^\times \times U(M) & \xrightarrow{\simeq} & (A/I[M])^\times. \end{array} \tag{2.4}$$

Since A/I and $A/I[M]$ are reduced (observed above) and A is local, it follows from [**2**, Proposition 4.20] that the lower horizontal arrow is an isomorphism. Since the left vertical arrow is clearly surjective (uses again that A is local), we conclude that the right vertical arrow must also be surjective, as desired. □

2.3. The main result. We are now ready to prove Theorem 0.1, which extends the assertion of Weibel's K-dimension conjecture from polynomial to monoid algebras. The following result is a refinement of [**22**, Proposition 6.1]. We do not use it here but include it because it may be useful in the generalization of Theorem 0.1 for non-affine schemes. We fix a field k.

LEMMA 2.4. *Let M be a monoid and $X \in \widetilde{\mathbf{Sch}}_k$. Assume that $C^M K_i(\mathcal{O}_{X,x}) = 0$ for all $i \leq -\dim(\mathcal{O}_{X,x})$ and all points $x \in X$. Then $C^M K_i(X) = 0$ for all $i \leq -\dim(X)$.*

PROOF. For $i < -\dim(X)$, this is stated in [**22**, Proposition 6.1] and proven in [**21**, Proposition 3]. However, the proof given there works in the modified case as well with no change. We briefly explain it.

Let d denote the dimension of X. Using the spectral sequence (2.2), it suffices to show that $H^p(X, C^M\mathcal{K}_{q,X}) = 0$ whenever $q - p \leq -d$. Suppose first that $q + d \leq 0$. In this case, it suffices to show that $C^M\mathcal{K}_{q,X} = 0$. But this follows from our assumption because $\dim(\mathcal{O}_{X,x}) \leq d \leq -q$ for all $x \in X$.

We now fix $0 \leq p \leq d$ and $q - p \leq -d$ (equivalently, $p > q + d - 1$) such that $q + d > 0$. By [**21**, Lemma 4] (where we take $r = q + d - 1$), it suffices to show that if $x \in X$ is a Zariski point with $\dim(\overline{\{x\}}) \geq q + d$, then $C^M K_q(\mathcal{O}_{X,x}) = 0$. But the condition $\dim(\overline{\{x\}}) \geq q + d$ implies that $\dim(\mathcal{O}_{X,x}) \leq -q$. Our hypothesis implies again that $C^M K_q(\mathcal{O}_{X,x}) = 0$. This finishes the proof. □

LEMMA 2.5. *Let M be a cancellative torsion-free semi-normal monoid and X a regular Noetherian scheme of dimension $d \geq 0$. Then $C^M K_i(X) = 0$ for $i \leq -d$.*

PROOF. Since X is regular, we can assume that it is connected. We first assume that X is affine. In this case, it follows from [**31**, Corollary 1.4] that $C^M K_0(X) = 0$. The same holds if we replace X by $X[T^{\pm 1}]$ because the latter scheme is also affine and regular.

On the other hand, the fundamental exact sequence of Bass yields a surjective homomorphism $C^M K_i(X[T^{\pm 1}]) \twoheadrightarrow C^M K_{i-1}(X)$. We conclude inductively that $C^M K_i(X) = 0$ for $i \leq 0$. Using the spectral sequence (2.2) (see the proof of Lemma 2.3), we now conclude that $C^M K_i(X) = 0$ for $i \leq -d$ if X is any regular Noetherian scheme of dimension d. □

LEMMA 2.6. *Let $\mathcal{C}_k$ be a subcategory of $\widetilde{\mathbf{Sch}}_k$ which admits weak resolution of singularities. Let $X \in \mathcal{C}_k$ be of dimension $d \geq 0$. Let M be a cancellative torsion-free semi-normal monoid. Then $K_i(X) \to K_i(X[M])$ is an isomorphism for all $i \leq -d$.*

PROOF. The lemma is equivalent to showing that $C^M K_i(X) = 0$ for $i \leq -d$. By Lemma 2.3, we can assume that X is reduced. Consequently, we can assume that $X[M]$ is reduced by [**2**, Theorem 4.19] as M is cancellative torsion-free. Note that $X[M] \in \mathcal{C}_k$ as M is finitely generated. We shall prove the lemma by induction on d. The $d = 0$ case follows from Lemma 2.5. So we assume that $d \geq 1$.

Due to the assumption of weak resolution of singularity, we have an abstract blow-up square (see [**3**, Introduction] for definition) as in (2.1). Adjoining our monoid, this yields another abstract blow-up square in $\mathcal{C}_k$:

$$\begin{array}{ccc} \widetilde{Y}[M] & \rightarrow & \widetilde{X}[M] \\ \downarrow & & \downarrow \\ Y[M] & \rightarrow & X[M]. \end{array} \tag{2.5}$$

For any integer $n \geq 1$, we let nY denote the infinitesimal thickening of Y inside X defined by the sheaf of ideals $\mathcal{I}_Y^n$, where $\mathcal{I}_Y$ is the sheaf of ideals on X defining Y. We define $n\widetilde{Y}$ analogously. Then it is easy to see that $nY[M]$ (resp. $n\widetilde{Y}[M]$) is an infinitesimal thickening of $Y[M]$ (resp. $\widetilde{Y}[M]$) inside $X[M]$ (resp. $\widetilde{X}[M]$).

Since $C^M K_*(X)$ is a natural direct factor of $K_*(X[M])$, we can apply the pro-cdh-descent theorem [**22**, Theorem A] to get an exact sequence of pro-abelian groups

$$\{C^M K_{i+1}(n\widetilde{Y})\} \to C^M K_i(X) \to \{C^M K_i(nY)\} \oplus C^M K_i(\widetilde{X}). \tag{2.6}$$

Since $Y \subset X$ and $\widetilde{Y} \subset \widetilde{X}$ are nowhere dense closed subsets, we see that $\dim(Y)$ and $\dim(\widetilde{Y})$ are less than d. Using induction on d and regularity of $\widetilde{X}$, we see that the end terms of this exact sequence vanish. We conclude that $C^M K_i(X) = 0$ for $i \leq -d$. □

Combining Theorem 2.1 and Lemma 2.6, we get:

COROLLARY 2.7. *Let M be a cancellative torsion-free semi-normal monoid and X a quasi-excellent scheme over $\mathbb{Q}$ of dimension $d \geq 0$. Then $K_i(X) \to K_i(X[M])$ is an isomorphism for all $i \leq -d$.*

We now state our main result on the extension of Weibel's conjecture to monoid algebras.

THEOREM 2.8. *Let M be a partially cancellative torsion-free monoid and let $\mathcal{C}_k$ be a subcategory of $\widetilde{\mathbf{Sch}}_k$ which admits weak resolution of singularities. Let $X \in \mathcal{C}_k$ be affine of dimension $d \geq 0$. Assume that one of the following holds.*

(1) $d \geq 1$.
(2) *M is cancellative and semi-normal.*

Then the map $K_i(X) \to K_i(X[M])$ is an isomorphism for all $i \leq -d$.

PROOF. In view of the main results of [**22**], the theorem is equivalent to proving that $C^M K_i(X) = 0$ for $i \leq -d$. Lemma 2.3 allows us to assume that X is reduced. By Lemma 1.8, we can assume that M is a cancellative torsion-free monoid. Using Lemma 1.7, we can further assume that M is positive. If $d \geq 1$, we can use Lemma 1.5 to assume that M is a cancellative torsion-free semi-normal monoid. If $d = 0$, then M is already given to be cancellative torsion-free semi-normal.

We have therefore reduced the proof of the theorem to the case where M is a cancellative torsion-free semi-normal monoid and $X \in \mathcal{C}_k$ a reduced affine scheme of dimension $d \geq 0$. We can therefore apply Lemma 2.6 to conclude the proof. □

Combining Corollary 2.7 and Theorem 2.8, we get:

COROLLARY 2.9. *Let M be a partially cancellative torsion-free monoid and let X be a quasi-excellent affine scheme of dimension $d \geq 0$ over $\mathbb{Q}$. Assume that one of the following holds.*

(1) $d \geq 1$.

(2) *M is cancellative and semi-normal.*

Then the map $K_i(X) \to K_i(X[M])$ is an isomorphism for all $i \leq -d$.

REMARK 2.10. One can easily check from the proof of Lemma 2.6 that the above proof of Theorem 2.8 remains valid (without any change) for all d-dimensional schemes over any ground field k which admits weak resolution of singularities for schemes of dimensions up to d. Since the resolution of singularities is known to hold in dimension up to three over any ground field (see [**8**]), we see that the assertion of Theorem 2.8 remains valid for affine schemes over any arbitrary field as long as $d \leq 3$.

We also remark that by the same reason as above, Theorem 2.8 is also valid if k is any field and X is either an affine normal crossing scheme or an affine toric variety over k. This is because X admits resolution of singularities in both cases.

2.4. Vanishing of $K_{<-d}(X[M])$. In this subsection, we shall prove a vanishing result for $K_{<-d}(X[M])$ if X is a quasi-excellent $\mathbb{Q}$-scheme of dimension d and M is a cancellative torsion-free monoid. We need the following extension of Lemma 2.5.

LEMMA 2.11. *Let M be a cancellative torsion free monoid and X a regular Noetherian scheme of dimension $d \geq 0$. Then $K_i(X[M]) = 0$ for $i < -d$.*

PROOF. Since X is regular, we can assume that it is connected. We first assume that X is affine. In this case, it follows from [**15**, Theorem 1.3] that $K_i(X[M]) = 0$ for $i \leq -1$. Suppose now that X is any Noetherian scheme of dimension d. Using [**22**, Theorem B], it suffices to show that $C^M K_i(X) = 0$ for $i < -d$. But this follows immediately from the spectral sequence (2.2) (see the proof of Lemma 2.3). □

THEOREM 2.12. *Let M be a cancellative torsion-free monoid and X a quasi-excellent $\mathbb{Q}$-scheme of dimension d. Then $K_i(X[M]) = 0$ for $i < -d$.*

PROOF. This proof is identical to that of Lemma 2.6. We give the sketch. We prove the theorem by induction on d. The $d = 0$ case follows from Lemma 2.11. So we assume that $d \geq 1$.

By Lemma 2.3, we can assume that X is reduced. Consequently, we can assume that $X[M]$ is reduced by [**2**, Theorem 4.19] as X is a $\mathbb{Q}$-scheme.

By Theorem 2.1, we have an abstract blow-up square as in (2.1). Adjoining our monoid, this yields another abstract blow-up square:

$$\begin{array}{ccc} \widetilde{Y}[M] & \longrightarrow & \widetilde{X}[M] \\ \downarrow & & \downarrow \\ Y[M] & \longrightarrow & X[M]. \end{array} \tag{2.7}$$

Now, as we did in the proof of Lemma 2.6, we can apply the pro-cdh-descent theorem [**22**, Theorem A] to get an exact sequence of pro-abelian groups

$$\{K_{i+1}(n\widetilde{Y}[M])\} \to K_i(X[M]) \to \{K_i(nY[M])\} \oplus K_i(\widetilde{X}[M]). \tag{2.8}$$

Since $Y \subset X$ and $\widetilde{Y} \subset \widetilde{X}$ are nowhere dense closed subsets, we see that $\dim(Y)$ and $\dim(\widetilde{Y})$ are less than d. Using induction on d and regularity of $\widetilde{X}$, we see that the end terms of this exact sequence vanish. We conclude that $K_i(X[M]) = 0$ for $i < -d$. $\square$

2.5. The positive characteristic case. The weak resolution of singularities in all dimensions is yet unknown if the ground field k has positive characteristic. Nevertheless, we can show that Theorem 2.8 is valid in this case too if we invert $\mathrm{char}(k)$. Using the reduction steps of § 1, this turns out to be actually an easy consequence of the homotopy invariance property of Weibel's homotopy K-theory. More precisely, we can prove the following.

THEOREM 2.13. *Let k be a field of characteristic $p > 0$. Let $X \in \widetilde{\mathbf{Sch}}_k$ be an affine scheme of dimension $d \geq 0$. Assume that one of the following holds.*

(1) $d \geq 1$.

(2) *M is cancellative and semi-normal.*

Then the map $K_i(X)[\frac{1}{p}] \to K_i(X[M])[\frac{1}{p}]$ is an isomorphism for all $i \leq -d$.

PROOF. In this proof, we shall work only with $\mathbb{Z}[\frac{1}{p}]$-modules. In particular, $K_i(X)$ will mean $K_i(X)[\frac{1}{p}]$ for simplicity of notation. In this case, we can replace the spectrum $K(X)$ by $KH(X)$ (see [**33**, Exercise 9.11(h)]), where the latter is Weibel's homotopy K-theory [**37**].

Now, we can assume M to be cancellative torsion-free and positive by using the KH-analogues of Lemmas 1.8 and 1.7. If we now write $X = \mathrm{Spec}\,(R)$, it follows that $R[M]$ is a positively graded R-algebra. Therefore, by the homotopy invariance property [**37**, Theorem 1.2], we get $KH_i(X) \cong KH_i(X[M])$ for $i \leq -d$. $\square$

3. Counterexample for non-affine schemes

Recall that Weibel's K_{-d}-regularity conjecture is true for all Noetherian separated schemes [**22**]. We saw in the previous section that this holds also for cancellative torsion-free semi-normal monoids. However, we shall show in this section that the monoid extension of K_{-d}-regularity conjecture is not valid for non-affine schemes if M is not semi-normal. This shows that the extension of Weibel's K_{-d}-regularity conjecture to monoids is a subtle question.

We let $M \subset \mathbb{Z}_+$ be the submonoid generated by $\{2, 3\}$. It is clear that M is cancellative torsion-free, but not semi-normal. It is also clear that the inclusion $R[M] \subset R[\mathbb{Z}_+]$ is same as the inclusion $R[x^2, x^3] \subset R[x]$ for any ring R. We fix a field k of characteristic zero and let $C = \mathrm{Spec}\,(k[M])$.

THEOREM 3.1. *Let X be a connected smooth projective curve over k of positive genus. Then the map $K_{-1}(X) \to K_{-1}(X[M])$ is not an isomorphism.*

PROOF. Since $K_{-1}(X) = 0$, the theorem is equivalent to showing that $K_{-1}(X[M]) \neq 0$.

We consider the conductor square

(3.1)
$$\begin{array}{ccccc} S & \xrightarrow{u} & T & \xrightarrow{\iota'} & \mathbb{A}^1_k \\ & \searrow\!\!\searrow & \downarrow{\scriptstyle f'} & & \downarrow{\scriptstyle f} \\ & & S & \xrightarrow{\iota} & C \end{array} \qquad \begin{array}{ccccc} S \times X & \xrightarrow{u} & T \times X & \xrightarrow{\iota'} & \mathbb{A}^1_k \times X \\ & \searrow\!\!\searrow & \downarrow{\scriptstyle f'} & & \downarrow{\scriptstyle f} \\ & & S \times X & \xrightarrow{\iota} & C \times X, \end{array}$$

where the square on the right is obtained by the one on the left by the base change via the map $X \to \operatorname{Spec}(k)$. The map f is the normalization map, $S \cong \operatorname{Spec}(k)$ and $T \cong \operatorname{Spec}(k[x]/(x^2))$. The inclusion $u : S \to T$ is induced by the augmentation $k[x]/(x^2) \twoheadrightarrow k$.

We write $X_Y = X \times Y$ for any k-scheme Y. Note that $X_S \cong X$. For any $n \geq 1$, we have a commutative diagram of relative K-theory exact sequences:

$$\begin{array}{ccccccccc} K_0(X_C) & \to & K_0(nX) & \to & K_{-1}(X_C, nX) & \xrightarrow{\alpha_n} & K_{-1}(X_C) & \to & 0 \\ \downarrow & & \downarrow & & \downarrow & & & & \\ K_0(X_{\mathbb{A}^1_k}) & \to & K_0(nX_T) & \xrightarrow{\beta_n} & K_{-1}(X_{\mathbb{A}^1_k}, nX_T) & \longrightarrow & 0, & & \end{array} \tag{3.2}$$

where the vertical arrows are induced by f. Note that for a closed immersion $W \subset Y$ defined by the sheaf of ideals $\mathcal{I}_W$, the subscheme $nW \subset Y$ is defined by $\mathcal{I}^n$. The map α_n is surjective because its cokernel will otherwise map injectively into $K_{-1}(nX)$. But this latter term is isomorphic to $K_{-1}(X)$ by Lemma 2.3 which is zero. The map β_n is surjective because $K_{-1}(X_{\mathbb{A}^1_k}) = 0$.

When $n = 1$, we have the situation

$$\begin{array}{ccccccc} K_0(X) & \xrightarrow{p^*} & K_0(X_C) & \xrightarrow{\iota^*} & K_0(X) & & \\ & \searrow^{\cong} & \downarrow & & \downarrow f'^* & & \\ & & K_0(X_{\mathbb{A}^1_k}) & \xrightarrow{\iota'^*} & K_0(X_T) & \overset{\beta_1}{\twoheadrightarrow} & K_{-1}(X_{\mathbb{A}^1_k}, X_T), \end{array} \tag{3.3}$$

where p^* is induced by the projection $X_C \to X$. It follows that the composite horizontal arrow on the top is identity. The indicated isomorphism is by the homotopy invariance. A diagram chase shows that ι^* is split surjective and $K_{-1}(X_{\mathbb{A}^1_k}, X_T) \cong \operatorname{Coker}(\iota'^*) = \operatorname{Coker}(f'^*)$. Using this, we also see that α_1 is an isomorphism. In particular, the theorem is equivalent to showing that $K_{-1}(X_C, X) \neq 0$.

We now consider the inclusion of inductive systems of closed pairs $\{(X_{\mathbb{A}^1_k}, X_T)\} \hookrightarrow \{(X_{\mathbb{A}^1_k}, nX_T)\}$. We make the following

CLAIM. The induced map of pro-abelian groups $\{K_{-1}(X_{\mathbb{A}^1_k}, nX_T)\} \to K_{-1}(X_{\mathbb{A}^1_k}, X_T)$ is surjective.

To prove the claim, it suffices to show that the map $K_{-1}(X_{\mathbb{A}^1_k}, nX_T) \to K_{-1}(X_{\mathbb{A}^1_k}, X_T)$ is surjective for all $n \geq 1$.

We fix $n \geq 1$ and consider the commutative diagram

$$\begin{array}{ccc} K_0(nX_T) & \twoheadrightarrow & K_{-1}(X_{\mathbb{A}^1_k}, nX_T) \\ \downarrow & & \downarrow \\ K_0(X_T) & \twoheadrightarrow & K_{-1}(X_{\mathbb{A}^1_k}, X_T). \end{array} \tag{3.4}$$

Because $K_{-1}(X_{\mathbb{A}^1_k}) = 0$, the relative K-theory exact sequence tells us that the horizontal arrows are surjective. It suffices therefore to show that the left vertical arrow is surjective.

Since $\dim(X) = 1$, it is easy to check using the Thomason-Trobaugh spectral sequence [**33**, Theorem 10.3] that there is a functorial split exact sequence

$$0 \to H^1(nX_T, \mathcal{O}^{\times}_{nX_T}) \to K_0(nX_T) \to H^0(X, \mathbb{Z}) \to 0. \tag{3.5}$$

To compute the left term, we consider the short exact sequence of sheaves

$$0 \to (1 + \mathcal{I}_X) \to \mathcal{O}^{\times}_{nX_T} \to \mathcal{O}^{\times}_X \to 0,$$

where $\mathcal{I}_X$ is the sheaf of ideals on nX_T defining X. Since $nX_T \cong X \times \mathrm{Spec}\,(k[x]/(x^{2n}))$ and $\mathrm{char}(k) = 0$, this sequence is split and the exponential map $\exp : \mathcal{O}_X \otimes_k (x)/(x^{2n}) \to (1 + \mathcal{I}_X)$ is an isomorphism of the sheaves of abelian groups. We thus have a commutative diagram of the split exact sequences of sheaves of abelian groups

$$\begin{array}{ccccccc} 0 \to \mathcal{O}_X \otimes_k (x)/(x^{2n}) & \to & \mathcal{O}^{\times}_{nX_T} & \to & \mathcal{O}^{\times}_X & \to 0 \\ \downarrow & & \downarrow & & \| & \\ 0 \to \mathcal{O}_X \otimes_k (x)/(x^2) & \to & \mathcal{O}^{\times}_{X_T} & \to & \mathcal{O}^{\times}_X & \to 0. \end{array} \tag{3.6}$$

This yields an associated commutative diagram of split exact sequences of the first cohomology groups. On the other hand, the left vertical arrow in (3.6) is split surjective because $(x)/(x^{2n}) \cong \overset{2n-1}{\underset{i=1}{\oplus}} (x^i)/(x^{i+1})$. It follows that the map $H^1(nX_T, \mathcal{O}^{\times}_{nX_T}) \to H^1(X_T, \mathcal{O}^{\times}_{X_T})$ is split surjective. Using (3.5), we conclude that the map $K_0(nX_T) \to K_0(X_T)$ is surjective. We have thus proven the claim.

Since $\mathcal{O}_X \cong \mathcal{O}_X \otimes_k (x)/(x^2)$, it also follows from (3.5) and (3.6) that $K_0(X_T) \cong H^1(X, \mathcal{O}_X) \oplus K_0(X)$. Using this in (3.3), we get

$$K_{-1}(X_{\mathbb{A}^1_k}, X_T) \cong \mathrm{Coker}(\iota'^*) \cong \mathrm{Coker}(f'^*) \cong H^1(X, \mathcal{O}_X). \tag{3.7}$$

In the final step, we consider the commutative diagram of pro-abelian groups

$$\begin{array}{ccc} \{K_{-1}(X_C, nX)\} & \xrightarrow{\psi} & K_{-1}(X_C, X) \\ {\scriptstyle f^*}\downarrow & & \downarrow{\scriptstyle f^*} \\ \{K_{-1}(X_{\mathbb{A}^1_k}, nX_T)\} & \xrightarrow{\psi'} & K_{-1}(X_{\mathbb{A}^1_k}, X_T), \end{array} \tag{3.8}$$

where ψ is induced by the inclusion $(X_C, X) \hookrightarrow (X_C, nX)$. The arrow ψ' is similarly defined. By the pro-cdh-descent theorem [**22**, Theorem A] (note that [**24**, Theorem 1.1] suffices for the present case), applied to the abstract blow-up square on the right of (3.1), we see that the left vertical arrow in (3.8) is an isomorphism. The bottom horizontal arrow is surjective by the above claim. It follows that $f^* \circ \psi = \psi' \circ f^*$ is surjective.

We conclude that the map $f^* : K_{-1}(X_C, X) \to K_{-1}(X_{\mathbb{A}^1_k}, X_T)$ is surjective. We can now apply (3.7) to see that $K_{-1}(X_C, X)$ can not be zero if the genus of X is positive. The proof of the theorem is complete. $\square$

4. SK_0 of monoid algebras: some reductions

Our goal in the next two sections is to prove Theorems 0.5 and 0.7. In this section, we establish some reduction steps which go into the proof of Theorem 0.7.

4.1. Structure of positive cancellative semi-normal monoids. Let M be a cancellative monoid with possible torsion. We shall denote the torsion part of $\mathrm{gp}(M)$ by $t(M)$. We let $\overline{M} := \mathrm{Image}(M \to \mathrm{gp}(M)/t(M))$. There is an identification $\mathrm{gp}(M) = \mathrm{gp}(\overline{M}) \times t(M)$.

Suppose now that M is a cancellative torsion-free monoid and let $\mathrm{gp}(M) \simeq \mathbb{Z}^r$. Recall from [**31**, § 5] that the interior of M is its subset consisting of all elements $a \in M$ such that for all $b \in M$, there is an integer $n > 0$ and $c \in M$ such that $na = b + c$. We denote this set by $\mathrm{Int}(M)$. Note that if M is generated by a finite set $\{x_1, \dots, x_n\}$, then $\sum_{i=1}^{n} x_i \in \mathrm{Int}(M)$. In particular, $\mathrm{Int}(M) \neq \emptyset$.

Since $\mathrm{gp}(M) \simeq \mathbb{Z}^r$, we can view M as the set of of integral points in the vector space $\mathbb{R}^r$. We let $\mathbb{R}_+M$ denote the set of non-negative $\mathbb{R}$-linear combinations of elements in M. In this case, we have $\mathrm{Int}(M) = \mathrm{Int}(\mathbb{R}_+M) \cap \mathbb{Z}^r$, where $\mathrm{Int}(\mathbb{R}_+M)$ is the topological interior of the cone $\mathbb{R}_+M$.

If M is positive cancellative but not necessarily torsion-free, then $\overline{M}$ is a positive cancellative torsion-free monoid. We shall let $F(M) := \{F_0, \dots, F_r\}$ denote the set of faces of the cone $\mathbb{R}_+\overline{M}$ including 0 and $\mathbb{R}_+\overline{M}$. We index $F(M)$ so that $\dim(F_i) < \dim(F_j)$ implies $i < j$ for all $F_i, F_j \in F(M)$. Let $\mathrm{rk}(M) = m$ so that $\dim(\mathbb{R}_+\overline{M}) = m$. Given two monoids L, N, we define $L \curlywedge N := L \times N \setminus \{(0, n) | \ n \in N \setminus \{0\}\}$.

We shall use the following description of the semi-normalization of M, due to Gubeladze [**18**, Lemma 9.1].

LEMMA 4.1. *Let M be a positive cancellative monoid (possibly with torsion). For every $F \in F(M)$, there is a subgroup $T_F \subset t(M)$ such that*

(1) $sn(M) = \cup_{F(M)} (\mathrm{Int}(n(\overline{M} \cap F)) \times T_F)$.
(2) $T_{\mathbb{R}_+\overline{M}} = t(M)$.
(3) $T_{F_1} \subset T_{F_2}$ *for* $F_1 \subset F_2$.

Recall from [**31**, § 5] that a submonoid E of any monoid M is called extremal if it is non-empty and $a + b \in E \Rightarrow a, b \in E$ for all $a, b \in M$. The following is from [**31**, Theorem 5.4].

LEMMA 4.2. *Let M be a cancellative monoid. Then a submonoid E of M is extremal if and only if there is a monoid homomorphism $\phi : M \to \mathbb{N}$ such that $\phi^{-1}(0) = E$.*

We note down some properties of the monoid $\overline{M}$ in the following lemma for later use.

LEMMA 4.3. *Let M be a positive cancellative semi-normal monoid. Then the following hold.*

(1) *$\overline{M}$ is a positive cancellative torsion-free semi-normal monoid.*
(2) *Each $\overline{M} \cap F_i$ is an extremal submonoid of $\overline{M}$.*
(3) *Each $\overline{M} \cap F_i$ is a cancellative torsion-free semi-normal monoid.*
(4) *If N is a submonoid of $\overline{M}$ such that $\mathrm{Int}(N) \cap (\overline{M} \cap F_i)$ is non-empty, then $N \subset \overline{M} \cap F_i$.*

PROOF. We have already observed that $\overline{M}$ is a positive cancellative torsion-free monoid. To see semi-normality, let $a \in gp(\overline{M})$ such that $2a, 3a \in \overline{M}$. Let

$sn(M)$ and $sn(\overline{M})$ denote the semi-normalizations of M and $\overline{M}$, respectively. Using Lemma 4.1, we get a commutative diagram

$$\begin{array}{ccc} M & \to & sn(M) = \cup_{F(M)} \left(\mathrm{Int}(n(\overline{M} \cap F)) \times T_F\right) \\ \phi \downarrow & & \downarrow \psi \\ \overline{M} & \longrightarrow & sn(\overline{M}) = \cup_{F(M)} \left(\mathrm{Int}(n(\overline{M} \cap F)\right), \end{array} \tag{4.1}$$

where ψ is induced from the projection on each face. Clearly, ψ is a surjective map.

Since $\overline{M}$ is cancellative, the map $\overline{M} \to gp(\overline{M})$ is injective. Hence the map $\overline{M} \to sn(\overline{M})$ is injective. Since $a, 2a, 3a \in sn(\overline{M})$ (see § 1.1), we can lift them to $sn(M)$ using (4.1)) and use the fact that $M = sn(M)$ to conclude that $a \in \overline{M}$. Hence, we have $\overline{M} = sn(\overline{M})$.

The part (2) follows from [**2**, Remark 2.6(c), Exercise 2.3(a)] (see also [**31**, § 5, Remark]). To prove (3), we only need to show that $\overline{M} \cap F_i$ is semi-normal. For this, let $a \in gp(\overline{M} \cap F_i)$ such that $2a, 3a \in \overline{M} \cap F_i$. Since $\overline{M}$ is semi-normal, we must have $a \in \overline{M}$. Since F_i is an extremal submonoid of $\overline{M}$ by (2) and $a + a \in F_i$, we must have $a \in F_i$. This proves (3).

For (4), let $z \in N$ and $x \in \mathrm{Int}(N) \cap (\overline{M} \cap F_i)$. By definition of $\mathrm{Int}(N)$, there exists an integer $n > 0$ such that $nx = z + y$ for some $y \in N$. Since F_i is extremal and $z+y = nx \in F_i$, it follows that $z, y \in F_i$. In particular, $z \in N \cap F_i \subset \overline{M} \cap F_i$. □

LEMMA 4.4. *Let M be a positive cancellative semi-normal monoid. With the above notations, let $I_k := \cup_{j \ge k} \left(\mathrm{Int}(n(\overline{M} \cap F_j))\right)_* \times T_{F_j}$ be subsets of M for $0 \le k \le r$. Then I_k is an ideal of M for each k.*

PROOF. Since M is semi-normal, Lemma 4.3(1) implies that $\overline{M}$ is also semi-normal. It follows from Lemma 4.3(3) that each $\overline{M} \cap F_j$ is also a cancellative torsion-free semi-normal monoid. Therefore, we deduce from [**31**, Lemma 6.6] that $\mathrm{Int}(n(\overline{M} \cap F_j)) = \mathrm{Int}(\overline{M} \cap F_j)$.

We need to show that $I_k + M \subset I_k$ for each k. For this, we note that $(\mathrm{Int}(\overline{M} \cap F_j))_* \times T_{F_j} = (\mathrm{Int}(\overline{M} \cap F_j) \times T_{F_j})_*$ for each j. Furthermore, $\mathbb{R}_+\overline{M} \cap \overline{M} = \amalg_j \mathrm{Int}(\overline{M} \cap F_j)$ is a disjoint union (see [**31**, Lemma 5.3]). Hence, Lemma 4.1 implies that $M = \amalg_j \mathrm{Int}(\overline{M} \cap F_j) \times T_F$ is a disjoint union, where we identify $gp(M) = gp(\overline{M}) \times t(M)$ and look at M as a submonoid of $gp(M)$.

We now let $a \in I_k$ and $b \in M$. Then $a = (a_1, a_2) \in \mathrm{Int}(\overline{M} \cap F_j) \times T_{F_j}$ for some $j \ge k$ and $b = (b_1, b_2) \in \mathrm{Int}(\overline{M} \cap F_i) \times T_{F_i}$ for some $0 \le i \le r$. Then $a + b = (a_1 + b_1, a_2 + b_2) \in M$ and hence it must belong to $\mathrm{Int}(\overline{M} \cap F_l) \times T_{F_l}$ for some l. We have to show that $l \ge k$ to finish the proof.

To show this, we note that $\overline{M} \cap F_l$ is an extremal submonoid of $\overline{M}$. It follows from this that $a_1, b_1 \in \overline{M} \cap F_l$. In particular, we get $a_1 \in \mathrm{Int}(\overline{M} \cap F_j) \cap \mathrm{Int}(\overline{M} \cap F_l)$. Lemma 4.3(4) therefore implies that $\overline{M} \cap F_j \subseteq \overline{M} \cap F_l$. Since each face of $\mathbb{R}_+\overline{M}$ is its intersection with finitely many hyperplanes, each of which must either be non-negative or non-positive on $\mathbb{R}_+\overline{M}$, it follows that $F_j \subseteq F_l$. This in turn implies that $l \ge j$. As $j \ge k$, we get $l \ge k$, as desired. □

4.2. Milnor square associated to positive torsion monoids. Let R be a ring and M a positive cancellative seminormal monoid as in § 4.1. We let $A_k =$

$R[M]/I_kR[M] = R[M/I_k]$ for $0 \leq k \leq r$. There is a sequence of surjective R-algebra homomorphisms $A_{r+1} := R[M] \twoheadrightarrow A_r \twoheadrightarrow \cdots \twoheadrightarrow A_k \twoheadrightarrow A_{k-1} \twoheadrightarrow \cdots \twoheadrightarrow A_1 \xrightarrow{\cong} A_0 = R$. Let $\phi_k : A_k \to A_{k-1}$ be the quotient map for $1 \leq k \leq r+1$.

LEMMA 4.5. *The following hold for* $1 \leq k \leq r+1$.

(1) $\mathrm{Ker}(\phi_k) = R(\mathrm{Int}(\overline{M} \cap F_{k-1}) \times T_{F_{k-1}})$.

(2) *There is a Milnor square*

$$\begin{array}{ccc} R[(\mathrm{Int}(\overline{M} \cap F_{k-1}) \times T_{F_{k-1}})_*] & \longrightarrow & A_k \\ \downarrow & & \downarrow \phi_k \\ R & \xrightarrow{\delta_{k-1}} & A_{k-1}. \end{array} \tag{4.2}$$

PROOF. Let I_{k-1}/I_k be the image of I_{k-1} in M/I_k under the quotient map $M \to M/I_k$. Then it is clear from Lemma 4.4 that I_{k-1}/I_k is an ideal in M/I_k and the monoid M/I_{k-1} is obtained from M/I_k by collapsing I_k/I_{k-1}. It follows that

$$0 \to R[I_{k-1}/I_k] \to R[M/I_k] \xrightarrow{\phi_k} R[M/I_{k-1}] \to 0 \tag{4.3}$$

is an exact sequence of R-modules (see § 1.1). The first assertion now follows as $R[I_{k-1}/I_k] = R[\mathrm{Int}(\overline{M} \cap F_{k-1}) \times T_{F_{k-1}}]$.

To prove (2), we first note that δ_k is the canonical inclusion $R \hookrightarrow A_k$. Moreover, there is a commutative diagram

(4.4)

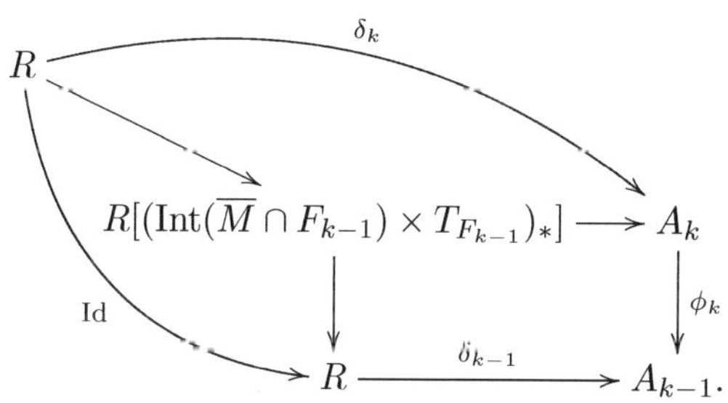

Using this diagram, we conclude immediately from (1) that (4.2) is Cartesian. Since ϕ_k is surjective, it follows that this is also a Milnor square. □

4.3. Reduction to positive semi-normal monoids. In this subsection, we shall prove some lemmas to reduce the proof of Theorem 0.7 to the case of positive semi-normal monoids. We begin by recalling the definition of SK_1 and SK_0 of rings.

Let R be a ring and let $H_0(R)$ denote the set of all continuous functions from Spec $(R) \to \mathbb{Z}$ with respect to the Zariski topology on Spec (R) and the discrete topology on $\mathbb{Z}$. It is easy to verify that this is a ring. There is a group homomorphism rk : $K_0(R) \to H_0(R)$ such that $\mathrm{rk}([P])(\mathfrak{p})$ is the rank of $P_{\mathfrak{p}}$ if P is a projective R-module. We define $\widetilde{K}_0(R) := \mathrm{Ker}(\mathrm{rk})$. There is a map det : $K_0(R) \to \mathrm{Pic}(R)$ which sends $[P]$ to $[\wedge^r(P)]$, where r is the rank of P. Its restriction yields a group homomorphism $\widetilde{\det} : \widetilde{K}_0(R) \to \mathrm{Pic}(R)$. We define $SK_0(R) := \mathrm{Ker}(\widetilde{\det})$. We let $SK_1(R) = SL(R)/E(R)$ so that there is a canonical decomposition $K_1(R) = SK_1(R) \oplus U(R)$. The following lemma is elementary.

LEMMA 4.6. $SK_0(R) = 0$ *if and only if for every projective R-module P, one has* $P \oplus R^s \cong \wedge^r(P) \oplus R^t$, *where* $r = \mathrm{rk}(P)$.

PROOF. Suppose first that $SK_0(R) = 0$. Let P be a projective R-module of rank $r \geq 1$. Then it is easy to see that $[P] = [\wedge^r P \oplus R^{r-1}]$ in $K_0(R)$. But this implies that $P \oplus R^s \cong \wedge^r P \oplus R^t$, as is well known. The converse is obvious. □

The following result which connects SK_1 with SK_0, is due to Bass [**1**, Corollary 5.12].

PROPOSITION 4.7. *If* (1.1) *is a Milnor square, then we have the* 6*-term exact sequence*

$$SK_1(A_1) \to SK_1(A_2) \oplus SK_1(B_1) \to SK_1(B_2) \to \\ SK_0(A_1) \to SK_0(A_2) \oplus SK_0(B_1) \to SK_0(B_2).$$

LEMMA 4.8. *Let R be an Artinian ring and M a cancellative torsion-free seminormal monoid. Then the following hold.*

(1) $SK_0(R[M]) = 0$.
(2) *If M is free and positive, then* $SK_1(R[M]) = 0$.

PROOF. By Lemma 1.4, we can assume that R is reduced. In this case, we have $R \cong k_1 \times \cdots \times k_m$, where each k_i is a field. The assertion (1) now follows [**15**, Theorem 1.3].

Similar to the case of $SK_0(R)$, we have $SK_1(R[M]) = SK_1(R_{\text{red}}[M])$ by [**1**, Chapter IX, Proposition 3.10, 3.11]). So we can assume R to be a field. Since $M \cong \mathbb{Z}_+^r$ by our assumption, we get $SK_1(R[M]) \cong SK_1(R) = 0$, where the first isomorphism is by the homotopy invariance. □

Recall from [**31**, § 14] that a ring extension $A \subset B$ is called an elementary subintegral extension if $B = A[x]$, where $x^2, x^3 \in A$. We say that $A \subset B$ is a subintegral extension if it is a filtered union of elementary subintegral extensions.

LEMMA 4.9. *Let $A \subset B$ be a subintegral extension. If $SK_0(B) = 0$, then $SK_0(A) = 0$.*

PROOF. By Lemma 4.6, we need to show that $P \oplus A^s \cong \wedge^r(P) \oplus A^t$ for any projective A-module P of rank $r \geq 1$. At any rate, our assumption and Lemma 4.6 together imply that $P_B := P \otimes_A B$ has the property that $P_B \oplus B^s \cong \wedge^r(P_B) \oplus B^t$. Equivalently, we have $(P \oplus A^s)_B \cong (\wedge^r(P) \oplus A^t)_B$. But this implies that $P \oplus A^s \cong \wedge^r(P) \oplus A^t$ by [**31**, Theorem 14.1]. □

LEMMA 4.10. *Let R be an Artinian ring and M any monoid. We let $N = M \setminus U(M)$, where $U(M)$ is the group of units of M. If $SK_0(R[N_*]) = 0$, then $SK_0(R[M]) = 0$.*

PROOF. Since M is finitely generated, we have $U(M) \simeq \mathbb{Z}^r \oplus G$ for some $r \geq 0$, where G is a finite abelian group. Under the hypothesis of the lemma, there is a Milnor square of rings as in (1.3) (see also [**31**, § 6]). Since $R[G]$ is an Artinian ring, we have $SK_0(R[U(M)]) \cong SK_0(R[G][\mathbb{Z}^r]) = 0$ by Lemma 4.8. As $SK_0(R[N_*]) = 0$, we conclude using Proposition 4.7 that $SK_0(R[M]) = 0$. □

5. SK_0 and the Levine–Weibel Chow group

In this final section, we shall first prove Theorem 0.7 and then deduce Theorem 0.5 using Theorem 0.7 and the affine Roitman torsion theorem [**25**] for the Levine-Weibel Chow group.

5.1. SK_0 of cancellative monoid algebras. We shall first prove our main result for the vanishing of SK_0 of monoids algebras when the underlying monoid is cancellative. More precisely, we prove:

LEMMA 5.1. *Let R be an Artinian ring and M a commutative cancellative monoid. Then $SK_0(R[M]) = 0$.*

PROOF. We can assume, using Lemmas 4.9 and 4.10, that M is semi-normal and positive. If we further assume that M is torsion-free, then $SK_0(R[M]) = 0$ by Lemma 4.8. So we can assume that M is positive, cancellative and semi-normal but not torsion-free.

We can write $M = sn(M) = \cup_{F(M)}(\mathrm{Int}(\overline{M} \cap F) \times T_F))$ by Lemma 4.1. Note that $\mathrm{Int}(n(\overline{M} \cap F)) = \mathrm{Int}(\overline{M} \cap F)$ by [**31**, Lemma 6.6]. We shall prove the lemma by applying Proposition 4.7 to (4.2). We note that $A_0 \cong A_1 \cong R$ and $A_{r+1} = R[M]$. Hence, it suffices to show using an induction argument that $SK_0(A_k) = 0$ for $1 \leq k \leq r+1$.

Since $SK_0(R) = 0$, the base case for induction is established. It suffices now to prove the lemma for A_k assuming it holds for A_{k-1} for $k \geq 2$. Using Lemma 4.5 and Proposition 4.7, it suffices to show that $SK_0(R[(\mathrm{Int}(\overline{M} \cap F_{k-1}) \times T_{F_{k-1}})_*]) = 0$.

Now, we observe that $R[(\mathrm{Int}(\overline{M} \cap F_{k-1}) \times T_{F_{k-1}})_*] = R[(\mathrm{Int}(\overline{M} \cap F_{k-1}))_* \leftthreetimes T_{F_{k-1}}]$. Letting $L = (\mathrm{Int}(\overline{M} \cap F_{k-1}))_*$, we see that L is a normal positive cancellative torsion-free monoid (see [**2**, Proposition 2.40]). Writing $F_{k-1} = \prod_{i=1}^{m} \mathbb{Z}/n_i$, it suffices to show by induction on m that

$$SK_0(R[L \leftthreetimes (\prod_{i=1}^{m} \mathbb{Z}/n_i)]) = 0. \tag{5.1}$$

If $m = 0$, then (5.1) is true by our assumption. In general, there is a Milnor square (see [**18**, Proof of Theorem 1.1, p. 214])

$$\begin{array}{ccc} \Lambda & \longrightarrow & R[L \leftthreetimes (\prod_{i=1}^{m} \mathbb{Z}/n_i)] \\ \downarrow & & \downarrow \\ R[\mathbb{Z}_+][L \leftthreetimes (\prod_{i=1}^{m-1} \mathbb{Z}/n_i)] & \stackrel{\pi}{\twoheadrightarrow} & R[\mathbb{Z}/n_m][L \leftthreetimes (\prod_{i=1}^{m-1} \mathbb{Z}/n_i)], \end{array} \tag{5.2}$$

where

(1) $\Lambda = A + B$,

(2) $A = R[\mathbb{Z}/n_m][(L \leftthreetimes \mathbb{Z}_+) \leftthreetimes (\prod_{i-1}^{m-1} \mathbb{Z}/n_i)]$,

(3) $B = R[t^{n_m}, t^{n_m+1} - t, \cdots, t^{2n_m-1} - t^{n_m-1}] \subset R[t] \simeq R[\mathbb{Z}_+]$ and

(4) π is induced by $t \mapsto x$ for some generator $x \in \mathbb{Z}/n_m$.

By induction on m and Proposition 4.7, it suffices to show that $SK_0(\Lambda) = 0$.

We now consider another Milnor square (see [**18**, Proof of Theorem 1.1])

$$\begin{array}{ccc} A & \longrightarrow & \Lambda \\ \downarrow & & \downarrow \\ R & \longrightarrow & B. \end{array}$$

Using this square and Proposition 4.7 again, it suffices to prove that $SK_0(A) = 0 = SK_0(B)$.

Since $L \rightthreetimes \mathbb{Z}_+$ is a finitely generated positive, cancellative torsion-free normal monoid, it follows by induction on m that $SK_0(A) = 0$. To prove the result for B, we can assume R is reduced by Lemma 1.4. We can further assume that R is a field. We now observe that there is a conductor square

$$\begin{array}{ccc} B & \longrightarrow & R[t] \\ \downarrow & & \downarrow \\ B/C & \longrightarrow & R[t]/C, \end{array}$$

where C is the conductor ideal of the extension $B \subset R[t]$. Being a subring of the integral domain $R[t]$, B is an integral domain. Since $t(t^{n_m} - 1) \in C$ is a non-zero divisor, we see that the height of C is positive. It follows that B/C and $R[t]/C$ are both Artinian rings. In particular, $SK_1(R[t]/C) = 0$. Since $SK_0(B/C) = SK_0(R[t]) = 0$, it follows from Proposition 4.7 that $SK_0(B) = 0$. □

5.2. The final step for Theorem 0.7. Let M be an arbitrary (finitely generated) monoid. Recall from [**31**, § 15] that an ideal $P \subset M$ is called prime if $P \neq M$ and if $x, y \in M$, $x + y \in P$ implies $x \in P$ or $y \in P$. This is equivalent to saying that $N = M \setminus P$ is a non-empty submonoid of M. In this case, there are monoid algebra morphisms $R[N] \to R[M] \to R[N] \cong R[M]/R[P]$ for any ring R whose composite is identity. It is also easy to check that if $\mathfrak{p}$ is a prime ideal of $R[M]$, them $\mathfrak{p} \cap M$ is a prime ideal of M.

An ideal $I \subset M$ is called a radical ideal if every element $x \in M$ with the property $nx \in I$ for some $n > 0$, belongs to I. If $I \subset M$ is an ideal, we let $\sqrt{I} = \{x \in M | nx \in I \text{ for some } n > 0\}$. We call this the radical of I. It is easy to check that $\sqrt{I}$ is a radical ideal of M. Our key step to finish the proof of Theorem 0.7 is the following result which generalizes [**31**, Lemma 15.6] to arbitrary monoids.

LEMMA 5.2. *Let $I \subset M$ be a proper radical ideal in a monoid M. Then there are prime ideals $\mathfrak{p}_1, \ldots, \mathfrak{p}_r$ in M such that $I = \bigcap_{i=1}^{r} \mathfrak{p}_i$.*

PROOF. In this proof, we shall use the multiplicative notation for the monoid operation on M. Since $\mathbb{Z}[I]$ is an ideal of the Noetherian ring $\mathbb{Z}[M]$ (note that M is finitely generated), we have the (irredundant) primary decomposition $\mathbb{Z}[I] = \bigcap_{i=1}^{r} Q_i$. We let P_i denote the unique associated prime of Q_i in $\mathbb{Z}[M]$. Recall that if we write the monoid operation of M multiplicatively, then there is a multiplicative monoid embedding $M \hookrightarrow \mathbb{Z}[M]$ which sends m to $(1 \cdot m)$. We let $\mathfrak{q}_i = Q_i \cap M$ and $\mathfrak{p}_i = P_i \cap M$ via this embedding. It is easy to check from the definition of $\mathbb{Z}[M]$ that $I = \mathbb{Z}[I] \cap M$. We therefore get $I = \mathbb{Z}[I] \cap M = \bigcap_{i=1}^{r} (Q_i \cap M) = \bigcap_{i=1}^{r} \mathfrak{q}_i$. In particular, we have $I = \bigcap_{i=1}^{r} \mathfrak{q}_i \subset \bigcap_{i=1}^{r} \mathfrak{p}_i$. Note that each $\mathfrak{p}_i$ is a prime ideal of M as we already observed earlier.

Suppose now that there is an element $x \in M$ which lies in $\bigcap_{i=1}^{r} \mathfrak{p}_i$. This implies that x lies in $\bigcap_{i=1}^{r} P_i$ inside $\mathbb{Z}[M]$. Since this intersection is same as $\sqrt{\mathbb{Z}[I]}$ in $\mathbb{Z}[M]$, it follows that $x^m \in \mathbb{Z}[I]$ for some $m > 0$. Since $x \in M$ and M is multiplicatively closed in $\mathbb{Z}[M]$, we also have $x^m \in M$. Consequently, we get $x^m \in \mathbb{Z}[I] \cap M = I$. But I is a radical ideal of M and hence we must have $x \in I$. We have therefore shown that $\bigcap_{i=1}^{r} \mathfrak{p}_i \subset I$. This proves the lemma. □

The following lemma is elementary.

LEMMA 5.3. *Let M be any monoid and $I \subset M$ any ideal. Then $R[\sqrt{I}] \subseteq \sqrt{R[I]}$ for any ring R.*

PROOF. As in the proof of Lemma 5.2, we shall use the multiplicative notation for the monoid structure of M. Let $u = a_1x_1 + \cdots + a_rx_r \in R[\sqrt{I}]$ with $a_i \in R$ and $x_i \in \sqrt{I}$ for each i. Since $I \subset M$ is an ideal, we can find $m_0 \gg 0$ such that $x_i^m \in I$ for all $m \geq m_0$ and all $i \geq 1$. It is then straightforward to check using the multinomial expansion of u^m that for all $m \geq rm_0$, all M-coefficients of u^m will lie in I. That is, if we write $u^m = b_1y_1 + \cdots + b_sy_s$, then each $y_i \in I$ for $1 \leq i \leq s$. But this implies that $u \in \sqrt{R[I]}$. □

Proof of Theorem 0.7: We can assume R to be reduced by Lemma 1.4. We can then further assume that R is a field. Let N be a cancellative monoid and $I \subset N$ an ideal such that $M = N/I$. If $I = N$, then $R[M] \cong R$ and the theorem is obvious. So we can assume that $I \subset N$ is a proper ideal. In particular, $I \cap U(N) = \emptyset$.

Since the image of $R[\sqrt{I}]$ in $R[M]$ is nilpotent by Lemma 5.3, we can assume that I is a radical ideal of N by Lemma 1.4. In this case, Lemma 5.2 says that we can write $I = \bigcap_{i=1}^{r} \mathfrak{p}_i$, where $\mathfrak{p}_1, \ldots, \mathfrak{p}_r$ are prime ideals in N.

We shall now prove the theorem by induction on $r \geq 1$. If $r = 1$, then I is a prime ideal of N so that $J = N \setminus I$ is a submonoid of N and $R[M] \cong R[J]$. Since N is cancellative, it follows that J is also cancellative. So we are done in this case by Lemma 5.1.

In general, we let $\mathfrak{p} = \mathfrak{p}_1$ and $\mathfrak{q} = \mathfrak{p}_2 \cap \cdots \cap \mathfrak{p}_r$ with $r \geq 2$. Then $I = \mathfrak{p} \cap \mathfrak{q}$ and $L = \mathfrak{p} \cup \mathfrak{q}$ is a proper ideal of N since $\mathfrak{p}, \mathfrak{q} \subset N \setminus U(N)$. Note here that the union of two ideals in a ring is generally not an ideal, but it is true for ideals in a monoid.

Since $\mathfrak{p}/I \xrightarrow{\cong} L/\mathfrak{q}$, it is easy to check using exact sequences of the type (4.3) that the diagram

$$\begin{array}{ccc} R[N]/R[I] & \to & R[N]/R[\mathfrak{q}] \\ \downarrow & & \downarrow \\ R[N]/R[\mathfrak{p}] & \to & R[N]/R[L] \end{array} \tag{5.3}$$

is Cartesian (see [**31**, Proof of Theorem 15.1]). Furthermore, if we let $N' = N \setminus \mathfrak{p}$ and $S = N' \cap L = N' \cap \mathfrak{q}$, then $R[N']/R[S] \xrightarrow{\cong} R[N]/R[L]$. On the other hand, the inclusion $N' \hookrightarrow N$ takes S into $\mathfrak{q}$ and induces a map $R[N']/R[S] \to R[N]/R[\mathfrak{q}]$. This shows that the right vertical arrow in (5.3) is a split surjection.

By induction on r, we see that $SK_0(R[N]/R[\mathfrak{p}]) = 0 = SK_0(R[N]/R[\mathfrak{q}])$. It follows from Proposition 4.7 that the sequence

$$SK_1(R[N]/R[\mathfrak{p}]) \oplus SK_1(R[N]/R[\mathfrak{q}]) \to SK_1(R[N]/R[L]) \to SK_0(R[N]/R[I]) \to 0$$

is exact. Since the right vertical arrow in (5.3) is a split surjective homomorphism of R-algebras, it follows that the first arrow from left in this exact sequence in surjective. But this implies that $SK_0(R[M]) = SK_0(R[N]/R[I]) = 0$. This finishes the proof of the theorem. □

5.3. The Levine-Weibel Chow group. Our goal in this subsection is to prove Theorem 0.5. As we explained in Introduction, our approach is to use the vanishing of SK_0 of the given monoid algebra and then use the affine Roitman torsion theorem for the Levine-Weibel Chow group from [**25**]. Before we give the details, we recall the definition of the Levine-Weibel Chow group from [**27**] for reader's reference.

Let k be an algebraically closed field of any characteristic. Let A be a finite type reduced k-algebra and let $X = \mathrm{Spec}\,(A)$ denote the spectrum of A. We shall say that a point $x \in X$ is regular if $\mathcal{O}_{X,x}$ is a regular local ring. We let $X_{\mathrm{reg}} \subset X$ denote the regular locus of X so that $x \in X_{\mathrm{reg}}$ if and only if it is a regular point. We let $X_{\mathrm{sing}} = X \setminus X_{\mathrm{reg}}$ denote the singular locus of X. A closed subscheme $C \subset X$ is called a Cartier curve if it is a scheme of pure dimension one such that the following hold.

(1) No irreducible component of C lies in X_{sing}.
(2) For every $x \in C \cap X_{\mathrm{sing}}$, the ideal $I_{C,x}$ of C in the local ring $\mathcal{O}_{X,x}$ is generated by a regular sequence.

For a Cartier curve C, let $k(C, X_{\mathrm{sing}})^{\times}$ denote the group of invertible elements in the ring of total quotients of C which are regular along $C \cap X_{\mathrm{sing}}$.

Let $\mathcal{Z}_0(X)$ denote the free abelian group on the set of regular closed points of X. Given a Cartier curve $C \subset X$ and $f \in k(C, X_{\mathrm{sing}})^{\times}$, we have the divisor $\mathrm{div}_C(f)$ of f in the sense of [**9**, § 1.2] (see also [**27**, § 1]). Since f is regular and invertible along X_{sing}, it follows that $\mathrm{div}_C(f) \in \mathcal{Z}_0(X)$. We let $\mathrm{CH}_0^{LW}(X)$ be the quotient of $\mathcal{Z}_0(X)$ by the subgroup $\mathcal{R}_0(X)$, generated by $\mathrm{div}_C(f)$, where C runs over all Cartier curves on X and $f \in k(C, X_{\mathrm{sing}})^{\times}$. We shall use the notations $\mathrm{CH}_0^{LW}(X)$ and $\mathrm{CH}_0^{LW}(A)$ interchangeably.

When X is regular, $\mathrm{CH}_0^{LW}(X)$ coincides with the classical Chow group of 0-cycles on X (see [**9**, Chapter 1]). This is however not the case when X has singularity. In this case, it is $\mathrm{CH}_0^{LW}(X)$ which is known to be the correct Chow group of 0-cycles and is supposed to constitute the 0-cycle part of the conjectural full theory of cohomological Chow groups of X. Furthermore, it is directly related to theory of vector bundles on X unlike the classical homological Chow group $\mathrm{CH}_0(X)$.

Since the structure sheaf of a regular closed point $x \in X$ has finite tor-dimension over X, it follows that this point has a class $cyc(x)$ in $K_0(X)$. We thus get a cycle class map $cyc : \mathcal{Z}_0(X) \to K_0(X)$. Furthermore, it is shown in [**27**, Proposition 2.1] that this map kills $\mathcal{R}_0(X)$ so that there is a well-defined cycle class map

$$cyc : \mathrm{CH}_0^{LW}(X) \to K_0(X). \tag{5.4}$$

We have the following result about this cycle class map which is supposed to be well known.

LEMMA 5.4. *Suppose that* $\dim(X) \geq 2$. *Then the cycle class map has the factorization*

$$cyc : \mathrm{CH}_0^{LW}(X) \to SK_0(X).$$

PROOF. Let $x \in X$ be a regular closed point and let $U = \mathrm{Spec}\,(X) \setminus \{x\}$. We then have the Thomason-Trobaugh localization exact sequence

$$K_0(\{x\}) \xrightarrow{\cong} K_0^{\{x\}}(X) \to K_0(X) \to K_0(U). \tag{5.5}$$

Since the image of the middle arrow is the subgroup generated by $cyc(x)$, it suffices to show that the maps $H^0(X, \mathbb{Z}) \to H^0(U, \mathbb{Z})$ and $\mathrm{Pic}(X) \to \mathrm{Pic}(U)$ are injective. The first assertion is obvious. So we need to prove the second assertion. For this, we let $S = \{x\}$. Using the isomorphism $H^1(X, \mathcal{O}_X^\times) \cong \mathrm{Pic}(X)$ and the exact sequence

$$H_S^1(X, \mathcal{O}_X^\times) \to H^1(X, \mathcal{O}_X^\times) \to H^1(U, \mathcal{O}_U^\times),$$

it suffices to show that $H_S^1(X, \mathcal{O}_X^\times) = 0$. Since x is a regular closed point, we have $H_S^1(X, \mathcal{O}_X^\times) \cong H_S^1(X_{\mathrm{reg}}, \mathcal{O}_{X_{\mathrm{reg}}}^\times)$ by excision. So we can assume X is regular.

In this case, we have a long exact sequence

$$H^0(X, \mathcal{O}_X^\times) \to H^0(U, \mathcal{O}_U^\times) \to H_S^1(X, \mathcal{O}_X^\times) \to H^1(X, \mathcal{O}_X^\times) \to H^1(U, \mathcal{O}_U^\times).$$

Since X is regular and codimension of S is at least two in X, it is well known that the map $H^i(X, \mathcal{O}_X^\times) \to H^i(U, \mathcal{O}_U^\times)$ is an isomorphism for $i \leq 1$. We are therefore done. □

The second key ingredient in the proof of Theorem 0.5 is the following affine Roitman torsion theorem for 0-cycles. This is an old conjecture of Murthy [**30**] and is now a theorem [**25**, Corollary 7.6].

THEOREM 5.5. *Let A be a reduced affine algebra over an algebraically closed field k. Then the cycle class map $cyc : \mathrm{CH}_0^{LW}(A) \to K_0(A)$ is injective.*

PROOF OF THEOREM 0.5. We let $X = \mathrm{Spec}\,(k[M])$. Using Theorem 0.7 and Lemma 5.4, it suffices to show that the map $cyc : \mathrm{CH}_0^{LW}(X) \to K_0(X)$ is injective. But this follows from Theorem 5.5. □

5.4. A different proof of Theorem 0.7 for pctf monoids. We end our discussion with another proof of Theorem 0.7 in the special case in which the underlying monoid is (partially cancellative and) torsion free. Note that Theorem 0.7 proves the vanishing of $SK_0(R[M])$ for all partially cancellative monoids which are not necessarily torsion-free. But we decided to include this different proof in the special case because it is more K-theoretic in nature and crucially uses negative K-theory. Our hope is that this K-theoretic approach may be helpful in future generalizations of Theorem 0.7 to more general monoid algebras.

We now begin the proof. We let M be a partially cancellative torsion-free monoid and R an Artinian ring. We want to show that $SK_0(R[M]) = 0$.

If M is cancellative, then the result follows from Lemma 5.1. We can therefore assume that M is torsion-free but only partially cancellative. As in the proof of Lemma 1.8, we can assume that $M = N/I$, where N is a cancellative torsion-free monoid. Associated to the Milnor-square (1.6), there exists a commutative diagram

$$\begin{array}{ccccc} SK_0(R) \oplus SK_0(R[N]) & \xrightarrow{\alpha} & SK_0(R[M]) & & \\ \downarrow & & \downarrow & & \\ K_0(R) \oplus K_0(R[N]) & \xrightarrow{\beta} & K_0(R[M]) & \to & K_{-1}(R[I_*]), \end{array} \tag{5.6}$$

where the bottom row is exact by Proposition 1.3. The vertical arrows are clearly injective. Here, the arrows α and β are the canonical maps.

Since R is Artinian and I_* is a cancellative torsion-free monoid, it follows from Lemmas 1.4 and 1.10 that the last term of the bottom exact sequence in (5.6) vanishes. A diagram chase shows that that $SK_0(R[M])$ lies in the image of β. Since $\widetilde{K}_0(R)$ is a canonical direct summand of $K_0(R)$, it follows that the image of $SK_0(R[M])$ in $\widetilde{K}_0(R[M])$ lies in the image of the map $\widetilde{K}_0(R) \oplus \widetilde{K}_0(R[N]) \to \widetilde{K}_0(R[M])$.

We now consider the commutative diagram

$$\begin{array}{ccccccc} & & SK_0(R) \oplus SK_0(R[N]) & \xrightarrow{\alpha} & SK_0(R[M]) & & \\ & & \downarrow & & \downarrow & & \\ \widetilde{K}_0(R[I_*]) & \longrightarrow & \widetilde{K}_0(R) \oplus \widetilde{K}_0(R[N]) & \xrightarrow{\beta} & \widetilde{K}_0(R[M]) & \longrightarrow & 0 \\ \downarrow & & \downarrow & & \downarrow & & \\ \mathrm{Pic}(R[I_*]) & \longrightarrow & \mathrm{Pic}(R) \oplus \mathrm{Pic}(R[N]) & \longrightarrow & \mathrm{Pic}(R[M]) & \longrightarrow & 0. \end{array} \tag{5.7}$$

If A is a ring and L is a projective A-module of rank one, then $[L]-[A] \in \widetilde{K}_0(A)$. Furthermore, the map det : $\widetilde{K}_0(A) \to \mathrm{Pic}(A)$ sends $[L] - [A]$ to $[L]$. It follows that the map $\widetilde{K}_0(A) \to \mathrm{Pic}(A)$ is surjective. We conclude that the lower vertical arrows in (5.7) are all surjective. Since the two columns on the right are exact and the two lower rows are also exact, a diagram chase shows that α is surjective. On the other hand, $SK_0(R) \oplus SK_0(R[N]) = 0$ by Lemma 5.1. It follows that $SK_0(R[M]) = 0$. This finishes the proof. □

Acknowledgments

This project was conceived and a part of the work was done when the authors were visiting Hausdorff Institute for Mathematics (HIM), Bonn as part of institute's K-theory program in 2017. The authors thank the institute and the organizers of the program for invitation and support. The second author would like to thank Tata Institute and University of Warwick for invitation and support. This paper was greatly inspired by several works of Gubeladze, Cortiñas, Haesemeyer, Walker and Weibel on K-theory of monoid algebras. The authors would like to thank the referee for carefully reading the paper and providing many valuable suggestions to improve its presentation.

References

[1] Hyman Bass, *Algebraic K-theory*, W. A. Benjamin, Inc., New York-Amsterdam, 1968. MR0249491

[2] Winfried Bruns and Joseph Gubeladze, *Polytopes, rings, and K-theory*, Springer Monographs in Mathematics, Springer, Dordrecht, 2009. MR2508056

[3] G. Cortiñas, C. Haesemeyer, M. Schlichting, and C. Weibel, *Cyclic homology, cdh-cohomology and negative K-theory*, Ann. of Math. (2) **167** (2008), no. 2, 549–573, DOI 10.4007/annals.2008.167.549. MR2415380

[4] G. Cortiñas, C. Haesemeyer, Mark E. Walker, and C. Weibel, *The K-theory of toric varieties*, Trans. Amer. Math. Soc. **361** (2009), no. 6, 3325–3341, DOI 10.1090/S0002-9947-08-04750-8. MR2485429

[5] G. Cortiñas, C. Haesemeyer, Mark E. Walker, and C. Weibel, *The K-theory of toric varieties in positive characteristic*, J. Topol. **7** (2014), no. 1, 247–286, DOI 10.1112/jtopol/jtt026. MR3180619

[6] Guillermo Cortiñas, Christian Haesemeyer, Mark E. Walker, and Charles Weibel, *Toric varieties, monoid schemes and cdh descent*, J. Reine Angew. Math. **698** (2015), 1–54, DOI 10.1515/crelle-2012-0123. MR3294649

[7] G. Cortiñas, C. Haesemeyer, M. E. Walker, and C. A. Weibel, *The K-theory of toric schemes over regular rings of mixed characteristic*, Singularities, algebraic geometry, commutative algebra, and related topics, Springer, Cham, 2018, pp. 455–479. MR3839807

[8] Vincent Cossart and Olivier Piltant, *Resolution of singularities of threefolds in positive characteristic. II*, J. Algebra **321** (2009), no. 7, 1836–1976, DOI 10.1016/j.jalgebra.2008.11.030. MR2494751

[9] William Fulton, *Intersection theory*, 2nd ed., Ergebnisse der Mathematik und ihrer Grenzgebiete. 3. Folge. A Series of Modern Surveys in Mathematics [Results in Mathematics and Related Areas. 3rd Series. A Series of Modern Surveys in Mathematics], vol. 2, Springer-Verlag, Berlin, 1998. MR1644323

[10] Thomas Geisser and Lars Hesselholt, *On the vanishing of negative K-groups*, Math. Ann. **348** (2010), no. 3, 707–736, DOI 10.1007/s00208-010-0500-z. MR2677901

[11] I. Dzh. Gubeladze, *The Anderson conjecture and a maximal class of monoids over which projective modules are free* (Russian), Mat. Sb. (N.S.) **135(177)** (1988), no. 2, 169–185, 271, DOI 10.1070/SM1989v063n01ABEH003266; English transl., Math. USSR-Sb. **63** (1989), no. 1, 165–180. MR937805

[12] Joseph Gubeladze, *Classical algebraic K-theory of monoid algebras*, K-theory and homological algebra (Tbilisi, 1987), Lecture Notes in Math., vol. 1437, Springer, Berlin, 1990, pp. 36–94, DOI 10.1007/BFb0086718. MR1079964

[13] J. Gubeladze, *Geometric and algebraic representations of commutative cancellative monoids* (English, with English and Georgian summaries), Proc. A. Razmadze Math. Inst. **113** (1995), 31–81. MR1414169

[14] Joseph Gubeladze, *Nontriviality of $SK_1(R[M])$*, J. Pure Appl. Algebra **104** (1995), no. 2, 169–190, DOI 10.1016/0022-4049(94)00130-1. MR1360174

[15] Joseph Gubeladze, *K-theory of affine toric varieties*, Homology Homotopy Appl. **1** (1999), 135–145, DOI 10.4310/hha.1999.v1.n1.a5. MR1693095

[16] Joseph Gubeladze, *The nilpotence conjecture in K-theory of toric varieties*, Invent. Math. **160** (2005), no. 1, 173–216, DOI 10.1007/s00222-004-0410-3. MR2129712

[17] Joseph Gubeladze, *Global coefficient ring in the nilpotence conjecture*, Proc. Amer. Math. Soc. **136** (2008), no. 2, 499–503, DOI 10.1090/S0002-9939-07-09106-X. MR2358489

[18] Joseph Gubeladze, *Unimodular rows over monoid rings*, Adv. Math. **337** (2018), 193–215, DOI 10.1016/j.aim.2018.08.011. MR3853049

[19] Christian Haesemeyer, *Descent properties of homotopy K-theory*, Duke Math. J. **125** (2004), no. 3, 589–620, DOI 10.1215/S0012-7094-04-12534-5. MR2166754

[20] Heisuke Hironaka, *Resolution of singularities of an algebraic variety over a field of characteristic zero. I, II*, Ann. of Math. (2) **79** (1964), 109–203; ibid. (2) **79** (1964), 205–326, DOI 10.2307/1970547. MR0199184

[21] Moritz Kerz and Florian Strunk, *On the vanishing of negative homotopy K-theory*, J. Pure Appl. Algebra **221** (2017), no. 7, 1641–1644, DOI 10.1016/j.jpaa.2016.12.021. MR3614971

[22] Moritz Kerz, Florian Strunk, and Georg Tamme, *Algebraic K-theory and descent for blow-ups*, Invent. Math. **211** (2018), no. 2, 523–577, DOI 10.1007/s00222-017-0752-2. MR3748313

[23] Amalendu Krishna, *On the negative K-theory of schemes in finite characteristic*, J. Algebra **322** (2009), no. 6, 2118–2130, DOI 10.1016/j.jalgebra.2009.05.038. MR2542834

[24] Amalendu Krishna, *An Artin-Rees theorem in K-theory and applications to zero cycles*, J. Algebraic Geom. **19** (2010), no. 3, 555–598, DOI 10.1090/S1056-3911-09-00521-9. MR2629600

[25] Amalendu Krishna, *Murthy's conjecture on 0-cycles*, Invent. Math. **217** (2019), no. 2, 549–602, DOI 10.1007/s00222-019-00871-8. MR3987177

[26] Amalendu Krishna and Husney Parvez Sarwar, *K-theory of Monoid Algebras and a Qestion of Gubeladze*, J. Inst. Math. Jussieu **18** (2019), no. 5, 1051–1085, DOI 10.1017/s1474748017000317. MR3995720

[27] Marc Levine and Chuck Weibel, *Zero cycles and complete intersections on singular varieties*, J. Reine Angew. Math. **359** (1985), 106–120. MR794801

[28] Hideyuki Matsumura, *Commutative ring theory*, Cambridge Studies in Advanced Mathematics, vol. 8, Cambridge University Press, Cambridge, 1986. Translated from the Japanese by M. Reid. MR879273

[29] John Milnor, *Introduction to algebraic K-theory*, Princeton University Press, Princeton, N.J.; University of Tokyo Press, Tokyo, 1971. Annals of Mathematics Studies, No. 72. MR0349811

[30] M. Pavaman Murthy, *Zero cycles and projective modules*, Ann. of Math. (2) **140** (1994), no. 2, 405–434, DOI 10.2307/2118605. MR1298718

[31] Richard G. Swan, *Gubeladze's proof of Anderson's conjecture*, Azumaya algebras, actions, and modules (Bloomington, IN, 1990), Contemp. Math., vol. 124, Amer. Math. Soc., Providence, RI, 1992, pp. 215–250, DOI 10.1090/conm/124/1144038. MR1144038

[32] Michael Temkin, *Desingularization of quasi-excellent schemes in characteristic zero*, Adv. Math. **219** (2008), no. 2, 488–522, DOI 10.1016/j.aim.2008.05.006. MR2435647

[33] R. W. Thomason and Thomas Trobaugh, *Higher algebraic K-theory of schemes and of derived categories*, The Grothendieck Festschrift, Vol. III, Progr. Math., vol. 88, Birkhäuser Boston, Boston, MA, 1990, pp. 247–435, DOI 10.1007/978-0-8176-4576-2_10. MR1106918

[34] L. N. Vaseršteĭn, *On the stabilization of the general linear group over a ring*, Math. USSR-Sb. **8** (1969), 383–400. MR0267009

[35] Ton Vorst, *The Serre problem for discrete Hodge algebras*, Math. Z. **184** (1983), no. 3, 425–433, DOI 10.1007/BF01163515. MR716288

[36] Charles A. Weibel, *K-theory and analytic isomorphisms*, Invent. Math. **61** (1980), no. 2, 177–197, DOI 10.1007/BF01390120. MR590161

[37] Charles A. Weibel, *Homotopy algebraic K-theory*, Algebraic K-theory and algebraic number theory (Honolulu, HI, 1987), Contemp. Math., vol. 83, Amer. Math. Soc., Providence, RI, 1989, pp. 461–488, DOI 10.1090/conm/083/991991. MR991991

[38] Charles Weibel, *The negative K-theory of normal surfaces*, Duke Math. J. **108** (2001), no. 1, 1–35, DOI 10.1215/S0012-7094-01-10811-9. MR1831819

[39] Charles A. Weibel, *The K-book*, Graduate Studies in Mathematics, vol. 145, American Mathematical Society, Providence, RI, 2013. An introduction to algebraic K-theory. MR3076731

School of Mathematics, Tata Institute of Fundamental Research, 1 Homi Bhabha Road, Colaba, Mumbai, India

Email address: amal@math.tifr.res.in

School of Mathematics, Tata Institute of Fundamental Research, 1 Homi Bhabha Road, Colaba, Mumbai, India

Email address: mathparvez@gmail.com

Contemporary Mathematics
Volume **749**, 2020
https://doi.org/10.1090/conm/749/15074

The principal-symbol index map for an algebra of pseudodifferential operators

Severino T. Melo

This paper is dedicated to the memory of Heinz-Otto Cordes (1925-2018)

ABSTRACT. Let B be a compact Riemannian manifold, let Ω denote the cylinder $\mathbb{R} \times B$, Δ_Ω its Laplace operator and $\Lambda = (1 - \Delta_\Omega)^{-1/2}$. Let $\mathfrak{A}$ denote the C*-algebra of bounded operators on $L^2(\mathbb{R} \times B)$ generated by all the classical pseudodifferential operators on $\mathbb{R} \times B$ of the form $L\Lambda^N$, N a nonnegative integer and L an N-th order differential operator whose (local) coefficients approach 2π-periodic functions at $+\infty$ and $-\infty$. Let $\mathfrak{E}$ denote the kernel of the continuous extension of the principal symbol to $\mathfrak{A}$.

The problem of computing the K-theory index map $\delta_1(K_1(\mathfrak{A}/\mathfrak{E})) \to K_0(\mathfrak{E}) \simeq \mathbb{Z}^2$ on an element of $K_1(\mathfrak{A}/\mathfrak{E})$ is reduced to the problem of computing the Fredholm indices of two elliptic operators on the compact manifold $S^1 \times B$.

For $B = S^1$, Hess went further and proved in her thesis that $K_0(\mathfrak{A}) \cong \mathbb{Z}^5$ and $K_1(\mathfrak{A}) \simeq \mathbb{Z}^4$

Introduction

More than fifty years after the Atiyah-Singer index theorem, people are still looking for generalizations of their theorem for classes of non-compact manifolds, for manifolds with boundary, corners or edges, or for manifolds with singularities in a more general sense. There are, in fact, a great variety of different classes of problems, and of different approaches or tecnhiques that could be used.

One of the approaches is to consider C*algebras generated by zero-order pseudodiferential operators which contain an *order-reducing device.* That means that there are linear isomorphisms between Sobolev spaces so that certain classes of differential operators multiplied on the right by them fall into the C*algebra. In the case of a compact manifold, the kernel of the continuous extension to the C*-algebra of the principal symbol map coincides with the ideal of compact operators, and the Fredholm index can be regarded as the K-theory index map associated to the exact sequence (1) below. For a noncompact manifold, one might argue that the relevant index is no longer the Fredholm index, but rather the K-theory index map taking values in K_0 of the kernel of the principal symbol.

2010 *Mathematics Subject Classification.* Primary 46L80, 19K56, 47G30.
Key words and phrases. K-Theory, C*-algebras, Index Theory, Pseudodifferential Operators.

Monthubert and Nistor followed that track in [**24, 25**], using algebras of pseudodifferential operators on groupoids which relate to (regular) algebras of pseudodifferential operators on a noncompact manifold. In their case, the relevant compactification of the original manifold is itself a manifold with corners. In this paper we consider an algebra of pseudodifferential operators on a cylinder $\mathbb{R} \times B$, where B is compact, which includes operators with periodic symbols. In our case the relevant compactification of the manifold is not even locally path connected.

The K-groups of our algebra, for the special cases of B equals a point and of B equals S^1, have been computed in [**23**] and [**13**], respectively.

The algebra considered here is an example of a *comparison algebra*, in the sense of Cordes, who pioneered [**4–6, 9, 10**] the use of Banach-algebra techniques in the study of the spectral theory of partial differential operators and singular integral operators on noncompact manifolds. Those algebras were defined in [**8**] as C*-algebras of bounded operators on $L^2(M)$ for a noncompact manifold M equipped with a smooth positive measure and a positive second-order elliptic differential operator H. Typically, M is equipped with a riemannian metric and its induced measure, and H is equal to the identity minus the Laplacian. One defines $\Lambda = H^{-1/2}$, where H also denotes the Friedrichs extension of the original operator defined on $C_c^\infty(M)$, which of course coincides with the unique self-adjoint extension in case the operator is essentially self-adjoint. Given a class of smooth functions $\mathcal{A}^\sharp$ and a class of first-order differential operators $\mathcal{D}^\sharp$ on M satisfying certain conditions, a comparison algebra is defined as the C*-subalgebra of the algebra of bounded operators on $L^2(M)$ generated by all operators of multiplication by $a \in \mathcal{A}^\sharp$ and by all $D\Lambda$, $D \in \mathcal{A}^\sharp$. In his monograph [**8**], Cordes develops general techniques to describe the ideal structure of a comparison algebra and how to find differential operators *within reach* of the algebra, meaning differential operators which, composed with an isomorphism from L^2 to a suitable Sobolev space, belong to the algebra.

In case the manifold is compact, there is only one possible choice of classes: $\mathcal{A}^\sharp = C^\infty(M)$ and $\mathcal{D}^\sharp$ equals to the set of all first order differential operators with smooth coefficients. In this case, the resulting comparison algebra coincides [**21**] with the C*algebra generated by all zero-order classical pseudodifferential operators on M, which we will denote by $\Psi(M)$. The principal symbol of operators, a priori a *-homomorphism defined on the subalgebra of classical pseudodifferential operators and taking values in smooth functions on the co-sphere bundle S^*M, extends to a surjective C*-algebra homomorphism $\sigma_M : \Psi(M) \to C(S^*M)$ with kernel equal to the ideal of compact operators $\mathfrak{K}_M$. In other words, we have the short exact sequence of C*-algebras

$$0 \longrightarrow \mathfrak{K}_M \hookrightarrow \Psi(M) \xrightarrow{\sigma_M} C(S^*(M)) \longrightarrow 0. \tag{1}$$

In Cordes' approach, the map σ_M arises as the Gelfand map for the commutative C*algebra $\Psi(M)/\mathfrak{K}_M$. Classically, (1) follows from the equality between the norm, modulo compact operators, of a singular integral operator on a compact manifold and the supremum norm of its symbol, proven by Gohberg [**15**] and Seeley [**29**]. Proofs of that result in the language of pseudodifferential operators appeared in [**16, 18**].

1. Preliminaries

In this section, we define the C*-algebra $\mathfrak{A}$ and review results about its structure that will be needed in the next and final section.

Let B denote a compact Riemannian manifold of dimension n without boundary. Consider the cylinder $\Omega = \mathbb{R} \times B$ and denote by $\Delta_\Omega = \Delta_\mathbb{R} + \Delta_B$ its Laplacian (we use the classical analysis convention sign for the Laplacian, $-\Delta_\Omega \geq 0$) Define $\mathfrak{A}$ as the C*-subalgebra of $\mathfrak{B}(L^2(\Omega))$ (we denote by $\mathfrak{B}(\mathcal{H})$ the C*-algebra of all bounded operators on a Hilbert space $\mathcal{H}$) generated by:

(i) all $A_1 = a(M_x)$, operators of multiplication by $a \in C^\infty(B)$,
$$[a(M_x)f](t,x) = a(x)f(t,x), \quad (t,x) \in \mathbb{R} \times B;$$

(ii) all $A_2 = b(M_t)$, operators of multiplication by $b \in C([-\infty, +\infty])$ (this denotes the set of all continuous functions on $\mathbb{R}$ which have limits at $-\infty$ and $+\infty$),
$$[b(M_t)f](t,x) = b(t)f(t,x), \quad (t,x) \in \mathbb{R} \times B;$$

(iii) all operators $A_3 = e^{ikM_t}$, $k \in \mathbb{Z}$, defined by
$$[e^{ikM_t} f](t,x) = e^{ikt} f(t,x), \quad (t,x) \in \mathbb{R} \times B;$$

(iv) $A_4 = \Lambda := (1 - \Delta_\Omega)^{-1/2}$;

(v) $A_5 = \frac{1}{i}\frac{\partial}{\partial t}\Lambda$;

(vi) $D_x\Lambda$, where $D_x = -i\sum_{k=1}^n c^k(x)\partial_{x_k}$ is a smooth vector field on B.

Since Ω is a complete manifold, its Laplacian, defined on $C_c^\infty(\Omega)$, has a unique extension (also denoted by Δ_Ω) as an unbounded self-adjoint operator on $L^2(\Omega)$ ([**14**], [**8**, Theorem IV.1.8]) and one can then define Λ by functional calculus . The operators in (v) and (vi), a priori defined on the dense subset $\Lambda^{-1}(C_c^\infty(\Omega))$, have continuous extensions to $L^2(\Omega)$ [**8**, Section V.1].

Moreover, if L denotes an M-th order differential operator on Ω whose (smooth) coefficients approach 2π-periodic functions as t tends to $-\infty$ and as t tends to $+\infty$, the operator $L\Lambda^M$ can be extended to a bounded operator on $L^2(\Omega)$, which belongs to $\mathfrak{A}$. These $L\Lambda^M$ are zero-order *classical* pseudodifferential operators (meaning that their local symbols have asymptotic expansions in homogeneous components).

Applying Cordes' comparison algebra techniques and using results from [**7**, **11**], the statements about the structure of $\mathfrak{A}$ contained in the rest of this Section were proven in [**20**] (see also [**19**]).

Let $\mathfrak{E}$ denote the commutator ideal of $\mathfrak{A}$, i.e, the smallest closed ideal containing all commutators $AB - BA$, A and B in $\mathfrak{A}$. Let $\overline{\Omega}$ denote the compactfication $[-\infty, +\infty] \times B$ of Ω, and let S^*M denote the cosphere bundle of any manifold M. We may then write
$$S^*\overline{\Omega} = \{(t,x,\tau,\xi);\ (t,\tau) \in T^*[-\infty,+\infty],\ (x,\xi) \in T^*B,\ |\tau|^2 + |\xi|^2 = 1\}.$$

THEOREM 1. *The spectrum of the commutative C^*-algebra $\mathfrak{A}/\mathfrak{E}$ is homeomorphic to the following subset of $(S^*\overline{\Omega}) \times S^1$:*
$$\mathbf{M} = \{(t,x,\tau,\xi,e^{i\theta});\ \theta \in \mathbb{R} \textit{ and } \theta = t \textit{ if } |t| < \infty\}.$$

Denoting by

$$\sigma : \mathfrak{A} \to C(\mathbf{M}) \tag{2}$$

the composition of the Gelfand isomorphism $\mathfrak{A}/\mathfrak{E} \simeq C(\mathbf{M})$ *with the projection* $\mathfrak{A} \to \mathfrak{A}/\mathfrak{E}$*, for each of the generators of* $\mathfrak{A}$ *listed above,* $\sigma(A_k)(t, x, \tau, \xi, e^{i\theta})$, $k = 1, \cdots, 6$, *is equal to, respectively,*

$$a(x),\ b(t),\ e^{ij\theta},\ 0,\ \tau,\ \sum_{k=1}^{n} c^j(x)\xi_j. \tag{3}$$

Since M is a compactification of $S^*\Omega$, we may regard $C(\mathbf{M})$ as a subspace of $C(S^*\Omega)$. Theorem 1 can then be rephrased by saying that the usual principal symbol of classical pseudodifferential operators, defined on a dense subalgebra of $\mathfrak{A}$, has a continuous extension σ to $\mathfrak{A}$ whose image is equal to $C(\mathbf{M})$ and whose kernel is $\mathfrak{E}$.

Let $F_d : L^2(S^1) \to \ell^2(\mathbb{Z})$ denote the discrete Fourier transform

$$(F_d u)_j = \int_0^1 u(e^{2\pi i\varphi})e^{-2\pi ij\varphi}d\varphi\ ,\ u \in L^2(\mathbb{S}^1) \text{ and } j \in \mathbb{Z}$$

(S^1 is equipped with the Lebesgue measure pushforwarded by $[0, 1) \ni \varphi \mapsto e^{2\pi i\varphi} \in S^1$; hence F_d is unitary). For each sequence $(b_j)_{j\in\mathbb{Z}}$ with limits as $j \to -\infty$ and as $j \to +\infty$, we define bounded operators $b(M_j)$ on $\ell^2(\mathbb{Z})$ and $b(D_\varphi)$ on $L^2(S^1)$ by

$$[b(M_j)u]((u_j)_{j\in\mathbb{Z}}) = (b_j u_j)_{j\in\mathbb{Z}} \quad \text{and} \quad b(D_\varphi) = F_d^{-1} b(M_j) F_d.$$

Given $a \in C(S^1)$, let $a(M_\varphi) \in \mathfrak{B}(L^2(S^1))$ denote the operator of multiplication by a. Finally let $\mathfrak{S}$ denote the C*-subalgebra of $\mathcal{L}(L^2(S^1))$ generated by the operators of the form $a(M_\varphi)$ and $b(D_\varphi)$, for $a \in C(S^1)$ and $b = (b_j)_{j\in\mathbb{Z}}$ a sequence as above.

In Cordes' language, $\mathfrak{S}$ is the only comparison algebra over S^1. The principal symbol, defined on the dense *-subalgebra of all classical zero-order pseudos, extends to a surjective C*-algebra homomorphism (see the Introduction)

$$\sigma_S : \mathfrak{S} \to C(S^1 \times \{-1, +1\}), \tag{4}$$

whose kernel is equal to the set $\mathfrak{K}_{S^1}$ of all compact operators on $L^2(S^1)$. On our set of generators, one has $[\sigma(a(M_\varphi))](e^{2\pi i\varphi}, \pm 1) = a(e^{2\pi i\varphi})$, for $a \in C(S^1)$, and $[\sigma(b(D_\varphi))](e^{2\pi i\varphi}, \pm 1) = b(\pm\infty)$, for $b = (b_j)_{j\in\mathbb{Z}}$ a sequence having limits as $j \to +\infty$ and as $j \to -\infty$. A detailed exposition of some of these facts can be found in [**1**].

For each $\varphi \in \mathbb{R}$, let U_φ be the operator on $L^2(S^1)$ given by $U_\varphi f(z) = z^{-\varphi} f(z)$, $z \in S^1$ (we take the principal branch of $z^{-\varphi} = e^{-\varphi \log z}$, $z \neq 0$). Let $Y_\varphi := F_d U_\varphi F_d^{-1}$. This defines a smooth family $\varphi \mapsto Y_\varphi$ of unitary operators on $\ell^2(\mathbb{Z})$ such that for all $k \in \mathbb{Z}$, $(Y_k u)_j = u_{j+k}$, and $Y_\varphi Y_\omega = Y_{\varphi+\omega}$. Given $u \in L^2(\mathbb{R})$, the sequence $\tilde{u}(\varphi) = (u(\varphi - j)_{j\in\mathbb{Z}})$ belongs to $\ell^2(\mathbb{Z})$ for almost every $\varphi \in [0, 1)$ and the map

$$\begin{array}{rcl} W : L^2(\mathbb{R}) & \longrightarrow & L^2(S^1) \otimes \ell^2(\mathbb{Z}) \simeq L^2(S^1; \ell^2(\mathbb{Z})) \\ u & \longmapsto & Wu, \qquad (Wu)(e^{2\pi i\varphi}) = Y_\varphi \tilde{u}(\varphi) \end{array}$$

is a Hilbert space isomorphism.

We have used $\otimes$ to denote Hilbert-space tensor product. Given $A \in \mathfrak{B}(\mathcal{H})$ and $B \in \mathfrak{B}(\mathcal{K})$, we also denote by $A \otimes B$ the bounded operator on $\mathcal{H} \otimes \mathcal{K}$ defined by $(A \otimes B)(u \otimes v) = Au \otimes Bv$. And also, given C*-algebras $\mathfrak{A}_H \subset \mathfrak{B}(\mathcal{H})$ and $\mathfrak{A}_K \subset \mathfrak{B}(\mathcal{K})$, we denote by $\mathfrak{A}_H \otimes \mathfrak{A}_K \subset \mathfrak{B}(\mathcal{H} \otimes \mathcal{K})$ the closed linear span of all $A \otimes B$, $A \in \mathfrak{A}_H$ and $B \in \mathfrak{A}_K$.

We will denote by $\mathfrak{K}_X$ denote the sets of all compact operators on $L^2(X)$, for $X = \Omega$, S^1, $\mathbb{Z}$, B, $\mathbb{Z} \times B$, $S^1 \times B$. We will also write W and F meaning $W \otimes I_B$

and $F \otimes I_B$, where I_B denotes the identity map on $L^2(B)$. Similarly we will also denote by σ_S the map $\sigma_S \otimes I_{\mathfrak{K}_\mathbb{Z}} \otimes I_{\mathfrak{K}_B}$ defined on $\mathfrak{S} \otimes \mathfrak{K}_\mathbb{Z} \otimes \mathfrak{K}_B$.

The *-isomorphism $\mathfrak{B}(L^2(\Omega)) \ni A \longmapsto WF^{-1}AFW^{-1} \in \mathfrak{B}(L^2(S^1) \otimes \ell^2(\mathbb{Z}) \otimes L^2(B))$ restricted to $\mathfrak{E}$ defines a C*-isomorphism

$$\mathfrak{E} \overset{\cong}{\longrightarrow} \mathfrak{S} \otimes \mathfrak{K}_\mathbb{Z} \otimes \mathfrak{K}_B. \tag{5}$$

The kernel of the surjective map $\sigma_S : \mathfrak{S} \otimes \mathfrak{K}_\mathbb{Z} \otimes \mathfrak{K}_B \to C(S^1 \times \{-1, +1\}) \otimes \mathfrak{K}_\mathbb{Z} \otimes \mathfrak{K}_B$ is equal to $\mathfrak{K}_{S^1} \otimes \mathfrak{K}_\mathbb{Z} \otimes \mathfrak{K}_B \cong \mathfrak{K}_\Omega$ and hence σ_S induces a C*-algebra isomorphism

$$\mathfrak{E}/\mathfrak{K}_\Omega \overset{\cong}{\longrightarrow} C(S^1 \times \{-1, +1\}) \otimes \mathfrak{K}_\mathbb{Z} \otimes \mathfrak{K}_B \;\cong\; C(S^1 \times \{-1, +1\};\, \mathfrak{K}_\mathbb{Z} \otimes \mathfrak{K}_B). \tag{6}$$

The composition of this isomorphism with the canonical projection $\mathfrak{E} \to \mathfrak{E}/\mathfrak{K}_\Omega$ defines a C*-algebra homomorphism

$$\gamma : \mathfrak{E} \longrightarrow C(S^1 \times \{-1, +1\};\, \mathfrak{K}(L^2(\mathbb{Z} \times B))), \tag{7}$$

which can be extended, via the left regular representation, to a map from $\mathfrak{A}$ to the algebra of bounded operators on the Banach space $C(S^1 \times \{-1, +1\};\, \mathfrak{K}(L^2(\mathbb{Z} \times B))$. The image of this Banach-algebra homomorphism is contained in $C(S^1 \times \{-1, +1\};\, \mathfrak{B}(L^2(\mathbb{Z} \times B))$ (an $f \in C(S^1 \times \{-1, +1\}; \mathfrak{B}(L^2(\mathbb{Z} \times B))$ is here regarded as a multiplication operator on $C(S^1 \times \{-1, +1\}; \mathfrak{K}(L^2(\mathbb{Z} \times B)))$ and this defines a C*-algebra homomorphism, which we also denote by γ,

$$\gamma : \mathfrak{A} \longrightarrow C(S^1 \times \{-1, +1\};\, \mathfrak{B}(L^2(\mathbb{Z} \times B))). \tag{8}$$

We define yet another *-homomorphism

$$\Gamma : \mathfrak{A} \longrightarrow C(S^1 \times \{-1, +1\};\, \mathfrak{B}(L^2(S^1 \times B))) \tag{9}$$

$$\Gamma_A(e^{2\pi i\varphi}, \pm 1) = F_d^{-1}\Gamma_A(e^{2\pi i\varphi}, \pm 1)F_d, \quad e^{2\pi i\varphi} \in S^1.$$

The image of Γ is contained in $C(S^1 \times \{-1, +1\};\, \Psi(S^1 \times B))$, with $\Psi(S^1 \times B)$ as defined at the end of the Introduction. It follows almost by definition that the image of $\mathfrak{E}$ under Γ equals $C(S^1 \times \{-1, +1\};\, \mathfrak{K}_{S^1 \times B})$.

In order to give a more explicit description of Γ, we need two more definitions. For each $\varphi \in \mathbb{R}$, let us denote by $e^{i\varphi M_\theta}$ the multiplication operator (by a possibly discontinuous function)

$$[e^{i\varphi M_\theta}u](e^{i\theta}) = e^{i\varphi\theta}u(e^{i\theta}), \quad -\pi < \theta \le \pi, \quad u \in L^2(S^1).$$

And let us denote by D_θ the densely defined operator on $L^2(S^1)$

$$(D_\theta u)(e^{i\theta}) = -i\tilde{u}'(\theta), \quad \tilde{u}(\theta) = u(e^{i\theta}), \quad \theta \in \mathbb{R}, \quad u \in C^\infty(S^1).$$

For each of the generators of $\mathfrak{A}$ defined on page 227, the value of $\Gamma_{A_k}(e^{2\pi i\varphi}, \pm 1)$, $1 \le k \le 6$, equals

$$\begin{aligned} &a(M_x), \quad b(\pm\infty)I, \quad e^{ijM_\theta}, \quad e^{-i\varphi M_\theta}[1 + (D_\theta - \varphi)^2 - \Delta_B]^{-1/2}e^{i\varphi M_\theta}, \\ &e^{-i\varphi M_\theta}(D_\theta - \varphi)[1 + (D_\theta - \varphi)^2 - \Delta_B]^{-1/2}e^{i\varphi M_\theta}, \\ &e^{-i\varphi M_\theta}D_x[1 + (D_\theta - \varphi)^2 - \Delta_B]^{-1/2}e^{i\varphi M_\theta}, \end{aligned} \tag{10}$$

respectively (the first three classes of generators are mapped to constant functions).

For any positive integer k, the homomorphisms σ and Γ defined in (2) and (9) canonically induce homomorphisms, which we also denote by σ and Γ, on the algebra $M_k(\mathfrak{A})$ of k-by-k matrices with entries in $\mathfrak{A}$. Regarding $M_k(\mathfrak{A})$ as a C*-subalgebra of $\mathfrak{B}(L^2(\Omega)^k)$, it turns out that an $A \in M_k(\mathfrak{A})$ is a Fredholm operator if and only if $\sigma(A)$ and $\Gamma(A)$ are invertible [**20**, Theorem 3.2]. Moreover, the quotient of $M_k(\mathfrak{A})$ by the ideal of compact operators on $L^2(\Omega)^k$ is isomorphic to the image of the *total symbol*

$$\sigma \oplus \Gamma : M_k(\mathfrak{A}) \longrightarrow M_k(C(\mathbf{M})) \oplus M_k(C(S^1 \times \{-1,+1\}; \Psi(S^1 \times B))). \tag{11}$$

2. The principal-symbol exact sequence

In this section we analyse the K-theory index map associated to the exact sequence of C*-algebras

$$0 \longrightarrow \mathfrak{E} \hookrightarrow \mathfrak{A} \xrightarrow{\sigma} C(\mathbf{M}) \longrightarrow 0, \tag{12}$$

induced by the continuous extension to $\mathfrak{A}$ of the usual principal symbol of pseudodifferential operators (see Theorem 1).

We first consider the exact sequence of C*-algebras associated to the map σ_S defined in (4),

$$0 \longrightarrow \mathfrak{K}_{S^1} \hookrightarrow \mathfrak{S} \xrightarrow{\sigma_S} C(S^1 \times \{-1,+1\}) \longrightarrow 0. \tag{13}$$

This sequence tells us that an $A \in \mathfrak{S}$ is a Fredholm operator if and only if $\sigma_S(A)$ never vanishes. A well-known index formula for zero-order pseudodifferential operators on the circle, which essentially goes back to Noether [**26**] (see also the example right after Theorem 19.2.4 in [**17**]), implies that the index of a Fredholm operator $A \in \mathfrak{S}$ is equal to the difference of the winding numbers of the restrictions of $\sigma_S(A)$ to the two copies of S^1 in $S^1 \times \{-1,+1\}$. A detailed proof of this fact without explicit mention of pseudodifferential or singular integral operators can be found in [**1**, Seção 10]. It then follows that there is a Fredholm operator of index 1 in $\mathfrak{S}$ and, therefore, the index map δ_1 in the standard exact sequence of K-groups associated to (13)

$$\begin{array}{ccccc}
\mathbb{Z} \cong K_0(\mathfrak{K}_{S^1}) & \longrightarrow & K_0(\mathfrak{S}) & \longrightarrow & K_0(C(S^1 \times \{-1,+1\})) \\
\delta_1 \uparrow & & & & \downarrow \delta_0 \\
K_1(C(S^1 \times \{-1,+1\})) & \longleftarrow & K_1(\mathfrak{S}) & \longleftarrow & K_1(\mathfrak{K}_{S^1}) = 0
\end{array}$$

is surjective. We then get $K_0(\mathfrak{S}) \cong \mathbb{Z}^2$ and $K_1(\mathfrak{S}) \cong \mathbb{Z}$.

PROPOSITION 1. *The canonical projection* $\pi : \mathfrak{E} \to \mathfrak{E}/\mathfrak{K}_\Omega$ *induces a* K_0*-isomorphism.*

PROOF. The previous considerations and the isomorphisms (5) and (6) imply that

$$K_0(\mathfrak{E}) \cong K_0(\mathfrak{E}/\mathfrak{K}_\Omega) \cong K_1(\mathfrak{A}/\mathfrak{E}) \cong \mathbb{Z}^2 \quad \text{and} \quad K_1(\mathfrak{E}) \cong \mathbb{Z}.$$

The standard exact sequence of K-groups associated to

$$0 \longrightarrow \mathfrak{K}_\Omega \hookrightarrow \mathfrak{E} \xrightarrow{\pi} \frac{\mathfrak{E}}{\mathfrak{K}_\Omega} \longrightarrow 0$$

then becomes

$$\begin{array}{ccccc}
\mathbb{Z} & \longrightarrow & \mathbb{Z}^2 & \longrightarrow & \mathbb{Z}^2 \\
\delta_1 \uparrow & & & & \downarrow \delta_0 \\
\mathbb{Z}^2 & \longleftarrow & \mathbb{Z} & \longleftarrow & 0
\end{array} \tag{14}$$

The index map δ_1 is non-zero (otherwise there would be an exact sequence $0 \to \mathbb{Z} \to \mathbb{Z}^2 \to \mathbb{Z}^2 \to 0$). If it were not surjective, the image of the upper left homomorphism would be a finite non-trivial group, which could not be a subgroup of $\mathbb{Z}^2$. The exactness of the sequence then provides the desired result. □

The following rephrasing of the fact that δ_1 in (14) is surjective is perhaps interesting in itself.

COROLLARY 1. *There exists $k > 0$ and a Fredholm operator of index 1 in $M_k(\mathfrak{E}) \subset \mathfrak{B}(L^2(\Omega)^k)$.*

We will need to use Banach algebra K-theory [**3**], in which K_0 consists of formal differences of classes of idempotents, while the elements of K_1 are classes of invertibles. The following lemma was used before in [**13, 23, 27**], see [**23**, Lemma 1] for a proof.

LEMMA 1. *Let $\mathcal{A}$ be a unital Banach algebra, $\mathcal{J}$ a closed ideal of $\mathcal{A}$, and let $\delta_1 : K_1(\mathcal{A}/\mathcal{J}) \to K_0(\mathcal{J})$ denote the index map associated to the exact sequence $0 \to \mathcal{J} \to \mathcal{A} \to \mathcal{A}/\mathcal{J} \to 0$. If $u \in M_k(\mathcal{A}/\mathcal{J})$ is invertible, $\pi(a) = u$ and $\pi(b) = u^{-1}$, then*

$$\delta_1([u]_1) = \left[\begin{pmatrix} 2ab - (ab)^2 & a(2 - ba)(1 - ba) \\ (1 - ba)b & (1 - ba)^2 \end{pmatrix}\right]_0 - \left[\begin{pmatrix} 1 & 0 \\ 0 & 0 \end{pmatrix}\right]_0$$

Next we consider the exact sequence obtained from (12) quotienting $\mathfrak{E}$ and $\mathfrak{A}$ by the compact ideal,

$$0 \longrightarrow \frac{\mathfrak{E}}{\mathfrak{K}_\Omega} \longrightarrow \frac{\mathfrak{A}}{\mathfrak{K}_\Omega} \xrightarrow{\pi} \frac{\mathfrak{A}}{\mathfrak{E}} \longrightarrow 0. \tag{15}$$

The following Proposition describes the connecting map $\delta_1 : K_1(\mathfrak{A}/\mathfrak{E}) \to K_0(\mathfrak{E}/\mathfrak{K}_\Omega)$ in terms of the index of Fredholm operators on $L^2(S^1 \times B)$. We recall that each element of $K_1(\mathfrak{A}/\mathfrak{E})$ is of the form $[[A]_\mathfrak{E}]_1$, where $A \in M_k(\mathfrak{A})$ is invertible modulo $M_k(\mathfrak{E})$ for some positive integer k and $[A]_\mathfrak{E}$ denotes its class in the quotient $M_k(\mathfrak{A})/M_k(\mathfrak{E})$.

PROPOSITION 2. *Let $A \in M_k(\mathfrak{A})$ be such that $[A]_\mathfrak{E}$ is invertible in $M_k(\mathfrak{A}/\mathfrak{E})$ for some positive integer k. Then $\Gamma_A(z, -1)$ and $\Gamma_A(z, +1)$ are Fredholm operators on $L^2(S^1 \times B)^k$ for every $z \in S^1$ and*

$$\delta_1([[A]_\mathfrak{E}]_1) = (\operatorname{ind} \Gamma_A(1, -1)\, [E]_0, \operatorname{ind} \Gamma_A(1, +1)\, [E]_0) \tag{16}$$

where E is a rank one projection on $L^2(S^1 \times B)$ and ind *denotes the Fredholm index (It will become clear in the proof how the left side of* (16) *can be regarded as an element of $K_1(\mathfrak{E}/\mathfrak{K}_\Omega)$).*

PROOF. Recall that the image of $\mathfrak{E}$ under Γ equals $C(S^1 \times \{-1, +1\}; \mathfrak{K}_{S^1 \times B})$.

Let $B \in M_k(\mathfrak{A})$ be such that $I - AB$ and $I - BA$ belong to $M_k(\mathfrak{E})$. Then, $I - \Gamma_A\Gamma_B$ and $I - \Gamma_B\Gamma_A$ belong to $C(S^1 \times \{-1, +1\}; M_k(\mathfrak{K}_{S^1 \times B}))$. This means that, at each $(z, \pm 1)$ in $S^1 \times \{-1, +1\}$, $\Gamma_A(z, \pm 1)$ is invertible modulo compacts, and hence it is a Fredholm operator.

Let $a = [A]_\mathfrak{K}$ and $b = [B]_\mathfrak{K}$. By Lemma 1,

$$\delta_1([[A]_\mathfrak{E}]_1) = \left[\begin{pmatrix} 2ab - (ab)^2 & a(2 - ba)(1 - ba) \\ (1 - ba)b & (1 - ba)^2 \end{pmatrix}\right]_0 - \left[\begin{pmatrix} 1 & 0 \\ 0 & 0 \end{pmatrix}\right]_0 \in K_0(\mathfrak{E}/\mathfrak{K}_\Omega) \tag{17}$$

(we are regarding $\mathfrak{E}/\mathfrak{K}_\Omega$ as an ideal of $\mathfrak{A}/\mathfrak{K}_\Omega$). Using the isomorphism (6) and the definitions of γ and Γ that follow (6), we see that the map

$$\mathfrak{E}/\mathfrak{K}_\Omega \ni [T] \quad \longmapsto \quad \Gamma_T \in C(S^1 \times \{-1,+1\};\, \mathfrak{K}_{S^1\times B})$$

is an isomorphism. Using this isomorphism as an identification, we may replace a and b by Γ_A and Γ_B in (17) and get $\delta_1([[A]_{\mathfrak{E}}]_1) =$

$$\left[\begin{pmatrix} 2\Gamma_A\Gamma_B - (\Gamma_A\Gamma_B)^2 & \Gamma_A(2-\Gamma_B\Gamma_A)(1-\Gamma_B\Gamma_A) \\ (1-\Gamma_B\Gamma_A)\Gamma_B & (1-\Gamma_B\Gamma_A)^2 \end{pmatrix}\right]_0 - \left[\begin{pmatrix} 1 & 0 \\ 0 & 0 \end{pmatrix}\right]_0$$

$$\in K_0(C(S^1 \times \{-1,+1\};\, \mathfrak{K}_{S^1\times B})) \cong K_0(C(S^1;\, \mathfrak{K}_{S^1\times B})) \oplus K_0(C(S^1;\, \mathfrak{K}_{S^1\times B}))$$

Let us consider now the map $p : C(S^1;\, \mathfrak{K}_{S^1\times B}) \to \mathfrak{K}_{S^1\times B}$, $p(f) = f(1)$. The map $q : \mathfrak{K}_{S^1\times B} \to C(S^1;\, \mathfrak{K}_{S^1\times B})$ that regards a compact operator as a constant function is a right inverse for p. Moreover, the kernel of p is the suspension of $\mathfrak{K}$, and then its K_0-group is 0. All that imply that p induces a K_0-isomorphism. Using this isomorphism as an identification, we may write $\delta_1([[A]_{\mathfrak{E}}]_1)$ as the element of $K_0(\mathfrak{K}_{S^1\times B}) \oplus K_0(\mathfrak{K}_{S^1\times B})$

$$\left(\left[\begin{pmatrix} 2A^-B^- - (A^-B^-)^2 & A^-(2-B^-A^-)(1-B^-A^-) \\ (1-B^-A^-)B^- & (1-B^-A^-)^2 \end{pmatrix}\right]_0 - \left[\begin{pmatrix} 1 & 0 \\ 0 & 0 \end{pmatrix}\right]_0,\right.$$

$$\left.\left[\begin{pmatrix} 2A^+B^+ - (A^+B^+)^2 & A^+(2-B^+A^+)(1-B^+A^+) \\ (1-B^+A^+)B^+ & (1-B^+A^+)^2 \end{pmatrix}\right]_0 - \left[\begin{pmatrix} 1 & 0 \\ 0 & 0 \end{pmatrix}\right]_0\right),$$

where we have denoted $\Gamma_A(1,\pm 1)$ by $A^\pm$ and analogously for $\Gamma_B(1,\pm 1)$.

If we apply Lemma 1 to $\mathcal{A} = \mathcal{B}(L^2(S^1 \times B))$, $\mathcal{J} = \mathfrak{K}_{S^1\times B}$ and $u = [A^\pm]_{\mathfrak{K}}$, recalling that the K-theory index map is a generalization of the Fredholm index [**28**], we conclude that this element of $K_0(\mathfrak{K}_{S^1\times B}) \oplus K_0(\mathfrak{K}_{S^1\times B})$ that we have just written above is equal to the left side of (16). □

Theorem 2 below follows immediately from Proposition 1 and Proposition 2. To simplify notation, in its statment we regard as identifications the canonical group isomorphisms

$$K_0(C(S^1 \times \{-1,+1\}; \mathfrak{K}_{S^1\times B})) \cong K_0(C(S^1; \mathfrak{K}_{S^1\times B})) \oplus K_0(C(S^1; \mathfrak{K}_{S^1\times B})) \cong \mathbb{Z} \oplus \mathbb{Z}.$$

THEOREM 2. *Let $\delta_1' : K_1(\mathfrak{A}/\mathfrak{K}) \to K_0(\mathfrak{E})$ denote the index map associated to the exact sequence $0 \to \mathfrak{E} \to \mathfrak{A} \to \mathfrak{A}/\mathfrak{E} \to 0$, let $\Gamma_* : K_0(\mathfrak{E}) \to \mathbb{Z} \oplus \mathbb{Z}$ denote the isomorphism induced by the surjective map $\Gamma : \mathfrak{E} \to C(S^1 \times \{-1,+1\}; \mathfrak{K}_{S^1\times B})$, and denote $\delta_1 = \Gamma_* \circ \delta_1'$. Given $A \in M_k(\mathfrak{A})$ such that $[A]_{\mathfrak{E}}$ is invertible in $M_k(\mathfrak{A}/\mathfrak{E})$ for some positive integer k, then $A^- = \Gamma_A(z,-1)$ and $A^+ = \Gamma_A(z,+1)$ are Fredholm operators on $L^2(S^1 \times B)^k$ and*

$$\delta_1([[A]_{\mathfrak{E}}]_1) = (\mathsf{ind}A^-, \mathsf{ind}A^+) \tag{18}$$

where ind *denotes the Fredholm index.*

The final goal should be to describe the index map associated to (12) in purely topological terms. That can now be easily achieved if we invoke the Atiyah-Singer Theorem.

In the following we adopt topological K-theory definitions and notation of [**3**, Chapter I]. We denote, as usual, by T^*M, B^*M and S^*M, respectively, the tangent bundle, the bundle of unit closed balls in the tangent bundle, and the co-sphere bundle of any manifold M.

THEOREM 3. *Let δ_1 denote the index map associated to (12), and let $f \in M_k(C(\mathbf{M}))$ be invertible for some positive k. The restrictions f_- and f_+ of f to the points of $\mathbf{M}$ with, respectively, $t = -\infty$ and $t = +\infty$ are invertible elements of $M_k(C(S^*(S^1 \times B)))$. Let*

$$x_\pm = [(E^k, E^k, f_\pm)] \in K(B^*(S^1 \times B), S^*(S^1 \times B)) \cong K(T^*(S^1 \times B))$$

denote the classes of the triples $(E^k, E^k, f_\pm)$, where E^k denotes the trivial bundle $(S^1 \times B) \times \mathbb{C}^k$. Then

$$\delta^1([f]_1) = (\mathsf{ind}_t(x_+), \mathsf{ind}_t(x_-)) \in \mathbb{Z} \oplus \mathbb{Z},$$

where $ind_t : K(T^*(S^1 \times B)) \to \mathbb{Z}$ denotes the topological index of Atiyah and Singer [**2**].

PROOF. Let $A \in M_k(\mathfrak{A})$ be such that $\sigma(A) = f$. Define $A^\pm = \Gamma_A(1, \pm 1)$. We claim that A^- and A^+ belong to $M_k(\Psi(S^1 \times B))$ and $\sigma_{S^1 \times B}(A^\pm) = f_\pm$. By definition, $f_\pm = \sigma(A)|_{t=\pm\infty}$. The maps $\mathfrak{A} \ni A \mapsto \sigma(A)|_{t=\pm\infty}$, $\mathfrak{A} \ni A \mapsto \Gamma_A(1, \pm 1))|_{t=\pm\infty}$ and $\Psi(M) \ni T \mapsto \sigma_{S^1 \times B}(T)$ are C*-algebra homomorphisms. So, it is enough to verify our claim on a set of generators of $\mathfrak{A}$. For each of the generators $A_1, A_2, \cdots, A_6$ defined on page 227, the claim follows immediately from (10).

Our theorem would already follow from the Atiyah-Singer Theorem [**2**] if A^+ and A^- were classical pseudodiferential operators. For the general case, we need to show that the topological index of the class $[(E^k, E^k, \sigma(T))] \in K(T^*(S^1 \times B))$ is equal to the Fredholm index of any Fredholm operator $T \in \Psi(S^1 \times B)$ (the Atiyah-Singer Theorem gives that for those T that belong to the dense subalgebra of all classical pseudodifferential operators).

In the rest of this proof, we will replace $S^1 \times B$ by an arbitrary compact manifold M.

Let δ^{tbs} denote index mapping for the exact sequence of C*algebras

$$0 \longrightarrow C_0(T^*M) \longrightarrow C(B^*M) \longrightarrow C(S^*M) \longrightarrow 0$$

and let $\iota : K_0(C_0(T^*M)) \mapsto K(T^*M)$ be the canonical isomorphism between these two groups. We then have

$$\iota \circ \delta^{tbs}([g]_1) = [(E^k, E^k, g)], \quad \text{for any invertible } g \in M_k(C(S^*M)). \tag{19}$$

For a proof of (19) in a slightly more general setting, see [**22**, Proposition 15], for example.

Now let $T \in M_k(\Psi(M))$ be a Fredholm operator. Its Fredholm index $\mathsf{ind}(T)$ coincides with $\delta^M([\sigma_M(T)]_1)$, where δ^M denotes the index map of (1). Because every $f \in C^\infty(S^*M)$ is the symbol of a zero-order classical pseudodifferential operator, there exists such a pseudo T_0 with invertible symbol so close to T that $[\sigma(T)]_1 = [\sigma(T_0)]_1$. Then

$$\mathsf{ind}(T) = \delta^M([\sigma_M(T)]_1) = \delta^M([\sigma_M(T_0)]_1) = \mathsf{ind}_t([(E^k, E^k, \sigma_M(T_0))]) =$$

$$\mathsf{ind}_t(\iota \circ \delta^{tbs}([\sigma_M(T_0)]_1)) = \mathsf{ind}_t(\iota \circ \delta^{tbs}([\sigma_M(T)]_1)) = \mathsf{ind}_t([(E^k, E^k, \sigma_M(T_0))]).$$

The Atiyah-Singer Theorem was used in the third of the above equalities. Twice we used (19). □

EXAMPLE 1. Some statements made here about operators in $\mathfrak{A}$ are proven in [**20**], using results from [**8**].

Let $L : C^\infty(\Omega)^k \to C^\infty(\Omega)^k$ be a differential operator of the form

$$L = \sum_{j+|\alpha|\leq N} a_{j,\alpha}(t,x)\left(\frac{1}{i}\frac{\partial}{\partial t}\right)^j \left(\frac{1}{i}\frac{\partial}{\partial x}\right)^\alpha,$$

where each $a_{j,\alpha} \in M_k(C^\infty(\Omega))$ has support contained in $U \times \mathbb{R}$, U the domain of a chart in B. Moreover, let us assume that $a_{j,\alpha}$ is *semi-periodic* in the following sense. There exist 2π-periodic (in t) $a^\pm_{j,\alpha} \in M_k(C^\infty(\Omega))$ and, for a choice of real-valued $\chi^\pm \in C^\infty(\mathbb{R})$ such that $\chi^+(t) + \chi^-(t) = 1$ for all t and $\chi_\pm(t) = 0$ for $\mp t > 1$, there exists $a^0_{j,\alpha} \in M_k(C^\infty(\Omega))$ vanishing at infinity such that

$$a_{j,\alpha}(t,x) = \chi^+(t)a^+_{j,\alpha}(t,x) + \chi^-(t)a^-_{j,\alpha}(t,x) + a^0_{j,\alpha}(t,x), \quad \text{for all} \quad (t,x) \in \Omega.$$

The operator $A = L\Lambda^N$, a priori defined on the dense subspace $[\Lambda^{-N}(C^\infty_c(\Omega))]^k$ of $L^2(\Omega)^k$, extends to a bounded operator $A \in M_k(\mathfrak{A})$ ([**8**, Sections VII-3 and IX-3], [**20**, Theorem 3.7]). Each of the generators $A_1, \cdots, A_5$ defined on page 227 is a particular example of such an A, with $k = 1$ and N equal to 0 or 1. And because B has a finite atlas, each A_6 is a finite sum of such A's.

The symbol σ_A coincides with the continuous extension to the compactification $\mathbf{M}$ of $S^*\Omega$ of the principal symbol of L,

$$\sigma(A)(t,x,\tau,\xi) = \sum_{j+|\alpha|=N} a_{j,\alpha}(t,x)\tau^j\xi^\alpha, \quad (t,x,\tau,\xi) \in S^*\Omega$$

It follows that A is invertible modulo $\mathfrak{E}$ if and only if L is *uniformily elliptic* in the sense that

$$\inf\{|\sum_{j+|\alpha|=N} a_{j,\alpha}(t,x)\tau^j\xi^\alpha|,\ (t,x,\tau,\xi) \in S^*\Omega\} > 0.$$

We also have

$$\Gamma_A(1,\pm 1) = \sum_{j+|\alpha|=N} a^\pm_{j,\alpha}(t,\theta)\left(\frac{1}{i}\frac{\partial}{\partial \theta}\right)^j \left(\frac{1}{i}\frac{\partial}{\partial x}\right)^\alpha \Lambda^N$$

(θ denotes the argument of a point in S^1). These are zero-order elliptic classical pseudodifferential operators on $S^1 \times B$ whose principal symbols are

$$\sigma(A)|_{t=\pm\infty}(e^{i\theta},x,\tau,\xi) = \sum_{j+|\alpha|=N} a^\pm_{j,\alpha}(\theta,x)\tau^j\xi^\alpha, \quad (e^{i\theta},x,\tau,\xi) \in S^*(S^1 \times B).$$

Having smooth symbols, their indices (and then $\delta_1([[A]_\mathfrak{E}]_1)$) can be more explicitly calculated as the integral of certain diferential forms on $S^*(S^1 \times B)$.

With the aid of a computer algebra system, Patrícia Hess [**13**] was able to follow this path and, using Fedosov's formula [**12**, formula (1)], find the index of a 2-by-2 pseudodifferential operator on $S^1 \times S^1$. That result was used in her proof that, for the case $B = S^1$, $K_0(\mathfrak{A}) \cong \mathbb{Z}^5$ and $K_1(\mathfrak{A}) \cong \mathbb{Z}^4$.

References

[1] R. F. DE ANDRADE. O Teorema do Índice para o Círculo. Master's Thesis, Universidade de São Paulo, 2008. http://www.teses.usp.br/teses/disponiveis/45/45131/tde-16102008-174340/pt-br.php

[2] M. F. Atiyah and I. M. Singer, *The index of elliptic operators. I*, Ann. of Math. (2) **87** (1968), 484–530, DOI 10.2307/1970715. MR236950

[3] Bruce Blackadar, *K-theory for operator algebras*, 2nd ed., Mathematical Sciences Research Institute Publications, vol. 5, Cambridge University Press, Cambridge, 1998. MR1656031

[4] M. Breuer and H. O. Cordes, *On Banach algebras with σ-symbol*, J. Math. Mech. **13** (1964), 313–323. MR0159239

[5] H. O. Cordes, *The algebra of singular integral operators in R^n*, J. Math. Mech. **14** (1965), 1007–1032. MR0181911

[6] H. O. Cordes, *Elliptic pseudodifferential operators—an abstract theory*, Lecture Notes in Mathematics, vol. 756, Springer, Berlin, 1979. MR551619

[7] H. O. Cordes, *On the two-fold symbol chain of a C^*-algebra of singular integral operators on a polycylinder*, Rev. Mat. Iberoamericana **2** (1986), no. 1-2, 215–234, DOI 10.4171/RMI/31. MR864659

[8] H. O. Cordes, *Spectral theory of linear differential operators and comparison algebras*, London Mathematical Society Lecture Note Series, vol. 76, Cambridge University Press, Cambridge, 1987. MR890743

[9] H. O. Cordes and E. A. Herman, *Singular integral operators on a half-line*, Proc. Nat. Acad. Sci. U.S.A. **56** (1966), 1668–1673, DOI 10.1073/pnas.56.6.1668. MR213918

[10] H. O. Cordes and E. A. Herman, *Gel′fand theory of pseudo differential operators*, Amer. J. Math. **90** (1968), 681–717, DOI 10.2307/2373478. MR454743

[11] H. O. Cordes and S. T. Melo, *An algebra of singular integral operators with kernels of bounded oscillation, and application to periodic differential operators*, J. Differential Equations **75** (1988), no. 2, 216–238, DOI 10.1016/0022-0396(88)90137-4. MR961154

[12] B. V. Fedosov, *The index of an elliptic system on a manifold* (Russian), Funkcional. Anal. i Priložcn. **4** (1970), no. 4, 57 67. MR0285034

[13] P. HESS. K-teoria de operadores pseudodiferenciais com símbolos semiperiódicos no cilindro. PhD Thesis, Universidade de São Paulo, 2008. http://www.teses.usp.br/teses/disponiveis/45/45132/tde-02072009-045613/pt-br.php

[14] Matthew P. Gaffney, *A special Stokes's theorem for complete Riemannian manifolds*, Ann. of Math. (2) **60** (1954), 140–145, DOI 10.2307/1969703. MR62490

[15] I. C. Gohberg, *On the theory of multidimensional singular integral equations*, Soviet Math. Dokl. **1** (1960), 960–963. MR0124704

[16] Lars Hörmander, *Pseudo-differential operators and hypoelliptic equations*, Singular integrals (Proc. Sympos. Pure Math., Vol. X, Chicago, Ill., 1966), Amer. Math. Soc., Providence, R.I., 1967, pp. 138–183. MR0383152

[17] Lars Hörmander, *The analysis of linear partial differential operators. III*, Classics in Mathematics, Springer, Berlin, 2007. Pseudo-differential operators; Reprint of the 1994 edition. MR2304165

[18] J. J. Kohn and L. Nirenberg, *An algebra of pseudo-differential operators*, Comm. Pure Appl. Math. **18** (1965), 269–305, DOI 10.1002/cpa.3160180121. MR0176362

[19] Severino Toscano do Rego Melo, *Comparison algebras with periodic symbols*, ProQuest LLC, Ann Arbor, MI, 1988. Thesis (Ph.D.)–University of California, Berkeley. MR2636906

[20] Severino T. Melo, *A comparison algebra on a cylinder with semi-periodic multiplications*, Pacific J. Math. **146** (1990), no. 2, 281–304. MR1078383

[21] Severino T. Melo, *Norm closure of classical pseudodifferential operators does not contain Hörmander's class*, Geometric analysis of PDE and several complex variables, Contemp. Math., vol. 368, Amer. Math. Soc., Providence, RI, 2005, pp. 329–336, DOI 10.1090/conm/368/06789. MR2126480

[22] S. T. Melo, R. Nest, and E. Schrohe, *C^*-structure and K-theory of Boutet de Monvel's algebra*, J. Reine Angew. Math. **561** (2003), 145–175, DOI 10.1515/crll.2003.064. MR1998610

[23] Severino T. Melo and Cíntia C. Silva, *K-theory of pseudodifferential operators with semi-periodic symbols*, *K*-Theory **37** (2006), no. 3, 235–248, DOI 10.1007/s10977-006-0018-z. MR2273458

[24] Bertrand Monthubert and Victor Nistor, *The K-groups and the index theory of certain comparison C^*-algebras*, Noncommutative geometry and global analysis, Contemp. Math., vol. 546, Amer. Math. Soc., Providence, RI, 2011, pp. 213–224, DOI 10.1090/conm/546/10791. MR2815137

[25] Bertrand Monthubert and Victor Nistor, *A topological index theorem for manifolds with corners*, Compos. Math. **148** (2012), no. 2, 640–668, DOI 10.1112/S0010437X11005458. MR2904199

[26] Fritz Noether, *Über eine Klasse singulärer Integralgleichungen* (German), Math. Ann. **82** (1920), no. 1-2, 42–63, DOI 10.1007/BF01457974. MR1511970

[27] Victor Nistor, *Higher index theorems and the boundary map in cyclic cohomology*, Doc. Math. **2** (1997), 263–295. MR1480038

[28] M. Rørdam, F. Larsen, and N. Laustsen, *An introduction to K-theory for C^*-algebras*, London Mathematical Society Student Texts, vol. 49, Cambridge University Press, Cambridge, 2000. MR1783408

[29] R. T. Seeley, *Integro-differential operators on vector bundles*, Trans. Amer. Math. Soc. **117** (1965), 167–204, DOI 10.2307/1994203. MR173174

Instituto de Matemática e Estatística, Universidade de São Paulo, Rua do Matão 1010, 05508-090 São Paulo, Brazil

Email address: toscano@ime.usp.br

Contemporary Mathematics
Volume **749**, 2020
https://doi.org/10.1090/conm/749/15075

A more general method to classify up to equivariant KK-equivalence II: Computing obstruction classes

Ralf Meyer

ABSTRACT. We describe Universal Coefficient Theorems for the equivariant Kasparov theory for C*-algebras with an action of the group of integers or over a unique path space, using KK-valued invariants. We compare the resulting classification up to equivariant KK-equivalence with the recent classification theorem involving a K-theoretic invariant together with an obstruction class in a certain Ext^2-group and with the classification by filtrated K-theory. This is based on a general theorem that computes these obstruction classes.

1. Introduction

Objects in a triangulated category, such as equivariant KK-categories, may be classified up to isomorphism using a primary homological invariant and a secondary "obstruction class" provided they have a projective resolution of length 2 in a suitable sense. This method was applied in [**4**] to objects in the circle-equivariant KK-category $\mathfrak{KK}^{\mathbb{T}}$, the equivariant KK-category $\mathfrak{KK}^X$ for C*-algebras over a finite, unique path space X, and graph C*-algebras with finite ideal lattice. Here we compute the obstruction classes that occur in this classification. This makes the classification for objects of $\mathfrak{KK}^{\mathbb{T}}$ and $\mathfrak{KK}^X$ more explicit.

The result suggests, in fact, a different invariant for objects in these categories that is fine enough to admit a Universal Coefficient Theorem. The price to pay is that the invariant uses bivariant K-theory instead of ordinary K-theory. We explain this for the equivariant bootstrap class in the category $\mathfrak{KK}^{\mathbb{Z}}$; the latter is equivalent to $\mathfrak{KK}^{\mathbb{T}}$ by Baaj–Skandalis duality.

An object of $\mathfrak{KK}^{\mathbb{Z}}$ is a C*-algebra A with a $\mathbb{Z}$-action, which is generated by a single automorphism α. The most obvious homological invariant on $\mathfrak{KK}^{\mathbb{Z}}$ maps this to the K-theory $\mathrm{K}_*(A)$ with the module structure over the group ring $\mathbb{Z}[x, x^{-1}]$ of $\mathbb{Z}$ that is induced by α. Let $\mathfrak{A}$ be the category of countable, $\mathbb{Z}/2$-graded modules over $\mathbb{Z}[x, x^{-1}]$. This is a stable Abelian category. The K-theory described above defines a stable homological functor $F_{\mathrm{K}}^{\mathbb{Z}}\colon \mathfrak{KK}^{\mathbb{Z}} \to \mathfrak{A}$. The category $\mathfrak{A}$ has cohomological dimension 2, that is, any object has a projective resolution of length 2. Let $\mathfrak{A}\delta$ be the additive category of pairs (A, δ) with $A \in\in \mathfrak{A}$ and $\delta \in \mathrm{Ext}^2_{\mathfrak{A}}(\Sigma A, A)$; morphisms from (A, δ) to (A', δ') in $\mathfrak{A}\delta$ are morphisms f from A to A' in $\mathfrak{A}$ with $\delta' f = f\delta$. It is shown in [**4**] that isomorphism classes of objects in the bootstrap class in $\mathfrak{KK}^{\mathbb{Z}}$ are

2010 *Mathematics Subject Classification.* Primary 19K35; Secondary 18E30, 46L35.

Supported by the DFG grant "Classification of non-simple purely infinite C*-algebras."

in bijection with isomorphism classes of objects in $\mathfrak{A}\delta$. In particular, any object A of $\mathfrak{A}$ lifts to an object in the bootstrap class in $\mathfrak{KK}^{\mathbb{Z}}$. The lifting is, however, not unique: different liftings of A are in bijection with $\mathrm{Ext}^2_{\mathfrak{A}}(\Sigma A, A)$; here two liftings B_1, B_2 are identified if there is an isomorphism $B_1 \cong B_2$ that induces the identity map on $A = F^{\mathbb{Z}}_{\mathrm{K}}(B_1) = F^{\mathbb{Z}}_{\mathrm{K}}(B_2)$.

In this article, we compute the obstruction class of an object of $\mathfrak{KK}^{\mathbb{Z}}$ explicitly. This question remained open in [**4**]. Let (A, α) be a C*-algebra with an automorphism α as above. The exact sequence in the Universal Coefficient Theorem splits, so that

$$\mathrm{KK}_0(A, A) \cong \mathrm{Hom}\big(\mathrm{K}_*(A), \mathrm{K}_*(A)\big) \oplus \mathrm{Ext}\big(\mathrm{K}_{*+1}(A), \mathrm{K}_*(A)\big).$$

While the splitting above is not natural, it is canonical enough to associate well defined elements $\alpha_0 \in \mathrm{Hom}\big(\mathrm{K}_*(A), \mathrm{K}_*(A)\big)$ and $\alpha_1 \in \mathrm{Ext}\big(\mathrm{K}_{*+1}(A), \mathrm{K}_*(A)\big)$ to α. Here α_0 is the action on $\mathrm{K}_*(A)$ induced by α, which is part of the $\mathbb{Z}[x, x^{-1}]$-module structure on $F^{\mathbb{Z}}_{\mathrm{K}}(A)$. We show how to compute the obstruction class from α_1. Namely, it is $\mu^*(\alpha_1)$ for a canonical map

$$\mu^*\colon \mathrm{Ext}\big(\mathrm{K}_{*+1}(A), \mathrm{K}_*(A)\big) \to \mathrm{Ext}^2_{\mathbb{Z}[x,x^{-1}]}\big(\mathrm{K}_{*+1}(A), \mathrm{K}_*(A)\big).$$

Thus an object (A, α) of $\mathfrak{KK}^{\mathbb{Z}}$ is determined uniquely up to isomorphism by A as an object of $\mathfrak{KK}$ together with the class $[\alpha]$ of α in $\mathrm{KK}_0(A, A)$. We treat the pair $(A, [\alpha])$ as an object of a certain exact category $\mathfrak{KK}[\mathbb{Z}]$ with a suspension automorphism. The forgetful functor $F^{\mathbb{Z}}\colon \mathfrak{KK}^{\mathbb{Z}} \to \mathfrak{KK}[\mathbb{Z}]$ is a stable homological functor. The fact that it gives a complete invariant for objects of $\mathfrak{KK}^{\mathbb{Z}}$ follows from a Universal Coefficient Theorem computing $\mathrm{KK}^{\mathbb{Z}}_*(A, B)$. This Universal Coefficient Theorem is already proven in [**11**], and it is shown there to be equivalent to a Pimsner–Voiculescu like exact sequence for $\mathrm{KK}^{\mathbb{Z}}_*(A, B)$.

The Universal Coefficient Theorem for the invariant $F^{\mathbb{Z}}\colon \mathfrak{KK}^{\mathbb{Z}} \to \mathfrak{KK}[\mathbb{Z}]$ is based on projective resolutions of length 1 for the stable homological ideal $\ker F^{\mathbb{Z}}$. We use these resolutions to compute the obstruction class related to the homological invariant $F^{\mathbb{Z}}_{\mathrm{K}}(A) := \mathrm{K}_*(A)$. Namely, let $0 \to B_1 \to B_0 \to A \to 0$ be a projective resolution of length 1 for the ideal $\ker F^{\mathbb{Z}}$ in $\mathfrak{KK}^{\mathbb{Z}}$. This remains exact with respect to the larger stable homological ideal $\ker F^{\mathbb{Z}}_{\mathrm{K}}$. The objects B_1 and B_0 are no longer projective for $\ker F^{\mathbb{Z}}_{\mathrm{K}}$, but they have projective resolutions of length 1. Thus $\mathrm{KK}^{\mathbb{Z}}_0(B_1, B_0)$ is computed by a Universal Coefficient Theorem, which splits non-naturally. So the arrow $B_1 \to B_0$ gives a class in $\mathrm{Ext}^1_{\mathbb{Z}[x,x^{-1}]}\big(\mathrm{K}_{*+1}(B_1), \mathrm{K}_*(B_0)\big)$. It must determine the obstruction class of A because A is $\mathrm{KK}^{\mathbb{Z}}$-equivalent to the cone of the map $B_1 \to B_0$. The result is the formula for the obstruction class of A asserted above. This computation of obstruction classes is carried out in Section 3 in the same generality as in [**4**], using a triangulated category with a stable homological ideal with enough projectives.

The Universal Coefficient Theorem for $\mathfrak{KK}^{\mathbb{Z}}$ is developed in Section 2.1. This is followed by an analogous treatment for the category of C*-algebras over a unique path space X in Section 2.2. In Section 4, we compare the classification theorems that our two invariants give for objects in the bootstrap classes in $\mathfrak{KK}^{\mathbb{Z}}$ and $\mathfrak{KK}^X$. In Section 5, we study $\mathfrak{KK}^X$ in the special case where X is totally ordered. Then filtrated K-theory is a third complete invariant for the same objects (see [**12**]), and we compare this invariant with the new ones. This is a rather long homological computation. In Section 5.2, the result of Section 5 is formulated more nicely if X has only two points, so that C*-algebras over X are extensions of C*-algebras.

2. Universal coefficient theorems with KK-valued functors

This section develops Universal Coefficient Theorems for the equivariant Kasparov categories $\mathfrak{KK}^{\mathbb{Z}}$ and $\mathfrak{KK}^X$ for a unique path space X, which are based on forgetful functors to $\mathfrak{KK}$. We shall use the general framework developed in [**11**] for doing relative homological algebra in triangulated categories. The starting point for this is a stable homological ideal $\mathfrak{I}$ in a triangulated category $\mathfrak{T}$. Being stable and homological means that there is a stable homological functor $F\colon \mathfrak{T} \to \mathfrak{A}$ to some stable Abelian category $\mathfrak{A}$ with

$$\mathfrak{I}(A,B) = \ker F(A,B) := \{g \in \mathfrak{T}(A,B) : F(g) = 0\}.$$

Here stability of $\mathfrak{A}$ and F means that $\mathfrak{A}$ is equipped with a suspension automorphism and that $F \circ \Sigma_{\mathfrak{T}} = \Sigma_{\mathfrak{A}} \circ F$.

A stable homological ideal $\mathfrak{I}$ allows to carry over various notions from homological algebra to $\mathfrak{T}$. In particular, there are $\mathfrak{I}$-exact chain complexes, $\mathfrak{I}$-projective objects and $\mathfrak{I}$-projective resolutions in $\mathfrak{T}$, which then allow to define $\mathfrak{I}$-derived functors. We shall only be interested in cases where there are enough $\mathfrak{I}$-projective objects. The thick subcategory $\langle \mathfrak{P}_{\mathfrak{I}} \rangle$ generated by the $\mathfrak{I}$-projective objects is an analogue of the “bootstrap class” in Kasparov theory. If $A \in\in \langle \mathfrak{P}_{\mathfrak{I}} \rangle$ has an $\mathfrak{I}$-projective resolution of length 1, then the graded group $\mathfrak{T}_*(A,B)$ for $B \in\in \mathfrak{T}$ may be computed by a Universal Coefficient Theorem (UCT). The Hom and Ext groups in this UCT are those in a certain Abelian category, namely, the target category of the universal $\mathfrak{I}$-exact stable homological functor F^u. The functor F^u strongly classifies objects of $\langle \mathfrak{P}_{\mathfrak{I}} \rangle$ with an $\mathfrak{I}$-projective resolution of length 1. That is, any isomorphism $F^u(A) \cong F^u(B)$ between such objects is induced by an isomorphism $A \cong B$ in $\mathfrak{T}$. In many cases of interest, the universal functor F^u is quite explicit. The following theorem allows to recognise it.

THEOREM 2.1 ([**11**, Theorem 57]). *Let $\mathfrak{T}$ be a triangulated category, let $\mathfrak{I} \subseteq \mathfrak{T}$ be a stable homological ideal, and let $F\colon \mathfrak{T} \to \mathfrak{A}$ be an $\mathfrak{I}$-exact stable homological functor into a stable Abelian category $\mathfrak{A}$. Let $\mathfrak{P}\mathfrak{A}$ be the class of projective objects in $\mathfrak{A}$. Suppose that idempotent morphisms in $\mathfrak{T}$ split. The functor F is the universal $\mathfrak{I}$-exact stable homological functor to a stable Abelian category and there are enough $\mathfrak{I}$-projective objects in $\mathfrak{T}$ if and only if*

(1) *$\mathfrak{A}$ has enough projective objects;*
(2) *the adjoint functor $F^{\vdash}$ of F is defined on $\mathfrak{P}\mathfrak{A}$;*
(3) *$F \circ F^{\vdash}(A) \cong A$ for all $A \in\in \mathfrak{P}\mathfrak{A}$.*

The ideals to be treated in this section are defined as the kernel on morphisms of a triangulated functor $F\colon \mathfrak{T} \to \mathfrak{S}$ to another triangulated category $\mathfrak{S}$. Ideals of this form are already treated in [**11**], using a construction of Freyd to embed $\mathfrak{S}$ into an Abelian category. Here we unify these two cases by allowing the functor F to take values in an *exact category* (see [**5, 7**]). An exact category is an additive category with a chosen class of “admissible” extensions. A typical example is a full, additive subcategory of an Abelian category that is closed under extensions. An Abelian category $\mathfrak{A}$ is exact where all extensions are admissible. And a triangulated category $\mathfrak{S}$ is exact where only split extensions are admissible. So exact categories contain the two cases treated in [**11**]. In many examples of classification results using a UCT, the range of the universal $\mathfrak{I}$-exact functor F^u_{old} is a certain exact subcategory of a module category, and only objects in this exact subcategory have projective resolutions of length 1. This holds for the UCTs in [**2, 3, 8, 12**] (see Remark 5.6 for one of these cases). The phenomenon is understood better in [**6**].

Quillen's Embedding Theorem allows to embed any exact category into an Abelian one in a fully faithful, fully exact way. So all results in the general theory carry over to the case where F takes values in an exact category. In particular, we may modify the definition of the universal $\mathfrak{I}$-exact stable homological functor by allowing exact categories as target. Let $F^u_{\text{old}}\colon \mathfrak{T} \to \mathfrak{A}$ be the universal $\mathfrak{I}$-exact functor into an Abelian category as in Theorem 2.1. Let $\mathfrak{E} \subseteq \mathfrak{A}$ be the exact subcategory of $\mathfrak{A}$ generated by the range of F^u_{old}. Let $F^u\colon \mathfrak{T} \to \mathfrak{E}$ be F^u_{old} viewed as a functor to $\mathfrak{E}$. This is the universal $\mathfrak{I}$-exact functor in the new sense. Its universal property follows from the one of F^u_{old} and Quillen's Embedding Theorem. If $\mathfrak{I}$ is defined by a forgetful functor $F\colon \mathfrak{T} \to \mathfrak{S}$ to another triangulated category, then the universal homological functor to an Abelian category typically uses Freyd's embedding of $\mathfrak{S}$ into an Abelian category. In contrast, we shall see that in many examples the universal $\ker F$-exact functor to an exact category gives an exact category that is very closely related to $\mathfrak{S}$.

Remark 2.2. Assume that all objects in $\mathfrak{T}$ have an $\mathfrak{I}$-projective resolution of finite length. Then the image of F^u_{old} is contained in the subcategory of objects in $\mathfrak{A}$ with a finite-length projective resolution. This subcategory is exact. And it is already generated as an exact category by the projective objects in $\mathfrak{A}$: this is proved by induction on the length of a resolution. Hence the image of F^u_{new} must be equal to the subcategory of objects with a finite-length projective resolution. So $F^u_{\text{new}} = F^u_{\text{old}}$ if and only if *all* objects of $\mathfrak{A}$ have a finite-length projective resolution.

2.1. Actions of the group of integers. We now treat a concrete example, namely, the case $\mathfrak{T} := \mathfrak{KK}^{\mathbb{Z}}$. This is a triangulated category. Its objects are pairs (A, α) with a separable C*-algebra A and $\alpha \in \operatorname{Aut}(A)$; the latter generates an action of $\mathbb{Z}$ on A by automorphisms. The arrows in $\mathfrak{KK}^{\mathbb{Z}}$ are the Kasparov groups $\mathrm{KK}^{\mathbb{Z}}_0(A, B)$, and the composition is the Kasparov product; we have dropped the automorphisms from our notation as usual to avoid clutter. The triangulated category structure on $\mathfrak{KK}^{\mathbb{Z}}$ is described in [**9**, Appendix]. The relative homological algebra in $\mathfrak{KK}^{\mathbb{Z}}$ is already studied in [**11**]. The main result is the following variant of the Pimsner–Voiculescu sequence for $\mathrm{KK}^{\mathbb{Z}}_*(A, B)$. Let A and B be C*-algebras with automorphisms α and β, respectively. Then there is an exact sequence

$$
\begin{array}{ccccc}
\mathrm{KK}_1(A,B) & \longrightarrow & \mathrm{KK}^{\mathbb{Z}}_0(A,B) & \xrightarrow{\text{forget}} & \mathrm{KK}_0(A,B) \\
{\scriptstyle \alpha^*\beta_*^{-1}-1}\uparrow & & & & \downarrow{\scriptstyle \alpha^*\beta_*^{-1}-1} \\
\mathrm{KK}_1(A,B) & \xleftarrow[\text{forget}]{} & \mathrm{KK}^{\mathbb{Z}}_1(A,B) & \longleftarrow & \mathrm{KK}_0(A,B)
\end{array}
\tag{2.3}
$$

(see [**11**, Section 5.1]); here $(\alpha^*\beta_*^{-1} - 1)(x) = \beta^{-1} \circ x \circ \alpha - x$ for all $x \in \mathrm{KK}_*(A, B)$. We may rewrite this as a pair of short exact sequences

$$
\begin{aligned}
\operatorname{coker}\bigl(\alpha^*\beta_*^{-1} - 1\colon \mathrm{KK}_{1+i}(A,B) \to \mathrm{KK}_{1+i}(A,B)\bigr) &\rightarrowtail \mathrm{KK}^{\mathbb{Z}}_i(A,B) \\
&\twoheadrightarrow \ker\bigl(\alpha^*\beta_*^{-1} - 1\colon \mathrm{KK}_i(A,B) \to \mathrm{KK}_i(A,B)\bigr)
\end{aligned}
$$

for $i = 0, 1$. We shall explain that the long exact sequence (2.3) is an instance of a Universal Coefficient Theorem as in [**11**, Theorem 66]. In particular, the cokernel and kernel in it are the Ext and Hom groups in a certain exact category. This is already proven in [**11**], but the consequences for classification are not explored there. We shall also treat other examples by the same method later.

The forgetful functor

$$R^{\mathbb{Z}}\colon \mathfrak{KK}^{\mathbb{Z}} \to \mathfrak{KK}, \qquad (A, \alpha) \mapsto A,$$

is triangulated. So its kernel on morphisms

$$\mathfrak{I}^{\mathbb{Z}}(A, B) := \{f \in \mathrm{KK}_0^{\mathbb{Z}}(A, B) : R^{\mathbb{Z}}(f) = 0 \text{ in } \mathrm{KK}_0(A, B)\}$$

is a stable homological ideal in $\mathfrak{KK}^{\mathbb{Z}}$. We now describe the universal $\mathfrak{I}^{\mathbb{Z}}$-exact functor. The main point here is to describe its target category. Let $\mathfrak{KK}[\mathbb{Z}]$ be the additive category of functors $\mathbb{Z} \to \mathfrak{KK}$ with natural transformations as arrows. Equivalently, an object of $\mathfrak{KK}[\mathbb{Z}]$ is an object $A \in\in \mathfrak{KK}$ with a group homomorphism $a\colon \mathbb{Z} \to \mathrm{KK}_0(A, A)^{\times}$, $n \mapsto a_n$, where $\mathrm{KK}_0(A, A)^{\times}$ denotes the multiplicative group of invertible elements in the ring $\mathrm{KK}_0(A, A)$. And an arrow $(A, a) \to (B, b)$ in $\mathfrak{KK}[\mathbb{Z}]$ is an arrow $f \in \mathrm{KK}_0(A, B)$ that satisfies $b_n \circ f = f \circ a_n$ for all $n \in \mathbb{Z}$. The homomorphism a is determined by its value at $1 \in \mathbb{Z}$, which may be any element $a_1 \in \mathrm{KK}_0(A, A)^{\times}$. So we also denote objects of $\mathfrak{KK}[\mathbb{Z}]$ as (A, a_1). An element $f \in \mathrm{KK}_0(A, B)$ is an arrow $(A, a) \to (B, b)$ in $\mathfrak{KK}[\mathbb{Z}]$ if and only if $b_1 \circ f = f \circ a_1$ or, equivalently, $b_1^{-1} \circ f \circ a_1 - f = 0$. Thus

$$\ker\bigl(\alpha^* \beta_*^{-1} - 1\colon \mathrm{KK}_0(A, B) \to \mathrm{KK}_0(A, B)\bigr) \cong \mathrm{Hom}_{\mathfrak{KK}[\mathbb{Z}]}\bigl((A, [\alpha]), (B, [\beta])\bigr).$$

Here we have implicitly used the functor

$$F^{\mathbb{Z}}\colon \mathfrak{KK}^{\mathbb{Z}} \to \mathfrak{KK}[\mathbb{Z}], \qquad (A, \alpha) \mapsto (A, [\alpha]),$$

where $[\alpha] \in \mathrm{KK}_0(A, A)^{\times}$ is the $\mathfrak{KK}$-class of the automorphism α. It has the same kernel on morphisms $\mathfrak{I}^{\mathbb{Z}}$ as the forgetful functor $R^{\mathbb{Z}}$.

We call a kernel–cokernel pair $K \rightarrowtail E \twoheadrightarrow Q$ in $\mathfrak{KK}[\mathbb{Z}]$ *admissible* if it splits in $\mathfrak{KK}$, that is, $E \cong K \oplus Q$ as objects of $\mathfrak{KK}$. This turns $\mathfrak{KK}[\mathbb{Z}]$ into an exact category (see [**5**, Exercise 5.3]). We use this exact structure to define projective objects and projective resolutions in $\mathfrak{KK}[\mathbb{Z}]$.

Given $A \in\in \mathfrak{KK}$ and a free Abelian group $G = \mathbb{Z}[I]$ on a countable set I, we define $A \otimes G \in\in \mathfrak{KK}$ by

$$A \otimes G := \bigoplus_{i \in I} A. \tag{2.4}$$

In particular, $A \otimes \mathbb{Z} := A$. This construction is an additive functor in G. Namely, let G and H be free Abelian groups and let $f\colon G \to H$ be a group homomorphism. The Universal Coefficient Theorem implies $\mathrm{KK}_0(\mathbb{C} \otimes G, \mathbb{C} \otimes H) \cong \mathrm{Hom}(G, H)$. So we get a functorial $f_{\mathbb{C}} \in \mathrm{KK}_0(\mathbb{C} \otimes G, \mathbb{C} \otimes H)$. Identifying $A \otimes G = A \otimes (\mathbb{C} \otimes G)$ (with the minimal tensor product of C*-algebras), we get a functorial

$$f_A := \mathrm{id}_A \otimes f_{\mathbb{C}} \in \mathrm{KK}_0(A \otimes G, A \otimes H).$$

Let $A \in\in \mathfrak{KK}$. Then $A \otimes \mathbb{Z}[x, x^{-1}]$ with the invertible element

$$x_A = \mathrm{id}_A \otimes x \in \mathrm{KK}_0\bigl(A \otimes \mathbb{Z}[x, x^{-1}], A \otimes \mathbb{Z}[x, x^{-1}]\bigr)$$

induced by multiplication with the invertible element $x \in \mathbb{Z}[x,x^{-1}]$ is an object of $\mathfrak{KK}[\mathbb{Z}]$. It behaves like a free module over A because

$$\mathrm{Hom}_{\mathfrak{KK}[\mathbb{Z}]}\big((A\otimes\mathbb{Z}[x,x^{-1}],x_A),(B,b)\big)\cong \mathrm{KK}_0(A,B).$$

In particular, $(A\otimes\mathbb{Z}[x,x^{-1}],x_A)$ is projective in the exact category $\mathfrak{KK}[\mathbb{Z}]$. The invertible element x_A in the KK-endomorphism ring of $(A\otimes\mathbb{Z}[x,x^{-1}],x_A)$ lifts to the shift automorphism

$$\tau\in\mathrm{Aut}(\mathrm{C}_0(\mathbb{Z},A)),\qquad (\tau f)(n):=f(n-1).$$

That is, $F^{\mathbb{Z}}(\mathrm{C}_0(\mathbb{Z},A),\tau)\cong(A\otimes\mathbb{Z}[x,x^{-1}],x_A)$. And

$$\mathrm{KK}_0^{\mathbb{Z}}(\mathrm{C}_0(\mathbb{Z},A),B)\cong\mathrm{KK}_0(A,B)\cong\mathrm{Hom}_{\mathfrak{KK}[\mathbb{Z}]}\big((A\otimes\mathbb{Z}[x,x^{-1}],x_A),B\big).\tag{2.5}$$

The first isomorphism here is [**9**, Equation (20)]. It applies the forgetful functor

$$\mathrm{KK}_0^{\mathbb{Z}}\big((\mathrm{C}_0(\mathbb{Z},A),\tau),(B,\beta)\big)\to\mathrm{KK}_0(\mathrm{C}_0(\mathbb{Z},A),B)$$

and then composes with the inclusion $*$-homomorphism $A\to\mathrm{C}_0(\mathbb{Z},A)$ that maps A identically onto the summand at $0\in\mathbb{Z}$. Equation (2.5) says that the partial adjoint of the functor $F^{\mathbb{Z}}$ is defined on $(A\otimes\mathbb{Z}[x,x^{-1}],x_A)$ and maps it to $(\mathrm{C}_0(\mathbb{Z},A),\tau)$.

LEMMA 2.6. *Any object of $\mathfrak{KK}[\mathbb{Z}]$ has a free resolution of length* 1.

PROOF. The trivial representation of the group $\mathbb{Z}$ on $\mathbb{Z}$ corresponds to $\mathbb{Z}$ made a module over $\mathbb{Z}[x,x^{-1}]$ by letting $x\in\mathbb{Z}[x,x^{-1}]$ act by 1. This module has the following free resolution of length 1:

$$0\to\mathbb{Z}[x,x^{-1}]\xrightarrow{\mathrm{mult}(x-1)}\mathbb{Z}[x,x^{-1}]\xrightarrow{\mathrm{ev}_1}\mathbb{Z}\to0,\tag{2.7}$$

where $\mathrm{ev}_1\colon\mathbb{Z}[x,x^{-1}]\to\mathbb{Z}$ is the homomorpism of evaluation at 1, so $\mathrm{ev}_1(x^n)=1$ for all $n\in\mathbb{Z}$. The extension in (2.7) splits as an extension of Abelian groups because $\mathbb{Z}$ is free as an Abelian group.

The resolution (2.7) induces a chain complex

$$0\to(A\otimes\mathbb{Z}[x,x^{-1}],[\alpha]\otimes x)\to(A\otimes\mathbb{Z}[x,x^{-1}],[\alpha]\otimes x)\to(A,[\alpha])\to0\tag{2.8}$$

in the additive category $\mathfrak{KK}[\mathbb{Z}]$. That is, the maps are arrows in $\mathfrak{KK}[\mathbb{Z}]$. The extension (2.7) splits by a group homomorphism, and the tensor product construction is additive. Hence (2.8) is exact, that is, it splits in $\mathfrak{KK}$. The arrow

$$A\otimes\mathbb{Z}[x,x^{-1}]=\bigoplus_{n\in\mathbb{Z}}A\xrightarrow{\bigoplus_{n\in\mathbb{Z}}[\alpha^n]}\bigoplus_{n\in\mathbb{Z}}A=A\otimes\mathbb{Z}[x,x^{-1}]$$

in $\mathfrak{KK}$ is an isomorphism

$$(A\otimes\mathbb{Z}[x,x^{-1}],\mathrm{id}_A\otimes x)\cong(A\otimes\mathbb{Z}[x,x^{-1}],[\alpha]\otimes x).$$

Thus (2.8) is isomorphic to a free resolution

$$0\to(A\otimes\mathbb{Z}[x,x^{-1}],1\otimes x)\to(A\otimes\mathbb{Z}[x,x^{-1}],1\otimes x)\to(A,[\alpha])\to0,\tag{2.9}$$

in $\mathfrak{KK}[\mathbb{Z}]$, where the boundary map on $(A\otimes\mathbb{Z}[x,x^{-1}],1\otimes x)$ has changed to $[\alpha]^{-1}\otimes\mathrm{mult}(x)-\mathrm{id}_{A\otimes\mathbb{Z}[x,x^{-1}]}$. □

Lemma 2.6 implies that the exact category $\mathfrak{KK}[\mathbb{Z}]$ has enough projective objects and that an object of $\mathfrak{KK}[\mathbb{Z}]$ is projective if and only if it is a direct summand of a free object. Since the partial adjoint of the functor $F^{\mathbb{Z}}$ is defined on all free objects, it is defined also on all projective objects of $\mathfrak{KK}[\mathbb{Z}]$, and it is inverse to $F^{\mathbb{Z}}$ on this subcategory. Idempotents in the category $\mathfrak{KK}^{\mathbb{Z}}$ split because it is triangulated and has countable direct sums.

PROPOSITION 2.10. *The functor* $F^{\mathbb{Z}}\colon \mathfrak{KK}^{\mathbb{Z}} \to \mathfrak{KK}[\mathbb{Z}]$ *is the universal* $\mathfrak{I}^{\mathbb{Z}}$*-exact stable homological functor from* $\mathfrak{KK}^{\mathbb{Z}}$ *to an exact category.*

PROOF. To prove this, we first embed $\mathfrak{KK}[\mathbb{Z}]$ into an Abelian category by Quillen's Embedding Theorem. One way to do this is to embed $\mathfrak{KK}$ into an Abelian category $\mathfrak{A}$ by Freyd's Embedding Theorem and then form $\mathfrak{A}[\mathbb{Z}]$. The Abelian category $\mathfrak{A}[\mathbb{Z}]$ has enough projective objects, and all its projective objects already belong to $\mathfrak{KK}[\mathbb{Z}]$. So Theorem 2.1 applies and shows that the functor to $\mathfrak{A}[\mathbb{Z}]$ is the universal $\mathfrak{I}$-exact functor to an Abelian category. The subcategory $\mathfrak{KK}[\mathbb{Z}] \subseteq \mathfrak{A}[\mathbb{Z}]$ is exact and contains the range of the functor. In fact, we shall show soon that the functor $\mathfrak{KK}^{\mathbb{Z}} \to \mathfrak{KK}[\mathbb{Z}]$ is surjective on objects (see Theorem 2.12). Taking this for granted, Remark 2.2 shows that the functor $\mathfrak{KK}^{\mathbb{Z}} \to \mathfrak{KK}[\mathbb{Z}]$ is the universal $\mathfrak{I}$-exact functor to an exact category. □

Equip $\mathrm{C}_0(\mathbb{Z}, A)$ with the $\mathbb{Z}$-action generated by the shift automorphism τ. We use (2.7) to lift the resolution (2.9) in $\mathfrak{KK}[\mathbb{Z}]$ to the following $\mathfrak{I}^{\mathbb{Z}}$-projective resolution of length 1 in $\mathfrak{KK}^{\mathbb{Z}}$:

$$0 \to \mathrm{C}_0(\mathbb{Z}) \otimes A \xrightarrow{\varphi} \mathrm{C}_0(\mathbb{Z}) \otimes A \xrightarrow{p} (A, [\alpha]) \to 0; \tag{2.11}$$

here $\varphi = [\tau] \otimes [\alpha]^{-1} - 1$, and p is the counit of the adjunction (2.5); that is, the first isomorphism in (2.5) maps p to the identity element in $\mathrm{KK}_0(A, A)$. In other words, when we forget the $\mathbb{Z}$-actions, then p restricts to the identity map at the 0th summand in $\mathrm{C}_0(\mathbb{Z}) \otimes A = \bigoplus_{n \in \mathbb{Z}} A$. When we use the resolution (2.11) to compute derived functors, we get

$$\mathrm{Ext}^1_{\mathfrak{KK}[\mathbb{Z}], \mathfrak{I}^{\mathbb{Z}}}\big((A, [\alpha]), (B, [\beta])\big) \cong \mathrm{coker}\big((\alpha^*)^{-1}\beta_* - 1\colon \mathrm{KK}_0(A, B) \to \mathrm{KK}_0(A, B)\big).$$

The map $(\alpha^*)^{-1}\beta_* - 1$, has the same cokernel as $\alpha^*\beta_*^{-1} - 1$.

A rather deep result says that the $\mathfrak{I}^{\mathbb{Z}}$-projective objects generate $\mathfrak{KK}^{\mathbb{Z}}$. Equivalently, if $R^{\mathbb{Z}}(A) \cong 0$ in $\mathfrak{KK}$, then already $A \cong 0$ in $\mathfrak{KK}^{\mathbb{Z}}$. This is related to the proof of the Baum–Connes conjecture for the group $\mathbb{Z}$. It follows immediately from the Pimsner–Voiculescu sequence (2.3).

THEOREM 2.12. *Let* (A, α) *and* (B, β) *be objects of* $\mathfrak{KK}^{\mathbb{Z}}$. *Any isomorphism* $F^{\mathbb{Z}}(A) \cong F^{\mathbb{Z}}(B)$ *in* $\mathfrak{KK}[\mathbb{Z}]$ *lifts to an isomorphism* $A \cong B$ *in* $\mathfrak{KK}^{\mathbb{Z}}$. *And any object of* $\mathfrak{KK}[\mathbb{Z}]$ *is isomorphic to* $F^{\mathbb{Z}}(A)$ *for some* $A \in\in \mathfrak{KK}^{\mathbb{Z}}$, *which is unique up to isomorphism.*

If $t \in \mathrm{KK}_0(A, B)^{\times}$ *is a* KK*-equivalence with* $t \circ [\alpha] = [\beta] \circ t$, *then there is an isomorphism in* $\mathrm{KK}_0^{\mathbb{Z}}(A, B)$ *that is mapped to* t *by the forgetful functor* $\mathfrak{KK}^{\mathbb{Z}} \to \mathfrak{KK}$.

PROOF. The Universal Coefficient Theorem for $\mathrm{KK}_0^{\mathbb{Z}}(A, B)$ and the ideal $\mathfrak{I}^{\mathbb{Z}}$ allows to lift any map $t\colon (A, [\alpha]) \to (B, [\beta])$ in $\mathfrak{KK}[\mathbb{Z}]$ to $\hat{t} \in \mathrm{KK}_0^{\mathbb{Z}}\big((A, \alpha), (B, \beta)\big)$. The naturality of the Universal Coefficient Theorem implies that the composition vanishes for two elements of the Ext-part of $\mathrm{KK}_0^{\mathbb{Z}}(A, B)$. Hence $\hat{t}$ is invertible if t is invertible. Any object of $\mathfrak{KK}[\mathbb{Z}]$ has a projective resolution of length 1. This allows to lift it to an object of $\mathfrak{KK}^{\mathbb{Z}}$ (see [**4**, Proposition 2.3]). This proves both assertions in the first paragraph. The second paragraph only describes isomorphisms in $\mathfrak{KK}[\mathbb{Z}]$ more concretely. □

We make the isomorphism criterion in Theorem 2.12 more explicit under the assumption that the C*-algebras A and B belong to the bootstrap class. Let $A^\pm$ be C*-algebras in the bootstrap class with

$$\mathrm{K}_0(A^+) = \mathrm{K}_0(A), \quad \mathrm{K}_1(A^+) = 0, \qquad \mathrm{K}_0(A^-) = 0, \quad \mathrm{K}_1(A^-) = \mathrm{K}_1(A).$$

Then $\mathrm{K}_*(A^+ \oplus A^-) \cong \mathrm{K}_*(A)$. So $A \cong A^+ \oplus A^-$ by the Universal Coefficient Theorem for KK. We use such a KK-equivalence to map α to an element of $\mathrm{KK}_0(A^+ \oplus A^-, A^+ \oplus A^-)$. We rewrite this as a 2×2-matrix

$$\begin{pmatrix} \alpha^{++} & \alpha^{+-} \\ \alpha^{-+} & \alpha^{--} \end{pmatrix}, \qquad \begin{aligned} \alpha^{++} &\in \mathrm{KK}_0(A^+, A^+) \cong \mathrm{Hom}\big(\mathrm{K}_0(A), \mathrm{K}_0(A)\big), \\ \alpha^{+-} &\in \mathrm{KK}_0(A^-, A^+) \cong \mathrm{Ext}\big(\mathrm{K}_1(A), \mathrm{K}_0(A)\big), \\ \alpha^{-+} &\in \mathrm{KK}_0(A^+, A^-) \cong \mathrm{Ext}\big(\mathrm{K}_0(A), \mathrm{K}_1(A)\big), \\ \alpha^{--} &\in \mathrm{KK}_0(A^-, A^-) \cong \mathrm{Hom}\big(\mathrm{K}_1(A), \mathrm{K}_1(A)\big). \end{aligned}$$

Here we have used the Universal Coefficient Theorem for the C*-algebras $A^\pm$. A similar decomposition $B \cong B^+ \oplus B^-$ allows us to map β to an element of $\mathrm{KK}_0(B^+ \oplus B^-, B^+ \oplus B^-)$, which we then rewrite as a 2×2-matrix

$$\begin{pmatrix} \beta^{++} & \beta^{+-} \\ \beta^{-+} & \beta^{--} \end{pmatrix}, \qquad \begin{aligned} \beta^{++} &\in \mathrm{KK}_0(B^+, B^+) \cong \mathrm{Hom}\big(\mathrm{K}_0(B), \mathrm{K}_0(B)\big), \\ \beta^{+-} &\in \mathrm{KK}_0(B^-, B^+) \cong \mathrm{Ext}\big(\mathrm{K}_1(B), \mathrm{K}_0(B)\big), \\ \beta^{-+} &\in \mathrm{KK}_0(B^+, B^-) \cong \mathrm{Ext}\big(\mathrm{K}_0(B), \mathrm{K}_1(B)\big), \\ \beta^{--} &\in \mathrm{KK}_0(B^-, B^-) \cong \mathrm{Hom}\big(\mathrm{K}_1(B), \mathrm{K}_1(B)\big). \end{aligned}$$

And we may also transfer an element $t \in \mathrm{KK}_0(A, B)$ to such a 2×2-matrix

$$\begin{pmatrix} t^{++} & t^{+-} \\ t^{-+} & t^{--} \end{pmatrix}, \qquad \begin{aligned} t^{++} &\in \mathrm{KK}_0(A^+, B^+) \cong \mathrm{Hom}\big(\mathrm{K}_0(A), \mathrm{K}_0(B)\big), \\ t^{+-} &\in \mathrm{KK}_0(A^-, B^+) \cong \mathrm{Ext}\big(\mathrm{K}_1(A), \mathrm{K}_0(B)\big), \\ t^{-+} &\in \mathrm{KK}_0(A^+, B^-) \cong \mathrm{Ext}\big(\mathrm{K}_0(A), \mathrm{K}_1(B)\big), \\ t^{--} &\in \mathrm{KK}_0(A^-, B^-) \cong \mathrm{Hom}\big(\mathrm{K}_1(A), \mathrm{K}_1(B)\big). \end{aligned}$$

The naturality of the exact sequence in the Universal Coefficient Theorem implies that the Kasparov product of an element of $\mathrm{Ext}\big(\mathrm{K}_{1+*}(A), \mathrm{K}_*(B)\big)$ with an element of $\mathrm{KK}_0(B, C)$ depends only on the image of the latter in $\mathrm{Hom}\big(\mathrm{K}_*(B), \mathrm{K}_*(C)\big)$. Thus the product of two Ext-terms always vanishes. Hence the condition $t[\alpha] = [\beta]t$ in Theorem 2.12 is equivalent to four equations

$$\begin{aligned} t^{++}\alpha^{++} &= \beta^{++}t^{++}, & t^{+-}\alpha^{--} + t^{++}\alpha^{+-} &= \beta^{+-}t^{--} + \beta^{++}t^{+-}, \\ t^{--}\alpha^{--} &= \beta^{--}t^{--}, & t^{-+}\alpha^{++} + t^{--}\alpha^{-+} &= \beta^{-+}t^{++} + \beta^{--}t^{-+}. \end{aligned}$$

The equations on the left assert that the two diagonal entries of $t[\alpha]$ and $[\beta]t$ are equal; those on the right assert equality of the two off-diagonal entries. Theorem 2.12 says that (A, α) and (B, β) are $\mathrm{KK}^{\mathbb{Z}}$-equivalent if and only if there is an invertible element t for which these equations hold. The equations on the left mean that the isomorphism $t_* = (t^{++}, t^{--}) \in \mathrm{Hom}\big(\mathrm{K}_*(A), \mathrm{K}_*(B)\big)$ induced by t intertwines the automorphisms $\alpha_* \in \mathrm{Aut}(\mathrm{K}_*(A))$ and $\beta_* \in \mathrm{Aut}(\mathrm{K}_*(B))$. Equivalently, t_* is an isomorphism of $\mathbb{Z}/2$-graded $\mathbb{Z}[x, x^{-1}]$-modules. The equations on the right mean that there is $t^1 := (t^{+-}, t^{-+}) \in \mathrm{Ext}\big(\mathrm{K}_{1+*}(A), \mathrm{K}_*(B)\big)$ with

$$t^1\alpha_* - \beta_*t^1 = \beta^1 t_* - t_*\alpha^1,$$

where we abbreviate $\alpha_* = (\alpha^{++}, \alpha^{--})$ and $\alpha^1 = (\alpha^{+-}, \alpha^{-+})$, and similarly for β. The choice of t^1 has no effect on the invertibility of t. So the criterion in Theorem 2.12 is whether the image of $\beta^1 t_* - t_*\alpha^1$ vanishes in the cokernel of

$$\text{(2.13)} \quad \mathrm{Ext}\big(\mathrm{K}_{1+*}(A), \mathrm{K}_*(B)\big) \to \mathrm{Ext}\big(\mathrm{K}_{1+*}(A), \mathrm{K}_*(B)\big), \qquad t^1 \mapsto t^1\alpha_* - \beta_* t^1.$$

We shall later identify this cokernel with the group $\mathrm{Ext}^2_{\mathbb{Z}[x,x^{-1}]}\big(\mathrm{K}_{1+*}(A), \mathrm{K}_*(B)\big)$ and show that the image of $\beta^1 t_* - t_*\alpha^1$ in this cokernel is the relative obstruction class for (A, α) and (B, β) and a $\mathbb{Z}[x, x^{-1}]$-module isomorphisms $t_* : \mathrm{K}_*(A) \cong \mathrm{K}_*(B)$. This gives the rule to translate between the classifying invariants in Theorem 2.12 and in [**4**].

REMARK 2.14. If $A = B$ is a unital Kirchberg algebra (separable, nuclear, unital, purely infinite and simple), then a much finer classification theorem for automorphisms is proved by Nakamura [**13**].

2.2. C*-algebras over unique path spaces. Now we prove a Universal Coefficient Theorem for $\mathfrak{KK}^X$ for a unique path space X. Let X be a countable set and let $\to$ be a relation on X, which says for which points $x, y \in X$ there is an edge $x \to y$. Equip X with the partial order generated by $\leftarrow$, that is, $x \preceq y$ if and only if there is a chain of edges $x = x_0 \leftarrow x_1 \leftarrow \cdots \leftarrow x_\ell = y$ with some $\ell \geq 0$ and $x_1, \ldots, x_{\ell-1} \in X$. Equip X with the Alexandrov topology generated by this partial order. We assume $(X, \to)$ to be a *unique path space*, that is, there is at most one chain of edges $x = x_0 \leftarrow x_1 \leftarrow \cdots \leftarrow x_\ell = y$ between any two points $x, y \in X$.

For $x \in X$, the subset $U_x := \{y \in X : x \preceq y\}$ is the minimal open subset containing x. A *C*-algebra over* X is equivalent to a C*-algebra A with fixed ideals $A(U_x) \triangleleft A$ for all $x \in X$, such that $A(U_x) \subseteq A(U_y)$ for all $x, y \in X$ with $x \to y$ or, equivalently, with $U_x \subseteq U_y$. The equivariant Kasparov category $\mathfrak{KK}^X$ for C*-algebras over X has separable C*-algebras over X as objects and the KK-groups $\mathrm{KK}^X_0(A, B)$ as arrows (see also [**10**], where this category is denoted $\mathfrak{KK}(X)$). The forgetful functor

$$R^X : \mathfrak{KK}^X \to \prod_{x \in X} \mathfrak{KK}, \qquad A \mapsto (A(U_x))_{x \in X},$$

is a triangulated functor between triangulated categories. Its kernel on morphisms

$$\mathfrak{I}^X(A, B) := \{f \in \mathrm{KK}^X_0(A, B) : f(U_x) = 0 \text{ in } \mathrm{KK}_0(A(U_x), B(U_x)) \text{ for all } x \in X\}$$

is a stable homological ideal. We now describe the universal $\mathfrak{I}^X$-exact stable homological functor as in Section 2.1.

Let $\mathfrak{KK}[X]$ be the category of functors $(X, \succeq) \to \mathfrak{KK}$ with natural transformations as arrows. Since X is a unique path space, the category associated to the partially ordered set $(X, \succ)$ is the path category of the directed graph $(X, \to)$. By the universal property of the path category, an object of $\mathfrak{KK}[X]$ is given by $A_x \in\in \mathfrak{KK}$ for $x \in X$ and $\alpha_{y,x} \in \mathrm{KK}_0(A_x, A_y)$ for $x, y \in X$ with $x \to y$, without any relations on the $\alpha_{y,x}$. This uniquely determines KK-classes $\alpha_{y,x} \in \mathrm{KK}_0(A_x, A_y)$ for $x, y \in X$ with $x \succeq y$ such that $\alpha_{x,x} = \mathrm{id}_{A_x}$ and $\alpha_{z,y} \circ \alpha_{y,x} = \alpha_{z,x}$ for all $x, y, z \in X$ with $x \succeq y \succeq z$. An arrow $(A_x, \alpha_{y,x}) \to (B_x, \beta_{y,x})$ is a family of arrows $f_x \in \mathrm{KK}_0(A_x, B_x)$ for $x \in X$ with $f_y\alpha_{y,x} = \beta_{y,x}f_x$ in $\mathrm{KK}_0(A_x, B_y)$ for all $x, y \in X$ with $x \to y$; then $f_y\alpha_{y,x} = \beta_{y,x}f_x$ holds for all $x, y \in X$ with $x \succeq y$. Define

$$F^X : \mathfrak{KK}^X \to \mathfrak{KK}[X]$$

by mapping a C*-algebra A over X to the object of $\mathfrak{KK}[X]$ where $A_x := A(U_x)$ and where $\alpha_{y,x} \in \mathrm{KK}_0(A_x, A_y)$ for $x, y \in X$ with $x \to y$ is the KK-class of the inclusion map $A(U_x) \hookrightarrow A(U_y)$. Then $\alpha_{y,x} \in \mathrm{KK}_0(A_x, A_y)$ for $x, y \in X$ with $x \succeq y$ is the KK-class of the inclusion map as well. A kernel–cokernel pair $K \to E \to Q$ in $\mathfrak{KK}[X]$ is called *admissible* if it splits pointwise, that is, $K_x \to E_x \to Q_x$ is a split extension in $\mathfrak{KK}$ for all $x \in X$; so $E_x \cong K_x \oplus Q_x$ in $\mathfrak{KK}$ for all $x \in X$, but the sections $Q_x \to E_x$ are not compatible with the structure maps $E_x \to E_y$ and $Q_x \to Q_y$ for $x \to y$. This turns $\mathfrak{KK}[X]$ into an exact category (see [**5**, Exercise 5.3]).

Let $z \in X$ and $A \in\in \mathfrak{KK}$. As in [**10**], let $i_z(A) \in\in \mathfrak{KK}^X$ be the C*-algebra A with

$$i_z(A)(U_x) := \begin{cases} A & \text{if } z \in U_x \text{, that is, } x \preceq z, \\ 0 & \text{otherwise.} \end{cases}$$

Then

$$\mathrm{KK}_0^X(i_z(A), B) \cong \mathrm{KK}_0\big(A, B(U_z)\big) \tag{2.15}$$

for all $B \in\in \mathfrak{KK}^X$ by [**10**, Proposition 3.13]. This isomorphism applies the restriction map $\mathrm{KK}_0^X(i_z(A), B) \to \mathrm{KK}_0(i_z(A)(U_z), B(U_z))$ and then identifies $i_z(A)(U_z) = A$. The object

$$j_z(A) := F^X(i_z(A)) \in\in \mathfrak{KK}[X]$$

has $j_z(A)_x = A$ for $x \preceq z$ and $j_z(A)_x = 0$ otherwise, and the map $j_z(A)_x \to j_z(A)_y$ for $x \succeq y$ is the identity map in $\mathrm{KK}_0(A, A)$ if $z \succeq x$ and the zero map in $\mathrm{KK}_0(0, j_z(A)_y)$ otherwise. We compute

$$\mathrm{Hom}_{\mathfrak{KK}[X]}\big(j_z(A), (B_x, \beta_{y,x})\big) \cong \mathrm{KK}_0(A, B_z). \tag{2.16}$$

Equations (2.15) and (2.16) imply

$$\mathrm{KK}_0^X(i_z(A), B) \cong \mathrm{KK}_0(A, B(U_z)) \cong \mathrm{Hom}_{\mathfrak{KK}[X]}\big(j_z(A), F^X(B)\big) \tag{2.17}$$

for all $B \in\in \mathfrak{KK}^X$.

THEOREM 2.18. *Let $(X, \to)$ be a countable unique path space. The objects of the exact category $\mathfrak{KK}[X]$ of the form $\bigoplus_{z\in X} j_z(A_z)$ for $A_z \in\in \mathfrak{KK}$ for $z \in X$ are projective. For any object $A = (A_x, \alpha_{x,y})$ of $\mathfrak{KK}[X]$, there is an admissible extension*

$$\bigoplus_{x\to y} j_y(A_x) \overset{\iota}{\rightarrowtail} \bigoplus_{x\in X} j_x(A_x) \overset{\pi}{\twoheadrightarrow} A, \tag{2.19}$$

which is a projective resolution of length 1. The exact category $\mathfrak{KK}[X]$ has enough projective objects, and the partial adjoint of F^X is defined on all projective objects of $\mathfrak{KK}[X]$ and is a section for F^X there. The functor $F^X\colon \mathfrak{KK}^X \to \mathfrak{KK}[X]$ is the universal $\mathfrak{I}^X$-exact stable homological functor to an exact category.

PROOF. The objects $j_z(A_z)$ are projective by (2.16). This is inherited by the direct sum $\bigoplus_{z\in X} j_z(A_z)$. The identity map on A_x has an adjunct $a_x\colon j_x(A_x) \to A$ by (2.16). These maps induce the map $\pi = (a_x)_{x\in X}\colon \bigoplus_{x\in X} j_x(A_x) \to A$. For each edge $x \to y$ in the directed graph $(X, \to)$, the map $(\mathrm{id}_{A_x}, -\alpha_{y,x})\colon A_x \to A_x \oplus A_y$ is adjunct to a map

$$a_{x\to y}\colon j_y(A_x) \to j_x(A_x) \oplus j_y(A_y) \subseteq \bigoplus_{x\in X} j_x(A_x)$$

by (2.16). It satisfies $\pi \circ a_{x\to y} = 0$. The maps $a_{x\to y}$ combine to a map

$$\iota\colon \bigoplus_{x\to y} j_y(A_x) \to \bigoplus_{x} j_x(A_x)$$

with $\pi \circ \iota = 0$.

Now we prove that the maps π and ι mapped to $\prod_{x\in X} \mathfrak{KK}$ by the forgetful functor form a split exact sequence. We consider the entries at a fixed $z \in X$. The entry of A at $z \in X$ is simply A_z. The entry of $\bigoplus_{x\in X} j_x(A_x)$ at $z \in X$ is the direct sum of A_x over all $x \in X$ with $x \succeq z$. The entry of $\bigoplus_{x\to y} j_y(A_x)$ at $z \in X$ is the direct sum of A_x for all edges $x \to y$ in X with $y \succeq z$. The entry of π at z is

$$(\alpha_{z,x})_{x\succeq z}\colon \bigoplus_{x\succeq z} A_x \to A_z.$$

This is split surjective with the canonical section that maps A_z identically onto the summand A_z for $x = z$. The entry of ι at z maps the summand A_x for $x \to y \succeq z$ to $A_x \oplus A_y \subseteq \bigoplus_{t\succeq z} A_t$ using $(\mathrm{id}_{A_x}, -\alpha_{y,x})$. We are going to define a map

$$s_1\colon \bigoplus_{t\succeq z} A_t \to \bigoplus_{x\to y\succeq z} A_x$$

which together with s_0 forms a contracting homotopy for the short chain complex formed by $\iota|_z$ and $\pi|_z$. By assumption, if $t \succeq z$ then there is a unique chain $t = x_0 \to x_1 \to \cdots \to x_\ell = z$. We let $(s_1)_z|_{A_t}$ for $t \succeq z$ map the summand A_t in $\bigoplus_{t\succeq z} A_t$ to the direct sum of A_{x_i} for the edges $x_i \to x_{i+1}$ for $i = 0, \dots, \ell - 1$, where we use $\alpha_{x_i,t}$ to map A_t to A_{x_i}. If $x \to y \succeq z$, then $y = x_1$ in the above chain. Therefore, $s_1 \circ \iota|_{A_x}$ is a map to $\bigoplus_{j=0}^{\ell} A_{x_j}$, where the entry at A_{x_j} is $\alpha_{x_j,x} - \alpha_{x_j,y} \circ \alpha_{y,x} = 0$ for $j = 1, \dots, \ell$, and the identity map for $j = 0$. So $s_1 \circ \iota$ is the identity map. Finally, we claim that $s_0 \circ \pi + \iota \circ s_1$ is the identity map on $\bigoplus_{t\succeq z} A_t$. This is checked on each summand A_t separately. Let $t = x_0 \to x_1 \to \cdots \to x_\ell = z$ be the unique chain as above. Then $\iota \circ s_1$ is a telescoping sum of $\pm\alpha_{x_j,t}$ for $j = 0, \dots, \ell$, where $\alpha_{t,t}$ occurs only with sign $+$ and $\alpha_{z,t}$ only with sign $-$. And $s_0 \circ \pi$ is the map $\alpha_{z,t}$. Thus $s_0 \circ \pi + \iota \circ s_1$ is the identity map on A_t. This finishes the proof that (2.19) is $\mathfrak{I}^X$-exact. Its entries are $\mathfrak{I}^X$-projective. So it is an $\mathfrak{I}^X$-projective resolution.

The projective resolution (2.19) implies that $\mathfrak{KK}[X]$ has enough projective objects and that an object is projective if and only if it is a direct summand of an object of the form $\bigoplus_{z\in X} j_z(B_z)$ for some separable C*-algebras B_z for $z \in X$. Equation (2.17) says that the adjoint functor to $F^X\colon \mathfrak{KK}^X \to \mathfrak{KK}[X]$ is defined on $j_z(A)$ and maps it to $i_z(A)$. Since the partial adjoint commutes with direct sums, it is also defined on $\bigoplus_{z\in X} j_z(B_z)$ for any $B_z \in\in \mathfrak{KK}$ and maps it to $\bigoplus_{z\in X} i_z(B_z)$. Idempotents in the category $\mathfrak{KK}^X$ split because it is triangulated and has countable direct sums. Therefore, the partial adjoint of F^X is defined on all projective objects of $\mathfrak{KK}[X]$ and is a section for F^X there. An argument as in the proof of Proposition 2.10 shows that F^X is the universal $\mathfrak{I}^X$-exact functor to an exact category. The target category is not smaller because the functor $\mathfrak{KK}^X \to \mathfrak{KK}[X]$ is surjective on objects by Corollary 2.20. □

Unlike in Section 2.1, the $\mathfrak{I}^X$-projective objects in $\mathfrak{KK}^X$ do not generate $\mathfrak{KK}^X$. The localising subcategory $\langle \mathfrak{P}_{\mathfrak{I}} \rangle$ generated by them is equal to the localising subcategory generated by objects of the form $i_x(A)$ for $x \in X$, $A \in\in \mathfrak{KK}$. If X is finite, then it is described in several equivalent ways in [**10**, Definition 4.7]. It contains all nuclear C*-algebras over X. This is the subcategory on which Theorem 2.18 implies a Universal Coefficient Theorem, using Hom and Ext groups in the category $\mathfrak{KK}[X]$. This implies the following classification theorem:

COROLLARY 2.20. *Let $(X,\to)$ be a countable unique path space and let A and B be C*-algebras over X that belong to the localising subcategory $\langle\mathfrak{P}_{\mathfrak{J}}\rangle$ generated by objects of the form $i_x(B)$ for $x\in X$, $B\in\in\mathfrak{KK}$. Any isomorphism between $F^X(A)$ and $F^X(B)$ in $\mathfrak{KK}[X]$ lifts to an isomorphism in $\mathfrak{KK}^X$. And any object of $\mathfrak{KK}[X]$ lifts to an object of $\langle\mathfrak{P}_{\mathfrak{J}}\rangle$. An isomorphism $F^X(A)\xrightarrow{\sim}F^X(B)$ is a family of invertible elements $t_x\in \mathrm{KK}_0(A(U_x),B(U_x))$ for $x\in X$ for which the diagrams*

$$\begin{array}{ccc} A(U_x) & \hookrightarrow & A(U_y) \\ \downarrow{\scriptstyle t_x} & & \downarrow{\scriptstyle t_y} \\ B(U_x) & \hookrightarrow & B(U_y) \end{array}$$

commute in $\mathfrak{KK}$ for all $x,y\in X$ with $x\to y$.

The criterion above may be made more explicit if, in addition, $A(U_x)$ and $B(U_x)$ belong to the bootstrap class for all $x\in X$. As in Section 2.1, we identify $A(U_x)\cong A(U_x)^+\oplus A(U_x)^-$ and rewrite the classes of the inclusion maps $\alpha_{y,x}\in \mathrm{KK}_0(A(U_x),A(U_y))$ and $\beta_{y,x}\in\mathrm{KK}_0(B(U_x),B(U_y))$ for $x\to y$ as 2×2-matrices, and similarly for the arrows $t_x\in\mathrm{KK}_0(A(U_x),B(U_x))$. As in the case of $\mathfrak{KK}^{\mathbb{Z}}$, the equality $\beta_{y,x}t_x=t_y\alpha_{y,x}$ for $x\to y$ in Corollary 2.20 may be rewritten as four equalities of matrix coefficients. The equality of the diagonal terms says that the diagrams $\big(\mathrm{K}_*(A(U_x)),\mathrm{K}_*(\alpha_{y,x})\big)$ and $\big(\mathrm{K}_*(B(U_x)),\mathrm{K}_*(\beta_{y,x})\big)$ of countable $\mathbb{Z}/2$-graded Abelian groups are isomorphic. That is, for all $x,y\in X$ with $x\to y$, the following diagram commutes:

$$\begin{array}{ccc} \mathrm{K}_*(A(U_x)) & \xrightarrow{\mathrm{K}_*(\alpha_{y,x})} & \mathrm{K}_*(A(U_y)) \\ \downarrow{\scriptstyle \mathrm{K}_*(t_x)} & & \downarrow{\scriptstyle \mathrm{K}_*(t_y)} \\ \mathrm{K}_*(B(U_x)) & \xrightarrow{\mathrm{K}_*(\beta_{y,x})} & \mathrm{K}_*(B(U_y)) \end{array}$$

The equality of the off-diagonal terms will be studied in Section 4.2.

3. Computation of the obstruction class

We recall the setup of [**4**]. Let $\mathfrak{T}$ be a triangulated category with countable direct sums. Let $\mathfrak{J}$ be a stable homological ideal in $\mathfrak{T}$ with enough projective objects. Let $F\colon\mathfrak{T}\to\mathfrak{A}$ be the universal $\mathfrak{J}$-exact stable homological functor (in this article, we allow $\mathfrak{A}$ to be exact). We assume that $\mathfrak{A}$ is paired as in [**4**, Definition 2.14], that is, $\mathfrak{A}=\mathfrak{A}_+\times\mathfrak{A}_-$ with $\Sigma\mathfrak{A}_+=\mathfrak{A}_-$ and $\Sigma\mathfrak{A}_-=\mathfrak{A}_+$. For instance, $\mathfrak{A}$ could be the category of $\mathbb{Z}$-graded or $\mathbb{Z}/2$-graded modules over some ring, with the suspension automorphism shifting the grading, and $\mathfrak{A}_+$ and $\mathfrak{A}_-$ may be taken to be the graded modules concentrated in even or odd degrees, respectively. We want to compute the obstruction class of an object $A\in\in\mathfrak{T}$. This is only meaningful if A is constructed from simpler ingredients.

We assume that there is an exact, $\mathfrak{J}$-exact triangle

$$B_1\xrightarrow{\varphi}B_0\xrightarrow{p}A\xrightarrow{i}\Sigma B_1, \tag{3.1}$$

where B_0 and B_1 are objects of $\langle\mathfrak{P}_{\mathfrak{J}}\rangle$ with projective resolutions of length 1. Then $A\in\in\langle\mathfrak{P}_{\mathfrak{J}}\rangle$ as well. The $\mathfrak{J}$-exactness assumption says that $i\in\mathfrak{J}$. Equivalently,

$F(i) = 0$, $F(\varphi)$ is monic, and $F(p)$ is epic. The objects B_0 and B_1 are uniquely determined up to isomorphism by $F(B_0)$ and $F(B_1)$ because they have projective resolutions of length 1 (compare [**4**, Proposition 2.3]). We may split $B_i \cong B_i^+ \oplus B_i^-$ with $F(B_i^\pm) \in \mathfrak{A}_\pm$ for $i = 0, 1$. Then φ becomes a 2×2-matrix

$$\varphi = \begin{pmatrix} \varphi_{++} & \varphi_{+-} \\ \varphi_{-+} & \varphi_{--} \end{pmatrix}$$

with $\varphi_{-+} \colon B_1^+ \to B_0^-$, and so on. The two diagonal entries give an element of

$$\mathfrak{T}(B_1^+, B_0^+) \oplus \mathfrak{T}(B_1^-, B_0^-) \cong \operatorname{Hom}_{\mathfrak{A}}\big(F(B_1), F(B_0)\big),$$

and the two off-diagonal entries give an element of

$$\mathfrak{T}(B_1^+, B_0^-) \oplus \mathfrak{T}(B_1^-, B_0^+) \cong \operatorname{Ext}^1_{\mathfrak{A}}\big(\Sigma F(B_1), F(B_0)\big);$$

here we have used the Universal Coefficient Theorem to compute $\mathfrak{T}(B_1^\pm, B_0^\pm)$. The results for these four groups only have a single Hom or a single Ext group because of the parity assumptions. Thus the splitting of φ into $\varphi^0 := \varphi_{++} + \varphi_{--}$ and $\varphi^1 := \varphi_{+-} + \varphi_{-+}$ splits the exact sequence

$$\operatorname{Ext}^1_{\mathfrak{A}}\big(\Sigma F(B_1), F(B_0)\big) \rightarrowtail \mathfrak{T}_*(B_1, B_0) \twoheadrightarrow \operatorname{Hom}_{\mathfrak{A}}\big(F(B_1), F(B_0)\big).$$

We shall compute the obstruction class of A in terms of φ^1.

By assumption, there is a short exact sequence

$$F(B_1) \xrightarrow{F(\varphi)} F(B_0) \xrightarrow{F(p)} F(A).$$

This induces a long exact sequence

$$0 \leftarrow \operatorname{Ext}^2_{\mathfrak{A}}\big(\Sigma F(A), F(A)\big) \xleftarrow{\partial} \operatorname{Ext}^1_{\mathfrak{A}}\big(\Sigma F(B_1), F(A)\big) \xleftarrow{F(\varphi)^*} \operatorname{Ext}^1_{\mathfrak{A}}\big(\Sigma F(B_0), F(A)\big) \xleftarrow{F(p)^*} \operatorname{Ext}^1_{\mathfrak{A}}\big(\Sigma F(A), F(A)\big) \leftarrow \cdots$$

because $\operatorname{Ext}^k_{\mathfrak{A}}\big(F(B_i), F(A)\big) = 0$ for $i = 0, 1$, $k \geq 2$.

THEOREM 3.2. *The obstruction class of A is*

$$\partial(p_*\varphi^1) = \partial(F(p) \circ \varphi^1) \in \operatorname{Ext}^2_{\mathfrak{A}}\big(\Sigma F(A), F(A)\big),$$

where $F(p) \in \operatorname{Hom}_{\mathfrak{A}}\big(\Sigma F(B_0), \Sigma F(A)\big)$ and $\varphi^1 \in \operatorname{Ext}^1_{\mathfrak{A}}\big(\Sigma F(B_1), F(B_0)\big)$.

PROOF. We shall recall the construction of obstruction classes in [**4**] along the way. It starts with an $\mathfrak{I}$-projective resolution of A of length 2. So first we have to construct this. We use the projective resolutions of $F(B_i)$ of length 1, which exist by assumption. They lift canonically to $\mathfrak{I}$-projective resolutions

$$0 \to P_{i1} \xrightarrow{d_{i1}} P_{i0} \xrightarrow{d_{i0}} B_i \tag{3.3}$$

in $\mathfrak{T}$ for $i = 0, 1$ (see [**11**, Theorem 59]). Since (3.3) is a resolution, d_{i1} is $\mathfrak{I}$-monic and d_{i0} is $\mathfrak{I}$-epic. The arrow $\varphi \in \mathfrak{T}(B_1, B_0)$ lifts to a chain map

$$\begin{array}{ccccc} P_{11} & \overset{d_{11}}{\rightarrowtail} & P_{10} & \overset{d_{10}}{\twoheadrightarrow} & B_1 \\ \downarrow{\scriptstyle \Phi_1} & & \downarrow{\scriptstyle \Phi_0} & & \downarrow{\scriptstyle \varphi} \\ P_{01} & \underset{d_{01}}{\rightarrowtail} & P_{00} & \underset{d_{00}}{\twoheadrightarrow} & B_0 \end{array} \tag{3.4}$$

between the $\mathfrak{J}$-projective resolutions (3.3) (see [**11**, Proposition 44]). We write $\rightarrowtail$ for $\mathfrak{J}$-monic and $\twoheadrightarrow$ for $\mathfrak{J}$-epic maps. We claim that

$$0 \to P_{11} \xrightarrow{(-d_{11},\Phi_1)} P_{10} \oplus P_{01} \xrightarrow{(\Phi_0,d_{01})} P_{00} \xrightarrow{p\circ d_{00}} A \to 0 \tag{3.5}$$

is an $\mathfrak{J}$-projective resolution of A of length 2. The entries are $\mathfrak{J}$-projective by construction. Next we prove that (3.5) is a resolution, that is, it becomes an exact chain complex when we apply F to it.

When we apply F to the diagram (3.4), the two rows become short exact sequences in $\mathfrak{A}$, and the vertical maps become a chain map between them. The mapping cone of this chain map is again an exact chain complex in $\mathfrak{A}$. It has the form

$$0 \to F(P_{11}) \to F(P_{10}) \oplus F(P_{01}) \to F(P_{00}) \oplus F(B_1) \to F(B_0) \to 0.$$

The map $F(\varphi)\colon F(B_1) \to F(B_0)$ is monic with cokernel $F(A)$. So the direct summand $F(B_1)$ and its image in $F(B_0)$ together form a contractible subcomplex. The quotient by it is again an exact chain complex in $\mathfrak{A}$. This is what we get by applying F to (3.5). So this is a resolution as asserted.

An axiom for triangulated categories provides an exact triangle

$$P_{11} \overset{(-d_{11},\Phi_1)}{\rightarrowtail} P_{10} \oplus P_{01} \twoheadrightarrow D \longrightarrow \Sigma P_{11}$$

containing $(-d_{11},\Phi_1)$. Similarly, the map $p\circ d_{00}\colon P_{00} \to A$ in (3.5) is part of an exact triangle

$$D' \overset{\gamma}{\rightarrowtail} P_{00} \overset{p\circ d_{00}}{\twoheadrightarrow} A \longrightarrow \Sigma D'. \tag{3.6}$$

The long exact sequences for F applied to these two exact triangles show that $F(D)$ is the cokernel of the monomorphism $F(-d_{11},\Phi_1)$, and that $F(D')$ is the kernel of the epimorphism $F(p\circ d_{00})$. The exactness of (3.5) implies $F(D) \cong F(D')$. Since B_0 and B_1 belong to $\langle\mathfrak{P}_{\mathfrak{J}}\rangle$, so do D and D'. And $F(D)$ has a projective resolution of length 1 by construction. Hence the Universal Coefficient Theorem applies to D and D'. Thus the isomorphism $F(D) \cong F(D')$ lifts to an isomorphism $D \cong D'$. We shall identify $D = D'$.

The Universal Coefficient Theorem for D gives a short exact sequence

$$\mathrm{Ext}^1_{\mathfrak{A}}\big(\Sigma F(D), F(P_{00})\big) \rightarrowtail \mathfrak{T}(D, P_{00}) \overset{F}{\twoheadrightarrow} \mathrm{Hom}_{\mathfrak{A}}\big(F(D), F(P_{00})\big). \tag{3.7}$$

We split $D = D^+ \oplus D^-$ and $P_{00} = P_{00}^+ \oplus P_{00}^-$ into objects of even and odd parity as in the construction of φ^1 above the theorem. Then $\mathfrak{T}(D, P_{00})$ splits accordingly as a 2×2-matrix. The sum of the diagonal terms $\mathfrak{T}(D^+, P_{00}^+) \oplus \mathfrak{T}(D^-, P_{00}^-)$ is isomorphic to $\mathrm{Hom}_{\mathfrak{A}}\big(F(D), F(P_{00})\big)$, whereas the sum of the off-diagonal terms $\mathfrak{T}(D^-, P_{00}^+) \oplus \mathfrak{T}(D^+, P_{00}^-)$ is isomorphic to $\mathrm{Ext}^1_{\mathfrak{A}}\big(\Sigma F(D), F(P_{00})\big)$. This is the (unnatural) splitting of the Universal Coefficient Theorem exact sequence (3.7) that follows because $\mathfrak{A}$ is paired. In particular, we decompose $\gamma = \gamma^0 + \gamma^1$ into its parity-preserving and parity-reversing parts.

Let A' be another object of $\langle\mathfrak{P}_{\mathfrak{J}}\rangle$ with an isomorphism $F(A') \cong F(A)$. The argument above shows that both A and A' are cones of some $\gamma, \gamma' \in \mathfrak{T}(D, P_{00})$ as in (3.6), which lift the inclusion map $F(D) \rightarrowtail F(P_{00})$ that we get from the resolution (3.5). The *relative obstruction class* is defined as follows: compose

$$\gamma - \gamma' \in \mathrm{Ext}^1_{\mathfrak{A}}\big(\Sigma F(D), F(P_{00})\big) \subseteq \mathfrak{T}(D, P_{00})$$

with the map $F(p \circ d_{00})\colon F(P_{00}) \to F(A)$ and apply the boundary map for the extension $F(D) \rightarrowtail F(P_{00}) \twoheadrightarrow F(A)$. That is, plug $\gamma - \gamma'$ into the map

$$\text{(3.8)} \quad \operatorname{Ext}^1_{\mathfrak{A}}(\Sigma F(D), F(P_{00})) \xrightarrow{F(p\circ d_{00})_*} \operatorname{Ext}^1_{\mathfrak{A}}(\Sigma F(D), F(A)) \xrightarrow{\partial_{DP_{00}A}} \operatorname{Ext}^2_{\mathfrak{A}}(\Sigma F(A), F(A)).$$

Let $\gamma^0\colon D \to P_{00}$ be the unique parity-preserving arrow that lifts the inclusion map $F(D) \rightarrowtail F(P_{00})$ and let A^0 be its cone. The *obstruction class* of A is the relative obstruction class for A and A^0. That is, we plug $\gamma^1 := \gamma - \gamma^0$ into (3.8).

Finally, we relate the obstruction class defined above to φ^1. The solid square in the following diagram commutes:

$$\text{(3.9)} \qquad \begin{array}{ccccccc} B_1 & \overset{\varphi}{\rightarrowtail} & B_0 & \overset{p}{\twoheadrightarrow} & A & \longrightarrow & \Sigma B_1 \\ \varepsilon\,\vdots\uparrow & & d_{00}\uparrow & & \| & & \Sigma\varepsilon\,\vdots\uparrow \\ D & \overset{\gamma}{\rightarrowtail} & P_{00} & \overset{p\circ d_{00}}{\twoheadrightarrow} & A & \longrightarrow & \Sigma D \end{array}$$

By the third axiom of triangulated categories, there is an arrow ε making all three squares commute. We shall only use the left square. Since D, B_1, B_0 and P_{00} have projective resolutions of length 1 and $\mathfrak{A}$ is paired, the Universal Coefficient Theorem allows us to split them into even and odd parts. Hence each of the arrows γ, φ, ε and d_{00} splits into a parity-preserving and a parity-reversing part. We have already used the splittings $\varphi = \varphi^0 + \varphi^1$ and $\gamma = \gamma^0 + \gamma^1$, and now we also split $\varepsilon = \varepsilon^0 + \varepsilon^1$. The arrow d_{00} is parity-preserving because $\mathfrak{T}(P_{00}, B_0) \cong \operatorname{Hom}_{\mathfrak{A}}(F(P_{00}), F(B_0))$ has no parity-reversing part. The left commuting square in (3.9) implies

$$\text{(3.10)} \qquad d_{00} \circ \gamma^1 = \varphi^0 \circ \varepsilon^1 + \varphi^1 \circ \varepsilon^0 = \varphi \circ \varepsilon^1 + \varphi^1 \circ \varepsilon.$$

The second step uses that the composite of two odd terms always vanishes, that is, the Ext-term in the Universal Coefficient Theorem is nilpotent.

Composing $\gamma^1 \in \operatorname{Ext}^1_{\mathfrak{A}}(\Sigma F(D), F(P_{00}))$ with $F(p \circ d_{00})$ as in (3.8) has the same effect as composing with $p \circ d_{00}$ because the exact sequence in the Universal Coefficient Theorem is natural. Thus the obstruction class of A is the image of $p \circ d_{00} \circ \gamma^1 \in \operatorname{Ext}^1_{\mathfrak{A}}(\Sigma F(D), F(A))$ under the boundary map

$$\partial_{DP_{00}A}\colon \operatorname{Ext}^1_{\mathfrak{A}}(\Sigma F(D), F(A)) \to \operatorname{Ext}^2_{\mathfrak{A}}(\Sigma F(A), F(A))$$

Equation (3.10) and $p \circ \varphi = 0$ imply

$$p \circ d_{00} \circ \gamma^1 = p \circ \varphi \circ \varepsilon^1 + p \circ \varphi^1 \circ \varepsilon = p \circ \varphi^1 \circ \varepsilon.$$

The composite $p \circ \varphi^1 = p_*(\varphi^1) \in \operatorname{Ext}^1_{\mathfrak{A}}(\Sigma F(B_1), F(A))$ also appears in the statement of the theorem. Composing with ε in $\mathfrak{T}$ has the same effect as composing with $F(\varepsilon)$ in the graded category $\operatorname{Ext}^*_{\mathfrak{A}}$. When we apply F to the morphism of exact triangles (3.9), we get the following morphism of extensions in $\mathfrak{A}$:

$$\begin{array}{ccccc} F(B_1) & \overset{F(\varphi)}{\rightarrowtail} & F(B_0) & \overset{F(p)}{\twoheadrightarrow} & F(A) \\ F(\varepsilon)\,\vdots\uparrow & & F(d_{00})\uparrow & & \| \\ F(D) & \overset{F(\gamma)}{\rightarrowtail} & F(P_{00}) & \overset{F(p)\circ F(d_{00})}{\twoheadrightarrow} & F(A) \end{array}$$

Since boundary maps in Ext-theory are natural, the boundary map

$$\partial\colon \operatorname{Ext}^1_{\mathfrak{A}}(\Sigma F(B_1), F(A)) \to \operatorname{Ext}^2_{\mathfrak{A}}(\Sigma F(A), F(A))$$

for the top row is equal to the composite of $F(\varepsilon)$ and the boundary map $\partial_{DP_{00}A}$. Thus the obstruction class of A is $\partial(F(p)\circ\varphi^1)$ as asserted. □

4. Comparison of classification theorems

In this section, we apply Theorem 3.2 in several cases to relate the classification theorem involving the obstruction class to other classification theorems. We first compare the classification for $\mathbb{Z}$-actions in Theorem 2.12 with the one obtained in [**4**]. Then we compare the classification for C*-algebras over a unique path space $(X,\to)$ in Corollary 2.20 with the one in [**4**]. In both cases, the invariants can be translated into each other rather directly.

4.1. Actions of the integers. The Universal Coefficient Theorem for actions of $\mathbb{Z}$ on C*-algebras in Section 2.1 is based on the stable homological ideal $\mathfrak{I}^{\mathbb{Z}}$ defined by the forgetful functor $\mathfrak{KK}^{\mathbb{Z}}\to\mathfrak{KK}$. Now we use another ideal $\mathfrak{I}^{\mathbb{Z}}_{\mathrm{K}}$. Let $\mathfrak{A}^{\mathbb{Z}}$ be the category of countable $\mathbb{Z}/2$-graded $\mathbb{Z}[x,x^{-1}]$-modules. Let

$$F^{\mathbb{Z}}_{\mathrm{K}}\colon \mathfrak{KK}^{\mathbb{Z}}\to\mathfrak{A}^{\mathbb{Z}},\qquad (A,\alpha)\mapsto\big(\mathrm{K}_*(A),\mathrm{K}_*(\alpha)\big),$$

that is, we map $(A,\alpha)\in\in\mathfrak{KK}^{\mathbb{Z}}$ to the $\mathbb{Z}/2$-graded Abelian group $\mathrm{K}_*(A)$ with the $\mathbb{Z}[x,x^{-1}]$-module structure given by the automorphism $\mathrm{K}_*(\alpha)$ of $\mathrm{K}_*(A)$. Let

$$\mathfrak{I}^{\mathbb{Z}}_{\mathrm{K}}(A,B) := \{\varphi\in\mathrm{KK}^{\mathbb{Z}}_0(A,B) : F^{\mathbb{Z}}_{\mathrm{K}}(\varphi)=0\}$$

be the kernel of $F^{\mathbb{Z}}_{\mathrm{K}}$ on morphisms. This example is treated in [**4**], but there $\mathfrak{KK}^{\mathbb{Z}}$ is disguised as $\mathfrak{KK}^{\mathbb{T}}$ for the circle group $\mathbb{T}$. These two categories are equivalent by Baaj–Skandalis duality (see [**1**]). The functor $F^{\mathbb{Z}}_{\mathrm{K}}$ corresponds to the functor on $\mathfrak{KK}^{\mathbb{T}}$ that maps a C*-algebra with a continuous $\mathbb{T}$-action to $\mathrm{K}^{\mathbb{T}}_*(A)$ with the canonical module structure over the representation ring $\mathbb{Z}[x,x^{-1}]$ of $\mathbb{T}$, which is used in [**4**].

If $B\in\in\mathrm{KK}^{\mathbb{Z}}$, then (2.5) implies

$$\mathrm{KK}^{\mathbb{Z}}_0(\mathrm{C}_0(\mathbb{Z}),B)\cong\mathrm{KK}_0(\mathbb{C},B)\cong\mathrm{K}_0(B)\cong\mathrm{Hom}_{\mathfrak{A}^{\mathbb{Z}}}\big(\mathbb{Z}[x,x^{-1}],F^{\mathbb{Z}}_{\mathrm{K}}(B)\big).$$

Thus the partial adjoint of $F^{\mathbb{Z}}_{\mathrm{K}}$ is defined on the rank-1 free module $\mathbb{Z}[x,x^{-1}]$ and maps it to $\mathrm{C}_0(\mathbb{Z})$. Then it is defined on all free modules, also in odd parity. Since idempotents in $\mathfrak{KK}^{\mathbb{Z}}$ split, the partial adjoint $(F^{\mathbb{Z}}_{\mathrm{K}})^{\vdash}$ of $F^{\mathbb{Z}}_{\mathrm{K}}$ is defined on all projective objects of $\mathfrak{A}^{\mathbb{Z}}$. Since $F^{\mathbb{Z}}_{\mathrm{K}}(\mathrm{C}_0(\mathbb{Z}))$ is the rank-1 free module again, we get $(F^{\mathbb{Z}}_{\mathrm{K}})\circ(F^{\mathbb{Z}}_{\mathrm{K}})^{\vdash}(P)\cong P$ for all projective objects of $\mathfrak{A}^{\mathbb{Z}}$. Any object in $\mathfrak{A}^{\mathbb{Z}}$ has a projective resolution of length 2. Hence it belongs to the image of $F^{\mathbb{Z}}_{\mathrm{K}}$ by [**4**, Lemma 2.4]. Remark 2.2 and Theorem 2.1 show that $F^{\mathbb{Z}}_{\mathrm{K}}$ is the universal $\mathfrak{I}^{\mathbb{Z}}_{\mathrm{K}}$-exact stable homological functor both to an exact and to an Abelian category.

The category $\mathfrak{A}^{\mathbb{Z}}$ is paired in an obvious way, using the subcategories $\mathfrak{A}^{\mathbb{Z}}_{\pm}$ of countable $\mathbb{Z}[x,x^{-1}]$-modules concentrated in degree 0 and 1, respectively. So the obstruction class is defined for any object of $\mathfrak{KK}^{\mathbb{Z}}$. Assume that $(A,\alpha)\in\in\mathfrak{KK}^{\mathbb{Z}}$ is such that A belongs to the bootstrap class in $\mathfrak{KK}$. Equivalently, (A,α) belongs to the localising subcategory of $\mathfrak{KK}^{\mathbb{Z}}$ generated by the $\mathfrak{I}^{\mathbb{Z}}_{\mathrm{K}}$-projective object $\mathrm{C}_0(\mathbb{Z})$. Then the main result of [**4**] shows that A is determined uniquely up to isomorphism by $\big(\mathrm{K}_*(A),\mathrm{K}_*(\alpha)\big)\in\in\mathfrak{A}^{\mathbb{Z}}$ and the obstruction class.

A chain complex in $\mathfrak{KK}^{\mathbb{Z}}$ that is $\mathfrak{I}^{\mathbb{Z}}$-exact is also $\mathfrak{I}^{\mathbb{Z}}_{\mathrm{K}}$-exact because $\mathfrak{I}^{\mathbb{Z}}\subseteq\mathfrak{I}^{\mathbb{Z}}_{\mathrm{K}}$. So the $\mathfrak{I}^{\mathbb{Z}}$-projective resolution in (2.11) is $\mathfrak{I}^{\mathbb{Z}}_{\mathrm{K}}$-exact. Its entries $\mathrm{C}_0(\mathbb{Z},A)$ are no longer $\mathfrak{I}^{\mathbb{Z}}_{\mathrm{K}}$-projective. We claim, however, that they have $\mathfrak{I}^{\mathbb{Z}}_{\mathrm{K}}$-projective resolutions of length 1. Let $P_1 \rightarrowtail P_0 \twoheadrightarrow \mathrm{K}_*(A)$ be a resolution of the $\mathbb{Z}/2$-graded Abelian group $\mathrm{K}_*(A)$ by countable free Abelian groups. Then

$$\mathbb{Z}[x,x^{-1}] \otimes P_1 \rightarrowtail \mathbb{Z}[x,x^{-1}] \otimes P_0 \twoheadrightarrow \mathbb{Z}[x,x^{-1}] \otimes \mathrm{K}_*(A)$$

is a projective resolution of

$$F_{\mathrm{K}}^{\mathbb{Z}}(\mathrm{C}_0(\mathbb{Z},A)) \cong \mathbb{Z}[x,x^{-1}] \otimes \mathrm{K}_*(A).$$

Here the tensor products with $\mathbb{Z}[x,x^{-1}]$ carry the module structure defined by multiplication with x in the tensor factor $\mathbb{Z}[x,x^{-1}]$. So we are in the situation of (3.1). Theorem 3.2 implies the following formula for the obstruction class of A:

THEOREM 4.1. *Let A be a C^*-algebra in the bootstrap class in $\mathfrak{KK}$ and $\alpha \in \mathrm{Aut}(A)$. Split $[\alpha] \in \mathrm{KK}_0(A,A)$ into parity-preserving and parity-reversing parts $\alpha^0 \in \mathrm{Hom}(\mathrm{K}_*(A),\mathrm{K}_*(A))$ and $\alpha^1 \in \mathrm{Ext}(\mathrm{K}_{1+*}(A),\mathrm{K}_*(A))$. Define*

$$\gamma\colon \mathrm{Ext}^1_{\mathbb{Z}}(\mathrm{K}_{1+*}(A),\mathrm{K}_*(A)) \to \mathrm{Ext}^1_{\mathbb{Z}}(\mathrm{K}_{1+*}(A),\mathrm{K}_*(A)), \qquad x \mapsto \alpha\circ x - x\circ\alpha.$$

Then $\mathrm{Ext}^2_{\mathbb{Z}[x,x^{-1}]}(\Sigma F_{\mathrm{K}}^{\mathbb{Z}}(A), F_{\mathrm{K}}^{\mathbb{Z}}(A)) \cong \operatorname{coker}\gamma$ and the obstruction class of A is the image of $-\alpha^1(\alpha^0)^{-1}$ in this cokernel.

PROOF. In our case, the map φ in (3.1) is the map $\mathrm{C}_0(\mathbb{Z}) \otimes A \to \mathrm{C}_0(\mathbb{Z}) \otimes A$ in (2.11). Split $A = A^+ \oplus A^-$ into its even and odd parts as before. Then

$$\mathrm{C}_0(\mathbb{Z}) \otimes A \cong \mathrm{C}_0(\mathbb{Z}) \otimes A^+ \oplus \mathrm{C}_0(\mathbb{Z}) \otimes A^-$$

is the parity decomposition of $\mathrm{C}_0(\mathbb{Z}) \otimes A$. The translation τ and the identity are parity-preserving. Decompose $[\alpha]$ and $[\alpha^{-1}]$ into their even and odd parts. The parity-reversing part of the map on $\mathrm{C}_0(\mathbb{Z}) \otimes A$ is $\varphi^1 = [\tau] \otimes [\alpha^{-1}]^1$. The map $p \in \mathrm{KK}_0^{\mathbb{Z}}(\mathrm{C}_0(\mathbb{Z}) \otimes A, A)$ satisfies $p(\tau\otimes \mathrm{id}_A) = [\alpha]p$ because it is $\mathbb{Z}$-equivariant, and it restricts to the identity map on the 0th summand of $\mathrm{C}_0(\mathbb{Z}) \otimes A$. The isomorphism $\mathrm{KK}_0^{\mathbb{Z}}((\mathrm{C}_0(\mathbb{Z},A),\tau),(B,\beta)) \cong \mathrm{KK}_0(A,B)$ in (2.5) forgets the $\mathbb{Z}$-action and then evaluates at 0. Therefore, the image of φ^1 in $\mathrm{KK}_0(A,A)$ is the restriction of $p\circ\varphi^1$ to the $0th$ summand. And this is $[\alpha][\alpha^{-1}]^1 \in \mathrm{KK}_0(A,A)$. Since $\alpha\alpha^{-1} = \mathrm{id}_A$ and products of two parity-reversing KK-classes always vanish, $[\alpha][\alpha^{-1}]^1 + [\alpha]^1[\alpha^{-1}] = [\mathrm{id}_A]^1 = 0$. Theorem 3.2 now says that the obstruction class for A is the image of

$$[\alpha][\alpha^{-1}]^1 = -[\alpha]^1[\alpha^{-1}] = -\alpha^1(\alpha^0)^{-1} \in \mathrm{Ext}^1_{\mathbb{Z}}(\mathrm{K}_{1+*}(A),\mathrm{K}_*(A)))$$

under the boundary map to $\mathrm{Ext}^2_{\mathbb{Z}[x,x^{-1}]}(\Sigma F_{\mathrm{K}}^{\mathbb{Z}}(A), F_{\mathrm{K}}^{\mathbb{Z}}(A))$. The description of the latter Ext group in the theorem follows when we compute it with the projective resolution in (3.5) defined by the length-1 resolutions of $\mathrm{C}_0(\mathbb{Z}) \otimes A$ above. This computation has already been done in [**4**, Section 3.2]. When the Ext groups are described in this way, the boundary map in Theorem 3.2 becomes a trivial map, mapping an element of $\mathrm{Ext}_{\mathbb{Z}}(\mathrm{K}_{1+*}(A),\mathrm{K}_*(A))$ to its image in $\operatorname{coker}\gamma$. □

The formula for the obstruction class depends, of course, on the isomorphism $\mathrm{Ext}^2_{\mathbb{Z}[x,x^{-1}]}(\Sigma F_{\mathrm{K}}^{\mathbb{Z}}(A), F_{\mathrm{K}}^{\mathbb{Z}}(A)) \cong \operatorname{coker}\gamma$, and this depends on the $\mathfrak{I}^{\mathbb{Z}}$-projective resolution from which it is obtained. Our theorem uses the most obvious resolution. One may apply the automorphism of $\mathrm{Ext}^2_{\mathbb{Z}[x,x^{-1}]}(\Sigma F_{\mathrm{K}}^{\mathbb{Z}}(A), F_{\mathrm{K}}^{\mathbb{Z}}(A))$ that composes with the automorphism $-\Sigma\alpha^0$ on $\Sigma F_{\mathrm{K}}^{\mathbb{Z}}(A)$. This replaces the obstruction class $-\alpha^1(\alpha^0)^{-1}$ by α^1. So the images of $-\alpha^1(\alpha^0)^{-1}$ and α^1 in $\mathrm{Ext}^2_{\mathbb{Z}[x,x^{-1}]}(\Sigma F_{\mathrm{K}}^{\mathbb{Z}}(A), F_{\mathrm{K}}^{\mathbb{Z}}(A))$ contain the same information.

Theorem 4.1 and the computations after Theorem 2.12 allow to deduce the classification by $F_{\mathrm{K}}^{\mathbb{Z}}$ and the obstruction class from the classification by the invariant $F^{\mathbb{Z}}$ in Section 2.1. Let (A, α) and (B, β) belong to the bootstrap class in $\mathfrak{KK}^{\mathbb{Z}}$. Assume that there is an isomorphism $t^0 \colon F_{\mathrm{K}}^{\mathbb{Z}}(A) \xrightarrow{\sim} F_{\mathrm{K}}^{\mathbb{Z}}(B)$ in $\mathfrak{A}^{\mathbb{Z}}$, that is, a grading-preserving $\mathbb{Z}[x, x^{-1}]$-module isomorphism $\mathrm{K}_*(A) \cong \mathrm{K}_*(B)$. We use this isomorphism to identify $\mathrm{K}_*(A) = \mathrm{K}_*(B)$. By Theorem 4.1, the relative obstruction class vanishes if and only if $\alpha^1 - \beta^1 \in \operatorname{Ext}(\mathrm{K}_{*+1}(A), \mathrm{K}_*(A))$ vanishes in $\operatorname{coker}(\gamma)$. Equivalently, there is $t^1 \in \operatorname{Ext}(\mathrm{K}_{1+*}(A), \mathrm{K}_*(B))$ for which $t = t^0 + t^1 \in \mathrm{KK}_0(A, B)$ satisfies $[\beta] \circ t = t \circ [\alpha]$. Then t is an isomorphism $F^{\mathbb{Z}}(A) \cong F^{\mathbb{Z}}(B)$ in $\mathfrak{KK}[\mathbb{Z}]$, and Theorem 2.12 shows that such an isomorphism lifts to an isomorphism in $\mathrm{KK}_0^{\mathbb{Z}}(A, B)$.

Recall that Baaj–Skandalis duality is an equivalence of triangulated categories $\mathfrak{KK}^{\mathbb{T}} \cong \mathfrak{KK}^{\mathbb{Z}}$. Hence everything said above about $\mathbb{Z}$-actions carries over to $\mathbb{T}$-actions. The functor $A \mapsto \mathrm{K}_*(A)$ becomes $B \mapsto \mathrm{K}_*^{\mathbb{T}}(B)$ on $\mathfrak{KK}^{\mathbb{T}}$, equipped with the natural module structure over the representation ring $\mathbb{Z}[x, x^{-1}]$ of $\mathbb{T}$. The automorphism α is replaced by an automorphism of $B \rtimes \mathbb{T}$, namely, the generator β of the dual action of $\mathbb{Z}$. Since $\mathrm{K}_*(B \rtimes \mathbb{T}) \cong \mathrm{K}_*^{\mathbb{T}}(B)$, the Universal Coefficient Theorem splits $\mathrm{KK}_0(B \rtimes \mathbb{T}, B \rtimes \mathbb{T})$ into the parity-preserving part $\operatorname{Hom}(\mathrm{K}_*^{\mathbb{T}}(B), \mathrm{K}_*^{\mathbb{T}}(B))$ and the parity-reversing part $\operatorname{Ext}(\mathrm{K}_{1+*}^{\mathbb{T}}(B), \mathrm{K}_*^{\mathbb{T}}(B))$. The obstruction class of $B \rtimes \mathbb{T}$ is the class of $-\beta^1(\beta^0)^{-1} \in \operatorname{Ext}(\mathrm{K}_{1+*}^{\mathbb{T}}(B), \mathrm{K}_*^{\mathbb{T}}(B))$ in the cokernel of the map

$$\gamma \colon \operatorname{Ext}(\mathrm{K}_{1+*}^{\mathbb{T}}(B), \mathrm{K}_*^{\mathbb{T}}(B)) \to \operatorname{Ext}(\mathrm{K}_{1+*}^{\mathbb{T}}(B), \mathrm{K}_*^{\mathbb{T}}(B)), \qquad x \mapsto \beta^0 x - x\beta^0.$$

As above, a grading-preserving $\mathbb{Z}[x, x^{-1}]$-module isomorphism $t \colon \mathrm{K}_*^{\mathbb{T}}(A) \to \mathrm{K}_*^{\mathbb{T}}(B)$ is compatible with the obstruction classes if and only if it lifts to an isomorphism between $A \rtimes \mathbb{T}$ and $B \rtimes \mathbb{T}$ in $\mathfrak{KK}[\mathbb{Z}]$, and such an isomorphism lifts further to an isomorphism between $A \rtimes \mathbb{T}$ and $B \rtimes \mathbb{T}$ in $\mathfrak{KK}^{\mathbb{Z}}$. By Baaj–Skandalis duality, the latter is equivalent to an invertible element in $\mathrm{KK}_0^{\mathbb{T}}(A, B)$.

4.2. C*-algebras over unique path spaces. Now we return to the setup of Section 2.2. So $(X, \to)$ is a countable directed graph with the unique path property. Let $\preceq$ be the partial order generated by $\leftarrow$ and equip X with the Alexandrov topology defined by $\preceq$. Let $\mathfrak{KK}^X$ be the category of C*-algebras over X. Objects in the appropriate bootstrap class in $\mathfrak{KK}^X$ are classified in [**4**] under the extra assumption that X be finite. Actually, the arguments in [**4**] work in the same way if the set X is countable. Here we treat this more general case right away.

Let $\mathfrak{A}^X$ be the category of all functors from X to the category of countable $\mathbb{Z}/2$-graded Abelian groups. Equivalently, an object of $\mathfrak{A}^X$ is a family of countable $\mathbb{Z}/2$-graded Abelian groups G_x for $x \in X$ with grading-preserving group homomorphisms $\gamma_{y,x} \colon G_x \to G_y$ for all $x, y \in X$ with $x \to y$. Morphisms $(G_x, \gamma_{y,x}) \to (H_x, \eta_{y,x})$ in $\mathfrak{A}^X$ are families of grading-preserving group homomorphisms $t_x \colon G_x \to H_x$ that satisfy $t_y \circ \gamma_{y,x} = \eta_{y,x} \circ t_x$ for all $x, y \in X$ with $x \to y$. We define the functor

$$F_{\mathrm{K}}^X : \mathfrak{KK}^X \to \mathfrak{A}^X$$

by mapping a C*-algebra over X to the diagram of $\mathbb{Z}/2$-graded Abelian groups $\mathrm{K}_*(A(U_x))$ for $x \in X$ with the maps induced by the inclusion maps $A(U_x) \triangleleft A(U_y)$ for $x \to y$. The target category $\mathfrak{A}^X$ is a paired, stable Abelian category, where $\mathfrak{A}_{\pm}^X \subseteq \mathfrak{A}^X$ are the subcategories of $\mathbb{Z}/2$-graded groups where the odd or even part vanishes, respectively. And F_{K}^X is a stable homological functor. Let $\mathfrak{I}_{\mathrm{K}}^X$ be its kernel on morphisms. Equation (2.17) implies

$$\mathrm{KK}_0^X(i_z(\mathbb{C}), B) \cong \mathrm{KK}_0\big(\mathbb{C}, B(U_z)\big) \cong \mathrm{K}_0\big(B(U_z)\big) \cong \mathrm{Hom}_{\mathfrak{A}^X}\big(j_z(\mathbb{C}), F_{\mathrm{K}}^X(B)\big).$$

Hence the partial adjoint $(F_{\mathrm{K}}^X)^{\vdash}$ of F_{K}^X is defined on $j_z(\mathbb{C})$ for all $z \in X$. As in Section 2.2, any object of $\mathfrak{A}^X$ is a quotient of a direct sum of objects of the form $j_z(\mathbb{C})$ for $z \in X$. Hence $(F_{\mathrm{K}}^X)^{\vdash}$ is defined on all projective objects of $\mathfrak{I}_{\mathrm{K}}^X$ and $F_{\mathrm{K}}^X \circ (F_{\mathrm{K}}^X)^{\vdash}(P) = P$ for all projective objects P of $\mathfrak{A}^X$; this is proved like the corresponding statement about $F_{\mathrm{K}}^{\mathbb{Z}}$ in Section 4.1. Therefore, the functor $F_{\mathrm{K}}^X : \mathfrak{KK}^X \to \mathfrak{A}^X$ is the universal $\mathfrak{I}_{\mathrm{K}}^X$-exact stable homological functor into an exact category or into an Abelian category by Theorem 2.1 and Remark 2.2. And F_{K}^X restricts to an equivalence of categories between the $\mathfrak{I}_{\mathrm{K}}^X$-projective objects in $\mathfrak{KK}^X$ and the projective objects in $\mathfrak{A}^X$. Let $\mathfrak{B}^X \subseteq \mathfrak{KK}^X$ be the localising subcategory generated by the $\mathfrak{I}_{\mathrm{K}}^X$-projective objects. This is the analogue of the bootstrap class in $\mathfrak{KK}^X$.

LEMMA 4.2. *Let $G = (G_x, \gamma_{y,x})$ and $H = (H_x, \eta_{y,x})$ be objects of $\mathfrak{A}^X$.*

(1) *There is a projective resolution for G of length 2 in $\mathfrak{A}^X$.*
(2) *There is $A \in \mathfrak{B}^X$ with $F_{\mathrm{K}}^X(A) \cong G$.*
(3) *Write* Ext *for the* Ext^1 *of Abelian groups. The group $\mathrm{Ext}^2_{\mathfrak{A}^X}(G, H)$ is naturally isomorphic to the cokernel of the map*

$$\prod_{x\in X} \mathrm{Ext}(G_x, H_x) \to \prod_{x\to y} \mathrm{Ext}(G_x, H_y), \qquad (t_x)_{x\in X} \mapsto (\eta_{y,x} \circ t_x - t_y \circ \gamma_{y,x})_{x\to y}, \tag{4.3}$$

PROOF. Since $\mathfrak{I}^X \subseteq \mathfrak{I}_{\mathrm{K}}^X$, any $\mathfrak{I}^X$-exact chain complex in $\mathfrak{KK}^X$ is also $\mathfrak{I}_{\mathrm{K}}^X$-exact. In particular, the $\mathfrak{I}^X$-projective resolution in (2.19) is $\mathfrak{I}_{\mathrm{K}}^X$-exact. We claim that its entries have $\mathfrak{I}_{\mathrm{K}}^X$-projective resolutions of length 1. Hence there is a projective resolution of $F_{\mathrm{K}}^X(A)$ of length 2 as in (3.5). To prove the same for all objects G of $\mathfrak{A}^X$, we carry over (2.19).

Let $J_z(B) \in\in \mathfrak{A}^{\mathbb{Z}}$ for a countable $\mathbb{Z}/2$-graded Abelian group B be the diagram with $J_z(B)_x = B$ if $z \succeq x$ and $J_z(B)_x = 0$ otherwise, with the identity map $J_z(B)_x \to J_z(B)_y$ for $z \succeq x \to y$ and the zero map otherwise. Then

$$\mathrm{Hom}_{\mathfrak{A}^X}\big(J_z(B), H\big) \cong \mathrm{Hom}\big(B, H_z\big), \tag{4.4}$$

where the isomorphism simply restricts a morphism in $\mathfrak{A}^X$ to the object $z \in X$. In particular, if $y \succeq z$, then $J_y(B)_z = B$ and so there is a canonical map $J_{y\succeq z}(B) \colon J_z(B) \to J_y(B)$ in $\mathfrak{A}^X$. Explicitly, this map is the identity map on $J_z(B)_x$ if $z \succeq x$ and the zero map on $J_z(B)_x = 0$ if $z \not\succeq x$. The proof that (2.19) is exact also proves the exactness of

$$0 \to \bigoplus_{x\to y} J_y(G_x) \xrightarrow{\psi} \bigoplus_{x\in X} J_x(G_x) \xrightarrow{q} G \to 0; \tag{4.5}$$

here ψ restricted to the summand $J_y(G_x)$ for $x \to y$ is the map $(J_{x\to y}(G_x), J_y(\gamma_{y,x}))$ to $J_x(G_x) \oplus J_y(G_y)$, and q maps $J_x(G_x)$ to G by the adjunct of the identity map on G_x under the adjunction in (4.4). Explicitly, the entry of $\bigoplus_{x\in X} J_x(G_x)$ at $z \in X$ is $\bigoplus_{x\succeq z} G_x$, which is mapped to G_z by $(\gamma_{z,x})_{x\succeq z}$. The proof that (4.5) is exact shows that the chain complexes of $\mathbb{Z}/2$-graded Abelian groups

$$0 \to \bigoplus_{x\to y} J_y(G_x)_z \xrightarrow{\psi_z} \bigoplus_{x\in X} J_x(G_x)_z \xrightarrow{q_z} G_z \to 0$$

are contractible for all $z \in X$.

For each $x \in X$, there is a resolution

$$P_{x,1} \overset{d_{x,1}}{\rightarrowtail} P_{x,0} \overset{d_{x,0}}{\twoheadrightarrow} G_x \tag{4.6}$$

of G_x of length 1 by countable free $\mathbb{Z}/2$-graded Abelian groups. The homomorphism $\gamma_{y,x}\colon G_x \to G_y$ for $x \to y$ lifts to a morphism of extensions

$$\begin{array}{ccccc} P_{x,1} & \overset{d_{x,1}}{\rightarrowtail} & P_{x,0} & \overset{d_{x,0}}{\twoheadrightarrow} & G_x \\ \downarrow{\scriptstyle\gamma^1_{y,x}} & & \downarrow{\scriptstyle\gamma^0_{y,x}} & & \downarrow{\scriptstyle\gamma_{y,x}} \\ P_{y,1} & \overset{d_{y,1}}{\rightarrowtail} & P_{y,0} & \overset{d_{y,0}}{\twoheadrightarrow} & G_y \end{array} \tag{4.7}$$

The construction J_z above is an exact functor, and it maps free Abelian groups to projective objects of $\mathfrak{A}^X$ by the adjunction (4.4). Hence

$$\bigoplus_{x\to y} J_y(P_{x,1}) \rightarrowtail \bigoplus_{x\to y} J_y(P_{x,0}) \twoheadrightarrow \bigoplus_{x\to y} J_y(G_x),$$
$$\bigoplus_{x\in X} J_x(P_{x,1}) \rightarrowtail \bigoplus_{x\in X} J_x(P_{x,0}) \twoheadrightarrow \bigoplus_{x\in X} J_sx(G_x)$$

are projective resolutions of length 1 in $\mathfrak{A}^X$. The resolution in (4.5) gives a projective resolution of G of length 2 as in (3.5). This proves the first assertion.

Now [**4**, Lemma 2.4] shows that there is $A \in\in \mathfrak{B}^X$ with $G = F^X_{\mathrm{K}}(A)$; the property that $\mathrm{KK}^X_0(A, B) = 0$ for all $\mathfrak{J}^X_{\mathrm{K}}$-contractible $B \in\in \mathfrak{KK}^X$ is equivalent to $A \in\in \mathfrak{B}^X$.

The projective resolution of G built above has the form

$$0 \to \bigoplus_{x\to y} J_y(P_{x,1}) \to \bigoplus_{x\to y} J_y(P_{x,0}) \oplus \bigoplus_{x\in X} J_x(P_{x,1}) \to \bigoplus_{x\in X} J_x(P_{x,0}) \to G,$$

where the maps $\bigoplus_{x\to y} J_y(P_{x,i}) \to \bigoplus_{x\in X} J_x(P_{x,i})$ for $i = 0, 1$ use $\gamma^i_{y,x}\colon P_{x,i} \to P_{y,i}$ in (4.7). We use this resolution to compute the group $\mathrm{Ext}^2_{\mathfrak{A}^X}(G, H)$. An element of $\mathrm{Ext}^2_{\mathfrak{A}^X}(G, H)$ is represented by a map $\bigoplus_{x\to y} J_y(P_{x,1}) \to H$. By the adjunction (4.4), this corresponds to a family of maps $f_{x\to y}\colon P_{x,1} \to H_y$. This family represents 0 in $\mathrm{Ext}^2_{\mathfrak{A}^X}(G, H)$ if and only if the corresponding map $\bigoplus_{x\to y} J_y(P_{x,1}) \to H$ factors through $\bigoplus_{x\to y} J_y(P_{x,0}) \oplus \bigoplus_x J_x(P_{x,1})$. Using the adjunction (4.4) again, a map on this direct sum corresponds to families of maps $g_{x\to y}\colon P_{x,0} \to H_y$ and $h_x\colon P_{x,1} \to H_x$. The resulting map $\bigoplus_{x\to y} J_y(P_{x,1}) \to H$ corresponds to the family of maps

$$-g_{x\to y} \circ d_{x,1} + \eta_{y,x} h_x - h_y \gamma^1_{y,x}\colon P_{x,1} \to H_y.$$

The resolutions (4.6) compute $\mathrm{Ext}(G_x, D)$ for any $\mathbb{Z}/2$-graded Abelian group D. So each $f_{x\to y}\colon P_{x,1} \to H_y$ represents an element of $\mathrm{Ext}(G_x, H_y)$, and it represents the zero element if and only if it is of the form $g_{x\to y} \circ d_{x,1}$ for some $g_{x\to y}\colon P_{x,0} \to H_y$. The elements h_x above represent elements of $\mathrm{Ext}(G_x, H_x)$. If they represent the zero element of $\mathrm{Ext}(G_x, H_x)$, then the term $\eta_{y,x} h_x - h_y \gamma^1_{y,x}$ above may be rewritten in the form $-g_{x\to y}\circ d_{x,1}$. Now we get the formula for $\mathrm{Ext}^2_{\mathfrak{A}^X}(G,H)$ in the third statement in the lemma. □

THEOREM 4.8. *Let A and B belong to $\mathfrak{B}^X$. An isomorphism $t\colon F_{\mathrm{K}}^X(A)\xrightarrow{\sim} F_{\mathrm{K}}^X(B)$ in $\mathfrak{A}^X$ lifts to an invertible element in $\mathrm{KK}_0^X(A,B)$ if and only if $t\delta_A=\delta_B t$ holds in $\mathrm{Ext}^2_{\mathfrak{A}^X}\big(\Sigma F_{\mathrm{K}}^X(A),F_{\mathrm{K}}^X(B)\big)$, where δ_A and δ_B are the obstruction classes of A and B.*

PROOF. The lemma verifies all the conditions to apply the classification method of [**4**]. □

The proof of the lemma also gives all the ingredients needed in Section 3. So we may now compute obstruction classes:

THEOREM 4.9. *Let B be an object of $\mathfrak{B}^X$. Let $B_x := B(U_x)$ for $x\in X$. Let $\beta_{y,x}\in \mathrm{KK}_0(B_x,B_y)$ be the KK-class of the inclusion map $B_x\hookrightarrow B_y$. Split $\beta_{y,x} = \beta^0_{y,x}+\beta^1_{y,x}$ with a parity-preserving part $\beta^0_{y,x}\in \mathrm{Hom}\big(\mathrm{K}_*(B_x),\mathrm{K}_*(B_y)\big)$ and a parity-reversing part $\beta^1_{y,x}\in \mathrm{Ext}\big(\mathrm{K}_{1+*}(B_x),\mathrm{K}_*(B_y)\big)$. The obstruction class of B is the class in the cokernel of the map in (4.3) that is represented by*

$$(\beta^1_{y,x})_{x\to y}\in \prod_{x\to y}\mathrm{Ext}\big(\mathrm{K}_{1+*}(B_x),\mathrm{K}_*(B_y)\big).$$

PROOF. The projective resolution (2.19) in $\mathfrak{KK}[X]$ lifts to an exact triangle

$$\bigoplus_{x\to y} i_y(B_x)\xrightarrow{\varphi}\bigoplus_{x\in X} i_x(B_x)\xrightarrow{p} B\to \Sigma\bigoplus_{x\to y} i_y(B_x)$$

in $\mathfrak{KK}^X$ by [**4**, Proposition 2.3]. This is how the Universal Coefficient Theorem in [**11**, Theorem 66] is proved. Here the map φ restricted to the summand $i_y(B_x)$ is the difference of two maps: the map $i_y(\beta_{y,x})$ to $i_y(B_y)$ and the canonical map $i_{x,y}(B_x)\colon i_y(B_x)\to i_x(B_x)$ that is the adjunct of the identity map $B_x\to i_x(B_x)_y = B_x$ under the adjunction (2.15). And the map p restricted to $i_x(B_x)$ is the adjunct of the identity map $B_x\to B_x$ under the adjunction (2.15).

Both $\bigoplus_{x\to y} i_y(B_x)$ and $\bigoplus_{x\in X} i_x(B_x)$ belong to $\mathfrak{B}^X$ and have $\mathfrak{I}^X_{\mathrm{K}}$-projective resolutions of length 1 (see (4.6)). So we are in the situation of (3.1). We split each $B_x := B(U_x)$ into its even and odd parity part $B_x = B_x^+\oplus B_x^-$. Then $i_y(B_x^+)$ and $i_y(B_x^-)$ are of even or odd parity, respectively. Split $\beta_{y,x} = \beta^0_{y,x}+\beta^1_{y,x}$ into a parity-preserving and a parity-reversing part. So $\beta^0_{y,x}\in \mathrm{Hom}\big(\mathrm{K}_*(B_x),\mathrm{K}_*(B_y)\big)$ and $\beta^1_{y,x}\in \mathrm{Ext}\big(\mathrm{K}_{1+*}(B_x),\mathrm{K}_*(B_y)\big)$ by the Universal Coefficient Theorem for KK, see the discussion after Theorem 2.12. The induced maps $i_y(\beta^0_{y,x})$ and $i_y(\beta^1_{y,x})$ are parity-preserving and parity-reversing, respectively. And the map $i_{x,y}(B_x)$ is parity-preserving. So the parity-reversing part φ^1 of φ is the map that restricts to $i_y(\beta^1_{y,x})\colon i_y(B_x)\to i_y(B_y)$ on the summand for $x\to y$. The composite $p\circ\varphi^1$ is the map $\bigoplus_{x\to y} i_y(B_x)\to B$ whose restriction to the summand $i_y(B_x)$ is adjunct to $\beta^1_{y,x}\colon B_x\to B_y$. These maps define an element of $\prod_{x\to y}\mathrm{Ext}\big(\mathrm{K}_{1+*}(B_x),\mathrm{K}_*(B_y)\big)$. The obstruction class of B is its image under the boundary map to $\mathrm{Ext}^2_{\mathfrak{A}^X}\big(\Sigma F^X_{\mathrm{K}}(B), F^X_{\mathrm{K}}(B)\big)$ by Theorem 3.2. We have described $\mathrm{Ext}^2_{\mathfrak{A}^X}(G,H)$ in Lemma 4.2 in such a way that this boundary map becomes tautological: it simply maps an element of $\prod_{x\to y}\mathrm{Ext}(G_x,H_y)$ to its class in the cokernel of the map in (4.3). In particular, the obstruction class of B is represented by $(\beta^1_{y,x})_{x\to y}\in\prod_{x\to y}\mathrm{Ext}\big(\mathrm{K}_{1+*}(B_x),\mathrm{K}_*(B_y)\big)$. □

Theorem 4.9 allows us to compare the classification theorems for C*-algebras in $\mathfrak{B}^X$ that use the invariant $F^X(B)$ or $F_{\mathrm{K}}^X(B)$ with the obstruction class. Let $A, B \in\in \mathfrak{B}^X$. An isomorphism

$$t^0 \colon F_{\mathrm{K}}^X(A) \xrightarrow{\sim} F_{\mathrm{K}}^X(B)$$

is equivalent to a family of isomorphisms

$$t_x^0 \colon \mathrm{K}_*(A(U_x)) \xrightarrow{\sim} \mathrm{K}_*(B(U_x))$$

that make the diagrams

$$\begin{array}{ccc} \mathrm{K}_*(A(U_x)) & \xrightarrow[\cong]{t_x^0} & \mathrm{K}_*(B(U_x)) \\ \downarrow{\scriptstyle \mathrm{K}_*(\alpha_{y,x})} & & \downarrow{\scriptstyle \mathrm{K}_*(\beta_{y,x})} \\ \mathrm{K}_*(A(U_y)) & \xrightarrow[\cong]{t_y^0} & \mathrm{K}_*(B(U_y)) \end{array}$$

commute for all edges $x \to y$. Here the maps $\mathrm{K}_*(\alpha_{y,x})$ and $\mathrm{K}_*(\beta_{y,x})$ are induced by the inclusion maps of our C*-algebras over X. The obstruction class for the isomorphism $(t_x^0)_{x\in X}$ vanishes if and only if there are $t_x^1 \in \mathrm{Ext}\big(\mathrm{K}_{1+*}(A(U_x)), \mathrm{K}_*(B(U_x))\big)$ for $x \in X$ such that

$$t_y^1 \circ \mathrm{K}_*(\alpha_{y,x}) - \mathrm{K}_*(\beta_{y,x}) \circ t_x^1 = \beta_{y,x}^1 \circ t_x^0 - t_y^0 \circ \alpha_{y,x}^1 \tag{4.10}$$

for all edges $x \to y$. Here

$$\alpha_{y,x}^1 \in \mathrm{Ext}\big(\mathrm{K}_{1+*}(A(U_x)), \mathrm{K}_*(A(U_y))\big), \qquad \beta_{y,x}^1 \in \mathrm{Ext}\big(\mathrm{K}_{1+*}(B(U_x)), \mathrm{K}_*(B(U_y))\big)$$

are the parity-reversing parts of the KK-classes of the inclusion maps. The elements t_x^0 and t_x^1 together form $t_x \in \mathrm{KK}_0(A(U_x), B(U_x))$. And (4.10) is equivalent to $t_y \circ \alpha_{y,x} = \beta_{y,x} \circ t_x$ in $\mathrm{KK}_0(A(U_x), B(U_y))$.

Corollary 2.20 says that any family of KK-equivalences $t_x \in \mathrm{KK}_0(A(U_x), B(U_x))$ with $t_y \circ \alpha_{y,x} = \beta_{y,x} \circ t_x$ for all edges $x \to y$ lifts to an invertible element in $\mathrm{KK}_0^X(A, B)$. The classification using the obstruction class says that a family of isomorphisms $t_x^0 \colon \mathrm{K}_*(A(U_x)) \xrightarrow{\sim} \mathrm{K}_*(B(U_x))$ lifts to an invertible element in $\mathrm{KK}_0^X(A, B)$ if and only if there exist t_x^1 satisfying (4.10). Corollary 2.20 makes a slightly stronger assertion because it says that any choice of elements t_x^1 satisfying (4.10) may be realised by an invertible element in $\mathrm{KK}_0^X(A, B)$.

The result in Corollary 2.20 about the existence of liftings is also slightly stronger than the corresponding result using the invariant F_{K}^X and the obstruction class because it does not require the C*-algebras $A(U_x)$ to belong to the bootstrap class.

5. Filtrated K-theory for totally ordered spaces

Now we consider the special unique path space

$$X = \{1 \leftarrow 2 \leftarrow \cdots \leftarrow n\}$$

for some $n \in \mathbb{N}_{\geq 2}$. We are going to compare the classification for C*-algebras over X that follows from the Universal Coefficient Theorem in [**12**] to the classifications in Section 4.2. The invariant used in [**12**] is filtrated K-theory. This is the diagram of K-theory groups formed by $\mathrm{K}_*(A(S))$ for all locally closed subsets $S \subseteq X$. Here a subset is locally closed if and only if it is of the form

$$[a, b] := \{a, a+1, \ldots, b-1, b\}, \qquad 1 \leq a \leq b \leq n.$$

The maps in the filtrated K-theory diagram are those that come from natural transformations. We are going to describe these below.

Let $\mathfrak{I}^X_{\text{fil}}$ be the kernel on morphisms of the filtrated K-theory functor, that is, $f \in \mathrm{KK}^X_0(A, B)$ belongs to $\mathfrak{I}^X_{\text{fil}}(A, B)$ if and only if it induces the zero map $\mathrm{K}_*(A[a,b]) \to \mathrm{K}_*(B[a,b])$ for all $1 \le a \le b \le n$. This is a stable homological ideal. It is shown in [**11**] that any object $A \in\in \mathfrak{B}^X$ has an $\mathfrak{I}^X_{\text{fil}}$-projective resolution

$$0 \to P_1 \xrightarrow{\varphi} P_0 \xrightarrow{f} A \tag{5.1}$$

of length 1. Then it follows that A is isomorphic to the cone of φ. And there is a Universal Coefficient Theorem for objects of $\mathfrak{B}^X$ based on Hom and Ext groups between their filtrated K-theory diagrams. Thus any isomorphism between the filtrated K-theory diagrams of $A, B \in\in \mathfrak{B}^X$ lifts to an equivalence in $\mathrm{KK}^X_0(A, B)$.

The minimal open subset U_x containing x is $[x, n]$ for each $x \in X$. Since $\mathrm{K}_*(A(U_x))$ is part of the filtrated K-theory of A, the ideal $\mathfrak{I}^X_{\text{fil}}$ is contained in the ideal $\mathfrak{I}^X_{\mathrm{K}}$ that is used above for a general unique path space. So any $\mathfrak{I}^X_{\text{fil}}$-exact chain complex is also $\mathfrak{I}^X_{\mathrm{K}}$-exact. If $P \in\in \mathfrak{B}^X$ is $\mathfrak{I}^X_{\text{fil}}$-projective, then the Abelian groups $\mathrm{K}_*(P(S))$ are free for all locally closed subsets $S \subseteq X$ (this follows from [**12**, Theorem 3.12] and will become manifest below). The computation of $\mathrm{Ext}^2_{\mathfrak{A}^X}$ in Lemma 4.2 shows that $\mathrm{Ext}^2_{\mathfrak{A}^X}(F^X_{\mathrm{K}}(P), D) = 0$ for all $D \in\in \mathfrak{A}^X$ if $P(U_x)$ is free for each $x \in X$. Hence P_1 and P_0 have $\mathfrak{I}^X_{\mathrm{K}}$-projective resolutions of length 1. So we are in the situation of (3.1) and Theorem 3.2 computes the obstruction class from the parity-reversing part of the map φ in (5.1). This computation is, however, quite non-trivial. We must first recall how the natural transformations in the filtrated K-theory diagram look like. Going beyond the results of [**12**], we then build an explicit $\mathfrak{I}^X_{\text{fil}}$-projective resolution. Next, we observe which parts of the map φ are parity-reversing. This gives a class in $\mathrm{Ext}^2_{\mathfrak{A}^X}$. To translate it into the setting of Theorem 4.9, we still have to compare the resolution used there with the one coming from (5.1). This requires most of the work.

We recall the description of the $\mathbb{Z}/2$-graded Abelian groups $\mathcal{NT}_*([a,b],[c,d])$ of natural transformations $\mathrm{K}_*(A([a,b])) \to \mathrm{K}_*(A([c,d]))$ in [**12**]. Let $1 \le a \le b \le n$. The functor $\mathfrak{KK}^X \to \mathfrak{Ab}^{\mathbb{Z}/2}$, $A \mapsto \mathrm{K}_*(A([a,b]))$, is represented by an object $\mathcal{R}_{[a,b]}$ of $\mathfrak{KK}^X$, which is described in [**12**]. Let $1 \le a \le b \le n$ and $1 \le c \le d \le n$. By the Yoneda Lemma,

$$\mathcal{NT}_*([a,b],[c,d]) \cong \mathrm{KK}^X_*(\mathcal{R}_{[c,d]}, \mathcal{R}_{[a,b]}) \cong \mathrm{K}_*(\mathcal{R}_{[a,b]}([c,d])).$$

These groups are computed in [**12**, Equation (3.1)]:

$$\mathcal{NT}_*([a,b],[c,d]) \cong \begin{cases} \mathbb{Z}_+ & \text{if } c \le a \le d \le b, \\ \mathbb{Z}_- & \text{if } a+1 \le c \le b+1 \le d, \\ 0 & \text{otherwise.} \end{cases} \tag{5.2}$$

We write $[a,b] \to [c,d]$ in the first two cases, that is, when there is a non-zero natural transformation in $\mathcal{NT}_*([a,b],[c,d])$. The groups $\mathcal{NT}_*([a,b],[c,d])$ form a $\mathbb{Z}/2$-graded ring $\mathcal{NT}$. The filtrated K-theory of a separable C*-algebra over X is a $\mathbb{Z}/2$-graded, countable module over it, which we denote by $FK(A)$.

The computations in [**12**] interpret the elements of $\mathcal{NT}_*([a,b],[c,d])$ as follows. Let A be a separable C*-algebra over X and let $M := FK(A)$. If $a < b \le c$, then $[b,c]$ is relatively open in $[a,c]$ with complement $[a,b-1]$. This induces the following natural six-term exact sequence in K-theory:

$$(5.3)\qquad \cdots \to M[b,c] \xrightarrow{i} M[a,c] \xrightarrow{r} M[a,b-1] \xrightarrow[\text{odd}]{\delta} M[b,c] \to \cdots$$

where the maps i, r preserve the $\mathbb{Z}/2$-grading and δ reverses it. Any natural transformation $M[a,b] \to M[c,d]$ is an integer multiple of a product of the maps i, r, δ above. More precisely, if $c \le a \le d \le b$, then there is a commuting square

$$(5.4)\qquad \begin{array}{ccc} M[a,b] & \xrightarrow{i} & M[c,b] \\ \downarrow r & \searrow \tau_{[a,b]}^{[c,d]} & \downarrow r \\ M[a,d] & \xrightarrow[i]{} & M[c,d], \end{array}$$

and its diagonal map $\tau_{[a,b]}^{[c,d]}$ generates $\mathcal{NT}_0([a,b],[c,d]) \cong \mathbb{Z}$. And if $a+1 \le c \le b+1 \le d$, then there is a commuting square

$$(5.5)\qquad \begin{array}{ccc} M[a,b] & \xrightarrow{\delta} & M[b+1,d] \\ \downarrow r & \searrow \tau_{[a,b]}^{[c,d]} & \downarrow i \\ M[a,c-1] & \xrightarrow[\delta]{} & M[c,d], \end{array}$$

and its diagonal map $\tau_{[a,b]}^{[c,d]}$ generates $\mathcal{NT}_1([a,b]),[c,d]) \cong \mathbb{Z}$. We have defined a generator $\tau_{[a,b]}^{[c,d]}$ for $\mathcal{NT}_*([a,b],[c,d])$ whenever $[a,b] \to [c,d]$, that is, whenever $\mathcal{NT}_*([a,b]),[c,d]) \neq 0$ by (5.2). We define $\tau_{[a,b]}^{[c,d]} = 0$ if $\mathcal{NT}_*([a,b]),[c,d]) = 0$. By the Yoneda Lemma, the natural transformations $\tau_{[a,b]}^{[c,d]}$ correspond to arrows

$$\left(\tau_{[a,b]}^{[c,d]}\right)^*: \mathcal{R}_{[c,d]} \to \mathcal{R}_{[a,b]}.$$

REMARK 5.6. An $\mathcal{NT}$-module is called *exact* if the sequences (5.3) are exact for all $a < b \le c$. The exact $\mathcal{NT}$-modules form a stable exact subcategory of the stable Abelian category of all $\mathcal{NT}$-modules, and the filtrated K-theory of any separable C*-algebra over X is exact as an $\mathcal{NT}$-module. The results in [**12**] imply that any exact $\mathcal{NT}$-module has a projective resolution of length 1. Hence it lifts to an object of the bootstrap class $\mathfrak{B}^X$. So the image of the filtrated K-theory functor is equal to the class of exact, countable $\mathcal{NT}$-modules. And the filtrated K-theory functor, viewed as a functor to the subcategory of exact, countable $\mathcal{NT}$-modules, is the universal $\mathfrak{I}^X_{\text{fil}}$-exact functor to an exact category. So $\mathfrak{I}^X_{\text{fil}}$ has the property that its universal exact functors to an Abelian and to an exact category are different.

Now we study the multiplication in $\mathcal{NT}$. We begin with decomposing the generators in $\mathcal{NT}$ further. We may rewrite the natural transformations $\tau_{[a,b]}^{[c,d]}$ defined above as products of the special natural transformations

$$\begin{aligned} i &= \tau_{[a+1,b]}^{[a,b]}: [a+1,b] \to [a,b], && a+1 \le b \le n, \\ r &= \tau_{[a,b+1]}^{[a,b]}: [a,b+1] \to [a,b], && a \le b \le n-1, \\ \delta &= \tau_{[1,a-1]}^{[a,n]}: [1,a-1] \to [a,n], && 2 \le a \le n. \end{aligned}$$

This is clear in the even case. In the odd case, we use the naturality of boundary maps to rewrite the boundary map $\delta\colon [a,b] \to [b+1,d]$ as the product of $i\colon [a,b] \to$

FIGURE 1. Natural transformations on filtrated K-theory for general n

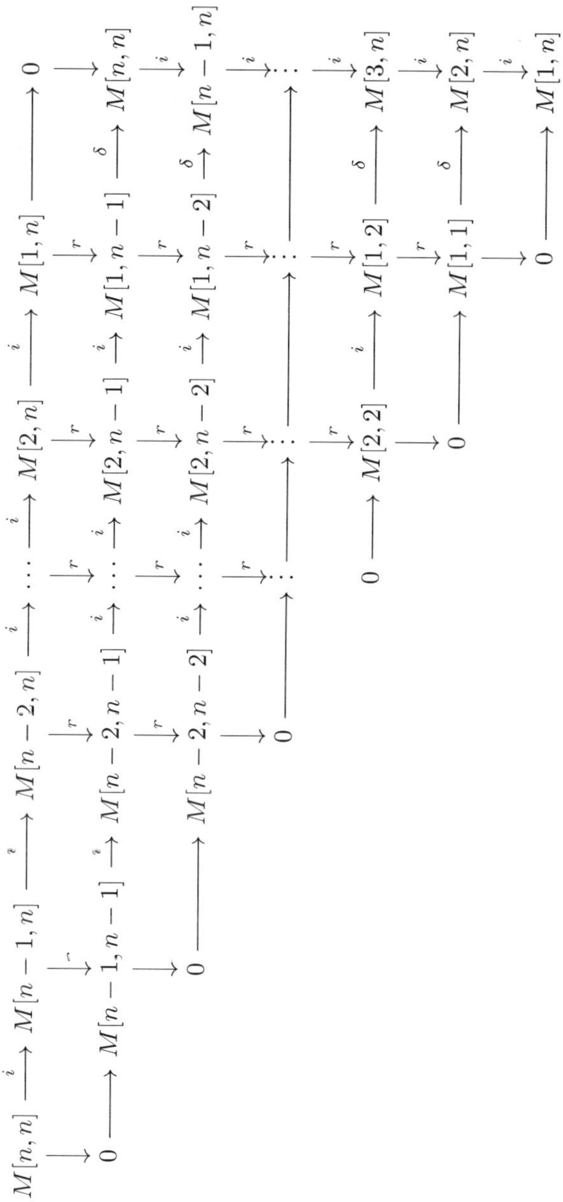

$[1, b]$, the boundary map $\delta\colon [1, b] \to [b + 1, n]$, and $r\colon [b + 1, n] \to [b + 1, d]$. The generating natural transformations defined above form a commuting diagram as in Figure 1. The last column and the first row in the diagram are the same. So the diagram repeats when we put a reflected copy of it next to it. Figure 2 shows the full diagram for $n = 3$. We claim that all relations among the generating natural transformations are given by this extended commuting diagram. In particular, a

FIGURE 2. Complete diagram of natural transformations on filtrated K-theory for $n = 3$

$$\begin{array}{cccccccccccc}
M[3,3] & \xrightarrow{i} & M[2,3] & \xrightarrow{i} & M[1,3] & \longrightarrow & 0 & & & & & \\
\downarrow & & \downarrow{\scriptstyle r} & & \downarrow{\scriptstyle r} & & \downarrow & & & & & \\
0 & \longrightarrow & M[2,2] & \xrightarrow{i} & M[1,2] & \xrightarrow{\delta} & M[3,3] & \longrightarrow & 0 & & & \\
 & & \downarrow & & \downarrow{\scriptstyle r} & & \downarrow{\scriptstyle i} & & \downarrow & & & \\
 & & 0 & \longrightarrow & M[1,1] & \xrightarrow{\delta} & M[2,3] & \xrightarrow{r} & M[2,2] & \longrightarrow & 0 & \\
 & & & & \downarrow & & \downarrow{\scriptstyle i} & & \downarrow{\scriptstyle i} & & \downarrow & \\
 & & & & 0 & \longrightarrow & M[1,3] & \xrightarrow{r} & M[1,2] & \xrightarrow{r} & M[1,1] &
\end{array}$$

composite of i, r, δ vanishes if and only if it factors through one of the objects 0 on the boundary of the extended diagram.

Let $1 \le a \le b \le n$ and $1 \le c \le d \le n$. A product of the generators of type i, r from $[a, b]$ to $[c, d]$ exists if and only if $c \le a$ and $d \le b$. Figure 1 shows that all such products are equal. Equation (5.2) shows that this product is 0 unless $a \le d$, so that $c \le a \le d \le b$. Then it is equal to $\tau_{[a,b]}^{[c,d]}$ by the definition in (5.4). As a consequence, $\tau_{[e,g]}^{[c,d]} \cdot \tau_{[a,b]}^{[e,g]} = \tau_{[a,b]}^{[c,d]}$ if $c \le e \le d \le g$ and $e \le a \le g \le b$; this is non-zero if and only if also $a \le d$ or, equivalently, $c \le e \le a \le d \le g \le b$.

Now consider a product of i, r, δ going from $[a, b]$ to $[c, d]$ and containing exactly one factor of δ. Using the diagram in Figure 1, we may rearrange this product in such a way that we first go right and then go down in the extended diagram as in Figure 2. If this goes through the zeros outside the drawn region, the product is 0. If not, we may combine consecutive i and consecutive r to bring the product into the following form:

$$[a, b] \xrightarrow{i} [1, b] \xrightarrow{\delta} [b + 1, n] \xrightarrow{r} [b + 1, d] \xrightarrow{i} [c, d].$$

The combination $r \circ \delta \circ i$ in the beginning is the boundary map $\delta\colon [a, b] \to [b+1, d]$. So we get $\tau_{[a,b]}^{[c,d]}$ if $c + 1 \le a \le d + 1 \le b$ and 0 otherwise by (5.5) and (5.2). Since we may rewrite all even $\tau_{[a,b]}^{[c,d]}$ in terms of i,r, we can now compute $\tau_{[e,g]}^{[c,d]} \cdot \tau_{[a,b]}^{[e,g]}$ if one of the transformations $\tau_{[e,g]}^{[c,d]}$ and $\tau_{[a,b]}^{[e,g]}$ is even and the other one is odd. Namely, the product is $\tau_{[a,b]}^{[c,d]}$ if $a+1 \le c \le b+1 \le d$, and 0 otherwise. In more detail, the assumption that exactly one of the transformations $\tau_{[e,g]}^{[c,d]}$ and $\tau_{[a,b]}^{[e,g]}$ is even means that $e+1 \le c \le g+1 \le d$ and $e \le a \le g \le b$, or $c \le e \le d \le g$ and $a+1 \le e \le b+1 \le g$. The assumption $a+1 \le c \le b+1 \le d$ becomes $e \le a < c \le g+1 \le b+1 \le d$ or $a < c \le e \le b+1 \le d \le g$ in these two cases, respectively.

Finally, any product with more than two factors δ vanishes because it may be deformed in the extended diagram in Figure 2 so as to factor through one of the zeros on the boundary. We sum up our results about the multiplication in $\mathcal{NT}$:

$$\tau_{[e,g]}^{[c,d]} \cdot \tau_{[a,b]}^{[e,g]} = \tau_{[a,b]}^{[c,d]} \neq 0 \Leftrightarrow \begin{cases} c \le e \le a \le d \le g \le b, \\ e \le a \le c-1 \le g \le b < d, \\ a < c \le e \le b+1 \le d \le g, \end{cases} \tag{5.7}$$

and $\tau_{[e,g]}^{[c,d]} \cdot \tau_{[a,b]}^{[e,g]} = 0$ otherwise. We write

$$[a,b] \to [e,g] \to [c,d] \Leftrightarrow \tau_{[e,g]}^{[c,d]} \cdot \tau_{[a,b]}^{[e,g]} = \tau_{[a,b]}^{[c,d]} \neq 0.$$

It can happen that $[a,b] \to [e,g]$, $[e,g] \to [c,d]$ and $[a,b] \to [c,d]$, but not $[a,b] \to [e,g] \to [c,d]$; that is, $\tau_{[e,g]}^{[c,d]} \cdot \tau_{[a,b]}^{[e,g]} = 0$ although $\tau_{[a,b]}^{[c,d]}, \tau_{[a,b]}^{[e,g]}, \tau_{[e,g]}^{[c,d]} \neq 0$.

Given $1 \le a \le b \le n$, there are one or two proper natural transformations to $M[a,b]$ that are shortest in the sense that all others factor through them. If $a < b < n$, these are $i\colon M[a+1,b] \to M[a,b]$ and $r\colon M[a,b+1] \to M[a,b]$. If $1 < a < b = n$, then r above is replaced by $\delta\colon M[1,a-1] \to M[a,n]$. One of these maps is missing if $a = b$ or if $(a,b) = (1,n)$, that is, on the two outer diagonals in the diagram in Figure 2 (now for general n). We define $M[a+1,a] := 0$ for $0 \le a \le n$ to make this a special case of the generic case.

Now we build an $\mathfrak{I}_{\text{fil}}^{X}$-projective resolution of A of length 1. This has not yet been done in [**12**], where only the existence of such a resolution is proven. For $1 \le a \le b \le n$, let $M[a,b]_{\text{ss}}$ be the quotient of $M[a,b]$ by the images of all proper natural transformations to $M[a,b]$ or, equivalently, by the images of the two shortest natural transformations:

$$M[a,b]_{\text{ss}} := \begin{cases} M[a,b] \,/\, (i(M[a+1,b]) + r(M[a,b+1])) & \text{if } b < n, \\ M[a,b] \,/\, (i(M[a+1,b]) + \delta(M[1,a-1])) & \text{if } b = n \end{cases}$$

(compare [**12**, Definition 3.7 and Lemma 3.8]). Choose a resolution

$$Q_1[a,b] \overset{d_1}{\rightarrowtail} Q_0[a,b] \overset{d_0}{\twoheadrightarrow} M[a,b]_{\text{ss}} \tag{5.8}$$

of $M[a,b]_{\text{ss}}$ by countable $\mathbb{Z}/2$-graded free Abelian groups. For $i = 0,1$, let

$$\hat{Q}_i[a,b] := \mathcal{R}_{[a,b]} \otimes_{\mathbb{Z}} Q_i[a,b],$$

where the tensor product is defined as in (2.4). Since $\mathcal{R}_{[a,b]} \otimes_{\mathbb{Z}} Q_i[a,b]$ is a direct sum of copies of suspensions of $\mathcal{R}_{[a,b]}$, the definition of $\mathcal{R}_{[a,b]}$ as a representing object implies

$$\mathrm{KK}_0^X(\hat{Q}_i[a,b], B) \cong \mathrm{Hom}\big(Q_i[a,b], \mathrm{K}_*(B[a,b])\big) \tag{5.9}$$

for all C*-algebras B over X; here Hom means grading-preserving group homomorphisms. This property characterises $\hat{Q}_i[a,b]$ uniquely up to isomorphism in $\mathfrak{KK}^X$. Equation (5.9) implies a similar description of the $\mathbb{Z}/2$-graded Abelian group $\mathrm{KK}_*^X(\hat{Q}_i[a,b],B)$, replacing Hom by group homomorphisms that need not respect the grading. Given a group homomorphism $g\colon Q_i[a,b]\to \mathrm{K}_*(B[a,b])$, let $g^\#$ denote the corresponding element of $\mathrm{KK}_*^X(\hat{Q}_i[a,b],B)$.

Since the $\mathbb{Z}/2$-graded Abelian group $Q_0[a,b]$ is free, the homomorphism d_0 in (5.8) lifts to a grading-preserving homomorphism

$$f[a,b]\colon Q_0[a,b]\to M[a,b]=\mathrm{K}_*(A[a,b]). \tag{5.10}$$

Let $f[a,b]^\# \in \mathrm{KK}_0^X(\hat{Q}_0[a,b],A)$ correspond to $f[a,b]$ by (5.9). Let

$$P_i := \bigoplus_{1\le a\le b\le n} \hat{Q}_i[a,b]$$

for $i=0,1$. The objects $\hat{Q}_i[a,b]$ and P_i for $i=0,1$ are $\mathfrak{I}^X_{\mathrm{fil}}$-projective because of (5.9). There is a unique element $f\in \mathrm{KK}_0^X(P_0,A)$ that restricts to $f[a,b]^\#$ on the summand $\hat{Q}_i[a,b]$.

LEMMA 5.11. *The map $FK(f)\colon FK(P_0)\to M:=FK(A)$ is surjective. Its kernel is isomorphic to $FK(P_1)$ as an $\mathcal{NT}$-module.*

PROOF. Let $f_*:=FK(f)$. The $\mathcal{NT}$-module M_{ss} is defined as the quotient $M/\mathcal{NT}_{\mathrm{nil}}\cdot M$ for a certain ideal $\mathcal{NT}_{\mathrm{nil}}$ in $\mathcal{NT}$. It follows that the functor $M\mapsto M_{\mathrm{ss}}$ is right exact. Even more, it is isomorphic to the tensor product functor with the right $\mathcal{NT}$-module $\mathcal{NT}_{\mathrm{ss}}$. This follows from the extension of $\mathcal{NT}$-modules $\mathcal{NT}_{\mathrm{nil}}\rightarrowtail \mathcal{NT}\twoheadrightarrow \mathcal{NT}_{\mathrm{ss}}$ and the right exactness of tensor product functors.

Right exactness implies $(\operatorname{coker} f_*)[a,b]_{\mathrm{ss}} = \operatorname{coker}(f_*[a,b]_{\mathrm{ss}})$ for all $1\le a\le b\le n$. The projective $\mathcal{NT}$-module $FK(P_0)$ has $FK(P_0)[a,b]_{\mathrm{ss}}\cong Q_0[a,b]$. Hence $(f_*)_{\mathrm{ss}}$ is surjective by construction of f, and the $\mathcal{NT}$-module $\operatorname{coker} f_*$ satisfies $(\operatorname{coker} f_*)_{\mathrm{ss}}=0$. This implies $\operatorname{coker} f_*=0$ by [**12**, Proposition 3.10]. That is, f_* is surjective. Let $N:=\ker FK(f)$. So there is an extension $N\rightarrowtail FK(P_0)\twoheadrightarrow M$ of $\mathcal{NT}$-modules. Since M and $FK(P_0)$ are exact $\mathcal{NT}$-modules, they satisfy $\mathrm{Tor}^1_{\mathcal{NT}}(\mathcal{NT}_{\mathrm{ss}},M)=0$ and $\mathrm{Tor}^1_{\mathcal{NT}}(\mathcal{NT}_{\mathrm{ss}},FK(P_0))=0$ by [**12**, Lemma 3.13]. Hence

$$N[a,b]_{\mathrm{ss}}\to FK(P_0)[a,b]_{\mathrm{ss}}\to M[a,b]_{\mathrm{ss}}$$

is a short exact sequence for all $1\le a\le b\le n$ and $\mathrm{Tor}^1_{\mathcal{NT}}(\mathcal{NT}_{\mathrm{ss}},N)=0$. Since $FK(P_0)[a,b]_{\mathrm{ss}}\cong Q_0[a,b]$, this implies $N[a,b]_{\mathrm{ss}}\cong Q_1[a,b]$. Now [**12**, Theorem 3.12] shows that N is a projective $\mathcal{NT}$-module. In fact, the proof of this theorem shows that $N\cong FK(P_1)$. More precisely, the quotient maps $N[a,b]\twoheadrightarrow N[a,b]_{\mathrm{ss}}$ split because $N[a,b]_{\mathrm{ss}}\cong Q_1[a,b]$ is free. Let

$$\varphi[a,b]\colon Q_1[a,b]\cong N[a,b]_{\mathrm{ss}}\to N[a,b]\subseteq FK(P_0)[a,b] \tag{5.12}$$

be sections. They induce an $\mathcal{NT}$-module homomorphism $FK(P_1)\to N$ by the universal property of the "free" $\mathcal{NT}$-module $FK(P_1)$. And the proof of [**12**, Theorem 3.12] shows that it is an isomorphism. $\square$

Disregarding the $\mathbb{Z}/2$-grading, we may write

$$FK(P_0)[a,b]=\bigoplus_{[c,d]\to[a,b]} Q_0[c,d],$$

that is, the sum runs over all $1 \le c \le d \le n$ with $a \le c \le b \le d$ or $c+1 \le a \le d+1 \le b$. So the map $\varphi[a,b]$ in (5.12) has components

$$\varphi_{[a,b]}^{[c,d]}\colon Q_1[a,b] \to Q_0[c,d]$$

for $[c,d] \to [a,b]$. Since $\varphi[a,b]$ is even, the map $\varphi_{[a,b]}^{[c,d]}$ has the same parity as $\tau_{[c,d]}^{[a,b]}$, that is, it is grading-preserving if $a \le c \le b \le d$ and grading-reversing if $c+1 \le a \le d+1 \le b$. The image of $\varphi[a,b]$ for $1 \le a \le b \le n$ is contained in $N = \ker FK(f)$, that is, $FK(f) \circ \varphi[a,b] = 0$ as a map $Q_1[a,b] \to \mathrm{K}_*(A[a,b])$. Unravelling the definition of $FK(f)$, this becomes

$$\sum_{[c,d]\to[a,b]} \tau_{[c,d]}^{[a,b]} \circ f[c,d] \circ \varphi_{[a,b]}^{[c,d]} = 0\colon Q_1[a,b] \to \mathrm{K}_*(A[a,b]). \tag{5.13}$$

The objects P_0 and P_1 are $\mathfrak{I}_{\mathrm{fil}}^X$-projective, and FK is fully faithful on $\mathfrak{I}_{\mathrm{fil}}^X$-projective objects. So the arrow $FK(P_1) \to FK(P_0)$ with components $\varphi_{[a,b]}^{[c,d]}$ for $[c,d] \to [a,b]$ lifts uniquely to an arrow $\varphi \in \mathrm{KK}_0^X(P_1, P_0)$. More precisely, the map φ is given by a matrix of maps $\mathcal{R}_{[c,d]} \otimes Q_1[c,d] \to \mathcal{R}_{[a,b]} \otimes Q_0[a,b]$ for all $1 \le c \le d \le n$ and $1 \le a \le b \le n$. The entries of this matrix are $\left((\tau_{[a,b]}^{[c,d]})^* \otimes \varphi_{[c,d]}^{[a,b]}\right)$, that is,

$$\varphi = \left((\tau_{[a,b]}^{[c,d]})^* \otimes \varphi_{[c,d]}^{[a,b]}\right)_{[c,d]\to[a,b]}\colon \bigoplus_{1\le c\le d\le n} \mathcal{R}_{[c,d]} \otimes Q_1[c,d] \to \bigoplus_{1\le a\le b\le n} \mathcal{R}_{[a,b]} \otimes Q_0[a,b]. \tag{5.14}$$

Here we use the convention that $\tau_{[a,b]}^{[c,d]} = 0$ if not $[a,b] \to [c,d]$.

Lemma 5.11 says that (5.1) with f and φ as above is $\mathfrak{I}_{\mathrm{fil}}^X$-exact, that is, FK applied to (5.1) is an exact sequence. The $\mathfrak{I}_{\mathrm{fil}}^X$-exactness of (5.1) says that the functor $B \mapsto \mathrm{K}_*(B[a,b])$ maps it to an exact sequence for each $1 \le a \le b \le n$. In fact, this gives projective resolutions. We write them down explicitly:

LEMMA 5.15. *Let $1 \le a \le b \le n$. Then*

$$\bigoplus_{[c,d]\to[a,b]} Q_1[c,d] \xrightarrow{\left(\varphi_{[c,d]}^{[e,g]}\right)_{[c,d]\to[e,g]\to[a,b]}} \bigoplus_{[e,g]\to[a,b]} Q_0[e,g] \xrightarrow{\left(\tau_{[e,g]}^{[a,b]} \circ f[e,g]\right)} \mathrm{K}_*(A[a,b]) \tag{5.16}$$

is a free resolution. Here $\left(\varphi_{[c,d]}^{[e,g]}\right)_{[c,d]\to[e,g]\to[a,b]}$ means that the matrix entry is $\varphi_{[c,d]}^{[e,g]}$ if $[c,d] \to [e,g] \to [a,b]$ as described in (5.7)*, and* 0 *otherwise.*

The boundary maps in (5.16) are inhomogeneous, that is, the matrix entries of the maps may have even or odd degree.

PROOF. Equation (5.2) computes the group

$$\mathrm{K}_*(\mathcal{R}_{[c,d]}([a,b])) \cong \mathcal{NT}_*([c,d],[a,b]).$$

Namely, it is $\mathbb{Z}$ in even or odd degree if $[c,d] \to [a,b]$ and 0 otherwise. Therefore, $P_i[a,b] \cong \bigoplus_{[c,d]\to[a,b]} Q_i[c,d]$ for $i = 0,1$, disregarding the grading. The map $\left((\tau_{[e,g]}^{[c,d]})^* \otimes \varphi_{[c,d]}^{[e,g]}\right)$ between the summands $\mathcal{R}_{[c,d]} \otimes Q_1[c,d]$ in P_1 and $\mathcal{R}_{[e,g]} \otimes Q_0[e,g]$ in P_0 induces the map $\varphi_{[c,d]}^{[e,g]}\colon Q_1[c,d] \to Q_0[e,g]$ if $[c,d] \to [e,g] \to [a,b]$, and 0 otherwise, compare (5.7). The map $\mathrm{K}_*(P_0[a,b]) \to \mathrm{K}_*(A[a,b])$ corresponds to a family of maps

$$\mathrm{K}_*(\mathcal{R}_{[c,d]}[a,b]) \otimes Q_0[c,d] \cong \mathrm{K}_*(\mathcal{R}_{[c,d]}[a,b] \otimes Q_0[c,d]) \to \mathrm{K}_*(A[a,b])$$

for $1 \le c \le d \le n$. As above, $\mathrm{K}_*(\mathcal{R}_{[c,d]}[a,b]) \neq 0$ only if $[c,d] \to [a,b]$, and then the map $Q_0[c,d] \to K_*(A[a,b])$ induced by $f\colon P_0 \to A$ is $\tau_{[c,d]}^{[a,b]} \circ f[c,d]$. $\square$

We have reached the first milestone in the computation of the obstruction class: the $\mathfrak{I}^X_{\mathrm{fil}}$-projective resolution (5.1). It is explicit enough to express the obstruction class that comes from filtrated K-theory in the terms of Theorem 4.9, namely, as being represented by a family of elements in $\mathrm{Ext}\big(\mathrm{K}_*(A[e+1,n]), \mathrm{K}_*(A[e,n])\big)$ for $e = 1, \dots, n-1$. We compute these Ext-groups with the resolutions in (5.16). So the obstruction class corresponds to a sequence of maps

$$\delta_e\colon \bigoplus_{[a,b]\to[e+1,n]} Q_1[a,b] \to \mathrm{K}_*(A[e,n]), \qquad e = 1, \dots, n-1.$$

In turn, each δ_e is given by maps $\delta_e^{[a,b]}\colon Q_1[a,b] \to \mathrm{K}_*(A[e,n])$ for all $1 \le a \le b \le n$ with $[a,b] \to [e+1,n]$. The following theorem computes these maps $\delta_e^{[a,b]}$. It is the main result of this section. Section 5.1 is dedicated to its proof.

THEOREM 5.17. *Let*

$$\delta_e^{[a,b]} := \begin{cases} \displaystyle\sum_{[c,d]\to[a,n]\to[a,b]} \tau_{[c,d]}^{[e,n]} \circ f[c,d] \circ \varphi_{[a,b]}^{[c,d]} & \textit{if } a = e \textit{ and } b < n, \\ 0 & \textit{otherwise.} \end{cases}$$

The resulting family of maps $(\delta_e)_{1\le e<n}$ lifts the obstruction class of A to an element of $\prod_{e=1}^{n-1} \mathrm{Ext}\big(\mathrm{K}_*(A[e+1,n]), \mathrm{K}_*(A[e,n])\big)$.

Any $\mathfrak{I}^X_{\mathrm{fil}}$-projective resolution of A is isomorphic to one of the form above because any $\mathfrak{I}^X_{\mathrm{fil}}$-epic map from an $\mathfrak{I}^X_{\mathrm{fil}}$-projective object to A is isomorphic to a map $f \in \mathrm{KK}_0^X(P_0, A)$ as above. Since $\mathfrak{I}^X_{\mathrm{fil}}$-projective resolutions of A are equivalent to projective resolutions of $FK(A)$, the maps $\delta_e^{[a,b]}$ may, in principle, be computed from $FK(A)$ by choosing a projective resolution. This gives the maps $f[c,d]$ and $\varphi_{[a,b]}^{[c,d]}$. Such a computation may, of course, be difficult in practice.

5.1. Proof of the obstruction class formula. First we examine the smaller invariant $F_{\mathrm{K}}^X\colon \mathfrak{KK}^X \to \mathfrak{A}^X$. This takes the part of filtrated K-theory consisting of $\mathrm{K}_*(A[a,n])$ for $1 \le a \le n$ with the maps i between them because the minimal open subset containing a is $U_a = [a,n]$. So the diagram $F_{\mathrm{K}}^X(A)$ is simply the first row in the diagram in Figure 1. We have $i_a\mathbb{C} \cong \mathcal{R}_{[a,n]}$ for $1 \le a \le n$ because both objects represent the same functor $A \mapsto \mathrm{K}_*(A[a,n])$. So $\mathcal{R}_{[a,n]}$ for $1 \le a \le n$ is $\mathfrak{I}^X_{\mathrm{K}}$-projective, and $F_{\mathrm{K}}^X(\mathcal{R}_{[a,n]})$ is the diagram

$$\text{(5.18)} \qquad \mathcal{P}_{[a,n]} := F_{\mathrm{K}}^X(\mathcal{R}_{[a,n]}) = \Big(\underbrace{\mathbb{Z} = \mathbb{Z} = \dots = \mathbb{Z}}_{a \text{ times}} \leftarrow \underbrace{0 = \dots = 0}_{n-a \text{ times}}\Big).$$

in $\mathfrak{A}^X$. If $1 \le a \le b \le n-1$, then $F_{\mathrm{K}}^X(\mathcal{R}_{[a,b]})$ is the diagram

$$\text{(5.19)} \qquad \mathcal{P}_{[a,b]} := F_{\mathrm{K}}^X(\mathcal{R}_{[a,b]}) = \Big(\underbrace{0 = \dots = 0}_{a \text{ times}} \leftarrow \underbrace{\mathbb{Z}_- = \dots = \mathbb{Z}_-}_{b+1-a \text{ times}} \leftarrow \underbrace{0 = \dots = 0}_{n-1-b \text{ times}}\Big)$$

because of the formula for $\mathrm{K}_*(\mathcal{R}_{[a,b]}([c,n]))$ in (5.2). Thus the objects $\mathcal{R}_{[a,b]}$ for $1 \le a \le b \le n$ are even if $b = n$ and odd if $b < n$.

The formula for the obstruction class in Theorem 3.2 uses the parity-reversing part of $\varphi \in \mathrm{KK}_0^X(P_1, P_0)$. This is described by the following lemma:

LEMMA 5.20. *The component* $(\tau_{[c,d]}^{[a,b]})^* \otimes \varphi_{[a,b]}^{[c,d]}$ *of* φ *is parity-reversing if and only if* $[c,d] \to [a,n] \to [a,b]$ *and* $b < n$.

PROOF. If $\tau_{[c,d]}^{[a,b]} \neq 0$, then either $a \le c \le b \le d$ or $c+1 \le a \le d+1 \le b$. The map $(\tau_{[c,d]}^{[a,b]})^* \otimes \varphi_{[a,b]}^{[c,d]}$ always belongs to $\mathrm{KK}_0^X(\hat{Q}_1[a,b], \hat{Q}_0[c,d])$. So $\varphi_{[a,b]}^{[c,d]} \colon Q_1[a,b] \to Q_0[c,d]$ is parity-preserving in the first case and parity-reversing in the second case. The object $\mathcal{R}_{[a,b]}$ is even if $b = n$ and odd if $b < n$.

First let $a \le c \le b \le d$. If $b = n$, then $d = n$ as well, so that $\mathcal{R}_{[a,b]}$ and $\mathcal{R}_{[c,d]}$ have the same parity, and $\varphi_{[a,b]}^{[c,d]}$ preserves parity. So we get a parity-preserving component of φ. For the same reasons, we get a parity-preserving component if $a \le c \le b \le d < n$, and a parity-reversing component if $a \le c \le b < d = n$.

Now let $c + 1 \le a \le d + 1 \le b$, so that $\varphi_{[a,b]}^{[c,d]}$ reverses parity. If $b = n$, then $d < n$, so that $\mathcal{R}_{[a,b]}$ and $\mathcal{R}_{[c,d]}$ have opposite parity. Hence $(\tau_{[c,d]}^{[a,b]})^* \otimes \varphi_{[a,b]}^{[c,d]}$ is parity-preserving altogether. If $b < n$, then $d < n$ and so $\mathcal{R}_{[a,b]}$ and $\mathcal{R}_{[c,d]}$ have the same parity. Thus $(\tau_{[c,d]}^{[a,b]})^* \otimes \varphi_{[a,b]}^{[c,d]}$ is parity-reversing. Inspection shows that the parity-reversing components are exactly those for which $[c,d] \to [a,n] \to [a,b]$ as in (5.7) and $b < n$. $\square$

Let $1 \le a \le b < n$. The exact triangle

$$\Sigma\mathcal{R}_{[a,n]} \xrightarrow{i^*} \Sigma\mathcal{R}_{[b+1,n]} \xrightarrow{\delta^*} \mathcal{R}_{[a,b]} \xrightarrow{r^*} \mathcal{R}_{[a,n]} \tag{5.21}$$

in $\mathfrak{KK}^X$ is $\mathfrak{I}_{\mathrm{K}}^X$-exact, that is, $F_{\mathrm{K}}^X(r^*) = 0$. Since $\Sigma\mathcal{R}_{[a,n]}$ and $\Sigma\mathcal{R}_{[b+1,n]}$ are $\mathfrak{I}_{\mathrm{K}}^X$-projective, it is an $\mathfrak{I}_{\mathrm{K}}^X$-projective resolution of $\mathcal{R}_{[a,b]}$ of length 1. Since we allow both odd and even arrows in diagrams, we may drop the suspensions in (5.21). For $1 \le a \le b \le n$, let

$$M(a,b) := \begin{cases} a & \text{if } b = n, \\ b+1 & \text{if } b < n. \end{cases}$$

Then $(\tau_{[a,b]}^{[M(a,b),n]})^* \colon \mathcal{R}_{[M(a,b),n]} \to \mathcal{R}_{[a,b]}$ is an $\mathfrak{I}_{\mathrm{K}}^X$-epimorphism both for $b = n$ and $b < n$. Its cone is $\mathcal{R}_{[a,n]}$ if $b < n$ and 0 if $b = n$. Now we write down $\mathfrak{I}_{\mathrm{K}}^X$-projective resolutions of P_i for $i = 0, 1$. Let

$$P_{i0} := \bigoplus_{1 \le a \le b \le n} \mathcal{R}_{[M(a,b),n]} \otimes Q_i[a,b], \tag{5.22}$$

$$P_{i1} := \bigoplus_{1 \le a \le b < n} \mathcal{R}_{[a,n]} \otimes Q_i[a,b]. \tag{5.23}$$

Then the following is an $\mathfrak{I}_{\mathrm{K}}^X$-projective resolution:

$$0 \to P_{i1} \xrightarrow{\bigoplus \tau^* \otimes \mathrm{id}} P_{i0} \xrightarrow{\bigoplus \tau^* \otimes \mathrm{id}} P_i \to 0. \tag{5.24}$$

Here $\bigoplus \tau^* \otimes \mathrm{id}$ means the direct sum of the maps $(\tau_{[z,w]}^{[x,y]})^* \otimes \mathrm{id}_{Q_i[a,b]}$ between the summands for fixed a, b, with the appropriate x, y, z, w. So (5.24) is the direct sum of the resolutions (5.21), tensored with $Q_i[a,b]$, over all $1 \le a \le b < n$, and the trivial resolutions $0 \to \mathcal{R}_{[a,n]} \to \mathcal{R}_{[a,n]}$, tensored with $Q_i[a,b]$, over all $1 \le a \le b = n$. The maps in (5.24) are inhomogeneous, that is, some components are in KK_0^X and others in KK_1^X.

The resolution (5.24) allows us to compute $\mathrm{KK}_0^X(P_1, P_0)$ with the Universal Coefficient Theorem for the invariant F_{K}^X. First, the long exact sequence for the direct sum of the exact triangles (5.21) implies a natural extension of Abelian groups

$$\operatorname{coker}\bigl(\mathrm{KK}_*^X(\Sigma P_{10}, P_0) \to \mathrm{KK}_*^X(\Sigma P_{11}, P_0)\bigr) \rightarrowtail \mathrm{KK}_*^X(P_1, P_0) \\ \twoheadrightarrow \ker\bigl(\mathrm{KK}_*^X(P_{10}, P_0) \to \mathrm{KK}_*^X(P_{11}, P_0)\bigr).$$

We may rewrite the kernel and cokernel here as Hom and Ext in $\mathfrak{A}^X$, using that (5.24) is an $\mathfrak{I}_{\mathrm{K}}^X$-projective resolution. The extension above splits non-naturally, giving the decomposition of φ into its parity-preserving and -reversing parts φ^+ and φ^-, respectively.

LEMMA 5.25. *The image of the parity-reversing part φ^- of φ in $\mathrm{Ext}^1_{\mathfrak{I}_{\mathrm{K}}^X}(P_1, P_0)$ is the map*

$$P_{11} = \bigoplus_{1 \le a \le b < n} \mathcal{R}_{[a,n]} \otimes Q_1[a,b] \to \bigoplus_{1 \le c \le d \le n} \mathcal{R}_{[c,d]} \otimes Q_0[c,d] = P_0$$

with matrix coefficients $\bigl(\tau_{[c,d]}^{[a,n]}\bigr)^* \otimes \varphi_{[a,b]}^{[c,d]}$ if $[c,d] \to [a,n] \to [a,b]$ and $b < n$, and 0 otherwise.

PROOF. In terms of the matrix description of φ, each matrix entry $\bigl(\tau_{[c,d]}^{[a,b]}\bigr)^* \otimes \varphi_{[a,b]}^{[c,d]}$ has even or odd parity and thus belongs to either Hom or Ext, respectively. By Lemma 5.20, the entry belongs to φ^- if and only if $\tau_{[c,d]}^{[a,b]}$ factors through $r^*\colon \mathcal{R}_{[a,b]} \to \mathcal{R}_{[a,n]}$. In this case, it factors as $\Bigl(\bigl(\tau_{[c,d]}^{[a,n]}\bigr)^* \otimes \varphi_{[a,b]}^{[c,d]}\Bigr) \circ (r^* \otimes \mathrm{id})$. Since r^* is the boundary map in (5.21), this exhibits a map $P_{11} \to P_0$. The map from $\mathrm{Ext}^1_{\mathfrak{A}^X}$ to KK_0^X in the UCT is defined by composing with the boundary map in the exact triangle that contains the given resolution. So the map $P_{11} \to P_0$ found above is the relevant component of φ. The formula in the lemma follows. □

According to the recipe in Theorem 3.2, the obstruction class in $\mathrm{Ext}^2_{\mathfrak{I}_{\mathrm{K}}^X}(\Sigma A, A)$ is the composite of the parity-reversing part of φ, viewed as an element of $\mathrm{Ext}^1_{\mathfrak{I}_{\mathrm{K}}^X}(P_1, P_0)$, with $f \in \mathrm{KK}_0^X(P_0, A)$ and with the class of the extension (5.1) in $\mathrm{Ext}^1_{\mathfrak{I}_{\mathrm{K}}^X}(A, P_1)$. Composing the two extensions gives a length-2 resolution

$$P_{11} \rightarrowtail P_{10} \to P_0 \twoheadrightarrow A. \tag{5.26}$$

The component $\mathcal{R}_{[a,n]} \otimes Q_1[a,b] \to A$ in the composite map $P_{11} \to P_0 \to A$ is the map $\mathcal{R}_{[a,n]} \otimes Q_1[a,b] \to A$ that corresponds to

$$\sum_{[c,d] \to [a,n] \to [a,b]} \tau_{[c,d]}^{[a,n]} \circ f[c,d] \circ \varphi_{[a,b]}^{[c,d]}\colon Q_1[a,b] \to \mathrm{K}_*(A[a,n]) \tag{5.27}$$

under the isomorphism (5.9). Here the sum runs only over those $[c,d]$ with $[c,d] \to [a,n] \to [a,b]$ as in Lemma 5.25. In contrast, the sum over all $[c,d]$ with $[c,d] \to [a,b]$ is 0 by (5.13).

In a sense, we have now computed the obstruction class. The length-2 resolution in (5.26) is, however, different from the one that is implicitly used in Theorem 4.9

to compute the relevant Ext^2-group and the obstruction class in it. To translate the formula for the obstruction class that we get from filtrated K-theory into the setting of Theorem 4.9, we must compare the underlying length-2 resolutions. First, we replace the resolution in (5.26) by one that is $\mathfrak{I}_{\mathrm{K}}^X$-projective.

The entries P_{10} and P_{11} are already $\mathfrak{I}_{\mathrm{K}}^X$-projective, and (5.24) is an $\mathfrak{I}_{\mathrm{K}}^X$-projective resolution of P_0. The objects P_i and P_{ij} are all sums over $1 \le a \le b \le n$, with summands of the form $\mathcal{R}_{[x,y]} \otimes Q_i[a,b]$ for suitable x, y depending on a, b; the summands in P_{i1} are 0 for $b = n$. Let $(\tau^* \otimes \varphi)$ denote the map between these sums for $i = 1$ to those for $i = 0$ with matrix entries

$$\bigl(\tau_{[z,w]}^{[x,y]}\bigr)^* \otimes \varphi_{[a,b]}^{[c,d]} \colon \mathcal{R}_{[x,y]} \otimes Q_1[a,b] \to \mathcal{R}_{[z,w]} \otimes Q_0[c,d].$$

As usual, $\tau_{[z,w]}^{[x,y]} = 0$ if $\mathcal{NT}_*([z,w],[x,y]) = 0$.

LEMMA 5.28. *There is a commuting diagram*

$$\begin{array}{ccccc} P_{11} & \xrightarrow{\bigoplus \tau^* \otimes \mathrm{id}} & P_{10} & \xrightarrow{\bigoplus \tau^* \otimes \mathrm{id}} & P_1 \\ \downarrow{\scriptstyle (\tau^*\otimes\varphi)} & & \downarrow{\scriptstyle (\tau^*\otimes\varphi)} & & \downarrow{\scriptstyle (\tau^*\otimes\varphi)} \\ P_{01} & \xrightarrow[\bigoplus \tau^* \otimes \mathrm{id}]{} & P_{00} & \xrightarrow[\bigoplus \tau^* \otimes \mathrm{id}]{} & P_0 \end{array}$$

PROOF. We compare maps between direct sums by comparing their matrix coefficients. For the two composite maps $P_{11} \to P_{00}$, these are maps $\mathcal{R}_{[a,n]} \otimes Q_1[a,b] \to \mathcal{R}_{[M(c,d),n]} \otimes Q_0[c,d]$ for $1 \le a \le b < n$ and $1 \le c \le d \le n$. The composite map through P_{10} is $\bigl(\tau_{[M(c,d),n]}^{[a,n]}\bigr)^* \otimes \varphi_{[a,b]}^{[c,d]}$ if $[a,n] \leftarrow [M(a,b),n] \leftarrow [M(c,d),n]$, and 0 otherwise; and the composite map through P_{01} is $\bigl(\tau_{[M(c,d),n]}^{[a,n]}\bigr)^* \otimes \varphi_{[a,b]}^{[c,d]}$ if $[a,n] \leftarrow [c,n] \leftarrow [M(c,d),n]$ and $d < n$, and 0 otherwise; the condition $d < n$ comes in because P_{01} contains only summands $\mathcal{R}_{[c,n]} \otimes Q_0[c,d]$ with $1 \le c \le d < n$. In both cases, the map vanishes unless $[c,d] \to [a,b]$ because of the factor $\varphi_{[a,b]}^{[c,d]}$. We claim that if $[c,d] \to [a,b]$ and $b < n$, then $[a,n] \leftarrow [M(a,b),n] \leftarrow [M(c,d),n]$ if and only if $[a,n] \leftarrow [c,n] \leftarrow [M(c,d),n]$ and $d < n$; here $M(a,b) = b+1$ because $b < n$. Indeed, if $d = n$, then $M(c,d) = c$, and $[a,n] \leftarrow [b+1,n] \leftarrow [c,n]$ means $a \le b+1 \le c$, which contradicts $[c,n] \to [a,b]$. If $d < n$, then $M(c,d) = d+1$. Then $[a,n] \leftarrow [c,n] \leftarrow [d+1,n]$ and $[a,n] \leftarrow [M(a,b),n] \leftarrow [M(c,d),n]$ are equivalent to $a \le c \le d+1$ and $a \le b+1 \le d+1$, respectively. If $[c,d] \to [a,b]$, both conditions say that we are in the case $a \le c \le b \le d$. The computations above show that the two maps $P_{11} \to P_{00}$ are equal.

Now consider the two maps $P_{10} \to P_0$. Its matrix coefficients are maps

$$\mathcal{R}_{[M(a,b),n]} \otimes Q_1[a,b] \to \mathcal{R}_{[c,d]} \otimes Q_0[c,d], \qquad 1 \le a \le b \le n,\ 1 \le c \le d \le n.$$

As above, the composite maps through P_{00} and P_1 are $\bigl(\tau_{[c,d]}^{[M(a,b),n]}\bigr)^* \otimes \varphi_{[a,b]}^{[c,d]}$ or 0. For the maps through P_1 and P_{00}, the former case occurs if $[M(a,b),n] \leftarrow [a,b] \leftarrow [c,d]$ or $[M(a,b),n] \leftarrow [M(c,d),n] \leftarrow [c,d]$, respectively. We may assume $[a,b] \leftarrow [c,d]$ and $[M(a,b),n] \leftarrow [c,d]$ because otherwise $\varphi_{[a,b]}^{[c,d]} = 0$ or $\tau_{[c,d]}^{[M(a,b),n]} = 0$. Under these assumptions, $[M(a,b),n] \leftarrow [a,b] \leftarrow [c,d]$ always holds by (5.7). And $[c,d] \to [M(a,b),n]$ implies $[M(a,b),n] \leftarrow [M(c,d),n] \leftarrow [c,d]$ because any natural transformation $\mathrm{K}_*(A[c,d]) \to \mathrm{K}_*(A[e,n])$ for some $1 \le e \le n$ factors through $\tau_{[c,d]}^{[M(c,d),n]}$. So the two maps $P_{10} \to P_0$ are equal as well. □

Using also the resolution (5.1) of A, we get the following $\mathfrak{I}^X_{\mathrm{K}}$-projective resolution of A:

$$P_{11} \xrightarrow{\begin{pmatrix} -(\tau^*\otimes\varphi) \\ \bigoplus \tau^*\otimes \mathrm{id}\end{pmatrix}} P_{01}\oplus P_{10} \xrightarrow{\left(\bigoplus \tau^*\otimes\mathrm{id} \quad (\tau^*\otimes\varphi)\right)} P_{00} \xrightarrow{\bigoplus f[a,b]^\#\circ(\tau^*\otimes\mathrm{id})} A.$$

The computation of the obstruction class in Theorem 4.8 starts with the following $\mathfrak{I}^X$-projective resolution of length 1 in $\mathfrak{KK}^X$:

$$\bigoplus_{b=1}^{n-1} \mathcal{R}_{[b,n]}\otimes A[b+1,n] \rightarrowtail \bigoplus_{b=1}^{n} \mathcal{R}_{[b,n]}\otimes A[b,n] \twoheadrightarrow A. \tag{5.29}$$

Since $\mathcal{R}_{[b,n]} = i_b(\mathbb{C})$ in the notation of Section 2.2, we have $\mathcal{R}_{[b,n]}\otimes A[b,n] \cong i_b(A[b,n])$. The restriction of the second map in (5.29) to this direct summand is the one that corresponds to the identity map on $A[b,n]$ under the isomorphism in (2.15). The first map in (5.29), restricted to the summand $\mathcal{R}_{[b,n]}\otimes A[b+1,n]$, is the difference of the two maps

$$\begin{aligned}(\tau^{[b,n]}_{[b+1,n]})^*\otimes\mathrm{id}&\colon \mathcal{R}_{[b,n]}\otimes A[b+1,n]\to\mathcal{R}_{[b+1,n]}\otimes A[b+1,n],\\ \mathrm{id}\otimes\tau^{[b,n]}_{[b+1,n]}&\colon \mathcal{R}_{[b,n]}\otimes A[b+1,n]\to\mathcal{R}_{[b,n]}\otimes A[b,n],\end{aligned}$$

where $\tau^{[b,n]}_{[b+1,n]}$ denotes the inclusion of $A[b+1,n]$ into $A[b,n]$; we could have written i for $\tau^{[b,n]}_{[b+1,n]}$ as in Figure 1. It is shown in Section 2.2 that this sequence is $\mathfrak{I}^X$-exact. And it is easy to prove this directly.

The projective resolutions of Abelian groups in (5.16) imply that there is an $\mathfrak{I}^X_{\mathrm{K}}$-projective resolution

$$\bigoplus_{[c,d]\to[b+1,n]} \mathcal{R}_{[b,n]}\otimes Q_1[c,d] \rightarrowtail \bigoplus_{[c,d]\to[b+1,n]} \mathcal{R}_{[b,n]}\otimes Q_0[c,d] \twoheadrightarrow \bigoplus_{b=1}^{n-1}\mathcal{R}_{[b,n]}\otimes A[b+1,n].$$

Splicing it with the resolution in (5.29) gives an $\mathfrak{I}^X_{\mathrm{K}}$-exact chain complex

$$W_1 \rightarrowtail W_0 \to \bigoplus_{b=1}^{n}\mathcal{R}_{[b,n]}\otimes A[b,n] \twoheadrightarrow A$$

with

$$W_i := \bigoplus_{b=1}^{n-1}\ \bigoplus_{[c,d]\to[b+1,n]} \mathcal{R}_{[b,n]}\otimes Q_i[c,d], \qquad i=0,1. \tag{5.30}$$

Next we are going to compare the two $\mathfrak{I}^X_{\mathrm{K}}$-exact chain complexes built above. We are going to build maps γ_{ij} and δ for $0\le i,j\le 1$ that make the following diagram commute, and such that δ gives the obstruction class:

$$\begin{array}{ccccccc} P_{11} & \rightarrowtail & P_{01}\oplus P_{10} & \longrightarrow & P_{00} & \twoheadrightarrow & A\\ \downarrow{\scriptstyle\gamma_{11}} & & \downarrow{\scriptstyle(\gamma_{01}\ \gamma_{10})} & & \downarrow{\scriptstyle\gamma_{00}} & & \|\\ W_1 & \rightarrowtail & W_0 & \twoheadrightarrow & \bigoplus_{e=1}^n \mathcal{R}_{[e,n]}\otimes A[e,n] & \longrightarrow & A\\ \downarrow{\scriptstyle\delta} & & & & & & \\ A & & & & & & \end{array} \tag{5.31}$$

We describe maps between direct sums through matrices of maps between the direct summands. Recall that W_i is defined in (5.30) and that

$$P_{i0} := \bigoplus_{1\le a\le b\le n} \mathcal{R}_{[M(a,b),n]} \otimes Q_i[a,b], \qquad P_{i1} := \bigoplus_{1\le a\le b<n} \mathcal{R}_{[a,n]} \otimes Q_i[a,b]$$

for $i = 0, 1$. The matrix coefficients of γ_{00} are maps

$$\gamma_{00}^{e,[a,b]} : \mathcal{R}_{[M(a,b),n]} \otimes Q_0[a,b] \to \mathcal{R}_{[e,n]} \otimes A[e,n]$$

for $1 \le a \le b \le n$ and $1 \le e \le n$. We let $\gamma_{00}^{e,[a,b]} = 0$ if $e \ne M(a,b)$. Let $e = M(a,b)$. Then $\gamma_{00}^{e,[a,b]}$ corresponds to a map

$$(\gamma_{00}^{e,[a,b]})^\flat : Q_0[a,b] \to \mathrm{K}_*(\mathcal{R}_{[e,n]}[e,n] \otimes A[e,n])$$

under the isomorphism (5.9). We have already used above that $\mathcal{R}_{[e,n]} \otimes A[e,n] \cong i_e(A[e,n])$; so

$$\mathcal{R}_{[e,n]}[e,n] \otimes A[e,n] \cong A[e,n]. \tag{5.32}$$

Using this isomorphism implicitly, we let

$$(\gamma_{00}^{e,[a,b]})^\flat := \tau_{[a,b]}^{[e,n]} \circ f[a,b] : Q_0[a,b] \xrightarrow{f[a,b]} \mathrm{K}_*(A[a,b]) \xrightarrow{\tau_{[a,b]}^{[e,n]}} \mathrm{K}_*(A[e,n]).$$

As usual, this is 0 unless $[a,b] \to [e,n]$. The map $\gamma_{01} : P_{01} \to W_0$ is given by a matrix of maps

$$\gamma_{01}^{e,[c,d],[a,b]} : \mathcal{R}_{[a,n]} \otimes Q_0[a,b] \to \mathcal{R}_{[e,n]} \otimes Q_0[c,d]$$

for $1 \le a \le b < n$, $1 \le e < n$, and $1 \le c \le d \le n$ with $[c,d] \to [e+1,n]$. We let

$$\gamma_{01}^{e,[c,d],[a,b]} := \begin{cases} \left(\tau_{[e,n]}^{[a,n]}\right)^* \otimes \mathrm{id}_{Q_0[a,b]} & \text{if } [a,b] = [c,d], \\ 0 & \text{otherwise.} \end{cases}$$

If $a = c$ and $b = d < n$, then $[c,d] \to [e+1,n]$ if and only if $a \le e \le b$, so that $\tau_{[e,n]}^{[a,n]} \ne 0$ in this formula.

The map $\gamma_{10} : P_{10} \to W_0$ is given by a matrix of maps

$$\gamma_{10}^{e,[c,d],[a,b]} : \mathcal{R}_{[M(a,b),n]} \otimes Q_1[a,b] \to \mathcal{R}_{[e,n]} \otimes Q_0[c,d]$$

for $1 \le a \le b \le n$, $1 \le e < n$, and $1 \le c \le d \le n$ with $[c,d] \to [e+1,n]$. We let

$$\gamma_{10}^{e,[c,d],[a,b]} := \begin{cases} \left(\tau_{[e,n]}^{[M(a,b),n]}\right)^* \otimes \varphi_{[a,b]}^{[c,d]} & \text{if } M(a,b) \le e < M(c,d), \\ 0 & \text{otherwise.} \end{cases}$$

The map $\gamma_{11} : P_{11} \to W_1$ is given by a matrix of maps

$$\gamma_{11}^{e,[c,d],[a,b]} : \mathcal{R}_{[a,n]} \otimes Q_1[a,b] \to \mathcal{R}_{[e,n]} \otimes Q_1[c,d]$$

for $1 \le a \le b < n$, $1 \le e < n$, and $1 \le c \le d \le n$ with $[c,d] \to [e+1,n]$. We let

$$\gamma_{11}^{e,[c,d],[a,b]} := \begin{cases} \left(\tau_{[e,n]}^{[a,n]}\right)^* \otimes \mathrm{id}_{Q_1[a,b]} & \text{if } a = c,\ b = d, \\ 0 & \text{otherwise.} \end{cases}$$

The map δ is given by a family of maps $\delta_{b,[c,d]} : \mathcal{R}_{[b,n]} \otimes Q_1[c,d] \to A$ for $1 \le b < n$ and $[c,d] \to [b+1,n]$. These correspond to maps $\delta_{b,[c,d]}^\flat : Q_1[c,d] \to \mathrm{K}_*(A[b,n])$ by the isomorphism (5.9). We define δ so that the maps $\delta_{b,[c,d]}^\flat$ are the maps denoted by $\delta_b^{[c,d]}$ in Theorem 5.17.

Now we must prove that the squares in the diagram commute. We begin on the right, comparing the two maps $P_{00} \to A$. Its restrictions $\mathcal{R}_{[M(a,b),n]} \otimes Q_0[a,b] \to A$ correspond to maps $Q_0[a,b] \to \mathrm{K}_*(A[M(a,b),n])$ under the isomorphism (5.9). This map is $\tau_{[a,b]}^{[M(a,b),n]} \circ f[a,b]$ both for the direct boundary map $P_{00} \to A$ and for the map through $\bigoplus_{e=1}^n \mathcal{R}_{[e,n]} \otimes A[e,n]$. So this square commutes.

Next we compare the two maps from $P_{10} \oplus P_{01}$ to $\bigoplus_{b=1}^n \mathcal{R}_{[b,n]} \otimes A[b,n]$. We first consider the restriction to P_{01}, then to P_{10}. The matrix coefficients of the map on P_{01} are maps $\mathcal{R}_{[a,n]} \otimes Q_0[a,b] \to \mathcal{R}_{[e,n]} \otimes A[e,n]$ for $1 \le a \le b < n$ and $1 \le e \le n$. Such maps correspond to group homomorphisms $Q_0[a,b] \to \mathrm{K}_*(\mathcal{R}_{[e,n]}[a,n] \otimes A[e,n])$. Recall that $\mathcal{R}_{[e,n]}[a,n] = \mathbb{C}$ if $a \le e$ and 0 otherwise. So we may assume without loss of generality that $a \le e$, and then we get corresponding maps $Q_0[a,b] \to \mathrm{K}_*(A[e,n])$. The map γ_{00} picks out the summand with $e = M(a,b) = b+1$, and $\big(\tau_{[M(a,b),n]}^{[a,n]}\big)^*$ induces the identity map

$$(\mathbb{Z} \cong \mathrm{K}_*\big(\mathcal{R}_{[a,n]}([b+1,n])\big) \to \mathrm{K}_*\big(\mathcal{R}_{[b+1,n]}([b+1,n])\big) \cong \mathbb{Z}).$$

Therefore, the map in the square through γ_{00} contributes the map

$$\delta_{e,b+1}\tau_{[a,b]}^{[e,n]} \circ f[a,b] \colon Q_0[a,b] \to \mathrm{K}_*(A[e,n]).$$

When we map through γ_{01} instead, then we first map $\mathcal{R}_{[a,n]} \otimes Q_0[a,b]$ to the direct sum of $\mathcal{R}_{[g,n]} \otimes Q_0[a,b]$ over all $g \in [a,b]$ using $\tau^* \otimes \mathrm{id}$ and then apply the boundary map on W_0. This gives a contribution in $\mathrm{K}_*(A[e,n])$ if $g = e$ or $g = e-1$, and these two contributions cancel each other for $a < e \le b$. For $e = b+1$, we get the same term as for the map that goes through P_{00}. And we get 0 for $e = a$ because $\tau_{[a+1,n]}^{[a,n]} \circ \tau_{[a,b]}^{[a+1,n]} = \tau_{[a,b]}^{[a,n]} = 0$. So the two maps are equal on P_{01}.

The matrix coefficients of the two maps on P_{10} are maps $\mathcal{R}_{[M(a,b),n]} \otimes Q_1[a,b] \to \mathcal{R}_{[e,n]} \otimes A[e,n]$ for $1 \le a \le b \le n$ and $1 \le e \le n$. As above, we may assume $M(a,b) \le e$ because otherwise any such map is zero. And then maps $\mathcal{R}_{[M(a,b),n]} \otimes Q_1[a,b] \to \mathcal{R}_{[e,n]} \otimes A[e,n]$ correspond to maps $Q_1[a,b] \to \mathrm{K}_*(A[e,n])$. We shall examine the difference of the map through P_{00} and the map through W_0. We first consider the map through P_{00}. It first applies the matrix $\tau^* \otimes \varphi$, going to the direct sum of $\mathcal{R}_{[M(c,d),n]} \otimes Q_0[c,d]$ for $1 \le c \le d \le n$. The map $Q_1[a,b] \to \mathrm{K}_*(A[e,n])$ for the composite map through $\mathcal{R}_{[M(c,d),n]} \otimes Q_0[c,d]$ is $\delta_{M(c,d),e}\tau_{[c,d]}^{[e,n]} \circ f[c,d] \circ \varphi_{[a,b]}^{[c,d]}$. So we get the sum of these terms over all $1 \le c \le d \le n$. When we apply $\gamma_{10} \colon P_{10} \to W_0$, then we apply the maps $\big(\tau_{[g,n]}^{[M(a,b),n]}\big)^* \otimes \varphi_{[a,b]}^{[c,d]}$ to the direct summands $\mathcal{R}_{[g,n]} \otimes Q_0[c,d]$ of W_0, where $1 \le c \le d \le n$ and $1 \le g < n$ are such that $[c,d] \to [g+1,n]$ and $M(a,b) \le g < M(c,d)$. The condition $[c,d] \to [g+1,n]$ is equivalent to $c > g$ if $d = n$ and $c \le g < d+1$ if $d < n$. So the set of g that are allowed is an interval $[x,y]$ or empty. The upper bound is always $y := M(c,d) - 1$. The lower bound x is $M(a,b)$ if $d = n$ or the maximum of c and $M(a,b)$ if $d < n$. By convention, we redefine $x := M(c,d)$ if the lower bound is bigger than $M(c,d)$. So g runs through the interval $[x,y]$ if $x \le y$, and otherwise $x = y+1 = M(c,d)$ and the set of possible g is empty.

We must compose γ_{10} with the boundary map on W_0. As above, this only contributes to the map $\mathcal{R}_{[M(a,b),n]} \otimes Q_1[a,b] \to \mathcal{R}_{[e,n]} \otimes A[e,n]$ if $g = e$ or $g = e-1$. And the contribution to the corresponding map $Q_1[a,b] \to \mathrm{K}_*(A[e,n])$ is $-\tau_{[c,d]}^{[e,n]} \circ f[c,d] \circ \varphi_{[a,b]}^{[c,d]}$ if $g = e$ and $+\tau_{[c,d]}^{[e,n]} \circ f[c,d] \circ \varphi_{[a,b]}^{[c,d]}$ if $g = e-1$. The contributions for $g = e$ and $g = e-1$ cancel if both occur. Therefore, when we sum over all g in the interval $[x, y]$ above, we get $-\tau_{[c,d]}^{[e,n]} \circ f[c,d] \circ \varphi_{[a,b]}^{[c,d]}$ if $e = x$, $\tau_{[c,d]}^{[e,n]} \circ f[c,d] \circ \varphi_{[a,b]}^{[c,d]}$ if $e = y+1$, and 0 otherwise. So we get the map

$$(\delta_{e,y+1} - \delta_{e,x}) \cdot \tau_{[c,d]}^{[e,n]} \circ f[c,d] \circ \varphi_{[a,b]}^{[c,d]} \colon Q_1[a,b] \to \mathrm{K}_*(A[e,n]).$$

This formula remains correct if no g is allowed because then $x = y+1$. Since $y+1 = M(c,d)$, the map involving $\delta_{e,y+1}$ is equal to the one that we get from the map through P_{00}. So when we take the difference of the two maps in the square, this term is cancelled. We remain with

$$\sum_{[c,d]\to[a,b]} \delta_{e,x} \cdot \tau_{[c,d]}^{[e,n]} \circ f[c,d] \circ \varphi_{[a,b]}^{[c,d]}, \tag{5.33}$$

where x depends on a, b, c, d as above. Recall that we only need the case $e \leq M(a,b)$. We are going to prove that the sum in (5.33) vanishes under this assumption. First we have to study the lower bound x for the different order relations among a, b, c, d. We may assume $[c,d] \to [a,b]$ because otherwise $\varphi_{[a,b]}^{[c,d]} = 0$. So either $a \leq c \leq b < d$ or $c+1 \leq a \leq d+1 \leq b$. It is also important whether $b = n$ or $d = n$. First assume $b = n$. So $M(a,b) = a$. If $a \leq c \leq b = d = n$, then $M(c,d) = c$ and so $x = a$. If $c+1 \leq a \leq d+1 \leq b = n$, then $d < n$. So $M(c,d) = d+1$ and $x = a$ as well. Therefore, $x = a$ and $[e,n] = [a,b]$ whenever $b = n$. In this case, the sum in (5.33) vanishes because of (5.13).

Now assume $b < n$, so $M(a,b) = b+1$. If $a \leq c \leq b < d = n$, then $M(c,d) = c < b+1$. So $x = M(c,d)$ and the summand in (5.33) vanishes because $e = x < M(a,b)$. And $\tau_{[a,b]}^{[b+1,n]} \cdot \tau_{[c,d]}^{[a,b]} = 0$ as well. If $a \leq c \leq b \leq d < n$, then $M(c,d) = d+1$ and $x = b+1$. In this case, we have $\tau_{[c,d]}^{[b+1,n]} = \tau_{[a,b]}^{[b+1,n]} \cdot \tau_{[c,d]}^{[a,b]} \neq 0$. So we may rewrite the sum in (5.33) as

$$\sum_{[c,d]\to[a,b]} \delta_{e,x} \cdot \tau_{[c,d]}^{[e,n]} \circ f[c,d] \circ \varphi_{[a,b]}^{[c,d]} = \sum_{[c,d]\to[a,b]} \tau_{[a,b]}^{[b+1,n]} \circ \tau_{[c,d]}^{[a,b]} \circ f[c,d] \circ \varphi_{[a,b]}^{[c,d]},$$

and this vanishes by (5.13). This proves the vanishing of (5.33) in all cases and finishes the proof that the square of maps $P_{01} \oplus P_{10} \to \bigoplus \mathcal{R}_{[e,n]} \otimes A[e,n]$ commutes.

Next, we consider the maps $P_{11} \to W_0$. We look at the matrix coefficient $\mathcal{R}_{[a,n]} \otimes Q_1[a,b] \to \mathcal{R}_{[e,n]} \otimes Q_0[c,d]$ of the maps $P_{11} \to W_0$ through W_1, P_{10} and P_{01} for fixed $1 \leq a \leq b < n$, $1 \leq e < n$, $[c,d] \to [e+1,n]$. In W_1, we have summands $\mathcal{R}_{[e,n]} \otimes Q_1[a,b]$ for those e with $[a,b] \to [e+1,n]$, which is equivalent to $a \leq e \leq b$ because $b < n$. The map γ_{11} maps the summands $\mathcal{R}_{[a,n]} \otimes Q_1[a,b]$ in P_{11} to each of these summands through $\left(\tau_{[e,n]}^{[a,n]}\right)^* \otimes \mathrm{id}$. The boundary map $W_1 \to W_0$ is obtained by tensoring the boundary map in (5.16) with $\mathrm{id}_{\mathcal{R}[e,n]}$. So we get the contribution $\left(\tau_{[e,n]}^{[a,n]}\right)^* \otimes \varphi_{[a,b]}^{[c,d]}$ to our matrix coefficient if $[c,d] \to [a,b] \to [e,n]$, and 0 otherwise. The boundary map to P_{10} maps the summands $\mathcal{R}_{[a,n]} \otimes Q_1[a,b]$ in P_{11} to the summand $\mathcal{R}_{[b+1,n]} \otimes Q_1[a,b]$ through $\left(\tau_{[b+1,n]}^{[a,n]}\right)^* \otimes \mathrm{id}$. When we continue with γ_{10}, we get the contribution $\left(\tau_{[e,n]}^{[a,n]}\right)^* \otimes \varphi_{[a,b]}^{[c,d]}$ to our matrix coefficient if and only if $M(a,b) \leq e < M(c,d)$. The map through P_{01} gives the contribution $-\left(\tau_{[e,n]}^{[a,n]}\right)^* \otimes \varphi_{[a,b]}^{[c,d]}$ to our matrix coefficient if $d < n$ and $a \leq c \leq e$ (recall that the summands $\mathcal{R}_{[c,n]} \otimes Q_0[c,d]$ of P_{01} only run over $1 \leq d < n$).

We must show that the sum of these terms is 0. Again we look at different cases regarding the order among a, b, c, d, e. We may assume $b < n$ and $[c, d] \to [e+1, n]$ because our matrix coefficient is only defined in this case. And we may assume $[e, n] \to [a, n]$ and $[c, d] \to [a, b]$ because otherwise $\big(\tau_{[e,n]}^{[a,n]}\big)^* \otimes \varphi_{[a,b]}^{[c,d]} = 0$. Besides $b < n$, these assumptions mean, first, that $e+1 \le c \le n = d$ or $c+1 \le e+1 \le d+1 \le n$ holds; secondly, $a \le e$; and, thirdly, $a \le c \le b \le d$ or $c+1 \le a \le d+1 \le b$. Assume first that $d = n$. Then our assumptions imply $a \le e < c \le b < d = n$. In this case, none of the three maps $P_{11} \to W_0$ give a non-zero contribution because $[c, d] \to [a, b] \to [e, n]$ is impossible, $M(a, b) = b + 1 > c = M(c, d)$ and $d = n$. So we may assume $d < n$ from now on. If $c + 1 \le a$, then it follows that $c+1 \le a \le e < d+1 \le b < n$. Again, none of the three maps $P_{11} \to W_0$ give a non-zero contribution in this case. So we may assume $a \le c$. Then either $a \le c \le e \le b \le d < n$ or $a \le c \le b < e \le d < n$. In the first case, the maps through W_1 and P_{01} give contributions that cancel each other, and the map through P_{10} gives no contribution because $e < M(a, b) = b+1$. In the second case, the maps through P_{10} and P_{01} give contributions that cancel each other, and the map through W_1 vanishes because $b < e$. Hence we get 0 in all cases, as needed. This finishes the proof that the square of maps $P_{11} \to W_0$ commutes.

Finally, we compute the composite map $\delta \circ \gamma_{11} \colon P_{11} \to A$ in our commuting diagram. Consider the restriction to $\mathcal{R}_{[a,n]} \otimes Q_1[a, b]$ for some $1 \le a \le b < n$. This map corresponds to a map $Q_1[a, b] \to \mathrm{K}_*(A[a, n])$ by (5.9). The map $\delta^b_{e,[c,d]} = \delta_e^{[c,d]}$ vanishes unless $e = c$, and the matrix coefficient $\gamma_{00}^{c,[c,d],[a,b]}$ vanishes for $[a, b] \ne [c, d]$ and is the identity map if $[a, b] = [c, d]$. So the composite map corresponds simply to the map

$$\sum_{[c,d]\to[a,n]\to[a,b]} \tau_{[c,d]}^{[a,n]} \circ f[c, d] \circ \varphi_{[a,b]}^{[c,d]} \colon Q_1[a, b] \to \mathrm{K}_*(A[a, n]).$$

This is exactly the formula for the obstruction class in (5.27). This finishes the proof of Theorem 5.17.

5.2. The case of extensions. We now specialise to the case $n = 2$. Then an object of $\mathfrak{KK}^X$ is equivalent to a C*-algebra extension

$$I \overset{i}{\rightarrowtail} A \overset{r}{\twoheadrightarrow} A/I,$$

where $I = A[2]$, $A = A[1, 2]$ and $A/I = A[1]$ and the maps are those in (5.3). Here we abbreviate $[1] = [1, 1]$ and $[2] = [2, 2]$. The filtrated K-theory is the six-periodic exact chain complex

$$\begin{array}{ccccc} \mathrm{K}_0(I) & \xrightarrow{i_*} & \mathrm{K}_0(A) & \xrightarrow{r_*} & \mathrm{K}_0(A/I) \\ {\scriptstyle\delta}\uparrow & & & & \downarrow{\scriptstyle\delta} \\ \mathrm{K}_1(A/I) & \xleftarrow[r_*]{} & \mathrm{K}_1(A) & \xleftarrow[i_*]{} & \mathrm{K}_1(I) \end{array} \tag{5.34}$$

The morphisms between the filtrated K-theory invariants are grading-preserving chain maps (morphisms of six-term exact sequences).

The invariant in Theorem 2.12 is the KK-class $[i] \in \mathrm{KK}_0(I, A)$. The invariant in Theorem 4.8 is the induced map $i_* \colon \mathrm{K}_*(I) \to \mathrm{K}_*(A)$, together with the obstruction class. To compute the latter, let $i^- \in \mathrm{Ext}\big(\mathrm{K}_{1+*}(I), \mathrm{K}_*(A)\big)$ be the parity-reversing part of $[i]$ in the Universal Coefficient Theorem for $\mathrm{KK}_0(I,A)$. The obstruction class is the image of i^- in the cokernel of the map

$$(5.35)\quad \mathrm{Ext}\big(\mathrm{K}_{1+*}(I),\mathrm{K}_*(I)\big)\oplus\mathrm{Ext}\big(\mathrm{K}_{1+*}(A),\mathrm{K}_*(A)\big)\to\mathrm{Ext}\big(\mathrm{K}_{1+*}(I),\mathrm{K}_*(A)\big),\\ (t_I,t_A)\mapsto i\circ t_I+t_A\circ i.$$

It follows from our theory that i_* and the image of i^- in the cokernel of (5.35) determine an object of $\mathfrak{B}^X$ uniquely up to KK^X-equivalence. The cokernel comes in because there are isomorphisms of C*-algebra extensions that act identically on $\mathrm{K}_*(I)$ and $\mathrm{K}_*(A)$, but have non-trivial components in $\mathrm{Ext}\big(\mathrm{K}_{1+*}(I),\mathrm{K}_*(I)\big)$ or $\mathrm{Ext}\big(\mathrm{K}_{1+*}(A),\mathrm{K}_*(A)\big)$. So the isomorphism class of an object in $\mathfrak{KK}^X$ does not determine i^- uniquely. Only its image in the cokernel of (5.35) is unique. And our theory shows that isomorphism classes of pairs consisting of $i_*\in\mathrm{Hom}\big(\mathrm{K}_*(I),\mathrm{K}_*(A)\big)$ and an element in the cokernel of (5.35) are in bijection with isomorphism classes of objects in the bootstrap class $\mathfrak{B}^X\subseteq\mathfrak{KK}^X$.

We are going to compare this classification result with the filtrated K-theory classification by the long exact sequences in (5.34). The long exact sequence in (5.34) contains i_* and the extension

$$(5.36)\quad \mathrm{coker}\big(i\colon \mathrm{K}_*(I)\to\mathrm{K}_*(A)\big)\rightarrowtail\mathrm{K}_*(A/I)\twoheadrightarrow\ker\big(i\colon\mathrm{K}_{*+1}(I)\to\mathrm{K}_{*+1}(A)\big),$$

and we may reconstruct the long exact sequence from these two pieces. Two extensions as in (5.36) have the same class in Ext if and only if the long exact sequences associated to them (for the same i_*) are isomorphic with an isomorphism that is the identity on $\mathrm{K}_*(I)$ and $\mathrm{K}_*(A)$. Therefore, the filtrated K-theory invariant is equivalent to the pair consisting of $i_*\colon\mathrm{K}_*(I)\to\mathrm{K}_*(A)$ and a class in $\mathrm{Ext}\big(\ker(i_*),\mathrm{coker}(i_*)\big)$. Now the following proposition clarifies the relationship between our different invariants:

PROPOSITION 5.37. *The cokernel in* (5.35) *is naturally isomorphic to the group*

$$\mathrm{Ext}^2_{\mathfrak{A}^X}(\Sigma A,A)\cong\mathrm{Ext}\Big(\ker\big(\mathrm{K}_{*+1}(I)\xrightarrow{i_*}\mathrm{K}_{*+1}(A)\big),\mathrm{coker}\big(\mathrm{K}_*(I)\xrightarrow{i_*}\mathrm{K}_*(A)\big)\Big).$$

And the obstruction class is the class that corresponds to minus the extension in (5.36).

PROOF. Let G be a $\mathbb{Z}/2$-graded Abelian group. By the long exact sequence for Hom and Ext, the restriction map $\mathrm{Ext}\big(\mathrm{K}_{*+1}(A),G\big)\to\mathrm{Ext}\big(i_*(\mathrm{K}_{*+1}(I)),G\big)$ is surjective and

$$\cdots\to\mathrm{Ext}\big(i_*(\mathrm{K}_{*+1}(I)),G\big)\to\mathrm{Ext}\big(\mathrm{K}_{*+1}(I),G\big)\\ \to\mathrm{Ext}\big(\ker(i_*\colon\mathrm{K}_{*+1}(I)\to\mathrm{K}_{*+1}(A)),G\big)\to 0$$

is exact. Hence the cokernel of the map $i^*\colon\mathrm{Ext}\big(\mathrm{K}_{*+1}(A),G\big)\to\mathrm{Ext}\big(\mathrm{K}_{*+1}(I),G\big)$ is $\mathrm{Ext}(\ker(i_*),G)$. If we let $G:=\mathrm{K}_*(A)$, then we may identify the cokernel of the map in (5.35) with the cokernel of the map

$$\mathrm{Ext}\big(\mathrm{K}_{*+1}(I),\mathrm{K}_*(I)\big)\xrightarrow{\text{restrict}}\mathrm{Ext}\big(\ker(i_*),\mathrm{K}_*(I)\big)\xrightarrow{i_*}\mathrm{Ext}\big(\ker(i_*),\mathrm{K}_*(A)\big).$$

Since the first map is surjective, this is equal to the cokernel of

$$i_*\colon\mathrm{Ext}\big(\ker(i_*),\mathrm{K}_*(I)\big)\to\mathrm{Ext}\big(\ker(i_*),\mathrm{K}_*(A)\big).$$

A variant of the proof above for the second variable identifies this cokernel with the group $\mathrm{Ext}\big(\ker(i_*),\mathrm{coker}(i_*)\big)$ as claimed. Given a class $\delta\in\mathrm{Ext}\big(\mathrm{K}_{1+*}(I),\mathrm{K}_*(A)\big)$, the map to $\operatorname{Ext}\bigl(\ker(i_*), \operatorname{coker}(i_*)\bigr)$ simply applies the bifunctoriality of Ext for the quotient map

$$\mathrm{K}_*(A) \to \operatorname{coker}\bigl(i_* \colon \mathrm{K}_*(I) \to \mathrm{K}_*(A)\bigr)$$

and the inclusion map

$$\ker\bigl(i_* \colon \mathrm{K}_{*+1}(I) \to \mathrm{K}_{*+1}(A)\bigr) \to \mathrm{K}_{*+1}(I).$$

It remains to compute the image of the obstruction class in Theorem 5.17 in $\operatorname{Ext}\bigl(\ker(i_*), \operatorname{coker}(i_*)\bigr)$. We recall the data used in Theorem 5.17 in our special case. The semi-simple part of the filtrated K-theory consists of

$$\begin{aligned}
\mathrm{K}_*(A[2])_{\mathrm{ss}} &\cong \frac{\mathrm{K}_*(I)}{\delta(\mathrm{K}_{*-1}(A/I))} \cong i_*(\mathrm{K}_*(I)) \subseteq \mathrm{K}_*(A),\\
\mathrm{K}_*(A[1,2])_{\mathrm{ss}} &\cong \frac{\mathrm{K}_*(A)}{i_*(\mathrm{K}_*(I))} = \operatorname{coker}(i_*),\\
\mathrm{K}_*(A[1])_{\mathrm{ss}} &\cong \frac{\mathrm{K}_*(A/I)}{r_*(\mathrm{K}_*(A))} \cong \delta_*(\mathrm{K}_*(A/I)) = \ker(i_*) \subseteq \mathrm{K}_{*-1}(I).
\end{aligned}$$

Our construction is based on free resolutions of these $\mathbb{Z}/2$-graded Abelian groups. In particular, we use a free resolution $Q_1[1] \rightarrowtail Q_0[1] \twoheadrightarrow \ker(i_*)$. We lift these resolutions to free resolutions

$$\begin{aligned}
Q_1[2] \oplus Q_1[1] &\rightarrowtail Q_0[2] \oplus Q_0[1] \twoheadrightarrow \mathrm{K}_*(A[2]),\\
Q_1[1,2] \oplus Q_1[2] &\rightarrowtail Q_0[1,2] \oplus Q_0[2] \twoheadrightarrow \mathrm{K}_*(A[1,2]),\\
Q_1[1] \oplus Q_1[1,2] &\rightarrowtail Q_0[1] \oplus Q_0[1,2] \twoheadrightarrow \mathrm{K}_*(A[1]),
\end{aligned}$$

which contain the maps $(\tau f[a,b])^\#$ and $\varphi_{[a,b]}^{[c,d]}$. We shall need the maps in the third extension and put them into a larger diagram, which commutes because of the construction of the maps $(\tau f[a,b])^\#$ and $\varphi_{[a,b]}^{[c,d]}$:

$$\begin{array}{ccccc}
Q_1[1] \oplus Q_1[1,2] & \xrightarrow{\begin{pmatrix} \varphi_{[1]}^{[1]} & 0 \\ \varphi_{[1]}^{[1,2]} & \varphi_{[1,2]}^{[1,2]} \end{pmatrix}} & Q_0[1] \oplus Q_0[1,2] & \xrightarrow{(f[1]\ r_*f[1,2])} & \mathrm{K}_*(A/I)\\
\downarrow{\scriptstyle \mathrm{pr}_1} & & \downarrow{\scriptstyle \mathrm{pr}_1} & & \downarrow{\scriptstyle \mathrm{can.}}\\
Q_1[1] & \xrightarrow{\varphi_{[1]}^{[1]}} & Q_0[1] & \xrightarrow{d_0} & \frac{\mathrm{K}_*(A/I)}{r_*(\mathrm{K}_*(A))}\\
\downarrow{\scriptstyle f[1]\circ\varphi_{[1]}^{[1]}} & & \downarrow{\scriptstyle f[1]} & & {\scriptstyle \cong}\downarrow{\scriptstyle \delta}\\
r_*(\mathrm{K}_*(A)) & \xrightarrow{\text{incl.}} & \mathrm{K}_*(A/I) & \xrightarrow{\delta_*} & \ker(i_*)
\end{array} \tag{5.38}$$

The matrix coefficient $\delta_e^{[a,b]}$ in Theorem 5.17 is defined only if $e = a \le b < n$. In our case $n = 2$, there is only one such matrix coefficient, namely, $\delta_1^{[1]} \colon Q_1[1] \to \mathrm{K}_*(A[1,2]) = \mathrm{K}_*(A)$. The sum defining it has only one summand, which is indexed by $[1,2] \to [1,2] \to [1]$. So

$$\delta_1^{[1]} = f[1,2] \circ \varphi_{[1]}^{[1,2]} \colon Q_1[1] \to \mathrm{K}_*(A).$$

We compose this map with the quotient map $\mathrm{K}_*(A) \twoheadrightarrow \operatorname{coker}(i_*)$. The exact sequence (5.34) shows that r_* induces an isomorphism from $\operatorname{coker}(i_*)$ onto $r_*(\mathrm{K}_*(A)) \subseteq \mathrm{K}_*(A/I)$. So we may as well compose with the map $r_* \colon \mathrm{K}_*(A) \to r_*(\mathrm{K}_*(A))$. The composite of the two maps in the top row in (5.38) is 0. Thus $r^* \circ \delta_1^{[1]} = -f[1] \circ \varphi_{[1]}^{[1]}$ (compare (5.13)). The commuting diagram (5.38) shows that the group extension of $\ker(i_*)$ in (5.36) belongs to the map $f[1] \circ \varphi_{[1]}^{[1]}$. The obstruction class belongs to the negative of this map. □

References

[1] Saad Baaj and Georges Skandalis, *C^*-algèbres de Hopf et théorie de Kasparov équivariante* (French, with English summary), K-Theory **2** (1989), no. 6, 683–721, DOI 10.1007/BF00538428. MR1010978

[2] Rasmus Bentmann, *Contributions to the structure theory of non-simple* C*-*algebrass*, Ph.D. Thesis, University of Copenhagen, 2013, `http://web.math.ku.dk/noter/filer/phd13rb.pdf`.

[3] Rasmus Bentmann and Manuel Köhler, *Universal coefficient theorems for* C*-*algebras over finite topological spaces* (2011), eprint. arXiv 1101.5702.

[4] Rasmus Bentmann and Ralf Meyer, *A more general method to classify up to equivariant KK-equivalence*, Doc. Math. **22** (2017), 423–454. MR3628788

[5] Theo Bühler, *Exact categories*, Expo. Math. **28** (2010), no. 1, 1–69, DOI 10.1016/j.exmath.2009.04.004. MR2606234

[6] Ivo Dell'Ambrogio, Greg Stevenson, and Jan Šťovíček, *Gorenstein homological algebra and universal coefficient theorems*, Math. Z. **287** (2017), no. 3-4, 1109–1155, DOI 10.1007/s00209-017-1862-7. MR3719530

[7] Bernhard Keller, *Derived categories and their uses*, Handbook of algebra, Vol. 1, Handb. Algebr., vol. 1, Elsevier/North-Holland, Amsterdam, 1996, pp. 671–701, DOI 10.1016/S1570-7954(96)80023-4. MR1421815

[8] Manuel Köhler, *Universal coefficient theorems in equivariant KK-theory*, Ph.D. Thesis, Georg-August-Universität Göttingen, 2010, `http://hdl.handle.net/11858/00-1735-0000-0006-B6A9-9`.

[9] Ralf Meyer and Ryszard Nest, *The Baum-Connes conjecture via localisation of categories*, Topology **45** (2006), no. 2, 209–259, DOI 10.1016/j.top.2005.07.001. MR2193334

[10] Ralf Meyer and Ryszard Nest, *C^*-algebras over topological spaces: the bootstrap class*, Münster J. Math. **2** (2009), 215–252. MR2545613

[11] Ralf Meyer and Ryszard Nest, *Homological algebra in bivariant K-theory and other triangulated categories. I*, Triangulated categories, London Math. Soc. Lecture Note Ser., vol. 375, Cambridge Univ. Press, Cambridge, 2010, pp. 236–289. MR2681710

[12] Ralf Meyer and Ryszard Nest, C*-*algebras over topological spaces: filtrated K-theory*, Canad. J. Math. **64** (2012), no. 2, 368–408, DOI 10.4153/CJM-2011-061-x. MR2953205

[13] Hideki Nakamura, *Aperiodic automorphisms of nuclear purely infinite simple C^*-algebras*, Ergodic Theory Dynam. Systems **20** (2000), no. 6, 1749–1765, DOI 10.1017/S0143385700000973. MR1804956

Mathematisches Institut, Georg-August Universität Göttingen, Bunsenstraße 3-5, 37073 Göttingen, Germany

Email address: `rmeyer2@uni-goettingen.de`

Contemporary Mathematics
Volume **749**, 2020
https://doi.org/10.1090/conm/749/15076

Grothendieck duality made simple

Amnon Neeman

ABSTRACT. It has long been accepted that the foundations of Grothendieck duality are complicated. This has changed recently.

By "Grothendieck duality" we mean what, in the old literature, used to go by the name "coherent duality". This isn't to be confused with what is nowadays called "Verdier duality", and used to pass as "ℓ-adic duality". (The prevailing current terminology—for duality in étale cohomology, that is "ℓ-adic duality"—is historically incorrect. The idea was originally due not to Verdier but to Grothendieck, see his work in SGA5 on what is nowadays called the formalism of the six operations. Since this survey is about coherent duality we elaborate no further.)

CONTENTS

0. Introduction

There are two classical paths to the foundations of Grothendieck duality: one due to Grothendieck and Hartshorne [**19**] and (much later) Conrad [**14**], and a second due to Deligne [**16**] and Verdier [**64**] and (much later) Lipman [**32**]. The consensus has been that both are unsatisfactory. If you listen to the detractors of the respective approaches: the first is a nightmare to set up, the second leads to a theory where you can't compute anything. While exaggerated, each criticism used to have a kernel of truth to it. Lipman summed it up more circumspectly and diplomatically some years ago, by saying there is no royal road to the subject.

2010 *Mathematics Subject Classification.* Primary 14F05, Secondary 13D09, 18G10.
Key words and phrases. Derived categories, Grothendieck duality.
The research was partly supported by the Australian Research Council.

And Lipman is probably the person who worked hardest on simplifying the foundations.

In passing let us mention that, while the two foundational avenues to setting up the subject must obviously be related, the details of this link are far from clear—in fact they haven't yet been fully worked out. And Lipman also happens to be the person who has tried hardest to understand this bridge.

Back to the history of the field: what happened is that in the mid 1970s the math community gradually started losing interest in the project, and by the mid 1980s the exodus was all but complete—most people had given up on improving the foundations and moved on. For three decades now there have been at most a dozen people actively working in the field—they break up into two groups: Lipman, his students and collaborators and Yekutieli, his students and collaborators. These two groups made regular, incremental progress, see for example Alonso, Jeremías and Lipman [**1**–**4**], Lipman [**29**–**31**], Lipman, Nayak and Sastry [**33**], Lipman and Neeman [**34**], Nayak [**39, 40**], Porta, Shaul and Yekutieli [**50**], Shaul [**56**–**58**], Sastry [**51, 52**], Yekutieli [**66**–**70**] and Yekutieli and Zhang [**72**–**77**].

And then, a couple of years ago, there was a seismic shift. In this article we attempt to describe the recent progress in a way accessible to the non-expert.

Let us stress that the term "non-expert" in the last paragraph is to be understood in the strong sense: the intended audience of this survey includes non-algebraic-geometers, and familiarity with derived categories isn't assumed. The reason for the inclusive exposition is that the questions opened up by the recent progress might well interest people in diverse fields, the most obvious being Hochschild homology and cohomology. The Hochschild experts might wish to start with Computation 3.2.11, as well as Problems 5.2.1, 5.2.2 and 5.3.3. Computation 3.2.11 spells out exactly where and how Hochschild homology and cohomology played a key role in the breakthrough we report. The three problems suggest obvious variants, explaining why each would be interesting to solve. To put it in a nutshell: up to the present time we—meaning the handful of algebraic geometers still working on the foundations of Grothendieck duality—have only been able to carry out the Hochschild homology computation of 3.2.11, which amounts to a very simple, special case of the general problem. Given how profoundly this baby computation has transformed our understanding of Grothendieck duality, we warmly invite mathematicians more adept and dextrous at using the Hochschild machinery to come to our aid.

1. Background

1.1. Reminder: Formally inverting morphisms. Let $\mathcal{C}$ be a category, and let $S \subset \mathrm{Mor}(\mathcal{C})$ be some collection of morphisms. It is a theorem of Gabriel and Zisman [**17**] that one may form a functor $F : \mathcal{C} \longrightarrow S^{-1}\mathcal{C}$ so that

(i) The functor F takes every element of S to an isomorphism.
(ii) If $H : \mathcal{C} \longrightarrow \mathcal{B}$ is a functor, taking every element of S to an isomorphism, then there exists a unique functor $G : S^{-1}\mathcal{C} \longrightarrow \mathcal{B}$ rendering commutative the triangle

$$\begin{array}{ccc} & & S^{-1}\mathcal{C} \\ & \overset{F}{\nearrow} & \vdots\ \exists ! G \\ \mathcal{C} & & \downarrow \\ & \underset{H}{\searrow} & \\ & & \mathcal{B} \end{array}$$

We call this construction *formally inverting the morphisms in* S.

REMARK 1.1.1. On objects the functor F is the identity: the objects of $S^{-1}\mathcal{C}$ are identical to those of $\mathcal{C}$. But the morphisms in $S^{-1}\mathcal{C}$ are complicated. Clearly any morphism of $\mathcal{C}$ must have an image in $S^{-1}\mathcal{C}$, but $S^{-1}\mathcal{C}$ must also contain inverses of the images of morphisms in S. And then we must be able to compose any finite string of these.

The morphisms of $S^{-1}\mathcal{C}$ are in fact equivalence classes of such finite strings. The problem becomes to figure out when two such strings are equivalent, that is which strings must have the same composite in $S^{-1}\mathcal{C}$. This is usually called *the calculus of fractions* of $S^{-1}\mathcal{C}$. And without an understanding of this calculus of fractions the category $S^{-1}\mathcal{C}$ is unwieldy.

The category $S^{-1}\mathcal{C}$ can be dreadful in general, for example: it may happen that $\mathcal{C}$ has small Hom-sets but $S^{-1}\mathcal{C}$ doesn't.

1.2. Reminder: Derived categories and derived functors. Let $\mathcal{A}$ be an abelian category. We will be looking at categories we will denote $\mathbf{D}_{\mathfrak{C}}(\mathcal{A})$. The category $\mathbf{D}_{\mathfrak{C}}(\mathcal{A})$ is as follows:

(i) The objects: an object in $\mathbf{D}_{\mathfrak{C}}(\mathcal{A})$ is a cochain complex of objects in $\mathcal{A}$, that is a diagram in $\mathcal{A}$

$$\cdots \longrightarrow A^{-2} \longrightarrow A^{-1} \longrightarrow A^{0} \longrightarrow A^{1} \longrightarrow A^{2} \longrightarrow \cdots$$

where the composite $A^i \longrightarrow A^{i+1} \longrightarrow A^{i+2}$ vanishes for every $i \in \mathbb{Z}$. The $\mathfrak{C}$ in $\mathbf{D}_{\mathfrak{C}}(\mathcal{A})$ stands for conditions: we may impose conditions on the objects, it may be that not every cochain complex belongs to $\mathbf{D}_{\mathfrak{C}}(\mathcal{A})$.

(ii) Morphisms: cochain maps are examples, that is commutative diagrams

$$\begin{array}{ccccccccccc}
\cdots \longrightarrow & A^{-2} & \longrightarrow & A^{-1} & \longrightarrow & A^{0} & \longrightarrow & A^{1} & \longrightarrow & A^{2} & \longrightarrow \cdots \\
& \downarrow & & \downarrow & & \downarrow & & \downarrow & & \downarrow & \\
\cdots \longrightarrow & B^{-2} & \longrightarrow & B^{-1} & \longrightarrow & B^{0} & \longrightarrow & B^{1} & \longrightarrow & B^{2} & \longrightarrow \cdots
\end{array}$$

where the rows are objects in $\mathbf{D}_{\mathfrak{C}}(\mathcal{A})$. But we also formally invert the cohomology isomorphisms.

REMARK 1.2.1. In the special case of $\mathbf{D}_{\mathfrak{C}}(\mathcal{A})$ the calculus of fractions is reasonably well understood, there's a rich literature about it—but we will not explain this here. This means that, whenever we tell the reader about some computation of morphisms, we will be asking the beginner to accept it on faith. The expert will notice that all the computations we mention are easy.

EXAMPLE 1.2.2. Let R be a commutative ring, and let $\mathcal{A} = R$–Mod be the abelian category of R–modules. Then the category $\mathbf{D}(R\text{–Mod})$ has for its objects all the cochain complexes of R–modules. The category $\mathbf{D}_{K\text{–Flat}}(R\text{–Mod})$ has for its objects all the K-flat complexes, the category $\mathbf{D}_{K\text{–Inj}}(R\text{–Mod})$ has for its objects all the K–injective complexes.

We remind the reader: a complex F^* is K–flat if, for all acyclic complexes A^*, the complex $A^* \otimes_R F^*$ is acyclic. A complex I^* is K–injective if, for all acyclic complexes A^*, the complex $\mathrm{Hom}_R(A^*, I^*)$ is acyclic.

Next we recall functors. If $F : \mathcal{A} \longrightarrow \mathcal{B}$ is an additive functor, we often want to extend F to derived categories. But the simple-minded approach does not in general work, you cannot simply apply F to the cochain complexes.

EXAMPLE 1.2.3. Let $f : R \longrightarrow S$ be a homomorphism of commutative rings, let $\mathcal{A} = R$–Mod and let $\mathcal{B} = S$–Mod. Fix an object $A \in S$–Mod. We wish to consider the functor $A \otimes_R (-)$, that is the functor that takes the object $B \in R$–Mod to the object $A \otimes_R B$ in S–Mod.

If we try to extend it to a functor $A \otimes_R (-) : \mathbf{D}(R\text{–Mod}) \longrightarrow \mathbf{D}(S\text{–Mod})$ we run into the following problem: it is entirely possible to have a cochain map $\widetilde{B}^* \longrightarrow B^*$ of R–modules, inducing an isomorphism in cohomology, but where $A \otimes_R \widetilde{B}^* \longrightarrow A \otimes_R B^*$ is *not* a cohomology isomorphism. The special case where $\widetilde{B}^* = 0$ is already problematic. A cochain map

$$\begin{array}{ccccccccccc}
\cdots \longrightarrow & 0 & \longrightarrow & 0 & \longrightarrow & 0 & \longrightarrow & 0 & \longrightarrow & 0 & \longrightarrow \cdots \\
& \downarrow & & \downarrow & & \downarrow & & \downarrow & & \downarrow & \\
\cdots \longrightarrow & B^{-2} & \longrightarrow & B^{-1} & \longrightarrow & B^0 & \longrightarrow & B^1 & \longrightarrow & B^2 & \longrightarrow \cdots
\end{array}$$

is a cohomology isomorphism if B^* is acyclic. But without some restrictions we would not expect $A \otimes_R B^*$ to be acyclic, meaning

$$\begin{array}{ccccccccccc}
\cdots \longrightarrow & 0 & \longrightarrow & 0 & \longrightarrow & 0 & \longrightarrow & 0 & \longrightarrow & 0 & \longrightarrow \cdots \\
& \downarrow & & \downarrow & & \downarrow & & \downarrow & & \downarrow & \\
\cdots \longrightarrow & A \otimes_R B^{-2} & \longrightarrow & A \otimes_R B^{-1} & \longrightarrow & A \otimes_R B^0 & \longrightarrow & A \otimes_R B^1 & \longrightarrow & A \otimes_R B^2 & \longrightarrow \cdots
\end{array}$$

will not be a cohomology isomorphism.

CONSTRUCTION 1.2.4. As above, suppose we are given an additive functor $F : \mathcal{A} \longrightarrow \mathcal{B}$. The remedy is to pass to "derived functors". The idea is as follows:

(i) Find a condition $\mathfrak{C}$ on the objects of $\mathbf{D}(\mathcal{A})$, so that if a cochain map $A^* \longrightarrow B^*$ between objects in $\mathbf{D}_{\mathfrak{C}}(\mathcal{A})$ is a cohomology isomorphism then so is $FA^* \longrightarrow FB^*$.

(ii) Prove that the natural functor $I : \mathbf{D}_{\mathfrak{C}}(\mathcal{A}) \longrightarrow \mathbf{D}(\mathcal{A})$ is an equivalence of categories.

Once we achieve (i) and (ii) above, we declare the derived functor of F to be the composite

$$\mathbf{D}(\mathcal{A}) \xrightarrow{\ I^{-1}\ } \mathbf{D}_{\mathfrak{C}}(\mathcal{A}) \xrightarrow{\ \ \ F\ \ \ } \mathbf{D}(B)$$

EXAMPLE 1.2.5. Let us return to the situation of Example 1.2.3: we are given a ring homomorphism $R \longrightarrow S$ and an S–module A. Then, while the functor $F(-) = A \otimes_R (-)$ does not respect general cochain maps inducing cohomology isomorphisms, it does respect them when the cochain complexes are K–flat as in Example 1.2.2. It turns out that the natural functor $I : \mathbf{D}_{K\text{-Flat}}(R\text{–Mod}) \longrightarrow \mathbf{D}(R\text{–Mod})$ is an equivalence of categories, and we define the functor $A \otimes^{\mathbf{L}}_R (-) : \mathbf{D}(R\text{–Mod}) \longrightarrow \mathbf{D}(S\text{–Mod})$ to be the composite

$$\mathbf{D}(R\text{–Mod}) \xrightarrow{\ I^{-1}\ } \mathbf{D}_{K\text{-Flat}}(R\text{–Mod}) \xrightarrow{\ \ A \otimes_R (-)\ \ } \mathbf{D}(S\text{–Mod})$$

Now consider the functor $\mathrm{Hom}_R(A, -) : R\text{–Mod} \longrightarrow S\text{–Mod}$. Once again this functor does not respect general cochain maps inducing cohomology isomorphisms. But it does respect them if the cochain complexes are K–injective as in Example 1.2.2, and the natural functor $I : \mathbf{D}_{K\text{-Inj}}(R\text{–Mod}) \longrightarrow \mathbf{D}(R\text{–Mod})$ is an equivalence of categories. We define the functor $\mathbf{R}\mathrm{Hom}_R(A,-) : \mathbf{D}(R\text{–Mod}) \longrightarrow \mathbf{D}(S\text{–Mod})$ to be the composite

$$\mathbf{D}(R\text{–Mod}) \xrightarrow{I^{-1}} \mathbf{D}_{K\text{–Inj}}(R\text{–Mod}) \xrightarrow{\mathrm{Hom}_R(A,-)} \mathbf{D}(S\text{–Mod})$$

REMARK 1.2.6. If $\mathcal{A}$ is an abelian category, the category $\mathcal{C}(\mathcal{A})$ has the same objects as $\mathbf{D}(\mathcal{A})$ but the only morphisms of $\mathcal{C}(\mathcal{A})$ are the genuine cochain maps. Inverses of cohomology isomorphisms are not allowed.

Generalizing the discussion of Example 1.2.5, we will allow ourselves to derive additive functors $\mathcal{C}(\mathcal{A}) \longrightarrow \mathcal{C}(\mathcal{B})$. For example: if A^* is an object of $\mathcal{C}(S\text{–Mod})$ there are standard functors

$$A^*\otimes_R(-) : \mathcal{C}(R\text{–Mod}) \longrightarrow \mathcal{C}(S\text{–Mod})\,, \quad \mathrm{Hom}_R(A^*,-) : \mathcal{C}(R\text{–Mod}) \longrightarrow \mathcal{C}(S\text{–Mod})\,.$$

The derived functors $A^* \otimes^{\mathbf{L}}_R (-)$ and $\mathbf{R}\mathrm{Hom}_R(A^*,-)$ are, respectively, the composites

$$\mathbf{D}(R\text{–Mod}) \xrightarrow{I^{-1}} \mathbf{D}_{K\text{–Flat}}(R\text{–Mod}) \xrightarrow{A^*\otimes_R(-)} \mathbf{D}(S\text{–Mod})$$

$$\mathbf{D}(R\text{–Mod}) \xrightarrow{I^{-1}} \mathbf{D}_{K\text{–Inj}}(R\text{–Mod}) \xrightarrow{\mathrm{Hom}_R(A^*,-)} \mathbf{D}(S\text{–Mod})$$

REMARK 1.2.7. As presented in Example 1.2.5 and Remark 1.2.6 the construction involves an arbitrary choice. More precisely: in Remark 1.2.6 the functor $A^* \otimes^{\mathbf{L}}_R (-)$ was defined by observing

(i) The natural functor $I : \mathbf{D}_{K\text{–Flat}}(R\text{–Mod}) \longrightarrow \mathbf{D}(R\text{–Mod})$ is an equivalence of categories.

(ii) On the category $\mathbf{D}_{K\text{–Flat}}(R\text{–Mod})$ the functor $A^* \otimes_R (-)$ is well-defined in the obvious way, meaning that when we restrict to K–flat cochain complexes the classical functor $A^* \otimes_R (-)$ respects cochain maps inducing cohomology isomorphisms.

This allowed us to form the functor $A^* \otimes^{\mathbf{L}}_R (-)$ as the composite

$$\mathbf{D}(R\text{–Mod}) \xrightarrow{I^{-1}} \mathbf{D}_{K\text{–Flat}}(R\text{–Mod}) \xrightarrow{A^*\otimes_R(-)} \mathbf{D}(S\text{–Mod})$$

But the observant reader will note that

(iii) The natural functor $I : \mathbf{D}_{K\text{–Inj}}(R\text{–Mod}) \longrightarrow \mathbf{D}(R\text{–Mod})$ is also an equivalence of categories.

(iv) On the category $\mathbf{D}_{K\text{–Inj}}(R\text{–Mod})$ the functor $A^*\otimes_R(-)$ is also well-defined in the obvious way.

Hence there is nothing to stop us from forming the composite

$$\mathbf{D}(R\text{–Mod}) \xrightarrow{I^{-1}} \mathbf{D}_{K\text{–Inj}}(R\text{–Mod}) \xrightarrow{A^*\otimes_R(-)} \mathbf{D}(S\text{–Mod})$$

And it turns out that the composite $A^* \otimes^{\mathbf{L}}_R (-)$, defined using (i) and (ii), *does not in general agree* with the composite defined using (iii) and (iv).

For a choice-free description one notes that, with the category $\mathcal{C}(R\text{–Mod})$ as in Remark 1.2.6 and with $F : \mathcal{C}(R\text{–Mod}) \longrightarrow D(R\text{–Mod})$ the Gabriel-Zisman quotient

map of Reminder 1.1, we have a triangle

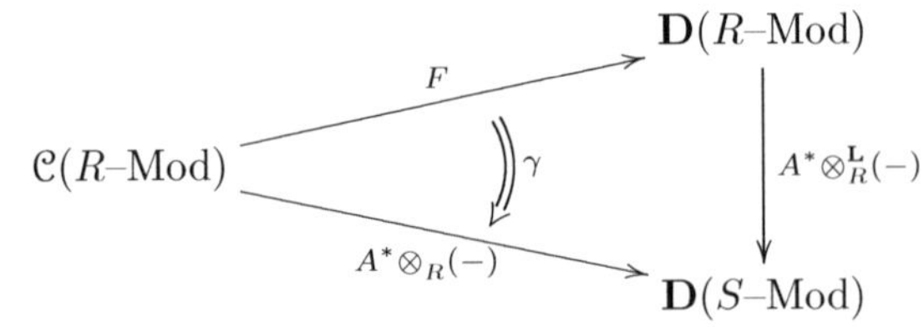

That is: there is a natural transformation from the composite functor $[A^*\otimes^{\mathbf{L}}_R(-)]\circ F$ to the functor $A^* \otimes_R (-)$. And what turns out to be true is that the triangle above has the obvious universal property: for any triangle

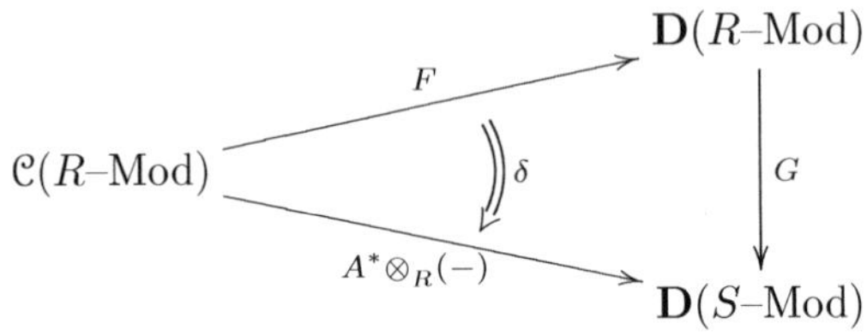

of functors and natural tranformation, there is a unique natural transformation $\varphi : G \Longrightarrow [A^* \otimes^{\mathbf{L}}_R (-)]$ such that $\delta = \gamma \circ (\varphi F)$. In category theoretic language: *the functor $A^* \otimes^{\mathbf{L}}_R (-)$ is the right Kan extension of $A^* \otimes_R (-)$ along F.*

This description as the right Kan extension is what earns the functor $A^* \otimes^{\mathbf{L}}_R (-)$ the title of the *left derived functor* of the functor $A^* \otimes_R (-)$, and is the reason for the $\mathbf{L}$ in the symbol. The functor $\mathbf{R}\mathrm{Hom}(A^*, -)$, which we met in Example 1.2.5 and Remark 1.2.6, turns out to be the *left Kan extension,* meaning the triangle of functors and natural transformation is

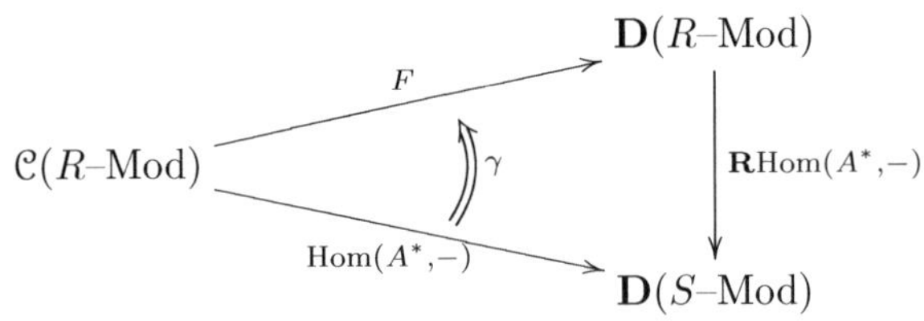

and the universality is with respect to all such triangles. Being the left Kan extension earns the functor $\mathbf{R}\mathrm{Hom}(A^*, -)$ the title of *right derived functor of* $\mathrm{Hom}(A^*, -)$, as well as the $\mathbf{R}$ in the symbol.

1.3. Conventions. Unless otherwise stated all rings are assumed commutative and noetherian, all schemes are assumed noetherian, and all morphisms of schemes are assumed of finite type. Since we will often deal with the derived category $\mathbf{D}(R\text{–Mod})$ we abbreviate it to $\mathbf{D}(R)$. If X is a scheme we will use $\mathbf{D}_{\mathbf{qc}}(X)$ as a shorthand for the category $\mathbf{D}_{\mathbf{qc}}(\mathcal{O}_X\text{–Mod})$. That is: the objects are cochain complexes of sheaves of $\mathcal{O}_X$–modules, and the condition we impose is that the cohomology sheaves are quasicoherent.

2. Statements of the main results

2.1. Generalities. We begin by setting up the framework.

REMINDER 2.1.1. Suppose $f : X \longrightarrow Y$ is a morphism of schemes. There are three induced functors[1] on the derived categories

$$\begin{array}{c} \mathbf{D_{qc}}(X) \\ \mathbf{L}f^* \uparrow \quad \downarrow \mathbf{R}f_* \quad \uparrow f^\times \\ \mathbf{D_{qc}}(Y) \end{array}$$

where each functor is left adjoint to the one to its right; in category theoretic notation we write $\mathbf{L}f^* \dashv \mathbf{R}f_* \dashv f^\times$. We remind the reader what these functors do.

(i) The functor $\mathbf{L}f^*$ is the left-derived pullback functor. We compute it as in Construction 1.2.4: let $\mathbf{D}_{\text{qc},K\text{–Flat}}(Y)$ be the derived category of complexes of $\mathcal{O}_Y$–modules, which are K–flat and have quasicoherent cohomology, and let $I : \mathbf{D}_{\text{qc},K\text{–Flat}}(Y) \longrightarrow \mathbf{D_{qc}}(Y)$ be the natural map. The functor I happens to be an equivalence. To evaluate $\mathbf{L}f^*$ on an object $C \in \mathbf{D_{qc}}(Y)$ you first form $I^{-1}(C) \in \mathbf{D}_{\text{qc},K\text{–Flat}}(Y)$, then pull back to obtain the complex $f^{-1}I^{-1}(C)$ on X, and finally form on X the tensor product $\mathbf{L}f^*C = \mathcal{O}_X \otimes_{f^{-1}\mathcal{O}_Y} f^{-1}I^{-1}(C)$.

(ii) The functor $\mathbf{R}f_*$ is the right-derived pushforward functor. Once again we compute it as in Construction 1.2.4: this time let $\mathbf{D}_{\text{qc},K\text{–Inj}}(X)$ be the derived category of complexes of $\mathcal{O}_X$–modules, which are K–injective and have quasicoherent cohomology, and let $I : \mathbf{D}_{\text{qc},K\text{–Inj}}(X) \longrightarrow \mathbf{D_{qc}}(X)$ be the natural map. This functor I also happens to be an equivalence. To evaluate $\mathbf{R}f_*$ on an object $D \in \mathbf{D_{qc}}(X)$ you first form $I^{-1}(D) \in \mathbf{D}_{\text{qc},K\text{–Inj}}(X)$, and then push forward to obtain the complex $\mathbf{R}f_*D = f_*I^{-1}(D)$ on Y.

(iii) The functor $f^\times$ is the mysterious one, Grothendieck duality is about understanding its properties. For a general f it turns out that $f^\times$ need not be the derived functor of any functor on abelian categories, with the notation as in Remark 1.2.7. The reader might wish to look at Remark 5.1.1 and at Appendix A for further discussion of how unusual $f^\times$ can be.

DISCUSSION 2.1.2. Since we're after an understanding of the functor $f^\times$, we need to agree what the word "understanding" will mean. Recall that the adjunction $\mathbf{R}f_* \dashv f^\times$ gives a natural isomorphism

$$\mathrm{Hom}(A, f^\times B) \xrightarrow{\varphi(A,B)} \mathrm{Hom}(\mathbf{R}f_*A, B)$$

Putting $A = f^\times B$ this specializes to a homomorphism

$$\mathrm{Hom}(f^\times B, f^\times B) \xrightarrow{\varphi(f^\times B,B)} \mathrm{Hom}(\mathbf{R}f_*f^\times B, B)$$

[1] Algebraic geometers might find the symbol $f^\times$ unfamiliar; the pre-2009 literature on Grothendieck duality talks almost exclusively about another functor $f^!$. The functors $f^\times$ and $f^!$ agree when f is proper, but not in general. There is a discussion of $f^!$ and its relation with $f^\times$ in Reminder 5.1.2, and a brief summary of the history in Remark 5.1.4. Until we reach that point, in this paper we will work exclusively with $f^\times$.

which sends $\mathrm{id} : f^\times B \longrightarrow f^\times B$ to the map $\varepsilon : \mathbf{R}f_* f^\times B \longrightarrow B$, the counit of adjunction. Consider

$$\mathrm{Hom}(A, f^\times B) \xrightarrow{\mathbf{R}f_*} \mathrm{Hom}(\mathbf{R}f_* A, \mathbf{R}f_* f^\times B) \xrightarrow{\mathrm{Hom}(-,\varepsilon)} \mathrm{Hom}(\mathbf{R}f_* A, B)$$

It is classical that naturality forces the composite to agree with $\varphi(A, B)$. Summarizing:

Conclusion 2.1.3. If we could compute, for every $B \in \mathbf{D_{qc}}(Y)$, the object $f^\times B$ and the morphism $\varepsilon : \mathbf{R}f_* f^\times B \longrightarrow B$, then we'd feel we understand the adjunction pretty well. After all the map $\varphi(A, B) : \mathrm{Hom}(A, f^\times B) \longrightarrow \mathrm{Hom}(\mathbf{R}f_* A, B)$ would be explicit: given an element $\alpha \in \mathrm{Hom}(A, f^\times B)$, that is a morphism $\alpha : A \longrightarrow f^\times B$, then the map $\varphi(A, B)$ would send α to the composite $\mathbf{R}f_* A \xrightarrow{\mathbf{R}f_*\alpha} \mathbf{R}f_* f^\times B \xrightarrow{\varepsilon} B$, which is an element $\varepsilon \circ \mathbf{R}f_*\alpha \in \mathrm{Hom}(\mathbf{R}f_* A, B)$. OK: it wouldn't be so clear how to go back, but classically people have been happy with understanding just this direction.

We will soon specialize to the case where f is smooth and proper, but the next result holds more generally and we state it in a strong form.

Theorem 2.1.4. *Assume $f : X \longrightarrow Y$ is a finite-type morphism of noetherian schemes. If $B \in \mathbf{D_{qc}}(Y)$ and $C \in \mathbf{D_{qc}}(X)$ then there is a canonical natural isomorphism $p_{B,C} : B \otimes^{\mathbf{L}} \mathbf{R}f_* C \longrightarrow \mathbf{R}f_*(\mathbf{L}f^* B \otimes^{\mathbf{L}} C)$ and a canonical natural transformation $\chi : \mathbf{L}f^* B \otimes^{\mathbf{L}} f^\times \mathcal{O}_Y \longrightarrow f^\times B$ such that the following pentagon commutes*

$$\begin{array}{ccccc}
\mathbf{R}f_*\big[\mathbf{L}f^* B \otimes^{\mathbf{L}} f^\times \mathcal{O}_Y\big] & \xrightarrow{p^{-1}_{B,f^\times\mathcal{O}_Y}} & B \otimes^{\mathbf{L}} \mathbf{R}f_* f^\times \mathcal{O}_Y & \xrightarrow{\mathrm{id}\otimes\varepsilon} & B \otimes^{\mathbf{L}} \mathcal{O}_Y \\
& \searrow^{\mathbf{R}f_*\chi} & & & \| \\
& & \mathbf{R}f_* f^\times B & \xrightarrow[\varepsilon]{} & B
\end{array}$$

Furthermore: the map χ is an isomorphism if and only if f is proper and of finite Tor-dimension.

The non-expert should view finite Tor-dimension as a technical condition that will be satisfied by all the f's we will consider. We will discuss properness in Remark 2.2.2.

Remark 2.1.5. Suppose f is proper and of finite Tor-dimension. Then we have an isomorphism $\mathbf{L}f^* B \otimes^{\mathbf{L}} f^\times \mathcal{O}_Y \longrightarrow f^\times B$, and the commutative pentagon of Theorem 2.1.4 makes precise the compatibility of this isomorphism with the counit ε of the adjunction $\mathbf{R}f_* \dashv f^\times$. Thus the import of Theorem 2.1.4 is that, as long as f is proper and of finite Tor-dimension, it suffices to compute $f^\times \mathcal{O}_Y$ and the counit of adjunction $\varepsilon : \mathbf{R}f_* f^\times \mathcal{O}_Y \longrightarrow \mathcal{O}_Y$. We are reduced to studying a single object, namely $\mathcal{O}_Y \in \mathbf{D_{qc}}(Y)$.

REMINDER 2.1.6. The next reduction comes from the observation that the category $\mathbf{D_{qc}}(X)$ has many endofunctors. There are many diagrams

$$\mathbf{D_{qc}}(X) \underset{\text{id}}{\overset{\Gamma_W}{\rightrightarrows}} \mathbf{D_{qc}}(X) \qquad c_W \Downarrow$$

That is: there are many choices of functors $\Gamma_W : \mathbf{D_{qc}}(X) \longrightarrow \mathbf{D_{qc}}(X)$, which come together with natural transformations $c_W : \Gamma_W \longrightarrow \text{id}$. The ones we have in mind are the Bousfield colocalizations. They come about as follows.

For every point $p \in X$ let $i_p : p \longrightarrow X$ be the inclusion, which we view as a morphism of schemes $i_p : \text{Spec}\big(k(p)\big) \longrightarrow X$. Suppose we are given a set of points $W \subset X$. The full subcategory $\mathbf{D}_{\mathbf{qc},W}(X) \subset \mathbf{D_{qc}}(X)$ will be the subcategory of all objects *supported on* W, we recall that this means

$$\mathbf{D}_{\mathbf{qc},W}(X) \quad = \quad \{E \in \mathbf{D_{qc}}(X) \mid \mathbf{L}i_p^*E = 0 \text{ for all } p \notin W\} \ .$$

Let $I_W : \mathbf{D}_{\mathbf{qc},W}(X) \longrightarrow \mathbf{D_{qc}}(X)$ be the inclusion. A straightforward generalization of a theorem of Bousfield [**13**] tells us that I_W has a right adjoint $J_W : \mathbf{D_{qc}}(X) \longrightarrow \mathbf{D}_{\mathbf{qc},W}(X)$, and the colocalizations we have in mind are the counits of adjunction $c_W : I_W J_W \longrightarrow \text{id}$.

We will give a concrete example later in this section.

REMARK 2.1.7. Let the notation be as in Reminder 2.1.6. It is customary to choose the set of points $W \subset X$ to be *closed under specialization*, meaning if $p \in W$, and if $q \in X$ belongs to the closure $\overline{\{p\}}$ of p, then $q \in W$. The advantage is that for such W the Bousfield colocalization can be computed locally. More precisely: let $u : U \longrightarrow X$ be an open immersion, let $\mathbf{D}_{\mathbf{qc},U\cap W}(U) \subset \mathbf{D_{qc}}(U)$ be the full subcategory of objects supported on $U \cap W$, let $I_{U\cap W} : \mathbf{D}_{\mathbf{qc},U\cap W}(U) \longrightarrow \mathbf{D_{qc}}(U)$ be the inclusion and $J_{U\cap W}$ its right adjoint, and let $c_{U\cap W} : \Gamma_{U\cap W} \longrightarrow \text{id}_{\mathbf{D_{qc}}(U)}$ be the counit of the adjunction $I_{U\cap W} \dashv J_{U\cap W}$. Then the relation with the $c_W : \Gamma_W \longrightarrow \text{id}_{\mathbf{D_{qc}}(X)}$ of Reminder 2.1.6 is simple: there is a canonical isomorphism $\mathbf{L}u^*\Gamma_W \cong \Gamma_{U\cap W}\mathbf{L}u^*$ making the triangle below commute

$$\begin{array}{ccc} \mathbf{L}u^*\Gamma_W & \xrightarrow{\mathbf{L}u^* c_W} & \\ \wr\downarrow & & \mathbf{L}u^* \\ \Gamma_{U\cap W}\mathbf{L}u^* & \xrightarrow{c_{U\cap W}\mathbf{L}u^*} & \end{array}$$

If W isn't specialization-closed this may fail, and the bottom line is that we feel infinitely more comfortable working with tools that lend themselves to local computations.

Given a $c_W : \Gamma_W \longrightarrow \text{id}$ as in Reminder 2.1.6, we can form the next gadget:

DEFINITION 2.1.8. We define ρ_W to be the composite

$$\mathbf{R}f_*\Gamma_W f^\times \mathcal{O}_Y \xrightarrow{\quad \mathbf{R}f_* c_W f^\times \quad} \mathbf{R}f_* f^\times \mathcal{O}_Y \xrightarrow{\quad \varepsilon \quad} \mathcal{O}_Y$$

REMARK 2.1.9. Ideally, we would aim to choose $c_W : \Gamma_W \longrightarrow \mathrm{id}$ in such a way that

(i) The composite ρ_W of Definition 2.1.8 is easy to compute.
(ii) From the computation of ρ_W we learn a lot about $\varepsilon : \mathbf{R}f_*f^\times \mathcal{O}_Y \longrightarrow \mathcal{O}_Y$.

Of course we could make a dumb choice of $c_W : \Gamma_W \longrightarrow \mathrm{id}$. For example: if we let $c_W : \Gamma_W \longrightarrow \mathrm{id}$ be the identity map $\mathrm{id} \longrightarrow \mathrm{id}$, then $\rho_W = \varepsilon$, we don't lose any information in passing from ε to ρ_W, but we also haven't simplified the computation. Or if we choose $\Gamma_W = 0$ then the computation of ρ_W becomes trivial, but worthless. The important thing is to choose $c_W : \Gamma_W \longrightarrow \mathrm{id}$ wisely.

2.2. If f is smooth and proper. In the most classical case of the theory we have the following results:

THEOREM 2.2.1. *Assume $f : X \longrightarrow Y$ is smooth and proper, of relative dimension n. Then there is a canonical isomorphism $\theta : \Omega^n_f[n] \xrightarrow{\sim} f^\times \mathcal{O}_Y$.*

REMARK 2.2.2. We should explain the theorem, starting with the hypotheses: if the non-expert tried to guess what it means for f to be smooth and proper, chances are she was right about proper but wrong about smooth. Let us elaborate.

It is customary to consider the following two conditions, which a continuous map $f : X \longrightarrow Y$ of topological spaces can satisfy:

(i) $f^{-1}(K)$ is compact whenever $K \subset Y$ is compact.
(ii) The map f is universally closed. This means that, if $f' : X' \longrightarrow Y'$ is some pullback of f, then $f'(K) \subset Y'$ is closed whenever $K \subset X'$ is closed.

In the category of locally compact Hausdorff spaces the two are equivalent, and a map satisfying these equivalent conditions is what's normally called proper. As it happens the topological spaces that come up in algebraic geometry are rarely Hausdorff, and in the category of schemes (i) and (ii) aren't equivalent. It turns out that the right way to define proper maps of schemes is to use (ii), this yields the theory one would intuitively expect.

But when it comes to smoothness algebraic geometers chose to be contrary. In differential geometry—and hence also in related topics like PDE—a smooth map of manifolds is defined to be a C^∞ map. With this definition alegbraic geometers never consider any map that's remotely non-smooth.

Even though the term "smooth map" was already in use in a well-established, clearly delineated context, algebraic geometers decided to steal the word and give it a different meaning. In this survey we follow the conventions of algebraic geometry: what we label a "smooth map" is what everyone else would dub a "submersion". In algebraic geometry, a morphism $f : X \longrightarrow Y$ of manifolds is *smooth* if, at every point $p \in X$, the derivative is a surjection $t_p : T_p \longrightarrow T_{f(p)}$. Here T_p is the tangent space at p and $T_{f(p)}$ is the tangent space at $f(p)$. The smooth map f has *relative dimension n* if the kernel of the linear map $t_p : T_p \longrightarrow T_{f(p)}$ is n–dimensional for every $p \in X$.

Assume f is a smooth map of relative dimension n in the sense above, and let Ω_X, Ω_Y be the cotangent bundles of X, Y respectively. The pullback $f^*\Omega_Y$ is naturally a subbundle of Ω_X, and the *relative cotangent bundle* is by definition the quotient $\Omega_f = \Omega_X / f^*\Omega_Y$, often written Ω^1_f. It is a vector bundle of rank n over X, whose top exterior power is what is usually written $\Omega^n_f = \wedge^n \Omega^1_f$. The line bundle Ω^n_f is called the *relative canonical bundle* of f. Thus Theorem 2.2.1 is the assertion that $f^\times \mathcal{O}_Y$ is canonically isomorphic to the object $\Omega^n_f[n] \in \mathbf{D_{qc}}(X)$, which is the cochain complex with only one nonvanishing term, namely the relative canonical bundle Ω^n_f in degree $-n$.

REMARK 2.2.3. With the notation as explained in Remark 2.2.2, Theorem 2.2.1 computes for us the object $f^\times \mathcal{O}_Y$. By Remark 2.1.6 our mission would be accomplished if we could also compute the counit of adjunction $\varepsilon : \mathbf{R}f_* f^\times \mathcal{O}_Y \longrightarrow \mathcal{O}_Y$. In Remark 2.1.9 we noted that it might prove expedient to take advantage of some Bousfield colocalization $c_W : \Gamma_W \longrightarrow \mathrm{id}$. The traditional choice, which happens to be well-suited for the current computation, is to take $c_W : \Gamma_W \longrightarrow \mathrm{id}$ to be the Bousfield colocalization of Reminder 2.1.6, where the set of points $W \subset X$ is the union of the irreducible closed subsets $Z \subset X$ such that the composite map $Z \longrightarrow X \longrightarrow Y$ is generically finite.

It should be noted that our Γ_W has been extensively studied and is very computable, the subject dealing with functors of this genre is called *local cohomology*. Most of what's written about Γ_W is in the commutative algebra literature.

Before the theorem it might help to illustrate the abstraction in a simple case.

EXAMPLE 2.2.4. Suppose $Y = \mathrm{Spec}(k)$ where k is field. Then X is smooth over the field k and n–dimensional. Take a minimal injective resolution for $\Omega^n_f[n] \cong f^\times \mathcal{O}_Y$: that is form a cochain complex

$$0 \longrightarrow I^{-n} \longrightarrow I^{-n+1} \longrightarrow \cdots \longrightarrow I^{-1} \longrightarrow I^0 \longrightarrow 0$$

where I^{-n} is the injective envelope of Ω^n_f, next I^{-n+1} is the injective envelope of I^{-n}/Ω^n_f, and so on.[2] Note that X is regular and n–dimensional, hence the injective dimension of Ω^n_f is $\leq n$. The minimal injective resolution must stop no later than I^0. The W of Remark 2.2.3 is the set of all closed points in X. The corresponding Bousfield colocalization $c_W : \Gamma_W \Omega^n_f[n] \longrightarrow \Omega^n_f[n]$ comes down to the cochain map

$$\begin{array}{ccccccccccccc}
0 & \longrightarrow & 0 & \longrightarrow & 0 & \longrightarrow & \cdots & \longrightarrow & 0 & \longrightarrow & I^0 & \longrightarrow & 0 \\
 & & \downarrow & & \downarrow & & & & \downarrow & & \| & & \\
0 & \longrightarrow & I^{-n} & \longrightarrow & I^{-n+1} & \longrightarrow & \cdots & \longrightarrow & I^{-1} & \longrightarrow & I^0 & \longrightarrow & 0
\end{array}$$

In the next theorem we will compute the composite ρ_W of Definition 2.1.8, that is we propose to compute the map $\mathbf{R}f_*\Gamma_W f^\times \mathcal{O}_Y \xrightarrow{\mathbf{R}f_* c_W} \mathbf{R}f_* f^\times \mathcal{O}_Y \xrightarrow{\varepsilon} \mathcal{O}_Y$. This map identifies, via the isomorphism $\Omega^n_f[n] \cong f^\times \mathcal{O}_Y$ of Theorem 2.2.1, with a composite

[2]The reader might recall that injective resolutions, even minimal ones, are only unique up to homotopy—hence it isn't obvious that the zeroth sheaf I^0 is functorial in anything—and the formula that's about to come, that is $I^0 = \Gamma_W \Omega^n_f[n]$, seems unreasonable at first sight.

Given any two injective resolutions of I^* and J^* of $\Omega^n_f[n]$ there are cochain maps $I^* \xrightarrow{\alpha} J^* \xrightarrow{\beta} I^*$, unique up to homotopy, so that $\alpha\beta$ and $\beta\alpha$ are homotopic to the identity. What is special here, because we are dealing with minimal injective resolutions of a line bundle Ω^n_f on a regular scheme X, is that any homotopy must vanish. Hence the minimal injective resolution I^* is unique up to canonical isomorphism, as is $I^0 = \Gamma_W \Omega^n_f[n]$.

For the experts in commutative algebra: the way one proves the vanishing of any homotopy is by noting that the injective sheaf I^{-j} is a direct sum of indecomposable injectives supported at points of dimension j, and there are no non-zero maps from an indecomposable injective supported at a point of dimension j to an indecomposable injective supported at a point of dimension $j+1$.

$\mathbf{R}f_*\Gamma_W\Omega^n_f[n] \overset{\mathbf{R}f_*c_W}{\longrightarrow} \mathbf{R}f_*\Omega^n_f[n] \overset{\varepsilon}{\longrightarrow} \mathcal{O}_Y$. Note that as $Y = \mathrm{Spec}(k)$ we have $\mathcal{O}_Y = k$. Recalling Reminder 2.1.1(ii), we compute $\mathbf{R}f_*$ by applying f_* to an injective resolution. That is, in our case ρ_W comes down to a composite cochain map

$$\begin{array}{ccccccccccccc}
0 & \longrightarrow & 0 & \longrightarrow & 0 & \longrightarrow & \cdots & \longrightarrow & 0 & \longrightarrow & f_*I^0 & \longrightarrow & 0 \\
 & & \downarrow & & \downarrow & & & & \downarrow & & \| & & \\
0 & \longrightarrow & f_*I^{-n} & \longrightarrow & f_*I^{-n+1} & \longrightarrow & \cdots & \longrightarrow & f_*I^{-1} & \longrightarrow & f_*I^0 & \longrightarrow & 0 \\
 & & \downarrow & & \downarrow & & & & \downarrow & & \downarrow{\scriptstyle \varepsilon^0} & & \\
0 & \longrightarrow & 0 & \longrightarrow & 0 & \longrightarrow & \cdots & \longrightarrow & 0 & \longrightarrow & k & \longrightarrow & 0
\end{array}$$

This makes it obvious why the $c_W : \Gamma_W f^\times \mathcal{O}_Y \longrightarrow f^\times \mathcal{O}_Y$ of Remark 2.2.3 is a reasonable choice: it certainly doesn't lose information, the map $\varepsilon^0 : f_*I^0 \longrightarrow k$ most definitely determines the cochain map $\varepsilon : \mathbf{R}f_*\Omega^n_f[n] = f_*I^* \longrightarrow k$.

Injective resolutions might be useful for proving a map is informative, but for computations one usually prefers other resolutions. Fortunately the functor Γ_W has other descriptions. For example there is a description in terms of local cohomology: it turns out that the $f_*I^0 = \mathbf{R}f_*\Gamma_W\Omega_f[n]$ above is isomorphic to

$$f_*I^0 \quad \cong \quad \bigoplus_{p\in X,\ \overline{\{p\}}=\{p\}} H^n_p(\Omega^n_f)\ .$$

That is, $f_*I^0 = \mathbf{R}f_*\Gamma_W\Omega^n_f[n]$ is the direct sum, over all closed points $p \in X$, of the nth local cohomology of the sheaf Ω^n_f at p. The standard computation of local cohomology, via Čech complexes, tells us that f_*I^0 may be written as the quotient of J, where J is the direct sum, over all closed points $p \in X$, of the vector space of meromorphic differential forms at p. A *meromorphic form at* p is the following list of data:

(i) The closed point $p \in X$.

For parts (ii) and (iii) below we write $\mathcal{O}_{X,p}$ for the stalk at p of the structure sheaf $\mathcal{O}_X$.

(ii) A system of coordinates $(x_1, \ldots, x_n)$ at p, that is $(x_1, \ldots, x_n)$ generate the maximal ideal of the local ring $\mathcal{O}_{X,p}$.

(iii) An expression

$$\frac{f dx_1 \wedge dx_2 \wedge \cdots \wedge dx_n}{x_1^N x_2^N \cdots x_n^N}$$

where $f \in \mathcal{O}_{X,p}$.

In Example 2.2.4 we simplified our life by assuming $Y = \mathrm{Spec}(k)$ with k a field. The general case is slightly more cumbersome to describe but similar. And now for the main result.

THEOREM 2.2.5. *Suppose $f : X \longrightarrow Y$ is a smooth and proper morphism of noetherian schemes, identify $\Omega^n_f[n] \cong f^\times\mathcal{O}_Y$ via the canonical isomorphism θ of Theorem 2.2.1, and let $c_W : \Gamma_W \longrightarrow \mathrm{id}$ be as chosen in Remark 2.2.3. Then the map $\rho_W : \mathbf{R}f_*\Gamma_W\Omega^n_f[n] \longrightarrow \mathcal{O}_Y$ of Definition 2.1.8 may be represented, in the notation of Example 2.2.4 [more accurately in the generalization of the notation to the case where Y is arbitrary], by the map $J \longrightarrow \mathcal{O}_Y$ taking a finite sum of meromorphic forms to the sum of their residues.*

EXAMPLE 2.2.6. Let us return to the situation of Example 2.2.4, where $Y = \mathrm{Spec}(k)$. But now assume further that $k = \overline{k}$ is an algebraically closed field. Then the residue of a meromorphic differential form is the obvious. In the expression

$$\frac{f dx_1 \wedge dx_2 \wedge \cdots \wedge dx_n}{x_1^N x_2^N \cdots x_n^N}$$

we may expand $f \in \mathcal{O}_{X,p}$ into a Taylor series, which we view as an element of the completion $\widehat{\mathcal{O}}_{X,p}$ of the ring $\mathcal{O}_{X,p}$. This gives an expansion of the entire meromorphic form into a Laurent series

$$\sum_{\text{all } k_i \leq N} \frac{a_{k_1,k_2,\ldots,k_n} dx_1 \wedge dx_2 \wedge \cdots \wedge dx_n}{x_1^{k_1} x_2^{k_2} \cdots x_n^{k_n}}$$

where the coefficients $a_{k_1,k_2,\ldots,k_n}$ belong to the field k. The map ρ_W takes the meromorphic form to $a_{1,1,\ldots,1}$, that is to the coefficient of

$$\frac{dx_1 \wedge dx_2 \wedge \cdots \wedge dx_n}{x_1 x_2 \cdots x_n}$$

in the Laurent series.

REMARK 2.2.7. The discussion of Example 2.2.6 should explain why we made the simplifying assumption that $Y = \mathrm{Spec}(k)$ with k algebraically closed. We run into subtleties already when $Y = \mathrm{Spec}(k)$ but we drop the hypothesis that $k = \overline{k}$. In this case the closed point p will not in general be k-rational, and we would not expect an element $f \in \mathcal{O}_{X,p} \subset \widehat{\mathcal{O}}_{X,p}$ to have a Taylor expansion with coefficients in k, in the generators $\{x_1, x_2, \ldots, x_n\}$ of the maximal ideal. The definition of the residue of a meromorphic form becomes subtler.

2.3. Application: Serre duality. Serre's classical duality theorem is a special case; let us recall precisely how. But first a reminder: back in Remark 1.2.1 we disclosed that the non-expert will be asked to accept, on faith, all computations of Hom-sets in derived categories—for example the ones she's about to witness.

Continue to assume that k is a field and $f : X \longrightarrow Y = \mathrm{Spec}(k)$ is smooth and proper, of relative dimension n. Let $\mathcal{V}$ be a vector bundle on X. Then the adjunction $\mathbf{R}f_* \dashv f^\times$ tells us that, in the commutative square below, the map φ is an isomorphism

$$\begin{array}{ccc}
\mathrm{Hom}_{\mathbf{D_{qc}}(X)}\big(\mathcal{V}[i], f^\times \mathcal{O}_Y\big) & \xrightarrow{\quad\varphi\quad} & \mathrm{Hom}_{\mathbf{D_{qc}}(Y)}\big(\mathbf{R}f_*\mathcal{V}[i], \mathcal{O}_Y\big) \\
\Big\| & & \Big\| \\
\mathrm{Hom}_{\mathbf{D_{qc}}(X)}\big(\mathcal{V}[i], \Omega^n_X[n]\big) & \xrightarrow{\qquad} & \mathrm{Hom}_{\mathbf{D_{qc}}(Y)}\big(\mathbf{R}f_*\mathcal{V}[i], k\big)
\end{array}$$

The bottom row simplifies to

$$H^{n-i}\big(\mathcal{H}om(\mathcal{V}, \Omega^n_X)\big) \quad \cong \quad \mathrm{Hom}\big(H^i(\mathcal{V}), k\big)$$

meaning there is a nondegenerate pairing $H^i(\mathcal{V}) \otimes H^{n-i}\big(\mathcal{H}om(\mathcal{V}, \Omega^n_X)\big) \longrightarrow k$. The pairing is explicit. It takes a morphism $\alpha : \mathcal{O}_X \longrightarrow \mathcal{V}[i]$ and a morphism $\beta : \mathcal{V} \longrightarrow \Omega^n_X[n-i]$ to the image of $1 \in \mathbf{R}f_*\mathcal{O}_X$ under the composite

$$\mathbf{R}f_*\mathcal{O}_X \xrightarrow{\ \mathbf{R}f_*\alpha\ } \mathbf{R}f_*\mathcal{V}[i] \xrightarrow{\ \mathbf{R}f_*\beta[i]\ } \mathbf{R}f_*\Omega^n_X[n] \xrightarrow{\quad\varepsilon\quad} k$$

After all: Discussion 2.1.2 and Conclusion 2.1.3 combine to tell us that the map $\beta[i] : \mathcal{V}[i] \longrightarrow \Omega^n_X[n] \cong f^\times \mathcal{O}_Y$ goes under the bijection φ to the composite

$$\mathbf{R}f_*\mathcal{V}[i] \xrightarrow{\mathbf{R}f_*\beta[i]} \mathbf{R}f_*\Omega^n_X[n] \xrightarrow{\quad\varepsilon\quad} k$$

and all we do is evaluate this composite at the element $\alpha : \mathcal{O}_X \longrightarrow \mathcal{V}[i]$ of the vector space $H^i(\mathcal{V}) = H^i\big(\mathcal{H}\mathrm{om}(\mathcal{O}_X, \mathcal{V})\big) = \mathrm{Hom}\big(\mathcal{O}_X, \mathcal{V}[i]\big)$.

REMARK 2.3.1. We can view Grothendieck duality as being Serre duality on steroids. Being macho, Grothendieck duality doesn't restrict the scheme Y to be the one-point space, doesn't assume the map f to be smooth and proper, and doesn't confine itself to only dealing with vector bundles.

3. The proofs

Modulo technicalities, the theorems of Section 2 are all contained in Hartshorne [**19**]. More precisely: the theorems in Hartshorne [**19**] aren't quite as clean or general, Section 2 is comprised of several technical improvements on those original assertions. However: with one inessential exception—which we will mention at the very end of Sketch 3.1.1—all the improvements had been obtained by the mid-1990s. In other words: none of the results in Section 2 is younger than two decades.

What is new is that we can now prove *every one of these statements* simply and directly, sidestepping the customary circuitous routes and long detours, and bypassing all of the traditional stopovers on distant planets. We will next discuss where the reader can find these simple, formal proofs. This naturally divides into two parts.

3.1. Simple proofs that have been around for decades. Let us begin with Reminder 2.1.1: we gave an explicit construction of the functors $\mathbf{L}f^*$ and $\mathbf{R}f_*$, and asserted the existence of a functor $f^\times$ right adjoint to $\mathbf{R}f_*$. The first short and formal proof of the existence of $f^\times$ may be found in Deligne's appendix [**16**] to Hartshorne's book [**19**]. In that proof the schemes are assumed noetherian and the derived categories are of bounded below complexes. The reader may find more general theorems in Balmer, Dell'Ambrogio and Sanders [**9**] and in [**43,46**], with the strongest theorem to date covering the case where f is any concentrated morphism of quasicompact, quasiseparated algebraic stacks. The modern proofs work by showing that $\mathbf{R}f_*$ respects coproducts and applying Brown representability.

Now we turn to Theorem 2.1.4, and for the reader's convenience we recall the statement

Theorem 2.1.4, reminder. *Assume $f : X \longrightarrow Y$ is a finite-type morphism of noetherian schemes. Then there is a canonical natural isomorphism $p_{A,B'} : A \otimes^{\mathbf{L}} \mathbf{R}f_*B' \longrightarrow \mathbf{R}f_*(\mathbf{L}f^*A \otimes^{\mathbf{L}} B')$ and a canonical natural transformation $\chi : \mathbf{L}f^*A \otimes^{\mathbf{L}} f^\times\mathcal{O}_Y \longrightarrow f^\times A$ such that the following pentagon commutes*

$$\begin{array}{ccccc}
\mathbf{R}f_*\big[\mathbf{L}f^*A \otimes^{\mathbf{L}} f^\times\mathcal{O}_Y\big] & \xrightarrow{p^{-1}_{A,f^\times\mathcal{O}_Y}} & A \otimes^{\mathbf{L}} \mathbf{R}f_*f^\times\mathcal{O}_Y & \xrightarrow{\mathrm{id}\otimes\varepsilon} & A \otimes^{\mathbf{L}} \mathcal{O}_Y \\
& \searrow^{\mathbf{R}f_*\chi} & & & \| \\
& & \mathbf{R}f_*f^\times A & \xrightarrow[\varepsilon]{\qquad} & A
\end{array}$$

Furthermore: the map χ is an isomorphism if and only if f is proper and of finite Tor-dimension.

The modern proof may be found in [**46**], and the reader might also wish to look at [**9**] for a generalization of Theorem 2.1.4 proved by the same techniques. To emphasize the formal nature of the argument we give an outline.

SKETCH 3.1.1. The functor $\mathbf{L}f^*$ is strong monoidal—it respects the tensor product, there is a natural isomorphism $\mathbf{L}f^*(A \otimes^{\mathbf{L}} B) \longrightarrow (\mathbf{L}f^*A) \otimes^{\mathbf{L}} (\mathbf{L}f^*B)$. If we put $B = \mathbf{R}f_*B'$ this gives the first map in the composite

$$\mathbf{L}f^*(A \otimes^{\mathbf{L}} \mathbf{R}f_*B') \longrightarrow (\mathbf{L}f^*A) \otimes^{\mathbf{L}} (\mathbf{L}f^*\mathbf{R}f_*B') \xrightarrow{\ \mathrm{id}\otimes\varepsilon'\ } (\mathbf{L}f^*A) \otimes^{\mathbf{L}} B' ,$$

where $\varepsilon' : \mathbf{L}f^*\mathbf{R}f_* \longrightarrow \mathrm{id}$ is the counit of the adjunction $\mathbf{L}f^* \dashv \mathbf{R}f_*$. Adjunction, applied to the highlighted composite above, gives a corresponding map $p_{A,B'}$: $A \otimes^{\mathbf{L}} \mathbf{R}f_*B' \longrightarrow \mathbf{R}f_*\big[(\mathbf{L}f^*A) \otimes^{\mathbf{L}} B'\big]$. The map p is an isomorphism, the so-called classical "projection formula". But since we are into seeing what part of the theory is formal, let us indicate the modern proof that p is an isomorphism.

If A is a perfect complex then it is "strongly dualizing"[3]—in particular there exists a dual complex $A^\vee$ and a canonical isomorphism $A \otimes^{\mathbf{L}} (-) \cong \mathcal{RH}\mathrm{om}(A^\vee, -)$. The following string of isomorphisms

$$\begin{aligned}
\mathrm{Hom}(C\ ,\ A \otimes^{\mathbf{L}} \mathbf{R}f_*B') &\cong \mathrm{Hom}(C\ ,\ \mathcal{RH}\mathrm{om}(A^\vee, \mathbf{R}f_*B')) \\
&\cong \mathrm{Hom}(C \otimes^{\mathbf{L}} A^\vee\ ,\ \mathbf{R}f_*B') \\
&\cong \mathrm{Hom}\big((\mathbf{L}f^*(C \otimes^{\mathbf{L}} A^\vee)\ ,\ B'\big) \\
&\cong \mathrm{Hom}\big((\mathbf{L}f^*C) \otimes^{\mathbf{L}} (\mathbf{L}f^*A^\vee)\ ,\ B'\big) \\
&\cong \mathrm{Hom}\big((\mathbf{L}f^*C) \otimes^{\mathbf{L}} (\mathbf{L}f^*A)^\vee\ ,\ B'\big) \\
&\cong \mathrm{Hom}\Big[(\mathbf{L}f^*C)\ ,\ \mathcal{RH}\mathrm{om}\big((\mathbf{L}f^*A)^\vee, B'\big)\Big] \\
&\cong \mathrm{Hom}\Big[C\ ,\ \mathbf{R}f_*\big((\mathbf{L}f^*A) \otimes^{\mathbf{L}} B'\big)\Big]
\end{aligned}$$

holds for every C and is a formal consequence of the definition of strongly dualizable objects—in particular the fifth isomorphism is formal, any strict monoidal functor must take strongly dualizable objects to strongly dualizable objects, and must respect duals. Yoneda gives that the isomorphism of Hom-sets must come from an isomorphism $A \otimes^{\mathbf{L}} \mathbf{R}f_*B' \cong \mathbf{R}f_*\big(\mathbf{L}f^*A) \otimes^{\mathbf{L}} B'\big)$, and it is an exercise in the definitions to check that this isomorphism is induced by the map $p_{A,B'}$. In other words: the map $p_{A,B'}$ induces an isomorphism as long as A is strongly dualizing—in the category $\mathbf{D_{qc}}(Y)$ this means as long as A is a perfect complex.

Thus the subcategory of all A's for which the map $p_{A,B'}$ induces an isomorphism, for every B', contains the perfect complexes and is closed under suspensions, triangles and coproducts. The closure under triangles and coproducts is because the functors $\mathbf{L}f^*$ and $\mathbf{R}f_*$ both respect triangles and coproducts. But then [**46**, Lemma 3.2] tells us that every object $A \in \mathbf{D_{qc}}(Y)$ belongs—the map $p_{A,B'}$ is an isomorphism for all pairs A, B'.

[3] Recall that an object A in a monoidal category is strongly dualizing if there exists a dual object $A^\vee$ and maps $A^\vee \otimes A \longrightarrow \mathbb{1}$ and $\mathbb{1} \longrightarrow A \otimes A^\vee$ so that the composites $A \longrightarrow A \otimes A^\vee \otimes A \longrightarrow A$ and $A^\vee \longrightarrow A^\vee \otimes A \longrightarrow A^\vee$ are both the identity. For the monoidal category $\mathbf{D_{qc}}(X)$ the strongly dualizing objects are the perfect complexes.

Once we know that the map p is an isomorphism we can repeat the idea. Put $B' = f^\times B''$, and then $p^{-1}_{A,f^\times B''}$ gives the first map in the composite

$$\mathbf{R}f_*\big[(\mathbf{L}f^*A)\otimes^{\mathbf{L}}(f^\times B'')\big] \xrightarrow{\;p^{-1}_{A,f^\times B''}\;} A\otimes^{\mathbf{L}}\mathbf{R}f_*f^\times B'' \xrightarrow{\;\mathrm{id}\otimes\varepsilon\;} A\otimes^{\mathbf{L}} B''$$

In this composite $\varepsilon : \mathbf{R}f_*f^\times B'' \longrightarrow B''$ is the counit of the adjunction $\mathbf{R}f_* \dashv f^\times$. Adjunction tells us that the composite corresponds to a map $\chi(A, B'') : (\mathbf{L}f^*A)\otimes^{\mathbf{L}} (f^\times B'') \longrightarrow f^\times(A\otimes^{\mathbf{L}} B'')$. The map $\chi = \chi_A$ of Theorem 2.1.4 is just the special case where $B'' = \mathcal{O}_Y$. The commutativity of the pentagon of Theorem 2.1.4 simply spells out what it means for the map χ_A to correspond, under the adjunction $\mathbf{R}f_* \dashv f^\times$, to the composite above—in other words the pentagon commutes by definition.

It remains to discuss when the map χ is an isomorphism.

One can easily write down a string of isomorphisms, much like those we used above to study $p_{A,B'}$, which combine to show that the map χ_A is an isomorphism as long as A is a perfect complex; see [**46**, top of page 228]. Consider the category of all A's so that the map χ_A is an isomorphism—it contains the perfect complexes, is closed under suspensions and triangles, and also closed under coproducts if $f^\times$ respects coproducts. Thus, as long as $f^\times$ respects coproducts, the map χ_A is an isomorphism for every A. In fact the condition that $f^\times$ respects coproducts is necessary and sufficient for the map χ to be an isomorphism.

Now [**46**, Lemma 5.1] comes to our aid: given a pair of adjoint triangulated functors $F \dashv G$, between compactly generated triangulated categories, the right adjoint G respects coproducts if and only if the left adjoint F respects compact objects.[4] Specializing to the pair of adjoint functors $\mathbf{R}f_* \dashv f^\times$, this tidbit of formal nonsense says that the functor $f^\times$ respects coproducts if and only if $\mathbf{R}f_*$ takes perfect complexes to perfect complexes. Thus we have tranformed a question about the right adjoint $f^\times$, which is mysterious, into a problem about its left adjoint $\mathbf{R}f_*$, which is explicit and computable. It is an old theorem of Illusie [**26**, Exposé III, Corollaire 4.3.1(a)] that if f is proper and of finite Tor-dimension then $\mathbf{R}f_*$ respects perfect complexes. The converse [which we don't use in this article], that is the theorem that if $\mathbf{R}f_*$ respects perfect complexes then f must be proper and of finite Tor-dimension, is much more recent. It may be found in [**34**].

This concludes our discussion of Theorem 2.1.4. Before proceeding to the remaining two theorems we include a brief interlude, recalling a couple of base-change maps.

REMINDER 3.1.2. Let us begin at the formal level. Suppose we are given four categories $\mathcal{W}$, $\mathcal{X}$, $\mathcal{Y}$ and $\mathcal{Z}$, as well as pairs of adjoint functors

$$\gamma : \mathcal{Y} \rightleftarrows \mathcal{W} : \Gamma \qquad\text{and}\qquad \beta : \mathcal{Z} \rightleftarrows \mathcal{X} : B$$

Recall: this is shorthand for the assertion that $\Gamma : \mathcal{W} \longrightarrow \mathcal{Y}$ is right adjoint to $\gamma : \mathcal{Y} \longrightarrow \mathcal{W}$, and $B : \mathcal{X} \longrightarrow \mathcal{Z}$ is right adjoint to $\beta : \mathcal{Z} \longrightarrow \mathcal{X}$. Assume further that we are given a pair of functors

$$\alpha : \mathcal{X} \longrightarrow \mathcal{W} \qquad\text{and}\qquad \delta : \mathcal{Z} \longrightarrow \mathcal{Y}$$

[4]Let $\mathcal{T}$ be a triangulated category with coproducts. An object $C \in \mathcal{T}$ is *compact* if $\mathrm{Hom}(C, -)$ respects coproducts—for this survey what's relevant is that the compact objects in $\mathbf{D_{qc}}(X)$ are the perfect complexes.

Then there is a canonical bijection between natural transformations

$$\begin{array}{ccc} \mathcal{W} & \xleftarrow{\ \alpha\ } & \mathcal{X} \\ {\scriptstyle\gamma}\uparrow & \overset{\rho}{\nearrow\!\!\!\nearrow} & \uparrow{\scriptstyle\beta} \\ \mathcal{Y} & \xleftarrow[\ \delta\ ]{} & \mathcal{Z} \end{array} \qquad \text{and} \qquad \begin{array}{ccc} \mathcal{W} & \xleftarrow{\ \alpha\ } & \mathcal{X} \\ {\scriptstyle\Gamma}\downarrow & \overset{\sigma}{\nwarrow\!\!\!\nwarrow} & \downarrow{\scriptstyle B} \\ \mathcal{Y} & \xleftarrow[\ \delta\ ]{} & \mathcal{Z} \end{array}$$

The formula is explicit: if $\eta(\gamma \dashv \Gamma)$ and $\varepsilon(\gamma \dashv \Gamma)$ are, respectively, the unit and counit of the adjunction $\gamma \dashv \Gamma$, while $\eta(\beta \dashv B)$ and $\varepsilon(\beta \dashv B)$ are, respectively, the unit and counit of the adjunction $\beta \dashv B$, then the formulas are that the bijection takes $\rho : \gamma\delta \longrightarrow \alpha\beta$ and $\sigma : \delta B \longrightarrow \Gamma\alpha$, respectively, to

$$\delta B \xrightarrow{\ \eta(\gamma\dashv\Gamma)\ } \Gamma\gamma\delta B \xrightarrow{\ \ \rho\ \ } \Gamma\alpha\beta B \xrightarrow{\ \varepsilon(\beta\dashv B)\ } \Gamma\alpha$$

$$\gamma\delta \xrightarrow{\ \eta(\beta\dashv B)\ } \gamma\delta B\beta \xrightarrow{\ \ \sigma\ \ } \gamma\Gamma\alpha\beta \xrightarrow{\ \varepsilon(\gamma\dashv\Gamma)\ } \alpha\beta$$

The standard terminology in category theory is that ρ and σ are *mates* of each other. More explicitly: ρ is the *left mate* of σ, and σ is the *right mate* of ρ.

As the reader may have guessed, this terminology was invented by Australian category theorists.

CONSTRUCTION 3.1.3. Any commutative square of schemes

$$\begin{array}{ccc} W & \xrightarrow{\ u\ } & X \\ {\scriptstyle f}\downarrow & & \downarrow{\scriptstyle g} \\ Y & \xrightarrow[\ v\ ]{} & Z \end{array}$$

gives rise to squares

$$\begin{array}{ccc} \mathbf{D_{qc}}(W) & \xleftarrow{\ \mathbf{L}u^*\ } & \mathbf{D_{qc}}(X) \\ {\scriptstyle \mathbf{L}f^*}\uparrow & \overset{\tau}{\nearrow\!\!\!\nearrow} & \uparrow{\scriptstyle \mathbf{L}g^*} \\ \mathbf{D_{qc}}(Y) & \xleftarrow[\ \mathbf{L}v^*\ ]{} & \mathbf{D_{qc}}(Z) \end{array} \qquad \text{and} \qquad \begin{array}{ccc} \mathbf{D_{qc}}(W) & \xleftarrow{\ \mathbf{L}u^*\ } & \mathbf{D_{qc}}(X) \\ {\scriptstyle \mathbf{R}f_*}\downarrow & & \downarrow{\scriptstyle \mathbf{R}g_*} \\ \mathbf{D_{qc}}(Y) & \xleftarrow[\ \mathbf{L}v^*\ ]{} & \mathbf{D_{qc}}(Z) \end{array}$$

where τ is the canonical isomorphism $\tau : \mathbf{L}f^*\mathbf{L}v^* \longrightarrow \mathbf{L}u^*\mathbf{L}g^*$. Reminder 3.1.2 yields a right mate for τ, producing a natural transformation we will call $\beta : \mathbf{L}v^*\mathbf{R}g_* \longrightarrow \mathbf{R}f_*\mathbf{L}u^*$. The classical flat base-change theorem tells us

3.1.3.1. *The map β is an isomorphism if the square is cartesian—meaning a pullback square in the category of schemes—and if v is flat.*

CONSTRUCTION 3.1.4. As in Construction 3.1.3 assume given a commutative square of schemes

$$\begin{array}{ccc} W & \xrightarrow{\ u\ } & X \\ {\scriptstyle f}\downarrow & \diamond & \downarrow{\scriptstyle g} \\ Y & \xrightarrow[\ v\ ]{} & Z \end{array}$$

If the base-change map $\mathbf{R}f_*\mathbf{L}u^* \xleftarrow{\beta} \mathbf{L}v^*\mathbf{R}g_*$ is an isomophism, for example if we are in the situation of 3.1.3.1, we may apply Reminder 3.1.2 to the squares

$$\begin{array}{ccc} \mathbf{D_{qc}}(W) & \xleftarrow{\mathbf{L}u^*} & \mathbf{D_{qc}}(X) \\ {\scriptstyle \mathbf{R}f_*}\big\downarrow & \overset{\beta^{-1}}{\Searrow} & \big\downarrow{\scriptstyle \mathbf{R}g_*} \\ \mathbf{D_{qc}}(Y) & \xleftarrow[\mathbf{L}v^*]{} & \mathbf{D_{qc}}(Z)\end{array} \qquad \text{and} \qquad \begin{array}{ccc} \mathbf{D_{qc}}(W) & \xleftarrow{\mathbf{L}u^*} & \mathbf{D_{qc}}(X) \\ {\scriptstyle f^\times}\big\uparrow & & \big\uparrow{\scriptstyle g^\times} \\ \mathbf{D_{qc}}(Y) & \xleftarrow[\mathbf{L}v^*]{} & \mathbf{D_{qc}}(Z)\end{array}$$

and produce a right mate for β^{-1}. We will write this base-change map as $\Phi = \Phi(\diamondsuit) : \mathbf{L}u^*g^\times \longrightarrow f^\times\mathbf{L}v^*$. And the relevant theorem tells us

3.1.4.1. *The map $\Phi(\diamondsuit)$ is an isomorphism if, in addition to the hypotheses of* 3.1.3.1 *already made to define $\Phi(\diamondsuit)$, we assume that g is proper and one of the following holds:*

(i) *f is of finite Tor-dimension.*
(ii) *We restrict the functors to $\mathbf{D}^+_{\mathbf{qc}}(Z) \subset \mathbf{D_{qc}}(Z)$; that is we only evaluate the map on bounded-below complexes.*

Remark 3.1.5. The assertion 3.1.3.1 goes all the way back to Grothendieck in the 1950s; it's an easy consequence of the fact that one can compute $\mathbf{R}f_*$ using Čech complexes.

There is a (complicated) proof of 3.1.4.1(ii) in Hartshorne [**19**], towards the end of the book. The first short and formal proof of 3.1.4.1(ii) is due to Verdier [**64**]. Over the years there have been technical improvements, and the strong version stated in 3.1.4.1(i) is recent. The reader is referred to [**43**, Lemma 5.20] for the proof.

For the applications in this article we do not need such refined forms of 3.1.4.1. Verdier's old theorem suffices.

It's time to move on to the remaining business: the proofs of Theorems 2.2.1 and 2.2.5. Both theorems are assertions involving a certain natural map θ; the first step is to construct this θ. We're about to do this to show that it can be done formally, using nothing more than the base-change map of Construction 3.1.4, the counit of some adjunction, and the Hochschild-Kostant-Rosenberg Theorem. The reader willing to skip the construction should proceed directly to Remark 3.1.9.

Construction 3.1.6. Suppose $f : X \longrightarrow Y$ is flat and proper. Consider the diagram

$$\begin{array}{ccccc} X & \xrightarrow{\delta} & X \times_Y X & \xrightarrow{\pi_1} & X \\ & & {\scriptstyle \pi_2}\big\downarrow & \diamondsuit & \big\downarrow{\scriptstyle f} \\ & & X & \xrightarrow[f]{} & Y \end{array}$$

where $\delta : X \longrightarrow X \times_Y X$ is the diagonal map. Since f is both flat and proper the hypotheses of 3.1.4.1 are satisfied, and the map $\Phi(\diamondsuit)$ is an isomorphism. We have the following composites, with the first two defining the maps α and γ that go into producing the third

$$\begin{array}{c}
f^\times \xrightarrow[\sim]{} \mathbf{L}\delta^*\mathbf{L}\pi_1^* f^\times \xrightarrow{\mathbf{L}\delta^*\Phi(\diamondsuit)} \mathbf{L}\delta^*\pi_2^\times \mathbf{L}f^* \qquad (\text{composite: } \alpha)\\[2ex]
\mathbf{R}\delta_* \xrightarrow{\sim} \mathbf{R}\delta_*\delta^\times\pi_2^\times \xrightarrow{\widetilde{\varepsilon}\pi_2^\times} \pi_2^\times \qquad (\text{composite: } \gamma)\\[2ex]
\mathbf{L}\delta^*\mathbf{R}\delta_*\mathbf{L}f^* \xrightarrow{\mathbf{L}\delta^*\gamma\mathbf{L}f^*} \mathbf{L}\delta^*\pi_2^\times\mathbf{L}f^* \xrightarrow{\alpha^{-1}} f^\times
\end{array}$$

The natural isomorphisms in the first two composites are because $\pi_1\delta = \pi_2\delta = \mathrm{id}$, hence $\mathbf{L}\delta^*\mathbf{L}\pi_1^* \cong \mathbf{L}(\mathrm{id})^* = \mathrm{id} = \mathrm{id}^\times \cong \delta^\times\pi_2^\times$. Since the map $\Phi(\diamondsuit)$ is an isomorphism the first row composes to an isomorphism, allowing us to form the third row. Evaluating the third row above at the object $\mathcal{O}_Y \in \mathbf{D_{qc}}(Y)$ we obtain the second and third maps in the composite below defining ζ

$$\mathbf{L}\delta^*\mathbf{R}\delta_*\mathcal{O}_X = \mathbf{L}\delta^*\mathbf{R}\delta_*\mathbf{L}f^*\mathcal{O}_Y \xrightarrow[\mathbf{L}\delta^*\gamma\mathbf{L}f^*]{} \mathbf{L}\delta^*\pi_2^\times\mathbf{L}f^*\mathcal{O}_Y \xrightarrow[\alpha^{-1}]{} f^\times\mathcal{O}_Y \qquad (\text{composite: } \zeta)$$

The object $\mathbf{L}\delta^*\mathbf{R}\delta_*\mathcal{O}_X$ is something we know and love—it is the derived category version of the tensor product of $\mathcal{O}_X$ with itself over $\mathcal{O}_{X\times_Y X}$. Formal nonsense tells us that $\mathbf{L}\delta^*\mathbf{R}\delta_*\mathcal{O}_X$ is a commutative monoid in the monoidal category $\mathbf{D_{qc}}(X)$, hence its sheaf cohomology $\mathcal{H}^*(\mathbf{L}\delta^*\mathbf{R}\delta_*\mathcal{O}_X)$ is a graded commutative ring. There is an obvious ring homomorphism

$$\wedge^*\mathcal{H}^{-1}[\mathbf{L}\delta^*\mathbf{R}\delta_*\mathcal{O}_X] = \wedge^*\Omega_f^1 \longrightarrow \mathcal{H}^*(\mathbf{L}\delta^*\mathbf{R}\delta_*\mathcal{O}_X)\ ,$$

from the exterior algebra on $\mathcal{H}^{-1}[\mathbf{L}\delta^*\mathbf{R}\delta_*\mathcal{O}_X] = \Omega_f^1$ to the ring $\mathcal{H}^*(\mathbf{L}\delta^*\mathbf{R}\delta_*\mathcal{O}_X)$. So far we have only assumed f flat and proper.

Now assume f is smooth and proper, of relative dimension n. The Hochschild-Kostant-Rosenberg Theorem [**21**] (see also Lipman [**30**, Proposition 4.6.3]) tells us that the homomorphism of graded rings of the paragraph above is an isomorphism. In particular we deduce

(i) The cohomology sheaves $\mathcal{H}^i(\mathbf{L}\delta^*\mathbf{R}\delta_*\mathcal{O}_X)$ vanish for $i < -n$.
(ii) We have constructed a natural map

$$\Omega_f^n \longrightarrow \wedge^n\mathcal{H}^{-1}[\mathbf{L}\delta^*\mathbf{R}\delta_*\mathcal{O}_X] \longrightarrow \mathcal{H}^{-n}[\mathbf{L}\delta^*\mathbf{R}\delta_*\mathcal{O}_X]$$

Formal nonsense, about t-structures in triangulated categories, tells us that the map of cohomology sheaves $\Omega_f^n \longrightarrow \mathcal{H}^{-n}[\mathbf{L}\delta^*\mathbf{R}\delta_*\mathcal{O}_X]$ can be realized as $\mathcal{H}^{-n}$ of a unique morphism $\psi : \Omega_f^n[n] \longrightarrow \mathbf{L}\delta^*\mathbf{R}\delta_*\mathcal{O}_X$ in the derived category $\mathbf{D_{qc}}(X)$. And the map θ of Theorem 2.2.1 is defined to be the composite

$$\Omega_f^n[n] \xrightarrow{\psi} \mathbf{L}\delta^*\mathbf{R}\delta_*\mathcal{O}_X \xrightarrow{\zeta} f^\times\mathcal{O}_Y$$

The precise version of Theorem 2.2.1 now says:

THEOREM 3.1.7. *If f is smooth and proper then the map θ of Construction* 3.1.6 *is an isomorphism.*

REMARK 3.1.8. In passing we mentioned that the Verdier version of 3.1.4.1 is sufficient for this paper—the reason is that in the proof we will only evaluate the base-change maps $\Phi(\diamondsuit)$ on the objects like $\mathcal{O}_Y$ or $\pi^\times \mathcal{O}_X$, which are bounded below. These are the only objects that come up in the definition of the map θ of Construction 3.1.6.

Lipman lectured about the approach to the map θ, presented in Construction 3.1.6, already in the 1980s. But it only appeared in print relatively recently, it may be found in Alonso, Jeremías and Lipman [**4**, Example 2.4 and Proposition 2.4.2]. The map θ has older avatars, for example in Verdier [**64**]—although the fact that Lipman's and Verdier's maps coincide was proved only recently, see [**35**].

REMARK 3.1.9. So far we have, up to one application of the Hochschild-Kostant-Rosenberg theorem, set up all the players entirely in formal nonsense fashion. And the historical asides along the way tell the reader that this much was known by the mid-1990s. Until now we haven't met anything younger than two decades.

It remains to discuss the proofs of Theorems 2.2.1 and 2.2.5, and this is where there has been major progress in the last few years. We open a new section for this.

3.2. The simple proofs discovered recently. Now that we have defined the maps occurring in Theorems 2.2.1 and 2.2.5, it remains to prove the claims of the theorems—we need to show that the maps do as the theorems assert they do. This should be a local problem, but for the longest time no one understood how to do this local computation simply and elegantly. Recall the first paragraph of the Introduction: there are two paths to the foundations of the subject, and in this article we've been following the one pioneered by Deligne and Verdier. The objection to this approach has long been that it leads to a theory where you can't compute anything. We've now reached the stage where a computation is in order—it should hardly come as a surprise that, until very recently, this was the point where the trail we have been on seemed to peter out, with the general direction appearing blocked and impenetrable.

We should probably explain, and for the purpose of clarity let us narrow our attention to just Theorem 2.2.1. Theorem 2.2.5 is similar but a touch more technical.

REMARK 3.2.1. Since we will say almost nothing about the proofs of Theorem 2.2.5, old or new, we should in passing acknowledge its long history and give references. The theorem was first sketched in Hartshorne [**19**, pp. 398–400]. The case of varieties over a perfect field was completely worked out in Lipman [**29**]. Lipman's main results were generalized in Hübl and Sastry [**25**]; see their Residue Theorem (iii) on p. 752 and its generalization (iii) on p. 785.

The results cited in the paragraph above all came with complicated proofs. The existence of a simple proof is a recent surprise, none of the experts—the dozen of us—expected such a thing, it is part of the exciting developments of the last few years. The reader can find the simple proof in [**48**, Section 2], and we will say a tiny bit more about this proof in §4.1.

STRATEGY 3.2.2. *Back to Theorem* 2.2.1*: in Construction* 3.1.6 *we defined a map* $\theta : \Omega^n_f[n] \longrightarrow f^\times \mathcal{O}_Y$*, and Theorem* 2.2.1 *asserts that* θ *is an isomorphism. Surely we should be able to prove this locally—a morphism in* $\mathbf{D_{qc}}(X)$ *is an isomorphism if and only if it restricts to an isomorphism on an open affine cover. Let* *us follow our noses and proceed the way such arguments usually work, to see where the obstacle lies, and to pinpoint the insight which uncovered an unconvoluted path circumventing this hurdle.*

LEMMA 3.2.3. *Suppose we are given a commutative square of schemes*

$$\begin{array}{ccc} U & \xrightarrow{\ u\ } & X \\ {\scriptstyle g}\downarrow & \clubsuit & \downarrow{\scriptstyle f} \\ V & \xrightarrow{\ v\ } & Y \end{array}$$

where the maps u and v are open immersions. Then there is a canonical natural isomorphism $\mathbf{L}u^*\Omega^n_f[n] \cong \Omega^n_g[n]$.

PROOF. Obvious. □

REMINDER 3.2.4. We are given a morphism $f : X \longrightarrow Y$, which we assume smooth and proper. We wish to show that the map

$$\Omega^n_f[n] \xrightarrow{\quad \theta \quad} f^\times \mathcal{O}_Y$$

of Construction 3.1.6 is an isomorphism, and the plan is to do this by studying it locally.

REDUCTION 3.2.5. Let us begin by showing that the problem is local in Y. Suppose that $v : V \longrightarrow Y$ is an open immersion with V affine. Next we

(i) Form the pullback square of schemes

$$\begin{array}{ccc} U & \xrightarrow{\ u\ } & X \\ {\scriptstyle g}\downarrow & \clubsuit & \downarrow{\scriptstyle f} \\ V & \xrightarrow{\ v\ } & Y \end{array}$$

It clearly suffices to show that, for every $v : V \longrightarrow Y$ as above, the map $\mathbf{L}u^*\theta$ is an isomorphism. Our reduction is about simplifying $\mathbf{L}u^*\theta$.

Let us pass from the square ($\clubsuit$) to the square of derived categories

$$\begin{array}{ccc} \mathbf{D_{qc}}(U) & \xleftarrow{\ \mathbf{L}u^*\ } & \mathbf{D_{qc}}(X) \\ {\scriptstyle g^\times}\uparrow & & \uparrow{\scriptstyle f^\times} \\ \mathbf{D_{qc}}(V) & \xleftarrow{\ \mathbf{L}v^*\ } & \mathbf{D_{qc}}(Y) \end{array}$$

The base-change map $\Phi(\clubsuit) : \mathbf{L}u^* f^\times \longrightarrow g^\times \mathbf{L}v^*$ is an isomorphism by 3.1.4.1. Applying to the object $\mathcal{O}_Y \in \mathbf{D_{qc}}(Y)$, and recalling that $\mathbf{L}v^*\mathcal{O}_Y = \mathcal{O}_V$, we deduce

(ii) We have produced an isomorphism

$$\mathbf{L}u^* f^\times \mathcal{O}_Y \longrightarrow g^\times \mathcal{O}_V$$

Next we apply Lemma 3.2.3 to the square ($\clubsuit$) in (i), obtaining

(iii) We have an isomorphism

$$\mathbf{L}u^*\Omega^n_f[n] \longrightarrow \Omega^n_g[n]$$

We are trying to show that the map

$$\Omega^n_f[n] \xrightarrow{\quad \theta \quad} f^\times \mathcal{O}_Y$$

is an isomorphism, and we have already agreed in (i) that it suffices to check that $\mathbf{L}u^*\theta$ is an isomorphism. By (ii) and (iii) above $\mathbf{L}u^*\theta$ rewrites as some map

$$\Omega^n_g[n] \longrightarrow g^\times \mathcal{O}_V$$

and it is an exercise in the definitions to check that this map is nothing other than the θ corresponding to $g : U \longrightarrow V$. In other words we are reduced to proving Theorem 2.2.1 in the case where Y is affine.

REDUCTION 3.2.6. Reduction 3.2.5 tells us that it suffices to prove Theorem 2.2.1 for smooth, proper morphisms $f : X \longrightarrow Y$ with Y affine. If we follow the usual yoga, the next step should be to reduce to the case where X is also affine. We want to prove that the map θ of Construction 3.1.6 an isomorphism, and it certainly suffices to show that $\mathbf{L}u^*\theta$ is an isomorphism for every open immersion $u : U \longrightarrow X$ with U affine. Choose therefore an open immersion $u : U \longrightarrow X$, and we would like to express $\mathbf{L}u^*\theta$ in some form that renders it easily computable. Now θ is a map

$$\Omega^n_f[n] \longrightarrow f^\times \mathcal{O}_Y$$

and we are embarking on a study of $\mathbf{L}u^*\theta$, which is a morphism

$$\mathbf{L}u^*\Omega^n_f[n] \longrightarrow \mathbf{L}u^* f^\times \mathcal{O}_Y$$

Let us begin by simplifying $\mathbf{L}u^*\Omega^n_f[n]$.

To this end we study the commutative square of schemes

$$\begin{array}{ccc} U & \xrightarrow{\quad u \quad} & X \\ {\scriptstyle fu}\big\downarrow & \clubsuit & \big\downarrow{\scriptstyle f} \\ Y & \xrightarrow{\quad \mathrm{id} \quad} & Y \end{array}$$

Applying Lemma 3.2.3 produces an isomorphism

$$\mathbf{L}u^*\Omega^n_f[n] \longrightarrow \Omega^n_{fu}[n]$$

In the usual jargon, our reduction so far tells us that the "object $\Omega^n_f[n]$ is local in X".

CAUTION 3.2.7. *The simple-minded way to proceed doesn't work, unfortunately the object $f^\times \mathcal{O}_Y$ isn't local in X. If we look at the composite $U \xrightarrow{u} X \xrightarrow{f} Y$, then $\mathbf{L}u^* f^\times \mathcal{O}_Y$ is not in general isomorphic to $(fu)^\times \mathcal{O}_Y$. See Sketch 4.1.1 for a counterexample.*

REDUCTION 3.2.8. Caution 3.2.7 tells us what doesn't work when we try to simplify $\mathbf{L}u^* f^\times$, and it's now time to present the way around the difficulty. To this end consider the following commutative diagram of schemes

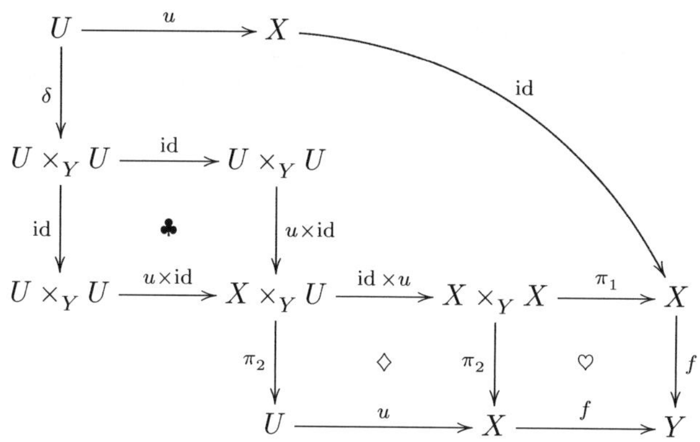

from which we obtain a diagram of derived categories

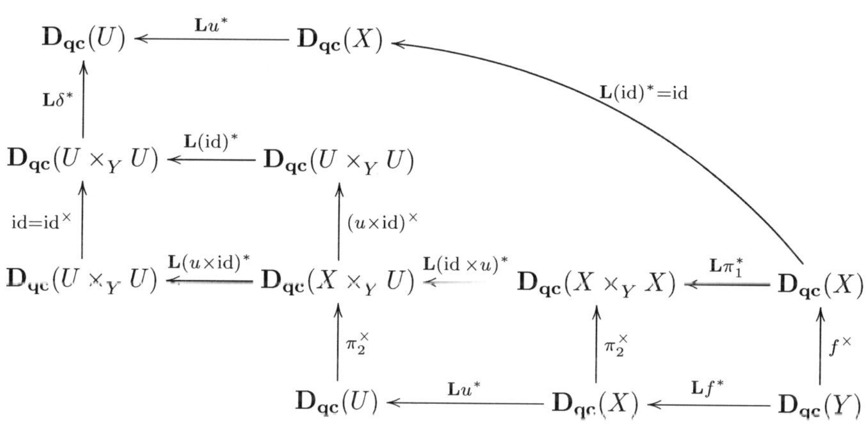

We would like to simplify the expression $\mathbf{L}u^* f^\times$, and in the diagram above we have a subdiagram which obviously commutes up to canonical natural isomorphism

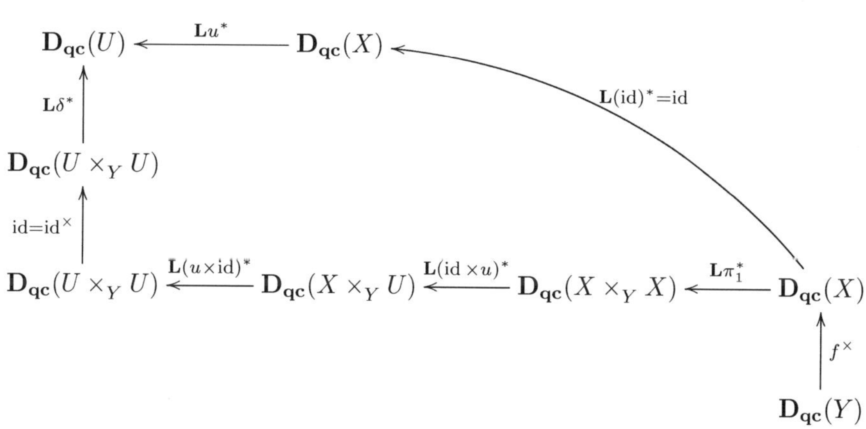

The squares $(\diamondsuit)$ and $(\heartsuit)$ in the commutative diagram of schemes satisfy the hypotheses of 3.1.4.1 and, up to the natural isomorphisms induced by $\Phi(\diamondsuit)$ and $\Phi(\heartsuit)$ composed with the canonical natural isomorphism of the diagram above, the

following subdiagram must also commute

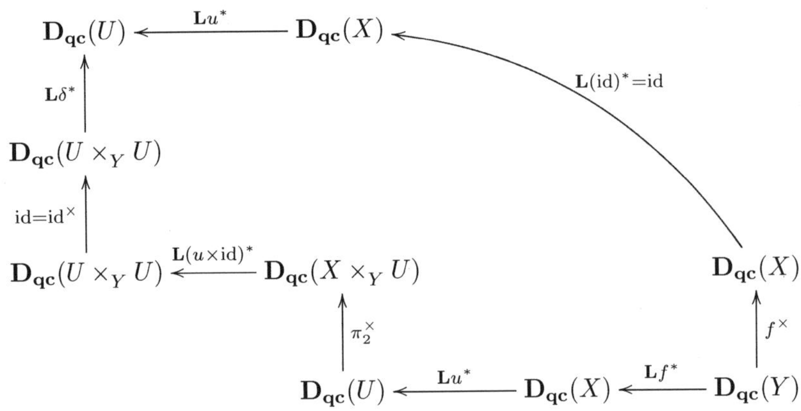

Up until now everything is entirely classical.

Of course there is nothing to stop us from looking at the base-change map of the square (♣) in the large diagram of schemes at the beginning of this Reduction. The square is cartesian, the horizontal map $(u \times \mathrm{id})$ is flat, and Construction 3.1.4 provides us with a morphism $\Phi(\clubsuit) : \mathbf{L}(\mathrm{id})^*(u \times \mathrm{id})^\times \longrightarrow \mathrm{id}^\times \mathbf{L}(u \times \mathrm{id})^*$, or more simply $\Phi(\clubsuit) : (u\times\mathrm{id})^\times \longrightarrow \mathbf{L}(u\times\mathrm{id})^*$. It might seem idiotic[5] to study $\Phi(\clubsuit)$, after all the hypotheses of 3.1.4.1 don't hold, the vertical map on the right decidedly isn't proper. And just in case the reader was wondering: it's not just that the hypotheses of 3.1.4.1 aren't satisfied—neither is the conclusion, the map $\Phi(\clubsuit)$ is known not to be an isomorphism in general.

The recent insight says

3.2.8.1. *Consider the following extract from our large diagram of derived categories*

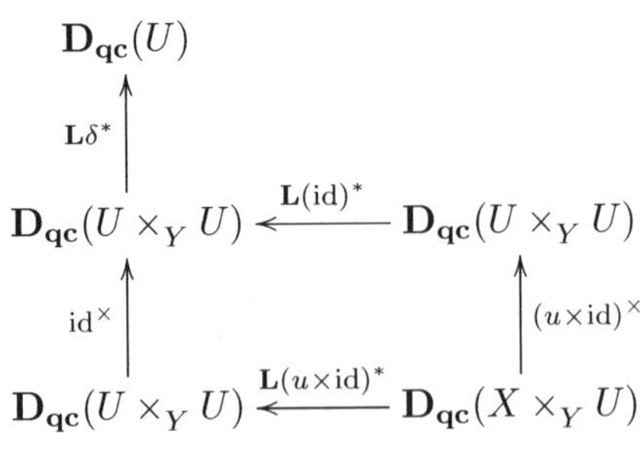

Then the composites from bottom right to top left agree, more precisely the map $\mathbf{L}\delta^*\Phi(\clubsuit)$ *gives an isomorphism* $\mathbf{L}\delta^*(u \times \mathrm{id})^\times \longrightarrow \mathbf{L}\delta^*\mathbf{L}(u \times \mathrm{id})^*$.

[5]In September 2016 the author presented the results (in a more technical incarnation) at a seminar at the University of Utah. The audience consisted of algebraic geometers and commutative algebraists. And this expert audience burst out laughing when told that the study of the base-change map $\Phi(\clubsuit)$ is what underpins the recent progress—to the experts this was hilarious.

With the aid of the isomorphism of 3.2.8.1 we deduce that, up to all the isomorphisms above, the diagram

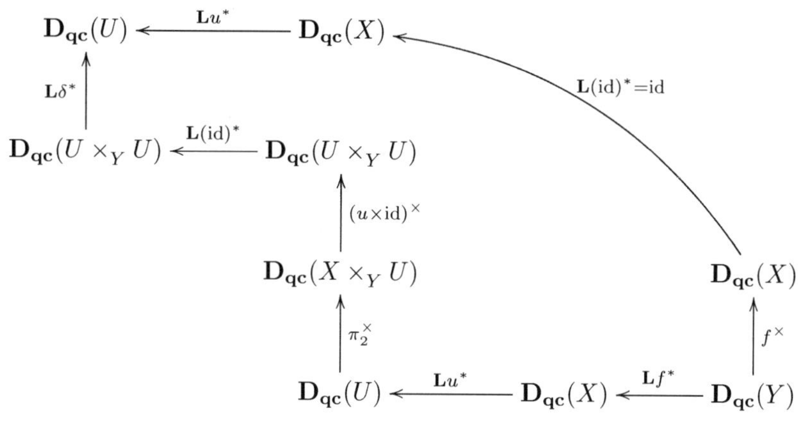

also commutes. This simplifies to

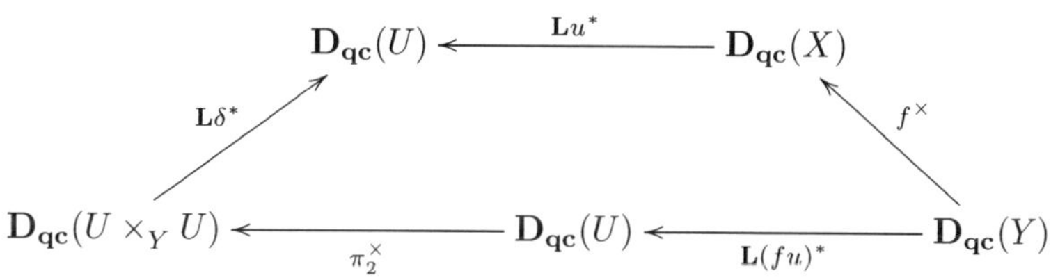

and the punchline is that we have found an isomorphism of the functor $\mathbf{L}u^*f^\times$ with the composite $\mathbf{L}\delta^*\pi_2^\times\mathbf{L}(fu)^*$, and in that composite *all the schemes are affine.*

REMARK 3.2.9. Reductions 3.2.6 and 3.2.8 transform the problem into one in which all the schemes are affine. It remains to show that

(i) The affine problem is tractable, in other words we can do the computation needed in the affine case. The reader can find in [**27**, Theorem 4.2.4] a working out of what needs to be computed, and in [**48**, Section 1] the computation is actually carried out. Below we present a rough outline, in Reminder 3.2.10 and Computation 3.2.11.

(ii) We should also say something about the proof of 3.2.8.1, after all this is the crux of what's new to the approach of the current document. The expert is referred to [**27**, (2.3.5.1)] for a general result, which implies 3.2.8.1 as a special case. In the interest of making the subject accessible to the non-expert we give a fairly complete, self-contained treatment in Sketch 3.2.12 below.

Let us first recall

REMINDER 3.2.10. If R is a ring let $\mathbf{D}(R)$ denote the (unbounded) derived category of cochain complexes of R–modules. From [**12**, Theorem 5.1] we learn that, for any ring R, the canonical functor $\mathbf{D}(R)\longrightarrow\mathbf{D_{qc}}\big[\mathrm{Spec}(R)\big]$ is an equivalence. Given a ring homomorphism $f:R\longrightarrow S$ we have an induced map of schemes, and by abuse of notation we will write it $f : \operatorname{Spec}(S) \longrightarrow \operatorname{Spec}(R)$. The diagram

$$\begin{array}{c} \mathbf{D_{qc}}\left[\operatorname{Spec}(S)\right] \\ \mathbf{L}f^* \Big\uparrow \;\Big\downarrow \mathbf{R}f_* \;\Big\uparrow f^\times \\ \mathbf{D_{qc}}\left[\operatorname{Spec}(R)\right] \end{array}$$

of Reminder 2.1.1 identifies with

$$\begin{array}{c} \mathbf{D}(S) \\ \mathbf{L}f^* \Big\uparrow \;\Big\downarrow f_* \;\Big\uparrow f^\times \\ \mathbf{D}(R) \end{array}$$

where $f_* : \mathbf{D}(S) \longrightarrow \mathbf{D}(R)$ is the forgetful functor—it takes an object of $\mathbf{D}(S)$, that is a cochain complex of S–modules, to itself viewed as a complex of R–modules. As f_* is exact there is no need to derive it, we have $\mathbf{R}f_* = f_*$. With the notation as in Example 1.2.5 the functor $\mathbf{L}f^*$, being the left adjoint of f_*, is given by the formula $\mathbf{L}f^*(-) = S \otimes_R^{\mathbf{L}} (-)$, while $f^\times$, being the right adjoint of f_*, is the functor $f^\times(-) = \mathbf{R}\mathrm{Hom}_R(S,-)$. And the units and counits of the adjunctions are all explicit.

Computation 3.2.11. *In view of Reminder* 3.2.10, *achieving Remark* 3.2.9*(i) has to be straightforward—it's just a matter of untangling the definitions and then doing a computation. We are given affine schemes U and Y, hence we may write* $Y = \operatorname{Spec}(R)$ *and* $U = \operatorname{Spec}(S)$. *The morphism of schemes $fu : U \longrightarrow Y$ corresponds to a ring homomorphism $\sigma : R \longrightarrow S$, and we have an isomorphism $U \times_Y U = Spec(S^{\mathrm{e}})$ with $S^{\mathrm{e}} = S \otimes_R S$. Reductions* 3.2.6 *and* 3.2.8 *produce isomophisms in the category $\mathbf{D_{qc}}(U)$*

$$\begin{aligned} \mathbf{L}u^*\Omega^n_f[n] &\cong \Omega^n_{fu}[n] \\ \mathbf{L}u^* f^\times \mathcal{O}_Y &\cong \mathbf{L}\delta^*\pi_2^\times \mathbf{L}(fu)^*\mathcal{O}_Y \end{aligned}$$

Using the descriptions on the right-hand-side, one checks that the equivalence $\mathbf{D_{qc}}(U) \cong \mathbf{D}(S)$ takes these objects, respectively, to $\mathrm{Tor}_n^{S^{\mathrm{e}}}(S,S)[n]$ *and* $S \otimes^{\mathbf{L}}_{S^{\mathrm{e}}} \mathbf{R}\mathrm{Hom}_R(S,S)$. *And the map $\mathbf{L}u^*\theta$ is nothing other than the composite*

$$\mathrm{Tor}_n^{S^{\mathrm{e}}}(S,S)[n] \longrightarrow S \otimes^{\mathbf{L}}_{S^{\mathrm{e}}} S \xrightarrow{\mathrm{id} \otimes^{\mathbf{L}}_{S^{\mathrm{e}}} I} S \otimes^{\mathbf{L}}_{S^{\mathrm{e}}} RHom_R(S,S)$$

where $I : S \longrightarrow RHom_R(S,S)$ is the obvious inclusion. It remains to show that, for S smooth over R of relative dimension n, the composite above is an isomorphism. The reader can find the computation in [**48**, Section 1].

Sketch 3.2.12. It remains to deliver on the promise of Remark 3.2.9(ii), we should say something about the proof of 3.2.8.1. The argument below is reasonably detailed—it may safely be skipped, the reader should feel free to proceed directly to Section 4.

We remind the reader: we consider the diagram

$$\begin{array}{ccccc}
\mathbf{D_{qc}}(U) & \xleftarrow{\ \mathbf{L}\delta^*\ } & \mathbf{D_{qc}}(U\times_Y U) & \xleftarrow{\ \mathbf{L}(\mathrm{id})^*\ } & \mathbf{D_{qc}}(U\times_Y U) \\
 & & \Big\uparrow{\scriptstyle \mathrm{id}^\times} & & \Big\uparrow{\scriptstyle (u\times\mathrm{id})^\times} \\
 & & \mathbf{D_{qc}}(U\times_Y U) & \xleftarrow{\ \mathbf{L}(u\times\mathrm{id})^*\ } & \mathbf{D_{qc}}(X\times_Y U)
\end{array}$$

and the assertion is that the functor $\mathbf{L}\delta^*$ takes the base-change map $\Phi(\clubsuit):(u\times\mathrm{id})^\times\longrightarrow\mathbf{L}(u\times\mathrm{id})^*$ to an isomorphism. To simplify the notation we will write v for the map $(u\times\mathrm{id}):U\times_Y U\longrightarrow X\times_Y U$.

Now the diagonal map $\delta:U\longrightarrow U\times_Y U$ is a closed immersion, hence the map $\mathbf{R}\delta_*=\delta_*$ is conservative—it suffices to prove that $\mathbf{R}\delta_*\mathbf{L}\delta^*$ takes $\Phi(\clubsuit)$ to an isomorphism. But we have isomorphisms

$$\mathbf{R}\delta_*\mathbf{L}\delta^*(-)\quad\cong\quad\mathbf{R}\delta_*\big[\mathbf{L}\delta^*(-)\otimes^{\mathbf{L}}\mathcal{O}_U\big]\quad\cong\quad(-)\otimes^{\mathbf{L}}\mathbf{R}\delta_*\mathcal{O}_U$$

where the second is the isomorphism p^{-1} of the projection formula, see Sketch 3.1.1. Consider therefore the two full subcategories of $\mathbf{D_{qc}}(U\times_Y U)$ given by

$$\mathbf{D}_{\mathbf{qc},\Delta}(U\times_Y U)\quad=\quad\left\{\mathfrak{s}\in\mathbf{D_{qc}}(U\times_Y U)\;\middle|\;\begin{array}{c}\mathbf{L}i^*\mathfrak{s}=0\text{ for all}\\ i:\mathrm{Spec}(K)\longrightarrow U\times_Y U-\Delta\\ \text{where }K\text{ is a field and where}\\ \Delta\subset U\times_Y U\text{ is the diagonal}\end{array}\right\}$$

$$\mathcal{S}\quad=\quad\left\{\mathfrak{s}\in\mathbf{D_{qc}}(U\times_Y U)\;\middle|\;\begin{array}{c}\text{the functor }(-)\otimes^{\mathbf{L}}\mathfrak{s}\\ \text{takes }\Phi(\clubsuit)\text{ to an isomorphism}\end{array}\right\}$$

The object $\mathbf{R}\delta_*\mathcal{O}_U$ clearly belongs to $\mathbf{D}_{\mathbf{qc},\Delta}(U\times_Y U)$, and we wish to show that it belongs to $\mathcal{S}$. It certainly suffices to prove that $\mathbf{D}_{\mathbf{qc},\Delta}(U\times_Y U)$ is contained in $\mathcal{S}$.

But both $\mathcal{S}$ and $\mathbf{D}_{\mathbf{qc},\Delta}(U\times_Y U)$ are localizing tensor ideals, and [**45**, Corollary 3.4] tells us that, as a localizing tensor ideal, $\mathbf{D}_{\mathbf{qc},\Delta}(U\times_Y U)$ is generated by the perfect complexes inside it[6]. It suffices to show that the perfect complexes in $\mathbf{D}_{\mathbf{qc},\Delta}(U\times_Y U)$ all belong to $\mathcal{S}$. In other words: it suffices to prove that, for every perfect complex P supported on the diagonal, the functor $(-)\otimes^{\mathbf{L}}P$ takes $\Phi(\clubsuit)$ to an isomorphism.

The morphism $v:U\times_Y U\longrightarrow X\times_Y U$ is an open immersion, hence the counit of adjunction $\varepsilon:\mathbf{L}v^*\mathbf{R}v_*\longrightarrow\mathrm{id}$ is an isomorphism. Consequently $\mathbf{R}v_*$ is fully faithful, and we have an isomorphism $\mathbf{L}v^*\mathbf{R}v_*P\cong P$. It therefore suffices to show that, for every perfect complex $P\in\mathbf{D_{qc}}(U\times_Y U)$ supported on the diagonal, the functor

$$\mathbf{R}v_*\big[(-)\otimes^{\mathbf{L}}P\big]\quad\cong\quad\mathbf{R}v_*\big[(-)\otimes^{\mathbf{L}}\mathbf{L}v^*\mathbf{R}v_*P\big]\quad\cong\quad\mathbf{R}v_*(-)\otimes^{\mathbf{L}}\mathbf{R}v_*P$$

takes $\Phi(\clubsuit)$ to an isomorphism, or to put it differently the functor $(-)\otimes^{\mathbf{L}}\mathbf{R}v_*P$ takes $\mathbf{R}v_*\Phi(\clubsuit)$ to an isomorphism. Now let $P\in\mathbf{D_{qc}}(U\times_Y U)$ be a perfect complex

[6]Note that $U\times_Y U$ is affine, so for us the old version in [**45**] suffices. We should mention that [**45**] builds on earlier papers by Hopkins [**22**] and Thomason and Trobaugh [**61**]. The reader can find later improvements in: the union of Thomason [**60**, Lemma 3.4] and Alonso, Jeremías and Souto [**5**, Corollary 4.11 and Theorem 4.12] generalize the result to all noetherian schemes, while Balmer and Favi [**10**] give a formal generalization to the world of tensor triangulated categories.

supported on the diagonal and let $\Gamma \subset X \times_Y U$ be the graph of the map $u : U \longrightarrow X$. The following is a cartesian square of open immersions

$$\begin{array}{ccc} U \times_Y U - \Delta & \xrightarrow{\quad\alpha\quad} & U \times_Y U \\ {\scriptstyle\beta}\downarrow & & \downarrow{\scriptstyle v} \\ X \times_Y U - \Gamma & \xrightarrow{\quad\gamma\quad} & X \times_Y U \end{array}$$

By 3.1.3.1 we have the first isomorphism below

$$\mathbf{L}\gamma^*\mathbf{R}v_*P \quad\cong\quad \mathbf{R}\beta_*\mathbf{L}\alpha^*P \quad\cong\quad 0\ ,$$

where the second isomorphism is because P is supported on Δ and hence $\mathbf{L}\alpha^*P = 0$. Thus both $\mathbf{L}\gamma^*\mathbf{R}v_*P = 0$ and $\mathbf{L}v^*\mathbf{R}v_*P \cong P$ are perfect complexes, and as $U \times_Y U$ and $X \times_Y U - \Gamma$ form an open cover for $X \times_Y U$ we deduce

(i) If $P \in \mathbf{D_{qc}}(U \times_Y U)$ is a perfect complex supported on the diagonal then $\mathbf{R}v_*P$ is a perfect complex on $X \times_Y U$, supported on the graph $\Gamma \subset X \times_Y U$ of the map $u : U \longrightarrow X$.

The reader can amuse herself by proving

(ii) The map $\Phi(\clubsuit) : v^\times \longrightarrow \mathbf{L}v^*$ is taken by $\mathbf{R}v_*$ to the composite

$$\mathbf{R}v_*v^\times \xrightarrow{\quad\widetilde{\varepsilon}\quad} \mathrm{id} \xrightarrow{\quad\eta\quad} \mathbf{R}v_*\mathbf{L}v^*$$

where $\widetilde{\varepsilon} : \mathbf{R}v_*v^\times \longrightarrow \mathrm{id}$ is the counit of the adjunction $\mathbf{R}v_* \dashv v^\times$, while $\eta : \mathrm{id} \longrightarrow \mathbf{R}v_*\mathbf{L}v^*$ is the unit of the adjunction $\mathbf{L}v^* \dashv \mathbf{R}v_*$.

Let $Q \in \mathbf{D_{qc}}(X \times_Y U)$ be a perfect complex supported on Γ. Then its dual $Q^\vee$ is also a perfect complex supported on Γ, and we have $(-) \otimes^{\mathbf{L}} Q \cong \mathcal{RH}\mathrm{om}(Q^\vee, -)$. From (i) and (ii) above it suffices to prove

(iii) For all perfect complexes $Q \in \mathbf{D_{qc}}(X \times_Y U)$ supported on Γ, the functor $(-) \otimes^{\mathbf{L}} Q$ takes the map η of (ii) to an isomorphism.

(iv) For all perfect complexes $Q \in \mathbf{D_{qc}}(X \times_Y U)$ supported on Γ, the functor $\mathcal{RH}\mathrm{om}(Q, -)$ takes the map $\widetilde{\varepsilon}$ of (ii) to an isomorphism.

To establish (iv) it suffices, by [**46**, Lemma 3.2], to show that for all pairs of perfect complexes $C, Q \in \mathbf{D_{qc}}(X \times_Y U)$, with Q supported on Γ, the functor $\mathrm{Hom}(C, -)$ takes $\mathcal{RH}\mathrm{om}(Q, \widetilde{\varepsilon})$ to an isomorphism. Now observe the isomorphism of functors

$$\mathrm{Hom}\big[C\ ,\ \mathcal{RH}\mathrm{om}(Q, -)\big] \quad\cong\quad \mathrm{Hom}(C \otimes^{\mathbf{L}} Q\ ,\ -)$$

As C and Q are both perfect and Q is supported on Γ, the complex $C \otimes^{\mathbf{L}} Q$ is perfect and is supported on Γ. Hence (iv) would follow from

(v) For all perfect complexes $Q \in \mathbf{D_{qc}}(X \times_Y U)$ supported on Γ, the functor $\mathrm{Hom}(Q, -)$ takes the map $\widetilde{\varepsilon}$ of (ii) to an isomorphism.

Let $Q \in \mathbf{D_{qc}}(X \times_Y U)$ be a perfect complex supported on Γ. Then $\mathbf{L}v^*Q$ is a perfect complex supported on Δ. The map $\eta : Q \longrightarrow \mathbf{R}v_*\mathbf{L}v^*Q$ is an isomorphism on the open set $U \times_Y U \subset X \times_Y U$, and is an isomorphism on $(X \times_Y U) - \Gamma$ because both Q and [by (i)] $\mathbf{R}v_*\mathbf{L}v^*Q$ vanish outside Γ. Since the open sets $U \times_Y U$ and $(X \times_Y U) - \Gamma$ cover $X \times_Y U$ it follows that $\eta : Q \longrightarrow \mathbf{R}v_*\mathbf{L}v^*Q$ is an isomorphism. Putting $A = \mathbf{L}v^*Q$ we deduce that (iii) and (v) would follow from

(vi) For all $A \in \mathbf{D_{qc}}(U \times_Y U)$, the functor $(-) \otimes^{\mathbf{L}} \mathbf{R}v_*A$ takes η to an isomorphism.

(vii) For all $A \in \mathbf{D_{qc}}(U \times_Y U)$, the functor $\mathrm{Hom}(\mathbf{R}v_* A\,,\, -)$ takes $\widetilde{\varepsilon}$ to an isomorphism.

To see (vi) observe the isomorphism of the projection formula

$$(-) \otimes^{\mathbf{L}} \mathbf{R}v_* A \quad \cong \quad \mathbf{R}v_* \big[\mathbf{L}v^*(-) \otimes^{\mathbf{L}} A\big]$$

Since the functor $\mathbf{L}v^*$ takes η to an ismorphism so does the right-hand-side above, and hence also the left-hand-side.

To see (vii) observe the commutative square

$$\begin{array}{ccc}
\mathrm{Hom}\big(\mathbf{R}v_*\mathbf{L}v^*(-)\,,\,?\big) & \xrightarrow{\ \sim\ } & \mathrm{Hom}\big(-\,,\,\mathbf{R}v_*v^\times(?)\big) \\
{\scriptstyle \mathrm{Hom}(\eta,?)}\big\downarrow & & \big\downarrow{\scriptstyle \mathrm{Hom}(-,\widetilde{\varepsilon})} \\
\mathrm{Hom}(-,?) & = & \mathrm{Hom}(-,?)
\end{array}$$

We wish to show that the vertical map on the right is an isomorphism when evaluated at $(-) = \mathbf{R}v_* A$, and the vertical map on the left makes it clear, after all $\eta\mathbf{R}v_* : \mathbf{R}v_* \longrightarrow \mathbf{R}v_*\mathbf{L}v^*\mathbf{R}v_*$ is an isomorphism.

4. Why did it take so long?

In some sense the ingredients of the proof were available already in the 1960s, but back then no one thought of applying the tools of homotopy theory—for example Brown representability—to problems in algebraic geometry. The methods employed in the classical proofs are fundamentally unsuited for the approach presented here. To mention just one facet: in the argument given here we relied heavily on the full power of the derived tensor product. The pre-1990 literature on Grothendieck duality all worked in the bounded-below derived category, where the derived tensor product exists only under strong restrictions and is next to useless.

That said, with the exception of 3.2.8.1 the ingredients of the argument were all available by the mid-1990s. And in Sketch 3.2.12 the reader learned that the proof of 3.2.8.1 could also have been given two decades ago. So why did we fail to see this?

4.1. The strangeness of the argument.

The reader should appreciate the bizarreness of looking at the base-change map in 3.2.8.1. It might help to elaborate a little by sketching a related and slightly easier computation.

SKETCH 4.1.1. Let k be a field, let $f : X \longrightarrow Y$ be the projection $f : \mathbb{P}^n_k \longrightarrow \mathrm{Spec}(k)$, and let $u : U \longrightarrow X$ be the open immersion $\mathbb{A}^n_k \longrightarrow \mathbb{P}^n_k$. In Caution 3.2.7 we warned the reader that the functor $u^\times f^\times$ is very different from $\mathbf{L}u^* f^\times$. Let us work out just how different by evaluating on $\mathcal{O}_Y$. Theorem 2.2.1 gives an isomorphism $\theta : \Omega^n_f[n] \longrightarrow f^\times \mathcal{O}_Y$, which means that $\mathbf{L}u^* f^\times \mathcal{O}_Y \cong \mathbf{L}u^*\Omega^n_f[n] \cong \Omega^n_{\mathbb{A}^n_k}[n] \cong \mathcal{O}_U[n]$, after all the canonical bundle of $U = \mathbb{A}^n_k$ is trivial. Under the equivalence $\mathbf{D}(S) \cong \mathbf{D_{qc}}(U)$ of Reminder 3.2.10 and Computation 3.2.11 the object $\mathbf{L}u^* f^\times \mathcal{O}_Y \in \mathbf{D_{qc}}(U)$ identifies with $S[n] \in \mathbf{D}(S)$, that is the cochain complex obtained by placing the S–module S in dimension $-n$.

Now let's compute $u^\times f^\times \mathcal{O}_Y \cong (fu)^\times \mathcal{O}_Y$. The morphism $fu : U \longrightarrow Y$ is a map of affine schemes, and in Reminder 3.2.10 we noted that the computation of $(fu)^\times$ can be carried over to $(fu)^\times : \mathbf{D}(R) \longrightarrow \mathbf{D}(S)$ and is given explicitly by the formula $(fu)^\times(-) = \mathbf{R}\mathrm{Hom}_R(S, -)$. In our case R is the field k, $S = k[x_1, \ldots, x_n]$ is the polynomial ring, and $(fu)^\times \mathcal{O}_Y$ computes to be $\mathbf{R}\mathrm{Hom}_k(S, k) = \mathrm{Hom}_k(S, k)$, which is a gigantic injective S–module placed in degree zero. The reader can check [**47**] to see just how gargatuan this injective module is. It turns out to depend on the cardinality of k.

Now consider the cartesian square

$$\begin{array}{ccc} \mathbb{A}^n_k & \xrightarrow{\mathrm{id}} & \mathbb{A}^n_k \\ {\scriptstyle \mathrm{id}}\downarrow & & \downarrow{\scriptstyle u} \\ \mathbb{A}^n_k & \xrightarrow{u} & \mathbb{P}^n_k \end{array}$$

The bottom horizontal map is flat, so there is a base-change map $\Phi : \mathbf{L}(\mathrm{id})^* u^\times \longrightarrow \mathrm{id}^\times \mathbf{L}u^*$; more simply we can write it as $\Phi : u^\times \longrightarrow \mathbf{L}u^*$. Since Φ is a natural tranformation between two functors $\mathbf{D_{qc}}(\mathbb{P}^n_k) \Longrightarrow \mathbf{D_{qc}}(\mathbb{A}^n_k)$ we may evaluate it at the object $f^\times \mathcal{O}_{\mathrm{Spec}(k)} \in \mathbf{D_{qc}}(\mathbb{P}^n_k)$, producing a morphism $\psi = \Phi_{f^\times \mathcal{O}_{\mathrm{Spec}(k)}} : u^\times f^\times \mathcal{O}_{\mathrm{Spec}(k)} \longrightarrow \mathbf{L}u^* f^\times \mathcal{O}_{\mathrm{Spec}(k)}$ in the category $\mathbf{D_{qc}}(\mathbb{A}^n_k)$. The equivalence $\mathbf{D_{qc}}(\mathbb{A}^n_k) \cong \mathbf{D}(S)$ must take it to a morphism

$$\mathrm{Hom}_k(S,k) \xrightarrow{\quad\psi\quad} S[n]$$

Doesn't it seem absurd to study this map?

But this is exactly what we do in the simple, recent proof of Theorem 2.2.5, which we haven't discussed in this article. The next paragraph gives a quick sketch—the non-experts may wish to skip ahead to §4.2.

Let $W \subset X$ be as in the statement of Theorem 2.2.5—that is W is the set of closed points in $X = \mathbb{P}^n_k$, and hence $U \cap W \subset U$ the set of closed points in $U = \mathbb{A}^n_k$. The recent proof of Theorem 2.2.5 hinges on the observation that the functor $\Gamma_{U\cap W}$ takes the map ψ above to an isomorphism. In Example 2.2.4 we told the reader how to compute $\Gamma_W f^\times \mathcal{O}_Y$, and the analogous recipe gives that $\Gamma_{U\cap W}\mathbf{L}u^* f^\times \mathcal{O}_Y \cong \mathbf{L}u^*\Gamma_W f^\times \mathcal{O}_Y$[7] can be computed by forming a minimal injective resolution for $S[n]$, that is the complex

$$0 \longrightarrow I^{-n} \longrightarrow I^{-n+1} \longrightarrow \cdots \longrightarrow I^{-1} \longrightarrow I^0 \longrightarrow 0$$

and putting $I^0 = \Gamma_{U\cap W}\mathbf{L}u^* f^\times \mathcal{O}_Y$. The morphism ψ in the category $\mathbf{D}(S)$ may be represented by a cochain map of injective resolutions

$$\begin{array}{ccccccccccccc} 0 & \longrightarrow & 0 & \longrightarrow & 0 & \longrightarrow & \cdots & \longrightarrow & 0 & \longrightarrow & \mathrm{Hom}_k(S,k) & \longrightarrow & 0 \\ & & \downarrow & & \downarrow & & & & \downarrow & & \downarrow{\scriptstyle \widetilde{\psi}} & & \\ 0 & \longrightarrow & I^{-n} & \longrightarrow & I^{-n+1} & \longrightarrow & \cdots & \longrightarrow & I^{-1} & \longrightarrow & I^0 & \longrightarrow & 0 \end{array}$$

and the functor $\Gamma_{U\cap W}$ takes this to a map $\Gamma_{U\cap W}(\widetilde{\psi}) : \Gamma_{U\cap W}\mathrm{Hom}_k(S,k) \longrightarrow I^0$. We haven't yet told the reader how to compute $\Gamma_{U\cap W}\mathrm{Hom}_k(S,k)$; the formula is that $\Gamma_{U\cap W}\mathrm{Hom}_k(S,k) \subset \mathrm{Hom}_k(S,k)$ is the submodule of all elements supported at the closed points. That is: an element $e \in \mathrm{Hom}_k(S,k)$ belongs to the submodule $\Gamma_{U\cap W}\mathrm{Hom}_k(S,k)$ if there exist a finite number of maximal ideals $\mathfrak{m}_1, \mathfrak{m}_2, \ldots, \mathfrak{m}_\ell \subset S$ and an integer $N > 0$ so that the ideal $\mathfrak{m}_1^N \mathfrak{m}_2^N \cdots \mathfrak{m}_\ell^N$ annihilates e. So the

[7]The isomorphism comes from Remark 2.1.7

assertion is that the obvious composite

$$\Gamma_{U\cap W}\mathrm{Hom}_k(S,k)\hookrightarrow \mathrm{Hom}_k(S,k) \xrightarrow{\ \psi\ } I^0$$

is an isomorphism of S–modules. In [**48**, Section 2] the reader can see how this is used in the proof of Theorem 2.2.5.

4.2. The historical block. The fact that the key new idea is so outlandish is only part of our excuse for taking so long. There were also the historical circumstances.

The foundations of Grothendieck duality was a lively, active field from about the mid-1960s until well into the 1970s. And then the interest gradually faded. The mathematical community accepted that the foundations were complicated, and stopped thinking about it. As an editor of one journal put it, when rejecting a recent paper of mine on the subject: "...your paper will be read by only the hardcore people...people have moved on...". It's pretty accurate to say that, in the last three decades, there have been two groups of people who have studied the foundations of Grothendieck duality, in the old-fashioned setting of classical, ordinary schemes: Lipman and his students and collaborators on one side, and Yekutieli and his students and collaborators on the other. It's not the whole story, one can point to exceptions such as Conrad [**14**], but it is a good approximation of the truth. In this I must count as one of Lipman's collaborators—every few years we run into each other, and he persuades me to return to the problem.

Lipman's approach has long been orthogonal to Yekutieli's: Yekutieli accepted the Grothendieck formalism, while Lipman has largely followed the approach of Deligne and Verdier. In this survey we've said next to nothing about the Grothendieck angle, making any comparison difficult—one could fairly say it takes an expert to recognize that the two subjects are truly one and the same. In fact: checking the details, that is verifying that the maps defined in the two theories agree, is often nontrivial. Just ask Lipman—he's probably the person who has tried hardest.

To put it in a nutshell: the story isn't that hundreds of people were feverishly working away at the problem, and this multitude overlooked the obvious for twenty-some years. What actually transpired is that at most a dozen of us were still studying foundational questions, half of us were exploring what turned out to be the wrong continent, and the solution, when it ultimately came, scored pretty high on the weirdness scale.

4.3. What finally woke us up. Back to the recent progress: the stimulus which, at long last, nudged us into probing in the right direction came from the work of Avramov and Iyengar [**7**], and later Avramov, Iyengar, Lipman and Nayak [**8**]. They discovered puzzling formulas, and since then there has been a proliferation—the reader can find increasingly general formulas in [**27**, **44**]. Let us present one example. Suppose we are given composable morphisms of schemes $U \xrightarrow{u} X \xrightarrow{f} Y$, and assume that $Y = \mathrm{Spec}(R)$ and $U = \mathrm{Spec}(S)$ are affine. Suppose the map u is an open immersion, the map f is proper and the composite $fu : U \longrightarrow Y$ is flat[8].

[8]The flatness of fu isn't crucial, it suffices for fu to be of finite Tor-dimension—the formula still holds, it is a special case of Avramov, Iyengar, Lipman and Nayak [**8**, (4.6.1)]. But in the finite-Tor-dimension generality we don't have a proof that's elementary. We will return to this in Problem 5.2.3.

The formula we happened to choose says that, for any $N \in \mathbf{D}(R)$,

$$\mathbf{L}u^* f^\times N \quad \cong \quad S \otimes^{\mathbf{L}}_{S^e} \mathbf{R}\mathrm{Hom}_R\big(S, S \otimes^{\mathbf{L}}_R N\big) \ ,$$

where $S^e = S \otimes_R S$. Where on earth did this come from?

The original presentation, in the articles [**7, 8**], touted the left-hand-side as a great simplification of the right-hand-side. Even the name given to the formulas—the "Reduction Formulas"—reflects this perspective. The formulas were first proved using the full strength of the existing theory of Grothendieck duality. It was not until [**27**] that we came up with an elementary proof of the formulas[9], and not until [**48**] that the right-hand-side became a computational tool for working out what's on the left.

What can I say: we were slow to see the light.

5. Future directions

For the last three decades Grothendieck duality has been a small, niche subject, with only a handful of dedicated practitioners. Granted that, the reader might well wonder what this section could possibly be about. What conceivable future can there be in a moribund, small field, long abandoned by the hordes?

There are two customary causes for the waning of a subject: it may die because the important questions have all been satisfactorily answered, or else because it hits a brick wall, and no one has any idea how to advance. In both cases a rebirth is possible, but it takes something of an earthquake. There needs to be a startling new development, a major new insight, or vital new questions that come up.

Grothendieck duality is an example of a subject that died because people got stuck—there were plenty of simple, natural questions left, but no really good ideas on how to tackle them. When we view the recent progress against this background, it can only be a matter of time before the field picks up again—if nothing else, the developments offer a radically new slant on the what's known. In this section we will sketch some of the obvious problems that now seem within reach. But before the open questions we need to brush up on facts that are known but haven't been covered yet—until now we have made a conscious effort to be minimal in our use of category theory, if a category or a functor was dispensable we omitted it.

5.1. Assorted background material. We have told the reader how to compute the functor $f^\times$ when f is smooth and proper. There is another classical situation in which the computation of $f^\times$ is understood, let us recall.

REMARK 5.1.1. The category $\mathbf{D_{qc}}(X)$ is monoidal, it has a tensor product. This tensor product has a right adjoint: there is a functor $\mathcal{RH}\mathrm{om}_{\mathbf{D_{qc}}(X)}(-,?)$ and an isomorphism, natural in everything in sight,

$$\mathrm{Hom}\big[A \otimes^{\mathbf{L}} B\,,\, C\big] \quad \cong \quad \mathrm{Hom}\big[A\,,\, \mathcal{RH}\mathrm{om}_{\mathbf{D_{qc}}(X)}(B,C)\big]$$

One way to construct it is to fix B and note that, since the functor $(-)\otimes^{\mathbf{L}} B$ respects coproducts, Brown representability gives a right adjoint $\mathcal{RH}\mathrm{om}_{\mathbf{D_{qc}}(X)}(B,-)$. Now

[9]As exposed in [**27**] the proof doesn't seem elementary—the article [**27**] was written for an expert audience. But the formula is a minor variant of Reduction 3.2.8 and, as presented in this document, the proof is manifestly elementary. And the truth is that, modulo peeling away the generality in [**27**] and dusting off superfluous fluff, the proof here is identical to the proof there.

let $f : X \longrightarrow Y$ be a morphism of schemes and suppose A, C are objects of $\mathbf{D_{qc}}(Y)$. We remind the reader of the following string of isomorphisms

$$\begin{aligned} \mathrm{Hom}\big[A\,,\,\mathbf{R}f_*f^\times C\big] &\cong \mathrm{Hom}\big[\mathbf{L}f^*A\,,\,f^\times C\big] \\ &\cong \mathrm{Hom}\big[\mathbf{L}f^*A\otimes^{\mathbf{L}}\mathcal{O}_X\,,\,f^\times C\big] \\ &\cong \mathrm{Hom}\big[\mathbf{R}f_*(\mathbf{L}f^*A\otimes^{\mathbf{L}}\mathcal{O}_X)\,,\,C\big] \\ &\cong \mathrm{Hom}\big[A\otimes^{\mathbf{L}}\mathbf{R}f_*\mathcal{O}_X\,,\,C\big] \\ &\cong \mathrm{Hom}\big[A\,,\,\mathcal{RH}om_{\mathbf{D_{qc}}(Y)}(\mathbf{R}f_*\mathcal{O}_X, C)\big] \end{aligned}$$

where the second isomorphism is because $\mathcal{O}_X$ is the unit of the tensor, the fourth comes from the projection formula, and the others are all by adjunction. Yoneda tells us that we have produced an isomorphism, natural in C,

$$\mathbf{R}f_*f^\times C \quad\cong\quad \mathcal{RH}om_{\mathbf{D_{qc}}(Y)}(\mathbf{R}f_*\mathcal{O}_X, C)$$

When f is an affine morphism this looks great—after all for an affine morphism f the functor $\mathbf{R}f_*$ is informative, especially when we view it as a functor from $\mathbf{D_{qc}}(X)$ to $\mathbf{D_{qc}}(f_*\mathcal{O}_X\text{–Mod})$.

The problem is that, in general, the expression $\mathcal{RH}om_{\mathbf{D_{qc}}(Y)}(\mathbf{R}f_*\mathcal{O}_X, C)$ isn't overly computable. What is frequently far more amenable to calculation is $\mathcal{RH}om_{\mathbf{D}(Y)}(\mathbf{R}f_*\mathcal{O}_X, C)$, which is related. That is: we compute the internal Hom not in the category $\mathbf{D_{qc}}(Y)$ but in the larger category $\mathbf{D}(Y)$, whose objects are all cochain complexes of sheaves of $\mathcal{O}_Y$–modules. This functor has the property that the cohomology sheaves $\mathcal{H}^i\big[\mathcal{RH}om_{\mathbf{D}(Y)}(\mathbf{R}f_*\mathcal{O}_X, C)\big]$ are what one would hope for. This means: for every open immersion $u : U \hookrightarrow Y$ consider the $\Gamma(U, \mathcal{O}_Y)$ module $\mathrm{Hom}_{\mathbf{D_{qc}}(U)}\big(\mathbf{L}u^*\mathbf{R}f_*\mathcal{O}_X, \mathbf{L}u^*C[i]\big)$. This comes with obvious restriction maps, elevating the construction to a presheaf of $\mathcal{O}_Y$–modules on Y. And the sheaf $\mathcal{H}^i\big[\mathcal{RH}om_{\mathbf{D}(Y)}(\mathbf{R}f_*\mathcal{O}_X, C)\big]$ turns out to be the sheafification of this presheaf. Unfortunately in order to pass, from the computable $\mathcal{RH}om_{\mathbf{D}(Y)}(\mathbf{R}f_*\mathcal{O}_X, C)$ to the desired $\mathcal{RH}om_{\mathbf{D_{qc}}(Y)}(\mathbf{R}f_*\mathcal{O}_X, C)\big] \cong \mathbf{R}f_*f^\times C$, one has to "quasicoherate"—not the world's most transparent process.[10]

There are cases where the computable Hom already has quasicoherent cohomology, in which case the two functors agree and the quasicoherator does nothing. Since we're making the assumption that f is an affine map [this is the case in which the computation of $\mathbf{R}f_*f^\times C$ will carry useful information about $f^\times C$], what is relevant for us is that $\mathbf{R}f_*f^\times C \cong \mathcal{RH}om_{\mathbf{D_{qc}}(Y)}(\mathbf{R}f_*\mathcal{O}_X, C)$ is given by the more computable expression if one of the conditions below holds:

(i) f is finite and C is bounded below.

(ii) f is finite and of finite Tor-dimension, and C is arbitrary.

In passing we note that, historically, the most widely used special case of the above has been where f is a closed immersion—possibly of finite Tor-dimension.

There isn't a whole lot more concrete computational knowledge about $f^\times$: in the body of the article we told the reader what is known when f is smooth and proper, and Remark 5.1.1 tells us some more when f is a finite map, possibly of finite Tor-dimension. Nevertheless one can use these (admittedly limited) pieces of

[10] In Appendix A we compute a simple example to illustrate the point.

information to deduce useful facts. But for this it helps to know another functor, a close cousin of $f^\times$. Before we introduce it, a reminder might help.

Reduction 3.2.8 had the following situation: we were given composable morphisms of schemes $U \xrightarrow{u} X \xrightarrow{f} Y$, with u an open immersion and f proper. The Reduction was all about computing the functor $\mathbf{L}u^* f^\times$, and until now we haven't mentioned that $\mathbf{L}u^* f^\times$ depends only on the composite map $U \xrightarrow{fu} Y$. That is:

Reminder 5.1.2. If $g : U \longrightarrow Y$ is any separated morphism of finite type, Nagata's theorem [**15**, **38**] allows us to choose a factorization of g as $U \xrightarrow{u} X \xrightarrow{f} Y$, with u an open immersion and f proper. We then define $g^! = \mathbf{L}u^* f^\times$, and it is a theorem that $g^!$ is independent of the factorization up to canonical isomorphism[11]. Classically the bulk of Grothendieck duality has been devoted to the study of the functors $g^!$. See Remark 5.1.4 below.

Having introduced $g^!$, we are now in a position to restate the recent progress in terms of it. Suppose we are given, as in the last paragraph, a map of schemes $g : U \longrightarrow Y$ which factorizes as $U \xrightarrow{u} X \xrightarrow{f} Y$, with u an open immersion and f proper. The square

$$\begin{array}{ccc} U & \xrightarrow{\ \mathrm{id}\ } & U \\ {\scriptstyle \mathrm{id}}\downarrow & & \downarrow{\scriptstyle u} \\ U & \xrightarrow{\ u\ } & X \end{array}$$

is cartesian, hence it has a base-change map $\Phi : u^\times \longrightarrow \mathbf{L}u^*$ as in Construction 3.1.4. The recent discoveries can be summarized as saying

(i) The induced map $\Phi f^\times : u^\times f^\times \longrightarrow \mathbf{L}u^* f^\times$ is independent of the factorization, and gives an unambiguous map $g^\times \longrightarrow g^!$. The construction taking g to $g^!$ yields a 2-functor we will denote $(-)^!$—in other words there is a compatibility with composition. And the map $g^\times \longrightarrow g^!$ is compatible too, it is a morphism of 2-functors $\psi : (-)^\times \longrightarrow (-)^!$ with many reasonable naturality properties. The reader can find this worked out in [**27**], and in greater generality in [**43**].

(ii) There are interesting situations in which some natural functor Γ takes $\psi(g) : g^\times \longrightarrow g^!$ to an isomorphism. We have encountered two examples, namely 3.2.8.1 and Sketch 4.1.1. For the general theory see [**27**, **43**].

Application 5.1.3. Suppose $f : X \longrightarrow Y$ is proper. Using the 2-functor $(-)^!$ one can prove the following

(i) If $G \in \mathbf{D}^+_{\mathbf{coh}}(Y)$ then $f^\times G \in \mathbf{D}^+_{\mathbf{coh}}(X)$. Here $\mathbf{D}^+_{\mathbf{coh}} \subset \mathbf{D}_{\mathbf{qc}}$ is the full subcategory of all objects with coherent cohomology sheaves, which vanish in sufficiently negative degrees.

[11] If one follows the Grothendieck-Hartshorne path to the subject, which we will discuss a little more in Remark 5.1.4, then the way to see the isomorphism $g^! \cong \mathbf{L}u^* f^\times$ is that the theory sets up a functor $g^!$, one shows that for open immersions u there is a natural isomorphism $u^! \cong \mathbf{L}u^*$, for proper maps f there is a natural isomorphism $f^! \cong f^\times$, and for composable maps there is a natural isomorphism $g^! = (fu)^! \cong u^! f^!$. But both Deligne [**16**] and Verdier [**64**] sketched an argument hinting how to prove directly the independence of factorization of $\mathbf{L}u^* f^\times$, using only 3.1.4.1 applied to suitable cartesian squares. There is some more detail in Lipman's book [**32**], and for a fuller argument, which works for algebraic stacks and hence must handle the 2-category technicalities more carefully, the reader is referred to [**43**].

Now suppose f is not only proper, but also of finite Tor-dimension. Then

(ii) If $G \in \mathbf{D}^b_{\mathbf{coh}}(Y)$ then $f^\times G \in \mathbf{D}^b_{\mathbf{coh}}(X)$. Here $\mathbf{D}^b_{\mathbf{coh}} \subset \mathbf{D_{qc}}$ is the full subcategory of complexes with bounded, coherent cohomology sheaves.

(iii) $G \in \mathbf{D}^-_{\mathbf{coh}}(Y)$ then $f^\times G \in \mathbf{D}^-_{\mathbf{coh}}(X)$. Here $\mathbf{D}^-_{\mathbf{coh}} \subset \mathbf{D_{qc}}$ is the full subcategory of complexes with coherent cohomology sheaves, vanishing in sufficiently positive degrees.

Thus if f is proper the adjoint pair $\mathbf{R}f_* : \mathbf{D_{qc}}(X) \rightleftarrows \mathbf{D_{qc}}(Y) : f^\times$ restricts to an adjunction $\mathbf{R}f_* : \mathbf{D}^+_{\mathbf{coh}}(X) \rightleftarrows \mathbf{D}^+_{\mathbf{coh}}(Y) : f^\times$. When f is not only proper but also of finite Tor-dimension, we also have two more adjoint pairs, namely $\mathbf{R}f_* : \mathbf{D}^-_{\mathbf{coh}}(X) \rightleftarrows \mathbf{D}^-_{\mathbf{coh}}(Y) : f^\times$ as well as $\mathbf{R}f_* : \mathbf{D}^b_{\mathbf{coh}}(X) \rightleftarrows \mathbf{D}^b_{\mathbf{coh}}(Y) : f^\times$.

And the proofs of (i), (ii) and (iii) go as follows: the assertions are local in X, hence it suffices to show that, for $u : U \longrightarrow X$ an open immersion from a (sufficiently small) open affine subset, the map $\mathbf{L}u^* f^\times = (fu)^!$ satisfies the properties. But $(-)^!$ is a 2-functor. We are allowed to factor fu in some other way, for example as a composite $fu = ghk$, and if we choose our factorization wisely then $(fu)^! = k^! h^! g^!$ might be more computable. For example g, h, k might fall into the classes where we understand $(-)^!$: open immersions, maps that are smooth and proper, and finite morphisms (possibly of finite Tor-dimension). For details the reader is referred to [**44**, Lemma 3.12 and its proof].

Thus the abstract nonsense approach does recover "coherent duality" as it was traditionally understood—meaning about complexes with coherent cohomology. The reader might also wonder about the relation of (for example) the categories $\mathbf{D}^b_{\mathbf{coh}}(X)$ and $\mathbf{D}^b(\mathbf{coh}/X)$. The relation is well-understood by now, but falls out side the scope of Grothendieck duality—it isn't about the functors $f^\times$ or $f^!$, it's a formal question about the interplay among the myriad derived categories one can associate to a single scheme X. We leave this out of the survey—we've barely started Section 5, and we've already been bombarded with a hail of new categories and functors.

REMARK 5.1.4. Now that we have introduced $(-)^!$ we can recall a historical point. The Grothendieck approach is entirely in terms of the 2-functor $(-)^!$, the 2-functor $(-)^\times$ never comes up. That is: starting with a separated, finite-type morphism of noetherian schemes $f : X \longrightarrow Y$, Grothendieck goes through an intricate, arduous procedure to arrive at a functor which turns out to agree with the $f^!$ of Reminder 5.1.2. In the Grothendieck setting it's a major theorem that, for proper f, the functor $f^!$ is right adjoint to $\mathbf{R}f_*$, that is $f^!$ satisfies the defining property of $f^\times$. Of course if you take Remark 5.1.2 as the definition of $f^!$ then this theorem becomes trivial: for a proper f we choose the Nagata factorization $X \xrightarrow{\mathrm{id}} X \xrightarrow{f} Y$, and by definition $f^! = \mathrm{id}^* f^\times = f^\times$.

We have already mentioned that, for general f, the existence of the right adjoint $f^\times$ for the functor $\mathbf{R}f_*$ was first proved in Deligne [**16**]. The article [**16**] notes that $f^\times$ agrees with $f^!$ for proper f, and for non-proper f Deligne dismisses $f^\times$ as too undeserving even to be graced with a name, a functor that doesn't lend itself to calculation. For non-proper f the functor $f^\times$ was deemed worthless and consigned to the trash heap of obscurity, until the 1990s it remained nameless.

Until the 1980s, the Deligne-Verdier approach to Grothendieck duality existed only as an offhand aside in Verdier [**64**], a remark saying such a theory should be possible. No one took the trouble to check the details. Lipman plunged into this project sometime in the late 1980s—it took him the best part of two decades, and the outcome was the book [**32**]. In his development of the subject Lipman chose not to ignore the right adjoint of $\mathbf{R}f_*$ when f isn't proper. He brushed off Deligne's disparaging appraisal of this functor—he christened[12] it $f^\times$, and then went on to study its properties.

As it turns out Lipman was right in resurrecting the maligned $f^\times$ from oblivion: the key to the new progress is to study $f^\times$ for non-proper maps. In particular the computability of $f^\times$, when f is a morphism of affine schemes, plays a key role—see Remark 3.2.10.

REMARK 5.1.5. Although none of the results presented in this survey is new, the published expositions of the recent work are all addressed to the expert—they are all couched in terms of $g^!$, its relation with $g^\times$, and how to use the relation for computations. Thus, even though the results give new and much simplified proofs of concrete, old theorems, you have to be an expert to discern this from the available manuscripts. This was inevitable: after all when we write research papers we'd like to have them published. When the subject happens to be out of fashion this adds an extra hurdle—the editor will naturally worry that the paper is likely to have only a miniscule readership and a meager impact (as measured by citations). But, even with the most sympathetic of editors, the paper will go to referees who will undoubtedly be experts. Thus the authors will invariably try to write to impress the experts. The focus will be on new results, not on simple proofs of old theorems—and even when such simplifications are present the fact may well be hidden, buried deep under a mountain of technicalities.

That said, the current survey is an attempt to open up the field to non-specialists. If the subject is to have a revival then it's imperative for the main points to be widely accessible. To keep this document as readable as possible we have, up to Section 5, avoided all but the most indispensable machinery. We are about to delve into areas where progress now seems within reach, and this is forcing us to first recall a little more background. We have to catch up a little, before the open questions can be stated clearly.

We have explained the 2-functor $(-)^!$. The next step is

REMINDER 5.1.6. We begin with a couple of old definitions

(i) Let X be a noetherian scheme. An object $D \in \mathbf{D}^b_{\mathbf{coh}}(X)$ is a *dualizing complex* if $\mathcal{RH}om(-, D)$ takes $\mathbf{D}^b_{\mathbf{coh}}(X)$ to itself[13], and furthermore the natural map $\mathcal{O}_X \longrightarrow \mathcal{RH}om(D, D)$ is an isomorphism. Equivalently: the functor $\mathcal{RH}om(-, D)$ induces an equivalence $\mathbf{D}^b_{\mathbf{coh}}(X)^{\mathrm{op}} \longrightarrow \mathbf{D}^b_{\mathbf{coh}}(X)$.

[12]The drawback of Lipman's notation is that, when handwritten, the symbol $f^\times$ is barely distinguishable from f^*. This renders it a calligraphic challenge to give blackboard talks in the subject.

[13]In Remark 5.1.1 we noted that there are two classical functors $\mathcal{RH}om$, namely the one with values in $\mathbf{D_{qc}}(X)$ and the one with values in $\mathbf{D}(X)$. Since we are only considering $\mathcal{RH}om(C, D)$ with $C \in \mathbf{D}^b_{\mathbf{coh}}(X)$ and $D \in \mathbf{D}^b_{\mathbf{coh}}(X) \subset \mathbf{D}^+_{\mathbf{qc}}(X)$, we are in a situation where the two are classically known to agree.

(ii) Let $f : X \longrightarrow Y$ be a flat, finite-type morphism of noetherian schemes, and consider the diagram

$$\begin{array}{ccccc} X & \xrightarrow{\ \delta\ } & X \times_Y X & \xrightarrow{\ \pi_1\ } & X \\ & & {\scriptstyle \pi_2}\downarrow & & \downarrow{\scriptstyle f} \\ & & X & \xrightarrow{\ f\ } & Y \end{array}$$

A dualizing complex $D \in \mathbf{D}^b_{\mathbf{coh}}(X)$ is Y*–rigid* if it comes together with an isomorphism $D \longrightarrow \delta^\times\big[\mathbf{L}\pi_1^* D \otimes^{\mathbf{L}} \mathbf{L}\pi_2^* D\big]$.

An old theorem tells us

(iii) Assume $X \xrightarrow{f} Y \xrightarrow{g} Z$ are finite-type, flat, separated morphisms of noetherian schemes. Then $f^! : \mathbf{D_{qc}}(Y) \longrightarrow \mathbf{D_{qc}}(X)$ takes Z–rigid dualizing complexes to Z–rigid dualizing complexes.

For the sake of historical accuracy: the fact that $f^!$ takes dualizing complexes to dualizing complexes goes all the way back to the dawn of the theory. The concept of rigid dualizing complexes started with Van den Bergh [**63**]. The formulation given here follows Lipman's reworking of Van den Bergh's result. And Yekutieli and Zhang [**74**–**77**] pursued this in depth, it was their contribution to simplifying and extending the Grothendieck approach to the subject.

In other words: in an alternative universe, a survey of the field would begin with rigid dualizing complexes and build up from there. It just so happens that the Deligne-Verdier angle on the subject was the first to achieve the status of satisfactorily cleaning up the foundations.

REMINDER 5.1.7. We should also remind the reader that there is an interesting noncommutative version. Since I'm not quite sure what a general noncommutative scheme should be, I will confine the discussion to affine noncommutative schemes.

Let R be a noetherian, commutative ring and let S be a flat, finitely generated, associative R–algebra which is right and left noetherian (but not necessarily commutative). Set $S^{\mathsf{e}} = S \otimes_R S^{\mathrm{op}}$. Following Yekutieli [**67**], a dualizing complex is an object $D \in \mathbf{D}(S^{\mathsf{e}})$ such that the functor $\mathbf{R}\mathrm{Hom}(\ \ , D)$ yields an equivalence of categories $\mathbf{D}^b(S\text{–mod})^{\mathrm{op}} \longrightarrow \mathbf{D}^b(S^{\mathrm{op}}\text{–mod})$. The paper [**67**] goes on to study the graded situation and produce some examples of dualizing complexes. If S is commutative then dualizing complexes in $\mathbf{D}(S^{\mathsf{e}})$, in Yekutieli's sense, can be shown to agree with dualizing complexes in $\mathbf{D}^b_{\mathbf{coh}}\big(\mathrm{Spec}(S)\big)$, as recalled in Reminder 5.1.6(i).

Following Van den Bergh [**63**], the dualizing complex $D \in \mathbf{D}(S^{\mathsf{e}})$ is R*–rigid* if it comes together with an isomorphism

$$D \quad \cong \quad \mathbf{R}\mathrm{Hom}_{S^{\mathsf{e}}}\big(S\ ,\ D \otimes^{\mathbf{L}}_R D\big)\ .$$

If R is a field, and S has a filtration whose associated graded ring is commutative and finitely generated as an R–algebra, then [**63**] cleverly shows that a rigid dualizing complex exists. When S is commutative then an R–rigid dualizing complexes in $\mathbf{D}(S^{\mathsf{e}})$, in Van den Bergh's sense, can be shown to agree with the $\mathrm{Spec}(R)$–rigid dualizing complex in $\mathbf{D}^b_{\mathbf{coh}}\big(\mathrm{Spec}(S)\big)$, as recalled in Reminder 5.1.6(ii).

There has been literature pursuing this further, for a couple of early papers the reader is referred to Yekutieli and Zhang [**72**, **73**]. But it's now high time to move on to the open problems.

5.2. Foundational questions. And now we are ready to state the first open question. This one is based not on the very recent work, it derives from the formulas of Avramov and Iyengar [**7**, **8**] that finally opened our eyes. Fittingly the first open question is about noncommutative algebraic geometry, which is unquestionably a hot field nowadays.

PROBLEM 5.2.1. Let $g : k \longrightarrow R$ be any finite-type, flat homomorphism of noetherian, commutative rings. Let the relation between the rings R and S be as in Reminder 5.1.7, meaning S is an R–algebra satisfying all the hypotheses of Reminder 5.1.7. Let $N \in \mathbf{D}(R\text{–mod}) \cong \mathbf{D}^b_{\mathbf{coh}}\big(\mathrm{Spec}(R)\big)$ be a k–rigid dualizing complex.

Question: is $S \otimes^{\mathbf{L}}_{S^{\mathrm{e}}} \mathbf{R}\mathrm{Hom}_R(S, S \otimes^{\mathbf{L}}_R N)$ a k–rigid dualizing complex in $\mathbf{D}(S^{\mathrm{e}}\text{–mod})$?

We should remark that, if S is commutative, this follows from the isomophism $f^! N \cong S \otimes^{\mathbf{L}}_{S^{\mathrm{e}}} \mathbf{R}\mathrm{Hom}_R(S, S \otimes^{\mathbf{L}}_R N)$ of the Reduction Formula, which we met in §4.3, coupled with Reminder 5.1.6(iii). But we don't yet understand the theory well enough to have a simple, direct proof of Reminder 5.1.6(iii) in the affine (commutative) case, hence have no idea if the result can be extended to the noncommutative context.

We should note that the case $k = R$ is already interesting. In fact let us confine ourselves to the case where the ring $k = R$ is Gorenstein; in this case it is known that $R \in \mathbf{D}(R\text{–mod})$ is an R–rigid dualizing complex, and the question specializes to: is $S \otimes^{\mathbf{L}}_{S^{\mathrm{e}}} \mathbf{R}\mathrm{Hom}_R(S, S)$ an R–rigid dualizing complex in $\mathbf{D}(S^{\mathrm{e}}\text{–mod})$? The reader should note that, in the noncommutative setting, R–rigid dualizing complexes are known to exist only when R is a field, in particular the existence results to date all assume equal characteristic.

PROBLEM 5.2.2. The work of Avramov, Iyengar and Lipman [**6**, Section 3] suggests an alternative definition for rigid dualizing complexes. With $k \longrightarrow R \longrightarrow S$ as in Problem 5.2.1, we can declare that a dualizing complex $D \in \mathbf{D}(S^{\mathrm{e}})$ is *AIL–k–rigid* if it comes with an isomorphism

$$D \quad\cong\quad \mathbf{R}\mathrm{Hom}_S\Big[\mathbf{R}\mathrm{Hom}_S\big(D\,,\, S \otimes^{\mathbf{L}}_{S^{\mathrm{e}}} \mathbf{R}\mathrm{Hom}_k(S, S)\big)\,,\, D\Big]$$

From the Reduction Formula of Avramov, Iyengar, Lipman and Nayak [**8**, (4.1.1)] we have that, as long as k is regular and finite dimensional and S is commutative, the two notions of rigidity agree.

Question 1: Do the two notions coincide when S isn't commutative?

Question 2, assuming the notions are different: Suppose $N \in \mathbf{D}(R)$ is an AIL–k–rigid dualizing complex. Is $S \otimes^{\mathbf{L}}_{S^{\mathrm{e}}} \mathbf{R}\mathrm{Hom}_R(S, S \otimes^{\mathbf{L}}_R N)$ also an AIL–k–rigid dualizing complex in $\mathbf{D}(S^{\mathrm{e}}\text{–mod})$?

The third foundational problem is about a derived stack version of the theory—as it happens derived stacks are also much in vogue nowadays.

PROBLEM 5.2.3. In many of the theorems we had to assume flatness, or at the very minimum finite Tor-dimension. The modern way to get around this is to work in the setting of derived algebraic geometry.

Question: is there an incarnation of the theory in derived algebraic geometry?

The Yekutieli school was the first to successfully employ DG methods in Grothendieck duality: see Yekutieli and Zhang [**71**, **76**, **77**], Yekutieli [**69**, survey], and more recently Shaul [**58**]. The Lipman school, inspired by the successes of the Yekutieli school, followed suit: the affine case of the Reduction Formulas, of Avramov and Iyengar, does extend from the flat case presented in §4.3 to the case where the map $R \longrightarrow S$ is of finite Tor-dimension. The proof given in Avramov, Iyengar, Lipman and Nayak [**8**, Section 4] goes by way of differential graded algebras. See also [**6**] for further instances, of the Lipman school exploiting the DG methods introduced by Yekutieli and Zhang.

For some planned future projects see Yekutieli [**65**]. Lipman is also interested in pursuing further the methods of derived algebraic geometry—he has been working his way through Lurie's book [**37**]—but I'm not aware of any manuscripts yet. In any case: at this point the subject is in its infancy, all are welcome to join in.

REMARK 5.2.4. The work in Hafiz Khusyairi 2017 PhD thesis might be relevant to Problem 5.2.3—the thesis is entirely in the setting of old-fashioned, ordinary, commutative schemes, but uses the formulas of Avramov and Iyengar as the starting point for setting up the theory. It then proceeds to develop the usual functoriality properties from there. Since the formulas have a DG analog Khusyairi's work might generalize.

5.3. Computational problems. In the previous section we sketched three foundational problems, about extending the theory to noncommutative and to derived algebraic geometry—both of which are "in" fields nowadays.

Let us now return to more classical problems. In the old, traditional world of ordinary, commutative algebraic geometry the foundations of Grothendieck duality have reached a reasonably satisfactory state. It is feasible to introduce the players and describe the relations among them in what could plausibly be called a short space, and it is possible to do so in such a way that the traditional computations become transparent and brief.

But the problem is that the traditional computations are limited. Let us assume $f : X \longrightarrow Y$ proper and of finite Tor-dimension, in which case Remark 2.1.6 allows us to reduce the problem to computing $f^{\times}\mathcal{O}_Y$ and the map $\varepsilon : \mathbf{R}f_* f^{\times}\mathcal{O}_Y \longrightarrow \mathcal{O}_Y$. The classical literature gives us a good understanding in the case where f is smooth, and an understanding of some sort in situations that are easily reduced to the smooth case, for example when f is Cohen-Macauley. Beyond that, what's known is not all that useful.

As it happens Nayak and Sastry are in the process of writing up a comprehensive account of what is known about the computations.[14] Before long there will be a manuscript containing everything that has been figured out so far—it will be valuable to have it all assembled in one place and the connections worked out. Once the document is ready the interested reader will be able to see, in print and in detail, just how paltry our understanding really is.

REMARK 5.3.1. The last paragraphs should not be interpreted as belittling the traditional case of Grothendieck duality, the special case where $f : X \longrightarrow Y$ is smooth and proper—this classical situation is already fascinating and has spawned a rich literature spanning many decades. Specializing further, assume that $Y = \mathrm{Spec}(k)$ is a point and $f : X \longrightarrow Y$ is smooth, proper and of relative dimension 1, and we find ourselves in ancient territory—we're in the well-understood case of duality for curves. The reader can find an excellent exposition of the old approaches

[14]Note added in proof: The manuscript now exists; see [**41**, **42**].

in Serre [**55**, pp. 25-34 and pp. 76-81], and [as far as the author knows] the cleverest, most recent idea is already five decades old, see Tate [**59**]. The case where $Y = \mathrm{Spec}(k)$ is still a point, $f : X \longrightarrow Y$ is still smooth and proper, but the relative dimension is arbitrary is classical Serre duality, the reader is referred to Serre [**54**], and also to the sketch presented in Section 2.3. In Beĭlinson [**11**] we learn how to generalize Tate's clever trick to higher dimension.

If $k = \mathbb{C}$ then X is a smooth, compact Kähler manifold, and the Hodge decomposition theorem identifies $H^n(\Omega^n_f) = \mathbf{R}f_*\Omega^n_f[n]$ with $H^{n,n}(X) \cong H^{2n}(X, \mathbb{C}) \cong \mathbb{C}$, where the last isomorphism is by Poincaré duality. We would therefore expect to be able to understand the residue map from this perspective too. The reader can find this explored in Harvey [**20**], Tong [**62**], and more recently in Sastry and Tong [**53**].

Now let us return to the generality of the relative case: that is $f : X \longrightarrow Y$ is assumed smooth and proper but Y is an arbitrary noetherian scheme. We know, from the results surveyed in this article, that $f^\times \mathcal{O}_Y$ is canonically isomorphic to $\Omega^n_f[n]$, and that the counit of adjunction $\varepsilon : \mathbf{R}f_* f^\times \mathcal{O}_Y \longrightarrow \mathcal{O}_Y$ is determined by the map taking a relative meromorphic n-form to its residue. In this article we presented a very recent approach to these theorems, we should say something about the older methods—after all understanding the relationship of the old tack with the new might well prove fruitful and illuminating.

The first issue is that all the data must be compatible with composition. That is: if $X \xrightarrow{f} Y \xrightarrow{g} Z$ are composable morphisms of schemes, both of which are smooth and proper, then the composite is smooth and proper and we have a string of canonical isomorphisms

$$\begin{aligned} \Omega^{m+n}_{gf}[m+n] \;\cong\; (gf)^\times \mathcal{O}_Z &\cong f^\times g^\times \mathcal{O}_Z \\ &\cong \mathbf{L}f^* g^\times \mathcal{O}_Z \otimes^{\mathbf{L}} f^\times \mathcal{O}_Y \\ &\cong \mathbf{L}f^* \Omega^m_g[m] \otimes^{\mathbf{L}} \Omega^n_f[n] \end{aligned}$$

The reader might wonder whether the composite is the obvious isomorphism—not sursprisingly the answer turns out to be Yes, see Lipman and Sastry [**36**]. Furthermore the counits of adjunction must be compatible. We have a counit of adjunction $\varepsilon(gf) : \mathbf{R}(gf)_*(gf)^\times \mathcal{O}_Z \longrightarrow \mathcal{O}_Z$, which must agree with the composite

$$\mathbf{R}g_*\mathbf{R}f_* f^\times g^\times \mathcal{O}_Z \xrightarrow{\quad\varepsilon(f)\quad} \mathbf{R}g_* g^\times \mathcal{O}_Z \xrightarrow{\quad\varepsilon(g)\quad} \mathcal{O}_Z\ .$$

Rewriting this in terms the string of canonical isomorphisms above yields a diagram which must commute

$$\begin{array}{ccccc} \mathbf{R}g_*\big(\Omega^m_g[m] \otimes^{\mathbf{L}} \mathbf{R}f_*\Omega^n_f[n]\big) & & \xrightarrow{\quad\cong\quad} & & \mathbf{R}(gf)_*\Omega^{m+n}_{gf}[m+n] \\ \Big\downarrow{\scriptstyle \varepsilon(f)} & & & & \Big\downarrow{\scriptstyle \varepsilon(gf)} \\ \mathbf{R}g_*\big(\Omega^m_g[m] \otimes^{\mathbf{L}} \mathcal{O}_Y\big) & = & \mathbf{R}g_*\Omega^m_g[m] & \xrightarrow{\quad\varepsilon(g)\quad} & \mathcal{O}_Z \end{array}$$

and the commutativity can be interpreted as a compatibility condition on the residue maps.

In the previous paragraph we learned that the counit of adjunction $\varepsilon : \mathbf{R}f_*\Omega^n_f[n] \longrightarrow \mathcal{O}_Y$, and hence the closely related residue map $\rho : \mathbf{R}f_*\Gamma_W\Omega^n_f[n] \longrightarrow \mathcal{O}_Y$, must be compatible with composition. It's even easier to see that ρ must be compatible with flat base change. For the purpose of computations, the compatibility with flat base change allows us to assume that $Y = \mathrm{Spec}(R)$ is affine—and if it helps we may even assume that R is a (strictly) henselian or even a complete local ring. Grothendieck's GFGA result [**18**, Théorème 5.1.4] allows us to replace X by its formal completion, and for some time now the experts have been pursuing the idea that doing so might lead to a better understanding of ρ. For a more extensive treatment the reader is referred to the forthcoming article by Nayak and Sastry; but see also Alonso, Jeremías and Lipman [**1, 2**], Lipman Nayak and Sastry [**33**], Nayak [**39**], Nayak and Sastry [**41, 42**] and Sastry [**52**].

REMARK 5.3.2. In the special case of smooth and proper morphisms $f : X \longrightarrow Y$, Remark 5.3.1 surveyed some of the work done in the quest for a better understanding of Grothendieck duality. The opening paragraphs of Section 5.3 dismissed what's known about more general f as being of limited computational value.

There is some literature: the reader might wish to look at Huang [**23, 24**], Kersken [**28**], Paršin [**49**] and Yekutieli [**68**] (see also the appendix by Sastry).

PROBLEM 5.3.3. Now put the recent results at center stage—they should allow us to go further with the computations. At least when f is flat, we have a simple and explicit formula for $f^!$. The generalization of Reduction 3.2.8 gives an isomorphism $f^! = \mathbf{L}\delta^*\pi^\times\mathbf{L}f^*$, where $\delta : X \longrightarrow X\times_Y X$ is the diagonal map and $\pi : X\times_Y X \longrightarrow X$ is the (second) projection. Any colocalization $c : \Gamma \longrightarrow \mathrm{id}$, where Γ takes the map $\psi : f^\times \longrightarrow f^!$ to an isomorphism, will permit us to form the composite

$$\mathbf{R}f_*\Gamma f^! \xrightarrow{\ \mathbf{R}f_*(\Gamma\psi)^{-1}\ } \mathbf{R}f_*\Gamma f^\times \xrightarrow{\ \mathbf{R}f_*cf^\times\ } \mathbf{R}f_*f^\times \xrightarrow{\ \ \varepsilon\ \ } \mathrm{id}$$

which should be computable, at least in the special case where X and Y are affine. For suitable choices, of the colocalization $c : \Gamma \longrightarrow \mathrm{id}$, the composite should deliver useful information about ε—and those of us competent to carry out the computations should be able to learn much more about the map $\varepsilon : \mathbf{R}f_*f^\times \longrightarrow \mathrm{id}$.

The computations will involve Hochschild homology and cohomology—terms like $S\otimes_{S^e}\mathbf{R}\mathrm{Hom}_R\big(S, S\otimes_R N\big)$ are bound to appear. Fortunately the world is full of experts in Hochschild homology and cohomology, and once they take an interest they will undoubtedly be able to move these computations much further than the handful of us, the few people who have been working on Grothendieck duality. Let's face it: in our tiny group none is adept at handling the Hochschild machinery. The Hochschild experts should feel invited to move right in.

Appendix A. A computation of the base-change map $u^\times \longrightarrow \mathbf{L}u^*$ when $u : U \longrightarrow C$ is an open immersion of curves

Let k be a field, let C be a complete algebraic curve smooth over k, and let $p \in C$ be a k-rational point. Put $U = C - \{p\}$ and let $u : U \longrightarrow C$ be the open immersion. The square

$$\begin{array}{ccc} U & = & U \\ \| & & \big\downarrow{\scriptstyle u} \\ U & \xrightarrow{\ u\ } & C \end{array}$$

is cartesian and the horizontal maps are flat, and Construction 3.1.4 yields a base-change map $\Phi : u^\times \longrightarrow \mathbf{L}u^*$. Let $\mathcal{L}$ be a line bundle on C; we propose to compute the map $\Phi(\mathcal{L}) : u^\times\mathcal{L} \longrightarrow \mathbf{L}u^*\mathcal{L}$. In Remark 5.1.1 we met the isomorphism

$$\mathbf{R}u_*u^\times\mathcal{L} \quad\cong\quad \mathcal{RH}om_{\mathbf{D_{qc}}(C)}(\mathbf{R}u_*\mathcal{O}_U, \mathcal{L})$$

One may check that the counit of adjunction $\varepsilon : \mathbf{R}u_*u^\times\mathcal{L} \longrightarrow \mathcal{L}$ is the map obtained by applying the functor $\mathcal{RH}om_{\mathbf{D_{qc}}(C)}(-,\mathcal{L})$ to the morphism $\mathcal{O}_C \longrightarrow \mathbf{R}u_*\mathcal{O}_U$. And the map $u^\times \longrightarrow \mathbf{L}u^*$ is just the composite $u^\times \stackrel{(\varepsilon')^{-1}u^\times}{\longrightarrow} \mathbf{L}u^*\mathbf{R}u_*u^\times \stackrel{\mathbf{L}u^*\varepsilon}{\longrightarrow} \mathbf{L}u^*$, where $\varepsilon' : \mathbf{L}u^*\mathbf{R}u_* \longrightarrow \mathrm{id}$ is the (invertible) counit of adjunction.

So much for abstract nonsense. Concretely we are reduced to computing what the functor $\mathcal{RH}om_{\mathbf{D_{qc}}(C)}(-,\mathcal{L})$ does to the morphism $\mathcal{O}_C \longrightarrow \mathbf{R}u_*\mathcal{O}_U$, after which we will apply $\mathbf{L}u^*$. Now the map $\mathcal{O}_C \longrightarrow \mathbf{R}u_*\mathcal{O}_U = u_*\mathcal{O}_U$ is the direct limit, as $n \to \infty$, of the maps $\mathcal{O}_C \longrightarrow \mathcal{O}_C(np)$. This means that, in the derived category $\mathbf{D_{qc}}(C)$, we need to compute the homotopy inverse limit of the sequence $\mathcal{L}(-np) \longrightarrow \mathcal{L}$. This is what we will now do.

Let $R = \mathcal{O}_{C,p}$, that is the stalk at p of the structure sheaf $\mathcal{O}_C$. Let $\mathfrak{m} \subset R$ be the maximal ideal. For each n we have a triangle

$$\mathcal{L}(-np) \longrightarrow \mathcal{L} \longrightarrow \mathcal{L} \otimes R/\mathfrak{m}^n$$

and taking homotopy inverse limits over n will yield a triangle. In general I find homotopy inverse limits difficult, but in the case of the inverse system $\mathcal{L} \otimes R/\mathfrak{m}^n$ it isn't so bad.

Let us first take the homotopy inverse limit in the category $\mathbf{D}(C)$, where we allow all complexes of sheaves of $\mathcal{O}_C$–modules—not only ones with quasicoherent cohomology, see Remark 5.1.1 for a discussion. Let $i : p \longrightarrow C$ be the inclusion of p; we turn it into a map of ringed spaces by giving p the structure sheaf R. The functor i_* (extension by zero) is exact, and has an exact left adjoint—the functor taking a sheaf to its stalk at p. Hence the induced functor $i_* : \mathbf{D}(p) \longrightarrow \mathbf{D}(C)$ respects products and therefore homotopy inverse limits. Thus the homotopy inverse limit of the system $\mathcal{L} \otimes R/\mathfrak{m}^n$ can be computed in $\mathbf{D}(p)$, and it comes down to the sheaf $\mathcal{L} \otimes i_*\widehat{R}$, the extension by zero of the completion of the stalk at p of $\mathcal{L}$.

This sheaf is manifestly not quasicoherent—to compute the homotopy inverse limit in the category $\mathbf{D_{qc}}(C)$ we need to derived quasicoherate. That is: we replace by an injective resolution and then quasicoherate. The injective resolution is easy enough: if K is the quotient field of R and $\widehat{K}$ its $\mathfrak{m}$-adic completion (i.e. the quotient field of $\widehat{R}$), then an injective resolution of $\widehat{R}$ as an R–module is given by $\widehat{K} \longrightarrow \widehat{K}/\widehat{R}$, and $i_*\widehat{K} \longrightarrow i_*[\widehat{K}/\widehat{R}]$ is an injective resolution of $i_*\widehat{R}$ in the category of sheaves of $\mathcal{O}_C$–modules. The module $i_*[\widehat{K}/\widehat{R}]$ is quasicoherent as it stands, and the quasicoherator takes $i_*\widehat{K}$ to the constant sheaf $\widehat{K}$. The morphism $\mathcal{L} \longrightarrow \underleftarrow{\mathrm{Holim}}\ (\mathcal{L} \otimes R/\mathfrak{m}^n)$ becomes identified with the cochain map

$$\begin{array}{ccccccc}
0 & \longrightarrow & \mathcal{L} & \longrightarrow & 0 & \longrightarrow & 0 \\
 & & \downarrow & & \downarrow & & \\
0 & \longrightarrow & \mathcal{L} \otimes \widehat{K} & \longrightarrow & \mathcal{L} \otimes i_*[\widehat{K}/\widehat{R}] & \longrightarrow & 0
\end{array}$$

Applying the functor $\mathbf{L}u^* = u^*$, that it restricting to $U \subset C$, kills the sheaf $i_*[\widehat{K}/\widehat{R}]$. We deduce the map of cochain complexes

$$\begin{array}{ccccccc}
0 & \longrightarrow & u^*\mathcal{L} & \longrightarrow & 0 & \longrightarrow & 0 \\
 & & \downarrow & & \downarrow & & \\
0 & \longrightarrow & u^*\mathcal{L} \otimes \widehat{K} & \longrightarrow & 0 & \longrightarrow & 0
\end{array}$$

And the map $u^\times\mathcal{L} \longrightarrow \mathbf{L}u^*\mathcal{L}$ is obtained by completing the triangle, it is the cochain map

$$\begin{array}{ccccccc} 0 & \longrightarrow & u^*\mathcal{L} & \longrightarrow & u^*\mathcal{L}\otimes\widehat{K} & \longrightarrow & 0 \\ & & \downarrow & & \downarrow & & \\ 0 & \longrightarrow & u^*\mathcal{L} & \longrightarrow & 0 & \longrightarrow & 0 \end{array}$$

To sum it up: the curve U is affine, let's say $U = \operatorname{Spec}(S)$. The line bundle $u^*\mathcal{L} \in \mathbf{D_{qc}}(U)$ corresponds, under the equivalence $\mathbf{D_{qc}}(U) \cong \mathbf{D}(S)$, to the rank–1 projective S–module $L = \Gamma(U,\mathcal{L})$. Consider the short exact sequence of S–modules

$$0 \longrightarrow L \longrightarrow L\otimes\widehat{K} \longrightarrow \tfrac{L\otimes\widehat{K}}{L} \longrightarrow 0$$

The morphism $u^\times\mathcal{L} \longrightarrow \mathbf{L}u^*\mathcal{L}$ in the derived category $\mathbf{D_{qc}}(U)$ corresponds, under the equivalence $\mathbf{D_{qc}}(U) \cong \mathbf{D}(S)$, to the map $\frac{L\otimes\widehat{K}}{L}[-1] \longrightarrow L$ that is represented by the short exact sequence.

Now there is an isomorphism of S–modules $\operatorname{Hom}_k(S,k) \cong \frac{L\otimes\widehat{K}}{L}$. The bad way to see this is as follows: both are injective S–modules, and the indecomposable injectives have the same multiplicity on both sides, see [**47**]. But in the case where $\mathcal{L}$ is the canonical bundle Ω^1_C we know there is a canonical isomorphism. If $f : C \longrightarrow \operatorname{Spec}(k)$ is the projection to a point, then $f^\times k \cong \Omega^1_C[1]$ canonically, hence

$$\operatorname{Hom}_k(S,k) \quad\cong\quad (fu)^\times k \quad\cong\quad u^\times f^\times k \quad\cong\quad u^\times\Omega^1_C[1] \quad\cong\quad \frac{\Omega^1_U\otimes\widehat{K}}{\Omega^1_U}$$

where the isomorphisms are all canonical.

Acknowledgments

The author would like to thank Leo Alonso, Asilata Bapat, Spencer Bloch, Jim Borger, Jesse Burke, Pierre Deligne, Anand Deopurkar, Luc Illusie, Ana Jeremías, Steve Lack, Joe Lipman, Bregje Pauwels, Pramath Sastry, Ross Street, Michel Van den Bergh, Amnon Yekutieli and an anonymous referee for questions, comments, corrections and improvements—based both on earlier versions of this manuscript, and on talks I've given presenting parts or all of the material. The contributions of the people listed have greatly influenced the exposition. Needless to say the flaws and mistakes that remain are entirely my fault.

References

[1] Leovigildo Alonso Tarrío, Ana Jeremías López, and Joseph Lipman, *Local homology and cohomology on schemes*, Ann. Sci. École Norm. Sup. (4) **30** (1997), no. 1, 1–39, DOI 10.1016/S0012-9593(97)89914-4. MR1422312

[2] Leovigildo Alonso Tarrío, Ana Jeremías López, and Joseph Lipman, *Studies in duality on Noetherian formal schemes and non-Noetherian ordinary schemes*, Contemporary Mathematics, vol. 244, American Mathematical Society, Providence, RI, 1999. MR1716706

[3] Leovigildo Alonso Tarrío, Ana Jeremías López, and Joseph Lipman, *Bivariance, Grothendieck duality and Hochschild homology I: Construction of a bivariant theory*, Asian J. Math. **15** (2011), no. 3, 451–497, DOI 10.4310/AJM.2011.v15.n3.a7. MR2838216

[4] Leovigildo Alonso Tarrío, Ana Jeremías López, and Joseph Lipman, *Bivariance, Grothendieck duality and Hochschild homology, II: The fundamental class of a flat scheme-map*, Adv. Math. **257** (2014), 365–461, DOI 10.1016/j.aim.2014.02.017. MR3187654

[5] Leovigildo Alonso Tarrío, Ana Jeremías López, and María José Souto Salorio, *Bousfield localization on formal schemes*, J. Algebra **278** (2004), no. 2, 585–610, DOI 10.1016/j.jalgebra.2004.02.030. MR2071654

[6] Luchezar Avramov, Srikanth B. Iyengar, and Joseph Lipman, *Reflexivity and rigidity for complexes, II: Schemes*, Algebra Number Theory **5** (2011), no. 3, 379–429, DOI 10.2140/ant.2011.5.379. MR2833796

[7] Luchezar L. Avramov and Srikanth B. Iyengar, *Gorenstein algebras and Hochschild cohomology*, Michigan Math. J. **57** (2008), 17–35, DOI 10.1307/mmj/1220879394. Special volume in honor of Melvin Hochster. MR2492438

[8] Luchezar L. Avramov, Srikanth B. Iyengar, Joseph Lipman, and Suresh Nayak, *Reduction of derived Hochschild functors over commutative algebras and schemes*, Adv. Math. **223** (2010), no. 2, 735–772, DOI 10.1016/j.aim.2009.09.002. MR2565548

[9] Paul Balmer, Ivo Dell'Ambrogio, and Beren Sanders, *Grothendieck-Neeman duality and the Wirthmüller isomorphism*, Compos. Math. **152** (2016), no. 8, 1740–1776, DOI 10.1112/S0010437X16007375. MR3542492

[10] Paul Balmer and Giordano Favi, *Generalized tensor idempotents and the telescope conjecture*, Proc. Lond. Math. Soc. (3) **102** (2011), no. 6, 1161–1185, DOI 10.1112/plms/pdq050. MR2806103

[11] A. A. Beĭlinson, *Residues and adèles* (Russian), Funktsional. Anal. i Prilozhen. **14** (1980), no. 1, 44–45. MR565095

[12] Marcel Bökstedt and Amnon Neeman, *Homotopy limits in triangulated categories*, Compositio Math. **86** (1993), no. 2, 209–234. MR1214458

[13] A. K. Bousfield, *The localization of spectra with respect to homology*, Topology **18** (1979), no. 4, 257–281, DOI 10.1016/0040-9383(79)90018-1. MR551009

[14] Brian Conrad, *Grothendieck duality and base change*, Lecture Notes in Mathematics, vol. 1750, Springer-Verlag, Berlin, 2000. MR1804902

[15] Brian Conrad, *Deligne's notes on Nagata compactifications*, J. Ramanujan Math. Soc. **22** (2007), no. 3, 205–257. MR2356346

[16] Pierre Deligne, *Cohomology à support propre en construction du foncteur $f^!$*, Residues and Duality, Lecture Notes in Mathematics, vol. 20, Springer–Verlag, 1966, pp. 404–421.

[17] P. Gabriel and M. Zisman, *Calculus of fractions and homotopy theory*, Ergebnisse der Mathematik und ihrer Grenzgebiete, Band 35, Springer-Verlag New York, Inc., New York, 1967. MR0210125

[18] Alexandre Grothendieck, *Éléments de géométrie algébrique III. Étude cohomologique des faisceaux cohérents I*, Inst. Hautes Études Sci. Publ. Math. (1961), no. 11, 425–511.

[19] Robin Hartshorne, *Residues and duality*, Lecture notes of a seminar on the work of A. Grothendieck, given at Harvard 1963/64. With an appendix by P. Deligne. Lecture Notes in Mathematics, No. 20, Springer-Verlag, Berlin-New York, 1966. MR0222093

[20] F. Reese Harvey, *Integral formulae connected by Dolbeault's isomorphism*, Rice Univ. Studies **56** (1970), no. 2, 77–97 (1971). MR0273067

[21] G. Hochschild, Bertram Kostant, and Alex Rosenberg, *Differential forms on regular affine algebras*, Trans. Amer. Math. Soc. **102** (1962), 383–408, DOI 10.2307/1993614. MR142598

[22] Michael J. Hopkins, *Global methods in homotopy theory*, Homotopy theory (Durham, 1985), London Math. Soc. Lecture Note Ser., vol. 117, Cambridge Univ. Press, Cambridge, 1987, pp. 73–96. MR932260

[23] I-Chiau Huang, *An explicit construction of residual complexes*, J. Algebra **225** (2000), no. 2, 698–739, DOI 10.1006/jabr.1999.8140. MR1741558

[24] I-Chiau Huang, *The residue theorem via an explicit construction of traces*, J. Algebra **245** (2001), no. 1, 310–354, DOI 10.1006/jabr.2001.8891. MR1868195

[25] Reinhold Hübl and Pramathanath Sastry, *Regular differential forms and relative duality*, Amer. J. Math. **115** (1993), no. 4, 749–787, DOI 10.2307/2375012. MR1231146

[26] Luc Illusie, *Conditions de finitude*, Théorie des intersections et théorème de Riemann-Roch, Springer-Verlag, Berlin, 1971, Séminaire de Géométrie Algébrique du Bois-Marie 1966–1967 (SGA 6, Exposé III), pp. 222–273. Lecture Notes in Mathematics, Vol. 225.

[27] Srikanth B. Iyengar, Joseph Lipman, and Amnon Neeman, *Relation between two twisted inverse image pseudofunctors in duality theory*, Compos. Math. **151** (2015), no. 4, 735–764, DOI 10.1112/S0010437X14007672. MR3334894

[28] Masumi Kersken, *Cousinkomplex und Nennersysteme* (German), Math. Z. **182** (1983), no. 3, 389–402, DOI 10.1007/BF01179758. MR696535

[29] Joseph Lipman, *Dualizing sheaves, differentials and residues on algebraic varieties* (English, with French summary), Astérisque **117** (1984), ii+138. MR759943

[30] Joseph Lipman, *Residues and traces of differential forms via Hochschild homology*, Contemporary Mathematics, vol. 61, American Mathematical Society, Providence, RI, 1987. MR868864

[31] Joseph Lipman, *Lectures on local cohomology and duality*, Local cohomology and its applications (Guanajuato, 1999), Lecture Notes in Pure and Appl. Math., vol. 226, Dekker, New York, 2002, pp. 39–89. MR1888195

[32] Joseph Lipman, *Notes on derived functors and Grothendieck duality*, Foundations of Grothendieck duality for diagrams of schemes, Lecture Notes in Math., vol. 1960, Springer, Berlin, 2009, pp. 1–259, DOI 10.1007/978-3-540-85420-3. MR2490557

[33] Joseph Lipman, Suresh Nayak, and Pramathanath Sastry, *Pseudofunctorial behavior of Cousin complexes on formal schemes*, Variance and duality for Cousin complexes on formal schemes, Contemp. Math., vol. 375, Amer. Math. Soc., Providence, RI, 2005, pp. 3–133, DOI 10.1090/conm/375/06934. MR2136143

[34] Joseph Lipman and Amnon Neeman, *Quasi-perfect scheme-maps and boundedness of the twisted inverse image functor*, Illinois J. Math. **51** (2007), no. 1, 209–236. MR2346195

[35] Joseph Lipman and Amnon Neeman, *On the fundamental class of an essentially smooth scheme-map*, Algebr. Geom. **5** (2018), no. 2, 131–159, DOI 10.14231/AG-2018-005. MR3769890

[36] Joseph Lipman and Pramathanath Sastry, *Regular differential and equidimensional scheme-maps*, J. Algebraic Geom. **1** (1992), no. 1, 101–130. MR1129841

[37] Jacob Lurie, *Higher topos theory*, Annals of Mathematics Studies, vol. 170, Princeton University Press, Princeton, NJ, 2009. MR2522659

[38] Masayoshi Nagata, *Imbedding of an abstract variety in a complete variety*, J. Math. Kyoto Univ. **2** (1962), 1–10, DOI 10.1215/kjm/1250524969. MR142549

[39] Suresh Nayak, *Pasting pseudofunctors*, Variance and duality for Cousin complexes on formal schemes, Contemp. Math., vol. 375, Amer. Math. Soc., Providence, RI, 2005, pp. 195–271, DOI 10.1090/conm/375/2136145. MR2136145

[40] Suresh Nayak, *Compactification for essentially finite-type maps*, Adv. Math. **222** (2009), no. 2, 527–546, DOI 10.1016/j.aim.2009.05.002. MR2538019

[41] Suresh Nayak and Pramathanath Sastry, *Grothendieck duality and transitivity I: Formal schemes*, arXiv:1903.01779 (2019).

[42] Suresh Nayak and Pramathanath Sastry, *Grothendieck duality and transitivity II: Traces and residues via Verdier's isomorphism*, arXiv:1903.01783 (2019).

[43] Amnon Neeman, *An improvement on the base-change theorem and the functor $f^!$*, https://arxiv.org/abs/1406.7599.

[44] Amnon Neeman, *The relation between Grothendieck duality and Hochschild homology*, K-Theory—Proceedings of the International Colloquium, Mumbai, 2016, Hindustan Book Agency, New Delhi, 2018, pp. 91–126. MR3930045

[45] Amnon Neeman, *The chromatic tower for $D(R)$*, Topology **31** (1992), no. 3, 519–532, DOI 10.1016/0040-9383(92)90047-L. With an appendix by Marcel Bökstedt. MR1174255

[46] Amnon Neeman, *The Grothendieck duality theorem via Bousfield's techniques and Brown representability*, J. Amer. Math. Soc. **9** (1996), no. 1, 205–236, DOI 10.1090/S0894-0347-96-00174-9. MR1308405

[47] Amnon Neeman, *The decomposition of $\mathrm{Hom}_k(\mathcal{S}, k)$ into indecomposable injectives*, Acta Math. Vietnam. **40** (2015), no. 2, 331–338, DOI 10.1007/s40306-014-0110-z. MR3366172

[48] Amnon Neeman, *Traces and residues*, Indiana Univ. Math. J. **64** (2015), no. 1, 217–229, DOI 10.1512/iumj.2015.64.5461. MR3320524

[49] A. N. Paršin, *On the arithmetic of two-dimensional schemes. I. Distributions and residues* (Russian), Izv. Akad. Nauk SSSR Ser. Mat. **40** (1976), no. 4, 736–773, 949. MR0419458

[50] Marco Porta, Liran Shaul, and Amnon Yekutieli, *On the homology of completion and torsion*, Algebr. Represent. Theory **17** (2014), no. 1, 31–67, DOI 10.1007/s10468-012-9385-8. MR3160712

[51] Pramathanath Sastry, *Base change and Grothendieck duality for Cohen-Macaulay maps*, Compos. Math. **140** (2004), no. 3, 729–777, DOI 10.1112/S0010437X03000654. MR2041779

[52] Pramathanath Sastry, *Duality for Cousin complexes*, Variance and duality for Cousin complexes on formal schemes, Contemp. Math., vol. 375, Amer. Math. Soc., Providence, RI, 2005, pp. 137–192, DOI 10.1090/conm/375/06936. MR2136144

[53] Pramathanath Sastry and Yue Lin L. Tong, *The Grothendieck trace and the de Rham integral*, Canad. Math. Bull. **46** (2003), no. 3, 429–440, DOI 10.4153/CMB-2003-043-3. MR1994868

[54] Jean-Pierre Serre, *Un théorème de dualité* (French), Comment. Math. Helv. **29** (1955), 9–26, DOI 10.1007/BF02564268. MR67489

[55] Jean-Pierre Serre, *Groupes algébriques et corps de classes* (French), Publications de l'institut de mathématique de l'université de Nancago, VII. Hermann, Paris, 1959. MR0103191

[56] Liran Shaul, *Reduction of Hochschild cohomology over algebras finite over their center*, J. Pure Appl. Algebra **219** (2015), no. 10, 4368–4377, DOI 10.1016/j.jpaa.2015.02.021. MR3346495

[57] Liran Shaul, *Relations between derived Hochschild functors via twisting*, Comm. Algebra **44** (2016), no. 7, 2898–2907, DOI 10.1080/00927872.2015.1065850. MR3507157

[58] Liran Shaul, *The twisted inverse image pseudofunctor over commutative DG rings and perfect base change*, Adv. Math. **320** (2017), 279–328, DOI 10.1016/j.aim.2017.08.041. MR3709107

[59] John Tate, *Residues of differentials on curves*, Ann. Sci. École Norm. Sup. (4) **1** (1968), 149–159. MR0227171

[60] R. W. Thomason, *The classification of triangulated subcategories*, Compositio Math. **105** (1997), no. 1, 1–27, DOI 10.1023/A:1017932514274. MR1436741

[61] R. W. Thomason and Thomas Trobaugh, *Higher algebraic K-theory of schemes and of derived categories*, The Grothendieck Festschrift, Vol. III, Progr. Math., vol. 88, Birkhäuser Boston, Boston, MA, 1990, pp. 247–435, DOI 10.1007/978-0-8176-4576-2_10. MR1106918

[62] Yue Lin L. Tong, *Integral representation formulae and Grothendieck residue symbol*, Amer. J. Math. **95** (1973), 904–917, DOI 10.2307/2373701. MR367255

[63] Michel van den Bergh, *Existence theorems for dualizing complexes over non-commutative graded and filtered rings*, J. Algebra **195** (1997), no. 2, 662–679, DOI 10.1006/jabr.1997.7052. MR1469646

[64] Jean-Louis Verdier, *Base change for twisted inverse image of coherent sheaves*, Algebraic Geometry (Internat. Colloq., Tata Inst. Fund. Res., Bombay, 1968), Oxford Univ. Press, London, 1969, pp. 393–408. MR0274464

[65] Amnon Yekutieli, *The derived category of sheaves of commutative dg rings (preview)*, arXiv:1312.6411.

[66] ———, *Duality and tilting for commutative dg rings*, arXiv:1312.6411.

[67] Amnon Yekutieli, *Dualizing complexes over noncommutative graded algebras*, J. Algebra **153** (1992), no. 1, 41–84, DOI 10.1016/0021-8693(92)90148-F. MR1195406

[68] Amnon Yekutieli, *An explicit construction of the Grothendieck residue complex* (English, with French summary), Astérisque **208** (1992), 127. With an appendix by Pramathanath Sastry. MR1213064

[69] Amnon Yekutieli, *Rigid dualizing complexes via differential graded algebras (survey)*, Triangulated categories, London Math. Soc. Lecture Note Ser., vol. 375, Cambridge Univ. Press, Cambridge, 2010, pp. 452–463. MR2681716

[70] Amnon Yekutieli, *The squaring operation for commutative DG rings*, J. Algebra **449** (2016), 50–107, DOI 10.1016/j.jalgebra.2015.09.038. MR3448166

[71] Amnon Yekutieli and James J. Zhang, *Rigid dualizing complexes on schemes*, arXiv math.AG/0405570.

[72] Amnon Yekutieli and James J. Zhang, *Serre duality for noncommutative projective schemes*, Proc. Amer. Math. Soc. **125** (1997), no. 3, 697–707, DOI 10.1090/S0002-9939-97-03782-9. MR1372045

[73] Amnon Yekutieli and James J. Zhang, *Rings with Auslander dualizing complexes*, J. Algebra **213** (1999), no. 1, 1–51, DOI 10.1006/jabr.1998.7657. MR1674648

[74] Amnon Yekutieli and James J. Zhang, *Dualizing complexes and perverse modules over differential algebras*, Compos. Math. **141** (2005), no. 3, 620–654, DOI 10.1112/S0010437X04001307. MR2135281

[75] Amnon Yekutieli and James J. Zhang, *Dualizing complexes and perverse sheaves on noncommutative ringed schemes*, Selecta Math. (N.S.) **12** (2006), no. 1, 137–177, DOI 10.1007/s00029-006-0022-4. MR2244264

[76] Amnon Yekutieli and James J. Zhang, *Rigid complexes via DG algebras*, Trans. Amer. Math. Soc. **360** (2008), no. 6, 3211–3248, DOI 10.1090/S0002-9947-08-04465-6. MR2379794

[77] Amnon Yekutieli and James J. Zhang, *Rigid dualizing complexes over commutative rings*, Algebr. Represent. Theory **12** (2009), no. 1, 19–52, DOI 10.1007/s10468-008-9102-9. MR2471449

Centre for Mathematics and its Applications, Mathematical Sciences Institute, Building 145, The Australian National University, Canberra, ACT 2601, Australia

Email address: `Amnon.Neeman@anu.edu.au`

Contemporary Mathematics
Volume **749**, 2020
https://doi.org/10.1090/conm/749/15077

Calculus of absolute Kähler forms and Milnor K-theory

Jinhyun Park

ABSTRACT. We develop a basic calculus for absolute Kähler forms over the formal power series rings of a field of characteristic 0. We show that for the single variable ring, one has a well-defined integral of a form in the relative Milnor K-groups. For the multivariable case, a partial answer is given.

Using these, we give the notion of Chen's iterated integrals for the absolute forms, and define the logarithms. The logarithms describe the projections from the Milnor K-groups to the relative Milnor K-groups.

1. Introduction

1.1. Overview. To some mathematicians, algebraic K-theory is often somewhat notorious for being abstract and esoteric. The difficulties one faces in studying algebraic K-theory are however real, and they often come from the large amount of topics one has to learn from various disciplines of mathematics, such as algebraic topology, algebraic geometry, homotopy theory, etc. The author hopes that, this paper can be seen as one of the papers that may "explain" certain basic aspects of algebraic K-theory from a very elementary point of view, such as calculus.

To experts of algebraic K-theory, probably that calculus has a place here may not be very surprising. For instance, one of the seminal results on the relative algebraic K-theory is the work of T. Goodwillie [**5**], that computes the relative algebraic K-groups in terms of the relative cyclic homology groups. The latter groups are often expressed in terms of de Rham complexes, and the de Rham complexes are in a sense the prime object of studies in calculus.

There is another place where the word calculus appears in algebraic K-theory. In [**1**], S. Bloch had constructed a map from the relative K_2 of an Artin local algebra R to the group of absolute Kähler differentials modulo exact forms. Under suitable assumptions on R, its modern generalization takes the form

$$B : K_n^M(R, I) \to \Omega^{n-1}_{R,I/\mathbb{Z}}/d\Omega^{n-1}_{R,I/\mathbb{Z}},$$

that one can call a *Bloch map*. This map B comes from the relative $d\log$ map

$$d\log : K_n^M(R, I) \to \Omega^n_{R,I/\mathbb{Z}},$$

2010 *Mathematics Subject Classification.* Primary 19D45; Secondary 13F25.

The author was partially supported by the National Research Foundation of Korea (NRF) grant (2018R1A2B6002287) funded by the Korean government (Ministry of Science and ICT) .

by suitably "integrating" $d\log$, see [**6**, §1]. In fact this *ibid.* even says "In other words, we integrate $d\log$ by taking a (co)limit, in the spirit of calculus." See the sentence just before [**6**, Remark 3.12].

The aim of this article is to show that, there indeed exists calculus in the sense of college freshmen, that is elementary and explicit, if one works with the formal power series rings over a field k of characteristic 0. We show that it is possible to describe a necessary and sufficient condition to be able to "integrate" a given absolute Kähler form of a formal power series using very elementary means of calculus plus some small epsilons, and such so-defined integrals are actually well-defined in the relative Milnor K-groups. Hence, in a sense, we provide a demystified way to do some K-theory from a very elementary point of view.

To the author, the initial motivation for this problem was not about the above Bloch map or the relative Milnor K-theory. It occurred when the author was working with S. Ünver on the papers [**10**] and [**11**]. In [**10**], we studied a cycle-theoretic description of the Milnor K-groups of the truncated polynomials $k_m := k[t]/(t^m)$ for integers $m \geq 1$ via certain "$d\log$" analogue functions on the algebraic cycle classes, called the regulators. In [**11**], we study the first off-Milnor range cycle-group $\mathrm{CH}^2(k_m, 3)$ via the "additive dilogarithm" of Ünver [**12**]. In that article, one persistent puzzling issue that kept bothering the author's mind was "Can Ünver's complicated additive dilogarithm formula of [**12**] possibly be understood in a more intuitive and systematic way, e.g. formal integrals or formal iterated integrals?"

The author's initial guess was that, like Euler's classical dilogarithm is obtained as the iterated integrals, maybe one might be able to obtain Ünver's additive dilogarithm by developing some notions of integrals and iterated integrals on the absolute Kähler forms over the formal power series $k[[t]]$. Unfortunately, except for $m = 2$ case, i.e. mod t^2, the author was not able to realize this (false) dream. However, instead the author found that it is actually possible to do some calculus using the absolute Kähler forms, and quite interestingly, it turned out the answer is related to the relative algebraic K-theory of $k[[t]]$ and the Bloch map B.

About the author's failure on answering the original question, the author now thinks that this could not be done because Ünver's additive dilogarithm is in a sense an analogue of the Bloch-Wigner dilogarithm function (see [**2**]) that satisfies the clean 5-term Abel's functional equation, and the Bloch-Wigner function is not obtained as an iterated integral. The Euler dilogarithm is an iterated integral, but it satisfies Abel's functional equation up to certain unclean lower terms. This is probably why the author couldn't express Ünver's additive dilogarithm in terms of iterated integrals developed in this paper.

The author hopes that this paper is an amusement to the experts, and an encouragement to the younger scholars because this paper is probably easy to read for them as well, while giving some easier perspectives to do algebraic K-theory.

1.2. Descriptions of results. In this paper, always we let k be a field of characteristic 0, although often assuming that the characteristic is $p > 0$ which is sufficiently large, some of the results may still work. We leave such matters to the reader. Consider the formal power series ring $k[[t]]$ in t with the coefficients in k.

1.2.1. *Calculus in* $k[[t]]$ *over* k. Doing calculus in $k[[t]]$ over k is probably every calculus student's dream: just knowing how to differentiate and integrate the polynomial functions is enough. In fancy terms, we can say that the de Rham complex of $k[[t]]$ over k is expressed simply as the 2-term complex $k[[t]] \xrightarrow{d} \Omega^1_{k[[t]]/k}$ with $\Omega^n_{k[[t]]/k} = 0$ for all $n \geq 2$, while $\ker d = k$. Here, d is surjective so that every form $\omega \in \Omega^1_{k[[t]]/k}$ has an integral $\int \omega \in k[[t]]$, defined up to translation by a constant in $\ker d = k$.

1.2.2. *Kähler forms over $\mathbb{Z}$.* However, when the de Rham complex is taken over the absolute base $\mathbb{Z}$, the situation is somewhat different and often rather confusing. Since this is a bit less common to use the Kähler forms with the absolute base $\mathbb{Z}$, let us recall a bit of basic facts on such absolute Kähler forms.

Recall (e.g. [**7**, Chapter II, §8, p.172]) that when R is a commutative ring with unity (which is automatically a $\mathbb{Z}$-algebra), the module $\Omega^1_{R/\mathbb{Z}}$ of Kähler differential 1-forms is an R-module with the $\mathbb{Z}$-derivation $d : R \to \Omega^1_{R/\mathbb{Z}}$ that satisfies the universal property: for any R-module M and any $\mathbb{Z}$-derivation $d' : R \to M$, there is a unique R-module homomorphism $f : \Omega^1_{R/\mathbb{Z}} \to M$ such that $d' = f \circ d$. It can be constructed also as the free R-module generated the symbols $\{db \mid b \in R\}$ modulo the relations $d(b+b') - db - db'$, $d(bb') - bdb' - b'db$ and da for $b, b' \in R$ and $a \in \mathbb{Z}$. By the universal property, this is unique up to an isomorphism of R-modules.

When k is a field of characteristic 0 and when R is a k-algebra, for $a, a' \in \mathbb{Z}$ with $a' \neq 0$, one checks that $d\left(\frac{a}{a'}\right) = \frac{a'da - ada'}{(a')^2}$. This implies the equality

$$\Omega^1_{R/\mathbb{Z}} = \Omega^1_{R/\mathbb{Q}}.$$

In fact, even when $R = k$, the differential $\Omega^1_{k/\mathbb{Z}}$ is not necessarily 0. In fact, when $c \in k$, we have $dc = 0$ if and only if c is algebraic over $\mathbb{Q}$.[1] See [**9**, Corollary 5.3-(c), p.74] for its proof.

In general, we take $\Omega^i_{R/\mathbb{Z}}$ to be the i-fold wedge product. Here, even when $R = k$ and $c_1, \cdots, c_r \in k$ are transcendental over $\mathbb{Q}$ and algebraically independent, then we could have $dc_1 \wedge \cdots \wedge dc_r \neq 0$, as well.

1.2.3. *Calculus of absolute Kähler forms.* As we saw, the absolute forms are relevant to the transcendency over $\mathbb{Q}$. To avoid trivial cases, suppose $\mathbb{Q} \subset k$ is a transcendental extension. Then the de Rham complex

$$k[[t]] \xrightarrow{d} \Omega^1_{k[[t]]/\mathbb{Z}} \xrightarrow{d} \Omega^2_{k[[t]]/\mathbb{Z}} \xrightarrow{d} \Omega^3_{k[[t]]/\mathbb{Z}} \xrightarrow{d} \cdots$$

does not necessarily terminate. Furthermore, the kernel of the first map $d : k[[t]] \to \Omega^1_{k[[t]]/\mathbb{Z}}$ is not equal to k, but the subgroup $\{c \in k \mid c \text{ is algebraic over } \mathbb{Q}\}$, although a virtue of the differential calculus is that the differentiation of a constant is zero.

At least the differential operators d tell us how to take derivatives of forms. What about integrals? One might hope to define the integral $\int$ of absolute Kähler forms in $\Omega^r_{k[[t]]/\mathbb{Z}}$ to be an "anti-derivative" in $\Omega^{r-1}_{k[[t]]/\mathbb{Z}}$ of the derivation $d : \Omega^{r-1}_{k[[t]]/\mathbb{Z}} \to \Omega^r_{k[[t]]/\mathbb{Z}}$ over $\mathbb{Z}$. However, this could be a bit more complicated than one can naïvely imagine. Even in the simplest case of $d : k[[t]] \to \Omega^1_{k[[t]]/\mathbb{Z}}$, this exterior derivation is not surjective. So, how do we know whether there is even a single anti-derivative of a 1-form in $\Omega^1_{k[[t]]/\mathbb{Z}}$?

[1] To illustrate part of this story, suppose $c \in k$ is algebraic over $\mathbb{Q}$. Let $p(x) \in \mathbb{Q}[x]$ be the minimal polynomial of c over $\mathbb{Q}$. Let $n := \deg_x p(x)$. From $p(c) = 0$, we then obtain $p'(c)dc = 0$ in $\Omega^1_{k/\mathbb{Z}} = \Omega^1_{k/\mathbb{Q}}$. By the minimality of the degree n, we must have $p'(c) \neq 0$ so that dividing by $p'(c)$, we have $dc = 0$.

If one insists to try anyway, for a nonzero constant $a \in k$, consider an 1-form $atdt \in \Omega^1_{k[[t]]/\mathbb{Z}}$. In the "usual" calculus (i.e. differentials forms are over k), one would like to say that $\int atdt = \frac{1}{2}at^2 + C$ for some constant $C \in k$. However, in the absolute base $\mathbb{Z}$ case, we note that for the derivation $d : k[[t]] \to \Omega^1_{k[[t]]/\mathbb{Z}}$,

$$d\left(\frac{1}{2}at^2 + C\right) = at^2 d\left(\frac{1}{2}\right) + \frac{1}{2}t^2 da + \frac{1}{2} \cdot 2atdt + dC = \frac{1}{2}t^2 da + atdt + dC.$$

But here neither $da = 0$ nor $dC = 0$ is guaranteed in $\Omega^1_{k[[t]]/\mathbb{Z}}$, unless a or C are algebraic over $\mathbb{Q}$. Hence $\frac{1}{2}at^2 + C$ is not necessarily an integral of $atdt$, in the strict sense.

However, one can observe the following: the newly introduced "garbage terms" $\frac{1}{2}t^2 da + dC$ do not involve dt, i.e. they are dt-free terms. Maybe, would it be always a good idea to ignore all these dt-free terms in the process?

1.2.4. *The results.* In this paper, we show that up to translating the given form by a "dt-free" form, we can define an integral of absolute Kähler forms, up to an exact form. Here is the story: we first show that for every $\omega \in \Omega^r_{k[[t]]/\mathbb{Z}}$, there exists a "$dt$-free companion" ω' in $\Omega^r_{k[[t]]/\mathbb{Z}}$ such that $\omega + \omega'$ is an exact form. Then we take any anti-derivative of the exact form $\omega + \omega'$, and call it *an* integral $\int \omega \in \Omega^{r-1}_{k[[t]]/\mathbb{Z}}$. In general, a different choice of a dt-free companion gives a different integral $\int \omega$. However, we have (see Theorems 2.10 and 2.11):

THEOREM 1.1. *Let $\omega \in \Omega^r_{k[[t]]/\mathbb{Z}}$ for $r \geq 1$. Then $\int \omega$ is well-defined up to translation by a form in $\Omega^{r-1}_{k/\mathbb{Z}}$ and an exact form in $d\Omega^{r-2}_{k[[t]]/\mathbb{Z}}$, where we let $\Omega^{-1}_{k[[t]]/\mathbb{Z}} := 0$ for notational consistency.*

More precisely, $\int \omega$ is well-defined in $\Omega^{r-1}_{k[[t]],(t)/\mathbb{Z}}/d\Omega^{r-2}_{k[[t]],(t)/\mathbb{Z}}$, where $\Omega^i_{k[[t]],(t)/\mathbb{Z}}$ $:= \ker(\Omega_{k[[t]]/\mathbb{Z}} \overset{\mathrm{ev}_{t=0}}{\to} \Omega^i_{k/\mathbb{Z}})$.

What is interesting is that the quotient group where $\int \omega$ belongs to is in fact the relative Milnor K-group $K^M_r(k[[t]], (t))$. See Corollary 2.13. Hence Theorem 1.1 can be rephrased as:

THEOREM 1.2. *Let $\omega \in \Omega^r_{k[[t]]/\mathbb{Z}}$ for $r \geq 1$. Then $\int \omega$ is well-defined as a class in $K^M_r(k[[t]], (t))$.*

While this relative $K^M_r(k[[t]], (t))$ is the natural place where the integrals live, it is often inconvenient to make explicit calculations. Here, one has a non-canonical decomposition

$$K^M_r(k[[t]], (t)) \simeq \Omega^{r-1}_{k[[t]],(t)/\mathbb{Z}}/d\Omega^{r-2}_{k[[t]],(t)/\mathbb{Z}} \simeq \bigoplus_{i \geq 1} t^i \Omega^{r-1}_{k/\mathbb{Z}} \tag{1.1}$$

into a direct sum that depends on the choice of a k-basis of the space $tk[[t]]$. The above decomposition uses $\{t, t^2, \cdots\}$. Taking this into considerations, we can understand the integral $\int w$ as an expansion

$$\int \omega = \sum_{i=1}^{\infty} t^i c_i,$$

where $c_i \in \Omega^{r-1}_{k/\mathbb{Z}}$. We can rephrase it as:

COROLLARY 1.3. *Let $\omega \in \Omega^r_{k[[t]]/\mathbb{Z}}$ for $r \geq 1$. Then under the identification (1.1), the coefficients of t^i of $\int \omega$ are well-defined in $\Omega^{r-1}_{k/\mathbb{Z}}$ for $i \geq 1$, but not for $i = 0$, in general.*

In summary, for the absolute Kähler forms of $k[[t]]$, up to translating by a dt-free form, every form has an anti-derivative, and its integral is well-defined in $K_n^M(k[[t]], (t))$. This admits a description as a power series expansion with positive t-degree terms, where the coefficient of t^i is in $\Omega^{r-1}_{k/\mathbb{Z}}$ for $i \geq 1$.

We remark that the above "pseudo-exactness" of every absolute Kähler form $\omega \in \Omega^r_{k[[t]]/\mathbb{Z}}$for $r > 0$ could possibly be proven in an elegant way using the Poincaré lemma $\mathrm{H}^r(\Omega^\bullet_{k[[t]],(t)/\mathbb{Z}}) = 0$ for $r > 0$, as one can deduce from [**14**, Corollary 9.9.3]. However, this approach via the Poincaré lemma does not immediately give an explicit expression of the integrals, so we believe this paper's method can be seen as a down-to-earth version of calculus for $k[[t]]$ over $\mathbb{Z}$.

Once we have the above notion of integrals, we can apply or extend them to a few directions. Here are several topics that apply or extend our theory of calculus discussed in this paper:

(1) Chen's iterated integrals for absolute Kähler forms over $k[[t]]$.
(2) Integrations of forms over the truncated polynomials $k_m := k[[t]]/(t^m)$ and the notion of significance ranges in t-adic approximation.
(3) The definition of $\log\{f_1, \cdots, f_n\} \in K_n^M(k[[t]], (t))$ for a Milnor symbol $\{f_1, \cdots, f_n\} \in K_n^M(k[[t]])$ as an integral of the form $\int \underline{\mathrm{dlog}}\{f_1, \cdots, f_n\} = \int d\log f_1 \wedge \cdots \wedge d\log f_n$. The notion of log for k_m as well.
(4) Multivariable calculus of absolute Kähler forms over $k[[\underline{t}]]{:=}k[[t_1, \cdots, t_r]]$, and the conditions for the existence of anti-derivatives.
(5) The notion of $\log\{f_1, \cdots, f_n\}$ for the symbol $\{f_1, \cdots, f_n\} \in K_n^M(k[[\underline{t}]])$.

So as not to make the introduction too long, we won't list all results obtained in this article. We encourage the reader to browse through the sections for more details. The author believes that the notion introduced in this article can be further extensively studied in follow-up works. We hope this paper would serve as a starting point.

2. Integrals of absolute Kähler forms

Let k be a field of characteristic 0. We study the notion of dt-free forms and integrals for absolute Kähler differential forms in $\Omega^r_{k[[t]]/\mathbb{Z}}$ for $r \geq 1$.

2.1. The dt-free parts, fake integrals, and dt-free companions. A key observation in developing calculus of absolute Kähler forms is the following notion:

DEFINITION 2.1. Let $\omega \in \Omega^r_{k[[t]]/\mathbb{Z}}$ for $r \geq 1$. This can be expressed as a formal sum of the terms of the form $ct^i da_1 \wedge \cdots \wedge da_r$ for $0 \leq i$ and $c, a_1, \cdots, a_r \in k$, and the terms of the form $ct^i da_1 \wedge \cdots \wedge da_{r-1} \wedge dt$ for $0 \leq i$ and $c, a_1, \cdots, a_{r-1} \in k$. Let $\omega_{\mathbb{Z}}$ be the sum of the former type of terms of ω, which we call, the *dt-free part of* ω, so that we have an expression of the form

$$\omega = \omega_{\mathbb{Z}} + \sum_{i=0}^{\infty} t^i \omega_i dt, \tag{2.1}$$

for some $\omega_i \in \Omega^{r-1}_{k/\mathbb{Z}}$. If ω is equal to its dt-free part, then we say ω is a dt-free form. Note that the subset $F^r_{k[[t]]/\mathbb{Z}} \subset \Omega^r_{k[[t]]/\mathbb{Z}}$ of dt-free forms is an abelian subgroup.

As an approximation to the correct definition of integrals of absolute Kähler forms, ignoring the dt-free parts we first define:

Definition 2.2. For a form $\omega \in \Omega^r_{k[[t]]/\mathbb{Z}}$, express ω as in (2.1). We define the *fake integral of* ω to be

$$\fint \omega := \sum_{i=0}^{\infty} \frac{t^{i+1}\omega_i}{i+1} \in \Omega^{r-1}_{k[[t]]/\mathbb{Z}}.$$

How far is $d \fint \omega$ from the given ω? We have the following useful result:

Proposition 2.3. *For every $\omega \in \Omega^r_{k[[t]]/\mathbb{Z}}$, the difference $d \fint \omega - \omega$ is dt-free. In other words, there exists a dt-free form $\omega' \in \Omega^r_{k[[t]]/\mathbb{Z}}$ such that $\omega + \omega'$ is exact for the derivation $d : \Omega^{r-1}_{k[[t]]/\mathbb{Z}} \to \Omega^r_{k[[t]]/\mathbb{Z}}$.*

Proof. Express ω as in (2.1). Then we have $d \fint w = \sum_{i=0}^{\infty}(t^i\omega_i dt + \frac{t^{i+1}}{i+1} d\omega_i) = (\omega - \omega_{\mathbb{Z}}) + \sum_{i=0}^{\infty} \frac{t^{i+1}}{i+1} d\omega_i \in \Omega^r_{k[[t]]/\mathbb{Z}}$. Hence for a dt-free form $\omega' := -\omega_{\mathbb{Z}} + \sum_{i=0}^{\infty} \frac{t^{i+1}}{i+1} d\omega_i$, we have $\omega + \omega' = d \fint \omega$. □

Definition 2.4. Let $\omega \in \Omega^r_{k[[t]]/\mathbb{Z}}$. A dt-free form $\omega' \in \Omega^r_{k[[t]]/\mathbb{Z}}$ such that $\omega+\omega'$ is exact is called *a dt-free companion of* ω.

The choice of dt-free companions is controlled up to exact forms:

Lemma 2.5. *Let $\omega \in \Omega^r_{k[[t]]/\mathbb{Z}}$ be a given form. Let $\omega', \omega'' \in \Omega^r_{k[[t]]/\mathbb{Z}}$ be two dt-free companions of ω, i.e. both $\omega + \omega'$ and $\omega + \omega''$ are exact. Then $\omega' - \omega''$ is an exact dt-free form. In particular, a dt-free companion $\omega' \in \Omega^r_{k[[t]]/\mathbb{Z}}$ gives a well-defined class in the quotient $\Omega^r_{k[[t]]/\mathbb{Z}}/d\Omega^{r-1}_{k[[t]]/\mathbb{Z}}$.*

Proof. By the given assumptions, there exist forms $\eta', \eta'' \in \Omega^{r-1}_{k[[t]]/\mathbb{Z}}$ such that $d\eta' = \omega + \omega'$ and $d\eta'' = \omega + \omega''$. Then $d(\eta' - \eta'') = \omega' - \omega''$ so that ω' and ω'' differ just by an exact form. Since $\omega' - \omega'' \in F^r_{k[[t]]/\mathbb{Z}}$, it is again dt-free. □

Remark 2.6. Note that in Lemma 2.5, we do not claim that every member of the coset $\omega' + d\Omega^{r-1}_{k[[t]]/\mathbb{Z}}$ is a dt-free companion of ω.

2.2. Integrals of absolute Kähler forms. We now define the integrals:

Definition 2.7. Let $\omega \in \Omega^r_{k[[t]]/\mathbb{Z}}$. Let $\omega' \in \Omega^r_{k[[t]]/\mathbb{Z}}$ be a dt-free companion of ω as in Definition 2.4. Define $\int \omega$ to be an anti-derivative of $\omega + \omega'$, called *an integral* of ω.

Example 2.8. Let $\omega \in \Omega^r_{k[[t]]/\mathbb{Z}}$.

(1) The fake integral $\fint \omega$ of Definition 2.2 is an integral of ω.
(2) If ω is a dt-free form, then $0 \in \Omega^{r-1}_{k[[t]]/\mathbb{Z}}$ is an integral of ω.
(3) Let $\eta \in \Omega^{r-1}_{k[[t]]/\mathbb{Z}}$ be an integral of ω and let $\epsilon \in \Omega^{r-2}_{k[[t]]/\mathbb{Z}}$. Then $\eta + d\epsilon$ is also an integral of ω. In particular, $\int \omega$ is at best defined in $\Omega^{r-1}_{k[[t]]/\mathbb{Z}}/d\Omega^{r-2}_{k[[t]]/\mathbb{Z}}$.

The natural question is whether $\int \omega$ is well-defined. We investigate it for $r = 1$ first. Observe:

LEMMA 2.9. *The collection of $f \in k[[t]]$ such that df is dt-free is precisely just $k \subset k[[t]]$.*

PROOF. Suppose $f = \sum_{i=0}^{\infty} a_i t^i$ with $a_i \in k$ satisfies the property that df is dt-free. Since $df = da_0 + \sum_{i=1}^{\infty}(t^i da_i + ia_i t^{i-1} dt)$, for this expression to be dt-free, we must have $ia_i t^{i-1} dt = 0$ for all $i \geq 1$. This holds if all $a_i = 0$ for $i \geq 1$. Then $df = da_0$. On the other hand, for any $a_0 \in k$, if $f = a_0$, then $df = da_0$ is dt-free. This proves the lemma. $\square$

Here is our first theorem:

THEOREM 2.10. *Let $\omega \in \Omega^1_{k[[t]]/\mathbb{Z}}$. Then the coefficients of t^i for $i \geq 1$ of an integral $\int \omega \in k[[t]]$ are independent of the choice of the integral $\int \omega$. In particular, the coefficients of t^i for $i \geq 1$ of any integral $\int \omega$ are all well-defined.*

PROOF. For a given $\omega \in \Omega^1_{k[[t]]/\mathbb{Z}}$, choose any two dt-free forms $\omega', \omega'' \in \Omega^1_{k[[t]]/\mathbb{Z}}$ such that both $\omega + \omega'$ and $\omega + \omega''$ are exact, so that there are $\eta', \eta'' \in k[[t]]$ satisfying $d\eta' = \omega + \omega'$ and $d\eta'' = \omega + \omega''$. This gives $d(\eta' - \eta'') = \omega' - \omega''$, which is exact and dt-free. Hence by Lemma 2.9, $\eta' - \eta''$ is just a member of k. But both η' and η'' are integrals of ω, so this shows that they differ just by a constant in k. Hence for each $\omega \in \Omega^1_{k[[t]]/\mathbb{Z}}$, the coefficients of t^i for $i \geq 1$ of any choice of $\int \omega$ are well-defined. $\square$

We now generalize Theorem 2.10 for $r \geq 2$. We remarked in Example 2.8-(3) that $\int \omega$ is at best well-defined modulo exact forms. We study its converse in a precise form:

THEOREM 2.11. *Let $\omega \in \Omega^r_{k[[t]]/\mathbb{Z}}$ with $r \geq 2$. Then any integral $\int \omega \in \Omega^{r-1}_{k[[t]]/\mathbb{Z}}$ is well defined up to an addition of a form in $\Omega^{r-1}_{k/\mathbb{Z}}$ and an exact form in $d\Omega^{r-2}_{k[[t]]/\mathbb{Z}}$.*

PROOF. For a given $\omega \in \Omega^r_{k[[t]]/\mathbb{Z}}$, let $\omega', \omega'' \in \Omega^r_{k[[t]]/\mathbb{Z}}$ be any two dt-free companions such that both $\omega + \omega'$ and $\omega + \omega''$ are exact. Thus there exist $\eta', \eta'' \in \Omega^{r-1}_{k[[t]]/\mathbb{Z}}$ such that $d\eta' = \omega + \omega'$ and $d\eta'' = \omega + \omega''$. This gives $d(\eta' - \eta'') = \omega' - \omega''$, which is exact and dt-free.

We may express $\eta' - \eta''$ as follows:

$$\eta' - \eta'' = \sum_{i \geq 0} \alpha_i t^i + \sum_{j \geq 0} \beta_j t^j dt, \tag{2.2}$$

where $\alpha_i \in \Omega^{r-1}_{k/\mathbb{Z}}$ and $\beta_j \in \Omega^{r-2}_{k/\mathbb{Z}}$. The derivative of $\eta' - \eta''$ is

$$d(\eta' - \eta'') = d\alpha_0 + \sum_{i \geq 1}(t^i d\alpha_i + (-1)^{r-1} i t^{i-1} \alpha_i \wedge dt) + d\beta_0 \wedge dt + \sum_{j \geq 1} t^j d\beta_j \wedge dt.$$

The part involving dt is $\left(d\beta_0 + \sum_{i \geq 1}(-1)^{r-1} i t^{i-1} \alpha_i + \sum_{j \geq 1} t^j d\beta_j\right) \wedge dt$

$$= \left((-1)^{r-1}\alpha_1 + d\beta_0 + \sum_{i \geq 1} t^i ((-1)^{r-1}(i+1)\alpha_{i+1} + d\beta_i) \right) \wedge dt. \tag{2.3}$$

Since $d(\eta' - \eta'') = \omega' - \omega''$ is dt-free, this dt-part should be zero. Hence from (2.3), we must have the equalities

$$\alpha_1 = -(-1)^{r-1} d\beta_0, \quad \alpha_{i+1} = -\frac{1}{i+1}(-1)^{r-1} d\beta_i, \quad i \geq 1$$

so that (2.2) becomes

$$\eta' - \eta'' = \alpha_0 + \sum_{i\geq 1}(\frac{1}{i}(-1)^{r-2}(d\beta_{i-1})t^i) + \sum_{j\geq 0}\beta_j t^j dt = \alpha_0 + \sum_{i\geq 1}\frac{1}{i}(-1)^{r-2} d(\beta_{i-1}t^i)$$

$$= \alpha_0 + d\left(\sum_{i\geq 1}\frac{1}{i}(-1)^{r-2}\beta_{i-1}t^i\right).$$

Since $\alpha_0 \in \Omega^{r-1}_{k/\mathbb{Z}}$, it does not involve any t at all. Hence the theorem holds. $\square$

Recall from [**10**, Lemma 5.4.3] (*cf.* [**8**, Lemma 4.4]) the following:

LEMMA 2.12. *Let k be a field of characteristic 0. Let $m \geq 1$ be an integer. Let $k_m := k[[t]]/(t^m) = k[t]/(t^m)$. Then for $n \geq 2$, we have*

$$K_n^M(k_m, (t)) \simeq \Omega^{n-1}_{k_m,(t)/\mathbb{Z}}/d\Omega^{n-2}_{k_m,(t)/\mathbb{Z}} \simeq tk_m \otimes_k \Omega^{n-1}_{k/\mathbb{Z}} = \bigoplus_{i=1}^{m-1} t^i \Omega^{n-1}_{k/\mathbb{Z}}.$$

Here $K_n^M(k_m,(t))$ is the relative Milnor K-group of k_m, and $\Omega^i_{k_m,(t)/\mathbb{Z}} := \ker(\mathrm{ev}_{t=0} : \Omega^i_{k_m/\mathbb{Z}} \to \Omega^i_{k/\mathbb{Z}})$.

Taking the inverse limit over $m \geq 1$ of Lemma 2.12, we deduce:

COROLLARY 2.13. *Let k be a field of characteristic 0. Then we have*

$$K_n^M(k[[t]], (t)) \simeq \Omega^{n-1}_{k[[t]],(t)/\mathbb{Z}}/d\Omega^{n-2}_{k[[t]],(t)/\mathbb{Z}} \simeq tk[[t]] \otimes_k \Omega^{n-1}_{k/\mathbb{Z}} = \bigoplus_{i=1}^{\infty} t^i \Omega^{n-1}_{k/\mathbb{Z}}.$$

Here, $\Omega^i_{k[[t]],(t)/\mathbb{Z}} := \ker(\mathrm{ev}_{t=0} : \Omega^i_{k[[t]]/\mathbb{Z}} \to \Omega^i_{k/\mathbb{Z}})$. When $n = 1$ we interpret $\Omega^{-1}_{k[[t]],(t)/\mathbb{Z}} := 0$ and $\Omega^0_{k[[t]],(t)\mathbb{Z}} := tk[[t]]$.

This is the place where integrals $\int \omega$ belong to. We remark that the second isomorphism depends on the choice of the k-basis $\{t, t^2, \cdots\}$ of $tk[[t]]$. Hence combining Theorems 2.10, 2.11 and Corollary 2.13, we deduce:

THEOREM 2.14. *Let k be a field of characteristic 0. Let $\omega \in \Omega^r_{k[[t]]/\mathbb{Z}}$ with $r \geq 1$. Let $\Omega^{-1}_{(k[[t]],(t))/\mathbb{Z}} := 0$ for notational conveniences. Then the following hold:*

1. *An integral $\int \omega$ of ω is well-defined in $K_n^M(k[[t]],(t)) \simeq \Omega^{r-1}_{k[[t]],(t)/\mathbb{Z}}/d\Omega^{r-2}_{k[[t]],(t)/\mathbb{Z}}$.*
2. *Furthermore, it gives a unique sequence of absolute Kähler forms $c_i \in \Omega^{r-1}_{k/\mathbb{Z}}$ for $i \geq 1$ such that $\int \omega$ is determined by the expression*

$$\int \omega = \sum_{i=1}^{\infty} t^i c_i.$$

2.3. Chen's iterated integrals. As an immediate application of the notion of integrals, we can readily define the following notion of the s-fold iterated integrals $\int_{C^s}$. Here, C stands for K.-T. Chen who devised this notion in [**4**].

Not to complicate the situation, we just consider the case when $\omega_1, \cdots, \omega_s \in \Omega^1_{k[[t]]/\mathbb{Z}}$, because for such forms their integrals are well-defined in $tk[[t]]$ by Theorem 2.10, so that they are easy to deal with.

When $s = 1$, we define $\int_{C^1} \omega_1 = \int \omega_1 \in tk[[t]]$ as before by forcibly abandoning the constant term. For $s = 2$, we define $\int_{C^2} \omega_1\omega_2 := \int (\int_{C^1} \omega_1)\omega_2$. Indeed, we have the well-defined integral $\int_{C^1} \omega_1 \in tk[[t]]$ so that $(\int_{C^1} \omega_1)\omega_2 \in t\Omega^1_{k[[t]]/\mathbb{Z}}$. Then its integral actually belongs to $t^2k[[t]]$, and by Theorem 2.10, this is well-defined. In general, for $s \geq 2$, inductively define

$$\int_{C^s} \omega_1 \cdots \omega_s := \int (\int_{C^{s-1}} \omega_1 \cdots \omega_{s-1})\omega_s \in t^s k[[t]].$$

LEMMA 2.15. *The iterated integral $\int_{C^s}$ defines an s-multi-linear map*

$$\int_{C^s} : \Omega^1_{k[[t]]/\mathbb{Z}} \otimes_{\mathbb{Z}} \cdots \otimes_{\mathbb{Z}} \Omega^1_{k[[t]]/\mathbb{Z}} \to t^s k[[t]].$$

PROOF. This immediately follows by induction from that $\int$ is additive. □

2.4. Significance ranges and significant figures. We proved in Theorem 2.10 that if $\omega \in \Omega^1_{k[[t]]/\mathbb{Z}}$ is given, then for $\int \omega \in k[[t]]$, only the coefficients of t^i for $i \geq 1$ are well-defined, but not the "constant" term. Due to this, we can simply truncate the constant term to have a clean well-defined integral $\int \omega \in tk[[t]]$. We will thus always take this truncation whenever we take the integral.

We would like to say informally that in this case, $[0, \infty) = \{0, 1, 2, \cdots\}$ is the significance range of $\int \omega$, and the order of $\int \omega$ (in t) is ≥ 1. We generalize this concept:

DEFINITION 2.16. (1) Let $\omega \in k[[t]] = \Omega^0_{k[[t]]/\mathbb{Z}}$. Express $\omega = \sum_{i \geq 0} \omega_i t^i$.

If the coefficients ω_i of t^i of ω are known with certainty for all $i \geq 0$, then we say that the *significance range* of ω is $[0, \infty)$. If the coefficients ω_i of t^i are known with certainty only when $i \in [0, r]$ for an integer $r \geq 0$, then we say that the *significance range* of ω is $[0, r]$.

The coefficients ω_i in the significance range are called the *significant figures* of ω.

(2) Suppose $r \geq 1$. Let $\omega \in \Omega^r_{k[[t]]/\mathbb{Z}}$, and express it as $\omega = \omega_{\mathbb{Z}} + \sum_{i=0}^{\infty} t^i \omega_i dt$ as in (2.1).

Putting aside the dt-free part, if the coefficients ω_i of t^i are known with certainty for all $i \geq 0$, then we say that the *significance range* of ω is $[0, \infty)$. Putting aside the dt-free part, the coefficients ω_i of t^i are known with certainty only when $i \in [0, r]$ for an integer $r \geq 0$, then we say that the *significance range* of ω is $[0, r]$. Similarly, the coefficients ω_i in the significance range are called the *significant figures* of ω.

(3) Let $r \geq 0$. Let $a \geq 0$ be an integer. We say that $\omega \in \Omega^r_{k[[t]]/\mathbb{Z}}$ is *of order* $\geq a$ *(in t)*, if the coefficients of t^i of ω are known to be 0 for all $0 \leq i < a$.

EXAMPLE 2.17. Let $\omega \in \Omega^r_{k[[t]]/\mathbb{Z}}/t^m\Omega^r_{k[[t]]/\mathbb{Z}} = \Omega^r_{k_m/\mathbb{Z}}$. Choose any lifting $\tilde{\omega} \in \Omega^r_{k[[t]]/\mathbb{Z}}$. Then the coefficients of t^i for the expansion of $\tilde{\omega}$ as in (2.1) for $i \geq m$, could be random. Hence the significance range of $\tilde{\omega}$ is contained in $[0, m-1]$. We say that the significance range of ω is the significance range of $\tilde{\omega}$.

We want to understand the behaviors of various operations on the significance ranges.

LEMMA 2.18. *Let* $\omega_1, \omega_2 \in \Omega^r_{k[[t]]/\mathbb{Z}}$, *and let* $R_{\omega_1}, R_{\omega_2} \subset [0, \infty)$ *be their respective significance ranges. Then the significance range of* $\omega_1 + \omega_2$ *is* $R_{\omega_1} \cap R_{\omega_2}$.

PROOF. Obvious because $\Omega^r_{k[[t]]/\mathbb{Z}}$ is an abelian group. □

LEMMA 2.19. *Let* $r \geq 1$. *Let* $f \in k[[t]]$ *and* $\omega \in \Omega^r_{k[[t]]/\mathbb{Z}}$ *such that their significance ranges are the subsets* $[0, b]$ *and* $[0, m]$ *of* $[0, \infty)$, *respectively. For an integer* $a \geq 0$, *suppose that* f *is of order* $\geq a$.

Then the significance range of $f\omega \in \Omega^r_{k[[t]]/\mathbb{Z}}$ *is the subset* $[0, \min\{b, a+m\}]$, *and it is of order* $\geq a$.

PROOF. Write $f = \sum_{i=0}^{\infty} c_i t^i$ with $c_i = 0$ for $i < a$, and $\omega = \omega_{\mathbb{Z}} + \sum_{j=0}^{\infty} t^j \omega_j dt$. Here, c_i for $i \in [0, b]$ and ω_j for $j \in [0, m]$ are only given with certainty. Then

$$f\omega = f\omega_{\mathbb{Z}} + \sum_{i=0}^{\infty}\sum_{j=0}^{\infty} t^{i+j} c_i \omega_j dt = f\omega_{\mathbb{Z}} + \sum_{\ell=0}^{\infty} t^{\ell} (\sum_{i=0}^{\ell} c_i \omega_{\ell-i}) dt$$

and $f\omega_{\mathbb{Z}}$ is dt-free. Here, we want to know when the coefficient

$$\sum_{i=0}^{\ell} c_i \omega_{\ell-i} dt = (c_0 \omega_\ell + c_1 \omega_{\ell-1} + \cdots + c_\ell \omega_0) dt \tag{2.4}$$

of t^ℓ is given with certainty. First of all, from the expression (2.4), all of $c_0, \cdots, c_\ell$ must be known with certainty, and this holds only when $\ell \in [0, b]$.

On the other hand, since $c_0 = \cdots = c_{a-1} = 0$, we do not necessarily need to know $\omega_\ell, \omega_{\ell-1}, \cdots, \omega_{\ell-(a-1)}$ with certainty, while we do need to know $\omega_{\ell-a}, \cdots, \omega_0$ with certainty. Hence we must have $\ell - a \in [0, m]$, which implies that $\ell \leq a + m$.

Combining the above two conditions, we thus need to have $0 \leq \ell \leq \min\{b, a + m\}$. But this is also sufficient, so that the significance range of $f\omega$ is $[0, \min\{b, a + m\}]$.

Finally, if $\ell < a$, then $c_i = 0$ for $i \in [0, \ell]$ so that the coefficient of t^ℓ is 0. Hence $f\omega$ is of order $\geq a$. □

2.5. Integrals of absolute forms over k_m.

LEMMA 2.20. *Let* $\omega \in \Omega^1_{k[[t]]/\mathbb{Z}}$. *Suppose its significance range is* $[0, b]$ *and it is of order* $\geq a$ *for an integer* $a \geq 0$. *Take an integral* $\int \omega \in tk[[t]]$, *forcibly killing the constant term by Theorem* 2.10. *Then its significance range is* $[0, b+1]$ *and it is of order* $\geq a + 1$.

PROOF. Express $\omega = \omega_{\mathbb{Z}} + \sum_{i=0}^{\infty} c_i t^i dt$ for some $a_i \in k$, where $\omega_{\mathbb{Z}}$ is the dt-free part of ω as in (2.1). Here, $c_0, \cdots, c_b$ are the significant figures. To prove the assertion on the significance range, we may take the fake integral $⨍ \omega$ by Theorem 2.10, with the constant term killed. Indeed, we have $⨍ \omega = \sum_{i=0}^{\infty} \frac{c_i}{i+1} t^{i+1}$. Since $c_0, \cdots, c_b$ are the significant figures of ω, the coefficients of $t^1, \cdots, t^{b+1}$ of $⨍ \tilde{\omega}$ of the integral are known with certainly. Its constant term is automatically 0. So, the significance range is $[0, b+1]$. On the other hand, $c_i = 0$ for $i < a$. Hence $⨍ \omega$ is of order $\geq a + 1$. □

COROLLARY 2.21. *Let $\omega \in \Omega^1_{k_m/\mathbb{Z}}$. Choose a lifting $\tilde{\omega} \in \Omega^1_{k[[t]]/\mathbb{Z}}$ so that its significance range is $[0, \cdots, m-1]$. Take an integral $\int \tilde{\omega} \in tk[[t]]$. Then its significance range is $[0, m]$ and it is of order ≥ 1. In particular, its image in tk_{m+1} is well-defined.*

This corollary encourages:

DEFINITION 2.22. For $\omega \in \Omega^1_{k_m/\mathbb{Z}}$, define $\int \omega \in tk_{m+1}$ to be the image of $\int \tilde{\omega} \in tk[[t]] \to tk_{m+1}$ for any lifting $\tilde{\omega} \in \Omega^1_{k[[t]]/\mathbb{Z}}$ of ω.

We can repeat the above for iterated integrals:

LEMMA 2.23. *Let $s \geq 1$ be an integer, and let $\omega_1, \cdots, \omega_s \in \Omega^1_{k[[t]]/\mathbb{Z}}$ be 1-forms whose significance ranges are all $[0, m-1]$ for an integer $m \geq 2$. Then the iterated integral $\int_{C^s} \omega_1 \cdots \omega_s \in k[[t]]$ has the significance range $[0, m+s-1]$ and it is of order $\geq s$.*

PROOF. We prove it by induction on s.

When $s = 1$, we know by Lemma 2.20 that $\int \omega_1 \in tk[[t]]$ has the significance range $[0, m]$ and of order ≥ 1.

For $s \geq 2$, suppose $\int_{C^{s-1}} \omega_1 \cdots \omega_{s-1}$ has the significance range $[0, m+(s-1)-1]$ and it is of order $\geq (s-1)$. Then the significance range of $(\int_{C^{s-1}} \omega_1 \cdots \omega_{s-1})\omega_s$ is $[0, \min\{m+(s-1)-1, (s-1)+m-1\}] = [0, m+(s-1)-1]$, and it is of order $\geq (s-1)$ by Lemma 2.19. Hence by integrating, the significance range of $\int_{C^s} \omega_1 \cdots \omega_s = \int ((\int_{C^{s-1}} \omega_1 \cdots \omega_{s-1})\omega_s)$ is $[0, m+s-1]$, and it is of order $\geq s$ by Lemma 2.20 again. This proves the lemma by induction. □

COROLLARY 2.24. *Let $s \geq 1$ be an integer, and let $\omega_1, \cdots, \omega_s \in \Omega^1_{k_m/\mathbb{Z}}$. For $1 \leq i \leq s$, choose liftings $\tilde{\omega}_i \in \Omega^1_{k[[t]]/\mathbb{Z}}$ so that their significance ranges are $[0, m-1]$. Then the iterated integral $\int_{C^s} \tilde{\omega}_1 \cdots \tilde{\omega}_s \in k[[t]]$ has a well-defined image in $t^s k_{m+s}$.*

This allows us to give:

DEFINITION 2.25. Let $s \geq 1$ be an integer, and let $\omega_1, \cdots, \omega_s \in \Omega^1_{k_m/\mathbb{Z}}$. For each $1 \leq i \leq s$, we choose a lifting $\tilde{\omega}_i \in \Omega^1_{k[[t]]/\mathbb{Z}}$ of ω_i. Define the iterated integral $\int_{C^s} \omega_1 \cdots \omega_s \in t^s k_{m+s}$ to be the image of $\int_{C^s} \tilde{\omega}_1 \cdots \tilde{\omega}_s$.

LEMMA 2.26. *The iterated integral $\int_{C^s}$ defines a multi-linear operation*

$$\int_{C^s} : \Omega^1_{k_m/\mathbb{Z}} \otimes_{\mathbb{Z}} \cdots \otimes_{\mathbb{Z}} \Omega^1_{k_m/\mathbb{Z}} \to t^s k_{m+s}.$$

PROOF. It follows immediately by induction from additivity of integral $\int$. □

DEFINITION 2.27. For the iterated integral operator $\int_{C^s}$ of Lemma 2.26, we can consider its anti-symmetrization

$$\mathrm{Alt}_s \int_{C_s} (\omega_1 \otimes \cdots \otimes \omega_s) := \frac{1}{s!} \sum_{\sigma \in \Sigma_n} (-1)^\sigma \int_{C_s} (\omega_{\sigma(1)} \cdots \omega_{\sigma(s)}).$$

This gives a map $\Omega^s_{k_m/\mathbb{Z}} \to t^s k_{m+s}$.

3. The logarithm and the additive logarithm

We apply the above machine of integrals to define and study the logarithm. We use it to discuss the notion of additive logarithm for each modulus $m \geq 2$.

3.1. The logarithm.

DEFINITION 3.1. Recall that we have a homomorphism $\underline{\mathrm{dlog}} : k[[t]]^\times \to \Omega^1_{k[[t]]/\mathbb{Z}}$ that sends $f \in k[[t]]^\times$ to $f^{-1}df$. For $\underline{\mathrm{dlog}}(f) = f^{-1}df \in \Omega^1_{k[[t]]/\mathbb{Z}}$, its integral $\int \underline{\mathrm{dlog}}(f) \in k[[t]]$ as in Theorem 2.10 projected to $tk[[t]]$, is denoted by $\log(f)$ and called the logarithm.

By definition, we have $d(\log(f)) = \underline{\mathrm{dlog}}(f)$ up to adding a dt-free form in $\Omega^1_{k[[t]]/\mathbb{Z}}$.

Recall that the underlying abelian group of the ring of big Witt vectors $\mathbb{W}(k)$ of k is isomorphic to $(1+tk[[t]])^\times$ as a group. Using this, we had another logarithm function defined as follows, which [**13**] calls $\log^0$:

$$\log^0 : k[[t]]^\times \to \mathbb{W}(k) \to tk[[t]], \tag{3.1}$$

where the first map sends f to $f/f(0) \in \mathbb{W}(k)$, and the second map sends $1+g \in \mathbb{W}(k)$ to $g - \frac{1}{2}g^2 + \frac{1}{3}g^3 - \cdots$.

LEMMA 3.2. *For $f \in k[[t]]^\times$, let $f_0 := f/f(0) \in \mathbb{W}(k)$. We have the equalities $\log(f) = \log(f_0) = \log^0(f)$ in $tk[[t]]$. In other words, the* log *defined in Definition* 3.1 *has an explicit description of* (3.1).

PROOF. Let $c := f(0) \in k^\times$ so that $f = cf_0$. We have $df = f_0dc + cdf_0$ so that $f^{-1}df = f^{-1}f_0dc + cf^{-1}df_0 = c^{-1}f_0^{-1}f_0dc + cc^{-1}f_0^{-1}df_0 = c^{-1}dc + f_0^{-1}df_0$. Since $c^{-1}dc$ is a dt-free form of $\Omega^1_{k[[t]]/\mathbb{Z}}$, we find that $\log(f) = \log(f_0)$ by definition.

It remains to prove that $\log(f_0) = \log^0(f_0)$. To check this, it is enough to check that $d\log(f_0) = d\log^0(f_0)$. Here, LHS is $f_0^{-1}df_0$. To compute the RHS, express $f_0 = 1+g$ for some $g \in tk[[t]]$. Since $\log^0(f_0) = g - \frac{1}{2}g^2 + \frac{1}{3}g^3 - \cdots$, we have $d\log^0(f_0) = dg - gdg + g^2dg - \cdots = (1-g+g^2-g^3+\cdots)dg = \frac{dg}{1+g} = \frac{dg}{f_0} =^\dagger f_0^{-1}df_0$, where the last equality $\dagger$ holds because $g = -1+f_0$ so that $dg = d(-1)+df_0 = df_0$. Hence we checked that $d\log(f_0) = d\log^0(f_0)$, and this proves the lemma. □

LEMMA 3.3. $\log : k[[t]]^\times \to tk[[t]]$ *is a group homomorphism. Furthermore, it induces a group isomorphism* $\log : \mathbb{W}(k) \to tk[[t]]$.

PROOF. Note that each $f \in k[[t]]^\times$ can be uniquely written as $f = f_0f(0)$ for some $f_0 \in \mathbb{W}(k)$, and we have $\log(f) = \log(f_0)$ by Lemma 3.2. So, to prove the lemma it is enough to prove that the restriction $\log : \mathbb{W}(k) \to tk[[t]]$ is an isomorphism of groups.

We first check that this is a group homomorphism. For this, it is enough to show that for every $n \in \mathbb{N}$, the composition $\log : (1+tk[[t]])^\times \to (tk[[t]])^+ \twoheadrightarrow (tk[[t]]/(t^n))$ is a group homomorphism, because the log is obtained by taking $tk[[t]] = \varprojlim_n tk[[t]]/(t^n)$.

Let $1+f, 1+g \in (1+tk[[t]])^\times$ be two elements. When $n=1$, the group $(tk[[t]]/(t^n))$ is 0, so there is nothing to prove. When $n=2$, we have $\log(1+f)(1+g) = \log(1+f+g+fg) = (f+g+fg) - \frac{1}{2}(f+g+fg)^2 + \cdots \equiv f+g \mod (t^2) \equiv \log(1+f) + \log(1+g) \mod (t^2)$. So, this is a group homomorphism.

Let $n \geq 2$. Suppose $\log(1+f)(1+g) \equiv \log(1+f) + \log(1+g) \mod (t^i)$ for all $1 \leq i \leq n$. To extend it to $n+1$, it is enough to check that $(-1)^n\frac{1}{n}(f+g+fg)^n \equiv (-1)^n\frac{1}{n}f^n + (-1)^n\frac{1}{n}g^n \mod (t^{n+1})$. But, since $f, g \in tk[[t]]$ so that $fg \in t^2k[[t]]$. Hence, the only terms in the trinomial expansion of $(f+g+fg)^n$ that may survive modulo (t^{n+1}) are f^n and g^n. Hence the assertion holds for $n+1$ as well. Hence by induction, we proved the first claim.

To show that $\log : \mathbb{W}(k) \to tk[[t]]$ is injective, for $1+f \in \mathbb{W}(k) = (1+tk[[t]])^\times$, note that $\log(1+f) = f \cdot (1-g)$ for some $g \in tk[[t]]$. So, $\log(1+f) = 0$ if and only if either $f = 0$ or $1-g = 0$, because $k[[t]]$ is an integral domain. But, $g \in tk[[t]]$ so that $1-g$ cannot be 0, thus $\log(1+f) = 0$ if and only if $f = 0$. This proves that log is injective.

Finally, to show that it is surjective, we define the exponential function

$$\exp : tk[[t]]^+ \to (1+tk[[t]])^\times \tag{3.2}$$

given by $\exp(f) = \sum_{i=0}^{\infty} \frac{f^i}{i!}$. This converges in the target space, and it is of the form $1+h$ for some $h \in tk[[t]]$. We need to check that $\log(\exp(f)) = f$. It is enough to prove it when $f = t$; the general case follows by plugging in f into t. Firstly, this composition makes sense because $\exp(0) = 0$ and the formal derivative $\partial_t \exp(t) = \exp(t)$ is invertible. Secondly, the formal derivative $\partial_t \log(\exp(t)) = 1$ and $\log(\exp 0) = 0$ while the formal power series t also satisfies $\partial_t(t) = 1$ and $t_{t=0} = 0$. Hence by comparing coefficients, we have the identity $\log(\exp(t)) = t$. This proves the surjectivity. □

REMARK 3.4. Recall that $K_1^M(k[[t]]) = k[[t]]^\times$. Its mod (t) reduction homomorphism $K_1^M(k[[t]]) \to K_1^M(k)$ is precisely the map $k[[t]]^\times \to k^\times$ given by $f \mapsto f(0)$, and its kernel is $K_1^M(k[[t]], (t)) = (1+tk[[t]])^\times = \mathbb{W}(k)$. By Lemma 3.3, the log of Definition 3.1 gives an isomorphism of $\mathbb{W}(k)$ with $tk[[t]]$. Hence one can view the log map as a description of the projection homomorphism $K_1^M(k[[t]]) \to K_1^M(k[[t]], (t))$ from the multiplicative world to its additive world.

3.2. The logarithm with modulus and weights. The above log function gives its reduction mod t^m version immediately:

LEMMA 3.5. *Let $m \geq 2$ be an integer. Then the homomorphism $\log : k[[t]]^\times \to tk[[t]]$ naturally induces a homomorphism $\log : k_m^\times \to tk[[t]]/(t^m) = tk_m$.*

PROOF. We have the natural surjective homomorphism $k_m^\times \to k_m^\times / k^\times = \mathbb{W}(k)/\mathbb{W}_m(k)$, where $\mathbb{W}_m(k) = (1+t^m k[[t]])^\times$. Hence we need to show that $\mathbb{W}_m(k) = (1+t^m k[[t]])^\times$ is mapped to $t^m k[[t]]$ under log. We can use the description of log by $\log^0$ by Lemma 3.2. Indeed, for $f \in k[[t]]$ we have $\log(1+t^m f) = t^m f - \frac{1}{2}(t^m f)^2 + \frac{1}{3}(t^m f)^3 - \cdots \equiv 0 \mod (t^m)$. □

REMARK 3.6. The homomorphism $\log : k_m^\times \to tk_m$ of Lemma 3.5 means that log sends the members with the significance range $[0, m-1]$ to the significance range $[0, m-1]$ and of order ≥ 1, because the constant term of log is forcibly chosen to be 0 in our definition.

REMARK 3.7. As in Remark 3.4, we can view the $\log : k_m^\times \to tk_m$ of Lemma 3.5 as $\log : K_1^M(k_m) \to K_1^M(k_m, (t))$.

REMARK 3.8. One could have defined the "first Bloch group" $B_1(R)$ of a ring R to be the free abelian group $\mathbb{Z}[\{[a] \mid a \in R^\times\}]$ modulo the relations $[ab]-[a]-[b]$ for each $a, b \in R^\times$. One can check that $B_1(R) \to R^\times$ given by $[a] \mapsto a$ is actually an isomorphism of groups. (cf. [**3**, p.1330])

Now, for a given integer $m \geq 2$, and $1 \leq w \leq m-1$, we let $\text{li}_{1,m,w} : B_1(k_m) = k_m^\times \to k$ to be the coefficient of t^w of log, thus

$$\log = \sum_{w=1}^{m-1} \text{li}_{1,m,w} t^w.$$

In the light of [**12**], one could have called the functions $\text{li}_{1,m,w}$ by the name *additive logarithm.*

3.3. The higher logarithm. We can extend the description of log of §3.1 to the higher level. More precisely, recall that we saw in Remarks 3.4 and 3.7 that the projections $K_1^M(k[[t]]) \to K_1^M(k[[t]],(t))$ and $K_1^M(k_m) \to K_1^M(k_m,(t))$ can be identified with the logarithm. We extend it to higher relative Milnor K-groups of $k[[t]]$ and k_m. The idea is simple.

Let R be a k-algebra. The map $\underline{\text{dlog}} : R^\times \to \Omega^1_{R/\mathbb{Z}}$, $f \mapsto f^{-1}df$ induces

$$\underline{\text{dlog}} = \underline{\text{dlog}}^{\wedge n} : \bigwedge^n R^\times \to \Omega^n_{R/\mathbb{Z}}, \quad f_1 \wedge \cdots \wedge f_n \mapsto \underline{\text{dlog}} f_1 \wedge \cdots \wedge \underline{\text{dlog}} f_n.$$

LEMMA 3.9. *The map* $\underline{\text{dlog}} : \bigwedge^n R^\times \to \Omega^n_{R/\mathbb{Z}}$ *descends to a map from* $K_n^M(R)$.

PROOF. This is well-known and easy: it is enough to show that the images of the Steinberg symbols vanish, i.e. $\underline{\text{dlog}}^{\wedge n}(f \wedge (1-f) \wedge f_3 \wedge \cdots \wedge f_n) = 0$ for $f, f_3, \cdots, f_n \in R^\times$ such that $1 - f \in R^\times$. It is enough to prove the statement for $n = 2$. In this case we have $\underline{\text{dlog}}(f) \wedge \underline{\text{dlog}}(1-f) = f^{-1}df \wedge (1-f)^{-1}d(1-f) = -f^{-1}(1-f)^{-1}df \wedge df = 0$. This proves the lemma. □

Now, by Theorem 2.14, for $f_1, \cdots, f_n \in k[[t]]^\times$ (resp. $\in k_m^\times$), the n-form

$$\underline{\text{dlog}}\{f_1, \cdots, \cdots f_n\} \in \Omega^n_{k[[t]]/\mathbb{Z}}$$

(resp. $\in \Omega^n_{k_m/\mathbb{Z}}$) has the integral

$$\int \underline{\text{dlog}}\{f_1, \cdots, f_n\} \in \Omega^{n-1}_{(k[[t]],(t))/\mathbb{Z}} / d\Omega^{n-2}_{(k[[t]],(t))/\mathbb{Z}} \quad (\text{resp. } \in \Omega^{n-1}_{(k_m,(t))/\mathbb{Z}} / d\Omega^{n-2}_{(k_m,(t))/\mathbb{Z}}),$$

but we have $\Omega^{n-1}_{(k[[t]],(t))/\mathbb{Z}} / d\Omega^{n-2}_{(k[[t]],(t))/\mathbb{Z}} = K_n^M(k[[t]],(t))$ by Corollary 2.13 (resp., $\Omega^{n-1}_{(k_m,(t))/\mathbb{Z}} / d\Omega^{n-2}_{(k_m,(t))/\mathbb{Z}} = K_n^M(k_m,(t))$ by Lemma 2.12). Hence:

THEOREM 3.10. *We have the log homomorphism*

$$\log := \int \underline{\text{dlog}} : K_n^M(k[[t]]) \to K_n^M(k[[t]],(t)), \quad \{f_1, \cdots, f_n\} \mapsto \int \underline{\text{dlog}}\{f_1, \cdots, f_n\}.$$

$$(\textit{resp. } \log := \int \underline{\text{dlog}} : K_n^M(k_m) \to K_n^M(k_m,(t)), \quad \{f_1, \cdots, f_n\} \mapsto \int \underline{\text{dlog}}\{f_1, \cdots, f_n\}.)$$

3.4. The polylogarithms. We can generalize the log to the higher ones, namely to the polylogarithms using the iterated integrals of absolute Kähler forms we discussed. Recall that we have the homomorphism $\underline{\text{dlog}} : k[[t]]^\times \to \Omega^1_{k[[t]]/\mathbb{Z}}$. Consider its s-fold tensor product. Then we can apply the s-fold iterated integral to define:

$$\text{li}_s := \int_{C^s} \circ (\underline{\text{dlog}} \otimes \cdots \otimes \underline{\text{dlog}}) : k[[t]]^\times \otimes_{\mathbb{Z}} \cdots \otimes_{\mathbb{Z}} k[[t]]^\times \to$$

$$\Omega^1_{k[[t]]/\mathbb{Z}} \otimes_{\mathbb{Z}} \cdots \otimes_{\mathbb{Z}} \Omega^1_{k[[t]]/\mathbb{Z}} \xrightarrow{\int_{C^s}} t^s k[[t]].$$

Its significance range is $[0, \infty)$, and it is of order $\geq s$.

EXAMPLE 3.11. Let's have a closer look at the case when $s = 2$. We begin with

$$\underline{\mathrm{dlog}} \otimes \underline{\mathrm{dlog}} : k[[t]]^\times \otimes k[[t]]^\times \to \Omega^1_{k[[t]]/\mathbb{Z}} \otimes \Omega^1_{k[[t]]/\mathbb{Z}}.$$

As part of an iterated integral, by integrating the first argument only, we then obtain a homomorphism

$$\log(-)\underline{\mathrm{dlog}}(-) : k[[t]]^\times \otimes_{\mathbb{Z}} k[[t]]^\times \to t\Omega^1_{k[[t]]/\mathbb{Z}},$$
$$f_1 \otimes f_2 \mapsto (\int \underline{\mathrm{dlog}}(f_1))\underline{\mathrm{dlog}}(f_2) = \log(f_1)\underline{\mathrm{dlog}}(f_2),$$

If we integrate it once again, then we obtain li_2. On the other hand, if we want to deduce a homomorphism $k[[t]]^\times \wedge k[[t]]^\times$, one can take the anti-symmetrizer, and it then gives

$$\mathrm{li}_2^{(1)} : k[[t]]^\times \wedge k[[t]]^\times \to t\Omega^1_{k[[t]]/\mathbb{Z}},$$
$$\mathrm{li}_2^{(1)} : f_1 \wedge f_2 \mapsto \frac{1}{2}(\log(f_1)\underline{\mathrm{dlog}}(f_2) - \log(f_2)\underline{\mathrm{dlog}}(f_1)).$$

This is $\frac{1}{2}$ times the differential "form" considered in [**13**].

3.5. Polylogarithms with modulus. We want to understand the significance range of li_s. We consider it for $s = 2$ first.

Recall that $\underline{\mathrm{dlog}} : k[[t]]^\times \to \Omega^1_{k[[t]]/\mathbb{Z}}$ is given by $f \mapsto f^{-1}df$. Since we proved that $d\log(f) = \underline{\mathrm{dlog}}(f)$ and log induces $k_m^\times \to tk_m$, we know that $\underline{\mathrm{dlog}}$ also induces $k_m^\times \to \Omega^1_{k_m/\mathbb{Z}}$. So, the significance range $[0, m-1]$ is sent to the range $[0, m-1]$. In other words, for any $f \in k_m^\times$, the significance range of $\underline{\mathrm{dlog}}(f)$ is $[0, m-1]$.

Given $f_1, f_2 \in k_m^\times$, by Lemma 2.23, the dilogarithm

$$\mathrm{li}_2(f_1 \otimes f_2) = \int_{C^2} \underline{\mathrm{dlog}}(f_1)\underline{\mathrm{dlog}}(f_2)$$

has the significance range $[0, m+1]$ and of order ≥ 2, so it is well-defined in $t^2 k_{m+2}$.

More generally, given $f_1, \cdots, f_s \in k_m^\times$, by Lemma 2.23, the s-logarithm $\mathrm{li}_r(f_1 \otimes \cdots \otimes f_s) = \int_{C^s} \underline{\mathrm{dlog}}(f_1) \cdots \underline{\mathrm{dlog}}(f_s)$ has the significance range $[0, m+s-1]$ and of order $\geq s$, so it is well-defined in $t^s k_{m+s}$.

4. Comparison with the regulators for k_m of Park-Ünver

In [**10**], Park and Ünver studied a cycle-theoretic description of $K_n^M(k_m)$ using higher Chow cycles over Spec $(k[[t]])$. Important parts of the discussion were the "regulator functions" Υ_i for $1 \leq i \leq m-1$, that take values in $\Omega^{n-1}_{k/\mathbb{Z}}$. In §4, we give their (collective) descriptions in terms of the logarithm of this article.

By the logarithm of Theorem 3.10, via Lemma 2.12, we have

$$\log : K_n^M(k_m) \to K_n^M(k_m, (t)) = \bigoplus_{i=1}^{m-1} t^i \Omega^{m-1}_{k/\mathbb{Z}} \simeq \bigoplus_{i=1}^{m-1} \Omega^{m-1}_{k/\mathbb{Z}}.$$

On the other hand, the regulator $\Upsilon_i(Z)$ for an integral cycle Z can be described explicitly when Z is the graph cycle $\Gamma_{(a_1,\cdots,a_n)}$ for $a_1, \cdots, a_n \in k_m^\times$ as follows: choose liftings $\tilde{a}_1, \cdots, \tilde{a}_n \in k[[t]]^\times$, then we had

$$\Upsilon_i(\Gamma_{(a_1,\cdots,a_n)}) = \mathrm{res}_{t=0} \frac{1}{t^i} \frac{d\tilde{a}_1}{\tilde{a}_1} \wedge \cdots \wedge \frac{d\tilde{a}_n}{\tilde{a}_n}.$$

Here, we can actually give a more systematic and collective description of all of Υ_i from a different angle. The main point is the following:

LEMMA 4.1. *Let k be a field of characteristic* 0.

(1) *Let $n = 0$ and let $f \in \Omega^0_{k[[t]]/\mathbb{Z}} = k[[t]]$. Express it as $f = \sum_{i=0}^{\infty} f_i t^i$, where $f_i \in k$ for each $0 \leq i$. Then we have*

$$f_i = \frac{1}{i}\mathrm{res}_{t=0}\frac{1}{t^i}df.$$

(2) *Let $n \geq 1$ be an integer. Let $\omega \in \Omega^n_{k[[t]]/\mathbb{Z}}$. Express it as $\omega = \omega_{\mathbb{Z}} + \sum_{i=0}^{\infty}\omega_i t^i dt =: \omega_{\mathbb{Z}} + \omega_t dt$, where $\omega_{\mathbb{Z}}$ is the dt-free part, and $\omega_i \in \Omega^{n-1}_{k/\mathbb{Z}}$ for each $0 \leq i$. Then we have*

$$\omega_i = \frac{1}{i}\mathrm{res}_{t=0}\frac{1}{t^i}d\omega_t.$$

PROOF. (1) Here $df = \sum_{j=1}^{\infty} j f_j t^{j-1} dt$ so that

$$\frac{1}{i}\mathrm{res}_{t=0}\frac{1}{t^i}df = \sum_{j=1}^{\infty}\frac{jf_j}{i}\mathrm{res}_{t=0}(t^{j-i-1}) =^{\dagger} \frac{if_i}{i} = f_i,$$

where † holds because $\mathrm{res}_{t=0}t^{j-i-1}dt$ is 0 if $j \neq i$, and 1 if $j = i$. This proves (1).

(2) is similar; just a bit more involved. Since $\omega_t = \sum_{i=0}^{\infty}\omega_i t^i$, we have $d\omega_t = \sum_{j=1}^{\infty}(t^j d\omega_j + jt^{j-1}\omega_j dt)$. Hence

$$\frac{1}{i}\mathrm{res}_{t=0}\frac{1}{t^i}d\omega_t = \sum_{j=1}^{\infty}\mathrm{res}_{t=0}\left(\frac{t^{j-i}d\omega_j}{i} + \frac{jt^{j-i-1}\omega_j}{i}dt\right) =^{\dagger} \sum_{j=1}^{\infty}\frac{jt^{j-i-1}\omega_j}{i}dt =^{\ddagger} \omega_j,$$

where † holds because the residue at $t = 0$ of a dt-free form is 0, while ‡ holds because $\mathrm{res}_{t=0}t^{j-i-1}dt$ is 0 if $j \neq i$ and 1 if $j = i$. This proves (2). □

The above lemma now gives the following description of Υ_i of [**10**]:

COROLLARY 4.2. *For $\{a_1, \cdots, a_n\} \in K_n^M(k_m)$, and $1 \leq i \leq m-1$, the coefficient of t^i in $\Omega^{m-1}_{k/\mathbb{Z}}$ of $\log\{a_1, \cdots, a_n\}$ is equal to $\frac{1}{i}\mathrm{res}_{t=0}\frac{1}{t^i}\underline{\mathrm{dlog}}\{\tilde{a}_1, \cdots, \tilde{a}_n\}$ for any choices $\tilde{a}_1, \cdots, \tilde{a}_n \in k[[t]]^{\times}$ of liftings of $a_i \in k_m^{\times}$. In particular, $\log\{a_1, \cdots, a_n\} = \sum_{i=1}^{m-1}\frac{1}{i}\Upsilon_i(\Gamma_{(a_1,\cdots,a_n)})$, where $\Gamma_{(a_1,\cdots,a_n)}$ is the graph cycle associated to the sequence $a_1, \cdots, a_n \in k_m^{\times}$.*

5. Additive dilogarithm of Ünver as an iterated integral

Recall that the Bloch group $B_2(R)$ of a ring R is defined to be the free abelian group $\mathbb{Z}[\{[a] \mid a \in R^{\times}$ such that $1 - a \in R^{\times}\}]$ on the set of units of R modulo the 5-term relations of Abel, namely $[a] - [b] + \left[\frac{b}{a}\right] - \left[\frac{1-a^{-1}}{1-b^{-1}}\right] + \left[\frac{1-a}{1-b}\right]$ for $a, b \in R^{\times}$ such that $1-a, 1-b, a-b \in R^{\times}$ as well. It has a natural map $\delta_R : B_2(R) \to R^{\times} \wedge_{\mathbb{Z}} R^{\times}$ given by $[a] \mapsto a \wedge (1-a)$ so that its cokernel is $K_2^M(R)$.

In Ünver's paper [**12**] on additive dilogarithms, the main question is, for a fixed $m \geq 2$, whether there is a map $\mathrm{li}_{2,m,w} : k_n^{\times} \wedge k_n^{\times} \to k$ of weight $m+1 \leq w \leq 2m-1$ of weight w for a sufficiently large $n \gg m$ such that the composite

$$B_2(k_n) \xrightarrow{\delta} k_n^{\times} \wedge k_n^{\times} \xrightarrow{\mathrm{li}_{2,m,w}} k$$

descends to $B_2(k_m)$. In [**12**], it is answered that there exists such a function, uniquely decided by the choice of m, w by lifting up to $n = 2m-1$.

These give a part of the power series $\sum_{w=m+1}^{2m-1} \mathrm{li}_{2,m,w}(f)t^w$, where $f \in k_n^\times$ is a generator of $B_2(k_n)$.

In general, the additive dilogarithms with modulus $m \geq 2$ of [**12**] are not expressed as iterated integrals. But, when $m = 2$, one has the following:

LEMMA 5.1. *Let $m = 2$, so that $w = 3$ is the only possible weight. Then for $f \in k[[t]]^\times$ such that $1 - f \in k[[t]]^\times$, we have the equality*

$$\mathrm{Li}_{2,2,3}(f) = \mathrm{res}_{t=0} \frac{1}{t^3} \left(\log^0(f) d\log(1-f) - \log^0(1-f) d\log(f) \right), \tag{5.1}$$

where $\mathrm{Li}_{2,2,3}(f) = \mathrm{li}_{2,(3,2),3}(f \wedge 1 - f)$ *is the dilogarithm of* [**12**, Proposition 2.2.3], *and the above value depends only on* $[f]_2$.

PROOF. We can uniquely express $f := s + s(1-s)a_1 t + s(1-s)a_2 t^2 + \cdots$ for some $s \in k \setminus \{0, 1\}$ and $a_i \in k$. Since the difference between $\log^0(f) d\log(1-f) - \log^0(1-f) d\log(f)$ and $\log^0(f) d\log^0(1-f) - \log^0(1-f) d\log^0(f)$ is dt-free in the sense of Definition 2.1, we may use the latter one to compute the residue of the right hand side of (5.1).

The formula in [**12**, Proposition 2.2.3] might look scary, but for $m = 2$ and $w = 3$ case, (following the notations there) only $(\alpha, \beta) = (1, 2)$ appears. In this case, $(s-1)^\alpha s^\beta - s^\alpha (s-1)^\beta = (s-1)^1 s^2 - s^1 (s-1)^2 = s^3 - s^2 - s^3 + 2s^2 - s = s^2 - s = -s(1-s)$. On the other hand, $C(1) = \{1\}, C(2) = \{1\}^{\times 2}$, and for $(a, b) \in C(1) \times C(2)$, we have $a = 1, b = (1, 1)$. Hence the value $\mathfrak{p}(f; (a, b)) = a_1 \cdot (a_1)^2$. Here $\mathfrak{w}(a) = 1$. Hence $\mathrm{Li}_{2,2,e}([f]) = -\frac{s(1-s)}{2} a_1^3$.

To compute the right hand side of (5.1), we note $f = s(1 + (1-s)z)$, $(1-f) = (1-s)(1-sz)$ where $z = a_1 t + a_2 t^2 + \cdots$. Hence

$$\log^0(f) = \log \frac{f}{s} = \log(1 + (1-s)z) = (1-s)z - \frac{1}{2}(1-s)^2 z^2 + \frac{1}{3}(1-s)^3 z^3 - \cdots,$$

$$\log^0(1-f) = \log \frac{1-f}{1-s} = \log(1 - sz) = -sz - \frac{1}{2}s^2 z^2 - \frac{1}{3}s^3 z^3 - \cdots.$$

Here, we keep in mind that we eventually want to compute the coefficient of $t^2 dt$, which is precisely the term that has the residue. Since $t | z$, for that purpose, we do not need terms with z^i, $i \geq 3$ in the above $\log^0(f)$ and $\log^0(1-f)$. Now, taking the derivative with respect to t, we have

$$d\log^0(f) = ((1-s)z' - (1-s)^2 zz' + \cdots)dt, d\log^0(1-f) = (-sz' - s^2 zz' - \cdots)dt.$$

(Here we ignore terms like dc for $c \in k$, because such absolute Kähler forms do not eventually produce the residues.) Hence we have $\log^0(f) d\log^0(1-f) - \log^0(1-f) d\log^0 f$

$$= (-s(1-s)zz' - s^2(1-s)z^2 z' + \frac{1}{2}s(1-s)^2 z^2 z' + \cdots)dt$$

$$-(-s(1-s)zz' + s(1-s)^2 z^2 z' - \frac{1}{2}s^2(1-s)z^2 z' + \cdots)dt.$$

Taking the coefficient of $t^2 dt$, we then precisely have $\frac{s(1-s)}{2} a_1^3 - s(1-s)a_1^3 = -\frac{s(1-s)}{2} a_1^3$. But this is what we computed on the left hand side. In particular, we observe that the value does not depend on a_i for $i \geq 2$. □

In particular, combined with Lemma 4.1, this Lemma 5.1 shows that $\mathrm{Li}_{2,2,3}$ is expressible as an iterated integral $\mathrm{li}_2^{(1)}$ of §3.4. It still remains to see in the future whether one can express $\mathrm{Li}_{2,m,w}$ in terms of certain combinations of iterated integrals with modulus.

6. Multivariable integral calculus

In §6, we study the multivariable analogue of what had been studied in the previous sections. Still k is a field of characteristic 0. In one variable case $k[[t]]$, an essence was that every form $\Omega^n_{k[[t]]/\mathbb{Z}}$ with $n > 0$ can be translated by a dt-free form into an exact form so that it has an anti-derivative. Furthermore, an anti-derivative is well-defined in $\Omega^{n-1}_{k[[t]],(t)/\mathbb{Z}}/d\Omega^{n-2}_{k[[t]],(t)/\mathbb{Z}} \simeq K^M_n(k[[t]],(t))$.

If we replace $k[[t]]$ by a formal power series ring $k[[\underline{t}]]$ with more variables $\underline{t} := \{t_1, \cdots, t_r\}$, however this kind of exactness modulo translation no longer holds. To see what necessary conditions are probably needed, consider first the two variable case of $r = 2$, and take a 1-form in $\Omega^1_{k[[t_1,t_2]]/k}$ of the form $\omega = Mdt_1 + Ndt_2$ over k. If this is exact, then it is necessarily closed. But this means that we must have $d\omega = M_{t_2}dt_2 \wedge dt_1 + N_{t_1}dt_1 \wedge dt_2 = (-M_{t_2} + N_{t_1})dt_1 \wedge dt_2 = 0$. This requires that $M_{t_2} = N_{t_1}$, similar to an elementary identity that we know well from freshmen calculus. So, we would need at least an absolute base $\mathbb{Z}$ version of this sort of equations. When the degree of the forms and the number of variables increase, we would have more complexities.

In §6, we propose a generalization of the notion of dt-free forms, and we identify a necessary and sufficient condition for exactness of an absolute Kähler form up to translation for some low degrees. Unfortunately, it was difficult to generalize it to all higher degrees in general at this moment, but we manage to get a sufficient condition that applies to a special class of higher degree forms, including the $d\log$ forms. This would be useful in extending the notion of log to the multivariable situation.

6.1. The dt-free forms. Let $r, n \geq 1$ be integers. Let $\underline{t} := \{t_1, \cdots, t_r\}$ be the set of r indeterminates.

DEFINITION 6.1. We have the following "geometric filtration"

$$F_0\Omega^n_{k[[\underline{t}]]/\mathbb{Z}} \subseteq F_1\Omega^n_{k[[\underline{t}]]/\mathbb{Z}} \subseteq \cdots \subseteq F_{\min\{r,n\}}\Omega^n_{k[[\underline{t}]]/\mathbb{Z}} = \Omega^n_{k[[\underline{t}]]/\mathbb{Z}},$$

where for each integer $0 \leq p \leq \min\{r, n\}$, F_p is the $k[[\underline{t}]]$-submodule generated by $dt_{i_1} \wedge \cdots \wedge dt_{i_{p'}}$ for any increasing sequence $1 \leq i_1 < \cdots < i_{p'} \leq r$ for all $0 \leq p' \leq p$.

A form in F_0 is called a *dt-free form*, while a form in $F_{\min\{r,n\}-1}$ is called *unsaturated.*

REMARK 6.2. For a given choice of parameters $\underline{t}$, we can let $\Omega^{p,q}_{k[[\underline{t}]]/\mathbb{Z}}$ with $q = n - p$ be the $k[[\underline{t}]]$-submodule of $\Omega^n_{k[[\underline{t}]]/\mathbb{Z}}$ generated by $dt_{i_1} \wedge \cdots \wedge dt_{t_p}$ for any increasing sequence $1 \leq i_1 < \cdots < i_p \leq r$ of integers of length p. A form in $\Omega^{p,q}_{k[[\underline{t}]]/\mathbb{Z}}$ is said to be of type (p, q). We have a non-canonical k-vector space decomposition

$$\Omega^n_{k[[\underline{t}]]/\mathbb{Z}} = \bigoplus_{0 \leq p \leq \min\{r,n\}} \Omega^{p,n-p}_{k[[\underline{t}]]/\mathbb{Z}}.$$

For a term of type (p, q), we may regard p as the "geometric degree", and q as the "arithmetic degree" of the form. We note that $d : \Omega^n_{k[[\underline{t}]]/\mathbb{Z}} \to \Omega^{n+1}_{k[[\underline{t}]]/\mathbb{Z}}$ maps $\Omega^{p,q}$ (with $q = n-p$) into the sum $\Omega^{p+1,q} \oplus \Omega^{p,q+1}$. One checks we can write $d = \partial^g + \partial^a$, and that $\partial^g \circ \partial^g = 0$ and $\partial^a \circ \partial^a = 0$. There is some similarity to the classical Hodge theory here. However, this plays no role for the rest of the paper.

EXAMPLE 6.3. When $r = 1$, we have $\min\{r, n\} = 1$ so that every term of a form $\omega \in \Omega^n_{k[[t]]/\mathbb{Z}}$ is always either a dt-free (type $(0,1)$) or a top dt-degree form (type $(1,0)$).

EXAMPLE 6.4. When $r = 2$, a 2-form ω in $\Omega^2_{k[[t_1,t_2]]/\mathbb{Z}}$ is in general written as

$$\omega = \omega_{\mathbb{Z}} + \omega_1 dt_1 + \omega_2 dt_2 + \omega_{12} dt_1 \wedge dt_2,$$

where $\omega_{\mathbb{Z}} \in k[[t_1,t_2]] \otimes_k \Omega^2_{k/\mathbb{Z}}$, $\omega_1, \omega_2 \in k[[t_1,t_2]] \otimes_k \Omega^1_{k/\mathbb{Z}}$ and $\omega_{12} \in k[[t_1,t_2]]$. Here, $\omega_{\mathbb{Z}}$ is a dt-free part, while $\omega_{12} dt_1 \wedge dt_2$ is a top dt-degree part.

6.2. Pseudo-closedness, pseudo-exactness and pseudo-totality.

6.2.1. *Pseudo-exactness and integrals.*

DEFINITION 6.5. Let $r, n \geq 1$ be integers. Let $\underline{t} := \{t_1, \cdots, t_r\}$ for r variables. Let $\omega \in \Omega^n_{k[[\underline{t}]]/\mathbb{Z}}$. We say ω is *pseudo-exact* if there exists an unsaturated form ω_0 such that $\omega - \omega_0$ is exact. If this happens we say that ω_0 is an *unsaturated companion* of ω. In this case, we have $\omega - \omega_0 = d\eta$ for some $\eta \in \Omega^{n-1}_{k[[\underline{t}]]/\mathbb{Z}}$. Define an integral $\int \omega$ of ω to be η.

The following generalizes Lemma 2.5:

LEMMA 6.6. *Let $\omega \in \Omega^n_{k[[\underline{t}]]/\mathbb{Z}}$ be a pseudo-exact form. Suppose for two unsaturated forms ω_0, ω_0', we have forms $\eta, \eta' \in \Omega^{n-1}_{k[[\underline{t}]]/\mathbb{Z}}$ such that $\omega - \omega_0 = d\eta$ and $\omega - \omega_0' = d\eta'$. Then $\omega_0 - \omega_0'$ is an exact unsaturated form. In particular, an unsaturated companion $\omega_0 \in \Omega^{n-1}_{k[[\underline{t}]]/\mathbb{Z}}$ gives a well-defined class in the quotient $\Omega^n_{k[[\underline{t}]]/\mathbb{Z}}/d\Omega^{n-1}_{k[[\underline{t}]]/\mathbb{Z}}$.*

PROOF. By the given assumptions, we have $\omega_0' - \omega_0 = d(\eta - \eta')$. Hence $\omega_0' - \omega_0$, which is obviously unsaturated, is also exact. □

The following generalizes Lemma 2.9:

LEMMA 6.7. *Let $f \in k[[\underline{t}]]$. Then for $d = d_{\mathbb{Z}}$, the collection of f such that df is dt-free is precisely just $k \subset k[[\underline{t}]]$.*

PROOF. Suppose $f = a_0 + \sum_I a_I t^I$ for some $a_0, a_I \in k$, where I is a multi-index of length r consisting of numbers in $\mathbb{N} \cup \{0\}$, such that $I \neq (0, \cdots, 0)$. Then $df = da_0 + \sum_I (t^I da_I + \text{extra})$, where extra is a sum whose terms are of the form $ca_I dt_i$ for some positive integer c and $1 \leq i \leq r$. Hence for this 1-form to be dt-free, we must have $a_I = 0$ for all multi-index I such that $I \neq (0, \cdots, 0)$. Hence $df = da_0$. On the other hand, for any $a_0 \in k$, and $f = a_0$, we have $df = da_0$ is dt-free. This proves the lemma. □

Here we have a multivariable generalization of Theorem 2.10:

THEOREM 6.8. *Let $\omega \in \Omega^1_{k[[\underline{t}]]/\mathbb{Z}}$. Suppose it is pseudo-exact. Then the coefficients of t^I for all multi-indices $I \neq (0, \cdots, 0)$ of $\int \omega \in k[[\underline{t}]]$ are independent of the choice of the integral $\int \omega$.*

PROOF. Suppose for two unsaturated (dt-free) forms ω_0, ω_0', we have $\eta, \eta' \in k[[\underline{t}]]$ such that $\omega - \omega_0 = d\eta$ and $\omega - \omega_0' = d\eta'$. Hence both η and η' are integrals of ω. Here, $d(\eta - \eta') = \omega_0 - \omega_0'$ is unsaturated (dt-free). Hence by Lemma 6.7, we have $\eta - \eta' \in k$. In particular, the coefficients of t^I is well-defined in k for every nonzero multi-index I. □

Of course, Theorem 6.8 is a bit limited because we do not know when a given 1-form ω is pseudo-exact. We will answer it soon. (See Theorem 6.15.) Before we study this issue, we remark that we would like to have a generalization of Theorem 6.8 to $n \geq 2$, but unfortunately due to its complexities, we were not yet able to prove it. We leave it as a guess:

CONJECTURE 6.9. *Let $\omega \in \Omega^n_{k[[\underline{t}]]/\mathbb{Z}}$ with $n \geq 2$. Suppose it is pseudo-exact. Then its integral $\int \omega \in \Omega^{n-1}_{k[[\underline{t}]]/\mathbb{Z}}$ is well-defined in the relative Milnor K-group $K_n^M(k[[\underline{t}]], (\underline{t}))$.*

6.2.2. *Pseudo-closedness.* We saw that if a form is pseudo-exact, then we can define its integrals. For $r = 1$, we already saw that all forms are pseudo-exact in §2. However, for $r \geq 2$ this is not true. We need a necessary condition, which we study now:

DEFINITION 6.10. Let $r, n \geq 1$ be integers. Let $\underline{t} := \{t_1, \cdots, t_r\}$ for r variables. Let $\omega \in \Omega^n_{k[[\underline{t}]]/\mathbb{Z}}$. We say ω is *pseudo-closed* if there exists an unsaturated form ω_0 such that $\omega - \omega_0$ is closed.

LEMMA 6.11. *Let $r, n \geq 1$ be integers. Let $\omega \in \Omega^n_{k[[\underline{t}]]/\mathbb{Z}}$. If it is pseudo-exact, then it is pseudo-closed.*

PROOF. Since ω is pseudo-exact, there exists an unsaturated form $\omega_0 \in \Omega^n_{k[[\underline{t}]]/\mathbb{Z}}$ and a form $\eta \in \Omega^{n-1}_{k[[\underline{t}]]/\mathbb{Z}}$ such that $\omega - \omega_0 = d\eta$. In particular $d(\omega - \omega_0) = d \circ d(\eta) = 0$, as desired. □

Is a pseudo-closed form always a pseudo-exact form? This question is important for us. For $r = 1$, we proved that every absolute Kähler form is pseudo-exact, in particular, every form is pseudo-closed as well.

6.2.3. *Pseudo-totality for $n = 1$.* For $r \geq 2$, however, not all forms are pseudo-exact, but at least whether the equivalence of pseudo-closedness and pseudo-exactness holds is an interesting question. We answer it positively for all $r \geq 1$ and $n = 1$. To do so, we give the following intermediate definition:

DEFINITION 6.12. Let $n = 1$ and $r \geq 1$. Let $\omega \in \Omega^1_{k[[\underline{t}]]/\mathbb{Z}}$. We say that ω is *pseudo-total* (derivative) if the (top) dt-degree part is of the form $\sum_{i=1}^r f_i dt_i$ for some $f_i \in k[[\underline{t}]]$ such that for each pair $1 \leq i, j \leq r$, we have

$$\frac{\partial f_i}{\partial t_j} = \frac{\partial f_j}{\partial t_i}.$$

DEFINITION 6.13. For any $\omega \in \Omega^1_{k[[\underline{t}]]/\mathbb{Z}}$, express it as $\omega = \omega_{\mathbb{Z}} + \sum_{i=1}^r f_i dt_i$ for some $f_i \in k[[\underline{t}]]$, and a dt-free $\omega_{\mathbb{Z}}$. Define its fake integral $⨍ \omega$ to be

$$⨍ \omega := \sum_{i=1}^r \int_{t_i} f_i dt_i,$$

where $\int_{t_i}(-)dt_i$ means we take the formal integration with respect to the variable t_i treating all other variables as constants.

LEMMA 6.14. *Let $\omega \in \Omega^1_{k[[\underline{t}]]/\mathbb{Z}}$. If ω is pseudo-total, then it is pseudo-exact.*

PROOF. We prove it by induction on $r \geq 1$. When $r = 1$, we already know the result by Theorem 2.10. Now let $r \geq 2$, and suppose that the lemma holds for $1, \cdots, r-1$.

Since ω is pseudo-total, for the dt-free part $\omega_{\mathbb{Z}}$, we have $\omega - \omega_{\mathbb{Z}} = \sum_i f_i dt_i$ with $\frac{\partial f_i}{\partial t_j} = \frac{\partial f_j}{\partial t_i}$ for each pair (i,j). Let $\int_{t_i} f_i dt_i =: G_i$.

Take the fake integral $⨍(\omega - \omega_{\mathbb{Z}}) = \sum_i \int_{t_i} f_i dt_i = \sum_i G_i$. Its total derivative is

$$d ⨍(\omega - \omega_{\mathbb{Z}}) = \sum_i dG_i = \sum_i \sum_j d_{t_j} G_i = \sum_i \left(f_i dt_i + \sum_{j \neq i} d_{t_j} G_i \right). \tag{6.1}$$

Here, for $j \neq i$, we have

$$\begin{aligned} d_{t_j} G_i &= \left(\frac{\partial}{\partial t_j} \int_{t_i} f_i dt_i \right) dt_j =^{\dagger} \left(\int_{t_i} \frac{\partial f_i}{\partial t_j} dt_i \right) dt_j =^{\ddagger} \left(\int_{t_i} \frac{\partial f_j}{d t_i} dt_i \right) dt_j \\ &= (f_j - f_j|_{t_i=0})\, dt_j, \end{aligned}$$

where the interchanging of $\int_{t_I}$ and $\frac{\partial}{\partial t_J}$ in $\dagger$ holds because the integral and the derivation are both formal ones and $t_i \neq t_j$, while the equality $\ddagger$ is the given pseudo-totality condition. Hence for a fixed i,

$$f_i dt_i + \sum_{j \neq i} d_{t_j} G_i = \sum_u f_u dt_u - \sum_{j \neq i} f_j|_{t_i=0} dt_j = (\omega - \omega_{\mathbb{Z}}) - \sum_{j \neq i} f_j|_{t_i=0} dt_j.$$

Putting this into (6.1), dividing by r, and rearranging the terms, we obtain

$$\omega - \omega_{\mathbb{Z}} = d\left(\frac{1}{r} ⨍(\omega - \omega_{\mathbb{Z}}) \right) + \frac{1}{r} \sum_{i \neq j} f_j|_{t_i=0} dt_j. \tag{6.2}$$

We claim that $A := \sum_{i \neq j} f_j|_{t_i=0} dt_j$ is exact. To illustrate the proof, we try $r = 2, 3$ case first, and the general case afterward.

When $r = 2$, we have $A = f_1(t_1, 0)dt_1 + f_2(0, t_2)dt_2$, and each term is purely a 1-form with a single t_i, so that it is exact by the case of 1-forms by Theorem 2.10. Hence (6.2) is exact for $r = 2$.

For $r = 3$, we have

$$A = (f_2|_{t_1=0} dt_2 + f_3|_{t_1=0} dt_3) + (f_1|_{t_2=0} dt_1 + f_3|_{t_2=0} dt_3) + (f_1|_{t_3=0} dt_1 + f_2|_{t_3=0} dt_2).$$

For simplicity, let $g_2 := f_2|_{t_1=0}, g_3 := f_3|_{t_1=0}$. Here $g_2, g_3 \in k[[t_2, t_3]]$ with $\frac{\partial g_2}{\partial t_3} = \frac{\partial g_3}{\partial t_2}$. Hence by the induction hypothesis, $f_2|_{t_1=0} dt_2 + f_3|_{t_1=0} dt_3$ is exact. Similarly, the two remaining terms of A are also exact.

In general, we write $A = \sum_{i=1}^r \sum_{j \neq i} f_j|_{t_i=0} dt_j$. For a fixed i, the sum $A_i := \sum_{j \neq i} f_j|_{t_i=0} dt_j$ is a 1-form with $\frac{\partial g_j}{\partial t_{j'}} = \frac{\partial g_{j'}}{\partial t_j}$ for each pair $j \neq j'$, distinct from i. Hence by the induction hypothesis, A_i is exact. Hence A is exact as well.

Hence we proved that for all $r \geq 1$, $\omega - \omega_{\mathbb{Z}}$ is an exact form. This proves that ω is a pseudo-exact 1-form. □

Now we have:

THEOREM 6.15. *Let* $\omega \in \Omega^1_{k[[\underline{t}]]/\mathbb{Z}}$. *Then the following are equivalent:*

(1) ω *is pseudo-total.*
(2) ω *is pseudo-exact.*
(3) ω *is pseudo-closed.*

PROOF. We proved that (1)⇒(2) in Lemma 6.14 and that (2)⇒(3) in Lemma 6.11.

It remains to prove (3)⇒(1). Since ω is pseudo-closed, there exists a dt-free form $\omega_0 \in \Omega^1_{k[[\underline{t}]]/\mathbb{Z}}$ such that $\omega - \omega_0$ satisfies $d(\omega - \omega_0) = 0$. Now express $\omega - \omega_0 = \omega'_{\mathbb{Z}} + \sum_{i=1}^r f_i dt_i$, where $\omega'_{\mathbb{Z}}$ is the dt-free part of $\omega - \omega_0$ and $f_i \in k[[\underline{t}]]$. Applying d, we obtain

$$0 = d\omega'_{\mathbb{Z}} + \sum_{i=1}^r df_i \wedge dt_i = d\omega'_{\mathbb{Z}} + \sum_{i=1}^r \sum_{j \neq i} \frac{\partial f_i}{\partial t_j} dt_j \wedge dt_i \tag{6.3}$$

$$= d\omega'_{\mathbb{Z}} + \sum_{i<j} \left(-\frac{\partial f_i}{\partial t_j} + \frac{\partial f_j}{\partial t_i} \right) dt_i \wedge dt_j.$$

When $\omega'_{\mathbb{Z}}$ is dt-free, its derivation $d\omega'_{\mathbb{Z}}$ is not necessarily dt-free but unsaturated, and at least we see that $d\omega'_{\mathbb{Z}}$ has no term involving the term of type $(2,0)$ like $dt_i \wedge dt_j$. Hence for the above equality (6.3) to hold, we must have $d\omega'_{\mathbb{Z}} = 0$ and $\frac{\partial f_i}{\partial t_j} = \frac{\partial f_j}{\partial t_i}$ for each pair $i \neq j$. In particular, we have $\omega = \omega_0 + \omega'_{\mathbb{Z}} + \sum_i f_i dt_i$, where $\omega_0 + \omega'_{\mathbb{Z}}$ is dt-free, and $\frac{\partial f_i}{\partial t_j} = \frac{\partial f_j}{\partial t_i}$. Hence ω is pseudo-total. □

6.3. The logarithm. We can apply Theorem 6.15 to define the logarithm for $k[[\underline{t}]]^\times$ as follows.

Let $f \in k[[\underline{t}]]^\times$. This defines $\underline{\text{dlog}} f := \frac{df}{f} \in \Omega^1_{k[[\underline{t}]]/\mathbb{Z}}$. We want to define $\log(f)$ to be its integral. To do so, we need to check that it is pseudo-exact, which is equivalent to pseudo-totality by Theorem 6.15:

LEMMA 6.16. *For* $f \in k[[\underline{t}]]^\times$, *the form* $\underline{\text{dlog}} f \in \Omega^1_{k[[\underline{t}]]/\mathbb{Z}}$ *is pseudo-total.*

PROOF. Up to removing a dt-free part (which we remove from all our calculations below), we have

$$\underline{\text{dlog}} f = \frac{df}{f} = \frac{\sum_i f_{t_i} dt_i}{f}, \text{ where } f_{t_i} := \frac{\partial f}{\partial t_i}.$$

Then we have

$$\frac{\partial \left(\frac{f_{t_i}}{f} \right)}{\partial t_j} = \frac{f_{t_i t_j} f - f_{t_i} f_{t_j}}{f^2}, \frac{\partial \left(\frac{f_{t_j}}{f} \right)}{\partial t_i} = \frac{f_{t_j t_i} f - f_{t_j} f_{t_i}}{f^2},$$

and these two are equal. Hence $\underline{\text{dlog}} f$ is pseudo-total. □

By Theorem 6.15 and Lemma 6.16, the form $\underline{\text{dlog}}(f)$ is pseudo-exact. Hence we have an integral $\int \underline{\text{dlog}}(f) \in k[[\underline{t}]]$, whose coefficients of t^I for nonzero indices I are all well-defined by Theorem 6.8. We thus have:

DEFINITION 6.17. For $f \in k[[\underline{t}]]^\times$, define $\log(f) := \int \underline{\text{dlog}}(f) \in k[[\underline{t}]]$. All of coefficients of t^I for multi-indices $I \neq (0, \cdots, 0)$ are well-defined.

LEMMA 6.18. *For* $\log^0(1+g) = g - \frac{1}{2}g^2 + \frac{1}{3}g^3 - \cdots$, *we have* $\log^0(1+g) = \log(1+g)$.

PROOF. The proof is similar to that of Lemma 3.2. We omit it. □

LEMMA 6.19. *We have* $\log(fg) = \log(f) + \log(g)$.

PROOF. The proof is similar to that part of Lemma 3.3. We omit it. □

6.4. Product of forms. In general, we would like to know when n-forms with $n \geq 2$ in the multivariable situation are pseudo-exact. It was not easy to answer yet. However, at least we can try the following, which is good enough for us to work with logarithms:

LEMMA 6.20. *Let* $n_1, n_2 \geq 1$ *be integers. Let* $\omega_1 \in \Omega^{n_1}_{k[[\underline{t}]]/\mathbb{Z}}$, $\omega_2 \in \Omega^{n_2}_{k[[\underline{t}]]/\mathbb{Z}}$. *Suppose* ω_1 *and* ω_2 *are pseudo-closed. Then* $\omega_1 \wedge \omega_2$ *is pseudo-closed as well.*

PROOF. Suppose for unsaturated n_i-forms ω_i' for $i = 1, 2$, we have $d(\omega_i - \omega_i') = 0$. Then $(\omega_1 - \omega_1') \wedge (\omega_2 - \omega_2')$ is closed by the Leibniz rule. But, this is equal to $\omega_1 \wedge \omega_2 - \omega_2 \wedge \omega_1' - \omega_1 \wedge \omega_2' + \omega_1' \wedge \omega_2'$, where all but the first term are unsaturated $(n_1 + n_2)$-forms. Hence $\omega_1 \wedge \omega_2$ is pseudo-closed. □

Similarly:

LEMMA 6.21. *Let* $n_1, n_2 \geq 1$ *be integers. Let* $\omega_1 \in \Omega^{n_1}_{k[[\underline{t}]]/\mathbb{Z}}$, $\omega_2 \in \Omega^{n_2}_{k[[\underline{t}]]/\mathbb{Z}}$. *Suppose* ω_1 *is pseudo-exact and* ω_2 *is pseudo-closed. Then* $\omega_1 \wedge \omega_2$ *is pseudo-exact.*

PROOF. Suppose for unsaturated n_i-forms ω_i' for $i = 1, 2$, we have $\omega_1 - \omega_1' = d\eta$ for some $\eta \in \Omega^{n_1 - 1}_{k[[\underline{t}]]/\mathbb{Z}}$ and $d(\omega_2 - \omega_2') = 0$. Since the wedge of an exact form and a closed form is exact by the Leibniz rule, the form $(\omega_1 - \omega_1') \wedge (\omega_2 - \omega_2')$ is exact. But, this is equal to $\omega_1 \wedge \omega_2 - \omega_2 \wedge \omega_1' - \omega_1 \wedge \omega_2' + \omega_1' \wedge \omega_2'$, where all but the first term are unsaturated $(n_1 + n_2)$-forms. Hence $\omega_1 \wedge \omega_2$ is pseudo-exact. □

COROLLARY 6.22. *Let* $n \geq 1$. *Suppose* $\omega_1 \in \Omega^1_{k[[\underline{t}]]/\mathbb{Z}}$ *is pseudo-exact, and for* $2 \leq i \leq n$, *the forms* $\omega_i \in \Omega^1_{k[[\underline{t}]]/\mathbb{Z}}$ *are pseudo-closed. Let* $\omega := \omega_1 \wedge \cdots \wedge \omega_n \in \Omega^n_{k[[\underline{t}]]/\mathbb{Z}}$. *Then there exists an integral* $\int \omega \in \Omega^{n-1}_{k[[\underline{t}]]/\mathbb{Z}}$.

PROOF. ω is pseudo-exact by Lemma 6.21, so it has an anti-derivative. □

REMARK 6.23. For $n = 1$, the integral of Corollary 6.22 is well-defined in $\sum_{i=1}^r t_i k[[\underline{t}]]$ by Theorem 6.8.

For $n \geq 2$, although we couldn't prove it in full generality yet, from some calculational inspections, it appears that the integral is actually well-defined in $\Omega^{n-1}_{k[[\underline{t}]],(\underline{t})/\mathbb{Z}} / d\Omega^{n-2}_{k[[\underline{t}]],(\underline{t})/\mathbb{Z}}$. Using the main theorem of [**6**] and taking an inverse limit, we have

$$\Omega^{n-1}_{k[[\underline{t}]],(\underline{t})/\mathbb{Z}} / d\Omega^{n-2}_{k[[\underline{t}]],(\underline{t})/\mathbb{Z}} \simeq K_n^M(k[[\underline{t}]], (\underline{t})).$$

This is consistent with the expectation of Conjecture 6.9.

DEFINITION 6.24. For integers $r, n \geq 1$, let $f_1, \cdots, f_n \in k[[\underline{t}]]^\times$. Consider the Milnor symbol $\{f_1, \cdots, f_n\} \in K_n^M(k[[\underline{t}]])$ and define $\underline{\mathrm{dlog}}\{f_1, \cdots, f_n\} \in \Omega^n_{k[[\underline{t}]]/\mathbb{Z}}$. Here, by Theorem 6.15 and Lemma 6.16, we know that each $\underline{\mathrm{dlog}} f_i \in \Omega^1_{k[[\underline{t}]]/\mathbb{Z}}$ is pseudo-exact. Now by Corollary 6.22, we have $\underline{\mathrm{dlog}}\{f_1, \cdots, f_n\} = \underline{\mathrm{dlog}} f_1 \wedge \cdots \wedge \underline{\mathrm{dlog}} f_n \in \Omega^n_{k[[\underline{t}]]/\mathbb{Z}}$ is pseudo-exact. Hence there exists an integral, which we define

$$\log\{f_1, \cdots, f_n\} := \int \underline{\mathrm{dlog}}\{f_1, \cdots, f_n\} \in \Omega^{n-1}_{k[[\underline{t}]]/\mathbb{Z}}.$$

In the spirit of Conjecture 6.9 and Remark 6.23, it seems the image of this integral is well-defined in $\Omega^{n-1}_{k[[\underline{t}]],(\underline{t})/\mathbb{Z}} / d\Omega^{n-2}_{k[[\underline{t}]],(\underline{t})/\mathbb{Z}} \simeq K^M_n(k[[\underline{t}]], (\underline{t}))$.

Under this, we may regard the above log as the realization of the projection $K^M_n(k[[\underline{t}]]) \to K^M_n(k[[\underline{t}]], (\underline{t}))$ to the relative Milnor K-group.

6.5. Logarithms for some local k-algebras. Let A be a local k-algebra that has a presentation $A = k[[\underline{t}]]/I$ for some ideal I such that $k = A/(\underline{t})A$ gives the residue field. For such rings, we can extend our construction of the log. In §6.5, we fix this A and the presentation.

LEMMA 6.25. *Let $f_1, f_2 \in A^\times$. The following are equivalent:*

(1) *$f_1 = f_2$ in $A^\times$.*
(2) *For any two liftings $\tilde{f}_1, \tilde{f}_2 \in k[[\underline{t}]]^\times$ of f_1, f_2, there exists $\alpha \in I$ such that $\tilde{f}_1 = \tilde{f}_2(1+\alpha)$.*

PROOF. For (1) $\Rightarrow$ (2), note that $f_2^{-1} f_1 \equiv 1 \mod I$. Hence $\tilde{f}_2^{-1}\tilde{f}_1 - 1 = \alpha$ for some $\alpha \in I$. This implies (2). (2) $\Rightarrow$ (1) is obvious. □

PROPOSITION 6.26. *Let $\tilde{f}_1, \tilde{f}_2 \in k[[\underline{t}]]^\times$. Suppose that $\tilde{f}_1 \equiv \tilde{f}_2$ in the local k-algebra $A = k[[\underline{t}]]/I$. Then $\log \tilde{f}_1 \equiv \log \tilde{f}_2$ in A.*

PROOF. By Lemma 6.25, we have $\tilde{f}_1 = \tilde{f}_2(1+\alpha)$. Hence $\log \tilde{f}_1 = \log \tilde{f}_2 + \log(1+\alpha)$ by Lemma 6.19. However, $\log(1+\alpha) = \log^0(1+\alpha)$ by Lemma 6.18, and the latter is in I by definition. This shows $\log \tilde{f}_1 \equiv \log \tilde{f}_2$ in A. □

This allows us to define $\log : A^\times \to \sum_{i=1}^r \bar{t}_i A$. This *a priori* depends on the choice of the presentation $A = k[[\underline{t}]]/I$, but we guess it is independent of the choice of the presentation, and it realizes the projection map $K^M_n(A) \to K^M_n(A \to k)$, where the latter is the relative Milnor K-group associated to the augmentation map associated to the residue field k of A. We do not yet have a good proof of it in general at this moment except for some special cases of A.

Acknowledgments

This article is prepared for the proceedings of the official ICM satellite conference on K-theory held at the University of Buenos Aires, Buenos Aires, Argentina, in July 2018. Taking this opportunity, the author deeply thanks the organizers of the great event, Professors Guillermo Cortiñas and Gisela Tartaglia. The author also wishes to express his gratitude to the referee whose comments significantly improved this article.

References

[1] Spencer Bloch, *K_2 of Artinian Q-algebras, with application to algebraic cycles*, Comm. Algebra **3** (1975), 405–428, DOI 10.1080/00927877508822053. MR0371891

[2] Spencer J. Bloch, *Higher regulators, algebraic K-theory, and zeta functions of elliptic curves*, CRM Monograph Series, vol. 11, American Mathematical Society, Providence, RI, 2000. MR1760901

[3] Jean-Louis Cathelineau, *Remarques sur les différentielles des polylogarithmes uniformes* (French, with English and French summaries), Ann. Inst. Fourier (Grenoble) **46** (1996), no. 5, 1327–1347. MR1427128

[4] Kuo Tsai Chen, *Iterated path integrals*, Bull. Amer. Math. Soc. **83** (1977), no. 5, 831–879, DOI 10.1090/S0002-9904-1977-14320-6. MR454968

[5] Thomas G. Goodwillie, *Relative algebraic K-theory and cyclic homology*, Ann. of Math. (2) **124** (1986), no. 2, 347–402, DOI 10.2307/1971283. MR855300

[6] S. O. Gorchinskiĭ and D. N. Tyurin, *Relative Milnor K-groups and differential forms of split nilpotent extensions* (Russian, with Russian summary), Izv. Ross. Akad. Nauk Ser. Mat. **82** (2018), no. 5, 23–60, DOI 10.4213/im8762; English transl., Izv. Math. **82** (2018), no. 5, 880–913. MR3859378

[7] Robin Hartshorne, *Algebraic geometry*, Springer-Verlag, New York-Heidelberg, 1977. Graduate Texts in Mathematics, No. 52. MR0463157

[8] R. Iwasa and W. Kai, *Chern classes with modulus*, Nagoya Math. J., **236** (2019), 84–1332. DOI 10.1017/nmj.2018.52

[9] Ernst Kunz, *Kähler differentials*, Advanced Lectures in Mathematics, Friedr. Vieweg & Sohn, Braunschweig, 1986. MR864975

[10] Jinhyun Park and Sinan Ünver, *Motivic cohomology of fat points in Milnor range*, Doc. Math. **23** (2018), 759–798. MR3861046

[11] J. Park and S. Ünver, *Algebraic cycles and the relative Bloch-Wigner sequence of higher moduli*, work in progress.

[12] Sinan Ünver, *On the additive dilogarithm*, Algebra Number Theory **3** (2009), no. 1, 1–34, DOI 10.2140/ant.2009.3.1. MR2491908

[13] S. Ünver, *Infinitesimal Chow Dilogarithm*, to appear in J. Algebraic Geom. DOI 10.1090/jag/746.

[14] Charles A. Weibel, *An introduction to homological algebra*, Cambridge Studies in Advanced Mathematics, vol. 38, Cambridge University Press, Cambridge, 1994. MR1269324

Department of Mathematical Sciences, KAIST, 291 Daehak-ro Yuseong-gu, Daejeon, 34141, Republic of Korea (South)

Email address: `jinhyun@mathsci.kaist.ac.kr; jinhyun@kaist.edu`

Contemporary Mathematics
Volume **749**, 2020
https://doi.org/10.1090/conm/749/15078

Noncommutative counterparts of celebrated conjectures

Gonçalo Tabuada

To Gabriel, for his long-awaited arrival.

ABSTRACT. In this survey, written for the proceedings of the *K-theory conference*, Buenos Aires and La Plata, Argentina (satellite event of the ICM 2018), we give a rigorous overview of the noncommutative counterparts of some celebrated conjectures of Grothendieck, Voevodsky, Beilinson, Weil, Tate, Parshin, Kimura, Schur, and others.

Introduction

Some celebrated conjectures of Grothendieck, Voevodsky, Beilinson, Weil, Tate, Parshin, Kimura, Schur, and others, were recently extended from the realm of algebraic geometry to the broad noncommutative setting of differential graded (=dg) categories. This noncommutative viewpoint led to a proof of these celebrated conjectures in several new cases. Moreover, it enabled a proof of the noncommutative counterparts of the celebrated conjectures in many interesting cases. The purpose of this survey, written for a broad mathematical audience, is to give a rigorous overview of these recent developments.

NOTATION. Given a perfect base field k of characteristic $p > 0$, we will write $W(k)$ for its ring of p-typical Witt vectors and $K := W(k)_{1/p}$ for the fraction field of $W(k)$. For example, when $k = \mathbb{F}_p$, we have $W(k) = \mathbb{Z}_p$ and $K = \mathbb{Q}_p$.

1. Celebrated conjectures

In this section, we briefly recall some celebrated conjectures of Grothendieck, Voevodsky, Beilinson, Weil, Tate, Parshin, and Kimura (concerning smooth proper schemes), as well as a conjecture of Schur (concerning smooth schemes).

1.1. Grothendieck standard conjecture of type C^+. Let k be a perfect base field of characteristic $p \geq 0$ and X a smooth proper k-scheme of dimension d. When $p = 0$, we will write $H^*_{\mathrm{dR}}(X)$ for the de Rham cohomology of X. In the same vein, when $p > 0$, we will write $H^*_{\mathrm{crys}}(X) := H^*_{\mathrm{crys}}(X/W(k)) \otimes_{W(k)} K$ for the crystalline cohomology of X. Given an integer $0 \leq i \leq 2d$, consider the associated

---

2010 *Mathematics Subject Classification.* Primary 54C40, 14E20; Secondary 46E25, 20C20.
Key words and phrases. Differential geometry, algebraic geometry.
The author was partially supported by the NSF CAREER Award #1350472 and by the Fundação para a Ciência e a Tecnologia (Portuguese Foundation for Science and Technology) through the project UID/MAT/00297/2019 (Centro de Matemática e Aplicações).

i^{th} Künneth projector $\pi^i\colon H^*_{\mathrm{dR}}(X) \to H^*_{\mathrm{dR}}(X)$, resp. $\pi^i\colon H^*_{\mathrm{crys}}(X) \to H^*_{\mathrm{crys}}(X)$, in de Rham cohomology, resp. in crystalline cohomology. In the sixties, Grothendieck [**17**] (see also [**30, 31**]) conjectured the following:

CONJECTURE ($\mathrm{C}^+(X)$). *The even Künneth projector $\pi^+ := \sum_{i\,\text{even}} \pi^i$ is algebraic*[1].

This conjecture is also usually called the "sign conjecture". It holds when $d \leq 2$, when X is an abelian variety (see Kleiman [**31**]), and also when k is a finite field (see Katz-Messing [**26**]). Besides these cases (and some other cases scattered in the literature), it remains wide open.

REMARK 1.1. Given smooth proper k-schemes X and Y, we have the implication of conjectures $\mathrm{C}^+(X) + \mathrm{C}^+(Y) \Rightarrow \mathrm{C}^+(X \times Y)$.

1.2. Grothendieck standard conjecture of type D. Let k be a perfect base field of characteristic $p \geq 0$ and X a smooth proper k-scheme of dimension d. Consider the graded $\mathbb{Q}$-vector space $\mathcal{Z}^*(X)_{\mathbb{Q}}/_{\sim \mathrm{hom}}$ of algebraic cycles on X up to homological equivalence (when $p = 0$, resp. $p > 0$, we make use of de Rham cohomology, resp. crystalline cohomology). Consider also the graded $\mathbb{Q}$-vector space $\mathcal{Z}^*(X)_{\mathbb{Q}}/_{\sim \mathrm{num}}$ of algebraic cycles on X up to numerical equivalence. In the sixties, Grothendieck [**17**] (see also [**30, 31**]) conjectured the following:

CONJECTURE ($\mathrm{D}(X)$). *The equality $\mathcal{Z}^*(X)_{\mathbb{Q}}/_{\sim \mathrm{hom}} = \mathcal{Z}^*(X)_{\mathbb{Q}}/_{\sim \mathrm{num}}$ holds.*

This conjecture holds when $d \leq 2$, when $d \leq 4$ and $p = 0$ (see Lieberman [**39**]), and also when X is an abelian variety and $p = 0$ (see Lieberman [**39**]). Besides these cases (and some other cases scattered in the literature), it remains wide open.

1.3. Voevodsky nilpotence conjecture. Let k be a base field of characteristic $p \geq 0$ and X a smooth proper k-scheme of dimension d. Following Voevodsky [**64**], consider the graded $\mathbb{Q}$-vector space $\mathcal{Z}^*(X)_{\mathbb{Q}}/_{\sim \mathrm{nil}}$ of algebraic cycles on X up to nilpotence equivalence. In the nineties, Voevodsky [**64**] conjectured the following:

CONJECTURE ($\mathrm{V}(X)$). *The equality $\mathcal{Z}^*(X)_{\mathbb{Q}}/_{\sim \mathrm{nil}} = \mathcal{Z}^*(X)_{\mathbb{Q}}/_{\sim \mathrm{num}}$ holds.*

This conjecture holds when $d \leq 2$ (see Voevodsky [**64**] and Voisin [**65**]), and also when X is an abelian threefold and $p = 0$ (see Kahn-Sebastian [**25**]). Besides these cases (and some other cases scattered in the literature), it remains wide open.

REMARK 1.2. Every algebraic cycle which is nilpotently trivial is also homologically trivial. Hence, we have the implication of conjectures $\mathrm{V}(X) \Rightarrow \mathrm{D}(X)$.

1.4. Beilinson conjecture. Let $k = \mathbb{F}_q$ be a finite field of characteristic p and X a smooth proper k-scheme of dimension d. Consider the graded $\mathbb{Q}$-vector space $\mathcal{Z}^*(X)_{\mathbb{Q}}/_{\sim \mathrm{rat}}$ of algebraic cycles on X up to rational equivalence. In the eighties, Beilinson [**4**] conjectured the following:

CONJECTURE ($\mathrm{B}(X)$). *The equality $\mathcal{Z}^*(X)_{\mathbb{Q}}/_{\sim \mathrm{rat}} = \mathcal{Z}^*(X)_{\mathbb{Q}}/_{\sim \mathrm{num}}$ holds.*

This conjecture holds when $d \leq 1$, and also when X is an abelian variety and $d \leq 3$ (see Kahn [**24**]). Besides these cases (and some other cases scattered in the literature), it remains wide open.

REMARK 1.3. Every algebraic cycle which is rationally trivial is also nilpotently trivial. Hence, in the case where k is a finite field, we have $\mathrm{B}(X) \Rightarrow \mathrm{V}(X)$.

[1] If π^+ is algebraic, then the odd Künneth projector $\pi^- := \sum_{i\,\text{odd}} \pi^i$ is also algebraic.

1.5. Weil conjecture. Let $k = \mathbb{F}_q$ be a finite field of characteristic p and X a smooth proper k-scheme of dimension d. Recall that the zeta function of X is defined as the formal power series $Z(X;t) := \exp(\sum_{n\geq 1} \#X(\mathbb{F}_{q^n})\frac{t^n}{n}) \in \mathbb{Q}[\![t]\!]$, where $\exp(t) := \sum_{n\geq 0} \frac{t^n}{n!}$. In the same vein, given an integer $0 \leq i \leq 2d$, consider the formal power series $Z_i(X;t) := \det(\mathrm{id} - t\,\mathrm{Fr}^i | H^i_{\mathrm{crys}}(X))^{-1} \in K[\![t]\!]$, where Fr stands for the Frobenius endomorphism of X and Fr^i for the induced automorphism of $H^i_{\mathrm{crys}}(X)$. Thanks to the Lefschetz trace formula established by Grothendieck and Berthelot (consult [**6**]), we have the following weight decomposition:

$$Z(X;t) = \frac{Z_0(X;t)Z_2(X;t)\cdots Z_{2d}(X;t)}{Z_1(X;t)Z_3(X;t)\cdots Z_{2d-1}(X;t)} \in K[\![t]\!]\,. \tag{1.1}$$

In the late forties, Weil [**66**] conjectured the following[2]:

CONJECTURE (W(X)). *The eigenvalues of the automorphism* Fr^i, *with* $0 \leq i \leq 2d$, *are algebraic numbers and all their complex conjugates have absolute value* $q^{\frac{i}{2}}$.

In the particular case of curves, this famous conjecture follows from Weil's pioneering work [**67**]. Later, in the seventies, it was proved in full generality by Deligne[3] [**12**]. In contrast with Weil's proof, which uses solely the classical intersection theory of divisors on surfaces, Deligne's proof makes use of several involved tools such as the theory of monodromy of Lefschetz pencils. The Weil conjecture has numerous applications. For example, when combined with the weight decomposition (1.1), it implies that the polynomials $\det(\mathrm{id} - t\,\mathrm{Fr}^i | H^i_{\mathrm{crys}}(X))$ have integer coefficients.

Recall that the Hasse-Weil zeta function of X is defined as the (convergent) infinite product $\zeta(X;s) := \prod_{x\in X^0}(1-(q^{\deg(x)})^{-s})^{-1}$, with $\mathrm{Re}(s) > d$, where X^0 stands for the set of closed points of X and $\deg(x)$ for the degree of the finite field extension $\kappa(x)/\mathbb{F}_q$. In the same vein, given an integer $0 \leq i \leq 2d$, consider the function $\zeta_i(X;s) := \det(\mathrm{id} - q^{-s}\,\mathrm{Fr}^i | H^i_{\mathrm{crys}}(X))^{-1}$. It follows from the Weil conjecture that $\zeta(X;s) = Z(X;q^{-s})$, with $\mathrm{Re}(s) > d$, and that $\zeta_i(X;s) = Z_i(X;q^{-s})$, with $\mathrm{Re}(s) > \frac{i}{2}$. Thanks to (1.1), we hence obtain the weight decomposition:

$$\zeta(X;s) = \frac{\zeta_0(X;s)\zeta_2(X;s)\cdots\zeta_{2d}(X;s)}{\zeta_1(X;s)\zeta_3(X;s)\cdots\zeta_{2d-1}(X;s)} \qquad \mathrm{Re}(s) > d\,. \tag{1.2}$$

Note that (1.2) implies automatically that the Hasse-Weil zeta function $\zeta(X;s)$ of X admits a (unique) meromorphic continuation to the entire complex plane.

REMARK 1.4 (Analogue of the Riemann hypothesis). The above conjecture W(X) is usually called the "analogue of the Riemann hypothesis" because it implies that if $z \in \mathbb{C}$ is a pole of $\zeta_i(X;s)$, then $\mathrm{Re}(z) = \frac{i}{2}$. Consequently, if $z \in \mathbb{C}$ is a pole, resp. zero, of $\zeta(X;s)$, then $\mathrm{Re}(z) \in \{0, 1, \ldots, d\}$, resp. $\mathrm{Re}(z) \in \{\frac{1}{2}, \frac{2}{3}, \ldots, \frac{2d-1}{2}\}$.

1.6. Tate conjecture. Let $k = \mathbb{F}_q$ be a finite field of characteristic p and X a smooth proper k-scheme of dimension d. Given a prime number $l \neq p$, consider the associated $\mathbb{Q}_l$-linear cycle class map with values in l-adic cohomology:

$$\mathcal{Z}^*(X)_{\mathbb{Q}_l}/_{\sim \mathrm{rat}} \longrightarrow H^{2*}_{l\text{-adic}}(X_{\overline{k}}, \mathbb{Q}_l(*))^{\mathrm{Gal}(\overline{k}/k)}\,. \tag{1.3}$$

[2]The above conjecture W(X) is a modern formulation of Weil's original conjecture; in the late forties crystalline cohomology was not yet developed.

[3]Deligne worked with étale cohomology instead. However, as explained by Katz-Messing in [**26**], Deligne's results hold similarly in crystalline cohomology. More recently, Kedlaya [**27**] gave an alternative proof of the Weil conjecture which uses solely p-adic techniques.

In the sixties, Tate [**60**] conjectured the following:

CONJECTURE ($\mathrm{T}^l(X)$). *The cycle class map* (1.3) *is surjective.*

This conjecture holds when $d \leq 1$, when X is an abelian variety and $d \leq 3$, and also when X is a $K3$-surface; consult Totaro's survey [**62**]. Besides these cases (and some other cases scattered in the literature), it remains wide open.

1.7. p-version of the Tate conjecture. Let $k = \mathbb{F}_q$ be a finite field of characteristic p and X a smooth proper k-scheme of dimension d. Consider the associated K-linear cycle class map with values in crystalline cohomology (see §1.5):

$$\mathcal{Z}^*(X)_K/_{\sim \mathrm{rat}} \longrightarrow H^{2*}_{\mathrm{crys}}(X)(*)^{\mathrm{Fr}^{2*}}. \tag{1.4}$$

Following Milne [**43**], the Tate conjecture admits the following p-version:

CONJECTURE ($\mathrm{T}^p(X)$). *The cycle class map* (1.4) *is surjective.*

This conjecture is equivalent to $\mathrm{T}^l(X)$ (for every $l \neq p$) when $d \leq 3$. Hence, it also holds in the cases mentioned in §1.6. Besides these cases (and some other cases scattered in the literature), it remains wide open.

REMARK 1.5. The p-version of the Tate conjecture can be alternatively formulated as follows: the $\mathbb{Q}_p$-linear cycle class map $\mathcal{Z}^*(X)_{\mathbb{Q}_p}/_{\sim \mathrm{rat}} \to H^{2*}_{\mathrm{crys}}(X)(*)^{\mathrm{Fr}_p^{2*}}$, where Fr_p^{2*} stands for the crystalline Frobenius, is surjective.

1.8. Strong form of the Tate conjecture. Let $k = \mathbb{F}_q$ be a finite field of characteristic p and X a smooth proper k-scheme of dimension d. Recall from §1.5 that the Hasse-Weil zeta function of X is defined as the (convergent) infinite product $\zeta(X;s) := \prod_{x \in X^0}(1-(q^{\deg(x)})^{-s})^{-1}$, with $\mathrm{Re}(s) > d$. Moreover, as mentioned in *loc. cit.*, $\zeta(X;s)$ admits a meromorphic continuation to the entire complex plane. In the sixties, Tate [**60**] also conjectured the following:

CONJECTURE ($\mathrm{ST}(X)$). *The order* $\mathrm{ord}_{s=j}\zeta(X;s)$ *of the Hasse-Weil zeta function* $\zeta(X;s)$ *at the pole* $s = j$, *with* $0 \leq j \leq d$, *is equal to* $-\dim_{\mathbb{Q}} \mathcal{Z}^j(X)_{\mathbb{Q}}/_{\sim \mathrm{num}}$.

REMARK 1.6. As proved by Tate in [**59**], resp. by Milne in [**43**], we have the equivalence of conjectures $\mathrm{ST}(X) \Leftrightarrow \mathrm{B}(X)+\mathrm{T}^l(X)$, resp. $\mathrm{ST}(X) \Leftrightarrow \mathrm{B}(X)+\mathrm{T}^p(X)$.

Thanks to Remark 1.6, the conjecture $\mathrm{ST}(X)$ holds when $d \leq 1$, and also when X is an abelian variety and $d \leq 3$. Besides these cases (and some other cases scattered in the literature), it remains wide open.

1.9. Parshin conjecture. Let $k = \mathbb{F}_q$ be a finite field of characteristic p and X a smooth proper k-scheme of dimension d. Consider the associated algebraic K-theory groups $K_n(X), n \geq 0$. In the eighties, Parshin conjectured the following:

CONJECTURE ($\mathrm{P}(X)$). *The groups* $K_n(X)$, *with* $n \geq 1$, *are torsion.*

This conjecture holds when $d \leq 1$ (see Quillen [**16**] and Harder [**21**]). Besides these cases (and some other cases scattered in the literature), it remains wide open.

REMARK 1.7. As proved by Geisser in [**15**], we have the implication of conjectures $\mathrm{B}(X) + \mathrm{ST}(X) \Rightarrow \mathrm{P}(X)$. This implies, in particular, that the conjecture $\mathrm{P}(X)$ also holds when X is an abelian variety and $d \leq 3$.

1.10. Kimura-finiteness conjecture. Let k be a base field of characteristic $p \geq 0$ and X a smooth proper k-scheme of dimension d. Consider the category of Chow motives $\mathrm{Chow}(k)_{\mathbb{Q}}$ introduced by Manin in [**40**]. By construction, this category is $\mathbb{Q}$-linear, idempotent complete, symmetric monoidal, and comes equipped with a symmetric monoidal functor $\mathfrak{h}(-)_{\mathbb{Q}}\colon \mathrm{SmProp}(k)^{\mathrm{op}} \to \mathrm{Chow}(k)_{\mathbb{Q}}$ defined on smooth proper k-schemes. A decade ago, Kimura [**29**] conjectured the following:

CONJECTURE (K(X)). *The Chow motive $\mathfrak{h}(X)_{\mathbb{Q}}$ is Kimura-finite*[4].

This conjecture holds when $d \leq 1$ and also when X is an abelian variety (see Kimura [**29**] and Shermenev [**46**]). Besides these cases[5] (and some other cases scattered in the literature), it remains wide open.

1.11. Schur-finiteness conjecture. Let k be a perfect base field of characteristic $p \geq 0$ and X a smooth k-scheme of dimension d. Consider the triangulated category of geometric mixed motives $\mathrm{DM}_{\mathrm{gm}}(k)_{\mathbb{Q}}$ introduced by Voevodsky in [**63**]. By construction, this category is $\mathbb{Q}$-linear, idempotent complete, symmetric monoidal, and comes equipped with a symmetric monoidal functor $M(-)_{\mathbb{Q}}\colon \mathrm{Sm}(k) \to \mathrm{DM}_{\mathrm{gm}}(k)_{\mathbb{Q}}$ defined on smooth k-schemes. Moreover, as proved in *loc. cit.*, the classical category of Chow motives $\mathrm{Chow}(k)_{\mathbb{Q}}$ may be embedded fully-faithfully into $\mathrm{DM}_{\mathrm{gm}}(k)_{\mathbb{Q}}$. An important conjecture in the theory of mixed motives is the following:

CONJECTURE (S(X)). *The mixed motive $M(X)_{\mathbb{Q}}$ is Schur-finite*[6].

This conjecture holds when $d \leq 1$ (see Guletskii [**19**] and Mazza [**42**]) and also when X is an abelian variety (see Kimura [**29**] and Shermenev [**46**]). Besides these cases (and some other cases scattered in the literature), it remains wide open.

REMARK 1.8. It is well-known that Kimura-finiteness implies Schur-finiteness. However, the converse does *not* holds. For example, O'Sullivan constructed a certain smooth surface X whose mixed motive $M(X)_{\mathbb{Q}}$ is Schur-finite but *not* Kimura-finite; consult [**42**]. An important open problem is the classification of all the Kimura-finite mixed motives and the computation of the corresponding Kimura-dimensions.

2. Noncommutative counterparts

In this section we describe the noncommutative counterparts of the celebrated conjectures of §1. We will assume some basic familiarity with the language of differential graded (=dg) categories; consult Keller's survey [**28**]. In particular, we

---

[4]Let $(\mathcal{C}, \otimes, \mathbf{1})$ be a $\mathbb{Q}$-linear, idempotent complete, symmetric monoidal category. Following Kimura [**29**], recall that an object $a \in \mathcal{C}$ is called *even-dimensional*, resp. *odd-dimensional*, if $\wedge^n(a) \simeq 0$, resp. $\mathrm{Sym}^n(a) \simeq 0$, for some $n \gg 0$. The biggest integer $\mathrm{kim}_+(a)$, resp. $\mathrm{kim}_-(a)$, for which $\wedge^{\mathrm{kim}_+(a)}(a) \not\simeq 0$, resp. $\mathrm{Sym}^{\mathrm{kim}_-(a)}(a) \not\simeq 0$, is called the *even Kimura-dimension*, resp. *odd Kimura-dimension*, of a. Recall also that an object $a \in \mathcal{C}$ is called *Kimura-finite* if $a \simeq a_+ \oplus a_-$, with a_+ even-dimensional and a_- odd-dimensional. The integer $\mathrm{kim}(a) := \mathrm{kim}_+(a_+) + \mathrm{kim}_-(a_-)$ is called the *Kimura-dimension* of a.

[5]In the particular case where $k = \overline{k}$, $p = 0$, and X is a surface with $p_g(X) = 0$, Guletskii and Pedrini proved in [**20**] that the conjecture K(X) is equivalent to a celebrated conjecture of Bloch [**7**] concerning the vanishing of the Albanese kernel.

[6]Let $(\mathcal{C}, \otimes, \mathbf{1})$ be a $\mathbb{Q}$-linear, idempotent complete, symmetric monoidal category. Following Deligne [**11**], every partition λ of an integer $n \geq 1$ gives rise to a Schur-functor $S_\lambda\colon \mathcal{C} \to \mathcal{C}$. For example, when $\lambda = (1, \ldots, 1)$, resp. $\lambda = (n)$, we have $S_{(1,\ldots,1)}(a) = \wedge^n(a)$, resp. $S_{(1)}(a) = \mathrm{Sym}^n(a)$. An object $a \in \mathcal{C}$ is called *Schur-finite* if $S_\lambda(a) \simeq 0$ for some partition λ.

will use freely the notion of *smooth proper* dg category in the sense of Kontsevich [**32**–**36**]. Examples include the finite-dimensional algebras of finite global dimension A (over a perfect base field) as well as the dg categories of perfect complexes $\mathrm{perf}_{\mathrm{dg}}(X)$ associated to smooth proper schemes X (or, more generally, to smooth proper algebraic stacks $\mathcal{X}$). In addition, we will make essential use of the recent theory of noncommutative motives; consult the book [**52**] and the survey [**48**].

2.1. Noncommutative Grothendieck standard conjecture of type C^+. Let k be a perfect base field of characteristic $p \geq 0$ and $\mathcal{A}$ a smooth proper k-linear dg category. In what follows, we will write $\mathrm{dgcat}_{\mathrm{sp}}(k)$ for the category of (essentially small) k-linear dg categories. Recall from [**41**, §9] that, when $p = 0$, periodic cyclic homology gives rise to a symmetric monoidal functor

$$HP_{\pm}(-)\colon \mathrm{dgcat}_{\mathrm{sp}}(k) \longrightarrow \mathrm{vect}_{\mathbb{Z}/2}(k) \tag{2.1}$$

with values in the category of finite-dimensional $\mathbb{Z}/2$-graded k-vector spaces. In the same vein, recall from [**47**, §2] that, when $p > 0$, topological periodic cyclic homology[7] gives rise to a symmetric monoidal functor

$$TP_{\pm}(-)_{1/p}\colon \mathrm{dgcat}_{\mathrm{sp}}(k) \longrightarrow \mathrm{vect}_{\mathbb{Z}/2}(K) \tag{2.2}$$

with values in the category of finite-dimensional $\mathbb{Z}/2$-graded K-vector spaces.

REMARK 2.1 (Relation with de Rham cohomology and crystalline cohomology). The above functor (2.1), resp. (2.2), may be understood as the noncommutative counterpart of de Rham cohomology, resp. crystalline cohomology. Concretely, given a smooth proper k-scheme X, we have the following natural isomorphisms of finite-dimensional $\mathbb{Z}/2$-graded vector spaces:

$$HP_{\pm}(\mathrm{perf}_{\mathrm{dg}}(X)) \simeq \Big(\bigoplus_{i\,\mathrm{even}} H^i_{\mathrm{dR}}(X), \bigoplus_{i\,\mathrm{odd}} H^i_{\mathrm{dR}}(X)\Big) \tag{2.3}$$

$$TP_{\pm}(\mathrm{perf}_{\mathrm{dg}}(X))_{1/p} \simeq \Big(\bigoplus_{i\,\mathrm{even}} H^i_{\mathrm{crys}}(X), \bigoplus_{i\,\mathrm{odd}} H^i_{\mathrm{crys}}(X)\Big). \tag{2.4}$$

On the one hand, (2.3) follows from the classical Hochschild-Kostant-Rosenberg theorem; see Feigin-Tsygan [**14**]. On the other hand, (2.4) follows from the recent work of Scholze on integral p-adic Hodge theory; consult [**13**][**57**, Thm. 5.2].

Recall from [**52**, §4.1] the definition of the category of noncommutative Chow motives $\mathrm{NChow}(k)_{\mathbb{Q}}$. By construction, this category is $\mathbb{Q}$-linear, idempotent complete, symmetric monoidal, and comes equipped with a symmetric monoidal functor $U(-)_{\mathbb{Q}}\colon \mathrm{dgcat}_{\mathrm{sp}}(k) \to \mathrm{NChow}(k)_{\mathbb{Q}}$. Moreover, we have a natural isomorphism

$$\mathrm{Hom}_{\mathrm{NChow}(k)_{\mathbb{Q}}}(U(k)_{\mathbb{Q}}, U(\mathcal{A})_{\mathbb{Q}}) \simeq K_0(\mathcal{D}_c(\mathcal{A}))_{\mathbb{Q}} =: K_0(\mathcal{A})_{\mathbb{Q}}\,, \tag{2.5}$$

where $\mathcal{D}(\mathcal{A})$ stands for the derived category of $\mathcal{A}$ and $\mathcal{D}_c(\mathcal{A})$ for its full triangulated subcategory of compact objects. As proved in [**41**, Thm. 9.2] when $p = 0$, resp. in [**47**, Thm. 2.3] when $p > 0$, the above functor (2.1), resp. (2.2), descends to the category of noncommutative Chow motives.

[7]Topological periodic cyclic homology is defined as the Tate cohomology of the circle group action on topological Hochschild homology; consult Hesselholt [**22**] and Nikolaus-Scholze [**45**].

Consider the even Künneth projector

$$\pi_+ \colon HP_\pm(\mathcal{A}) \to HP_\pm(\mathcal{A}) \qquad \text{resp. } \pi_+ \colon TP_\pm(\mathcal{A})_{1/p} \to TP_\pm(\mathcal{A})_{1/p}$$

in periodic cyclic homology, resp. in topological periodic cyclic homology. This projector is *algebraic* if there exists an endomorphism $\underline{\pi}_+ \colon U(\mathcal{A})_\mathbb{Q} \to U(\mathcal{A})_\mathbb{Q}$ such that $HP_\pm(\underline{\pi}_+) = \pi_+$, resp. $TP_\pm(\underline{\pi}_+)_{1/p} = \pi_+$. Under these definitions, the Grothendieck standard conjecture of type C^+ admits the following noncommutative counterpart:

CONJECTURE ($\mathrm{C}^+_{\mathrm{nc}}(\mathcal{A})$). *The even Künneth projector π_+ is algebraic*[8].

REMARK 2.2. Similarly to Remark 1.1, given smooth proper k-linear dg categories $\mathcal{A}$ and $\mathcal{B}$, we have the implication $\mathrm{C}^+_{\mathrm{nc}}(\mathcal{A}) + \mathrm{C}^+_{\mathrm{nc}}(\mathcal{B}) \Rightarrow \mathrm{C}^+_{\mathrm{nc}}(\mathcal{A} \otimes \mathcal{B})$.

The next result, whose proof makes use of the above isomorphisms (2.3)-(2.4), relates this conjecture with Grothendieck's original conjecture:

THEOREM 2.3. *([**50**, Thm. 1.1] and [**57**, Thm. 1.1]) Given a smooth proper k-scheme X, we have the equivalence of conjectures* $\mathrm{C}^+(X) \Leftrightarrow \mathrm{C}^+_{\mathrm{nc}}(\mathrm{perf}_{\mathrm{dg}}(X))$.

2.2. Noncommutative Grothendieck standard conjecture of type D. Let k be a perfect base field of characteristic $p \geq 0$ and $\mathcal{A}$ a smooth proper k-linear dg category. Note that by combining the above isomorphism (2.5) with the functor (2.1), resp. (2.2), we obtain an induced $\mathbb{Q}$-linear homomorphism:

$$K_0(\mathcal{A})_\mathbb{Q} \longrightarrow HP_+(\mathcal{A}) \qquad \text{resp. } K_0(\mathcal{A})_\mathbb{Q} \longrightarrow TP_+(\mathcal{A})_{1/p}\,. \tag{2.6}$$

The homomorphism (2.6) may be understood as the noncommutative counterpart of the cycle class map. In what follows, we will write $K_0(\mathcal{A})_\mathbb{Q}/_{\sim \mathrm{hom}}$ for the quotient of $K_0(\mathcal{A})_\mathbb{Q}$ by the kernel of (2.6). Consider also the Euler bilinear pairing:

$$\chi \colon K_0(\mathcal{A}) \times K_0(\mathcal{A}) \longrightarrow \mathbb{Z} \qquad ([M],[N]) \mapsto \sum_{n\in\mathbb{Z}} (-1)^n \mathrm{dim}_k \mathrm{Hom}_{\mathcal{D}_c(\mathcal{A})}(M, N[n])\,.$$

This pairing is not symmetric neither skew-symmetric. Nevertheless, as proved in [**52**, Prop. 4.24], the left and right kernels of χ agree. In what follows, we will write $K_0(\mathcal{A})/_{\sim \mathrm{num}}$ for the quotient of $K_0(\mathcal{A})$ by the kernel of χ and $K_0(\mathcal{A})_\mathbb{Q}/_{\sim \mathrm{num}}$ for the $\mathbb{Q}$-vector space[9] $K_0(\mathcal{A})/_{\sim \mathrm{num}} \otimes_\mathbb{Z} \mathbb{Q}$. Under these definitions, the Grothendieck standard conjecture of type D admits the following noncommutative counterpart:

CONJECTURE ($\mathrm{D}_{\mathrm{nc}}(\mathcal{A})$). *The equality $K_0(\mathcal{A})_\mathbb{Q}/_{\sim \mathrm{hom}} = K_0(\mathcal{A})_\mathbb{Q}/_{\sim \mathrm{num}}$ holds.*

The next result, whose proof makes use of the above isomorphisms (2.3)-(2.4), relates this conjecture with Grothendieck's original conjecture:

THEOREM 2.4. *([**50**, Thm. 1.1] and [**57**, Thm. 1.1]) Given a smooth proper k-scheme X, we have the equivalence of conjectures* $\mathrm{D}(X) \Leftrightarrow \mathrm{D}_{\mathrm{nc}}(\mathrm{perf}_{\mathrm{dg}}(X))$.

[8] If π_+ is algebraic, then the odd Künneth projector π_- is also algebraic.

[9] As proved in [**47**, Thm. 5.1], $K_0(\mathcal{A})/_{\sim \mathrm{num}}$ is a finitely generated free abelian group. Consequently, $K_0(\mathcal{A})_\mathbb{Q}/_{\sim \mathrm{num}}$ is a finite-dimensional $\mathbb{Q}$-vector space.

2.3. Noncommutative Voevodsky nilpotence conjecture. Let k be a base field of characteristic $p \geq 0$ and $\mathcal{A}$ a smooth proper k-linear dg category. Similarly to Voevodsky's definition of the nilpotence equivalence relation, an element α of the Grothendieck group $K_0(\mathcal{A})_\mathbb{Q}$ is called *nilpotently trivial* if there exists an integer $n \gg 0$ such that the associated element $\alpha^{\otimes n}$ of the Grothendieck group $K_0(\mathcal{A}^{\otimes n})_\mathbb{Q}$ is equal to zero. In what follows, we will write $K_0(\mathcal{A})_\mathbb{Q}/_{\sim \mathrm{nil}}$ for the quotient of $K_0(\mathcal{A})_\mathbb{Q}$ by the nilpotently trivial elements. Under these definitions, the Voevodsky nilpotence conjecture admits the noncommutative counterpart:

CONJECTURE ($\mathrm{V}_{\mathrm{nc}}(\mathcal{A})$). *The equality* $K_0(\mathcal{A})_\mathbb{Q}/_{\sim \mathrm{nil}} = K_0(\mathcal{A})_\mathbb{Q}/_{\sim \mathrm{num}}$ *holds.*

REMARK 2.5. The image of a nilpotently trivial element $\alpha \in K_0(\mathcal{A})_\mathbb{Q}$ under the above $\mathbb{Q}$-linear homomorphism (2.6) is equal to zero. Consequently, similarly to Remark 1.2, we have the implication of conjectures $\mathrm{V}_{\mathrm{nc}}(\mathcal{A}) \Rightarrow \mathrm{D}_{\mathrm{nc}}(\mathcal{A})$.

The next result, whose proof makes use of the theory of orbit categories, relates this conjecture with Voevodsky's original conjecture:

THEOREM 2.6. ([**5**, Thm. 1.1]) *Given a smooth proper k-scheme X, we have the equivalence of conjectures* $\mathrm{V}(X) \Leftrightarrow \mathrm{V}_{\mathrm{nc}}(\mathrm{perf}_{\mathrm{dg}}(X))$.

2.4. Noncommutative Beilinson conjecture. Let $k = \mathbb{F}_q$ be a finite field of characteristic p and $\mathcal{A}$ a smooth proper k-linear dg category. The Beilinson conjecture admits the following noncommutative counterpart:

CONJECTURE ($\mathrm{B}_{\mathrm{nc}}(\mathcal{A})$). *The equality* $K_0(\mathcal{A})_\mathbb{Q} = K_0(\mathcal{A})_\mathbb{Q}/_{\sim \mathrm{num}}$ *holds.*

REMARK 2.7. Similarly to Remark 1.3, note that in the case where k is a finite field, we have the implication of conjectures $\mathrm{B}_{\mathrm{nc}}(\mathcal{A}) \Rightarrow \mathrm{V}_{\mathrm{nc}}(\mathcal{A})$.

The next result, whose proof makes use of the Hirzebruch-Riemann-Roch theorem, relates this conjecture with Beilinson's original conjecture:

THEOREM 2.8. ([**55**, Thm. 1.3]) *Given a smooth proper k-scheme X, we have the equivalence of conjectures* $\mathrm{B}(X) \Leftrightarrow \mathrm{B}_{\mathrm{nc}}(\mathrm{perf}_{\mathrm{dg}}(X))$.

2.5. Noncommutative Weil conjecture. Let $k = \mathbb{F}_q$ be a finite field of characteristic p and $\mathcal{A}$ a smooth proper k-linear dg category. As explained in [**54**, §6], the topological periodic cyclic homology group $TP_0(\mathcal{A})_{1/p}$, resp. $TP_1(\mathcal{A})_{1/p}$, comes equipped with an automorphism F_0, resp. F_1, called the "cyclotomic Frobenius"[10]. Hence, we define the *even/odd zeta function of $\mathcal{A}$* as the formal power series:

$$\begin{aligned} Z_{\mathrm{even}}(\mathcal{A};t) &:= \det(\mathrm{id} - t\,\mathrm{F}_0 | TP_0(\mathcal{A})_{1/p})^{-1} \in K[\![t]\!] \\ Z_{\mathrm{odd}}(\mathcal{A};t) &:= \det(\mathrm{id} - t\,\mathrm{F}_1 | TP_1(\mathcal{A})_{1/p})^{-1} \in K[\![t]\!] . \end{aligned}$$

Under these definitions, Weil's conjecture admits the noncommutative counterpart:

CONJECTURE ($\mathrm{W}_{\mathrm{nc}}(\mathcal{A})$). *The eigenvalues of the automorphism* F_0, *resp.* F_1, *are algebraic numbers and all their complex conjugates have absolute value* 1, *resp.* $q^{\frac{1}{2}}$.

[10]The cyclotomic Frobenius is not compatible with the $\mathbb{Z}/2$-graded structure of $TP_*(\mathcal{A})_{1/p}$. Instead, we have canonical isomorphisms $\mathrm{F}_n \simeq q \cdot \mathrm{F}_{n+2}$ for every $n \in \mathbb{Z}$.

In contrast with the commutative world, the cyclotomic Frobenius is not induced from an endomorphism[11] of $\mathcal{A}$. Consequently, in contrast with the commutative world, it is not known if the polynomials $\det(\mathrm{id} - t\,\mathrm{F}_0 | TP_0(\mathcal{A})_{1/p})$ and $\det(\mathrm{id} - t\,\mathrm{F}_1 | TP_1(\mathcal{A})_{1/p})$ have integer coefficients (or rational coefficients). Nevertheless, after choosing an embedding $\iota\colon K \hookrightarrow \mathbb{C}$, we define the *even/odd Hasse-Weil zeta function of* $\mathcal{A}$ as follows:

$$\begin{aligned}\zeta_{\mathrm{even}}(\mathcal{A};s) &:= \det(\mathrm{id} - q^{-s}(\mathrm{F}_0 \otimes_{K,\iota} \mathbb{C})\,|\,TP_0(\mathcal{A})_{1/p} \otimes_{K,\iota} \mathbb{C})^{-1} \\ \zeta_{\mathrm{odd}}(\mathcal{A};s) &:= \det(\mathrm{id} - q^{-s}(\mathrm{F}_1 \otimes_{K,\iota} \mathbb{C})\,|\,TP_1(\mathcal{A})_{1/p} \otimes_{K,\iota} \mathbb{C})^{-1}.\end{aligned}$$

REMARK 2.9 (Analogue of the noncommutative Riemann hypothesis). Similarly to Remark 1.4, the conjecture $\mathrm{W}_{\mathrm{nc}}(\mathcal{A})$ may be called the "analogue of the noncommutative Riemann hypothesis" because it implies that if $z \in \mathbb{C}$ is a pole of $\zeta_{\mathrm{even}}(\mathcal{A};s)$, resp. $\zeta_{\mathrm{odd}}(\mathcal{A};s)$, then $\mathrm{Re}(z) = 0$, resp. $\mathrm{Re}(z) = \frac{1}{2}$ (independently of the chosen embedding ι).

The next result, whose proof makes use of the above isomorphism (2.4), relates the above conjecture with Weil's original conjecture:

THEOREM 2.10. *([**54**, Thm. 1.5]) Given a smooth proper k-scheme X, we have the equivalence of conjectures* $\mathrm{W}(X) \Leftrightarrow \mathrm{W}_{\mathrm{nc}}(\mathrm{perf}_{\mathrm{dg}}(X))$.

2.6. Noncommutative Tate conjecture. Let $k = \mathbb{F}_q$ be a finite field of characteristic p and $\mathcal{A}$ a smooth proper k-linear dg category. Given a prime number $l \neq p$, consider the following abelian groups

$$\mathrm{Hom}(\mathbb{Z}(l^\infty), \pi_{-1}(L_{KU}(K(\mathcal{A} \otimes_k \mathbb{F}_{q^n})))) \qquad n \geq 1\,, \tag{2.7}$$

where $\mathbb{Z}(l^\infty)$ stands for the Prüfer l-group[12], $K(\mathcal{A} \otimes_k \mathbb{F}_{q^n})$ for the algebraic K-theory spectrum of the dg category $\mathcal{A} \otimes_k \mathbb{F}_{q^n}$, and $L_{KU}(-)$ for the Bousfield localization functor with respect to topological complex K-theory KU. Under these notations, the Tate conjecture admits the following noncommutative counterpart:

CONJECTURE ($\mathrm{T}^l_{\mathrm{nc}}(\mathcal{A})$). *The abelian groups* (2.7) *are trivial.*

REMARK 2.11. Note that the conjecture $\mathrm{T}^l_{\mathrm{nc}}(\mathcal{A})$ holds, for example, whenever the abelian groups $\pi_{-1}(L_{KU}(K(\mathcal{A} \otimes_k \mathbb{F}_{q^n}))), n \geq 1$, are finitely generated.

The next result, whose proof is obtained by leveraging the pioneering work of Thomason [**61**], relates this conjecture with Tate's original conjecture:

THEOREM 2.12. *([**55**, Thm. 1.3]) Given a smooth proper k-scheme X, we have the equivalence of conjectures* $\mathrm{T}^l(X) \Leftrightarrow \mathrm{T}^l_{\mathrm{nc}}(\mathrm{perf}_{\mathrm{dg}}(X))$.

2.7. Noncommutative p-version of the Tate conjecture. Let $k = \mathbb{F}_q$ be a finite field of characteristic p and $\mathcal{A}$ a smooth proper k-linear dg category. Recall from §2.5 that the K-vector space $TP_0(\mathcal{A})_{1/p}$ comes equipped with an automorphism F_0 called the "cyclotomic Frobenius". Moreover, as explained in [**55**, §3], the right-hand side of (2.6) gives rise to a K-linear homomorphism:

$$K_0(\mathcal{A})_K \longrightarrow TP_0(\mathcal{A})_{1/p}^{\mathrm{F}_0}. \tag{2.8}$$

[11]Note that in the particular case where $\mathcal{A}$ is a k-algebra A the Frobenius map $a \mapsto a^q$ is a k-algebra endomorphism if and only if A is commutative.

[12]The functor $\mathrm{Hom}(\mathbb{Z}(l^\infty), -)$ agrees with the classical l-adic Tate module functor.

Under these notations, the p-version of the Tate conjecture admits the following noncommutative counterpart:

CONJECTURE ($\mathrm{T}^p_{\mathrm{nc}}(\mathcal{A})$). *The homomorphism* (2.8) *is surjective.*

The next result, whose proof makes use of the above isomorphism (2.4), relates this conjecture with the original conjecture:

THEOREM 2.13. *(*[**55**, Thm. 1.3]*) Given a smooth proper k-scheme X, we have the equivalence of conjectures* $\mathrm{T}^p(X) \Leftrightarrow \mathrm{T}^p_{\mathrm{nc}}(\mathrm{perf}_{\mathrm{dg}}(X))$.

2.8. Noncommutative strong form of the Tate conjecture. Let $k = \mathbb{F}_q$ be a finite field of characteristic p and $\mathcal{A}$ a smooth proper k-linear dg category. Recall from §2.5 the definition of the even Hasse-Weil zeta function $\zeta_{\mathrm{even}}(\mathcal{A}; s)$ of $\mathcal{A}$. Under these notations, the strong of the Tate conjecture admits the following noncommutative counterpart:

CONJECTURE ($\mathrm{ST}_{\mathrm{nc}}(\mathcal{A})$). *The order* $\mathrm{ord}_{s=0}\zeta_{\mathrm{even}}(\mathcal{A}; s)$ *of the even Hasse-Weil zeta function* $\zeta_{\mathrm{even}}(\mathcal{A}; s)$ *at the pole* $s = 0$ *is equal to* $-\mathrm{dim}_{\mathbb{Q}} K_0(\mathcal{A})_{\mathbb{Q}}/_{\sim \mathrm{num}}$.

REMARK 2.14 (Alternative formulation). By definition of the even Hasse-Weil zeta function of $\mathcal{A}$, the integer $-\mathrm{ord}_{s=0}\zeta_{\mathrm{even}}(\mathcal{A}; s)$ agrees with the algebraic multiplicity of the eigenvalue $q^0 = 1$ of the automorphism $\mathrm{F}_0 \otimes_{K,\iota} \mathbb{C}$ (or, equivalently, of F_0). Hence, the conjecture $\mathrm{ST}_{\mathrm{nc}}(\mathcal{A})$ may be alternatively formulated as follows: *the algebraic multiplicity of the eigenvalue* 1 *of* F_0 *agrees with* $\mathrm{dim}_{\mathbb{Q}} K_0(\mathcal{A})_{\mathbb{Q}}/_{\sim \mathrm{num}}$. This shows, in particular, that the integer $\mathrm{ord}_{s=0}\zeta_{\mathrm{even}}(\mathcal{A}; s)$ is independent of the embedding $\iota\colon K \hookrightarrow \mathbb{C}$ used in the definition of $\zeta_{\mathrm{even}}(\mathcal{A}; s)$.

REMARK 2.15. Similarly to Remark 1.6, as proved in [**54**, Thm. 9.3], we have the equivalence of conjectures $\mathrm{ST}_{\mathrm{nc}}(\mathcal{A}) \Leftrightarrow \mathrm{B}_{\mathrm{nc}}(\mathcal{A}) + \mathrm{T}^p_{\mathrm{nc}}(\mathcal{A})$.

The next result, whose proof is obtained by combining Remarks 1.6 and 2.15 with Theorems 2.8 and 2.13, relates this conjecture with Tate's original conjecture:

THEOREM 2.16. *(*[**54**, Thm. 1.17]*) Given a smooth proper k-scheme X, we have the equivalence of conjectures* $\mathrm{ST}(X) \Leftrightarrow \mathrm{ST}_{\mathrm{nc}}(\mathrm{perf}_{\mathrm{dg}}(X))$.

2.9. Noncommutative Parshin conjecture. Let $k = \mathbb{F}_q$ be a finite field of characteristic p and $\mathcal{A}$ a smooth proper k-linear dg category. The Parshin conjecture admits the following noncommutative counterpart:

CONJECTURE ($\mathrm{P}_{\mathrm{nc}}(\mathcal{A})$). *The groups $K_n(\mathcal{A})$, with $n \geq 1$, are torsion.*

The next result, whose proof is clear, relates this conjecture with Parshin's original conjecture:

THEOREM 2.17. *(*[**55**, Thm. 1.3]*) Given a smooth proper k-scheme X, we have the equivalence of conjectures* $\mathrm{P}(X) \Leftrightarrow \mathrm{P}_{\mathrm{nc}}(\mathrm{perf}_{\mathrm{dg}}(X))$.

2.10. Noncommutative Kimura-finiteness conjecture. Let k be a base field of characteristic $p \geq 0$ and $\mathcal{A}$ a smooth proper k-linear dg category. Recall from §2.1 that, by construction, the category of noncommutative Chow motives $\mathrm{NChow}(k)_{\mathbb{Q}}$ is $\mathbb{Q}$-linear, idempotent complete and symmetric monoidal. Hence, the Kimura-finiteness conjecture admits the following noncommutative counterpart:

CONJECTURE ($\mathrm{K}_{\mathrm{nc}}(\mathcal{A})$). *The noncommutative Chow motive $U(\mathcal{A})_{\mathbb{Q}}$ is Kimura-finite.*

The next result, whose proof makes use of a precise bridge between the category of Chow motives and the category of noncommutative Chow motives, relates this conjecture with Kimura's original conjecture:

THEOREM 2.18. *(*[**53**, Thm. 2.1]*) Given a smooth proper k-scheme X, we have the implication of conjectures* $\mathrm{K}(X) \Rightarrow \mathrm{K}_{\mathrm{nc}}(\mathrm{perf}_{\mathrm{dg}}(X))$.

2.11. Noncommutative Schur-finiteness conjecture. Let k be a perfect base field of characteristic $p \geq 0$ and $\mathcal{A}$ a smooth k-linear dg category. Recall from [**52**, §8-§9] the definition of the triangulated category of noncommutative mixed motives $\mathrm{NMot}(k)_{\mathbb{Q}}$ (denoted by $\mathrm{Nmot}^{\mathbb{A}^1}_{\mathrm{loc}}(k)_{\mathbb{Q}}$ in *loc. cit.*). By construction, this category is $\mathbb{Q}$-linear, idempotent complete, symmetric monoidal, and comes equipped with a symmetric monoidal functor $\mathrm{U}(-)_{\mathbb{Q}}\colon \mathrm{dgcat}_{\mathrm{s}}(k) \to \mathrm{NMot}(k)_{\mathbb{Q}}$ defined on smooth dg categories. Under these notations, the Schur-finiteness conjecture admits the following noncommutative counterpart:

CONJECTURE ($\mathrm{S}_{\mathrm{nc}}(\mathcal{A})$). *The noncommutative mixed motive* $\mathrm{U}(\mathcal{A})_{\mathbb{Q}}$ *is Schur-finite.*

The next result, whose proof makes use of a precise bridge between the category of Chow motives and the category of noncommutative Chow motives, relates this conjecture with Schur's original conjecture:

THEOREM 2.19. *(*[**52**, Prop. 9.17]*) Given a smooth k-scheme X, we have the equivalence of conjectures* $\mathrm{S}(X) \Leftrightarrow \mathrm{S}_{\mathrm{nc}}(\mathrm{perf}_{\mathrm{dg}}(X))$.

3. Applications to commutative geometry

Morally speaking, the theorems of §2 show that the celebrated conjectures of Grothendieck, Voevodsky, Beilinson, Weil, Tate, Parshin, and Schur, belong not only to the realm of algebraic geometry but also to the broad setting of dg categories. This noncommutative viewpoint, where one studies a scheme via its dg category of perfect complexes, led to a proof[13] of these celebrated conjectures in several new cases. In this section, we describe some of these new cases.

NOTATION 3.1. In order to simplify the exposition, we will often use the letter C to denote one of the celebrated conjectures $\{\mathrm{C}^+, \mathrm{D}, \mathrm{V}, \mathrm{B}, \mathrm{W}, \mathrm{T}^l, \mathrm{T}^p, \mathrm{ST}, \mathrm{P}, \mathrm{K}, \mathrm{S}\}$.

3.1. Derived invariance. Let k be a base field of characteristic $p \geq 0$. Note that the theorems of §2 imply automatically the following result:

COROLLARY 3.2. *Let X and Y be two smooth proper k-schemes (in the case of conjecture* S *we assume solely that X and Y are smooth) with (Fourier-Mukai) equivalent categories of perfect complexes* $\mathrm{perf}(X)$ *and* $\mathrm{perf}(Y)$. *Under these assumptions, we have the following equivalences of conjectures:*

$$\mathrm{C}(X) \Leftrightarrow \mathrm{C}(Y) \quad \text{with} \quad \mathrm{C} \in \{\mathrm{C}^+, \mathrm{D}, \mathrm{V}, \mathrm{B}, \mathrm{W}, \mathrm{T}^l, \mathrm{T}^p, \mathrm{ST}, \mathrm{P}, \mathrm{S}\}.$$

Roughly speaking, Corollary 3.2 shows that the celebrated conjectures of §1 are invariant under derived equivalence. This flexibility is very useful and is often used in the proofs of some theorems below.

[13]In what concerns the Weil conjecture (and the Grothendieck standard conjecture of type C^+ over a finite field), the noncommutative viewpoint led to an alternative proof of this celebrated conjecture in several new cases, which avoids all the involved tools used by Deligne.

3.2. Quadric fibrations. Let k be a perfect base field of characteristic $p \geq 0$, B a smooth proper k-scheme of dimension d (in the case of conjecture S we assume solely that B is smooth), and $q\colon Q \to B$ a flat quadric fibration of relative dimension d_q.

THEOREM 3.3. *Assume that all the fibers of q are quadrics of corank ≤ 1 and that the locus $Z \hookrightarrow B$ of critical values of q is smooth.*

(i) *When d_q is even, we have the following equivalences of conjectures*

$$\mathrm{C}(B) + \mathrm{C}(\widetilde{B}) \Leftrightarrow \mathrm{C}(Q) \quad \textit{with} \quad \mathrm{C} \in \{\mathrm{C}^+, \mathrm{D}, \mathrm{V}, \mathrm{B}, \mathrm{W}, \mathrm{T}^l\,(l \neq 2), \mathrm{T}^p, \mathrm{ST}, \mathrm{P}, \mathrm{S}\},$$

where $\widetilde{B}$ stands for the discriminant twofold cover of B (ramified over Z).

(ii) *When d_q is odd, $p \neq 2$, $d \leq 1$, and k is algebraically closed or a finite field, we have the following equivalences of conjectures:*

$$\mathrm{C}(B) + \mathrm{C}(Z) \Leftrightarrow \mathrm{C}(Q) \quad \textit{with} \quad \mathrm{C} \in \{\mathrm{C}^+, \mathrm{D}, \mathrm{V}, \mathrm{B}, \mathrm{W}, \mathrm{T}^l, \mathrm{T}^p, \mathrm{ST}, \mathrm{P}, \mathrm{S}\}.$$

(iii) *When d_q is odd and $p \neq 2$, we have the following implication of conjectures*

$$\{\mathrm{S}(U_j)\} + \{\mathrm{S}(\widetilde{Z}_j)\} \Rightarrow \mathrm{S}(Q),$$

where U_j is an affine open subscheme of B and $\widetilde{Z}_j$ is a certain Galois twofold cover of $Z_j := Z \cap U_j$ induced by the restriction of q to Z_j.

Roughly speaking, Theorem 3.3 relates the celebrated conjectures for the total space Q with the celebrated conjectures for the base B. Items (i)-(ii) were proved in [**5**, Thm. 1.2][**56**, Thm. 1.1(i)] in the case of the conjectures V and S. The proof of the other cases is similar. Item (iii) was proved in [**56**, Thm. 1.1(ii)].

COROLLARY 3.4 (Low-dimensional bases). *Let Q be as in Theorem 3.3.*

(i) *When d_q is even and $d \leq 1$, the following conjectures hold:*

$$\mathrm{C}(Q) \quad \text{with} \quad \mathrm{C} \in \{\mathrm{C}^+, \mathrm{D}, \mathrm{V}, \mathrm{B}, \mathrm{W}, \mathrm{T}^l\,(l \neq 2), \mathrm{T}^p, \mathrm{ST}, \mathrm{P}, \mathrm{S}\}.$$

Moreover, $\mathrm{C}^+(Q)$ holds when $d \leq 2$, $\mathrm{D}(Q)$ holds when $d \leq 2$ or when $d \leq 4$ and $p = 0$, and $\mathrm{V}(Q)$ holds when $d \leq 2$.

(ii) *When d_q is odd, $p \neq 2$, $d \leq 1$, and k is algebraically closed or a finite field, the following conjectures hold:*

$$\mathrm{C}(Q) \quad \text{with} \quad \mathrm{C} \in \{\mathrm{C}^+, \mathrm{D}, \mathrm{V}, \mathrm{B}, \mathrm{W}, \mathrm{T}^l, \mathrm{T}^p, \mathrm{ST}, \mathrm{P}, \mathrm{S}\}.$$

(iii) *When d_q is odd, $p \neq 2$, and $d \leq 1$, the conjecture $\mathrm{S}(Q)$ holds. Moreover, when $d \leq 2$, we have the implication of conjectures $\mathrm{S}(B) \Rightarrow \mathrm{S}(Q)$.*

REMARK 3.5 (Kimura-finiteness). Assume that B is a smooth k-curve.

(i) When d_q is even, it follows from Corollary 3.4(i) that the mixed motive $M(Q)_\mathbb{Q}$ is Schur-finite. As proved in [**51**, Thm. 1.1(i)], $M(Q)_\mathbb{Q}$ is moreover Kimura-finite and $\mathrm{kim}(M(Q)_\mathbb{Q}) = d_q \cdot \mathrm{kim}(M(B)_\mathbb{Q}) + \mathrm{kim}(M(\widetilde{B})_\mathbb{Q})$.

(ii) When d_q is odd, $p \neq 2$, and k is algebraically closed or a finite field, it follows from Corollary 3.4(ii) that the mixed motive $M(Q)_\mathbb{Q}$ is Schur-finite. As proved in [**51**, Thm. 1.1(ii)], the mixed motive $M(Q)_\mathbb{Q}$ is moreover Kimura-finite and $\mathrm{kim}(M(Q)_\mathbb{Q}) = (d_q + 1) \cdot \mathrm{kim}(M(B)_\mathbb{Q}) + \#Z$.

REMARK 3.6 (Bass-finiteness conjecture). Let $k = \mathbb{F}_q$ be a finite field and X a smooth k-scheme of finite type. In the seventies, Bass [**3**] conjectured that the algebraic K-theory groups $K_n(X), n \geq 0$, are finitely generated. In the same vein, we can consider the mod 2-torsion Bass-finiteness, where $K_n(X)$ is replaced by $K_n(X)_{1/2}$. As proved in [**56**, Thm. 1.10], the above Theorem 3.3 (items (i)-(iii)) holds similarly for the mod 2-torsion Bass-finiteness conjecture.

3.3. Intersections of quadrics. Let k be a perfect base field of characteristic $p \geq 0$ and X a smooth complete intersection of m quadric hypersurfaces in $\mathbb{P}^n$. The linear span of these quadric hypersurfaces give rise to a flat quadric fibration $q\colon Q \to \mathbb{P}^{m-1}$ of relative dimension $n-1$.

THEOREM 3.7. *Assume that all the fibers of q are quadrics of corank ≤ 1 and that the locus $Z \hookrightarrow \mathbb{P}^{m-1}$ of critical values of q is smooth. Under these assumptions, we have the following equivalences/implications of conjectures:*

$$\begin{cases} \mathrm{C}(Q) \Leftrightarrow \mathrm{C}(X) & 2m \leq n-1 \\ \mathrm{C}(Q) \Rightarrow \mathrm{C}(X) & 2m > n-1 \end{cases} \quad \text{with} \quad \mathrm{C} \in \{\mathrm{C}^+, \mathrm{D}, \mathrm{V}, \mathrm{B}, \mathrm{W}, \mathrm{T}^l, \mathrm{T}^p, \mathrm{ST}, \mathrm{P}, \mathrm{S}\}.$$

Intuitively speaking, Theorem 3.7 shows that in order to solve the celebrated conjectures for intersections of quadrics, it suffices to solve the celebrated conjectures from quadric fibrations (and vice-versa). This result was proved in [**56**, Thm. 1.5] in the case of the conjecture S. The proof of the other cases is similar.

COROLLARY 3.8 (Intersections of up-to-five quadrics). *Let X (and the associated fibration q) be as in Theorem 3.7.*

(i) *When n is odd and $m \leq 2$, the following conjectures hold:*

$$\mathrm{C}(X) \quad \text{with} \quad \mathrm{C} \in \{\mathrm{C}^+, \mathrm{D}, \mathrm{V}, \mathrm{B}, \mathrm{W}, \mathrm{T}^l\,(l \neq 2), \mathrm{T}^p, \mathrm{ST}, \mathrm{P}, \mathrm{S}\}.$$

Moreover, $\mathrm{C}^+(X)$ holds when $m \leq 3$, $\mathrm{D}(X)$ holds when $m \leq 3$ or when $m \leq 5$ and $p = 0$, and $\mathrm{V}(X)$ holds when $m \leq 3$.

(ii) *When n is even, $p \neq 2$, $m \leq 2$, and k is algebraically closed or a finite field, the following conjectures hold:*

$$\mathrm{C}(X) \quad \text{with} \quad \mathrm{C} \in \{\mathrm{C}^+, \mathrm{D}, \mathrm{V}, \mathrm{B}, \mathrm{W}, \mathrm{T}^l, \mathrm{T}^p, \mathrm{ST}, \mathrm{P}, \mathrm{S}\}.$$

(iii) *When n is even, $p \neq 2$, and $m \leq 3$, the conjecture $\mathrm{S}(X)$ holds.*

3.4. Families of sextic du Val del Pezzo surfaces. Let k be a perfect base field of characteristic $p \geq 0$, B a smooth proper k-scheme of dimension d (in the case of conjecture S we assume solely that B is smooth), and $f\colon X \to B$ a *family of sextic du Val del Pezzo surfaces*, i.e., a flat morphism such that for every geometric point $b \in B$ the associated fiber X_b is a sextic du Val del Pezzo surface[14]. Following Kuznetsov [**38**, §5], let $\mathcal{M}_2$, resp. $\mathcal{M}_3$, be the relative moduli stack of semistable sheaves on fibers of X over B with Hilbert polynomial $h_2(t) := (3t+2)(t+1)$, resp. $h_3(t) := (3t+3)(t+1)$, and Z_2, resp. Z_3, the coarse moduli space of $\mathcal{M}_2$, resp. $\mathcal{M}_3$.

THEOREM 3.9. *Assume that $p \neq 2, 3$ and that X is smooth. Under these assumptions, we have the following equivalences of conjectures:*

$$\mathrm{C}(B) + \mathrm{C}(Z_2) + \mathrm{C}(Z_3) \Leftrightarrow \mathrm{C}(X) \quad \textit{with} \quad \mathrm{C} \in \{\mathrm{C}^+, \mathrm{D}, \mathrm{V}, \mathrm{B}, \mathrm{W}, \mathrm{T}^l, \mathrm{T}^p, \mathrm{ST}, \mathrm{P}, \mathrm{S}\}.$$

Roughly speaking, Theorem 3.9 relates the celebrated conjectures for the total space X with the celebrated conjectures for the base B. Theorem 3.9 was proved in [**56**, Thm. 1.7] in the case of the conjecture S. The proof of the other cases is similar.

---

[14]Recall that a *sextic du Val del Pezzo surface* is a projective surface S with at worst du Val singularities and whose ample anticanonical class K_S is such that $K_S^2 = 6$.

COROLLARY 3.10 (Low-dimensional bases). *Let X be as in Theorem 3.9. When $d \leq 1$, the following conjectures hold:*

$$\mathrm{C}(X) \quad \text{with} \quad \mathrm{C} \in \{\mathrm{C}^+, \mathrm{D}, \mathrm{V}, \mathrm{B}, \mathrm{W}, \mathrm{T}^l, \mathrm{T}^p, \mathrm{ST}, \mathrm{P}, \mathrm{S}\}.$$

Moreover, $\mathrm{C}^+(X)$ holds when $d \leq 2$, $\mathrm{D}(X)$ holds when $d \leq 2$ or when $d \leq 4$ and $p = 0$, and $\mathrm{V}(X)$ holds when $d \leq 2$.

3.5. Linear sections of Grassmannians. Let k be a base field of characteristic $p = 0$, W a k-vector space of dimension 6 or 7, $X := \mathrm{Gr}(2, W)$ the Grassmannian variety of 2-dimensional subspaces equipped with the Plücker embedding $\mathrm{Gr}(2, W) \hookrightarrow \mathbb{P}(\wedge^2(W))$, and Y the Pfaffian variety $\mathrm{Pf}(4, W^*) \subset \mathbb{P}(\wedge^2(W^*))$. Given a linear subspace $L \subset \wedge^2(W^*)$, consider the associated linear sections

$$X_L := X \times_{\mathbb{P}(\wedge^2(W))} \mathbb{P}(L^\perp) \qquad Y_L := Y \times_{\mathbb{P}(\wedge^2(W^*))} \mathbb{P}(L),$$

where $L^\perp$ stands for the kernel of the induced homomorphism $\wedge^2(W) \twoheadrightarrow L^*$.

THEOREM 3.11. *Assume that X_L and Y_L are smooth*[15]*, and that $\mathrm{codim}(X_L) = \dim(L)$ and $\mathrm{codim}(Y_L) = \dim(L^\perp)$. Under these assumptions (which hold for a generic choice of L), we have the following equivalences of conjectures:*

$$\mathrm{C}(X_L) \Leftrightarrow \mathrm{C}(Y_L) \quad \textit{with} \quad \mathrm{C} \in \{\mathrm{C}^+, \mathrm{D}, \mathrm{V}, \mathrm{S}\}.$$

Intuitively speaking, Theorem 3.11 shows that in order to solve the celebrated conjectures for the linear section X_L, it suffices to solve the celebrated conjectures for the linear section Y_L (and vice-versa). Theorem 3.11 was proved in [**5**, Thm. 1.7] in the case of the conjecture V. The proof of the other cases is similar.

When $\dim(W) = 6$, we have $\dim(X_L) = 8 - \dim(L)$ and $\dim(Y_L) = \dim(L) - 2$. Moreover, in the case where $\dim(L) = 5$, resp. $\dim(L) = 6$, X_L is a Fano threefold, resp. $K3$-surface, and Y_L is a cubic threefold, resp. cubic fourfold.

When $\dim(W) = 7$, we have $\dim(X_L) = 10 - \dim(L)$ and $\dim(Y_L) = \dim(L) - 4$. Moreover, in the case where $\dim(L) = 5$, resp. $\dim(L) = 6$, X_L is a Fano fivefold, resp. Fano fourfold, and Y_L is a curve of degree 42, resp. surface of degree 42. Furthermore, in the case where $\dim(L) = 7$, X_L and Y_L are derived equivalent Calabi-Yau threefolds. In this particular latter case, Theorem 3.11 follows then from the above Corollary 3.2.

COROLLARY 3.12 (Low-dimensional sections). *Let X_L be as in Theorem 3.11.*

(i) *When $\dim(W) = 6$ and $\dim(L) \leq 3$, the following conjectures $\mathrm{C}(X_L)$, with $\mathrm{C} \in \{\mathrm{C}^+, \mathrm{D}, \mathrm{V}, \mathrm{S}\}$, hold. Moreover, $\mathrm{C}^+(X_L)$ holds when $\dim(L) \leq 4$, $\mathrm{D}(X_L)$ holds when $\dim(L) \leq 6$, and $\mathrm{V}(X_L)$ holds when $\dim(L) \leq 4$.*
(ii) *When $\dim(W) = 7$ and $\dim(L) \leq 5$, the following conjectures $\mathrm{C}(X_L)$, with $\mathrm{C} \in \{\mathrm{C}^+, \mathrm{D}, \mathrm{V}, \mathrm{S}\}$, hold. Moreover, $\mathrm{C}^+(X_L)$ holds when $\dim(L) \leq 6$, $\mathrm{D}(X_L)$ holds when $\dim(L) \leq 8$, and $\mathrm{V}(X_L)$ holds when $\dim(L) \leq 6$.*

REMARK 3.13 (Kimura-finiteness conjecture). Let X_L be as in Theorem 3.11. When $\dim(W) = 6$ and $\dim(L) \leq 3$ or when $\dim(W) = 7$ and $\dim(L) \leq 5$, it follows from Corollary 3.12 that the Chow motive $\mathfrak{h}(X_L)_\mathbb{Q}$ is Schur-finite. As proved in [**49**, Thm. 1.10], the Chow motive $\mathfrak{h}(X_L)_\mathbb{Q}$ is moreover Kimura-finite.

[15]The linear section X_L is smooth if and only if the linear section Y_L is smooth.

3.6. Linear sections of Lagrangian Grassmannians. Let k be a base field of characteristic $p = 0$, W a k-vector space of dimension 6 equipped with a symplectic form ω, and $X := \mathrm{LGr}(3, W)$ the associated Lagrangian Grassmannian of 3-dimensional subspaces. The natural representation of the symplectic group $\mathrm{Sp}(\omega)$ on $\wedge^3(W)$ decomposes into a direct sum $W \oplus V$. Moreover, the classical Plücker embedding $\mathrm{Gr}(3, W) \hookrightarrow \mathbb{P}(\wedge^3(W))$ restricts to an embedding $\mathrm{LGr}(3, W) \hookrightarrow \mathbb{P}(V)$ of the Lagrangian Grassmannian. Consider also the classical projective dual variety $\mathrm{LGr}(3, 6)^\vee \subseteq \mathbb{P}(V^*)$. This is a quartic hypersurface which is singular along a closed subvariety Z of dimension 9. Let us denote by Y the open dense subset $\mathrm{LGr}(3, 6)^\vee \backslash Z$. Given a linear subspace $L \subseteq V^*$ such that $\mathbb{P}(L) \cap Z = \emptyset$, consider the associated smooth linear sections $X_L := X \times_{\mathbb{P}(V)} \mathbb{P}(L^\perp)$ and $Y_L := Y \times_{\mathbb{P}(V^*)} \mathbb{P}(L)$.

THEOREM 3.14. *Assume that* $\mathrm{codim}(X_L) = \dim(L)$ *and* $\mathrm{codim}(Y_L) = \dim(L^\perp)$. *Under these assumption (which hold for a generic choice of L), we have the following equivalences of conjectures:*

$$\mathrm{C}(X_L) \Leftrightarrow \mathrm{C}(Y_L) \quad \textit{with} \quad \mathrm{C} \in \{\mathrm{C}^+, \mathrm{D}, \mathrm{V}, \mathrm{S}\}.$$

Intuitively speaking, Theorem 3.14 shows that in order to solve the celebrated conjectures for the linear section X_L, it suffices to solve the celebrated conjectures for the linear section Y_L (and vice-versa). Theorem 3.14 was proved in [**49**, Thm. 1.5] in the case of the conjecture S. The proof of the other cases is similar.

We have $\dim(X_L) = 6 - \dim(L)$ and $\dim(Y_L) = \dim(L) - 2$. Moreover, in the case where $\dim(L) = 3$, resp. $\dim(L) = 4$, X_L is a Fano threefold, resp. $K3$-surface of degree 16, and Y_L is a plane quartic, resp. $K3$-surface of degree 4.

COROLLARY 3.15 (Low-dimensional sections). *Let* X_L *be as in Theorem* 3.14. *When* $\dim(L) \leq 3$, *the following conjectures* $\mathrm{C}(X_L)$, *with* $\mathrm{C} \in \{\mathrm{C}^+, \mathrm{D}, \mathrm{V}, \mathrm{S}\}$, *hold. Moreover,* $\mathrm{C}^+(X_L)$ *holds when* $\dim(L) < 4$, $\mathrm{D}(X_L)$ *holds when* $\dim(L) \leq 6$, *and* $\mathrm{V}(X_L)$ *holds when* $\dim(L) \leq 4$.

REMARK 3.16 (Kimura-finiteness conjecture). Let X_L be as in Theorem 3.14. When $\dim(L) \leq 3$, it follows from Corollary 3.15 that the Chow motive $\mathfrak{h}(X_L)_\mathbb{Q}$ is Schur-finite. As proved in [**49**, Thm. 1.10], $\mathfrak{h}(X_L)_\mathbb{Q}$ is moreover Kimura-finite.

3.7. Linear sections of spinor varieties. Let k be a base field of characteristic $p = 0$, W a k-vector space of dimension 10 equipped with a nondegenerate quadratic form $q \in \mathrm{Sym}^2(W^*)$, and $X := \mathrm{OGr}_+(5, W)$ and $Y := \mathrm{OGr}_-(5, W)$ the connected components of the orthogonal Grassmannian of 5-dimensional subspaces. These are called the *spinor varieties.* By construction, we have a canonical embedding $\mathrm{OGr}_+(5, W) \hookrightarrow \mathbb{P}(V)$, where V stands for the corresponding half-spinor representation of the spin-group $\mathrm{Spin}(W)$. In the same vein, making use of the isomorphism $\mathbb{P}(V) \simeq \mathbb{P}(V^*)$ induced by the nondegenerate quadratic form q, we have the embedding $\mathrm{OGr}_-(5, W) \hookrightarrow \mathbb{P}(V^*)$. Given a linear subspace $L \subseteq V^*$, consider the associated linear sections $X_L := X \times_{\mathbb{P}(V)} \mathbb{P}(L^\perp)$ and $Y_L := Y \times_{\mathbb{P}(V^*)} \mathbb{P}(L)$.

THEOREM 3.17. *Assume that* X_L *and* Y_L *are smooth, and that* $\mathrm{codim}(X_L) = \dim(L)$ *and* $\mathrm{codim}(Y_L) = \dim(L^\perp)$. *Under these assumptions (which hold for a generic choice of L), we have the following equivalences of conjectures:*

$$\mathrm{C}(X_L) \Leftrightarrow \mathrm{C}(Y_L) \quad \textit{with} \quad \mathrm{C} \in \{\mathrm{C}^+, \mathrm{D}, \mathrm{V}, \mathrm{S}\}.$$

Intuitively speaking, Theorem 3.14 shows that in order to solve the celebrated conjectures for the linear section X_L, it suffices to solve the celebrated conjectures for the linear section Y_L (and vice-versa). Theorem 3.17 was proved in [**49**, Thm. 1.6] in the case of the conjecture S. The proof of the other cases is similar.

We have $\dim(X_L) = 10 - \dim(L)$ and $\dim(Y_L) = \dim(L) - 6$. Moreover, in the case where $\dim(L) = 7$, X_L is a Fano threefold and Y_L is a curve of genus 7. Furthermore, in the case where $\dim(L) = 8$, X_L and Y_L are derived equivalent $K3$-surfaces of degree 12. In this particular latter case, Theorem 3.17 follows then from the above Corollary 3.2.

COROLLARY 3.18 (Low-dimensional sections). *Let X_L be as in Theorem 3.17. When $\dim(L) \leq 7$, the following conjectures $\mathrm{C}(X_L)$, with $\mathrm{C} \in \{\mathrm{C}^+, \mathrm{D}, \mathrm{V}, \mathrm{S}\}$, hold. Moreover, $\mathrm{C}^+(X_L)$ holds when $\dim(L) \leq 8$, $\mathrm{D}(X_L)$ holds when $\dim(L) \leq 10$, and $\mathrm{V}(X_L)$ holds when $\dim(L) \leq 8$.*

REMARK 3.19 (Kimura-finiteness conjecture). Let X_L be as in Theorem 3.17. When $\dim(L) \leq 7$, it follows from Corollary 3.18 that the Chow motive $\mathfrak{h}(X_L)_{\mathbb{Q}}$ is Schur-finite. As proved in [**49**, Thm. 1.10], $\mathfrak{h}(X_L)_{\mathbb{Q}}$ is moreover Kimura-finite.

3.8. Linear sections of determinantal varieties. Let k be a perfect base field of characteristic $p \geq 0$, U_1 and U_2 two finite-dimensional k-vector spaces of dimensions d_1 and d_2, respectively, $V := U_1 \otimes U_2$, and $0 < r < d_1$ an integer. Consider the determinantal variety $\mathcal{Z}^r_{d_1,d_2} \subset \mathbb{P}(V)$ defined as the locus of those matrices $U_2 \to U_1^*$ with rank $\leq r$; recall that this condition can be described as the vanishing of the $(r+1)$-minors of the matrix of indeterminates:

$$\begin{pmatrix} x_{1,1} & \cdots & x_{1,d_2} \\ \vdots & \ddots & \vdots \\ x_{d_1,1} & \cdots & x_{d_1,d_2} \end{pmatrix}.$$

EXAMPLE 3.20 (Segre varieties). In the particular case where $r = 1$, the determinantal varieties reduce to the classical Segre varieties. Concretely, $\mathcal{Z}^1_{d_1,d_2}$ reduces to the image of Segre homomorphism $\mathbb{P}(U_1) \times \mathbb{P}(U_2) \to \mathbb{P}(V)$. For example, $\mathcal{Z}^1_{2,2}$ is the classical quadric hypersurface:

$$\{[x_{1,1} : x_{1,2} : x_{2,1} : x_{2,2}] \,|\, \det \begin{pmatrix} x_{1,1} & x_{1,2} \\ x_{2,1} & x_{2,2} \end{pmatrix} = 0\} \subset \mathbb{P}^3 .$$

In contrast with the Segre varieties, the determinantal varieties $\mathcal{Z}^r_{d_1,d_2}$, with $r \geq 2$, are not smooth. The singular locus of these latter varieties $\mathcal{Z}^r_{d_1,d_2}$ consists of those matrices $U_2 \to U_1^*$ with rank $< r$, *i.e.* it agrees with the closed subvariety $\mathcal{Z}^{r-1}_{d_1,d_2}$. Nevertheless, it is well-known that $\mathcal{Z}^r_{d_1,d_2}$ admits a canonical Springer resolution of singularities $X := \mathcal{X}^r_{d_1,d_2} \to \mathcal{Z}^r_{d_1,d_2}$. Dually, consider the variety $\mathcal{W}^r_{d_1,d_2} \subset \mathbb{P}(V^*)$, defined as the locus of those matrices $U_2^* \to U_1$ with corank $\geq r$, and the associated canonical Springer resolution of singularities $Y := \mathcal{Y}^r_{d_1,d_2} \to \mathcal{W}^r_{d_1,d_2}$. Given a linear subspace $L \subseteq V^*$, consider the associated linear sections $X_L := X \times_{\mathbb{P}(V)} \mathbb{P}(L^\perp)$ and $Y_L := Y \times_{\mathbb{P}(V^*)} \mathbb{P}(L)$. Note that whenever $\mathbb{P}(L^\perp)$ does not intersects the singular locus of $\mathcal{Z}^r_{d_1,d_2}$, we have $X_L = \mathbb{P}(L^\perp) \cap \mathcal{Z}^r_{d_1,d_2}$, i.e., X_L is a linear section of the determinantal variety $\mathcal{Z}^r_{d_1,d_2}$.

THEOREM 3.21. *Assume that* X_L *and* Y_L *are smooth, and that* $\mathrm{codim}(X_L) = \dim(L)$ *and* $\mathrm{codim}(Y_L) = \dim(L^\perp)$. *Under these assumptions (which hold for a generic choice of L), we have the following equivalences of conjectures:*

$$\mathrm{C}(X_L) \Leftrightarrow \mathrm{C}(Y_L) \quad \text{with} \quad \mathrm{C} \in \{\mathrm{C}^+, \mathrm{D}, \mathrm{V}, \mathrm{B}, \mathrm{W}, \mathrm{T}^l, \mathrm{T}^p, \mathrm{ST}, \mathrm{P}, \mathrm{S}\}\,.$$

Intuitively speaking, Theorem 3.14 shows that in order to solve the celebrated conjectures for the linear section X_L, it suffices to solve the celebrated conjectures for the linear section Y_L (and vice-versa). Theorem 3.21 was proved in [**57**, Cor. 2.4], resp. in [**55**, Cor. 1.7], in the case of the conjectures C^+ and D, resp. in the case of the conjectures B, T^l, T^p, and P. The proof of the other cases is similar.

By construction, we have the following equalities:

$$\dim(X_L) = r(d_1 + d_2 - r) - 1 - \dim(L) \quad \dim(Y_L) = r(d_1 - d_2 - r) - 1 + \dim(L)\,.$$

Moreover, in the case where $\dim(L) = d_2 r$, the linear sections X_L and Y_L are derived equivalent Calabi-Yau varieties. In this particular latter case, Theorem 3.21 follows then from the above Corollary 3.2.

COROLLARY 3.22 (High-dimensional sections). *Let* X_L *be as in Theorem* 3.21. *When* $\dim(L) \leq 2 - r(d_1 - d_2 - r)$, *the following conjectures hold:*

$$\mathrm{C}(X_L) \quad \text{with} \quad \mathrm{C} \in \{\mathrm{C}^+, \mathrm{D}, \mathrm{V}, \mathrm{B}, \mathrm{W}, \mathrm{T}^l, \mathrm{T}^p, \mathrm{ST}, \mathrm{P}, \mathrm{S}\}\,. \tag{3.1}$$

Moreover, $\mathrm{C}^+(X_L)$ *holds when* $\dim(L) \leq 3 - r(d_1 - d_2 - r)$, $\mathrm{D}(X_L)$ *holds when* $\dim(L) \leq 3 - r(d_1 - d_2 - r)$ *or when* $\dim(L) \leq 5 - r(d_1 - d_2 - r)$ *and* $p = 0$, *and* $\mathrm{V}(X_L)$ *holds when* $\dim(L) \leq 3 - r(d_1 - d_2 - r)$.

EXAMPLE 3.23 (Segre varieties). Let $r = 1$. Thanks to Corollary 3.22, when $\dim(L) = 3 - d_1 + d_2$, the following conjectures hold:

$$\mathrm{C}(X_L) \quad \text{with} \quad \mathrm{C} \in \{\mathrm{C}^+, \mathrm{D}, \mathrm{V}, \mathrm{B}, \mathrm{W}, \mathrm{T}^l, \mathrm{T}^p, \mathrm{ST}, \mathrm{P}, \mathrm{S}\}\,.$$

In all these cases, X_L is a smooth linear section of the Segre variety $\mathcal{Z}^1_{d_1,d_2}$. Moreover, X_L is Fano if and only if $\dim(L) < d_1$. Furthermore, $\dim(X_L) = 2d_1 - 5$. Therefore, by letting $d_1 \to \infty$ and by keeping $\dim(L)$ fixed, we obtain infinitely many examples of smooth proper k-schemes X_L, of arbitrary high dimension, satisfying the celebrated conjectures of §1. Note that in the particular case of the Weil conjecture (and in the case of the Grothendieck standard conjecture of type C^+ over a finite field) this proof avoids all the technical tools used by Deligne.

EXAMPLE 3.24 (Rational normal scrolls). Let $r = 1$, $d_1 = 4$ and $d_2 = 2$. In this particular case, the Segre variety $\mathcal{Z}^1_{4,2} \subset \mathbb{P}^7$ agrees with the rational normal 4-fold scroll $S_{1,1,1,1}$. Choose a linear subspace $L \subseteq V^*$ of dimension 1 such that the hyperplane $\mathbb{P}(L^\perp) \subset \mathbb{P}^7$ does not contains any 3-plane of the rulling of $S_{1,1,1,1}$; this condition holds for a generic choice of L. In this case, the linear section X_L agrees with the 3-fold scroll $S_{1,1,2}$. Hence, thanks to Example 3.23, we conclude that the following conjectures hold:

$$\mathrm{C}(S_{1,1,2}) \quad \text{with} \quad \mathrm{C} \in \{\mathrm{C}^+, \mathrm{D}, \mathrm{V}, \mathrm{B}, \mathrm{W}, \mathrm{T}^l, \mathrm{T}^p, \mathrm{ST}, \mathrm{P}, \mathrm{S}\}\,.$$

EXAMPLE 3.25 (Square matrices). Let $d_1 = d_2$. Thanks to Corollary 3.22, when $\dim(L) = 2 + r^2$, the following conjectures hold:

$$\mathrm{C}(X_L) \quad \text{with} \quad \mathrm{C} \in \{\mathrm{C}^+, \mathrm{D}, \mathrm{V}, \mathrm{B}, \mathrm{W}, \mathrm{T}^l, \mathrm{T}^p, \mathrm{ST}, \mathrm{P}, \mathrm{S}\}\,.$$

In all these cases, we have $\dim(X_L) = 2r(d_1 - r) - 3$. Therefore, by letting $d_1 \to \infty$ and by keeping $\dim(L)$ fixed, we obtain infinitely many new examples of smooth proper k-schemes X_L, of arbitrary high dimension, satisfying the celebrated conjectures of §1. Similarly to Example 3.23, note that in the particular case of the Weil conjecture (and in the case of the Grothendieck standard conjecture of type C^+ over a finite field) this proof avoids all the technical tools used by Deligne.

REMARK 3.26 (Kimura-finiteness conjecture). Let X_L be as in Theorem 3.21. When $\dim(L) = 2 - r(d_1 - d_2 - r)$, it follows from Corollary 3.22 that the Chow motive $\mathfrak{h}(X_L)_{\mathbb{Q}}$ is Schur-finite. A similar proof shows that $\mathfrak{h}(X_L)_{\mathbb{Q}}$ is Kimura-finite.

4. Applications to noncommutative geometry

The theorems of §2 enabled also a proof of the noncommutative counterparts of the celebrated conjectures of Grothendieck, Voevodsky, Beilinson, Weil, Tate, Kimura, and Schur, in many interesting cases. In this section, we describe some of these interesting cases. Similarly to §3, we will often use the letter C to denote one of the celebrated conjectures $\{\mathrm{C}^+, \mathrm{D}, \mathrm{V}, \mathrm{B}, \mathrm{W}, \mathrm{T}^l, \mathrm{T}^p, \mathrm{ST}, \mathrm{P}, \mathrm{K}, \mathrm{S}\}$. Moreover, given a smooth (proper) algebraic stack $\mathcal{X}$, we will write $\mathrm{C}_{\mathrm{nc}}(\mathcal{X})$ instead of $\mathrm{C}_{\mathrm{nc}}(\mathrm{perf}_{\mathrm{dg}}(\mathcal{X}))$.

4.1. Finite-dimensional algebras of finite global dimension. Let k be a perfect base field of characteristic $p \geq 0$ and A a finite-dimensional k-algebra of finite global dimension. Examples include path algebras of finite quivers without oriented cycles as well as their quotients by admissible ideals.

THEOREM 4.1. *The following conjectures hold:*

$$\mathrm{C}_{\mathrm{nc}}(A) \quad \textit{with} \quad \mathrm{C} \in \{\mathrm{C}^+, \mathrm{D}, \mathrm{V}, \mathrm{B}, \mathrm{W}, \mathrm{T}^p, \mathrm{ST}, \mathrm{P}, \mathrm{K}, \mathrm{S}\}.$$

Moreover, when $k = \overline{k}$, the conjecture $\mathrm{T}^l_{\mathrm{nc}}(A)$ also holds.

Theorem 4.1 was proved in [**54**, Thm. 3.1] in the case of the conjectures W and ST. The proof of the other cases is similar.

4.2. Semi-orthogonal decompositions. Let k be a base field of characteristic $p \geq 0$ and $\mathcal{B}, \mathcal{C} \subseteq \mathcal{A}$ smooth proper k-linear dg categories inducing a semi-orthogonal decomposition $\mathrm{H}^0(\mathcal{A}) = \langle \mathrm{H}^0(\mathcal{B}), \mathrm{H}^0(\mathcal{C})\rangle$ in the sense of Bondal-Orlov [**8**].

THEOREM 4.2. *We have the following equivalences of conjectures*

$$\mathrm{C}_{\mathrm{nc}}(\mathcal{B}) + \mathrm{C}_{\mathrm{nc}}(\mathcal{C}) \Leftrightarrow \mathrm{C}_{\mathrm{nc}}(\mathcal{A}) \quad \text{with} \quad \mathrm{C} \in \{\mathrm{C}^+, \mathrm{D}, \mathrm{V}, \mathrm{B}, \mathrm{W}, \mathrm{T}^l, \mathrm{T}^p, \mathrm{ST}, \mathrm{P}, \mathrm{K}, \mathrm{S}\}.$$

Intuitively speaking, Theorem 4.2 shows that the noncommutative counterparts of the celebrated conjectures are additive with respect to semi-orthogonal decompositions. Theorem 4.2 was proved in [**54**, Thm. 3.2] in the case of the conjectures W and ST. The proof of the other cases is similar.

4.3. Calabi-Yau dg categories associated to hypersurfaces. Let k be a base field of characteristic $p \geq 0$ and $X \subset \mathbb{P}^n$ a smooth hypersurface of degree $\deg(X) \leq n+1$. Following Kuznetsov [**37**], we have the following semi-orthogonal decomposition ($\mathcal{T}(X)$ stands for the orthogonal of the exceptional objects):

$$\mathrm{perf}(X) = \langle \mathcal{T}(X), \mathcal{O}_X, \ldots, \mathcal{O}_X(n - \deg(X))\rangle .$$

Moreover, the associated k-linear dg category $\mathcal{T}_{\mathrm{dg}}(X)$, defined as the dg enhancement of $\mathcal{T}(X)$ induced from $\mathrm{perf}_{\mathrm{dg}}(X)$, is a smooth proper Calabi-Yau dg category[16] of fractional dimension $\frac{(n+1)(\deg(X)-2)}{\deg(X)}$. By combining Theorem 4.2 with the Theorems of §2, we hence obtain the following result:

COROLLARY 4.3. *We have the equivalences of conjectures*

$$\mathrm{C}(X) \Leftrightarrow \mathrm{C}_{\mathrm{nc}}(\mathcal{T}_{\mathrm{dg}}(X)) \quad \textit{with} \quad \mathrm{C} \in \{\mathrm{C}^+, \mathrm{D}, \mathrm{V}, \mathrm{B}, \mathrm{W}, \mathrm{T}^l, \mathrm{T}^p, \mathrm{ST}, \mathrm{P}, \mathrm{S}\}$$

as well as the implication $\mathrm{K}(X) \Rightarrow \mathrm{K}_{\mathrm{nc}}(\mathcal{T}_{\mathrm{dg}}(X))$.

Roughly speaking, Corollary 4.3 shows that in what concerns the celebrated conjectures, there is no difference between the hypersurface X and the associated Calabi-Yau dg category $\mathcal{T}_{\mathrm{dg}}(X)$.

4.4. Root stacks. Let k be a base field of characteristic $p \geq 0$, X a smooth proper k-scheme of dimension d (in the case of conjecture S we assume solely that X is smooth), $\mathcal{L}$ a line bundle on X, $\varsigma \in \Gamma(X, \mathcal{L})$ a global section, and $n \geq 1$ an integer. Following Cadman [**9**, §2.2], the associated *root stack* is defined as the fiber-product

$$\begin{array}{ccc} \mathcal{X} := \sqrt[n]{(\mathcal{L},\varsigma)/X} & \longrightarrow & [\mathbb{A}^1/\mathbb{G}_m] \\ {\scriptstyle f}\downarrow & & \downarrow{\scriptstyle \theta_n} \\ X & \xrightarrow[(\mathcal{L},\varsigma)]{} & [\mathbb{A}^1/\mathbb{G}_m]\,, \end{array}$$

where θ_n stands for the morphism induced by the nth power map on $\mathbb{A}^1$ and $\mathbb{G}_m$. As proved by Ishii-Ueda in [**23**, Thm. 1.6], whenever the zero locus $Z \hookrightarrow X$ of ς is smooth, we have a semi-orthogonal decomposition

$$\mathrm{perf}(\mathcal{X}) = \langle \mathrm{perf}(Z)_{n-1}, \ldots, \mathrm{perf}(Z)_1, f^*(\mathrm{perf}(X))\rangle\,,$$

where all the categories $\mathrm{perf}(Z)_j$ are (Fourier-Mukai) equivalent to $\mathrm{perf}(Z)$. Hence, by combining Theorem 4.2 with the Theorems of §2, we obtain the following results:

COROLLARY 4.4. *Assume that the zero locus $Z \hookrightarrow X$ of the global section ς is smooth. Under this assumption, we have the equivalences of conjectures*

$$\mathrm{C}(X) + \mathrm{C}(Z) \Leftrightarrow \mathrm{C}_{\mathrm{nc}}(\mathcal{X}) \quad \textit{with} \quad \mathrm{C} \in \{\mathrm{C}^+, \mathrm{D}, \mathrm{V}, \mathrm{B}, \mathrm{W}, \mathrm{T}^l, \mathrm{T}^p, \mathrm{ST}, \mathrm{P}, \mathrm{S}\}$$

as well as the implication $\mathrm{K}(X) + \mathrm{K}(Z) \Rightarrow \mathrm{K}_{\mathrm{nc}}(\mathcal{X})$.

COROLLARY 4.5 (Low-dimensional root stacks). *Let $\mathcal{X}$ be as in Corollary 4.4. When $d \leq 1$, the following conjectures hold:*

$$\mathrm{C}_{\mathrm{nc}}(\mathcal{X}) \quad \text{with} \quad \mathrm{C} \in \{\mathrm{C}^+, \mathrm{D}, \mathrm{V}, \mathrm{B}, \mathrm{W}, \mathrm{T}^l, \mathrm{T}^p, \mathrm{ST}, \mathrm{P}, \mathrm{K}, \mathrm{S}\}\,.$$

Moreover, $\mathrm{C}^+_{\mathrm{nc}}(\mathcal{X})$ holds when $d \leq 2$, $\mathrm{D}_{\mathrm{nc}}(\mathcal{X})$ holds when $d \leq 2$ or when $d \leq 4$ and $p = 0$, and $\mathrm{V}_{\mathrm{nc}}(\mathcal{X})$ holds when $d \leq 2$.

[16] In the particular case where $n = 5$ and $\deg(X) = 3$, the dg categories $\mathcal{T}_{\mathrm{dg}}(X)$ obtained in this way are usually called "noncommutative $K3$-surfaces" because they share many of the key properties of the dg categories of perfect complexes of the classical $K3$-surfaces.

4.5. Global orbifolds. Let k be a base field of characteristic $p \geq 0$, G a finite group of order n, X a smooth proper k-scheme of dimension d equipped with a G-action (in the case of conjecture S we assume solely that X is smooth), and $\mathcal{X} := [X/G]$ the associated global orbifold.

THEOREM 4.6. *Assume that $p \nmid n$ and that k contains the nth roots of unity. Under these assumptions, we have the following implications of conjectures*

$$\sum_{\sigma \subseteq G} \mathrm{C}(X^\sigma) \Rightarrow \mathrm{C}_{\mathrm{nc}}(\mathcal{X}) \quad \text{with} \quad \mathrm{C} \in \{\mathrm{C}^+, \mathrm{D}, \mathrm{V}, \mathrm{B}, \mathrm{W}, \mathrm{T}^l\,(l \nmid n), \mathrm{T}^p, \mathrm{ST}, \mathrm{P}, \mathrm{K}, \mathrm{S}\}\,,$$

where σ is a cyclic subgroup of G.

Intuitively speaking, Theorem 4.6 shows that in order to solve the noncommutative counterparts of the celebrated conjectures for the global orbifold $\mathcal{X}$, it suffices to solve the celebrated conjectures for the underlying scheme X. Theorem 4.6 was proved in [**58**, Thm. 9.2][**57**, Thm. 3.1], resp. in [**55**, Thm. 1.16], in the case of the conjectures C^+, D, and V, resp. in the case of the conjectures B, T^l $(l \nmid n)$, T^p, and P. The proof of the other cases is similar.

COROLLARY 4.7 (Low-dimensional global orbifolds). *Let $\mathcal{X}$ be as in Theorem 4.6. When $d \leq 1$, the following conjectures hold:*

$$\mathrm{C}_{\mathrm{nc}}(\mathcal{X}) \quad \text{with} \quad \mathrm{C} \in \{\mathrm{C}^+, \mathrm{D}, \mathrm{V}, \mathrm{B}, \mathrm{W}, \mathrm{T}^l\,(l \nmid n), \mathrm{T}^p, \mathrm{ST}, \mathrm{P}, \mathrm{K}, \mathrm{S}\}\,.$$

Moreover, $\mathrm{C}^+_{\mathrm{nc}}$ holds when $d \leq 2$, $\mathrm{D}_{\mathrm{nc}}(\mathcal{X})$ holds when $d \leq 2$ or when $d \leq 4$ and $p = 0$, $\mathrm{V}_{\mathrm{nc}}(\mathcal{X})$ holds when $d \leq 2$ or when X is an abelian 3-fold and $p = 0$, and $\mathrm{T}^l_{\mathrm{nc}}(\mathcal{X})$ (with $l \nmid n$) and $\mathrm{T}^p_{\mathrm{nc}}(\mathcal{X})$ hold when X is a K3-surface.

COROLLARY 4.8 (Abelian G-varieties). *Let $\mathcal{X}$ be as in Theorem 4.6. When X is an abelian variety and G acts by group homomorphisms*[17]*, the conjectures $\mathrm{C}_{\mathrm{nc}}(\mathcal{X})$, with $\mathrm{C} \in \{\mathrm{C}^+, \mathrm{D}, \mathrm{W}, \mathrm{K}, \mathrm{S}\}$, hold. Moreover, when $d \leq 3$, the conjectures $\mathrm{C}_{\mathrm{nc}}(\mathcal{X})$, with $\mathrm{C} \in \{\mathrm{B}, \mathrm{T}^l\,(l \nmid n), \mathrm{T}^p, \mathrm{ST}, \mathrm{P}\}$, also hold.*

4.6. Twisted global orbifolds. Let k, G, X (of dimension d), and $\mathcal{X} := [X/G]$, be as in §4.5. In this subsection we consider the case where the global orbifold $\mathcal{X}$ is equipped with a sheaf of Azumaya algebras $\mathcal{F}$ of rank r. In other words, $\mathcal{F}$ is a G-equivariant sheaf of Azumaya algebras of rank r over X. Similarly to the dg category $\mathrm{perf}_{\mathrm{dg}}(\mathcal{X})$, we can also consider the dg category $\mathrm{perf}_{\mathrm{dg}}(\mathcal{X}; \mathcal{F})$ of perfect complexes of $\mathcal{F}$-modules. In what follows, we will write $\mathrm{C}_{\mathrm{nc}}(\mathcal{X}; \mathcal{F})$ instead of $\mathrm{C}_{\mathrm{nc}}(\mathrm{perf}_{\mathrm{dg}}(\mathcal{X}; \mathcal{F}))$. The next result is the "twisted" version of the Theorem 4.6:

THEOREM 4.9. *Assume that $p \nmid nr$ and that k contains the nth roots of unity. Under these assumptions, we have the following implications of conjectures*

$$\sum_{\sigma \subseteq G} \mathrm{C}(Y_\sigma) \Rightarrow \mathrm{C}_{\mathrm{nc}}(\mathcal{X}; \mathcal{F}) \quad \textit{with} \quad \mathrm{C} \in \{\mathrm{C}^+, \mathrm{D}, \mathrm{V}, \mathrm{B}, \mathrm{W}, \mathrm{T}^l\,(l \nmid nr), \mathrm{T}^p, \mathrm{ST}, \mathrm{P}, \mathrm{K}, \mathrm{S}\}\,,$$

where σ is a cyclic subgroup of G and Y_σ is a certain $\sigma^\vee$-Galois cover of X^σ induced by the restriction of $\mathcal{F}$ to X^σ.

Intuitively speaking, Theorem 4.9 shows that in order to solve the noncommutative counterparts of the celebrated conjectures for the twisted global orbifold $(\mathcal{X}; \mathcal{F})$, it suffices to solve the celebrated conjectures for certain Galois covers of

[17]For example, in the case where $G = \mathbb{Z}/2$, we can consider the canonical involution $a \mapsto -a$.

the underlying scheme X. Theorem 4.9 was proved in [**55**, Thm. 1.23] in the case of the conjectures B, T^l $(l \nmid nr)$, T^p, and P. The proof of the other cases is similar.

COROLLARY 4.10 (Low-dimensional twisted global orbifolds). *Let $\mathcal{X}$ and $\mathcal{F}$ be as in Theorem 4.9. When $d \leq 1$, the following conjectures hold:*

$$\mathrm{C}_{\mathrm{nc}}(\mathcal{X};\mathcal{F}) \quad \text{with} \quad \mathrm{C} \in \{\mathrm{C}^+, \mathrm{D}, \mathrm{V}, \mathrm{B}, \mathrm{W}, \mathrm{T}^l\,(l \nmid nr), \mathrm{T}^p, \mathrm{ST}, \mathrm{P}, \mathrm{K}, \mathrm{S}\}\,.$$

Moreover, $\mathrm{C}^+_{\mathrm{nc}}(\mathcal{X};\mathcal{F})$ holds when $d \leq 2$, $\mathrm{D}_{\mathrm{nc}}(\mathcal{X};\mathcal{F})$ holds when $d \leq 2$ or when $d \leq 4$ and $p = 0$, and $\mathrm{V}_{\mathrm{nc}}(\mathcal{X};\mathcal{F})$ holds when $d \leq 2$.

4.7. Intersections of bilinear divisors. Let k be a base field of characteristic $p \geq 0$ and V a finite-dimensional k-vector space. Consider the canonical action of $\mathbb{Z}/2$ on $\mathbb{P}(V)\times\mathbb{P}(V)$ and the associated global orbifold $\mathcal{X} := [(\mathbb{P}(V)\times\mathbb{P}(V))/(\mathbb{Z}/2)]$. Note that by construction we have the following morphism:

$$f\colon \mathcal{X} \longrightarrow \mathbb{P}(\mathrm{Sym}^2(V)) \qquad ([v_1],[v_2]) \mapsto [v_1 \otimes v_2 + v_2 \otimes v_1]\,.$$

Given a linear subspace $L \subset \mathrm{Sym}^2(V^*)$ of dimension ≤ 3 when $\dim(V)$ is even, resp. of dimension ≤ 6 when $\dim(V)$ is odd, consider the associated linear section $\mathcal{X}_L := f^{-1}(\mathbb{P}(L^\perp))$. Note that such a linear section corresponds to the intersection of $\dim(L)$ bilinear divisors in $\mathcal{X}$ parametrized by L.

THEOREM 4.11. *Assume that $\mathrm{codim}(\mathcal{X}_L) = \dim(L)$. Under this assumption (which holds for a generic choice of L), we have the implications of conjectures*

$$\mathrm{C}(Y) \Rightarrow \mathrm{C}_{\mathrm{nc}}(\mathcal{X}_L) \quad \textit{with} \quad \mathrm{C} \in \{\mathrm{C}^+, \mathrm{D}, \mathrm{V}, \mathrm{B}, \mathrm{W}, \mathrm{T}^l, \mathrm{T}^p, \mathrm{ST}, \mathrm{K}, \mathrm{S}\}\,,$$

where Y is a certain double cover of $\mathbb{P}(L)$ induced by f.

Intuitively speaking, Theorem 4.11 shows that in order to solve the noncommutative counterparts of the celebrated conjectures for intersections of bilinear divisors, it suffices to solve the celebrated conjectures for a certain double cover of the projective space $\mathbb{P}(L)$. Theorem 4.11 was proved in [**50**, Thm. 1.13] in the case of the conjectures C^+ and D. The proof of the other cases is similar.

COROLLARY 4.12 (Intersections of up-to-five bilinear divisors). *Let $\mathcal{X}_L$ be as in Theorem 4.11. When $\dim(L) \leq 2$, the following conjectures hold:*

$$\mathrm{C}_{\mathrm{nc}}(\mathcal{X}_L) \quad \text{with} \quad \mathrm{C} \in \{\mathrm{C}^+, \mathrm{D}, \mathrm{V}, \mathrm{B}, \mathrm{W}, \mathrm{T}^l, \mathrm{T}^p, \mathrm{ST}, \mathrm{P}, \mathrm{K}, \mathrm{S}\}\,.$$

Moreover, $\mathrm{C}_{\mathrm{nc}}(\mathcal{X}_L)$ holds when $\dim(L) \leq 3$, $\mathrm{D}_{\mathrm{nc}}(\mathcal{X}_L)$ holds when $\dim(L) \leq 3$ or when $\dim(L) \leq 5$ and $p = 0$, and $\mathrm{V}_{\mathrm{nc}}(\mathcal{X}_L)$ holds when $\dim(L) \leq 3$.

4.8. Moishezon manifolds associated to quartic double solids. Let $k = \mathbb{C}$ be the field of complex numbers and $X \to \mathbb{P}^2$ one of the quartic double solids introduced by Artin-Mumford in [**2**]. These are examples of unirational, but not rational, conic bundles. Thanks to the work of Cossec [**10**], these conic bundles can be alternatively described as those singular double coverings $X \to \mathbb{P}^3$ which are ramified over a quartic symmetroid Z. On the one hand, we can consider the Enriques surface S_Z obtained as the quotient of a natural involution (acting without fixed points) on the blow-up of Z. On the other hand, we can consider a small resolution of singularities $\mathcal{X} \to X$. Such a resolution is not an algebraic variety, but rather a Moishezon manifold[18].

THEOREM 4.13. *We have the equivalences of conjectures*

$$\mathrm{C}(S_Z) \Leftrightarrow \mathrm{C}_{\mathrm{nc}}(\mathcal{X}) \quad \textit{with} \quad \mathrm{C} \in \{\mathrm{C}^+, \mathrm{D}, \mathrm{V}, \mathrm{S}\}$$

as well as the implication $\mathrm{K}(S_Z) \Rightarrow \mathrm{K}_{\mathrm{nc}}(\mathcal{X})$.

Roughly speaking, Theorem 4.13 shows that in what concerns the celebrated conjectures, there is no difference between Enriques surfaces and Moishezon manifolds. Theorem 4.13 was proved in [**5**, Thm. 1.14] in the case of the conjecture V. The proof of the other cases is similar.

COROLLARY 4.14. *The conjectures* $\mathrm{C}_{\mathrm{nc}}(\mathcal{X})$, *with* $\mathrm{C} \in \{\mathrm{C}^+, \mathrm{D}, \mathrm{V}\}$, *hold.*

References

[1] M. Artin, *Algebraization of formal moduli. II. Existence of modifications*, Ann. of Math. (2) **91** (1970), 88–135, DOI 10.2307/1970602. MR260747

[2] M. Artin and D. Mumford, *Some elementary examples of unirational varieties which are not rational*, Proc. London Math. Soc. (3) **25** (1972), 75–95, DOI 10.1112/plms/s3-25.1.75. MR0321934

[3] Hyman Bass, *Some problems in "classical" algebraic K-theory*, Algebraic K-theory, II: "Classical" algebraic K-theory and connections with arithmetic (Proc. Conf., Battelle Memorial Inst., Seattle, Wash., 1972), Springer, Berlin, 1973, pp. 3–73. Lecture Notes in Math., Vol. 342. MR0409606

[4] A. A. Beĭlinson, *Height pairing between algebraic cycles*, K-theory, arithmetic and geometry (Moscow, 1984), Lecture Notes in Math., vol. 1289, Springer, Berlin, 1987, pp. 1–25, DOI 10.1007/BFb0078364. MR923131

[5] Marcello Bernardara, Matilde Marcolli, and Gonçalo Tabuada, *Some remarks concerning Voevodsky's nilpotence conjecture*, J. Reine Angew. Math. **738** (2018), 299–312, DOI 10.1515/crelle-2015-0068. MR3794895

[6] Pierre Berthelot, *Cohomologie cristalline des schémas de caractéristique* $p > 0$ (French), Lecture Notes in Mathematics, Vol. 407, Springer-Verlag, Berlin-New York, 1974. MR0384804

[7] Spencer Bloch, *Lectures on algebraic cycles*, Duke University Mathematics Series, IV, Duke University, Mathematics Department, Durham, N.C., 1980. MR558224

[8] A. Bondal and D. Orlov, *Semiorthogonal decomposition for algebraic varieties.* Available at arXiv:alg-geom/9506012.

[9] Charles Cadman, *Using stacks to impose tangency conditions on curves*, Amer. J. Math. **129** (2007), no. 2, 405–427, DOI 10.1353/ajm.2007.0007. MR2306040

[10] François R. Cossec, *Reye congruences*, Trans. Amer. Math. Soc. **280** (1983), no. 2, 737–751, DOI 10.2307/1999644. MR716848

[11] P. Deligne, *Catégories tensorielles* (French, with English summary), Mosc. Math. J. **2** (2002), no. 2, 227–248, DOI 10.17323/1609-4514-2002-2-2-227-248. Dedicated to Yuri I. Manin on the occasion of his 65th birthday. MR1944506

[12] Pierre Deligne, *La conjecture de Weil. I* (French), Inst. Hautes Études Sci. Publ. Math. **43** (1974), 273–307. MR0340258

[13] E. Elmanto, *Topological periodic cyclic homology of smooth* $\mathbb{F}_p$*-algebras.* Talk at the Arbeitsgemeinschaft 2018. Available at `https://www.mfo.de/occasion/1814/www_view`.

[14] B. L. Feĭgin and B. L. Tsygan, *Additive K-theory*, K-theory, arithmetic and geometry (Moscow, 1984), Lecture Notes in Math., vol. 1289, Springer, Berlin, 1987, pp. 67–209, DOI 10.1007/BFb0078368. MR923136

[18]Recall that a *Moishezon manifold* $\mathcal{X}$ is a compact complex manifold whose field of meromorphic functions on each component has transcendence degree equal to the dimension of the connected component. As proved by Moishezon in [**44**], $\mathcal{X}$ is a smooth projective $\mathbb{C}$-scheme if and only if it admits a Kähler metric. In the remaining cases, as proved by Artin in [**1**], $\mathcal{X}$ is a proper algebraic space over $\mathbb{C}$.

[15] Thomas Geisser, *Tate's conjecture, algebraic cycles and rational K-theory in characteristic p*, K-Theory **13** (1998), no. 2, 109–122, DOI 10.1023/A:1007709804443. MR1611623
[16] Daniel R. Grayson, *Finite generation of K-groups of a curve over a finite field (after Daniel Quillen)*, Algebraic K-theory, Part I (Oberwolfach, 1980), Lecture Notes in Math., vol. 966, Springer, Berlin-New York, 1982, pp. 69–90. MR689367
[17] A. Grothendieck, *Standard conjectures on algebraic cycles*, Algebraic Geometry (Internat. Colloq., Tata Inst. Fund. Res., Bombay, 1968), Oxford Univ. Press, London, 1969, pp. 193–199. MR0268189
[18] Alexander Grothendieck, *Formule de Lefschetz et rationalité des fonctions L* (French), Séminaire Bourbaki, Vol. 9, Soc. Math. France, Paris, 1995, pp. Exp. No. 279, 41–55. MR1608788
[19] Vladimir Guletskiĭ, *Finite-dimensional objects in distinguished triangles*, J. Number Theory **119** (2006), no. 1, 99–127, DOI 10.1016/j.jnt.2005.10.008. MR2228952
[20] V. Guletskiĭ and C. Pedrini, *Finite-dimensional motives and the conjectures of Beilinson and Murre*, K-Theory **30** (2003), no. 3, 243–263, DOI 10.1023/B:KTHE.0000019787.69435.89. Special issue in honor of Hyman Bass on his seventieth birthday. Part III. MR2064241
[21] G. Harder, *Die Kohomologie S-arithmetischer Gruppen über Funktionenkörpern* (German), Invent. Math. **42** (1977), 135–175, DOI 10.1007/BF01389786. MR473102
[22] Lars Hesselholt, *Topological Hochschild homology and the Hasse-Weil zeta function*, An alpine bouquet of algebraic topology, Contemp. Math., vol. 708, Amer. Math. Soc., Providence, RI, 2018, pp. 157–180, DOI 10.1090/conm/708/14264. MR3807755
[23] Akira Ishii and Kazushi Ueda, *The special McKay correspondence and exceptional collections*, Tohoku Math. J. (2) **67** (2015), no. 4, 585–609, DOI 10.2748/tmj/1450798075. MR3436544
[24] Bruno Kahn, *Équivalences rationnelle et numérique sur certaines variétés de type abélien sur un corps fini* (French, with English and French summaries), Ann. Sci. École Norm. Sup. (4) **36** (2003), no. 6, 977–1002 (2004), DOI 10.1016/j.ansens.2003.02.002. MR2032532
[25] Bruno Kahn and Ronnie Sebastian, *Smash-nilpotent cycles on abelian 3-folds*, Math. Res. Lett. **16** (2009), no. 6, 1007–1010, DOI 10.4310/MRL.2009.v16.n6.a8. MR2576688
[26] Nicholas M. Katz and William Messing, *Some consequences of the Riemann hypothesis for varieties over finite fields*, Invent. Math. **23** (1974), 73–77, DOI 10.1007/BF01405203. MR332791
[27] Kiran S. Kedlaya, *Fourier transforms and p-adic 'Weil II'*, Compos. Math. **142** (2006), no. 6, 1426–1450, DOI 10.1112/S0010437X06002338. MR2278753
[28] Bernhard Keller, *On differential graded categories*, International Congress of Mathematicians. Vol. II, Eur. Math. Soc., Zürich, 2006, pp. 151–190. MR2275593
[29] Shun-Ichi Kimura, *Chow groups are finite dimensional, in some sense*, Math. Ann. **331** (2005), no. 1, 173–201, DOI 10.1007/s00208-004-0577-3. MR2107443
[30] Steven L. Kleiman, *The standard conjectures*, Motives (Seattle, WA, 1991), Proc. Sympos. Pure Math., vol. 55, Amer. Math. Soc., Providence, RI, 1994, pp. 3–20, DOI 10.1090/pspum/055.1/1265519. MR1265519
[31] S. L. Kleiman, *Algebraic cycles and the Weil conjectures*, Dix exposés sur la cohomologie des schémas, Adv. Stud. Pure Math., vol. 3, North-Holland, Amsterdam, 1968, pp. 359–386. MR292838
[32] Maxim Kontsevich, *Mixed noncommutative motives*. Talk at the Workshop on Homological Mirror Symmetry, Miami, 2010. Available at `www-math.mit.edu/auroux/frg/miami10-notes`.
[33] Maxim Kontsevich, *Notes on motives in finite characteristic*, Algebra, arithmetic, and geometry: in honor of Yu. I. Manin. Vol. II, Progr. Math., vol. 270, Birkhäuser Boston, Inc., Boston, MA, 2009, pp. 213–247, DOI 10.1007/978-0-8176-4747-6_7. MR2641191
[34] Maxim Kontsevich, *XI Solomon Lefschetz Memorial Lecture series: Hodge structures in non-commutative geometry*, Non-commutative geometry in mathematics and physics, Contemp. Math., vol. 462, Amer. Math. Soc., Providence, RI, 2008, pp. 1–21, DOI 10.1090/conm/462/09058. Notes by Ernesto Lupercio. MR2444365
[35] ———, *Categorification, NC Motives, Geometric Langlands, and Lattice Models*. Talk at the Geometric Langlands Seminar, University of Chicago, 2006. Notes available at the webpage `https://www.ma.utexas.edu/users/benzvi/notes.html`.
[36] ———, *Noncommutative motives*. Talk at the IAS on the occasion of the 61st birthday of Pierre Deligne (2005). Available at `http://video.ias.edu/Geometry-and-Arithmetic`.

[37] Alexander Kuznetsov, *Calabi-Yau and fractional Calabi-Yau categories*, J. Reine Angew. Math. **753** (2019), 239–267, DOI 10.1515/crelle-2017-0004. MR3987870

[38] A. Kuznetsov, *Derived categories of families of sextic del Pezzo surfaces*. Available at arXiv:1708.00522.

[39] David I. Lieberman, *Numerical and homological equivalence of algebraic cycles on Hodge manifolds*, Amer. J. Math. **90** (1968), 366–374, DOI 10.2307/2373533. MR230336

[40] Ju. I. Manin, *Correspondences, motifs and monoidal transformations* (Russian), Mat. Sb. (N.S.) **77 (119)** (1968), 475–507. MR0258836

[41] Matilde Marcolli and Gonçalo Tabuada, *Noncommutative numerical motives, Tannakian structures, and motivic Galois groups*, J. Eur. Math. Soc. (JEMS) **18** (2016), no. 3, 623–655, DOI 10.4171/JEMS/598. MR3463419

[42] Carlo Mazza, *Schur functors and motives*, K-Theory **33** (2004), no. 2, 89–106, DOI 10.1007/s10977-004-6468-2. MR2131746

[43] J. Milne, *The Tate conjecture over finite fields*. AIM talk. Available at Milne's webpage `http://www.jmilne.org/math/articles/2007e.pdf`.

[44] B. G. Moĭšezon, *On n-dimensional compact complex manifolds having n algebraically independent meromorphic functions. I* (Russian), Izv. Akad. Nauk SSSR Ser. Mat. **30** (1966), 133–174. MR0216522

[45] Thomas Nikolaus and Peter Scholze, *On topological cyclic homology*, Acta Math. **221** (2018), no. 2, 203–409, DOI 10.4310/ACTA.2018.v221.n2.a1. MR3904731

[46] A. M. Šermenev, *Motif of an Abelian variety* (Russian), Funckcional. Anal. i Priložen. **8** (1974), no. 1, 55–61. MR0335523

[47] Gonçalo Tabuada, *Noncommutative motives in positive characteristic and their applications*, Adv. Math. **349** (2019), 648–681, DOI 10.1016/j.aim.2019.04.020. MR3942181

[48] Gonçalo Tabuada, *Recent developments on noncommutative motives*, New directions in homotopy theory, Contemp. Math., vol. 707, Amer. Math. Soc., Providence, RI, 2018, pp. 143–173, DOI 10.1090/conm/707/14258. MR3807746

[49] Gonçalo Tabuada, *A note on the Schur-finiteness of linear sections*, Math. Res. Lett. **25** (2018), no. 1, 237–253, DOI 10.4310/mrl.2018.v25.n1.a10. MR3818621

[50] Gonçalo Tabuada, *A note on Grothendieck's standard conjectures of type C^+ and D*, Proc. Amer. Math. Soc. **146** (2018), no. 4, 1389–1399, DOI 10.1090/proc/13955. MR3754327

[51] Gonçalo Tabuada, *Kimura-finiteness of quadric fibrations over smooth curves* (English, with English and French summaries), C. R. Math. Acad. Sci. Paris **355** (2017), no. 6, 628–632, DOI 10.1016/j.crma.2017.05.006. MR3661545

[52] Gonçalo Tabuada, *Noncommutative motives*, University Lecture Series, vol. 63, American Mathematical Society, Providence, RI, 2015. With a preface by Yuri I. Manin. MR3379910

[53] Gonçalo Tabuada, *Chow motives versus noncommutative motives*, J. Noncommut. Geom. **7** (2013), no. 3, 767–786, DOI 10.4171/JNCG/134. MR3108695

[54] ———, *Noncommutative Weil conjecture*. Available at arXiv:1808.00950.

[55] ———, *HPD-invariance of the Tate, Beilinson and Parshin conjectures*. Available at arXiv:1712.05397.

[56] ———, *Schur-finiteness (and Bass-finiteness) conjecture for quadric fibrations and for families of sextic du Val del Pezzo surfaces*. Available at arXiv:1708.05382.

[57] Gonçalo Tabuada, *A note on Grothendieck's standard conjectures of type* C^+ *and* D *in positive characteristic*, Proc. Amer. Math. Soc. **147** (2019), no. 12, 5039–5054, DOI 10.1090/proc/14768. MR4021067

[58] Gonçalo Tabuada and Michel Van den Bergh, *Additive invariants of orbifolds*, Geom. Topol. **22** (2018), no. 5, 3003–3048, DOI 10.2140/gt.2018.22.3003. MR3811776

[59] John Tate, *Conjectures on algebraic cycles in l-adic cohomology*, Motives (Seattle, WA, 1991), Proc. Sympos. Pure Math., vol. 55, Amer. Math. Soc., Providence, RI, 1994, pp. 71–83, DOI 10.1090/pspum/055.1/1265523. MR1265523

[60] John T. Tate, *Algebraic cycles and poles of zeta functions*, Arithmetical Algebraic Geometry (Proc. Conf. Purdue Univ., 1963), Harper & Row, New York, 1965, pp. 93–110. MR0225778

[61] R. W. Thomason, *A finiteness condition equivalent to the Tate conjecture over $\mathbf{F}_q$*, Algebraic K-theory and algebraic number theory (Honolulu, HI, 1987), Contemp. Math., vol. 83, Amer. Math. Soc., Providence, RI, 1989, pp. 385–392, DOI 10.1090/conm/083/991987. MR991987

[62] Burt Totaro, *Recent progress on the Tate conjecture*, Bull. Amer. Math. Soc. (N.S.) **54** (2017), no. 4, 575–590, DOI 10.1090/bull/1588. MR3683625

[63] Vladimir Voevodsky, *Triangulated categories of motives over a field*, Cycles, transfers, and motivic homology theories, Ann. of Math. Stud., vol. 143, Princeton Univ. Press, Princeton, NJ, 2000, pp. 188–238. MR1764202

[64] V. Voevodsky, *A nilpotence theorem for cycles algebraically equivalent to zero*, Internat. Math. Res. Notices **4** (1995), 187–198, DOI 10.1155/S1073792895000158. MR1326064

[65] Claire Voisin, *Remarks on zero-cycles of self-products of varieties*, Moduli of vector bundles (Sanda, 1994; Kyoto, 1994), Lecture Notes in Pure and Appl. Math., vol. 179, Dekker, New York, 1996, pp. 265–285. MR1397993

[66] André Weil, *Variétés abéliennes et courbes algébriques* (French), Actualités Sci. Ind., no. 1064 = Publ. Inst. Math. Univ. Strasbourg 8 (1946), Hermann & Cie., Paris, 1948. MR0029522

[67] André Weil, *Sur les courbes algébriques et les variétés qui s'en déduisent* (French), Actualités Sci. Ind., no. 1041 = Publ. Inst. Math. Univ. Strasbourg **7** (1945), Hermann et Cie., Paris, 1948. MR0027151

Department of Mathematics, Massachusetts Institute of Technology, Cambridge, Massachusetts 02139; Departamento de Matemática, FCT, UNL, Portugal; Centro de Matemática e Aplicações (CMA), FCT, UNL, Portugal.

Email address: tabuada@math.mit.edu

Contemporary Mathematics
Volume **749**, 2020
https://doi.org/10.1090/conm/749/15079

Bott periodicity and almost commuting matrices

Rufus Willett

Abstract. We give a proof of the Bott periodicity theorem for topological K-theory of C^*-algebras based on Loring's treatment of Voiculescu's almost commuting matrices and Atiyah's rotation trick. We also explain how this relates to the Dirac operator on the circle; this uses Yu's localization algebra and an associated explicit formula for the pairing between the first K-homology and first K-theory groups of a (separable) C^*-algebra.

1. Introduction

The aim of this note is to give a proof of Bott periodicity using Voiculescu's famous example [**8**] of 'almost commuting' unitary matrices that cannot be approximated by exactly commuting unitary matrices. Indeed, in his thesis [**4**, Chapter I] Loring explained how the properties of Voiculescu's example can be seen as arising from Bott periodicity; this note explains how one can go 'backwards' and use Loring's computations combined with Atiyah's rotation trick [**1**, Section 1] to prove Bott periodicity. Thus the existence of matrices with the properties in Voiculescu's example is in some sense equivalent to Bott periodicity.

A secondary aim is to explain how to interpret the above in terms of Yu's localization algebra [**10**] and the Dirac operator on the circle. The localization algebra of a topological space X is a C^*-algebra $C_L^*(X)$ with the property that the K-theory of $C_L^*(X)$ is canonically isomorphic to the K-homology of X. We explain how Voiculescu's matrices, and the computations we need to do with them, arise naturally from the Dirac and Bott operators on the circle using the localisation algebra. A key ingredient is a new explicit formula for the pairing between K-homology and K-theory in terms of Loring's machinery.

We do not claim any real technical originality in this note: as will be obvious, our approach to Bott periodicity is heavily dependent on Loring's work in particular. However, we think this approach is particularly attractive and concrete – it essentially reduces the proof of the Bott periodicity theorem to some formal observations and a finite dimensional matrix computation – and hope that this exposition is worthwhile from that point of view. We also hope it is interesting insofar as it bridges a gap between different parts of the subject. A secondary motivation comes from possible applications in the setting of controlled K-theory; this will be explored elsewhere, however.

2010 *Mathematics Subject Classification.* Primary 46L80.

The author was partly supported by NSF grant DMS 1564281.

2. Preliminaries and Atiyah's rotation trick

In this section, we set up notation and recall Atiyah's rotation trick.

Let $S^1 := \{z \in \mathbb{C} \mid |z| = 1\}$. Let S denote the C^*-algebra $C_0(S^1\backslash\{1\})$, so $S \cong C_0(\mathbb{R})$. For a C^*-algebra A, let SA denote the suspension $S \otimes A \cong C_0(S^1\backslash\{1\}, A)$. Let $b \in K_1(S)$ denote the Bott generator, i.e. b is the class of the unitary $z \mapsto z^{-1}$ in the unitization of S. Recall moreover that for a unital C^*-algebra A, the *Bott map* is defined on the class of a projection $p \in M_n(A)$ by

$$\beta_A : K_0(A) \to K_1(SA), \quad [p] \mapsto [bp + (1-p)],$$

where the right hand side is understood to be the class in $K_1(SA)$ of the unitary $z \mapsto z^{-1}p+1-p$ in the unitization of $C(S^1\backslash\{1\}, M_n(A)) \cong M_n(SA)$. This identifies with an element in the unitary group of the $n \times n$ matrices over the unitization of SA. One checks directly that the process above β_A indeed gives a well-defined map on K-theory that induces a map in the non-unital case in the usual way, and moreover that the collection of maps is natural in A in the sense that for any $*$-homomorphism $\phi : A \to B$, the corresponding diagram

$$\begin{array}{ccc} K_0(A) & \xrightarrow{\beta_A} & K_1(SA) \\ \downarrow{\scriptstyle \phi_*} & & \downarrow{\scriptstyle (\mathrm{id}_S \otimes \phi)_*} \\ K_0(B) & \xrightarrow{\beta_B} & K_1(SB) \end{array}$$

commutes (see for example [**7**, Section 11.1.1]).

THEOREM 2.1 (Bott periodicity theorem). *For every C^*-algebra A, β_A is an isomorphism.*

An important ingredient in our approach to this will be *Atiyah's rotation trick* [**1**, Section 1]: this allows one to reduce the proof of Bott periodicity to constructing a homomorphism $\alpha_A : K_1(SA) \to K_0(A)$ for each C^*-algebra A, so that the collection $\{\alpha_A \mid A \text{ a } C^*\text{-algebra}\}$ satisfies two natural axioms. As well as the fact that it simplifies the proof of Bott periodicity, a key virtue the rotation trick (as already pointed out in [**1**, Section 7]) is that the axioms satisfied by the α_A are easily checked for several different natural analytic and geometric constructions of a collection of homomorphisms α_A; this paper essentially codifies the observation that Voiculescu's almost commuting matrices give yet another way of constructing an appropriate family $\{\alpha_A\}$.

In order to give a precise statement of the axioms, recall from [**3**, Section 4.7] that for unital C^*-algebras A and B, the formulas

$$\times : K_0(A) \otimes K_0(B) \to K_0(A \otimes B), \quad [p] \otimes [q] \mapsto [p \otimes q]$$

and

$$\times : K_1(A) \otimes K_0(B) \to K_1(A \otimes B), \quad [u] \otimes [p] \mapsto [u \otimes p + 1 \otimes (1-p)]$$

induce canonical external products on K-theory; moreover, applying these products to unitizations and restricting, these extend to well-defined products in the non-unital case too.

Here then is a precise statement of the reduction that Atiyah's rotation trick allows: see the exposition in [**3**, Section 4.9] for a proof of the result as stated here.

PROPOSITION 2.2. *Assume that for each C^*-algebra A there is homomorphism $\alpha_A : K_1(SA) \to K_0(A)$ satisfying the following conditions.*

(i) $\alpha_{\mathbb{C}}(b) = 1$.

(ii) *For any C^*-algebras A and B, the diagram below*

$$\begin{array}{ccc} K_1(SA) \otimes K_0(B) & \xrightarrow{\times} & K_1(S(A \otimes B)) \\ \Big\downarrow{\scriptstyle \alpha_A \otimes 1} & & \Big\downarrow{\scriptstyle \alpha_{A \otimes B}} \\ K_0(A) \otimes K_0(B) & \xrightarrow{\times} & K_0(A \otimes B) \end{array}$$

commutes; here the horizontal arrows are the canonical external products in K-theory discussed above, and we have used the canonical identifications

$$SA \otimes B = (S \otimes A) \otimes B = S \otimes (A \otimes B) = S(A \otimes B)$$

to make sense of thc top horizontal arrow.

Then α_A and β_A are mutually inverse for all C^*-algebras A. □

3. Almost commuting matrices

Our goal in this section is to construct homomorphisms $\alpha_A : K_1(SA) \to K_0(A)$ with the properties in Proposition 2.2 above, and thus prove Bott periodicity. To motivate our construction, we start by recalling Voiculescu's almost commuting matrices.

Let $\{\delta_0, ..., \delta_{n-1}\}$ be the canonical orthonormal basis for $\mathbb{C}^n$, and define unitaries by

$$u_n : \delta_k \mapsto e^{2\pi ik/n}\delta_k \quad \text{and} \quad v_n : \delta_k \mapsto \delta_{k+1}, \tag{1}$$

where the '$k+1$' in the subscript above should be interpreted modulo n, or in other words $v_n : \delta_{n-1} \mapsto \delta_0$. It is straightforward to check that $\|[u_n, v_n]\| \to 0$ as $n \to \infty$ (the norm here and throughout is the operator norm). Voiculescu [**8**] proved the following result.

THEOREM 3.1. *There exists $\epsilon > 0$ such that if u_n', v_n' are $n \times n$ unitary matrices that actually commute, then we must have $\max\{\|u_n - u_n'\|, \|v_n - v_n'\|\} \geqslant \epsilon$ for all n.* □

In words, the sequence of pairs (u_n, v_n) cannot be well-approximated by pairs that actually commute. Voiculescu's original proof of this fact uses non-quasidiagonality of the unilateral shift; there is a close connection of this to K-theory and Bott periodicity, but this is not obvious from the original proof. A more directly K-theoretic argument is due to Loring from his thesis [**4**]. Fix smooth functions $f, g, h : S^1 \to [0, 1]$ with the properties in [**4**, pages 10-11]: the most salient of these are that

$$f^2 + g^2 + h^2 = f, \quad f(1) = 1, \quad g(1) = h(1) = 0, \quad \text{and} \quad gh = 0.$$

For a pair of unitaries u, v in a unital C^*-algebra B, define the *Loring element*

$$e(u, v) := \begin{pmatrix} f(u) & g(u) + h(u)v \\ v^*h(u) + g(u) & 1 - f(u) \end{pmatrix}, \tag{2}$$

which in general is a self-adjoint element of $M_2(B)$. For later use, note that the conditions on f, g, h above imply the following formula, where for brevity we write $e = e(u, v)$, $f = f(u)$, $g = g(u)$, $h = h(u)$,

$$e^2 = e + \begin{pmatrix} hvg + gv^*h & [f, hv] \\ [v^*h, f] & v^*h^2v - h^2 \end{pmatrix}; \tag{3}$$

in particular, if u and v happen to commute, this shows that e is an idempotent. Similarly, one can show that if $\|uv - vu\|$ is suitably small, then $\|e^2 - e\|$ is also small, so in particular the spectrum of e misses $1/2$: indeed, this is done qualitatively in [**4**, Proposition 3.5], while a quantitative result for a specific choice of f, g, and h can be found in [**5**, Theorem 3.5]; the latter could be used to make the conditions on t that are implicit in our results more explicit. Thus if χ is the characteristic function of $[1/2, \infty)$, then χ is continuous on the spectrum of e, and so $\chi(e)$ is a well-defined projection in B. Loring shows that if $e_n \in M_{2n}(\mathbb{C})$ is the Loring element associated to the matrices $u_n, v_n \in M_n(\mathbb{C})$ as in line (1), then for all suitably large n,

$$\operatorname{rank}(e_n) - n = 1.$$

However, it is not difficult to see that if u_n and v_n were well-approximated by pairs of actually commuting matrices in the sense that the conclusion of Theorem 3.1 fails, then for all suitably large n one would have that $\operatorname{rank}(e_n) = n$. Thus Loring's work in particular reproves Theorem 3.1.

Now, let us get back to constructing a family of homomorphisms $\alpha_A : K_1(SA) \to K_0(A)$ satisfying the conditions of Proposition 2.2, and thus to prove Bott periodicity. For this, it will be convenient to have a continuously parametrised versions of the sequence (u_n), which we now build.

Let then $\delta_n : S^1 \to \mathbb{C}$ be defined by $\delta_n(x) = \frac{1}{\sqrt{2\pi}} z^n$, so the collection $\{\delta_n\}_{n \in \mathbb{Z}}$ is the canonical 'Fourier orthonormal basis' for $L^2(S^1) \cong \ell^2(\mathbb{Z})$. For each $t \in [1, \infty)$, define a unitary operator $u_t : L^2(S^1) \to L^2(S^1)$ to be diagonal with respect to this basis, and given by

$$u_t \delta_n := \begin{cases} e^{2\pi i n t^{-1}} \delta_n & 0 \leqslant t^{-1} n \leqslant 1 \\ \delta_n & \text{otherwise} \end{cases}.$$

Thus u_t agrees with the operator of rotation by $2\pi t^{-1}$ radians on $\operatorname{span}\{\delta_0, ..., \delta_n \mid n \leqslant t\}$, and with the identity elsewhere. Let A be a unital C^*-algebra faithfully represented on some Hilbert space H, and represent $C(S^1)$ on $L^2(S^1)$ by multiplication operators in the canonical way. Represent $SA = \{f \in C(S^1, A) \mid f(1) = 0\}$ faithfully on $L^2(S^1, H)$ as multiplication operators. Let χ be the characteristic function of $\{z \in \mathbb{C} \mid \operatorname{Re}(z) \geqslant \frac{1}{2}\}$.

Note that the Bott element b acts on $L^2(S^1)$ via the (backwards) bilateral shift. When compressed to the subspace $\operatorname{span}\{\delta_0, ..., \delta_n \mid n \leqslant t\}$ of $L^2(S^1)$, the operators u_t and b are thus slight variants of Voiculescu's almost commuting unitaries from line (1).

LEMMA 3.2. *Let A be a unital C^*-algebra, let $\widetilde{SA}$ be the unitization of its suspension, and let $v \in \widetilde{SA}$ be a unitary operator. Then with notation as in the above discussion:*

(i) *the spectrum of the Loring element $e(u_t \otimes 1_A, v)$ (cf. line (2)) does not contain $1/2$ for all large t;*

(ii) *the difference*

$$\chi(e(u_t \otimes 1_A, v)) - \begin{pmatrix} 1 & 0 \\ 0 & 0 \end{pmatrix}$$

is in $\mathcal{K}(L^2(S^1)) \otimes M_2(A)$ *for all* t*;*

(iii) *the function*

$$t \mapsto \chi(e(u_t \otimes 1_A, v))$$

is operator norm continuous for all large t.

PROOF. For part (i), we first claim that for any $f \in C(S^1)$, $[v_t, f] \to 0$ as $t \to \infty$. Indeed, it suffices to check this for the Bott element $b(z) = z^{-1}$ as this function generates $C(S^1)$ as a C^*-algebra. With respect to the orthonormal basis $\{\delta_n : n \in \mathbb{Z}\}$ we have that b acts by

$$b : \delta_n \mapsto \delta_{n-1},$$

i.e. b is the inverse of the usual bilateral shift. On the other hand, we have that

$$\|[v_t, b]\| = \|(v_t - bv_tb^*)b\| = \|v_t - bv_tb^*\|,$$

and one computes directly that $v_t - bv_tb^*$ is a multiplication operator by an element of $\ell^\infty(\mathbb{Z})$ that tends to zero as t tends to infinity, completing the proof of the claim.

It follows from the claim that $[v_t \otimes 1, f \otimes a] \to 0$ as $t \to \infty$ for any $f \in C(S^1)$, and any $a \in A$. Hence $[k(v_t) \otimes 1, a] \to 0$ for any $a \in C(S^1, A)$ and any $k \in C(S^1)$ (as $k(v_t)$ is in the C^*-algebra generated by v_t). Part (i) follows from this and the formula in line (3), plus the fact that $hg = gh = 0$.

For part (ii), consider

$$e(u_t \otimes 1_A, v) = \begin{pmatrix} f(u_t) \otimes 1_A & g(u_t) \otimes 1_A + (h(u_t) \otimes 1_A)v \\ g(u_t) \otimes 1_A + v^*(h(u_t) \otimes 1_A) & (1 - f(u_t)) \otimes 1_A \end{pmatrix}.$$

As e is norm continuous in the 'input unitaries', we may assume that $v \in C(S^1, A)$ is of the form $z \mapsto \sum_{n=-M}^{M} z^n a_n$ for some finite $M \in \mathbb{N}$. It follows from this, the formula for v_t, and the facts that $h(1) = 0 = g(1)$ and $f(1) = 1$, that there exists $N \in \mathbb{N}$ (depending on M and t) such that the operator $e(u_t \otimes 1_A, v)$ leaves some subspace of $(L^2(S^1) \otimes H)^{\oplus 2}$ of the form

$$(\text{span}\{\delta_{-N}, ..., \delta_N\} \otimes H)^{\oplus 2}$$

invariant, and moreover that it agrees with the operator $\begin{pmatrix} 1 & 0 \\ 0 & 0 \end{pmatrix}$ on the orthogonal complement of this subspace. It follows from this that

$$\chi(e(u_t \otimes 1_A, v)) - \begin{pmatrix} 1 & 0 \\ 0 & 0 \end{pmatrix}$$

is also zero on the orthogonal complement of

$$(\text{span}\{\delta_{-N}, ..., \delta_N\} \otimes H)^{\oplus 2}$$

and we are done.

For part (iii), it is straightforward to check that the functions $t \mapsto h(u_t)$, $t \mapsto f(u_t)$ and $t \mapsto g(u_t)$ are continuous, as over a compact interval in the t variable, they only involve continuous changes on the span of finitely many of the eigenvectors $\{\delta_n\}_{n \in \mathbb{Z}}$. It follows from this and the formula for $e(u, v)$ that the function

$$t \mapsto e(u_t \otimes 1_A, v)$$

is norm continuous. The claim follows from this, the fact that χ is continuous on the spectrum $e(u_t \otimes 1_A, v)$ for large t, and continuity of the functional calculus in the appropriate sense (see for example [**7**, Lemma 1.2.5]). □

COROLLARY 3.3. *With notation as above, provisionally define*

$$\alpha_A : K_1(SA) \to K_0(A \otimes \mathcal{K})$$

by the formula for $u \in M_n(\widetilde{A})$

$$[u] \mapsto [\chi(e(v_t \otimes 1_{M_n(A)}, u))] - \begin{bmatrix} 1 & 0 \\ 0 & 0 \end{bmatrix},$$

where t *is chosen sufficiently large (depending on* u*) so that all the conditions in Lemma 3.2 hold. Then this is a well-defined homomorphism for any* C^**-algebra* A.

PROOF. This is straightforward to check from the formulas involved together with the universal property of K_1 as exposited in [**7**, Proposition 8.1.5], for example. □

Abusing notation slightly, we identity $K_0(A \otimes \mathcal{K})$ with $K_0(A)$ via the canonical stabilization isomorphism, and thus treat α_A as a homomorphism from $K_1(SA)$ to $K_0(A)$. To complete our proof of Bott periodicity, it remains to check that these homomorphisms α_A have the properties from Proposition 2.2. The second of these properties is almost immediate; we leave it to the reader to check.

The first property, that $\alpha_{\mathbb{C}}(b) = 1$, is more substantial, and we give the proof here following computations in Loring's thesis.

PROPOSITION 3.4. *With notation as above,* $\alpha_{\mathbb{C}}(b) = 1$.

PROOF. We must compute the element

$$[\chi(e(u_t, b))] - \begin{bmatrix} 1 & 0 \\ 0 & 0 \end{bmatrix} \in K_0(\mathcal{K}) \cong \mathbb{Z}$$

for suitably large t, and show that it is one. We will work with an integer value $t = N$ for N suitably large. Note that with respect to the canonical basis $\{\delta_n\}_{n \in \mathbb{Z}}$ of $L^2(S^1)$, the element b acts as the (inverse of the) unilateral shift. On the other hand, on this basis $u_N(\delta_n) = \delta_n$ for all $n \notin (0, N)$. Define

$$H_N := \operatorname{span}\{\delta_n \mid 1 \leqslant n \leqslant N\},$$

It follows from the above observations, the fact that $f(1) = 1$ and $h(1) = g(1) = 0$, and a direct computation, that $H_N^{\oplus 2}$ is an invariant subspace of $L^2(S^1)^{\oplus 2}$ for both $e(u_N, b)$, and for $\begin{pmatrix} 1 & 0 \\ 0 & 0 \end{pmatrix}$. Moreover, these two operators agree on the orthogonal complement $(H_N^{\oplus 2})^{\perp}$. On $H_N^{\oplus 2}$, $e(u_N, b)$ agrees with the operator $e(u_N, b_N)$, where we abuse notation by writing u_N also for the restriction of u_N to H_N, and where $b_N : H_N \to H_N$ is the permutation operator defined by

$$b_N : \delta_n \mapsto \delta_{n+1 \bmod n}$$

(i.e. b_N is the cyclic shift of the canonical basis). From these computations, we have that if we identify $K_0(\mathcal{K}(L^2(S^1))) \cong \mathbb{Z}$ and $K_0(\mathcal{B}(H_N)) \cong \mathbb{Z}$ via the canonical inclusion $\mathcal{B}(H_N) \to \mathcal{K}(L^2(S^1))$ (which induces an isomorphism on K-theory) then

$$[\chi(e(u_N, b))] - \begin{bmatrix} 1 & 0 \\ 0 & 0 \end{bmatrix} = [\chi(e(u_N, b_N))] - \begin{bmatrix} 1 & 0 \\ 0 & 0 \end{bmatrix} \in \mathbb{Z}.$$

Thus we have reduced the proposition (and therefore the proof of Bott periodicity) to a finite-dimensional matrix computation for N large: we must show that if $e_N := \chi(e(u_N, b_N))$, then the trace of the $2N \times 2N$ matrix

$$\chi(e_N) - \begin{pmatrix} 1 & 0 \\ 0 & 0 \end{pmatrix}$$

is 1 for all large N, or equivalently, that the trace of the $2N \times 2N$ matrix e_N is $N-1$ for all large N.

This can be computing directly, following Loring. The computation is elementary, although slightly involved; we proceed as follows.

STEP 1 ($\|\chi(e_N) - e_N\| = O(1/N)$ for all N). For notational convenience, we fix N, drop the subscript $_N$, and write f for $f(v_N) = f(v)$ and similarly for g and h. We have that $|\chi(x) - x| \leqslant 2|x^2 - x|$ for all $x \in \mathbb{R}$, whence from the functional calculus $\|\chi(e) - e\| \leqslant 2\|e - e^2\|$. Using the formula in line (3) above, it will thus suffice to show that the norm of

$$\begin{pmatrix} hbg + gb^*h & [f, bv] \\ [b^*h, f] & b^*h^2b - h^2 \end{pmatrix}$$

is bounded by C/N, where $C > 0$ is an absolute constant not depending on N. Note that if a function $k : S^1 \to \mathbb{C}$ is Lipschitz with Lipschitz constant $\mathrm{Lip}(k)$, we have that

$$\|[k, b]\| = \|bkb^* - k\| = \sup_{x \in [0,1]} |k(e^{2\pi ix}) - k(2\pi i(x + 1/N))| \leqslant \frac{2\pi \mathrm{Lip}(k)}{N}.$$

This, combined with the fact that $\|h\| \leqslant 1$ implies that

$$\left\| \begin{pmatrix} h[b, g] + [g, b^*]h & h[f, b] \\ [u^*, f]h & u^*(h[h, b] + [h, b]h) \end{pmatrix} \right\| \leqslant \frac{4\pi}{N}(\mathrm{Lip}(f) + \mathrm{Lip}(g) + \mathrm{Lip}(h))$$

and we are done with this step.

STEP 2 ($\mathrm{tr}(\chi(e_N)) - \mathrm{tr}(3e_N^2 - 2e_N^3) \to 0$ as $N \to \infty$). The result of step one says that there is a constant $C > 0$ such that the eigenvalues of e_N are all within C/N of either one or zero for all large N. The function $p(x) = 3x^2 - 2x^3$ has the property that $p(0) = 0$, $p(1) = 1$, and $p'(0) = p'(1) = 0$. Hence there is a constant D such that the eigenvalues of $p(e_N)$ are all within D/N^2 of either 0 or 1. It follows that $\|\chi(e_N) - p(e_N)\| \leqslant D/N^2$. Hence

$$|\mathrm{tr}(\chi(e_N)) - \mathrm{tr}(p(e_N))| = |\mathrm{tr}(\chi(e_N) - p(e_N))| \leqslant 2N\|\chi(e_N) - p(e_N)\| \leqslant \frac{2D}{N}.$$

This tends to zero as $N \to \infty$, completing the proof of step 2.

STEP 3 ($\mathrm{tr}(e_N^2) = N$ for all N). Using the formula in line (3) (with the same notational conventions used there) and rearranging a little, we get

$$\mathrm{tr}(e^2) = \mathrm{tr}(e) + \mathrm{tr}(hbg + gb^*h) + \mathrm{tr}(b^*h^2b - h^2).$$

From the formula for e, we get that $\mathrm{tr}(e) = N$. The second two terms are both zero using the trace property, and that $gh = 0$.

STEP 4 ($\mathrm{tr}(e_N^3) - (N - \frac{1}{2}) \to 0$ as $N \to \infty$). Again, we use the formula in line (3), multiplied by $e = e_N$ to see that

$$\mathrm{tr}(e^3) = \mathrm{tr}(e^2) + \mathrm{tr}\left(e \begin{pmatrix} hbg + gb^*h & [f, hb] \\ [b^*h, f] & b^*hb - h^2 \end{pmatrix}\right).$$

The first term is N using step 3. Multiplying the second term out, simplifying using the trace properties and that $gh = 0$, we see that the trace of the second term equals

$$\operatorname{tr}(3h^2(f - bfb^*)) = 3\sum_{k=0}^{N-1} h(e^{2\pi ik/N})^2(f(e^{2\pi ik/N}) - f(e^{2\pi i(k-1)/N})).$$

Assuming (as we may) that f is differentiable, this converges as N tends to infinity to

$$3\int_0^1 h(x)^2 f'(x)dx.$$

Using that $h = 0$ on $[0, 1/2]$, and that $h = f - f^2$ on $[1/2, 1]$ (plus the precise form for f in [**4**, pages 10-11]) we get that

$$3\int_{\frac{1}{2}}^1 (f(e^{2\pi ix}) - f(e^{2\pi ix})^2)f'(e^{2\pi ix})dx = 3\int_0^1 \lambda - \lambda^2 d\lambda = \frac{1}{2}.$$

This completes the proof of step 4.

Combining steps 2, 3, and 4 completes the argument and we are done. □

4. Connection with the localization algebra

Homomorphisms α_A with the properties required by Proposition 2.2 are maybe more usually defined using differential, or Toeplitz, operators associated to the circle. There is a direct connection between this picture and our unitaries u_t, which we now explain; we will not give complete proofs here as this would substantially increase the length of the paper, but at least explain the key ideas and connections.

The following definition is inspired by work of Yu [**10**] in the case that A is commutative. It agrees with Yu's original definition for unital commutative C^*-algebras.

Definition 4.1. Let A be a C^*-algebra, and assume that A is represented nondegenerately and essentially[1] on some Hilbert space H. Let $C_{ub}([1,\infty), \mathcal{B}(H))$ denote the C^*-algebra of bounded uniformly continuous functions from $[1,\infty)$ to $\mathcal{B}(H)$. The *localization algebra* of A is defined to be

$$C_L^*(A) := \left\{ f \in C_{ub}([1,\infty), \mathcal{B}(H)) \;\middle|\; \begin{array}{c} f(t)a \in \mathcal{K}(H) \text{ for all } t \in [1,\infty), a \in A \\ \text{and } \|[f(t), a]\| \to 0 \text{ for all } a \in A \end{array} \right\}.$$

If $A = C_0(X)$ is commutative, we will write $C_L^*(X)$ instead of $C_L^*(C_0(X))$.

The localization algebra of a C^*-algebra A does not depend on the choice of essential representation H up to non-canonical isomorphism, and its K-theory does not depend on H up to canonical isomorphism (these remarks follow from [**2**, Theorem 2.7]); thus we say 'the' localization algebra of A, even thought this involves a slight abuse of terminology. Moreover, building on work of Yu [**10**] and Qiao-Roe [**6**] in the commutative case, [**2**, Theorem 4.5] gives a canonical isomorphism

$$K^*(A) \to K_*(C_L^*(A))$$

from the K-homology groups of A to the K-theory groups of $C_L^*(A)$ (at least when A is separable).

[1]This means that no non-zero element of A acts as a compact operator, so in particular, such a representation is faithful.

One can define a pairing

$$K_i(C_L^*(A)) \otimes K_j(A) \mapsto \mathbb{Z}$$

between the K-theory of the localization algebra (i.e. K-homology) and the K-theory of a C^*-algebra A. The most complicated case, and also the one relevant to the current discussion of this occurs when $i = j = 1$, so let us focus on this. Let $(u_t)_{t\in[1,\infty)}$ be a unitary in the unitization of $C_L^*(A)$ and v in the unitization of A be another unitary (the construction also works with matrix algebras involved in exactly the same way, but we ignore this for notational simplicity). Let H be the Hilbert space used in the definition of the localization algebra, and for $t \in [1, \infty)$, let

$$e(u_t, v) \in M_2(\mathcal{B}(H))$$

be the Loring element of line (2). One can check that for all large t, the spectrum of $e(u_t, v)$ does not contain $1/2$. Hence if χ be the characteristic function of $[1/2, \infty)$, we get a difference

$$\chi(e(u_t, v)) - \begin{pmatrix} 1 & 0 \\ 0 & 0 \end{pmatrix} \in M_2(\mathcal{B}(H))$$

of projections for all suitably large t. It is moreover not difficult to check that this difference is in $M_2(\mathcal{K}(H))$, and thus defines an element in $K_0(\mathcal{K}(H)) \cong \mathbb{Z}$, which does not depend on t for t suitably large. We may thus define

$$\langle [u_t], [v] \rangle := [e(u_t, v)] - \begin{bmatrix} 1 & 0 \\ 0 & 0 \end{bmatrix} \in K_0(\mathcal{K}) \cong \mathbb{Z}$$

for any suitably large t. One checks that this formula gives a well-defined pairing between $K_1(A)$ and $K^1(A)$. More substantially, one can check that it agrees with the canonical pairing between K-homology and K-theory (at least up to sign conventions).

Let us go back to the case of interest for Bott periodicity. In terms of elliptic operators, one standard model for the canonical generator of the K-homology group $K_1(S^1)$ of the circle is the class of the Dirac operator $D = \frac{-i}{2\pi}\frac{d}{d\theta}$, where θ is 'the' angular coordinate. We consider D as an unbounded operator on $L^2(S^1)$ with domain the smooth functions $C^\infty(S^1)$. Let $\chi : \mathbb{R} \to [-1, 1]$ be any continuous function such that $\lim_{t\to\pm\infty} \chi(t) = \pm 1$, and for $t \in [1, \infty)$, define

$$F_t := \chi(t^{-1}D)$$

using the unbounded functional calculus. Concretely, each F_t is diagonal with respect to the canonical orthonormal basis $\{\delta_n \mid n \in \mathbb{Z}\}$, acting by

$$F_t : \delta_n \mapsto \chi(t^{-1}n)\delta_n \ ;$$

this follows as $D : \delta_n \mapsto n\delta_n$ for each n.

Using the above concrete description of the eigenspace decomposition of F_t, it is not too difficult to show that the function $t \mapsto F_t$ defines an element of the multiplier algebra $M(C_L^*(S^1))$ of the localization algebra $C_L^*(S^1)$. Moreover, one checks similarly that the function $t \mapsto \frac{1}{2}(F_t + 1)$ maps to a projection in $M(C_L^*(S^1))/C_L^*(S^1)$, and thus defines a class $[D]_0 \in K_0(M(C_L^*(S^1))/C_L^*(S^1))$. Yu defines the K-homology class associated to this operator to be the image $[D]$ of $[D]_0$ under the boundary map

$$\partial : K_0(M(C_L^*(S^1))/C_L^*(S^1)) \to K_1(C_L^*(S^1))$$

(all this is part of a very general machine for turning elliptic differential operators into elements of $K_*(C_L^*(S^1))$: see [**9**, Chapter 8]). This boundary map is explicitly computable (compare for example [**7**, Section 12.2]): the image of $[D]_0$ under this map is the class of the unitary

$$e^{2\pi i \frac{1}{2}(F_t+1)} = -e^{\pi i \chi(t^{-1}D)}.$$

Choosing χ to be the function which is negative one on $(-\infty, 0]$, one on $[1, \infty)$, and that satisfies $\chi(t) = 2t - 1$ on $(0, 1)$, we see that

$$-e^{\pi i \chi(t^{-1}D)} = u_t \tag{4}$$

i.e. the canonical generator of the K-homology group $K_1(S^1)$ is given precisely by the class of u_t in $K_1(C_L^*(S^1))$.

The formula for α_A is then a specialization of the general formula above for the pairing, using the element of line (4) (more precisely, its amplification to $C(S^1, A)$ in an obvious sense). Thus our proof of Bott periodicity using almost commuting matrices fits very naturally into the localization picture of K-homology.

Acknowledgment

Thank you to the anonymous referee for several helpful comments.

References

[1] M. F. Atiyah, *Bott periodicity and the index of elliptic operators*, Quart. J. Math. Oxford Ser. (2) **19** (1968), 113–140, DOI 10.1093/qmath/19.1.113. MR228000

[2] M. Dadarlat, R. Willett, and J. Wu, *Localization C^*-algebras and K-theoretic duality*, Ann. K-Theory **3** (2018), no. 4, 615–630, DOI 10.2140/akt.2018.3.615. MR3892961

[3] N. Higson and J. Roe, *Analytic K-homology*, Oxford Mathematical Monographs, Oxford University Press, Oxford, 2000. Oxford Science Publications. MR1817560

[4] T. Loring, *The torus and noncommutative topology (K-theory C^*-algebras, shape theory, cohomology)*, ProQuest LLC, Ann Arbor, MI, 1986. Thesis (Ph.D.)–University of California, Berkeley. MR2635051

[5] T. A. Loring, *Quantitative K-theory related to spin Chern numbers*, SIGMA Symmetry Integrability Geom. Methods Appl. **10** (2014), Paper 077, 25, DOI 10.3842/SIGMA.2014.077. MR3261867

[6] Y. Qiao and J. Roe, *On the localization algebra of Guoliang Yu*, Forum Math. **22** (2010), no. 4, 657–665, DOI 10.1515/FORUM.2010.036. MR2661442

[7] M. Rørdam, F. Larsen, and N. Laustsen, *An introduction to K-theory for C^*-algebras*, London Mathematical Society Student Texts, vol. 49, Cambridge University Press, Cambridge, 2000. MR1783408

[8] D. Voiculescu, *Asymptotically commuting finite rank unitary operators without commuting approximants*, Acta Sci. Math. (Szeged) **45** (1983), no. 1-4, 429–431. MR708811

[9] R. Willett and G. Yu, *Higher index theory*. Available at `https://math.hawaii.edu/~rufus/higherindextheory`, 2018.

[10] G. Yu, *Localization algebras and the coarse Baum-Connes conjecture*, K-Theory **11** (1997), no. 4, 307–318, DOI 10.1023/a:1007766031161. MR1451759

University of Hawai'i at Mānoa

Email address: `rufus@math.hawaii.edu`

Selected Published Titles in This Series

749 **Guillermo Cortiñas and Charles A. Weibel, Editors,** K-theory in Algebra, Analysis and Topology, 2020

746 **Erica Flapan and Helen Wong, Editors,** Topology and Geometry of Biopolymers, 2020

745 **Federico Binda, Marc Levine, Manh Toan Nguyen, and Oliver Röndigs, Editors,** Motivic Homotopy Theory and Refined Enumerative Geometry, 2020

744 **Pieter Moree, Anke Pohl, L'ubomír Snoha, and Tom Ward, Editors,** Dynamics: Topology and Numbers, 2020

743 **H. Garth Dales, Dmitry Khavinson, and Javad Mashreghi, Editors,** Complex Analysis and Spectral Theory, 2020

742 **Francisco-Jesús Castro-Jiménez, David Bradley Massey, Bernard Teissier, and Meral Tosun, Editors,** A Panorama of Singularities, 2020

741 **Houssam Abdul-Rahman, Robert Sims, and Amanda Young, Editors,** Analytic Trends in Mathematical Physics, 2020

740 **Alina Bucur and David Zureick-Brown, Editors,** Analytic Methods in Arithmetic Geometry, 2019

739 **Yaiza Canzani, Linan Chen, and Dmitry Jakobson, Editors,** Probabilistic Methods in Geometry, Topology and Spectral Theory, 2019

738 **Shrikrishna G. Dani, Surender K. Jain, Jugal K. Verma, and Meenakshi P. Wasadikar, Editors,** Contributions in Algebra and Algebraic Geometry, 2019

737 **Fernanda Botelho, Editor,** Recent Trends in Operator Theory and Applications, 2019

736 **Jane Hawkins, Rachel L. Rossetti, and Jim Wiseman, Editors,** Dynamical Systems and Random Processes, 2019

735 **Yanir A. Rubinstein and Bernard Shiffman, Editors,** Advances in Complex Geometry, 2019

734 **Peter Kuchment and Evgeny Semenov, Editors,** Differential Equations, Mathematical Physics, and Applications, 2019

733 **Peter Kuchment and Evgeny Semenov, Editors,** Functional Analysis and Geometry, 2019

732 **Samuele Anni, Jay Jorgenson, Lejla Smajlović, and Lynne Walling, Editors,** Automorphic Forms and Related Topics, 2019

731 **Robert G. Niemeyer, Erin P. J. Pearse, John A. Rock, and Tony Samuel, Editors,** Horizons of Fractal Geometry and Complex Dimensions, 2019

730 **Alberto Facchini, Lorna Gregory, Sonia L'Innocente, and Marcus Tressl, Editors,** Model Theory of Modules, Algebras and Categories, 2019

729 **Daniel G. Davis, Hans-Werner Henn, J. F. Jardine, Mark W. Johnson, and Charles Rezk, Editors,** Homotopy Theory: Tools and Applications, 2019

728 **Nicolás Andruskiewitsch and Dmitri Nikshych, Editors,** Tensor Categories and Hopf Algebras, 2019

727 **André Leroy, Christian Lomp, Sergio López-Permouth, and Frédérique Oggier, Editors,** Rings, Modules and Codes, 2019

726 **Eugene Plotkin, Editor,** Groups, Algebras and Identities, 2019

725 **Shijun Zheng, Marius Beceanu, Jerry Bona, Geng Chen, Tuoc Van Phan, and Avy Soffer, Editors,** Nonlinear Dispersive Waves and Fluids, 2019

724 **Lubjana Beshaj and Tony Shaska, Editors,** Algebraic Curves and Their Applications, 2019

723 **Donatella Danielli, Arshak Petrosyan, and Camelia A. Pop, Editors,** New Developments in the Analysis of Nonlocal Operators, 2019

For a complete list of titles in this series, visit the AMS Bookstore at **www.ams.org/bookstore/conmseries/**.